Mathematics for Economics and Finance

The aim of this book is to bring students of economics and finance who have only an introductory background in mathematics up to a quite advanced level in the subject, thus preparing them for the core mathematical demands of econometrics, economic theory, quantitative finance and mathematical economics, which they are likely to encounter in their final-year courses and beyond. The level of the book will also be useful for those embarking on the first year of their graduate studies in Business, Economics or Finance.

The book also serves as an introduction to quantitative economics and finance for mathematics students at undergraduate level and above. In recent years, mathematics graduates have been increasingly expected to have skills in practical subjects such as economics and finance, just as economics graduates have been expected to have an increasingly strong grounding in mathematics.

The authors avoid the pitfalls of many texts that become too theoretical. The use of mathematical methods in the real world is never lost sight of and quantitative analysis is brought to bear on a variety of topics including foreign exchange rates and other macro level issues. This makes for a comprehensive volume which should be particularly useful for advanced undergraduates, for postgraduates interested in quantitative economics and finance, and for practitioners in these fields.

Michael Harrison is Emeritus Senior Lecturer and Fellow of Trinity College Dublin, where he lectured from 1969 to 2009. He currently lectures in the School of Economics at University College Dublin.

Patrick Waldron is a graduate of the Universities of Dublin and Pennsylvania and a Research Associate in the Department of Economics at Trinity College Dublin.

Mathematics for Economics and Finance

**Michael Harrison and
Patrick Waldron**

Routledge
Taylor & Francis Group

LONDON AND NEW YORK

First published 2011
by Routledge
2 Park Square, Milton Park, Abingdon, Oxon, OX14 4RN

Simultaneously published in the USA and Canada
by Routledge
711 Third Avenue, New York, NY 10017

Routledge is an imprint of the Taylor & Francis Group, an informa business

Typeset in Times New Roman by Sunrise Setting Ltd, Devon, United Kingdom
Printed and bound in Great Britain by TJ International Ltd, Padstow, Cornwall

British Library Cataloguing in Publication Data
A catalogue record for this book is available from the British Library

Library of Congress Cataloging in Publication Data
A catalog record for this book has been requested

ISBN 978-0-415-57303-0 (hbk)
ISBN 978-0-415-57304-7 (pbk)
ISBN 978-0-203-82999-8 (ebk)

Contents

List of figures

List of tables

Foreword

This book has two parts, the first labelled MATHEMATICS, the second APPLICATIONS. Specifically, the applications are to economics and finance. In their Preface, the authors, Michael Harrison and Patrick Waldron, advise that the book is written for "advanced undergraduates of either mathematics or economics". However, the advanced undergraduate in economics who does not have a substantial exposure to mathematics will find Part I of this book challenging. The advanced undergraduate in mathematics, on the other hand, should find this book quite suitable for: reviewing branches of mathematics used in theoretical economics and finance; learning some aspects of these branches of mathematics, perhaps not covered in their mathematics courses, but useful in economic and financial applications; and then being presented with a comprehensive survey of economic and financial theory that draws upon this mathematics.

The advanced undergraduate in economics can use Part I as a checklist as to what mathematical background is expected in the applications in Part II. If the student has enough background to work through the presentation of the subject, then Part I is sufficient in that area. But as the authors note, their "development is often rather rapid". If the student finds the material too compact in an area, he or she can seek a suitable alternative text, which explains the subject less rapidly, then return to the Harrison and Waldron presentation.

In general, then, the two parts of this volume represent, first, an orderly and comprehensive assemblage of the mathematics needed for a great deal of important economic and financial theory; and, second, a presentation of the theory itself. The diligent student who makes his or her way through this volume will have gained a great deal of mathematical power, and knowledge of much economic and financial theory.

Any volume must have its inclusions and exclusions. One topic that has quite limited space in this volume is optimization subject to inequality constraints, such as the problem of tracing out mean–variance efficient sets subject to any system of linear equality, and/or (weak) inequality constraints in which some or all variables may be required to be non-negative. Markowitz and Todd (2000) is devoted to this topic. If the reader decides to read the latter after reading the present book, he or she will find that the present book's extensive coverage of vectors and matrices will be a big help. More generally, the mathematics background presented here will allow the student to move comfortably in the area of "computational finance" as well as financial theory.

Harry M. Markowitz

San Diego, California

October 2010

Preface

This book provides a course in intermediate mathematics for students of economics and finance. It originates from a series of lectures given to third-year undergraduates in the Faculty of Business, Economic and Social Studies and the School of Mathematics at the University of Dublin, Trinity College. The prime aim is to prepare students for some of the mathematical demands of courses in econometrics, economic theory, quantitative finance and mathematical economics, which they may encounter in their final-year programmes. The presentation may also be useful to those embarking on the first year of their postgraduate studies. In addition, it serves as an introduction to economics and finance for mathematics students.

In recent decades, mathematics graduates have been increasingly expected to have skills in practical subjects such as economics and finance, while economics graduates have been expected to have an increasingly strong grounding in mathematics. The growing need for those working in economics and finance to have a strong foundation in mathematics has been highlighted by such layman's texts as Ridley (1993), Kelly (1994), Davidson (1996), Bass (1999), Poundstone (2005) and Bernstein (2007). The present book is, in part, a response to these trends, offering advanced undergraduates of either mathematics or economics the opportunity to branch into the other subject.

There are many good texts on mathematics for economists at the introductory and advanced levels. However, there appears to be a lack of material at what we call the intermediate level, a level that takes much for granted and is concerned with proofs as well as rigorous applications, but that does not seek the abstraction and rigour of the sort of treatments that may be found in pure mathematics textbooks. Indeed, we have been unable to find a single book that covers the material herein at the level and in the way that our teaching required. Hence the present volume.

The book does not aim to be comprehensive. Rather, it contains material that we feel is particularly important to have covered before final-year work begins, and that can be covered comfortably in a standard academic year of three terms or two semesters. Other more specialized topics can be covered later, perhaps as part of a student's final year, as is possible in our programme. Thus we focus in Part I on the mathematics of linear algebra, difference equations, vector calculus and optimization. The mathematical treatment of these topics is quite detailed, but numerous worked examples are also provided. Part II of the book is devoted to a selection of applications from economics and finance. These include deterministic and stochastic dynamic macroeconomic models, input–output analysis, some probability, statistics, quadratic programming and econometric methodology, single- and multi-period choice under certainty, including general equilibrium and the term structure of interest rates, choice under uncertainty, and topics from portfolio theory, such as the capital asset pricing model.

It is assumed that most readers will have already completed good introductory courses in mathematics, economics and/or finance. The section on notation and preliminaries (pp. xix–xxiii) lists many of the basic ideas from these areas with which it is assumed that most readers will have some familiarity.

These are not rigid prerequisites, as the material presented is reasonably self-contained. However, the development is often rather rapid and the discussion fairly advanced at times. Therefore, most students, we feel, will find the book quite challenging but, we hope, correspondingly rewarding as the contents are mastered. Those who have solid preparations in mathematics, economics and finance should not be lulled into a false sense of security by the familiarity of the early material in some of the chapters; and those who have not taken such preliminary courses should be prepared for an amount of additional background reading from time to time. The exercises at the end of each chapter provide additional insight into some of the proofs, and practice in the application of the mathematical methods discussed. A solution manual will be available.

Economics may be defined as the scientific study of optimal decision-making under resource constraints and uncertainty. It is less about forecasting – however much that may be the popular perception – than it is about reacting optimally to the best available forecasts. Macroeconomics is concerned with the study of the economy as a whole. It seeks to explain such things as the determinants of the level of aggregate output (national income), the rate of growth of aggregate output, the general level of prices, and inflation, i.e. the rate of growth of prices. Microeconomics is fundamentally about the allocation of wealth or expenditure among different goods or services, which via the interaction of consumers and producers determines relative prices. Basic finance, or financial economics, is about the allocation of expenditure across two or more time periods, which gives the term structure of interest rates. A further problem in finance is the allocation of expenditure across (a finite number or a continuum of) states of nature, which yields random variables called rates of return on risky assets. Clearly, we might try to combine the concerns of microeconomics and finance to produce a rather complex problem, the solution of which points up the crucial role of mathematics.

It is probably fair to say that economics, particularly financial economics, has in recent years become as important an application of mathematics as theoretical physics. Some would say it is just another branch of applied mathematics. In mathematics departments that have traditionally taught linear algebra courses with illustrations from physics but have also followed the modern trend of offering joint programmes with economics or finance departments, the present work could be used as the basis for a general course in linear algebra for all students. The successful application of techniques from both mathematics and physics to the study of economics and finance has been outlined in the popular literature by authors such as those listed at the start of this preface. It is hoped that this book goes some way to providing students with the wherewithal to do successful mathematical work in economics and finance.

Michael Harrison and Patrick Waldron
School of Economics, University College Dublin
Dublin 4, Ireland
and
Department of Economics, Trinity College Dublin
Dublin 2, Ireland

November 2010

Acknowledgements

We would each like to acknowledge our mathematics and quantitative economics and finance teachers at (for Harrison) the University of Lancaster, particularly Alan Airth, Anna Koutsoyiannis and Tin Nguyen, and (for Waldron) Trinity College Dublin and the University of Pennsylvania, particularly Adrian Raftery, David Simms, John Woods, Angel de la Fuente, Bob Litzenberger and Krishna Ramaswamy.

Our thanks are also due to several cohorts of undergraduate students at Trinity College Dublin, some of whom drew attention to various typographical errors in previous versions of parts of the book used as course notes (they know who they are); and to Vahagn Galstyan for a number of useful comments on some of the linear algebra chapters.

We are grateful to three anonymous reviewers who read an earlier version of our manuscript and provided many helpful comments and suggestions; in particular to the reviewer whose detailed observations have greatly improved our treatment of Stein's lemma (Theorem 13.7.1) and Siegel's paradox (Theorem 13.10.2). We remain entirely responsible for the current version of the book, of course.

For biographical notes not otherwise attributed we have drawn on the official website of the Nobel Prize (http://nobelprize.org/nobel_prizes/economics/laureates); the MacTutor History of Mathematics archive at the University of St. Andrews (http://www-history.mcs.st-andrews.ac.uk/); the Earliest Known Uses of Some of the Words of Mathematics website (http://jeff560.tripod.com/mathword.html) and the Mathematics Genealogy Project (http://genealogy.math.ndsu.nodak.edu/).

We thank those at Routledge who have helped to bring this project to fruition, especially Simon Holt, Robert Langham, Lisa Salonen, Emily Senior and Thomas Sutton. We also thank Geoff Amor, our copyeditor, Gareth Toye, our cover designer, and Jessica Stock and her team at Sunrise Setting Ltd.

We are indebted to Donald Knuth for the TeX typesetting system and to other members of the worldwide TeX and LaTeX community, especially to Leslie Lamport for LaTeX, to Till Tantau for the TikZ and PGF packages, and to Christian Feuersänger for PGFPLOTS.

We are especially grateful to Professor Harry M. Markowitz, Nobel laureate, for kindly contributing the Foreword.

Last, but not least, thanks to June Harrison for all the dinners.

List of abbreviations

AR	autoregressive
BLUE	best linear unbiased estimator
CAPM	capital asset pricing model
CARA	constant absolute risk aversion
cdf	cumulative distribution function
CES	constant elasticity of substitution
CRRA	constant relative risk aversion
DARA	decreasing absolute risk aversion
DCF	discounted cash flow
DRRA	decreasing relative risk aversion
EMH	efficient markets hypothesis
EU	European Union
EUR	euro
GBP	pound sterling
GLS	generalized least squares
HARA	hyperbolic absolute risk aversion
$I(0)$	integrated to order zero
IARA	increasing absolute risk aversion
iff	if and only if
iid	independent and identically distributed
IRR	internal rate of return
IRRA	increasing relative risk aversion
IS	investment and saving equilibrium
ISO	International Organization for Standardization
LM	liquidity preference and money supply equilibrium
m	million
MVN	multivariate normal
NPV	net present value
OLS	ordinary least squares
pdf	probability density function
RLS	restricted least squares
rv	random variable *or* random vector
s.t.	so that *or* subject to *or* such that
UK	United Kingdom
US(A)	United States (of America)
VAR	vector autoregressive
VNM	von Neumann–Morgenstern
WLS	weighted least squares

Notation and preliminaries

We assume familiarity with basic mathematics from Pythagoras of Samos ($c.569$–$c.475$BC) and his famous theorem onwards. Basic knowledge of probability and statistics, as well as micro- and macroeconomics and especially financial economics, would also be advantageous. The material on probability and statistics in Chapter 13 aims to be completely self-contained, but will represent quite a steep learning curve for readers who do not have some preliminary training in these areas. Similarly, some prior exposure to economics and finance will enable readers to proceed more rapidly through the other applications chapters.

For the sake of clarity, we gather together in these preliminary pages some of the more important mathematical and statistical concepts with which we assume prior familiarity, concentrating especially on those for which notational conventions unfortunately vary from one textbook to another.

Mathematical and technical terms appear in boldface where they are first introduced or defined; the corresponding page number in the index is also in boldface.

Economics students who are new to formal mathematics should be aware of common pitfalls of flawed logic, in particular with the importance of presenting the parts of a definition in the correct order and with the process of proving a theorem by arguing from the assumptions to the conclusions. Familiarity with various approaches to proofs is assumed, though the principles of proof by contradiction, proof by contrapositive and proof by induction are described when these methods are first used.[1] Similarly, mathematics students, who may be familiar with many of the mathematics topics covered, should think about the nature, subject matter and scientific methodology of economics before starting to work through the book.

Readers should be familiar with the expressions "such that" and "subject to" (both often abbreviated "s.t.") and "if and only if" (abbreviated as "iff"), and also with their meanings and use. The symbol $\forall$ is mathematical shorthand for "for all" and $\exists$ is mathematical shorthand for "there exists". The expression iff signifies a necessary and sufficient condition, or equivalence. Briefly, Q is **necessary** for P if P implies Q; and similarly P is **sufficient** for Q if P implies Q. Furthermore, P implies Q if and only if the **contrapositive**, "not Q" implies "not P", is true. We shall sometimes use an alternative symbol for a necessary and sufficient condition, namely $\Leftrightarrow$, which signifies that the truth of the left-hand side implies the truth of the right-hand side and vice versa, and also that the falsity of the left-hand side implies the falsity of the right-hand side and vice versa. Other logical symbols used are $\Rightarrow$, which means "implies", and $\Leftarrow$, which means "is implied by" or "follows from".

We will make frequent use of the identity symbol, $\equiv$, particularly in definitions; of $\approx$, which denotes "approximately equal"; and of $\sum_{i=1}^{n} x_i$ and $\prod_{j=1}^{n} x_j$ to denote the sum of, and the product of, n numbers $x_1, x_2, \ldots, x_n$, respectively; i.e.

$$\sum_{i=1}^{n} x_i = x_1 + x_2 + \cdots + x_n$$

and

$$\prod_{j=1}^{n} x_j = x_1 \times x_2 \times \cdots \times x_n$$

When the upper and lower limits are clear from the context, we occasionally just write $\sum_i$ or $\prod_j$.

The sum of the first n terms of a **geometric series** or **geometric progression** with first term a and **common ratio** ϕ is

$$\sum_{i=1}^{n} a\phi^{i-1} = a + a\phi + a\phi^2 + \cdots + a\phi^{n-1} = \begin{cases} \dfrac{a(1-\phi^n)}{1-\phi} = \dfrac{a(\phi^n - 1)}{\phi - 1} & \text{if } \phi \neq 1 \\ n \times a & \text{if } \phi = 1 \end{cases}$$

If $-1 < \phi < 1$, then the sum to infinity of the series is $a/(1-\phi)$. If $\phi \leq -1$ (and $a \neq 0$), then the sum oscillates without converging as $n \to \infty$. If $\phi \geq 1$, then the sum goes to $\pm\infty$ as $n \to \infty$, depending on the sign of a.

The expression $n!$, referred to as n **factorial**, denotes the product of the integers from 1 to n, inclusive, and 0! is defined to be unity; i.e. $n! \equiv \prod_{i=1}^{n} i = 1 \times 2 \times \cdots \times n$ and $0! \equiv 1$.

Readers are assumed to be familiar with basic set notation and Venn diagrams.[2] If X is the **universal set** and $B \subseteq X$, i.e. B is a **subset** of X, then $X \setminus B$ denotes the **complement** of B or $X \setminus B \equiv \{x \in X : x \notin B\}$. We say that sets B and C are **disjoint** if $B \cap C = \{\ \}$, i.e. if the intersection of B and C is the **null set** or **empty set**. The **Cartesian product** of the n sets $X_1, X_2, \ldots, X_n$ is the set of **ordered n-tuples**, $(x_1, x_2, \ldots, x_n)$, where the ith component, x_i, of each n-tuple is an element of the ith set, X_i.

We assume knowledge of the sets and use of natural numbers, $\mathbb{N}$, integers, $\mathbb{Z}$, and real and complex numbers, $\mathbb{R}$ and $\mathbb{C}$. Throughout the book, italic letters such as x denote specific numbers in $\mathbb{R}$. The Cartesian product $\mathbb{R} \times \mathbb{R} \times \cdots \times \mathbb{R}$, denoted by $\mathbb{R}^n = \{(x_1, x_2, \ldots, x_n) \mid x_1, x_2, \ldots, x_n \in \mathbb{R}\}$, is called **(Euclidean) n-space**. Points in $\mathbb{R}^n$ (and sometimes in an arbitrary vector or metric space X) are denoted by lower-case boldface letters, such as $\mathbf{x}$, while an upper-case boldface letter, such as $\mathbf{X}$, will generally denote a matrix. Any $\mathbf{x} \in \mathbb{R}^n$ can also be written as the n-tuple $(x_1, x_2, \ldots, x_n)$, where $x_1, x_2, \ldots, x_n$ are referred to as the **(Cartesian) coordinates**[3] of $\mathbf{x}$. A tilde over a symbol will be used to denote a random variable (e.g. $\tilde{x}$) or a random vector (e.g. $\tilde{\mathbf{x}}$). The notation $\mathbb{R}^n_+ \equiv \{\mathbf{x} \in \mathbb{R}^n : x_i \geq 0, i = 1, 2, \ldots, n\}$ denotes the **non-negative orthant** of $\mathbb{R}^n$, and $\mathbb{R}^n_{++} \equiv \{\mathbf{x} \in \mathbb{R}^n : x_i > 0, i = 1, \ldots, n\}$ denotes the **positive orthant**.

The interval $[a, b] \equiv \{x \in \mathbb{R} : a \leq x \leq b\}$ is called a **closed interval** and $(a, b) \equiv \{x \in \mathbb{R} : a < x < b\}$ is called an **open interval**. The context will generally allow readers to distinguish between the 2-tuple $(a, b) \in \mathbb{R}^2$ and the open interval $(a, b) \subset \mathbb{R}$.

The most important result on complex numbers relied on is de Moivre's theorem,[4] which allows us to write $(\cos\theta + i\sin\theta)^t$ as $\cos t\theta + i\sin t\theta$ and $(\cos\theta - i\sin\theta)^t$ as $\cos t\theta - i\sin t\theta$, where $i \equiv \sqrt{-1}$.

Recall also that the **conjugate** of the complex number $z = a + ib$ is $\bar{z} = a - ib$, and that the conjugate of a sum is the sum of the conjugates and the conjugate of a product is the product of the conjugates. Also, the **modulus** of z is the positive square root $|z| = \sqrt{a^2 + b^2} = \sqrt{z\bar{z}}$.

The fundamental theorem of algebra states that a polynomial of degree n with real (or complex) coefficients has exactly n, possibly complex, roots, with the complex roots coming in conjugate pairs and allowing for the possibility of several roots having the same value.

The following definitions relating to functions and relations are important.

DEFINITION 0.0.1 A **function** (or **map** or **mapping**) $f: X \to Y: x \mapsto f(x)$ from a **domain** X to a **co-domain** Y is a rule that assigns to each element of the set X a unique element $f(x)$ of the set Y called the **image** of x.

DEFINITION 0.0.2 If $f: X \to Y$ and $g: Y \to Z$, then the **composition of functions** or **composite function** or **function of a function** $g \circ f: X \to Z$ is defined by $g \circ f(x) = g(f(x))$.

DEFINITION 0.0.3 A **correspondence** $f: X \to Y$ from a **domain** X to a **co-domain** Y is a rule that assigns to each element of X a non-empty subset of Y.

DEFINITION 0.0.4 The **range** of the function $f: X \to Y$ is the set $f(X) = \{f(x) \in Y: x \in X\}$.

DEFINITION 0.0.5 The function $f: X \to Y$ is **injective** (one-to-one) if and only if $f(x) = f(x') \Rightarrow x = x'$.

DEFINITION 0.0.6 The function $f: X \to Y$ is **surjective** (onto) if and only if $f(X) = Y$.

DEFINITION 0.0.7 The function $f: X \to Y$ is **bijective** (or **invertible**) if and only if it is both injective and surjective.

An invertible function $f: X \to Y$ has a well-defined inverse function $f^{-1}: Y \to X$ with $f(f^{-1}(y)) = y$ for all $y \in Y$ and $f^{-1}(f(x)) = x$ for all $x \in X$.

For any function $f: X \to Y$, if $A \subseteq X$, then

$$f(A) \equiv \{f(x): x \in A\} \subseteq Y$$

and if $B \subseteq Y$, the notation f^{-1} is also used to denote

$$f^{-1}(B) \equiv \{x \in X: f(x) \in B\} \subseteq X$$

If f is invertible and $y \in Y$, then $f^{-1}(\{y\}) = \{f^{-1}(y)\}$. If f is not invertible, then $f^{-1}(\{y\})$ can be empty or have more than one element, but $f^{-1}: f(X) \to X$ still defines a correspondence.

DEFINITION 0.0.8 If $f: X \to Y$ is a differentiable function $(X, Y \subseteq \mathbb{R})$, then $f': X \to \mathbb{R}$ denotes the **derivative** of f, i.e. $f'(x)$ is the derivative of f at x, also occasionally denoted $\frac{df}{dx}(x)$ or dy/dx if it is known that $y = f(x)$.

DEFINITION 0.0.9 The function $f: X \to Y$ is **homogeneous of degree** k ($k \in \mathbb{R}$) if and only if $f(\theta x) = \theta^k f(x)$ for all $\theta \in \mathbb{R}$.

When $k = 1$ and f is homogeneous of degree one, the function is sometimes called **linearly homogeneous**.

DEFINITION 0.0.10 A **binary relation** R on the set X is a subset R of $X \times X$ or a collection of pairs (x, y) where $x \in X$ and $y \in X$.

If $(x, y) \in R$, we usually write $x R y$.[5]

DEFINITION 0.0.11 The following properties of a binary relation R on a set X are often of interest:

(a) A relation R is **reflexive** if and only if $x R x$ for all $x \in X$.
(b) A relation R is **symmetric** if and only if $x R y \Rightarrow y R x$.
(c) A relation R is **transitive** if and only if $x R y$ and $y R z \Rightarrow x R z$.
(d) A relation R is **complete** if and only if, for all $x, y \in X$, either $x R y$ or $y R x$ (or both); in other words, a complete relation orders the whole set.
(e) An **equivalence relation** is a relation that is reflexive, symmetric and transitive. An equivalence relation partitions X in a natural way into disjoint equivalence classes.

In consumer theory, we will consider the weak preference relation $\succeq$, where $\mathbf{x} \succeq \mathbf{y}$ means that either the consumption bundle $\mathbf{x}$ is preferred to $\mathbf{y}$ or the consumer is indifferent between the two, i.e. that $\mathbf{x}$ is at least as good as $\mathbf{y}$.

We expect readers to have a sound knowledge of basic calculus, including the taking of limits and single-variable differentiation and integration. We assume familiarity with the definition of a derivative in terms of a limit, and with the single-variable versions of the chain rule and the product rule. We also assume knowledge of l'Hôpital's rule,[6] which states that, if the limits of the numerator and denominator in a fraction are both zero or both infinite, then the limit of the original ratio equals the limit of the ratio of the derivative of the numerator to the derivative of the denominator.

Familiarity with integration by substitution and integration by parts and with the standard rules for differentiation and integration of scalar-valued functions, in particular polynomial and trigonometric functions, is assumed.[7]

Among the trigonometric identities used later are

- the cosine rule

$$a^2 = b^2 + c^2 - 2bc \cos A$$

- the double-angle formula

$$\cos 2A = 2 \cos^2 A - 1$$

- the fundamental identity

$$\cos^2 A + \sin^2 A = 1$$

An ordered arrangement of r objects from a set of n objects is called a **permutation**, and the number of different permutations of r objects that can be chosen from a set of n objects, denoted ^nP_r, is given by

$$^nP_r = \frac{n!}{(n-r)!}$$

A selection of r objects from a set of n objects without regard for their order is called a **combination**, and the number of different combinations of r objects that can be chosen from a set of n objects, denoted ^nC_r, is given by

$$^nC_r = \frac{^nP_r}{r!} = \frac{n!}{(n-r)!\,r!}$$

We expect students to be comfortable with the properties of the exponential function $e: \mathbb{R} \to \mathbb{R}_{++}: x \mapsto e^x$, where $e \approx 2.7182\ldots$; and with its inverse, the natural logarithm function $\ln: \mathbb{R}_{++} \to \mathbb{R}: x \mapsto \ln x$; and also with the use of logarithms to any base. In particular, we rely on the fact that

$$\lim_{n \to \infty} \left(1 + \frac{r}{n}\right)^n = e^r$$

This is sometimes used as the definition of e, but others[8] prefer to start with

$$e^r \equiv 1 + r + \frac{r^2}{2!} + \frac{r^3}{3!} + \cdots = \sum_{j=0}^{\infty} \frac{r^j}{j!}$$

The notation $|X|$ denotes the number of elements in the set X, or the **cardinality** of X, while $|z|$ denotes the modulus of the (complex) number z and $|\mathbf{X}|$ denotes the determinant of the matrix $\mathbf{X}$, more often denoted $\det(\mathbf{X})$. The modulus is just the absolute value when z is real rather than complex. There is obviously some potential for confusion from use of the same symbol for three different concepts, but the context and notation within the symbol will almost always make the distinctions clear.

The collection of all possible subsets of the set X, or the **power set** of X, is denoted by 2^X. Note that $|2^X| = 2^{|X|}$.

The least upper bound or **supremum** of a set, X, of real numbers, denoted $\sup(X)$, is the smallest real number that is greater than or equal to every number in the set. For example, $\sup\{1, 2, 3, 4\} = 4$ and $\sup\{x \in \mathbb{R}^n: 0 < x < 1\} = 1$. The second of these examples indicates that the supremum is not necessarily the maximum real number in the set.

The greatest lower bound or **infimum** of a set, X, of real numbers, denoted $\inf(X)$, is the largest real number that is less than or equal to every number in the set. For example, $\inf\{1, 2, 3, 4\} = 1$, $\inf\{x \in \mathbb{R}^n: 0 \leq x \leq 1\} = 0$ and $\inf\{x \in \mathbb{R}^n: x^3 > 2\} = 2^{1/3}$. The infimum is not necessarily the minimum real number in a set.

Part I
MATHEMATICS

Introduction

Part I of the book is devoted to mathematics. It begins with a quite extensive treatment of linear algebra, dealing in Chapters 1 to 4 with matrices, determinants, eigenvalues and eigenvectors, and quadratic forms and definiteness, before going on to the somewhat more theoretical topics of vector spaces and linear transformations in Chapters 5 and 6. Chapter 7 provides the foundations for vector calculus, including such things as affine and convex combinations, sets and functions, basic topology, and limits and continuity. Some of this material (on limits) is referred to in the following quite lengthy Chapter 8 on dynamic modelling using difference equations, but most is intended as preparation for Chapter 9, which covers important subjects in vector differentiation and multiple integration. Finally, Chapter 10 discusses the topics of convexity and concavity, unconstrained and constrained optimization, and duality.

As mentioned in the Preface, the discussion of topics is reasonably self-contained but it progresses quite rapidly and is rather advanced at times. Moreover, the order in which chapters are read is important, as much of the material in earlier chapters is essential for a proper understanding of the mathematics covered in the later chapters. Proofs are given for the vast majority of the theorems introduced. In the few cases in which no proof is provided, a suitable reference to a proof is given. There are many worked numerical examples to help with the understanding of concepts, methods and results, as well as a few economic examples and illustrations intended, primarily, to motivate the mathematics covered, but also to introduce issues for generalization and further study in Part II.

1 Systems of linear equations and matrices

1.1 Introduction

This chapter focuses on matrices. It begins by discussing linear relationships and systems of linear equations, and then introduces the matrix concept as a tool for helping to handle and analyse such systems. Several examples of how matrices might arise in specific economic applications are given to motivate the mathematical detail that follows. These examples will be used again and further developed later in the book. The mathematical material that follows the examples comprises discussions of matrix operations, the rules of matrix algebra, and a taxonomy of special types of matrix encountered in economic and financial applications.

1.2 Linear equations and examples

Linear algebra is a body of mathematics that helps us to handle, analyse and solve systems of linear relationships. A great deal of economics and finance makes use of such linear relationships. A linear relationship may be represented by an equation of the form

$$z = \alpha x + \beta y \tag{1.1}$$

where x, y and z are variables and α and β are constants. Such relationships have several nice properties. One is that they are homogeneous of degree one, or linearly homogeneous, i.e. if all variables on the right-hand side are scaled (multiplied) by a constant, θ, then the left-hand side is scaled in the same way. Specifically, using (1.1), we have

$$z^* = \alpha(\theta x) + \beta(\theta y) = \theta(\alpha x + \beta y) = \theta z \tag{1.2}$$

Another property of linear relationships is that, for different sets of values for their variables, they are additive and their sum is also linear. Suppose we have the two equations $z_1 = \alpha x_1 + \beta y_1$ and $z_2 = \alpha x_2 + \beta y_2$, then

$$z_1 + z_2 = \alpha(x_1 + x_2) + \beta(y_1 + y_2) \tag{1.3}$$

after slight rearrangement, which may be written as

$$Z = \alpha X + \beta Y \tag{1.4}$$

where $X = x_1 + x_2$, $Y = y_1 + y_2$ and $Z = z_1 + z_2$. The result, equation (1.4), is a linear equation in the sums of the respective variables. The generalization to the case of n equations is straightforward and has $X = \sum_{i=1}^{n} x_i$, $Y = \sum_{i=1}^{n} y_i$ and $Z = \sum_{i=1}^{n} z_i$.

These simple properties constitute one reason why linear relationships are so widely used in economics and finance, and particularly when relationships, such as demand and supply curves, are first introduced to students.[1] Other reasons are that linear equations are easy to represent graphically, at least when they involve two and three variables. With two variables, they can be depicted as straight lines in two-dimensional diagrams; and with three variables, they can be represented as planes in three dimensions. With more than three variables, such as in much of our analysis below, we use the concept of a hyperplane; see Section 7.4.1. Further, linear equations are very adaptable.

We may note three types of adaptability. First, linear equations may be used to deal with certain nonlinear relationships by appropriate redefinition of variables. For example, the non-linear equation $z = \alpha x + \beta y^2$ may be written as $z = \alpha x + \beta w$, where $w = y^2$. Second, by some transformation and redefinition of variables, other nonlinear equations may be expressed as linear equations. For example, taking the logarithms of both sides of the equation $z = x^\alpha y^\beta$ yields $\log z = \alpha \log x + \beta \log y$, which may be written as $z^* = \alpha x^* + \beta y^*$, where $x^* = \log x$, $y^* = \log y$ and $z^* = \log z$. Third, linear equations may be used as approximations to more complex nonlinear relationships that do not linearize by either of the above methods. Such approximations are likely to be useful only locally rather than globally. Thus some nonlinear equations can be handled as if they were simpler linear equations.

Finally, linear equation systems, unlike systems of nonlinear equations, are usually easy to solve. The solution and general manipulation of systems of linear equations is facilitated by the use of matrices, and matrix algebra is an important part of linear algebra. A basic definition, therefore, is that of a **matrix**.

DEFINITION 1.2.1 A **matrix** is a rectangular array of **scalars** called **elements** (or **entries**). In this book, scalars will generally be real numbers.

Such an array is exemplified in the following general notation for a matrix.

NOTATION 1.2.1 *A general matrix* **A** *can be written as*

$$
\mathbf{A} = \begin{bmatrix} a_{11} & a_{12} & \ldots & a_{1n} \\ a_{21} & a_{22} & \ldots & a_{2n} \\ \vdots & \vdots & & \vdots \\ a_{m1} & a_{m2} & \ldots & a_{mn} \end{bmatrix} = [a_{ij}]
$$

Notice the use of a bold upper-case Roman letter for the matrix and the corresponding lower-case letter for its scalar elements. This will be our usual convention when referring to matrices and scalars. Note also the use of subscripts denoting the row and the column to which each element belongs. The number of rows and the number of columns determine the **order** or **dimension** of the matrix; here the order is $m \times n$, signifying m rows and n columns and hence mn elements in all. Sometimes it may be useful to make the order explicit by means of a subscript: $\mathbf{A}_{m \times n}$. Note further that it is often very useful to denote a matrix by means of its typical element, $[a_{ij}]$. Thus an efficient way of representing a matrix is to write $[a_{ij}]_{m \times n}$. When $m = n$, and the number of rows and columns is the same, we say that the matrix is a **square matrix**.

A matrix for which $m = 1$ is a single row of n elements; this is often called a **row vector**. Similarly, an $m \times 1$ matrix may be called a **column vector**. We shall also sometimes refer to $m \times 1$ column vectors as m-**vectors**, especially from Chapter 5 onwards. Our notation for row and column vectors, respectively, is as follows.

NOTATION 1.2.2 *A general row vector* **r** *can be written as*

$$\mathbf{r} = \begin{bmatrix} r_1 & r_2 & \cdots & r_n \end{bmatrix}$$

NOTATION 1.2.3 *A general column vector* **c** *can be written as*

$$\mathbf{c} = \begin{bmatrix} c_1 \\ c_2 \\ \vdots \\ c_m \end{bmatrix}$$

or $(c_1, c_2, \ldots, c_m)$ *to economize on space.*

A further row notation for a column vector will be introduced later, when the concept of the transpose of a matrix has been defined. When it is clear from the context whether we are dealing with row vectors or column vectors, we will use the term **vector** on its own. The use of bold lower-case Roman letters, as here, is a common means of denoting vectors.

One of our first tasks will be to define the operations that can be performed with matrices and to investigate the rules that govern their algebraic manipulation. However, before we turn to these matters, we shall first consider some examples that illustrate how matrices arise in economics. Later on, when we have developed a sufficient body of mathematics, we shall return to generalizations of these examples to analyse them more carefully.

1.2.1   Single-equation econometric model

Consider the problem of estimating the demand function for a product. Economic theory suggests that a demand relationship may be of the form $Q = f(P, Y)$, where Q is quantity demanded per period, P is the price of the product and Y is income. If it is assumed that the demand function is linear, then we have the theoretical model

$$Q = f(P, Y) = \alpha + \beta P + \gamma Y \tag{1.5}$$

where the Greek letters are (constant) parameters. Of course, estimation of this demand function, which means estimation of the unknown parameters, requires data. If we have T time-series observations on each of the variables Q, P and Y, then we may write

$$\tilde{Q}_t = \alpha + \beta P_t + \gamma Y_t + \tilde{u}_t, \quad t = 1, 2, \ldots, T \tag{1.6}$$

where t denotes time and $\tilde{u}_t$ is included as a random disturbance because the data do not conform to the theoretical linear relationship exactly, possibly because Q can be measured only with error or because Q is influenced in a lesser way by variables not included on the right-hand side of (1.5). In practice, the linear relationship (1.5) is only an approximation to the **stochastic** relationship (1.6). We will return to these topics in Chapter 13 and again in Section 14.2.

Equation (1.6) actually represents a system of T equations, one for each time period:

$$\tilde{Q}_1 = \alpha + \beta P_1 + \gamma Y_1 + \tilde{u}_1$$
$$\tilde{Q}_2 = \alpha + \beta P_2 + \gamma Y_2 + \tilde{u}_2$$
$$\vdots$$
$$\tilde{Q}_T = \alpha + \beta P_T + \gamma Y_T + \tilde{u}_T \qquad (1.7)$$

It is convenient to define matrices

$$\tilde{\mathbf{y}} = \begin{bmatrix} \tilde{Q}_1 \\ \tilde{Q}_2 \\ \vdots \\ \tilde{Q}_T \end{bmatrix}_{T \times 1} \quad \text{and} \quad \mathbf{X} = \begin{bmatrix} 1 & P_1 & Y_1 \\ 1 & P_2 & Y_2 \\ \vdots & \vdots & \vdots \\ 1 & P_T & Y_T \end{bmatrix}_{T \times 3} \qquad (1.8)$$

and matrices of parameters and random disturbances

$$\boldsymbol{\beta} = \begin{bmatrix} \alpha \\ \beta \\ \gamma \end{bmatrix}_{3 \times 1} \quad \text{and} \quad \tilde{\mathbf{u}} = \begin{bmatrix} \tilde{u}_1 \\ \tilde{u}_2 \\ \vdots \\ \tilde{u}_T \end{bmatrix}_{T \times 1} \qquad (1.9)$$

for then the entire system of equations (1.7) may be written simply as

$$\tilde{\mathbf{y}} = \mathbf{X}\boldsymbol{\beta} + \tilde{\mathbf{u}} \qquad (1.10)$$

In this equation we have followed standard practice by using bold lower-case Roman letters for the matrices consisting of a single column of the T values of the **dependent variable**, $\tilde{Q}_t$, and of the T unobservable random disturbances, $\tilde{u}_t$, while a bold lower-case Greek letter is used for the column of three unknown parameters. The first column of $\mathbf{X}$, consisting entirely of ones, is associated with the intercept α that appears in the specification of the demand equation. Of course, the interpretation and manipulation of matrix equation (1.10) presupposes an understanding of matrix equality, matrix addition and matrix multiplication, as well as the rules of matrix arithmetic. We shall turn to these matters shortly. The statistical aspects of estimation of parameters will be discussed briefly in Chapter 13 and referred to again in Chapter 14.

1.2.2   Static macroeconomic model of a closed economy

Consider now a simple linear (**Keynesian**) macroeconomic model of a closed economy that includes aggregate consumption and investment functions and a national income accounting identity:[2]

$$C = f(Y) = \alpha_1 + \alpha_2 Y \qquad (1.11)$$
$$I = g(Y, R) = \beta_1 + \beta_2 Y + \beta_3 R \qquad (1.12)$$
$$Y = C + I + G \qquad (1.13)$$

where C, I and Y are the **endogenous** variables, consumption, investment and national income, respectively, whose values are determined by the operation of the system, and R and G are the **exogenous** variables, rate of interest and government expenditure, respectively, determined outside the system, say, by the government. The coefficients in this system have interpretations that will be familiar to economics students. For example, $\alpha_2 = dC/dY$ is the rate of change of consumption with respect to national income and is known as the **marginal propensity to consume** ($1 - \alpha_2$ is the **marginal propensity to save**); and similarly for β_2 and β_3. These equations may be rearranged and written as

$$C + 0 - \alpha_2 Y = \alpha_1 + 0 + 0 \tag{1.14}$$

$$0 + I - \beta_2 Y = \beta_1 + \beta_3 R + 0 \tag{1.15}$$

$$-C - I + Y = 0 + 0 + G \tag{1.16}$$

suggesting the parameter matrices

$$\mathbf{A} = \begin{bmatrix} 1 & 0 & -\alpha_2 \\ 0 & 1 & -\beta_2 \\ -1 & -1 & 1 \end{bmatrix} \quad \text{and} \quad \mathbf{B} = \begin{bmatrix} \alpha_1 & 0 & 0 \\ \beta_1 & \beta_3 & 0 \\ 0 & 0 & 1 \end{bmatrix} \tag{1.17}$$

and the matrices of variables

$$\mathbf{x} = \begin{bmatrix} C \\ I \\ Y \end{bmatrix} \quad \text{and} \quad \mathbf{z} = \begin{bmatrix} 1 \\ R \\ G \end{bmatrix} \tag{1.18}$$

The first element of $\mathbf{z}$, the number one, is not a variable at all, of course, but it is useful to include this along with the exogenous variables proper whenever an intercept appears in an equation, as is the case in the consumption and investment relationships. Thus we might present the model much more compactly in matrix notation as

$$\mathbf{Ax} = \mathbf{Bz} \tag{1.19}$$

One may be interested in estimating the parameters of a system such as (1.19), which is another econometric exercise and would require further stochastic specification and data on the endogenous and exogenous variables.[3] Such data could be accommodated in two $3 \times n$ matrices, assuming n observations were available on all of the variables.

Another concern is the solution of the system for the endogenous variables in terms of the exogenous variables; or, more generally, the solution for $\mathbf{x}$ in the system $\mathbf{Ax} = \mathbf{b}$, where

$$\mathbf{A} = \begin{bmatrix} a_{11} & a_{12} & \cdots & a_{1n} \\ a_{21} & a_{22} & \cdots & a_{2n} \\ \vdots & \vdots & & \vdots \\ a_{m1} & a_{m2} & \cdots & a_{mn} \end{bmatrix} = [a_{ij}]_{m \times n}$$

$$\mathbf{x} = \begin{bmatrix} x_1 \\ x_2 \\ \vdots \\ x_n \end{bmatrix} = [x_j]_{n \times 1} \quad \text{and} \quad \mathbf{b} = \begin{bmatrix} b_1 \\ b_2 \\ \vdots \\ b_m \end{bmatrix} = [b_i]_{m \times 1} \tag{1.20}$$

i.e. a general linear simultaneous equation system. Note that, when $b_i = 0$ for all i, we describe such a system as **homogeneous**; otherwise it is **non-homogeneous**. Also note that, if a solution or solutions exist(s) for a homogeneous or non-homogeneous system of linear equations, we say the system is **consistent**; otherwise it is **inconsistent**. We return to the practical question of how a solution may be obtained for a certain kind of linear equation system in Chapter 2, while a more general result on the solution of homogeneous linear systems is provided in Chapter 6.

1.2.3   Static input–output model

A simple alternative way of describing and analysing an economy to that provided by the Keynesian macroeconomic model of the previous example is by means of the Leontief static input–output model.[4] This assumes that in the economy there are n industries, each producing one particular commodity, and that net outputs of these goods are required for use by consumers. We may denote these **final demands** as $f_1, f_2, \ldots, f_n$, where $f_i \geq 0$ for all i. However, there are other demands for the goods, because, to produce any given good, other goods are required as factors of production, i.e. as industrial inputs, in fixed proportions in the process of production of the good in question. The neoclassical theory of production deals with a more flexible class of production functions, which allow continuous substitution among inputs to the production process. In the two-input case, such production functions can be represented by **isoquant maps**, an **isoquant** being a level set of production possibilities showing all combinations of the inputs that yield the same quantity of output; in the special case of the Leontief production function, the isoquants are L-shaped, as will be mentioned again in Section 11.3.

Letting the fixed number of units of input i required to produce one unit of good j be a_{ij} (≥ 0), the total amount of good i required in production is then

$$a_{i1}x_1 + a_{i2}x_2 + \cdots + a_{in}x_n, \quad i = 1, 2, \ldots, n \tag{1.21}$$

where the a_{ij} are technological coefficients called **input–output coefficients**, and the x_j ($j = 1, 2, \ldots, n$) are the total outputs required of the n goods. The amounts $a_{ij}x_j$ ($i, j = 1, 2, \ldots, n$) are the **intermediate demands** for commodity i by industry j. If we add the final consumer demand for good i (f_i) to all intermediate industrial demands for good i, then we may write the total amount demanded (and produced) of good i as

$$f_i + a_{i1}x_1 + a_{i2}x_2 + \cdots + a_{in}x_n = x_i, \quad i = 1, 2, \ldots, n \tag{1.22}$$

For all goods simultaneously, we may make use of matrices and write

$$\mathbf{f} + \mathbf{A}\mathbf{x} = \mathbf{x} \tag{1.23}$$

where

$$\mathbf{f} = [f_i]_{n \times 1}, \quad \mathbf{A} = [a_{ij}]_{n \times n} \quad \text{and} \quad \mathbf{x} = [x_i]_{n \times 1} \tag{1.24}$$

using our more economical notation for a matrix. As with the previous examples of matrix expressions, before we can properly interpret this equation and proceed to its analysis and solution, we require a knowledge of the matrix operations used and of the rules of matrix algebra.

1.3 Matrix operations

In describing matrix operations, we simply state the basic definitions, making use of the typical term.

1.3.1 Matrix equality

DEFINITION 1.3.1 We have $\mathbf{A} = \mathbf{B}$ if and only if $\mathbf{A}$ and $\mathbf{B}$ are of the same order and $a_{ij} = b_{ij}$ for all i, j.

The meaning of "if and only if" (often written in shorthand form as "iff") in this definition is explained in the preliminary section on notation. We refer to the requirement that the matrices be of the same order as **conformability in equality**. The reader should have no difficulty in writing down two matrices to illustrate this straightforward concept.

1.3.2 Matrix addition

DEFINITION 1.3.2 We have $\mathbf{A} + \mathbf{B} = \mathbf{C}$ if and only if $\mathbf{A}$, $\mathbf{B}$ and $\mathbf{C}$ are of the same order and $c_{ij} = a_{ij} + b_{ij}$ for all i, j.

Note that the order requirement for **conformability in addition** is the same as that for conformability in equality.

EXAMPLE 1.3.1 If

$$\mathbf{A} = \begin{bmatrix} 1 & 2 \\ 3 & 4 \end{bmatrix} \quad \text{and} \quad \mathbf{B} = \begin{bmatrix} 2 & -2 \\ -4 & 4 \end{bmatrix} \tag{1.25}$$

then

$$\mathbf{C} = \mathbf{A} + \mathbf{B} = \begin{bmatrix} 3 & 0 \\ -1 & 8 \end{bmatrix} \tag{1.26}$$

This example makes clear that each element in $\mathbf{C}$ is just the sum of the corresponding elements in $\mathbf{A}$ and $\mathbf{B}$. Thus $c_{11} = a_{11} + b_{11} = 1 + 2 = 3$, $c_{12} = a_{12} + b_{12} = 2 + (-2) = 2 - 2 = 0$, and similarly for the other two elements of $\mathbf{C}$. $\qquad \diamond$

1.3.3 Multiplication of a matrix by a scalar

DEFINITION 1.3.3 We have $\mu\mathbf{A} = [\mu a_{ij}] = [a_{ij}\mu] = \mathbf{A}\mu$.

The concise matrix notation here indicates that multiplication of a matrix by a scalar, μ, involves multiplying every element of the matrix by that scalar. Note the use of the law of commutativity of multiplication of scalars in moving from the second to the third term in this definition.

When the scalar μ is of the form $1/k$, we occasionally employ the notation

$$\mathbf{A}/k \equiv \frac{\mathbf{A}}{k} \equiv \frac{1}{k}\mathbf{A} \tag{1.27}$$

EXAMPLE 1.3.2 Using the matrix **A** defined in Example 1.3.2, and setting $\mu = 5$,

$$\mu \mathbf{A} = 5 \begin{bmatrix} 1 & 2 \\ 3 & 4 \end{bmatrix} = \begin{bmatrix} 5 \times 1 & 5 \times 2 \\ 5 \times 3 & 5 \times 4 \end{bmatrix} = \begin{bmatrix} 5 & 10 \\ 15 & 20 \end{bmatrix} \tag{1.28}$$

and

$$\mathbf{A}\mu = \begin{bmatrix} 1 & 2 \\ 3 & 4 \end{bmatrix} 5 = \begin{bmatrix} 1 \times 5 & 2 \times 5 \\ 3 \times 5 & 4 \times 5 \end{bmatrix} = \begin{bmatrix} 5 & 10 \\ 15 & 20 \end{bmatrix} \tag{1.29}$$

$\diamond$

1.3.4 Matrix subtraction

DEFINITION 1.3.4 We have $\mathbf{A} - \mathbf{B} = \mathbf{D}$ if and only if **A**, **B** and **D** are conformable and $d_{ij} = a_{ij} - b_{ij}$ for all i, j.

This definition follows from the preceding definitions, using $\mu = -1$ in the expression $\mathbf{A} + \mu \mathbf{B}$. The conformability requirement is, again, that the matrices are of the same order.

EXAMPLE 1.3.3 Using matrices **A** and **B** from Example 1.3.2,

$$\mathbf{D} = \mathbf{A} - \mathbf{B} = \begin{bmatrix} -1 & 4 \\ 7 & 0 \end{bmatrix} \tag{1.30}$$

while

$$\mathbf{D}^* = \mathbf{B} - \mathbf{A} = \begin{bmatrix} 1 & -4 \\ -7 & 0 \end{bmatrix} \tag{1.31}$$

In this example, each element of **D** is just the difference between corresponding elements in **A** and **B**. Thus $d_{11} = a_{11} - b_{11} = 1 - 2 = -1$, and $d_{12} = a_{12} - b_{12} = 2 - (-2) = 2 + 2 = 4$, etc., and similarly for the elements of **D***. Note, too, that $\mathbf{D}^* = (-1)\mathbf{D}$ or $\mathbf{A} - \mathbf{B} = (-1)(\mathbf{B} - \mathbf{A})$. $\diamond$

1.3.5 Matrix multiplication

DEFINITION 1.3.5 We have $\mathbf{AB} = \mathbf{E}$ if and only if **A** has the same number of columns as **B** has rows (say, **A** is of order $m \times r$ and **B** is of order $r \times n$) and $e_{ij} = \sum_{k=1}^{r} a_{ik} b_{kj}$ for all i, j.

The equation in this definition simply says that the element in a given row and column of the matrix product, **E**, is determined as the sum of the products of successive elements in the corresponding row of the first matrix, **A**, and the corresponding column of the second matrix, **B**. It also implies that the order of the product is $m \times n$.

EXAMPLE 1.3.4 If

$$\mathbf{A} = \begin{bmatrix} 2 & 1 & 0 \\ 1 & 3 & 1 \end{bmatrix} \quad \text{and} \quad \mathbf{B} = \begin{bmatrix} 2 \\ 3 \\ 4 \end{bmatrix} \tag{1.32}$$

then

$$\mathbf{AB} = \mathbf{E} = \begin{bmatrix} 7 \\ 15 \end{bmatrix} \tag{1.33}$$

where the calculations are, for the element in the first row and first column of $\mathbf{AB} = \mathbf{E}$,

$$e_{11} = \sum_{k=1}^{3} a_{1k}b_{k1} = a_{11}b_{11} + a_{12}b_{21} + a_{13}b_{31} = (2 \times 2) + (1 \times 3) + (0 \times 4) = 7 \quad (1.34)$$

and for the element in the second row and first column of the product,

$$e_{21} = \sum_{k=1}^{3} a_{2k}b_{k1} = a_{21}b_{11} + a_{22}b_{21} + a_{23}b_{31} = (1 \times 2) + (3 \times 3) + (1 \times 4) = 15 \quad (1.35)$$

$$\diamond$$

Note carefully the need for **conformability in multiplication** in the sense that the number of columns in $\mathbf{A}$ and the number of rows in $\mathbf{B}$ must be the same for the matrix product to be defined. It follows that, in general, $\mathbf{AB} \neq \mathbf{BA}$. Indeed, the product $\mathbf{BA}$ may not be defined, even if $\mathbf{AB}$ is, as in the case of the simple matrices in Example 1.3.5. It is therefore useful to distinguish between **pre-** and **post-multiplication** of one matrix by another when referring to the formation of matrix products. In the numerical example, $\mathbf{B}$ is pre-multiplied by $\mathbf{A}$ (or, equivalently, $\mathbf{A}$ is post-multiplied by $\mathbf{B}$) to form the product $\mathbf{E}$.

Finally, it may be noted that now we have clear definitions of these basic matrix operations, it becomes easy to see how the matrix sums, products and equalities that appear in our three examples in Sections 1.2.1, 1.2.2 and 1.2.3 reproduce the original scalar versions of the equations in the systems considered. For instance, substituting for $\mathbf{A}$, $\mathbf{x}$ and $\mathbf{b}$ from (1.20) into $\mathbf{Ax} = \mathbf{b}$, and using matrix multiplication and matrix equality, will generate the full set of equations in a general linear simultaneous equation system. The following example uses a three-equation system as an illustration.

EXAMPLE 1.3.5 Let

$$\mathbf{A} = \begin{bmatrix} a_{11} & a_{12} & a_{13} \\ a_{21} & a_{22} & a_{23} \\ a_{31} & a_{32} & a_{33} \end{bmatrix}, \quad \mathbf{x} = \begin{bmatrix} x_1 \\ x_2 \\ x_3 \end{bmatrix} \quad \text{and} \quad \mathbf{b} = \begin{bmatrix} b_1 \\ b_2 \\ b_3 \end{bmatrix} \quad (1.36)$$

Then, substituting in

$$\mathbf{Ax} = \mathbf{b} \quad (1.37)$$

we have

$$\begin{bmatrix} a_{11} & a_{12} & a_{13} \\ a_{21} & a_{22} & a_{23} \\ a_{31} & a_{32} & a_{33} \end{bmatrix} \begin{bmatrix} x_1 \\ x_2 \\ x_3 \end{bmatrix} = \begin{bmatrix} b_1 \\ b_2 \\ b_3 \end{bmatrix} \quad (1.38)$$

We may refer to this form of the system as being in matrix notation, from which, using the operation of matrix multiplication, we get the vector notation

$$\begin{bmatrix} a_{11}x_1 + a_{12}x_2 + a_{13}x_3 \\ a_{21}x_1 + a_{22}x_2 + a_{23}x_3 \\ a_{31}x_1 + a_{32}x_2 + a_{33}x_3 \end{bmatrix} = \begin{bmatrix} b_1 \\ b_2 \\ b_3 \end{bmatrix} \quad (1.39)$$

and thence the scalar representation of the linear system of equations

$$a_{11}x_1 + a_{12}x_2 + a_{13}x_3 = b_1 \qquad (1.40)$$

$$a_{21}x_1 + a_{22}x_2 + a_{23}x_3 = b_2 \qquad (1.41)$$

$$a_{31}x_1 + a_{32}x_2 + a_{33}x_3 = b_3 \qquad (1.42)$$

$$\diamond$$

We thus motivate the initial need for matrix algebra as a shorthand for writing systems of linear equations, such as those considered above; see also Chiang and Wainwright (2005, Section 4.2). The matter of the solution of such systems is deferred until Chapter 2, following discussion of the rules of matrix algebra, some special types of matrix and the concept of determinant.

1.4   Rules of matrix algebra

There are only a few rules for the algebraic manipulation of matrices using the operations just defined. Some of these are similar to the familiar rules of scalar algebra but some are different. It is important to note that, unlike the matrix operations, which are simply definitions, these various rules are theorems, i.e. they are results that follow logically from the previous definitions and the rules of scalar algebra, which we take for granted. Therefore, all of the rules to be stated may be proved. It is not intended to prove them all here, but one or two proofs will be instructive as they illustrate how some matrix theorems may be proved simply by reference to appropriate typical terms and appeal to the relevant definitions.

1.4.1   Rules of matrix addition

Matrix addition is both **commutative** and **associative**, i.e.

$$\mathbf{A} + \mathbf{B} = \mathbf{B} + \mathbf{A} \qquad (1.43)$$

and

$$(\mathbf{A} + \mathbf{B}) + \mathbf{C} = \mathbf{A} + (\mathbf{B} + \mathbf{C}) \qquad (1.44)$$

assuming that the matrices are conformable in addition. The commutativity rule, for example, may be stated and proved formally as follows.

THEOREM 1.4.1 *Let* $\mathbf{A}$ *and* $\mathbf{B}$ *be matrices of order* $m \times n$, *then* $\mathbf{A} + \mathbf{B} = \mathbf{B} + \mathbf{A}$.

Proof: $\mathbf{A} + \mathbf{B} = [a_{ij}] + [b_{ij}] = [a_{ij} + b_{ij}]$, by definition.
Now $[a_{ij} + b_{ij}] = [b_{ij} + a_{ij}]$, using the commutative rule of scalar algebra.
Therefore $\mathbf{A} + \mathbf{B} = [b_{ij} + a_{ij}] = [b_{ij}] + [a_{ij}] = \mathbf{B} + \mathbf{A}$, again by appeal to the definition of matrix addition. $\qquad \square$

Proof of the associativity rule of matrix addition is left as an exercise; see Exercise 1.11.

1.4.2 Rule of matrix multiplication

Matrix multiplication is associative, i.e. $(\mathbf{AB})\mathbf{C} = \mathbf{A}(\mathbf{BC})$. However, matrix multiplication is not, in general, commutative; recall Definition 1.3.5, Example 1.3.5 and the subsequent comments on matrix multiplication. The proof of the associativity rule of matrix multiplication uses a typical term involving double summation; see Exercise 1.11 again.

We note, though it is fairly trivial, that $\mu(\theta\mathbf{A}) = (\mu\theta)\mathbf{A} = \theta(\mu\mathbf{A})$, where μ and θ are scalars, because scalars are commutative in multiplication.

1.4.3 Distributive law of matrix multiplication over matrix addition

Matrix multiplication is **distributive** over matrix addition, i.e. $\mathbf{A}(\mathbf{B} \pm \mathbf{C}) = \mathbf{AB} \pm \mathbf{AC}$ and $(\mathbf{A} \pm \mathbf{B})\mathbf{D} = \mathbf{AD} \pm \mathbf{BD}$, provided that the matrices involved are appropriately conformable. It is left as an exercise to determine the dimensions of the matrices in these two equations; recall the definitions of the relevant matrix operations, and see Exercise 1.12.

We note that $\mu(\mathbf{B} \pm \mathbf{C}) = \mu\mathbf{B} \pm \mu\mathbf{C}$ and $(\mu \pm \theta)\mathbf{A} = \mu\mathbf{A} \pm \theta\mathbf{A} = \mathbf{A}(\mu \pm \theta)$. An instance of this last rule may be formalized as follows.

THEOREM 1.4.2 *Let $\mathbf{A}$ be a matrix of order $m \times n$, and μ and θ be scalars, then $(\mu - \theta)\mathbf{A} = \mu\mathbf{A} - \theta\mathbf{A}$.*

Proof: $(\mu - \theta)\mathbf{A} = [(\mu - \theta)a_{ij}]$, by definition.

Now $[(\mu - \theta)a_{ij}] = [\mu a_{ij} - \theta a_{ij}]$, using the distributive rule of scalar algebra, and then $[\mu a_{ij} - \theta a_{ij}] = [\mu a_{ij} + (-\theta)a_{ij}] = [\mu a_{ij}] + [(-\theta)a_{ij}] = \mu\mathbf{A} + (-\theta)\mathbf{A} = \mu\mathbf{A} - \theta\mathbf{A}$, using scalar algebra and appropriate matrix definitions. □

All the rules that have been stated may be proved in similar fashion to those that have been presented, and the proofs are left as exercises.

1.5 Some special types of matrix and associated rules

In our work with matrices, and in particular in our various economic and financial applications later, we shall encounter a number of special kinds of matrix. The following subsections contain definitions for the main special matrices that we will use, and state a number of properties (theorems) associated with them. Again, most proofs will be left as exercises, but a few proofs will be provided where they give additional insight into the methods of deriving useful matrix results.

1.5.1 Zero matrix

NOTATION 1.5.1 *The zero matrix is $\mathbf{0} = [a_{ij}]$, where $a_{ij} = 0$ for all i, j.*

All of the elements of a **zero matrix** are scalar zeros; it is unnecessary to write out such a matrix in full. Note that the order of a zero matrix is arbitrary and will depend on the precise context. In cases of ambiguity, it is desirable to indicate the order of a zero matrix by a subscript, such as $\mathbf{0}_{m \times n}$, $\mathbf{0}_{1 \times n}$ or $\mathbf{0}_{m \times 1}$. Where the matrices are row and column vectors of zeros, a single subscript may be used, as long as there is no possibility of confusion. The zero matrix is sometimes referred to as the additive identity matrix, for the reason stated in the first of the following properties.

PROPERTY 1 *We have $\mathbf{A} \pm \mathbf{0} = \mathbf{A}$, where $\mathbf{0}$ and $\mathbf{A}$ are conformable in addition.*

PROPERTY 2 *We have* $0\mathbf{A} = \mathbf{0}$ *for all* $\mathbf{A}$, *where the zero on the left-hand side is a scalar and the zero matrix on the right-hand side is the same dimension as* $\mathbf{A}$.

PROPERTY 3 *We have* $\mathbf{A} - \mathbf{A} = \mathbf{0}$, *where* $\mathbf{0}$ *is of the same dimension as* $\mathbf{A}$.

PROPERTY 4 *If* $\mathbf{A}$ *is* $m \times n$, *then* $\mathbf{0}_{p \times m} \mathbf{A} = \mathbf{0}_{p \times n}$ *and* $\mathbf{A} \mathbf{0}_{n \times r} = \mathbf{0}_{m \times r}$.

It should be noted that the equation $\mathbf{AB} = \mathbf{0}$ does not necessarily imply that either $\mathbf{A}$ or $\mathbf{B}$ is a zero matrix, as a similar expression in scalar algebra would suggest.

EXAMPLE 1.5.1 Let

$$\mathbf{A} = \begin{bmatrix} 1 & 1 \end{bmatrix} \quad \text{and} \quad \mathbf{B} = \begin{bmatrix} 1 & -1 \\ -1 & 1 \end{bmatrix} \tag{1.45}$$

Then

$$\mathbf{AB} = \begin{bmatrix} 1 & 1 \end{bmatrix} \begin{bmatrix} 1 & -1 \\ -1 & 1 \end{bmatrix} = \begin{bmatrix} 0 & 0 \end{bmatrix} = \mathbf{0}_{1 \times 2} \tag{1.46}$$

The product $\mathbf{BA}$ is undefined, as the two matrices are not conformable in multiplication in this order. ◇

This example illustrates the danger of transferring results from scalar algebra to matrix algebra.

The proofs of these few properties of the zero matrix follow simply from the definitions and basic matrix operations, and are left as exercises; see Exercise 1.13.

1.5.2   Identity matrix

The **identity matrix** of dimension n is the $n \times n$ square matrix that contains scalar ones on the diagonal from top left to bottom right and zeros elsewhere, and is denoted $\mathbf{I}_n = [\delta_{ij}]_{n \times n}$, where $\delta_{ij} = 1$ for $i = j$ and $\delta_{ij} = 0$ for $i \neq j$.

The **Kronecker delta**[5] is the name given to δ_{ij} as used in the notation for the identity matrix. This symbol will be used from time to time below.

PROPERTY 1 *We have* $\mathbf{AI}_n = \mathbf{I}_m \mathbf{A} = \mathbf{A}$, *where* $\mathbf{A}$ *is any* $m \times n$ *matrix*.

Thus the identity matrix serves in matrix algebra much as unity does in scalar algebra, and is also referred to as the multiplicative identity, but note carefully the need for conformability in multiplication when the identity matrix is used.

A simple appeal to the definitions of the identity matrix and matrix multiplication will allow the reader to establish the above property.

1.5.3   Trace of a matrix

The diagonal on which the ones lie in the identity matrix, and the corresponding diagonal in any square matrix, is known as the **principal diagonal** or **main diagonal** or **leading diagonal** or **primary diagonal**.

The sum of the elements on the principal diagonal is called the **trace** of the matrix and denoted tr($\cdot$). The trace of the identity matrix of order n is tr($\mathbf{I}_n$) $= \sum_{i=1}^{n} 1 = n$.

PROPERTY 1 *We have* tr($\mathbf{AB}$) $=$ tr($\mathbf{BA}$), *where* $\mathbf{A}$ *is an* $m \times n$ *matrix and* $\mathbf{B}$ *is an* $n \times m$ *matrix.*

The proof of this result constitutes a useful exercise in the manipulation of typical terms, but note that it will involve the use of double summation and hence may be a little tricky; see Exercise 1.14.

1.5.4 Inverse matrix

If $\mathbf{A}$, $\mathbf{B}$ and $\mathbf{C}$ are square matrices of the same order, say, $n \times n$, and $\mathbf{AC} = \mathbf{I} = \mathbf{BA}$, then $\mathbf{C}$ is called a **right inverse** of $\mathbf{A}$ and $\mathbf{B}$ is called a **left inverse** of $\mathbf{A}$.

In fact, $\mathbf{B} = \mathbf{BI} = \mathbf{B(AC)} = \mathbf{(BA)C} = \mathbf{IC} = \mathbf{C}$, so the right inverse must equal the left inverse, and they can be called simply the **inverse** of $\mathbf{A}$ (and, vice versa, $\mathbf{A}$ is called the inverse of $\mathbf{B}$). Matrices $\mathbf{A}$ and $\mathbf{B}$ are said to be **invertible**.

EXAMPLE 1.5.2 Let

$$\mathbf{A} = \begin{bmatrix} 1 & 2 \\ 2 & 3 \end{bmatrix} \quad \text{and} \quad \mathbf{B} = \begin{bmatrix} -3 & 2 \\ 2 & -1 \end{bmatrix} \tag{1.47}$$

Using the operation of matrix multiplication, we have that

$$\mathbf{AB} = \begin{bmatrix} 1 & 2 \\ 2 & 3 \end{bmatrix} \begin{bmatrix} -3 & 2 \\ 2 & -1 \end{bmatrix} = \begin{bmatrix} 1 & 0 \\ 0 & 1 \end{bmatrix} = \mathbf{I}_2 \tag{1.48}$$

and

$$\mathbf{BA} = \begin{bmatrix} -3 & 2 \\ 2 & -1 \end{bmatrix} \begin{bmatrix} 1 & 2 \\ 2 & 3 \end{bmatrix} = \begin{bmatrix} 1 & 0 \\ 0 & 1 \end{bmatrix} = \mathbf{I}_2 \tag{1.49}$$

By definition, then, $\mathbf{A}$ and $\mathbf{B}$ are both invertible: $\mathbf{A}$ is the inverse of $\mathbf{B}$, and $\mathbf{B}$ is the inverse of $\mathbf{A}$. $\diamond$

Not all square matrices will have inverses associated with them, but we will leave the question of what property of any particular square matrix guarantees the existence of an inverse until Chapter 2.

The concept of the inverse of a matrix, which, when it exists, serves in matrix algebra much like the reciprocal of a number does in scalar algebra, is very important, and there are several useful properties relating to it. We will state the main properties of matrix inverses rather formally as theorems and consider their proofs; it is instructive to do so and provides useful illustrations of matrix manipulation using the matrix operations, the rules of matrix algebra and the identity matrix and its property presented above.

The first property is established using a **proof by contradiction**. This form of proof establishes the truth of a proposition by showing that the falsity of the proposition implies a contradiction. Since a proposition must be either true or false, and its falsity is shown to be impossible by dint of the contradiction, it follows that the proposition must be true. In other words, to prove by contradiction that P, it is shown that "not P" implies Q *and* "not Q". Then, since "not P" implies a contradiction, conclude P. This method of proof will be used in other contexts later.

PROPERTY 1 *Let* **A** *be a square invertible matrix; then the inverse of* **A** *is unique.*

Proof: Let **B** be the inverse of **A**, and suppose that another inverse, $\mathbf{B}^* \neq \mathbf{B}$, exists.
Then $\mathbf{AB}^* = \mathbf{I}$, and $\mathbf{BAB}^* = \mathbf{BI} = \mathbf{B}$, pre-multiplying by the inverse **B**.
But $\mathbf{BA} = \mathbf{I}$, since **B** is the inverse of **A**, so $\mathbf{BAB}^* = \mathbf{IB}^* = \mathbf{B}^*$.
Therefore $\mathbf{B}^* = \mathbf{B}$, which is a contradiction.   □

Because of this uniqueness result, a special symbol, involving superscript -1, has been adopted to signify a matrix inverse. Thus we write $\mathbf{A}^{-1}$ to denote the unique inverse of the invertible matrix **A**. The approach of assuming non-uniqueness to actually establish uniqueness is common and will be used again later.

PROPERTY 2 *Let* **A** *be a square invertible matrix; then* $(\mathbf{A}^{-1})^{-1} = \mathbf{A}$.

Proof: If **A** is invertible, hence $\mathbf{AA}^{-1} = \mathbf{I}$, then $\mathbf{A} = (\mathbf{A}^{-1})^{-1}$ by direct appeal to our definition of a matrix inverse. However, consider the following alternative.
We have $(\mathbf{A}^{-1})^{-1}\mathbf{A}^{-1} = \mathbf{I}$, by definition.
Therefore $(\mathbf{A}^{-1})^{-1}\mathbf{A}^{-1}\mathbf{A} = \mathbf{IA}$, post-multiplying by **A**.
Therefore $(\mathbf{A}^{-1})^{-1}\mathbf{I} = \mathbf{IA}$ and so $(\mathbf{A}^{-1})^{-1} = \mathbf{A}$.   □

PROPERTY 3 *Let* **A** *and* **B** *be invertible matrices of the same order; then* $(\mathbf{AB})^{-1} = \mathbf{B}^{-1}\mathbf{A}^{-1}$.

Proof: We have $\mathbf{AB}(\mathbf{AB})^{-1} = \mathbf{I}$, by definition.
Therefore $\mathbf{A}^{-1}\mathbf{AB}(\mathbf{AB})^{-1} = \mathbf{A}^{-1}\mathbf{I}$, pre-multiplying by $\mathbf{A}^{-1}$, and so $\mathbf{B}(\mathbf{AB})^{-1} = \mathbf{A}^{-1}$.
Therefore $\mathbf{B}^{-1}\mathbf{B}(\mathbf{AB})^{-1} = \mathbf{B}^{-1}\mathbf{A}^{-1}$, pre-multiplying by $\mathbf{B}^{-1}$, and $\mathbf{I}(\mathbf{AB})^{-1} = \mathbf{B}^{-1}\mathbf{A}^{-1}$, which gives the result: $(\mathbf{AB})^{-1} = \mathbf{B}^{-1}\mathbf{A}^{-1}$.   □

This last theorem on the inverse of a matrix product is an important result that we shall make a good deal of use of later. It is also used in the proof of the following theorem, which is left as an exercise; see Exercise 1.16.

PROPERTY 4 *Let* **A** *be a square invertible matrix, and define*

$$\mathbf{A}^r = \underbrace{\mathbf{AA}\ldots\mathbf{A}}_{r \text{ times}}, \ \mathbf{A}^0 = \mathbf{I}, \ \mathbf{A}^p\mathbf{A}^q = \mathbf{A}^{p+q} \ and \ \mathbf{A}^{-r} = (\mathbf{A}^{-1})^r = \underbrace{\mathbf{A}^{-1}\mathbf{A}^{-1}\ldots\mathbf{A}^{-1}}_{r \text{ times}}$$

Then $\mathbf{A}^r$ *is invertible and* $(\mathbf{A}^r)^{-1} = (\mathbf{A}^{-1})^r = \mathbf{A}^{-r}$.

Finally, again without proof, which involves a straightforward appeal to the original definition of a matrix inverse (see Exercise 1.17), we get the following property.

PROPERTY 5 *Let* **A** *be a square invertible matrix and* μ *be a scalar; then* $\mu\mathbf{A}$ *is invertible and*

$$(\mu\mathbf{A})^{-1} = \frac{1}{\mu}\mathbf{A}^{-1} = \frac{\mathbf{A}^{-1}}{\mu}$$

1.5.5   Scalar matrix

The paradoxical sounding term, **scalar matrix**, denotes a square matrix of the form $\mathbf{S} = \mu \mathbf{I}_n = [\mu \delta_{ij}]$, i.e. a matrix whose principal diagonal elements are the non-zero number μ and whose remaining elements are all zero. This is sometimes denoted $\text{diag}[\mu]$ (provided that the dimension is clear from the context). The identity matrix is the scalar matrix that results when $\mu = 1$. Illustration of this concept is very easy and is left to the reader.

PROPERTY 1 *We have* $\text{tr}(\mathbf{S}) = n\mu.$

PROPERTY 2 *We have*

$$\mathbf{S}^{-1} = \frac{1}{\mu} \mathbf{I}_n = \left[\frac{1}{\mu} \delta_{ij} \right]$$

It is easy to establish these properties, so, again, the proofs are left as exercises. Property 2 for scalar matrices, which may be obtained by direct consideration of the product $\mathbf{S}^{-1}\mathbf{S}$, is a special case of Property 5 for matrix inverses; see Exercise 1.17.

1.5.6   Diagonal matrix

Just as the scalar matrix may be viewed as a simple generalization of the identity matrix, so the **diagonal matrix** may be viewed as a further generalization of the scalar matrix. A diagonal matrix is a square matrix of the form $\mathbf{D} = [d_i \delta_{ij}]_{n \times n}$, which is sometimes also written as $\text{diag}[d_i]$, where the subscript i now indicates that the diagonal elements are not necessarily all equal. Thus all of the elements of a diagonal matrix are zero except those on the principal diagonal, which are arbitrary scalars. Once again, such a matrix is simple to visualize.

PROPERTY 1 *We have* $\text{tr}(\mathbf{D}) = \sum_{i=1}^{n} d_i.$

PROPERTY 2 *We have*

$$\mathbf{D}^{-1} = \left[\frac{1}{d_i} \delta_{ij} \right] = \text{diag} \left[\frac{1}{d_i} \right]$$

provided that $d_i \neq 0$ *for all* i.

The proofs of these properties are easy. For the latter one, it will suffice to note that the operation of matrix multiplication gives that

$$\mathbf{D}\mathbf{D}^{-1} = [d_i \delta_{ij}] \left[\frac{1}{d_i} \delta_{ij} \right] = [\delta_{ij}]_{n \times n} = \mathbf{I}_n \tag{1.50}$$

Therefore, by the definition of an inverse and its uniqueness, the result is established.

1.5.7   Transpose of a matrix

If $\mathbf{A}$ is a matrix of order $m \times n$, then the **transpose** of $\mathbf{A}$, denoted $\mathbf{A}^\top$, is the $n \times m$ matrix formed by interchanging the rows and columns of $\mathbf{A}$. We may write this definition symbolically as

$$\mathbf{A}^\top = [a_{ij}^\top]_{n \times m} = [a_{ji}]_{n \times m} \tag{1.51}$$

where $a_{ij}^\top$ is the typical element of the transpose. The transpose of a matrix is a much-used concept and we note five useful properties. Three of these will be proved; the others are left as exercises.

PROPERTY 1 *We have* $(\mathbf{A}^\top)^\top = \mathbf{A}$, *which follows simply from the definition of transposition.*

PROPERTY 2 *Let* $\mathbf{A}$ *and* $\mathbf{B}$ *be* $m \times n$ *matrices; then* $(\mathbf{A} \pm \mathbf{B})^\top = \mathbf{A}^\top \pm \mathbf{B}^\top$, *i.e. the transpose of a matrix sum or difference is the sum or difference of the individual matrix transposes.*

Proof: We have $(\mathbf{A} \pm \mathbf{B})^\top = [a_{ij} \pm b_{ij}]^\top = [c_{ij}]^\top$, say.
Now $[c_{ij}]^\top = [c_{ij}^\top] = [c_{ji}]$ by the definition of a transpose, and $[c_{ji}] = [a_{ji} \pm b_{ji}] = [a_{ji}] \pm [b_{ji}] = [a_{ij}^\top] \pm [b_{ij}^\top] = \mathbf{A}^\top \pm \mathbf{B}^\top$ by definition of matrix addition and subtraction. $\square$

PROPERTY 3 *Let* $\mathbf{A}$ *be an* $m \times n$ *matrix and* $\mathbf{B}$ *be an* $n \times m$ *matrix; then* $(\mathbf{AB})^\top = \mathbf{B}^\top \mathbf{A}^\top$, *i.e. the transpose of a matrix product is the product of the individual matrix transposes, but in reverse order. Note that the orders of* $\mathbf{A}$ *and* $\mathbf{B}$ *ensure conformability in the case of both of the products.*

Proof:

$$(\mathbf{AB})^\top = \left[\sum_{k=1}^{n} a_{ik} b_{kj} \right]^\top , \text{ using the typical term of the product}$$

$$= \left[\sum_{k=1}^{n} a_{jk} b_{ki} \right] , \text{ by definition of transposition}$$

$$= \left[\sum_{k=1}^{n} b_{ki} a_{jk} \right] , \text{ by the commutative law of multiplication of scalars}$$

$$= \left[\sum_{k=1}^{n} b_{ik}^\top a_{kj}^\top \right] , \text{ by definition of transposition}$$

$$= \mathbf{B}^\top \mathbf{A}^\top , \text{ by the definition of matrix multiplication} \tag{1.52}$$

$\square$

PROPERTY 4 *We have* $\text{tr}(\mathbf{A}^\top) = \text{tr}(\mathbf{A})$, *where* $\mathbf{A}$ *is a square matrix.*

PROPERTY 5 *If* $\mathbf{A}$ *is a square invertible matrix; then* $(\mathbf{A}^\top)^{-1} = (\mathbf{A}^{-1})^\top$.

Proof: We have $\mathbf{A}\mathbf{A}^{-1} = \mathbf{I}$ by definition.

Therefore $(\mathbf{A}^{-1})^\top \mathbf{A}^\top = \mathbf{I}^\top = \mathbf{I}$, transposing both sides and using the result on the transpose of a product.

Therefore $(\mathbf{A}^{-1})^\top \mathbf{A}^\top (\mathbf{A}^\top)^{-1} = \mathbf{I}(\mathbf{A}^\top)^{-1}$, post-multiplying by $(\mathbf{A}^\top)^{-1}$.

Therefore $(\mathbf{A}^{-1})^\top \mathbf{I} = (\mathbf{A}^\top)^{-1}$.

Therefore $(\mathbf{A}^{-1})^\top = (\mathbf{A}^\top)^{-1}$.   □

1.5.8   Symmetric matrix

A square matrix, $\mathbf{A}$, is **symmetric** if and only if $\mathbf{A}^\top = \mathbf{A}$, so that $a_{ij} = a_{ji}$ for all i, j. Symmetric matrices will feature importantly in later sections; they are easy to identify and exemplify. There are several useful properties associated with symmetric matrices, but we shall develop these later when more material has been covered. A useful result on an invertible symmetric matrix is given in the following lemma.

LEMMA 1.5.1 *Let $\mathbf{A}$ be symmetric and invertible; then $(\mathbf{A}^{-1})^\top = \mathbf{A}^{-1}$.*

Proof: We have $\mathbf{A}\mathbf{A}^{-1} = \mathbf{I}$ by definition.

Therefore $(\mathbf{A}\mathbf{A}^{-1})^\top = \mathbf{I}^\top = \mathbf{I}$, and $(\mathbf{A}\mathbf{A}^{-1})^\top = (\mathbf{A}^{-1})^\top \mathbf{A}^\top = (\mathbf{A}^{-1})^\top \mathbf{A}$, using the rule for transposition of matrix products and the fact that $\mathbf{A}$ is symmetric. Multiplying on the right by $\mathbf{A}^{-1}$ yields

$$(\mathbf{A}^{-1})^\top \mathbf{A}\mathbf{A}^{-1} = \mathbf{I}\mathbf{A}^{-1}.$$

Hence, $(\mathbf{A}^{-1})^\top = \mathbf{A}^{-1}$ and the inverse of a symmetric matrix is also symmetric.   □

So, if a matrix is symmetric and invertible, its inverse is also a symmetric matrix; see again Example 1.5.4, which concerns such a matrix.

1.5.9   Orthogonal matrix

A square matrix, $\mathbf{A}$, is **orthogonal** if and only if $\mathbf{A}^\top = \mathbf{A}^{-1}$. We shall meet orthogonal matrices in Chapters 3 and 6. They play an important role in certain theoretical results and in some applications, as we shall see later.

EXAMPLE 1.5.3 The (symmetric) matrix

$$\mathbf{A} = \begin{bmatrix} \frac{2}{\sqrt{5}} & \frac{1}{\sqrt{5}} \\ \frac{1}{\sqrt{5}} & \frac{-2}{\sqrt{5}} \end{bmatrix} \tag{1.53}$$

is orthogonal, since

$$\mathbf{A}\mathbf{A}^\top = \mathbf{A}^\top \mathbf{A} = \begin{bmatrix} \frac{2}{\sqrt{5}} & \frac{1}{\sqrt{5}} \\ \frac{1}{\sqrt{5}} & \frac{-2}{\sqrt{5}} \end{bmatrix} \begin{bmatrix} \frac{2}{\sqrt{5}} & \frac{1}{\sqrt{5}} \\ \frac{1}{\sqrt{5}} & \frac{-2}{\sqrt{5}} \end{bmatrix} = \begin{bmatrix} 1 & 0 \\ 0 & 1 \end{bmatrix} = \mathbf{I}_2 \tag{1.54}$$

implying that $\mathbf{A}^\top = \mathbf{A}^{-1}$.   ◇

The fact that $\mathbf{A}$ is symmetric in this example means that $\mathbf{A}^2 = \mathbf{I}$; hence, the matrix is its own inverse.

1.5.10   Idempotent matrix

A square matrix, $\mathbf{A}$, is **idempotent** if and only if $\mathbf{A}^2 = \mathbf{A}$. This implies that $\mathbf{A}^r = \mathbf{A}$, where r is any positive integer. It may be noted that the identity matrix is idempotent. Like symmetric matrices, idempotent matrices are important in certain applications and they have some important properties. Indeed, we shall encounter matrices that are both symmetric and idempotent. However, as with the symmetric matrix, we defer consideration of the properties of idempotent matrices until later.

EXAMPLE 1.5.4 It is easily verified that the following 2×2 matrices are all idempotent:

$$\begin{bmatrix} 1 & 0 \\ 0 & 0 \end{bmatrix}, \quad \begin{bmatrix} 1 & 1 \\ 0 & 0 \end{bmatrix} \quad \text{and} \quad \begin{bmatrix} 1 & 0 \\ 1 & 0 \end{bmatrix} \tag{1.55}$$

$\diamond$

1.5.11   Triangular matrix

Triangular matrices are square matrices whose elements on one side of the principal diagonal are all zero, and whose remaining elements are arbitrary. In the case of all elements above the principal diagonal being zero, we have a **lower triangular matrix**. In the case of all elements below the principal diagonal being zero, we have an **upper triangular matrix**. As will be seen in Corollary 4.4.17, certain square matrices, $\mathbf{A}$, can be factorized or decomposed as the product of a lower and an upper triangular matrix, $\mathbf{L}$ and $\mathbf{U}$, respectively, i.e. we may write $\mathbf{A} = \mathbf{LU}$. This decomposition facilitates the solution of certain types of systems of equations.

1.5.12   Band matrix

A **band matrix** is a square matrix that contains a given non-zero scalar value in every position on the principal diagonal and another (possibly different) *non-zero* scalar in every position on the immediate off-diagonals above and below the principal diagonal, and so on. These bands may or may not extend across the entire matrix. Where they do not, zeros appear in all remaining cells of the matrix. Thus, for example,

$$\mathbf{B} = \begin{bmatrix} b_0 & b_1 & 0 & 0 & 0 \\ b_1 & b_0 & b_1 & 0 & 0 \\ 0 & b_1 & b_0 & b_1 & 0 \\ 0 & 0 & b_1 & b_0 & b_1 \\ 0 & 0 & 0 & b_1 & b_0 \end{bmatrix} \quad \text{and} \quad \mathbf{G} = \begin{bmatrix} \gamma_0 & \gamma_1 & \gamma_2 & \gamma_3 & \gamma_4 \\ \gamma_1 & \gamma_0 & \gamma_1 & \gamma_2 & \gamma_3 \\ \gamma_2 & \gamma_1 & \gamma_0 & \gamma_1 & \gamma_2 \\ \gamma_3 & \gamma_2 & \gamma_1 & \gamma_0 & \gamma_1 \\ \gamma_4 & \gamma_3 & \gamma_2 & \gamma_1 & \gamma_0 \end{bmatrix} \tag{1.56}$$

are two 5×5 band matrices, but a scalar matrix is not a band matrix. The first is also called a **tri-diagonal matrix**, for obvious reasons. Band matrices arise in Section 14.4.1.

Most of the above special types of matrix are square. The exceptions are some zero matrices and some of the matrices whose transposes have been examined. The remaining types are not (necessarily) square.

1.5.13   Vector of ones

NOTATION 1.5.2 *The vector of ones is*

$$\mathbf{1}_n = \begin{bmatrix} 1 \\ 1 \\ \vdots \\ 1 \end{bmatrix}$$

This vector is often useful in representing summation operations in matrix notation; see Exercise 1.20. Note also that $\mathbf{1}_n \mathbf{1}_m^\top$ is an $n \times m$ matrix all of whose elements are unity.

1.5.14 Partitioned matrix

A **partitioned matrix** is a matrix whose rows and/or columns have been delineated so as to form several submatrices. For instance, given the $m \times n$ matrix $\mathbf{A}$, the first p rows may be distinguished from the remaining $m - p$ rows, and the first q columns from the remaining $n - q$ columns. Thus the matrix may be written as

$$\mathbf{A} = \begin{bmatrix} \mathbf{A}_{11} & \mathbf{A}_{12} \\ \mathbf{A}_{21} & \mathbf{A}_{22} \end{bmatrix} \tag{1.57}$$

where $\mathbf{A}_{11}$ is of order $p \times q$, $\mathbf{A}_{12}$ is $p \times (n - q)$, $\mathbf{A}_{21}$ is $(m - p) \times q$ and $\mathbf{A}_{22}$ is $(m - p) \times (n - q)$.

As long as the partitioning is such that the submatrices are appropriately conformable, the operations of equality, addition, subtraction and multiplication of partitioned matrices are precisely as defined previously, treating the submatrices as if they were individual elements. Hence, given matrix $\mathbf{A}$ in (1.57) and conformable

$$\mathbf{B} = \begin{bmatrix} \mathbf{B}_{11} & \mathbf{B}_{12} \\ \mathbf{B}_{21} & \mathbf{B}_{22} \end{bmatrix} \tag{1.58}$$

we have that

$$\mathbf{A} + \mathbf{B} = \begin{bmatrix} \mathbf{A}_{11} + \mathbf{B}_{11} & \mathbf{A}_{12} + \mathbf{B}_{12} \\ \mathbf{A}_{21} + \mathbf{B}_{21} & \mathbf{A}_{22} + \mathbf{B}_{22} \end{bmatrix} \tag{1.59}$$

as long as $\mathbf{B}$ is partitioned in the same way as $\mathbf{A}$.

Now, consider the product of partitioned matrix $\mathbf{A}$, from (1.57), and

$$\mathbf{C} = \begin{bmatrix} \mathbf{C}_{11} & \mathbf{C}_{12} & \mathbf{C}_{13} \\ \mathbf{C}_{21} & \mathbf{C}_{22} & \mathbf{C}_{23} \end{bmatrix} \tag{1.60}$$

Recalling the definition of matrix multiplication, the product may be written as

$$\mathbf{AC} = \begin{bmatrix} \mathbf{A}_{11}\mathbf{C}_{11} + \mathbf{A}_{12}\mathbf{C}_{21} & \mathbf{A}_{11}\mathbf{C}_{12} + \mathbf{A}_{12}\mathbf{C}_{22} & \mathbf{A}_{11}\mathbf{C}_{13} + \mathbf{A}_{12}\mathbf{C}_{23} \\ \mathbf{A}_{21}\mathbf{C}_{11} + \mathbf{A}_{22}\mathbf{C}_{21} & \mathbf{A}_{21}\mathbf{C}_{12} + \mathbf{A}_{22}\mathbf{C}_{22} & \mathbf{A}_{21}\mathbf{C}_{13} + \mathbf{A}_{22}\mathbf{C}_{23} \end{bmatrix} \tag{1.61}$$

as long as $\mathbf{C}$ is conformably partitioned. As an exercise, examine what the dimensions of the individual submatrices of $\mathbf{C}$ must be for this product to be well defined; see Exercise 1.22.

The inverse of a partitioned square matrix is of interest, assuming that it exists. In principle, it is possible to deduce the form of the partitioned inverse using only basic matrix operations and a condition on the submatrices $\mathbf{A}_{11}$ and $\mathbf{A}_{22}$. Let $\mathbf{A}$ be as defined in (1.57) but with $m = n$ and $p = q$, let $\mathbf{A}_{11}$ and $\mathbf{A}_{22}$ be square invertible matrices of dimensions $p \times p$ and $(n - p) \times (n - p)$, respectively, and let

$$\begin{bmatrix} \mathbf{A}_{11} & \mathbf{A}_{12} \\ \mathbf{A}_{21} & \mathbf{A}_{22} \end{bmatrix}^{-1} = \begin{bmatrix} \mathbf{B} & \mathbf{C} \\ \mathbf{D} & \mathbf{E} \end{bmatrix} \tag{1.62}$$

The dimensions of $\mathbf{A}_{12}$ and $\mathbf{A}_{21}$, as well as of all the partitions on the right-hand side of (1.62), are implied by the dimensions of $\mathbf{A}_{11}$ and $\mathbf{A}_{22}$. It follows that

$$\begin{bmatrix} \mathbf{A}_{11} & \mathbf{A}_{12} \\ \mathbf{A}_{21} & \mathbf{A}_{22} \end{bmatrix} \begin{bmatrix} \mathbf{B} & \mathbf{C} \\ \mathbf{D} & \mathbf{E} \end{bmatrix} = \begin{bmatrix} \mathbf{I}_p & \mathbf{0}_{p \times (n-p)} \\ \mathbf{0}_{(n-p) \times p} & \mathbf{I}_{n-p} \end{bmatrix} \tag{1.63}$$

Multiplying out the left-hand side of (1.63) and equating corresponding partitions will generate four matrix equations, the solution of which will yield the values of $\mathbf{B}, \mathbf{C}, \mathbf{D}$ and $\mathbf{E}$ in terms of the submatrices $\mathbf{A}_{ij}, i, j = 1, 2$. The details of this derivation are left as an exercise (see Exercise 1.23) but the result is as follows:[6]

$$\begin{bmatrix} \mathbf{A}_{11} & \mathbf{A}_{12} \\ \mathbf{A}_{21} & \mathbf{A}_{22} \end{bmatrix}^{-1} = \begin{bmatrix} \mathbf{B} & \mathbf{C} \\ \mathbf{D} & \mathbf{E} \end{bmatrix} = \begin{bmatrix} \mathbf{F} & -\mathbf{F}\mathbf{A}_{12}\mathbf{A}_{22}^{-1} \\ -\mathbf{A}_{22}^{-1}\mathbf{A}_{21}\mathbf{F} & \mathbf{A}_{22}^{-1} + \mathbf{A}_{22}^{-1}\mathbf{A}_{21}\mathbf{F}\mathbf{A}_{12}\mathbf{A}_{22}^{-1} \end{bmatrix} \tag{1.64}$$

where $\mathbf{F} = (\mathbf{A}_{11} - \mathbf{A}_{12}\mathbf{A}_{22}^{-1}\mathbf{A}_{21})^{-1}$.

A special case of this result, and the alternative result referred to in note 6, arises when $\mathbf{A}$ is **block symmetric**, in the sense that $\mathbf{A}_{21} = \mathbf{A}_{12}^\top$. This special case will be used in Chapter 14, and the derivation of its details is left as an exercise in that chapter; see Exercise 14.10.

Another special case of (1.64) is for the so-called **block diagonal matrix**, which arises when $\mathbf{A}_{12} = \mathbf{0}_{p \times (n-p)}$ and $\mathbf{A}_{21} = \mathbf{0}_{(n-p) \times p}$. Note that, unless $p = n/2$, these two zero matrices are not equal. The partitioned inverse in this case is

$$\begin{bmatrix} \mathbf{A}_{11} & \mathbf{0} \\ \mathbf{0} & \mathbf{A}_{22} \end{bmatrix}^{-1} = \begin{bmatrix} \mathbf{A}_{11}^{-1} & \mathbf{0} \\ \mathbf{0} & \mathbf{A}_{22}^{-1} \end{bmatrix} \tag{1.65}$$

still assuming that $\mathbf{A}_{11}$ and $\mathbf{A}_{22}$ are invertible. This useful result is easy to verify by multiplication of the partitioned inverse on the right-hand side of (1.65) by the original block diagonal matrix to produce the (partitioned) identity matrix

$$\begin{bmatrix} \mathbf{A}_{11}^{-1} & \mathbf{0} \\ \mathbf{0} & \mathbf{A}_{22}^{-1} \end{bmatrix} \begin{bmatrix} \mathbf{A}_{11} & \mathbf{0} \\ \mathbf{0} & \mathbf{A}_{22} \end{bmatrix} = \begin{bmatrix} \mathbf{A}_{11}^{-1}\mathbf{A}_{11} + \mathbf{0}\mathbf{0} & \mathbf{A}_{11}^{-1}\mathbf{0} + \mathbf{0}\mathbf{A}_{22} \\ \mathbf{0}\mathbf{A}_{11} + \mathbf{A}_{22}^{-1}\mathbf{0} & \mathbf{0}\mathbf{0} + \mathbf{A}_{22}^{-1}\mathbf{A}_{22} \end{bmatrix}$$

$$= \begin{bmatrix} \mathbf{I} + \mathbf{0} & \mathbf{0} + \mathbf{0} \\ \mathbf{0} + \mathbf{0} & \mathbf{0} + \mathbf{I} \end{bmatrix} = \begin{bmatrix} \mathbf{I} & \mathbf{0} \\ \mathbf{0} & \mathbf{I} \end{bmatrix} = \mathbf{I} \tag{1.66}$$

where we have omitted the cumbersome dimension subscripts on the various zero and identity matrices. The result is also an obvious generalization of the inverse of a diagonal matrix that we encountered in Section 1.5.6.

1.5.15   Kronecker product of matrices

DEFINITION 1.5.1 The **Kronecker product** or **direct product** of two matrices $\mathbf{A}$ and $\mathbf{B}$, where $\mathbf{A}$ is of order $m \times n$ and $\mathbf{B}$ is of order $p \times q$, is the partitioned matrix of order $mp \times nq$ given by

$$\mathbf{A} \otimes \mathbf{B} \equiv [a_{ij}\mathbf{B}] = \begin{bmatrix} a_{11}\mathbf{B} & a_{12}\mathbf{B} & \cdots & a_{1n}\mathbf{B} \\ a_{21}\mathbf{B} & a_{22}\mathbf{B} & \cdots & a_{2n}\mathbf{B} \\ \vdots & \vdots & & \vdots \\ a_{m1}\mathbf{B} & a_{m2}\mathbf{B} & \cdots & a_{mn}\mathbf{B} \end{bmatrix}_{mp \times nq} \tag{1.67}$$

In other words, the Kronecker product of two matrices is formed by multiplying the second matrix by each (scalar) element of the first matrix in turn and then arranging the matrices so formed in a partitioned matrix, as shown in (1.67). For conciseness, the middle expression in (1.67) extends the idea of the typical term to the typical submatrix of the partitioned matrix. It is possible to represent a great deal of information as a Kronecker product; note its dimensions.

EXAMPLE 1.5.5 Using the matrices $\mathbf{A}$ and $\mathbf{B}$ defined in Example 1.3.2, we have the Kronecker product

$$
\mathbf{A} \otimes \mathbf{B} = \begin{bmatrix} 1 & 2 \\ 3 & 4 \end{bmatrix} \otimes \begin{bmatrix} 2 & -2 \\ -4 & 4 \end{bmatrix}
$$

$$
= \begin{bmatrix} 1 \begin{bmatrix} 2 & -2 \\ -4 & 4 \end{bmatrix} & 2 \begin{bmatrix} 2 & -2 \\ -4 & 4 \end{bmatrix} \\ 3 \begin{bmatrix} 2 & -2 \\ -4 & 4 \end{bmatrix} & 4 \begin{bmatrix} 2 & -2 \\ -4 & 4 \end{bmatrix} \end{bmatrix} \tag{1.68}
$$

or, writing the product in its most detailed form,

$$
\mathbf{A} \otimes \mathbf{B} = \begin{bmatrix} 2 & -2 & 4 & -4 \\ -4 & 4 & -8 & 8 \\ 6 & -6 & 8 & -8 \\ -12 & 12 & -16 & 16 \end{bmatrix}. \tag{1.69}
$$

$\diamond$

The manipulation of Kronecker products is facilitated by a number of theorems. The following results are particularly useful.

PROPERTY 1 *Let $\mathbf{A}$, $\mathbf{B}$, $\mathbf{C}$ and $\mathbf{D}$ be matrices of orders $m \times n$, $p \times q$, $n \times r$ and $q \times s$, respectively; then $(\mathbf{A} \otimes \mathbf{B})(\mathbf{C} \otimes \mathbf{D}) = \mathbf{AC} \otimes \mathbf{BD}$.*

Proof: We have $\mathbf{A} \otimes \mathbf{B} = [a_{ij}\mathbf{B}]$ and $\mathbf{C} \otimes \mathbf{D} = [c_{ij}\mathbf{D}]$ by definition of Kronecker product. By the definition of (partitioned) matrix multiplication,

$$
(\mathbf{A} \otimes \mathbf{B})(\mathbf{C} \otimes \mathbf{D}) = \left[\sum_{k=1}^{n} a_{ik} \mathbf{B} c_{kj} \mathbf{D} \right] \tag{1.70}
$$

Using the properties of multiplication of matrices by scalars, it can be seen that the typical (ijth, ($p \times s$)-dimensional) partition of this matrix is just $\left(\sum_{k=1}^{n} a_{ik}c_{kj} \right) \mathbf{BD}$. By the definitions of matrix multiplication and Kronecker product, this is just the typical term of

$$
(\mathbf{A} \otimes \mathbf{B})(\mathbf{C} \otimes \mathbf{D}) = \mathbf{AC} \otimes \mathbf{BD} \tag{1.71}
$$

$\square$

As an exercise, you might determine the order of $(\mathbf{A} \otimes \mathbf{B})(\mathbf{C} \otimes \mathbf{D})$; see Exercise 1.24.

PROPERTY 2 *Let* **A** *and* **B** *be square invertible matrices of dimensions* $m \times m$ *and* $n \times n$, *respectively; then* $(\mathbf{A} \otimes \mathbf{B})^{-1} = \mathbf{A}^{-1} \otimes \mathbf{B}^{-1}$.

Proof: It follows immediately from Property 1 that

$$(\mathbf{A} \otimes \mathbf{B})(\mathbf{A}^{-1} \otimes \mathbf{B}^{-1}) = \mathbf{A}\mathbf{A}^{-1} \otimes \mathbf{B}\mathbf{B}^{-1} = \mathbf{I}_m \otimes \mathbf{I}_n = \mathbf{I}_{mn} \qquad (1.72)$$

which establishes this useful result.   □

As will become clear later, neither **B** nor $\mathbf{A} \otimes \mathbf{B}$ in Example 1.5.15 is invertible.

PROPERTY 3 *Let* **A** *and* **B** *be square matrices. Then* $\mathrm{tr}(\mathbf{A} \otimes \mathbf{B}) = \mathrm{tr}(\mathbf{A})\mathrm{tr}(\mathbf{B})$.

Proof: The proof of this final result on Kronecker products is left as an exercise; see Exercise 1.26.   □

EXERCISES

1.1 If $\mathbf{A} = \begin{bmatrix} 2 & 4 \\ -1 & 6 \end{bmatrix}$ and $\mathbf{B} = \begin{bmatrix} 0 & 1 \\ 2 & 2 \end{bmatrix}$, show that $\mathbf{AB} \neq \mathbf{BA}$.

1.2 Given $\mathbf{A} = \begin{bmatrix} 1 & 0 & 3 \\ 2 & -1 & 1 \end{bmatrix}$, $\mathbf{B} = \begin{bmatrix} 3 & 4 & 1 \\ 0 & -1 & 5 \\ 1 & 2 & -2 \end{bmatrix}$ and $\mathbf{C} = \begin{bmatrix} 2 \\ -1 \\ 4 \end{bmatrix}$, calculate $(\mathbf{AB})^{\top}, \mathbf{B}^{\top}\mathbf{A}^{\top}$, $(\mathbf{AC})^{\top}$ and $\mathbf{C}^{\top}\mathbf{A}^{\top}$, and comment on your results.

1.3 Find all matrices **B** obeying the equation $\begin{bmatrix} 0 & 1 \\ 0 & 2 \end{bmatrix} \mathbf{B} = \begin{bmatrix} 0 & 0 & 1 \\ 0 & 0 & 2 \end{bmatrix}$.

1.4 Find all matrices **B** that commute with $\mathbf{A} = \begin{bmatrix} 0 & 1 \\ 0 & 2 \end{bmatrix}$ to give $\mathbf{AB} = \mathbf{BA}$.

1.5 Let $\mathbf{A} = \begin{bmatrix} 0 & -3 & 6 & -10 \\ 3 & 0 & 9 & \frac{1}{2} \\ -6 & -9 & 0 & 1 \\ 10 & -\frac{1}{2} & -1 & 0 \end{bmatrix}$. Calculate $\mathbf{A} - \mathbf{A}^{\top}$ and comment on the result.

1.6 Let $\mathbf{A} = \begin{bmatrix} 1 & 1 & 1 \\ 2 & -1 & 2 \\ -1 & -1 & 0 \end{bmatrix}$ and $\mathbf{B} = \begin{bmatrix} -\frac{2}{3} & \frac{1}{3} & -1 \\ \frac{2}{3} & -\frac{1}{3} & 0 \\ 1 & 0 & 1 \end{bmatrix}$. Find the products **AB** and **BA**.

Show that the pre-multiplication of any three-element column vector, **y**, by the product **BA** leaves the vector **y** unchanged; i.e. show that $\mathbf{BAy} = \mathbf{y}$. Hence solve the system of equations

$$x_1 + x_2 + x_3 = 6$$
$$2x_1 - x_2 + 2x_3 = 6$$
$$-x_1 - x_2 = -5$$

(Hint: recall the identity matrix and the inverse of a matrix from Section 1.5.2 and Section 1.5.4, respectively.)

1.7 Let $A = \begin{bmatrix} 1 & 2 \\ 2 & 4 \end{bmatrix}$. Find a matrix $B \neq 0$ such that $AB = \begin{bmatrix} 0 & 0 \\ 0 & 0 \end{bmatrix}$, verifying that $AB = 0$ does not, in general, imply that either $A = 0$ or $B = 0$.

1.8 Given the matrix $A = \begin{bmatrix} 2 & 1 \\ 5 & 3 \end{bmatrix}$, find the matrix B such that $AB = \begin{bmatrix} 1 & 0 \\ 0 & 1 \end{bmatrix}$, using only the matrix operations of equality and multiplication.

1.9 If x is a column vector with n elements, and y is a column vector with m elements, what is the product $xy^{\top}$, which is called an **outer product**? Compare $x^{\top}y$ for the case $m = n$; this is an example of an inner product or scalar product; see Definition 5.4.12.

When x is fixed, the expression $x^{\top}y$ is called a **linear form** in y. In Definition 4.3.1, we will also encounter quadratic forms. In Chapter 14, both linear forms and quadratic forms appear in our more detailed treatment of the single-equation econometric model.

1.10 A large firm controls five factories. The input and output levels of each factory are given by numbers in one column of the table

Factory	a	b	c	d	e
Good 1	1	-2	0	1	1
Good 2	0	3	$-\frac{1}{2}$	1	-2
Good 3	$-\frac{1}{2}$	1	2	-2	-1

where inputs are negative and outputs are positive. If the prices of the three goods are given by the respective elements of $[3 \quad 4 \quad 6]^{\top}$, find the total profit of the firm. Write down the matrix formula that would effect this calculation.

1.11 Specifying the general dimensions of the matrices involved:

(a) prove that $(A + B) + C = A + (B + C)$, the associative rule of matrix addition; and
(b) prove that $(AB)C = A(BC)$, the associative rule of matrix multiplication.

(Hint: as the proof of the second part of this question uses a typical term involving double summation, take care with the subscripts and ranges of summation.)

1.12 Expand $(A + B)(A - B)$ and $(A - B)(A + B)$, where A and B are suitably conformable matrices.

(a) What are the dimensions of the two matrices, in general?
(b) What theorem of matrix algebra did you use to expand the expressions? Prove this theorem using an argument based on the use of typical terms.
(c) Are the two expansions the same? If not, why not? How many terms are there in each expansion?

1.13 Let $\mathbf{A}$ be an $m \times n$ matrix $(m \neq n)$.

(a) Prove that $0\mathbf{A} = \mathbf{0}$ (note that the first zero is scalar and that the second is a matrix).
(b) Prove that $\mathbf{A} + \mathbf{0} = \mathbf{A}$. What is the dimension of the zero in this equation?
(c) Explain why, in general, $\mathbf{0A} \neq \mathbf{A0}$, where each $\mathbf{0}$ is a suitably conformable zero matrix. Under what circumstances is $\mathbf{0A} = \mathbf{A0}$?

1.14 Define the trace, $\mathrm{tr}(\mathbf{A})$, of a square matrix, $\mathbf{A}$. Prove that, if $\mathbf{A}$ and $\mathbf{B}$ are such that both $\mathbf{AB}$ and $\mathbf{BA}$ exist, then $\mathrm{tr}(\mathbf{AB}) = \mathrm{tr}(\mathbf{BA})$. (Hint: as in Exercise 1.11, take care with the subscripts and ranges of summation.)

1.15 Prove that diagonal matrices of the same order are commutative in multiplication with each other.

1.16 Let $\mathbf{A}$ be a square invertible matrix, and define

$$\mathbf{A}^r = \underbrace{\mathbf{AA}\ldots\mathbf{A}}_{r \text{ times}}, \quad \mathbf{A}^0 = \mathbf{I}, \quad \mathbf{A}^p\mathbf{A}^q = \mathbf{A}^{p+q} \text{ and } \mathbf{A}^{-r} = (\mathbf{A}^{-1})^r = \underbrace{\mathbf{A}^{-1}\mathbf{A}^{-1}\ldots\mathbf{A}^{-1}}_{r \text{ times}}$$

Prove that $\mathbf{A}^r$ is invertible and $(\mathbf{A}^r)^{-1} = (\mathbf{A}^{-1})^r = \mathbf{A}^{-r}$.

1.17 Let $\mathbf{A}$ be a square invertible matrix.

(a) Show that $\mathbf{A}^2 = \mathbf{AA}$ is invertible, with $(\mathbf{A}^2)^{-1} = \mathbf{A}^{-1}\mathbf{A}^{-1} = (\mathbf{A}^{-1})^2$.
(b) Prove that $\left(\frac{1}{4}\mathbf{A}\right)^{-1} = 4\mathbf{A}^{-1}$.
(c) Suppose that $\mathbf{A} = 0.5\mathbf{I}_{15}$. Find $\mathbf{A}^{-1}$ and $\mathrm{tr}(\mathbf{A})$.

1.18 Prove that $\mathrm{tr}(\mathbf{A}) = \mathrm{tr}(\mathbf{A}^\top)$ for any square matrix $\mathbf{A}$.

1.19 Find an orthogonal matrix other than the identity matrix that is symmetric.

1.20 If $\mathbf{1}^\top = [1 \quad 1 \quad 1]$ and $\mathbf{A}$ is a 3×3 matrix, what is the relationship between:

(a) the rows of $\mathbf{A}$ and the column vector $\mathbf{A1}$; and
(b) the columns of $\mathbf{A}$ and the row vector $\mathbf{1}^\top\mathbf{A}$?

Comment on the interpretation of $\frac{1}{3}\mathbf{A1}$ and $\frac{1}{3}\mathbf{1}^\top\mathbf{A}$; and also on the nature of $\mathbf{11}^\top\mathbf{A}$, $\mathbf{A11}^\top$ and $\mathbf{1}^\top\mathbf{1A}$.

1.21 Let $\mathbf{x} = [x_i]$ be an n-vector. Recalling Exercise 1.20, find a square matrix $\mathbf{A}$, which, when pre-multiplying $\mathbf{x}$, will yield the n-vector $[x_i - \bar{x}]$, where $\bar{x} = (1/n)\sum_{i=1}^{n} x_i$ is the arithmetic mean of the elements of $\mathbf{x}$.

1.22 Consider the matrix $\mathbf{A}$ in equation (1.57) and matrix $\mathbf{C}$ in equation (1.60). What are the restrictions on the dimensions of the individual submatrices, $\mathbf{C}_{ij}$, $i = 1, 2$, $j = 1, 2, 3$, for the product $\mathbf{AC}$ in equation (1.61) to be well defined?

1.23 For the partitioned matrix in equation (1.57), derive the partitioned inverse in equation (1.64) assuming that $m = n$ and that $\mathbf{A}_{11}$ and $\mathbf{A}_{22}$ are invertible. Show that an alternative form of the partitioned inverse is

$$\begin{bmatrix} \mathbf{A}_{11} & \mathbf{A}_{12} \\ \mathbf{A}_{21} & \mathbf{A}_{22} \end{bmatrix}^{-1} = \begin{bmatrix} \mathbf{A}_{11}^{-1} + \mathbf{A}_{11}^{-1}\mathbf{A}_{12}\mathbf{G}\mathbf{A}_{21}\mathbf{A}_{11}^{-1} & -\mathbf{A}_{11}^{-1}\mathbf{A}_{12}\mathbf{G} \\ -\mathbf{G}\mathbf{A}_{21}\mathbf{A}_{11}^{-1} & \mathbf{G} \end{bmatrix}$$

where $\mathbf{G} = (\mathbf{A}_{22} - \mathbf{A}_{21}\mathbf{A}_{11}^{-1}\mathbf{A}_{12})^{-1}$.

1.24 What is the order of $(\mathbf{A} \otimes \mathbf{B})(\mathbf{C} \otimes \mathbf{D})$ in Property 1 of Kronecker products?

1.25 Prove that $(\mathbf{A} \otimes \mathbf{B})^{-1} = \mathbf{A}^{-1} \otimes \mathbf{B}^{-1}$ without appeal to Property 1 of Kronecker products.

1.26 Provide a proof of Property 3 of Kronecker products, i.e. that $\mathrm{tr}(\mathbf{A} \otimes \mathbf{B}) = \mathrm{tr}(\mathbf{A})\,\mathrm{tr}(\mathbf{B})$, where $\mathbf{A}$ and $\mathbf{B}$ are arbitrary square matrices.

2   Determinants

2.1   Introduction

In Chapter 1 we defined a matrix, saw a few examples of how systems of linear equations and matrices arise in economic and financial applications, and presented the basic operations and rules of matrix algebra, paying attention to the proofs of many of the matrix properties that were stated. We also catalogued a number of special types of matrix that we will encounter later. In this chapter, we introduce the concept of a determinant, which is important in the theory and solution of systems of linear equations. The determinant also provides us with a means of obtaining an explicit formula for evaluating the inverse of a matrix.

At a basic level, we can describe a determinant as a real-valued function of a square matrix variable, i.e. a function that associates a real number $f(\mathbf{A})$ with a square matrix $\mathbf{A}$. Formally, we may write $f : M \to \mathbb{R}$, where M is the set of square matrices and $\mathbb{R}$ is the set of real numbers. However, we require a more precise definition that will also allow us to calculate the numerical value of $f(\mathbf{A})$. Before we give this, we need to be clear on a few preliminary ideas.

2.2   Preliminaries

2.2.1   Permutation

The idea of a **permutation** of some number, r, of integers from the set of integers $\{1, 2, \ldots, n\}$ is fundamental. As discussed on p. xxiii, a permutation is an ordered arrangement of r of the first n integers. The number of such different permutations, $^{n}P_{r}$, is given by the general result

$$^{n}P_{r} = \frac{n!}{(n-r)!} \tag{2.1}$$

defined on p. xxiii, from which it follows that $^{n}P_{n}$, the number of ordered arrangements of all n integers, is $n!$, since $(n-n)! = 0! \equiv 1$. Of these $n!$ permutations, the typical permutation, i.e. the jth, may be denoted by $j_1 j_2 \ldots j_n$.

2.2.2   Inversion

An **inversion** in a permutation of n integers occurs whenever a larger integer precedes a smaller one in the permutation. For example, if $j_1 > j_2$, we have one inversion; if $j_1 > j_3$ also, we have another inversion; and so on.

Table 2.1 Classification of permutations of the first three integers

Permutation	Number of inversions	Classification
123	0	Even
132	1	Odd
⋮	⋮	⋮
312	2	Even
321	3	Odd

2.2.3 Odd and even permutations

A permutation is **odd** if it contains an odd number of inversions; and a permutation is **even** if it contains an even number of inversions.

To make these first three ideas concrete, consider the set $\{1, 2, 3\}$. In this case, $n = 3$ and $^n P_n = 3! = 6$. The individual permutations, together with the number of inversions they contain and their classification as odd or even, are given in Table 2.1, except for two permutations, which have been omitted so that the reader can fill in the missing information as an exercise; see Exercise 2.1.

2.2.4 Elementary product

An **elementary product**, formed from an $n \times n$ matrix $\mathbf{A}$, is any product of n elements from $\mathbf{A}$, no two elements of which come from the same row or same column. The number of elementary products is $n!$. For example, if

$$\mathbf{A} = \begin{bmatrix} a_{11} & a_{12} \\ a_{21} & a_{22} \end{bmatrix} \tag{2.2}$$

then the $2! = 2$ elementary products are $a_{11}a_{22}$ and $a_{12}a_{21}$. Products such as $a_{11}a_{12}$ and $a_{21}a_{22}$ are not elementary products because in each of these cases both elements in the product are from the same row. Similarly, the products $a_{11}a_{21}$ and $a_{12}a_{22}$ are not elementary products. (Why?) If $\mathbf{A} = [a_{ij}]_{n \times n}$, then the elementary products are, typically, $a_{1j_1}a_{2j_2} \ldots a_{nj_n}$, where $j_1 j_2 \ldots j_n$, relating to the columns from which elements come, is the typical permutation of the integers in the set $\{1, 2, \ldots, n\}$.

2.2.5 Signing convention

By convention, we sign elementary products as **positive** if the $j_1 j_2 \ldots j_n$ is an even permutation and **negative** if the $j_1 j_2 \ldots j_n$ is an odd permutation. Thus, for example, in the 2×2 case, we have $+a_{11}a_{22}$ and $-a_{12}a_{21}$; and in the 3×3 case, we have $+a_{11}a_{22}a_{33}, -a_{11}a_{23}a_{32}, -a_{12}a_{21}a_{33}$; and so on. The determination of the remaining three elementary products from $\mathbf{A} = [a_{ij}]_{3 \times 3}$ is left as an exercise; see Exercise 2.1.

2.3 Definition and properties

Making use of the preliminary ideas from the previous section, we can now formally state the definition of the determinant function and proceed to consider its properties.

2.3.1 Definition

DEFINITION 2.3.1 The **determinant**, det($\mathbf{A}$) or $|\mathbf{A}|$, of the $n \times n$ matrix, $\mathbf{A}$, is the sum of all signed elementary products of $\mathbf{A}$: $\det(\mathbf{A}) = \sum \pm a_{1j_1} a_{2j_2} \ldots a_{nj_n}$, where $\sum$ indicates that the terms are to be summed over all $n!$ permutations $(j_1 j_2 \ldots j_n)$.

The det($\cdot$) notation is generally used for an abstract matrix, e.g. det($\mathbf{A}$) or det($\mathbf{B}$); the $|\cdot|$ notation is generally used – in place of the square brackets – when the matrix is written out in full, e.g.

$$\begin{vmatrix} 1 & 0 & 2 \\ 2 & 1 & 0 \\ 2 & 3 & 1 \end{vmatrix} \tag{2.3}$$

Thus, for the case of $n = 2$, $\det(\mathbf{A}) = a_{11}a_{22} - a_{12}a_{21}$. For the case of $n = 3$,

$$\det(\mathbf{A}) = a_{11}a_{22}a_{33} - a_{11}a_{23}a_{32} - a_{12}a_{21}a_{33} + a_{12}a_{23}a_{31} + a_{13}a_{21}a_{32} - a_{13}a_{22}a_{31} \tag{2.4}$$

When n is small, as in these two cases, and the numerical values of the a_{ij} are known, det($\mathbf{A}$) is easy to evaluate using this definitional formula.

EXAMPLE 2.3.1 Consider the 3×3 matrix

$$\mathbf{A} = \begin{bmatrix} 1 & 0 & 2 \\ 2 & 1 & 0 \\ 2 & 3 & 1 \end{bmatrix} \tag{2.5}$$

Then

$$\begin{aligned} \det(\mathbf{A}) &= a_{11}a_{22}a_{33} - a_{11}a_{23}a_{32} - a_{12}a_{21}a_{33} + a_{12}a_{23}a_{31} + a_{13}a_{21}a_{32} - a_{13}a_{22}a_{31} \\ &= (1 \times 1 \times 1) - (1 \times 0 \times 3) - (0 \times 2 \times 1) + (0 \times 0 \times 2) \\ &\quad + (2 \times 2 \times 3) - (2 \times 1 \times 2) \\ &= 1 - 0 - 0 + 0 + 12 - 4 = 9 \end{aligned} \tag{2.6}$$

$\diamond$

When n is large, however, the definitional formula proves cumbersome and we use a more efficient computational approach, developed in Section 2.4. In the meantime, we consider the properties of the determinant concept.

2.3.2 Properties

The statement of all of the following properties assumes that $\mathbf{A}$ is an $n \times n$ matrix.

PROPERTY 1 *If $\mathbf{A}$ contains a row of zeros or a column of zeros, then $\det(\mathbf{A}) = 0$.*

PROPERTY 2 *If $\mathbf{A}$ is diagonal, then $\det(\mathbf{A}) = a_{11}a_{22} \ldots a_{nn} = \prod_{i=1}^{n} a_{ii}$.*

PROPERTY 3 *If $\mathbf{A}$ is triangular, then $\det(\mathbf{A}) = \prod_{i=1}^{n} a_{ii}$.*

PROPERTY 4 *If every element in a single row of* **A** *is multiplied by a constant, k, to form* **B***, say, then* $\det(\mathbf{B}) = k \det(\mathbf{A})$.

PROPERTY 5 *We have* $\det(k\mathbf{A}) = k^n \det(\mathbf{A})$.

PROPERTY 6 *If any two rows of* **A** *are interchanged to form* **B***, say, then* $\det(\mathbf{B}) = -\det(\mathbf{A})$.

PROPERTY 7 *If* **A** *has two identical rows, then* $\det(\mathbf{A}) = 0$.

PROPERTY 8 *If a multiple of one row of* **A** *is added to another row to form* **B***, say, then* $\det(\mathbf{B}) = \det(\mathbf{A})$.

PROPERTY 9 *If* **A** *has two rows that are proportional, then* $\det(\mathbf{A}) = 0$.

PROPERTY 10 *We have* $\det(\mathbf{A}^\top) = \det(\mathbf{A})$.

PROPERTY 11 *We have* $\det(\mathbf{AB}) = \det(\mathbf{A}) \det(\mathbf{B})$.

The reader should consider the proofs of at least the first ten of these properties of determinants and attempt them as exercises. The proofs of the first five properties are straightforward and involve direct appeal to the definition of a determinant. To take the case of Property 5, for example, we prove it in the following theorem.

THEOREM 2.3.1 *Let* **A** *be an* $n \times n$ *matrix and let* $k \in \mathbb{R}$*; then* $\det(k\mathbf{A}) = k^n \det(\mathbf{A})$.

Proof: We have $k\mathbf{A} = [ka_{ij}]$ by definition. Therefore, also by definition,

$$\det(k\mathbf{A}) = \sum ka_{1j_1} ka_{2j_2} \ldots ka_{nj_n}$$
$$= k^n \sum a_{1j_1} a_{j_2} \ldots a_{nj_n}$$
$$= k^n \det(\mathbf{A}) \tag{2.7}$$

□

The proof of Property 6 is likely to pose more difficulty; it hinges on the fact that the row interchange changes the sign of all permutations of the column subscripts, hence the sign of all of the elementary products of **A**. The proof of Property 7 follows simply from Property 6. The proof of Property 8 is also rather tricky but may be obtained by computing the determinants involved and then verifying the equality. Properties 1 and 8 are sufficient to establish Property 9, while Property 10 is straightforward to establish. The tenth result is important in that it facilitates showing that almost every theorem on determinants that refers to rows is also true when re-cast to refer to columns. For example, consider the column version of Property 6, proved in the following theorem.

THEOREM 2.3.2 *Let* **A** *be an* $n \times n$ *matrix and* $\mathbf{B}^*$ *be the matrix formed by the interchange of any two columns of* **A***; then* $\det(\mathbf{B}^*) = -\det(\mathbf{A})$.

Proof: From Property 10, we have

$$\det(\mathbf{B}^*) = \det(\mathbf{B}^{*\top})$$
$$= -\det(\mathbf{A}^{\top}) \tag{2.8}$$

by Property 6, since $\mathbf{B}^{*\top}$ is just $\mathbf{A}^{\top}$ with two rows interchanged; therefore

$$\det(\mathbf{B}^*) = -\det(\mathbf{A}) \tag{2.9}$$

by Property 10.   □

Property 3 will be proved later; see Theorem 2.4.2. The proof of Property 11 is longer and requires concepts that have not yet been introduced, so it is not given; for a proof see Anton and Rorres (2011, p. 108).

2.3.3   Singularity and non-singularity

DEFINITION 2.3.2 Square matrices that have zero determinants are said to be **singular**; and those with non-zero determinants are said to be **non-singular**.

Examples of singular matrices include those referred to in the first, seventh and ninth properties listed in Section 2.3.2. The following theorem establishes the importance of non-singularity.

THEOREM 2.3.3 *If a square matrix,* $\mathbf{A}$*, is invertible, then* $\det(\mathbf{A}) \neq 0$*, i.e.* $\mathbf{A}$ *is non-singular.*

Proof: If $\mathbf{A}$ is invertible, then $\mathbf{A}\mathbf{A}^{-1} = \mathbf{I}$, by definition,
Therefore $\det(\mathbf{A}\mathbf{A}^{-1}) = \det(\mathbf{I}) = 1$, using Property 2.
Therefore $\det(\mathbf{A})\det(\mathbf{A}^{-1}) = 1$, using Property 11.
Therefore $\det(\mathbf{A}) \neq 0$, hence $\mathbf{A}$ is non-singular by definition.   □

The following is a useful corollary of this proof.

COROLLARY 2.3.4 *We have*

$$\det(\mathbf{A}^{-1}) = [\det(\mathbf{A})]^{-1} = \frac{1}{\det(\mathbf{A})} \tag{2.10}$$

The theorem can be strengthened to yield a necessary and sufficient condition stating that $\mathbf{A}$ is invertible if and only if $\det(\mathbf{A}) \neq 0$, i.e. if and only if $\mathbf{A}$ is non-singular. The first part of this modified theorem ($\Rightarrow$) is identical to the weaker theorem. However, the proof of the second part of the modified theorem ($\Leftarrow$) requires concepts that have not yet been introduced, so is not given here; for a proof see Anton and Rorres (2010, p. 109).

2.4   Co-factor expansions of determinants

The definition of a determinant given in the previous section is not computationally efficient. A more efficient and more usual method of calculation makes use of the idea of a **co-factor expansion** of a determinant. Such expansions are explained in this section. These expansions also provide means of deriving an explicit formula for the calculation of matrix inverses. Before we see these important results, however, we need to introduce two basic concepts.

DEFINITION 2.4.1 The **minor** of matrix element a_{ij}, denoted M_{ij}, is the determinant of the submatrix that remains when the ith row and jth column are deleted from a square matrix **A**.

There are n^2 minors defined for an $n \times n$ matrix, one corresponding to every element or every pairing of row and column; hence, the use of the ij subscripts and the association of M_{ij} with the typical element a_{ij}.

DEFINITION 2.4.2 The **co-factor** of matrix element a_{ij} is $C_{ij} \equiv (-1)^{i+j} M_{ij}$.

A co-factor, like a minor, is associated with a particular row and column and, therefore, with a particular element, a_{ij}, of a matrix; and it differs from a minor only in its sign, i.e. $C_{ij} = \pm M_{ij}$, depending on the row and column numbers. If $i + j$ is an even number, the sign of the ijth co-factor is the same as that of the corresponding minor; if $i + j$ is odd, the co-factor and minor are of opposite sign.

Now, for the 3×3 case,

$$\mathbf{A} = \begin{bmatrix} a_{11} & a_{12} & a_{13} \\ a_{21} & a_{22} & a_{23} \\ a_{31} & a_{32} & a_{33} \end{bmatrix} \tag{2.11}$$

and, as in (2.4), the determinant is

$$\det(\mathbf{A}) = a_{11}a_{22}a_{33} - a_{11}a_{23}a_{32} - a_{12}a_{21}a_{33} + a_{12}a_{23}a_{31}$$
$$+ a_{13}a_{21}a_{32} - a_{13}a_{22}a_{31} \tag{2.12}$$

Identifying certain common factors in these six terms, this can be written as

$$\det(\mathbf{A}) = a_{11}(a_{22}a_{33} - a_{23}a_{32}) - a_{12}(a_{21}a_{33} - a_{23}a_{31}) + a_{13}(a_{21}a_{32} - a_{22}a_{31})$$
$$= a_{11}M_{11} - a_{12}M_{12} + a_{13}M_{13}$$
$$= a_{11}C_{11} + a_{12}C_{12} + a_{13}C_{13} \tag{2.13}$$

This is referred to as the co-factor expansion of $\det(\mathbf{A})$ along the first row, because the elements in the first row have been removed from pairs of the six terms as common factors, leaving the corresponding co-factors multiplying them. We could have removed elements from any row, or indeed any column, to obtain different co-factor expansions. There are $2n$ co-factor expansions in all. Thus, for an $n \times n$ matrix **A**, $\det(\mathbf{A})$ may be computed by expansion using the elements of any row and their associated co-factors, i.e.

$$\det(\mathbf{A}) = a_{i1}C_{i1} + a_{i2}C_{i2} + \cdots + a_{in}C_{in}, \quad i = 1, 2, \ldots, n \tag{2.14}$$

or expansion using the elements of any column and their associated co-factors, i.e.

$$\det(\mathbf{A}) = a_{1j}C_{1j} + a_{2j}C_{2j} + \cdots + a_{nj}C_{nj}, \quad j = 1, 2, \ldots, n \tag{2.15}$$

THEOREM 2.4.1 *Expansions using the elements of one row (or column) and co-factors from a different row (or column) always sum to zero, i.e.*

$$a_{i1}C_{j1} + a_{i2}C_{j2} + \cdots + a_{in}C_{jn} = 0, \quad i, j \, (i \neq j) = 1, 2, \ldots, n \tag{2.16}$$

and

$$a_{1i}C_{1j} + a_{2i}C_{2j} + \cdots + a_{ni}C_{nj} = 0, \quad i, j \, (i \neq j) = 1, 2, \ldots, n \tag{2.17}$$

Proof: See Anton and Rorres (2011, p. 110). □

Expansions such as those in (2.16) and (2.17) are called expansions involving **alien co-factors**.

An important matrix in the formula for an inverse is the following one.

DEFINITION 2.4.3 The **adjoint** or **adjugate** of a square matrix **A**, denoted adj(**A**), is the transpose of the matrix formed by replacing the elements of **A** by their co-factors, i.e. $\text{adj}(\mathbf{A}) = [C_{ij}]^{\top}$.

Note that the adjoint of a 2×2 matrix can be found by swapping the principal diagonal elements and reversing the sign of the off-diagonal elements. For the 3×3 case, consider the following example.

EXAMPLE 2.4.1 Recall the matrix given in (2.5):

$$\mathbf{A} = \begin{bmatrix} 1 & 0 & 2 \\ 2 & 1 & 0 \\ 2 & 3 & 1 \end{bmatrix} \tag{2.18}$$

The minors and co-factors associated with the elements of **A** are easily determined. For instance, the co-factors corresponding to a_{11} and a_{12} are obtained as

$$C_{11} = (-1)^{1+1}M_{11} = \begin{vmatrix} 1 & 0 \\ 3 & 1 \end{vmatrix} = 1 \tag{2.19}$$

and

$$C_{12} = (-1)^{1+2}M_{12} = -\begin{vmatrix} 2 & 0 \\ 2 & 1 \end{vmatrix} = -2 \tag{2.20}$$

Completion of these co-factor calculations yields the matrix of co-factors

$$\begin{bmatrix} C_{11} & C_{12} & C_{13} \\ C_{21} & C_{22} & C_{23} \\ C_{31} & C_{32} & C_{33} \end{bmatrix} = \begin{bmatrix} 1 & -2 & 4 \\ 6 & -3 & -3 \\ -2 & 4 & 1 \end{bmatrix} \tag{2.21}$$

and the adjoint of **A** follows simply as the transpose of (2.21), namely

$$\text{adj}(\mathbf{A}) = \begin{bmatrix} 1 & 6 & -2 \\ -2 & -3 & 4 \\ 4 & -3 & 1 \end{bmatrix} \tag{2.22}$$

Notice that the elements in, for example, the first **column** of adj(A) are the co-factors in the first **row** of the co-factor matrix associated with the elements in the first **row** of **A**. It is therefore easy to find the determinant of **A** using the co-factor expansion along the first row of **A**. This is obtained by multiplying the elements in the first row of **A** either by their associated co-factors in the first row of the co-factor matrix or, alternatively, by the corresponding elements in the first column of adj(A):

$$\det(\mathbf{A}) = a_{11}C_{11} + a_{12}C_{12} + a_{13}C_{13} = (1 \times 1) + (0 \times -2) + (2 \times 4) = 9 \tag{2.23}$$

in agreement with the original calculation of this determinant in Example 2.3.1. ◇

It would be a useful exercise to repeat the calculation of the determinant of the matrix in Example 2.4 using, say, the co-factor expansion along the third row and the co-factor expansion down the second column of **A**, selecting the appropriate co-factors in each case from the elements of the co-factor matrix in (2.21) or from adj(A) in (2.22); see Exercise 2.2. In practice, computational efficiency is gained by choosing the co-factor expansion corresponding to the row (or column) of a matrix in which the most zeros appear.

An extreme example of the efficiency gains associated with choosing the most appropriate co-factor expansion is provided by the calculation of the determinant of a triangular matrix. Consider the case of a 3×3 upper triangular matrix

$$\mathbf{U} = \begin{bmatrix} a_{11} & a_{12} & a_{13} \\ 0 & a_{22} & a_{23} \\ 0 & 0 & a_{33} \end{bmatrix} \tag{2.24}$$

Then, as in (2.23), det(**U**) may be calculated using a co-factor expansion along the first row:

$$\det(\mathbf{U}) = a_{11}C_{11} + a_{12}C_{12} + a_{13}C_{13}$$
$$= a_{11} \begin{vmatrix} a_{22} & a_{23} \\ 0 & a_{33} \end{vmatrix} - a_{12} \begin{vmatrix} 0 & a_{23} \\ 0 & a_{33} \end{vmatrix} + a_{13} \begin{vmatrix} 0 & a_{22} \\ 0 & 0 \end{vmatrix}$$
$$= a_{11}(a_{22}a_{33} - 0) - a_{12}(0 - 0) + a_{13}(0 - 0)$$
$$= a_{11}a_{22}a_{33} \tag{2.25}$$

A better approach, however, is to use a co-factor expansion down the first column, which contains two zeros, taking the co-factors from the first column of the co-factor matrix or first row of adj(**U**). This yields the alternative calculation

$$\det(\mathbf{U}) = a_{11}C_{11} + a_{21}C_{21} + a_{31}C_{31}$$
$$= a_{11} \begin{vmatrix} a_{22} & a_{23} \\ 0 & a_{33} \end{vmatrix} + 0C_{21} + 0C_{31}$$
$$= a_{11}(a_{22}a_{33} - 0) + 0 + 0$$
$$= a_{11}a_{22}a_{33} \tag{2.26}$$

The difference in the amount of computation involved in the two approaches is not great in this 3×3 example, but it would be considerable if the dimension of **U** were higher.

For then, n co-factors would need to be evaluated in the first approach, whereas only one would still be required in the second approach.

Consideration of the calculation of the determinant of a triangular matrix leads to the following theorem.

THEOREM 2.4.2 *Let* **A** *be an* $n \times n$ *triangular matrix; then*

$$\det(\mathbf{A}) = \prod_{i=1}^{n} a_{ii} = a_{11}a_{22}\dots a_{nn} \tag{2.27}$$

the product of the elements on the principal diagonal of **A**.

The proof of this theorem may be obtained by consideration of the definition of a determinant and use of the fact that all elementary products except $a_{11}a_{22}\dots a_{nn}$ will contain at least one zero element and so be zero themselves. Hence, $\det(\mathbf{A}) = \sum \pm a_{1j_1}a_{2j_2}\dots a_{nj_n} = a_{11}a_{22}\dots a_{nn}$.

An alternative **proof by induction** is given here, however, as this method of proof will be used again later. The principle of proof by induction is simply that, if the truth of a statement for the positive integer n implies that the statement is true for $n+1$, and if the statement is also true for $n = 1$, then the statement is true for all positive integers.[1]

Proof: Let $P(n)$ be the proposition that an $n \times n$ upper triangular matrix satisfies (2.27). First, we check the truth of $P(1)$.

$P(1)$ is just the trivial statement that the determinant of the 1×1 matrix $[a_{11}]$ is the number a_{11}, which is clearly true.

Now, we assume the truth of $P(n)$ and endeavour to derive the truth of $P(n+1)$.

Partition the $(n+1) \times (n+1)$ upper triangular matrix as

$$\mathbf{A} \equiv \begin{bmatrix} \mathbf{A}_{11} & \mathbf{a} \\ \mathbf{0} & a_{(n+1)(n+1)} \end{bmatrix} \tag{2.28}$$

where $\mathbf{A}_{11}$ is $n \times n$, $\mathbf{0}$ is $1 \times n$, $\mathbf{a}$ is $n \times 1$ and $a_{(n+1)(n+1)}$ is a scalar.

Using a co-factor expansion across the last row, the determinant of **A** is

$$\det(\mathbf{A}) = a_{(n+1)1}C_{(n+1)1} + a_{(n+1)2}C_{(n+1)2} + \dots + a_{(n+1)(n+1)}C_{(n+1)(n+1)}$$

$$= a_{(n+1)(n+1)}C_{(n+1)(n+1)} \tag{2.29}$$

since $a_{(n+1)i} = 0$, $i = 1, 2, \dots, n$. Now $C_{(n+1)(n+1)}$ is the determinant of the $n \times n$ upper triangular matrix $\mathbf{A}_{11}$, so by $P(n)$ we have that $C_{(n+1)(n+1)} = a_{11}a_{22}\dots a_{nn}$. It then follows from (2.29) that

$$\det(\mathbf{A}) = a_{(n+1)(n+1)}C_{(n+1)(n+1)} = C_{(n+1)(n+1)}a_{(n+1)(n+1)}$$

$$= a_{11}a_{22}\dots a_{nn}a_{(n+1)(n+1)} \tag{2.30}$$

which is $P(n+1)$.

As we have shown that the truth of $P(n)$ implies the truth of $P(n+1)$, our proof by induction is complete. $\square$

The case of $n = 2$ provides a very simple illustration. Given

$$\mathbf{A} = \begin{bmatrix} a_{11} & a_{12} \\ 0 & a_{22} \end{bmatrix} \tag{2.31}$$

we have from the sum of the signed elementary products that $\det(\mathbf{A}) = a_{11}a_{22} - 0 = a_{11}a_{22}$, the product of the two diagonal elements, in accordance with Theorem 2.4.2.

A similar proof, based on co-factor expansions of determinants along their first row, applies in the case of lower triangular matrices; see Exercise 2.5.

Returning to the adjoint matrix, the essential role of the adjoint in the computation of matrix inverses is made clear in the following theorem.

THEOREM 2.4.3 *If an $n \times n$ matrix, $\mathbf{A}$, is invertible, then $\mathbf{A}^{-1} = \mathrm{adj}(\mathbf{A})/\det(\mathbf{A})$.*

Proof: Consider the product $\mathbf{A}\,\mathrm{adj}(\mathbf{A})$, whose typical element is

$$\sum_{k=1}^{n} a_{ik} C_{jk} = a_{i1} C_{j1} + a_{i2} C_{j2} + \cdots + a_{in} C_{jn} \tag{2.32}$$

When $i = j$, this is the co-factor expansion of $\det(\mathbf{A})$ along the ith row; when $i \neq j$, the expansion is zero because it involves alien co-factors.

Therefore $\mathbf{A}\,\mathrm{adj}(\mathbf{A}) = \mathrm{diag}[\det(\mathbf{A})] = \det(\mathbf{A})\mathbf{I}$.

Since $\det(\mathbf{A}) \neq 0$ by Theorem 2.3.3, then

$$\frac{\mathbf{A}\,\mathrm{adj}(\mathbf{A})}{\det(\mathbf{A})} = \mathbf{A}\frac{\mathrm{adj}(\mathbf{A})}{\det(\mathbf{A})} = \mathbf{I} \tag{2.33}$$

Therefore $\mathrm{adj}(\mathbf{A})/\det(\mathbf{A})$ is the inverse of $\mathbf{A}$ by definition and uniqueness of a matrix inverse. □

2.5 Solution of systems of equations

2.5.1 Cramer's rule

Cramer's theorem and the rule to which it gives rise provide a well-known application of determinants to the solution of certain square systems of equations, i.e. systems containing n equations in n unknowns. The theorem may be formally stated as follows, using the alternative notation for a determinant.[2]

THEOREM 2.5.1 (CRAMER'S THEOREM). *If $\mathbf{Ax} = \mathbf{b}$ is a square system of equations in which the matrix $\mathbf{A}$ is $n \times n$ and non-singular, and $\mathbf{x}$ (the vector of unknowns) and $\mathbf{b}$ are n-vectors, then the system has a unique solution given by $\mathbf{x} = (1/|\mathbf{A}|)[|\mathbf{A}_j|]$, where $\mathbf{A}_j$ ($j = 1, 2, \ldots, n$) is the matrix obtained by replacing the jth column of $\mathbf{A}$ by $\mathbf{b}$.*

Thus the solution for the individual elements of $\mathbf{x}$ is

$$x_1 = \frac{|\mathbf{A}_1|}{|\mathbf{A}|}, \quad x_2 = \frac{|\mathbf{A}_2|}{|\mathbf{A}|}, \quad \ldots, \quad x_n = \frac{|\mathbf{A}_n|}{|\mathbf{A}|} \tag{2.34}$$

Proof:

(a) Uniqueness. Since $\mathbf{A}$ is non-singular, its inverse, $\mathbf{A}^{-1}$, exists and is unique by Property 1 on p. 18. Pre-multiplication of both sides of $\mathbf{Ax} = \mathbf{b}$ by $\mathbf{A}^{-1}$ yields a solution $\mathbf{x} = \mathbf{A}^{-1}\mathbf{b}$. Let $\mathbf{s}$ also be a solution of the system so that $\mathbf{As} = \mathbf{b}$. It follows, by pre-multiplying both sides by $\mathbf{A}^{-1}$, that $\mathbf{A}^{-1}\mathbf{As} = \mathbf{A}^{-1}\mathbf{b}$. Therefore, $\mathbf{s} = \mathbf{A}^{-1}\mathbf{b} = \mathbf{x}$ and the solution is unique.

(b) The solution may be written as $\mathbf{x} = \mathbf{A}^{-1}\mathbf{b} = (1/|\mathbf{A}|)\,\mathrm{adj}(\mathbf{A})\mathbf{b}$, substituting for $\mathbf{A}^{-1}$. Thus

$$\mathbf{x} = \left[\frac{1}{|\mathbf{A}|}(b_1 C_{1j} + b_2 C_{2j} + \cdots + b_n C_{nj}) \right] \tag{2.35}$$

using the typical (jth) element of the vector on the right-hand side of the equation. The expression in round brackets is recognizable as a co-factor expansion down the jth column of an $n \times n$ matrix, where that jth column contains the elements of $\mathbf{b}$ and every other element of the matrix is identical to the corresponding element in $\mathbf{A}$. It is expedient to manufacture such a matrix, $\mathbf{A}_j$, and therefore be able to write $x_j = |\mathbf{A}_j|/|\mathbf{A}|$. □

EXAMPLE 2.5.1 Recall the example in Section 1.2.2 and consider the simplified macroeconomic model

$$C = \alpha_1 + \alpha_2 Y \tag{2.36}$$

$$Y = C + Z \tag{2.37}$$

where it is required to solve for C (consumption) and Y (national income) in terms of Z (autonomous expenditure, comprising investment, I, and government expenditure, G), α_1 and α_2. We may write this simple square system in the form $\mathbf{Ax} = \mathbf{b}$, where

$$\mathbf{A} = \begin{bmatrix} 1 & -\alpha_2 \\ -1 & 1 \end{bmatrix}, \quad \mathbf{x} = \begin{bmatrix} C \\ Y \end{bmatrix} \quad \text{and} \quad \mathbf{b} = \begin{bmatrix} \alpha_1 \\ Z \end{bmatrix} \tag{2.38}$$

The determinant of $\mathbf{A}$ is $a_{11}a_{22} - a_{12}a_{21} = 1 - \alpha_2$; hence, $\mathbf{A}$ is non-singular if $\alpha_2 \neq 1$. Economic theory suggests that the marginal propensity to consume is greater than zero and less than unity, so $0 < \alpha_2 < 1$; we may therefore suppose that $|\mathbf{A}| \neq 0$ and conclude that Cramer's theorem applies. It follows that

$$|\mathbf{A}_1| = \begin{vmatrix} \alpha_1 & -\alpha_2 \\ Z & 1 \end{vmatrix} \quad \text{and} \quad |\mathbf{A}_2| = \begin{vmatrix} 1 & \alpha_1 \\ -1 & Z \end{vmatrix} \tag{2.39}$$

$$x_1 = C = \frac{|\mathbf{A}_1|}{|\mathbf{A}|} = \frac{\alpha_1 + \alpha_2 Z}{1 - \alpha_2} = \frac{\alpha_1}{1 - \alpha_2} + \frac{\alpha_2}{1 - \alpha_2}Z \tag{2.40}$$

$$x_2 = Y = \frac{|\mathbf{A}_2|}{|\mathbf{A}|} = \frac{\alpha_1 + Z}{1 - \alpha_2} = \frac{\alpha_1}{1 - \alpha_2} + \frac{1}{1 - \alpha_2}Z \tag{2.41}$$

◇

This solution, though very simple, is not without interest. In particular, the coefficients of Z in the two equations are the so-called multipliers that measure the overall impact of a change in Z on C and Y, respectively. The second of these, $1/(1 - \alpha_2)$, or the reciprocal of the marginal propensity to save, is the elementary Keynesian multiplier from introductory macroeconomics.

EXAMPLE 2.5.2 As a second example, let us use Cramer's rule to solve the following system of equations for x_i, $i = 1, 2, 3$:

$$2x_1 + 2x_3 = 1 \tag{2.42}$$

$$x_1 - 2x_2 + x_3 = 2 \tag{2.43}$$

$$3x_1 + x_2 = 3 \tag{2.44}$$

This system is square, with three equations and three unknowns. Writing it in the form $\mathbf{Ax} = \mathbf{b}$, we have

$$\begin{bmatrix} 2 & 0 & 2 \\ 1 & -2 & 1 \\ 3 & 1 & 0 \end{bmatrix} \begin{bmatrix} x_1 \\ x_2 \\ x_3 \end{bmatrix} = \begin{bmatrix} 1 \\ 2 \\ 3 \end{bmatrix} \tag{2.45}$$

Using the co-factor expansion along the first row to exploit the zero value of a_{12}, we have

$$|\mathbf{A}| = a_{11}C_{11} + a_{12}C_{12} + a_{13}C_{13} = 2 \begin{vmatrix} -2 & 1 \\ 1 & 0 \end{vmatrix} + 0 + 2 \begin{vmatrix} 1 & -2 \\ 3 & 1 \end{vmatrix} = 12 \tag{2.46}$$

so $\mathbf{A}$ is non-singular. Proceeding with Cramer's rule, using appropriate co-factor expansions, we have

$$|\mathbf{A}_1| = \begin{vmatrix} 1 & 0 & 2 \\ 2 & -2 & 1 \\ 3 & 1 & 0 \end{vmatrix} = 15, \quad |\mathbf{A}_2| = \begin{vmatrix} 2 & 1 & 2 \\ 1 & 2 & 1 \\ 3 & 3 & 0 \end{vmatrix} = -9 \tag{2.47}$$

and

$$|\mathbf{A}_3| = \begin{vmatrix} 2 & 0 & 1 \\ 1 & -2 & 2 \\ 3 & 1 & 3 \end{vmatrix} = -9 \tag{2.48}$$

Therefore,

$$x_1 = \frac{|\mathbf{A}_1|}{|\mathbf{A}|} = \frac{15}{12} = \frac{5}{4} \tag{2.49}$$

$$x_2 = \frac{|\mathbf{A}_2|}{|\mathbf{A}|} = \frac{-9}{12} = -\frac{3}{4} \tag{2.50}$$

and, since $|\mathbf{A}_3| = |\mathbf{A}_2|$,

$$x_3 = x_2 = -\frac{3}{4} \tag{2.51}$$

$\diamond$

The solution obtained in Example 2.5.1 may be verified as correct by substituting the numerical values of the x_i into the original equations.

2.5.2   Scalar methods

Cramer's theorem makes it clear that having the same number of equations as unknowns is not a sufficient condition for the existence of a unique solution for a system of linear equations. In fact, this counting rule of thumb is neither necessary nor sufficient for the existence of a unique solution, as the following cases indicate.

1. The same number of equations and unknowns, but no unique solution:

$$x + y = 1 \tag{2.52}$$
$$2x + 2y = 2 \tag{2.53}$$

This is a case in which the equations are in fact coincident and the matrix of coefficients is singular. There are infinitely many solutions of the form $y = 1 - x$, where x can be any real number.

2. The same number of equations and unknowns, but no solution:

$$x + y = 1 \tag{2.54}$$
$$x + y = 2 \tag{2.55}$$

This is a case in which the equations do not coincide but the matrix of coefficients is still singular. This system is actually inconsistent in the sense described in Chapter 1. We will have more to say about consistency and inconsistency of systems of linear equations in Chapter 5; see, especially, Theorem 5.4.9.

3. More equations than unknowns, but a unique solution:

$$x = y \tag{2.56}$$
$$x + y = 2 \tag{2.57}$$
$$x - 2y + 1 = 0 \tag{2.58}$$

The solution in this case is $x = 1$, $y = 1$ and comes about due to one of the three equations being redundant. The properties of the matrix of coefficients in cases like this will be discussed further in Chapters 3 and 5.

4. In systems of nonlinear equations, it is possible to have fewer equations than unknowns, but a unique solution; e.g. $x^2 + y^2 = 0$, the solution for which is $x = 0$, $y = 0$.

In terms of the geometric representation of the simultaneous equation problem, in both the generic and linear cases, two curves in the coordinate plane can intersect in zero, one or more points; two surfaces in three-dimensional coordinate space typically intersect in a curve; and three surfaces in three-dimensional coordinate space can intersect in zero, one or more points. The need for a more precise solution theory and methodology is clear.

An approach to solving simultaneous equations that can be applied to both linear and nonlinear problems involves the following procedure.

1. Solution of one equation (say, the first) for a given variable in terms of the other variables.
2. Elimination of the given variable in all other equations by substitution using the solution from the previous step.

3. Repetition of step 1 to obtain a solution for one equation of the reduced system from step 2 for a given variable in terms of the other variables.
4. Repetition of step 2 for all remaining equations by appropriate substitution.
5. And so on, until only one equation and one variable remain.
6. Back-substitution to find numerical values for each variable.
7. Checking of the solution by substitution in the original system, which is a strongly recommended practice for any method of solving simultaneous equations.

This method will often help in finding the solution of systems of simultaneous nonlinear equations that arise as the first-order conditions in optimization problems; see Chapter 10.

EXAMPLE 2.5.3 Solve the equations

$$x + y = 2 \tag{2.59}$$
$$2y - x = 7 \tag{2.60}$$

for x and y.

Using the above approach, and solving the first equation for x in terms of y, we have

$$x = 2 - y \tag{2.61}$$

Elimination of x from the remaining equation gives

$$2y - (2 - y) = 7 \tag{2.62}$$

from which we have $3y = 9$ and therefore $y = 3$. Substitution in (2.61) may be used to determine x:

$$x = 2 - y = 2 - 3 = -1 \tag{2.63}$$

$\diamond$

It is a useful exercise to draw a diagram showing the two equations in this example and the solutions corresponding to their point of intersection; see Exercise 2.8.

EXAMPLE 2.5.4 Solve the equations

$$x + 2y + 3z = 6 \tag{2.64}$$
$$4x + 5y + 6z = 15 \tag{2.65}$$
$$7x + 8y + 10z = 25 \tag{2.66}$$

for x, y and z.

Here, we first solve one equation, equation (2.64), for x in terms of y and z:

$$x = 6 - 2y - 3z \tag{2.67}$$

Then we use this to eliminate x from the other two equations, to give

$$4(6 - 2y - 3z) + 5y + 6z = 15 \tag{2.68}$$
$$7(6 - 2y - 3z) + 8y + 10z = 25 \tag{2.69}$$

What remains is a 2×2 system:

$$-3y - 6z = -9 \tag{2.70}$$
$$-6y - 11z = -17 \tag{2.71}$$

Solving the first equation for y, we have

$$y = 3 - 2z \tag{2.72}$$

which allows us to eliminate y from the last equation, setting

$$-6(3 - 2z) - 11z = -17 \tag{2.73}$$

The solution for z is easily found from this, namely,

$$z = 1 \tag{2.74}$$

and substituting back appropriately gives $y = 1$ and $x = 1$. $\diamond$

2.5.3   *Elementary row operations*

Another alternative approach to that provided by Cramer's theorem involves **elementary row operations**, which can be performed on a system of simultaneous equations without changing the solution(s). There are three types of elementary row operation:

1. addition or subtraction of a multiple of one equation to or from another equation;
2. multiplication of a particular equation by a non-zero constant; and
3. interchange of two equations.

Note that each of these operations is reversible (invertible).

One algorithm for solving simultaneous equation systems using elementary row operations involves the following steps:

1. (a) Elimination of the first variable from all except the first equation.
   (b) Elimination of the second variable from all except the first two equations.
   (c) Elimination of the third variable from all except the first three equations.
   (d) And so on.
2. (a) Dividing the first equation by the coefficient of the first variable.
   (b) Dividing the second equation by the coefficient of the second variable.
   (c) Dividing the third equation by the coefficient of the third variable.
   (d) And so on.
3. Solution of the last equation, which is easy as it now includes only one variable.

4. Substitution of this solution in the second last equation to solve for the second last variable.
5. And so on, recursively.

Instead of finding the solution by recursive substitution, we could use further elementary row operations to remove the last variable from all the equations except the last, to remove the second last variable from all the equations except the second last, and so on, producing a system with only one variable remaining in each equation. The second type of elementary row operation could then be used on each equation in turn to find the final solution.

EXAMPLE 2.5.5 Let us repeat Example 2.5.2 using elementary row operations.
Adding the first equation to the second gives

$$x + y = 2 \tag{2.75}$$
$$3y = 9 \tag{2.76}$$

Thus, from the second equation we have $y = 3$. Substitution of this value into the first equation then yields $x = -1$. ◇

EXAMPLE 2.5.6 Let us repeat Example 2.5.2 using elementary row operations.

$$x + 2y + 3z = 6 \tag{2.77}$$
$$4x + 5y + 6z = 15 \tag{2.78}$$
$$7x + 8y + 10z = 25 \tag{2.79}$$

We can add -4 times (2.77) to (2.78) and -7 times (2.77) to (2.79) to obtain

$$x + 2y + 3z = 6 \tag{2.80}$$
$$-3y - 6z = -9 \tag{2.81}$$
$$-6y - 11z = -17 \tag{2.82}$$

Now we can add -2 times (2.81) to (2.82) to obtain

$$x + 2y + 3z = 6 \tag{2.83}$$
$$-3y - 6z = -9 \tag{2.84}$$
$$z = 1 \tag{2.85}$$

This gives us the solution for z, and back-substitution or further elementary row operations will give the solutions for x and y. ◇

2.5.4 Matrix representation of elementary row operations

The elementary row operations performed in the previous section are effectively operations on the matrix of coefficients in the system of simultaneous linear equations. If we write the numbers that appear in the 3×3 system in Example 2.5.2 as a 3×4 matrix,

$$\begin{bmatrix} 1 & 2 & 3 & 6 \\ 4 & 5 & 6 & 15 \\ 7 & 8 & 10 & 25 \end{bmatrix} \tag{2.86}$$

we can perform exactly the same elementary row operations on this **augmented matrix**, with no need to write out "x", "y", "z", "$\pm$" and "$=$" repeatedly at each stage.

EXAMPLE 2.5.7 The augmented matrices corresponding to the two steps in Example 2.5.3 are

$$\begin{bmatrix} 1 & 2 & 3 & 6 \\ 0 & -3 & -6 & -9 \\ 0 & -6 & -11 & -17 \end{bmatrix} \quad \text{and} \quad \begin{bmatrix} 1 & 2 & 3 & 6 \\ 0 & -3 & -6 & -9 \\ 0 & 0 & 1 & 1 \end{bmatrix} \tag{2.87}$$

$\diamond$

We can use the augmented matrix method to solve several systems of equations with the same matrix on the left-hand side all at once. The matrix equations $\mathbf{Ax} = \mathbf{b}_1$, $\mathbf{Ax} = \mathbf{b}_2$ and $\mathbf{Ax} = \mathbf{b}_3$ can be combined into the single augmented matrix $[\mathbf{A} \ \mathbf{b}_1 \ \mathbf{b}_2 \ \mathbf{b}_3]$. This is equivalent to solving the matrix equation $\mathbf{AX} = \mathbf{B}$, where $\mathbf{b}_1$, $\mathbf{b}_2$ and $\mathbf{b}_3$ are the three columns of the matrix $\mathbf{B}$.

A particular example of this type of matrix equation has $\mathbf{X} = \mathbf{A}^{-1}$ and $\mathbf{B} = \mathbf{I}$, i.e. $\mathbf{AA}^{-1} = \mathbf{I}$. So another method of finding the inverse of an $n \times n$ matrix $\mathbf{A}$ is to apply elementary row operations to the $n \times 2n$ augmented matrix $[\mathbf{A} \ \mathbf{I}]$.

EXAMPLE 2.5.8 Using the coefficient matrix from Example 2.5.2 again, we can start with the augmented matrix

$$\begin{bmatrix} 1 & 2 & 3 & 1 & 0 & 0 \\ 4 & 5 & 6 & 0 & 1 & 0 \\ 7 & 8 & 10 & 0 & 0 & 1 \end{bmatrix} \tag{2.88}$$

Repeating the elementary row operations carried out in Examples 2.5.3 and 2.5.4 turns this into

$$\begin{bmatrix} 1 & 2 & 3 & 1 & 0 & 0 \\ 0 & -3 & -6 & -4 & 1 & 0 \\ 0 & -6 & -11 & -7 & 0 & 1 \end{bmatrix} \tag{2.89}$$

and then

$$\begin{bmatrix} 1 & 2 & 3 & 1 & 0 & 0 \\ 0 & -3 & -6 & -4 & 1 & 0 \\ 0 & 0 & 1 & 1 & -2 & 1 \end{bmatrix} \tag{2.90}$$

A few more operations are required to turn the left-hand partition of the augmented matrix into the 3×3 identity matrix. Multiplying the second row by $-1/3$ produces

$$\begin{bmatrix} 1 & 2 & 3 & 1 & 0 & 0 \\ 0 & 1 & 2 & \frac{4}{3} & -\frac{1}{3} & 0 \\ 0 & 0 & 1 & 1 & -2 & 1 \end{bmatrix} \tag{2.91}$$

Subtracting twice row 3 from row 2 and three times row 3 from row 1 produces

$$\begin{bmatrix} 1 & 2 & 0 & -2 & 6 & -3 \\ 0 & 1 & 0 & -\frac{2}{3} & \frac{11}{3} & -2 \\ 0 & 0 & 1 & 1 & -2 & 1 \end{bmatrix} \tag{2.92}$$

Finally, subtracting twice row 2 from row 1 produces

$$\begin{bmatrix} 1 & 0 & 0 & -\frac{2}{3} & -\frac{4}{3} & 1 \\ 0 & 1 & 0 & -\frac{2}{3} & \frac{11}{3} & -2 \\ 0 & 0 & 1 & 1 & -2 & 1 \end{bmatrix} \tag{2.93}$$

This tells us that

$$\begin{bmatrix} 1 & 2 & 3 \\ 4 & 5 & 6 \\ 7 & 8 & 10 \end{bmatrix}^{-1} = \begin{bmatrix} -\frac{2}{3} & -\frac{4}{3} & 1 \\ -\frac{2}{3} & \frac{11}{3} & -2 \\ 1 & -2 & 1 \end{bmatrix} \tag{2.94}$$

which is easily verified by multiplying the two matrices.   ◇

2.5.5 Elementary matrices

So far, we have seen that elementary row operations may be applied to scalar equations or to the rows of an augmented matrix to obtain a solution for a set of linear simultaneous equations or the inverse of a matrix. The concept of an elementary matrix may be used to summarize these operations.

An **elementary matrix** is a square matrix derived from an identity matrix by performing a single elementary row operation. For example, consider the following two matrices:

$$\mathbf{E} = \begin{bmatrix} 0 & 1 & 0 \\ 1 & 0 & 0 \\ 0 & 0 & 1 \end{bmatrix} \quad \text{and} \quad \mathbf{E}^* = \begin{bmatrix} 0 & 0 & 1 \\ 0 & 1 & 0 \\ 1 & 0 & 0 \end{bmatrix} \tag{2.95}$$

The matrix $\mathbf{E}$ is the elementary matrix formed by interchanging the first and the second rows (or columns) of $\mathbf{I}_3$, while $\mathbf{E}^*$ is the elementary matrix formed by interchanging the first and the third rows (or columns) of $\mathbf{I}_3$.

It is easily seen that pre-multiplying any matrix $\mathbf{A}$ by a conformable elementary matrix $\mathbf{E}$ has the same effect on $\mathbf{A}$ as the elementary row operation that was performed on the identity matrix to form $\mathbf{E}$.

To confirm this, the reader might investigate the nature of the products $\mathbf{AE}$, $\mathbf{EA}$, $\mathbf{AE}^*$ and $\mathbf{E}^*\mathbf{A}$, using matrix multiplication and a 3×3 matrix, $\mathbf{A}$, of his or her choice; see Exercise 2.9.

Similarly, the 3×3 matrix that corresponds to the elementary row operation of adding twice row 2 to row 1 and the 4×4 matrix that corresponds to the elementary row operation of multiplying row 4 by -3 are respectively

$$
\begin{bmatrix} 1 & 2 & 0 \\ 0 & 1 & 0 \\ 0 & 0 & 1 \end{bmatrix} \quad \text{and} \quad \begin{bmatrix} 1 & 0 & 0 & 0 \\ 0 & 1 & 0 & 0 \\ 0 & 0 & 1 & 0 \\ 0 & 0 & 0 & -3 \end{bmatrix} \tag{2.96}
$$

In the examples above, we have sometimes performed two elementary row operations in a single step. For example, going from (2.91) to (2.92), we implicitly pre-multiplied the augmented matrix by two elementary matrices

$$
\mathbf{E}_1 \equiv \begin{bmatrix} 1 & 0 & 0 \\ 0 & 1 & -2 \\ 0 & 0 & 1 \end{bmatrix} \quad \text{and} \quad \mathbf{E}_2 \equiv \begin{bmatrix} 1 & 0 & -3 \\ 0 & 1 & 0 \\ 0 & 0 & 1 \end{bmatrix} \tag{2.97}
$$

Note that these elementary matrices commute, so that

$$
\mathbf{E}_1\mathbf{E}_2 = \mathbf{E}_2\mathbf{E}_1 = \begin{bmatrix} 1 & 0 & -3 \\ 0 & 1 & -2 \\ 0 & 0 & 1 \end{bmatrix} \tag{2.98}
$$

2.5.6   Row-echelon forms

Applying elementary row operations to the augmented matrix formed from an invertible matrix and the identity matrix or pre-multiplying the augmented matrix by the corresponding elementary matrices will produce the inverse of the original matrix. Applying the same techniques to a matrix that is not invertible (and not necessarily square) can reduce it to either of two forms, which resemble the identity matrix, and which we now define.

DEFINITION 2.5.1

(a) A matrix is in **row-echelon form** if:

(i) every row either consists entirely of zeros or else has non-zero entries that begin with a **leading 1**;

(ii) any rows consisting entirely of zeros are grouped together at the bottom of the matrix; and

(iii) the leading 1 in any non-zero row is to the right of any leading 1 in a row nearer the top of the matrix.

(b) A matrix is in **reduced row-echelon form** if it is in row-echelon form and each of its columns that contains a leading 1 has zeros on every other row.

An obvious example of a matrix in row-echelon form is an upper triangular matrix with ones on the leading diagonal. An obvious example of a matrix in reduced row-echelon form is an identity matrix. A partitioned matrix with an identity matrix in the top left corner, anything in the top right corner, and zeros in the bottom left and bottom right corners is also in reduced row-echelon form.

Elementary row operations can be used to reduce any matrix (typically an augmented matrix) to row-echelon form. This procedure is known as **Gaussian elimination**. Further elementary row operations can then be used to reduce this row-echelon form to reduced row-echelon form. This procedure is known as **Gauss–Jordan elimination**.[3]

The difference between the Gaussian method and the Gauss–Jordan method when applied to solving simultaneous linear equations is that the former reduces a non-singular system to a form where the last variable is determined and the other variables can be found by back-substitution, while the latter reduces a non-singular system to a form where the entire solution can be seen immediately.

The two solution methods described in Section 2.5.3 are effectively Gaussian elimination and Gauss–Jordan elimination.

If the matrix $\mathbf{A}$ is not invertible (including the case where $\mathbf{A}$ is not square), then it will be found that Gaussian elimination reduces it to a form in which the last row(s) or last column(s) consist entirely of zeros.

Exercise 2.11 asks the reader to solve a system of simultaneous linear equations by Gaussian elimination.

Each step in Gauss–Jordan elimination amounts to pre-multiplying the augmented matrix by an elementary matrix, say six of them,

$$\mathbf{E}_6\mathbf{E}_5\mathbf{E}_4\mathbf{E}_3\mathbf{E}_2\mathbf{E}_1[\mathbf{A}\ \ \mathbf{b}] = [\mathbf{I}\ \ \mathbf{x}] \tag{2.99}$$

It follows that

$$\mathbf{E}_6\mathbf{E}_5\mathbf{E}_4\mathbf{E}_3\mathbf{E}_2\mathbf{E}_1\mathbf{A} = \mathbf{I} \tag{2.100}$$

and

$$\mathbf{E}_6\mathbf{E}_5\mathbf{E}_4\mathbf{E}_3\mathbf{E}_2\mathbf{E}_1\mathbf{b} = \mathbf{x} \tag{2.101}$$

Equation (2.100) implies that

$$\mathbf{A}^{-1} = \mathbf{E}_6\mathbf{E}_5\mathbf{E}_4\mathbf{E}_3\mathbf{E}_2\mathbf{E}_1 \tag{2.102}$$

This confirms that Gauss–Jordan elimination may be used to solve for each of the columns of the inverse, or to solve for the whole thing at once, and that the solution may be written as

$$\mathbf{x} = \mathbf{E}_6\mathbf{E}_5\mathbf{E}_4\mathbf{E}_3\mathbf{E}_2\mathbf{E}_1\mathbf{b} = \mathbf{A}^{-1}\mathbf{b} \tag{2.103}$$

We will use products of elementary matrices again in the proof of Theorem 4.4.16. This approach can be extended to provide a proof that the determinant of a product is the product of the determinants, and a proof that invertibility (existence of $\mathbf{A}^{-1}$) is a necessary as well as a sufficient condition for non-singularity ($\det(\mathbf{A}) \neq 0$); this is the approach used in the previously cited proofs by Anton and Rorres (2010, pp. 108–9).

EXERCISES

2.1 Complete the contents of Table 2.1, and hence determine all of the signed elementary products that may be formed from the matrix $A = [a_{ij}]_{3 \times 3}$.

2.2 Given the matrix in (2.18), calculate det(A) using the co-factor expansions along each row and each column, and confirm that your six answers are identical.

2.3 Write down a matrix of order 3×3 with numerical elements. Find first its square and then its cube, checking the latter by using the two processes $A(A^2)$ and $(A^2)A$. Compute $|A|$, $|A^2|$ and $|A^3|$, and comment on your findings.

2.4 Given $A = \begin{bmatrix} 1 & 3 & 2 \\ 2 & 6 & 9 \\ 7 & 6 & 1 \end{bmatrix}$ and $E = \begin{bmatrix} 0 & 1 & 0 \\ 1 & 0 & 0 \\ 0 & 0 & 1 \end{bmatrix}$, calculate det(A), det(E) and det(B), where $B = EA$. Use a co-factor expansion down the third column for at least one of your calculations, and a direct method using the elementary products of one of the 3×3 matrices for another. Verify that $\det(B) = \det(E)\det(A)$.

2.5 Let L be an $n \times n$ lower triangular matrix. Prove that $\det(L) = l_{11}l_{22}\ldots l_{nn}$, where l_{ii} is the typical element on the principal diagonal of L.

2.6 Use Cramer's rule to solve the system of equations

$$x_1 + x_2 + x_3 = 6$$
$$2x_1 - x_2 + 2x_3 = 6$$
$$-x_1 - x_2 = -5$$

2.7 Use Cramer's rule to solve the macroeconomic model

$$C = f(Y, Y_{[-1]}) = \alpha_1 + \alpha_2 Y + \alpha_3 Y_{[-1]}$$
$$I = g(Y, R) = \beta_1 + \beta_2 Y + \beta_3 R$$
$$Y = C + I + G$$

for the endogenous variables C, I and Y in terms of the predetermined variables R, G and $Y_{[-1]}$, where $Y_{[-1]}$ denotes a lagged value of Y, and the parameters α_i and β_i ($i = 1, 2, 3$). Comment on the economic interpretation of your results, given $0 < \alpha_2 < 1$, $0 < \alpha_3 < \alpha_2$, $0 < \beta_2 < 1$ and $\beta_3 < 0$.

2.8 Solve the following equation system using an approach other than Cramer's rule, and illustrate your solution graphically:

$$x + y = 2$$
$$2y - x = 7$$

2.9 Let $\mathbf{A} = \begin{bmatrix} 1 & 2 & 3 \\ 4 & 5 & 6 \\ 7 & 8 & 9 \end{bmatrix}$ and define the matrices

$$\mathbf{E}_1 = \begin{bmatrix} 1 & 0 & 0 \\ 0 & 0 & 1 \\ 0 & 1 & 0 \end{bmatrix}, \quad \mathbf{E}_2 = \begin{bmatrix} 0 & 0 & 1 \\ 0 & 1 & 0 \\ 1 & 0 & 0 \end{bmatrix} \quad \text{and} \quad \mathbf{E}_3 = \begin{bmatrix} 0 & 0 & 1 \\ 0 & 0 & 0 \\ 1 & 0 & 0 \end{bmatrix}$$

Examine the matrix products $\mathbf{AE}_i$, $\mathbf{AE}_i^\top$, $\mathbf{E}_i\mathbf{A}$ and $\mathbf{E}_i^\top\mathbf{A}$ for $i = 1, 2, 3$, and comment on your findings.

2.10 Let $\mathbf{V} = \begin{bmatrix} 0 & 1 & 0 \\ 0 & 0 & 1 \\ 0 & 0 & 0 \end{bmatrix}$. Find $\mathbf{V}^2$ and $\mathbf{V}^3$; and examine the products $\mathbf{VA}$, $\mathbf{V}^2\mathbf{A}$ and $\mathbf{V}^\top\mathbf{A}$, where $\mathbf{A}$ is a square matrix of your choice and $\top$ denotes matrix transposition. Examine $|\mathbf{V}|$, $|\mathbf{V}^2|$, $|\mathbf{V}^3|$, $|\mathbf{VA}|$, $|\mathbf{V}^2\mathbf{A}|$ and $|\mathbf{V}^3\mathbf{A}|$, and comment on your findings.

2.11 Use Gaussian elimination to solve the following system of equations for x, y and z:

$$x + 2y + 3z = 6$$
$$4x + 5y + 6z = 15$$
$$7x + 8y + 10z = 25$$

2.12 Prove that the determinant of an orthogonal matrix has the value $+1$ or -1.

2.13 A square matrix, $\mathbf{A}$, is said to be **skew-symmetric** iff $\mathbf{A}^\top = -\mathbf{A}$.

(a) Prove that the elements on the principal diagonal of a skew-symmetric matrix are all zero.
(b) What is the trace of a skew-symmetric matrix?
(c) Prove that the determinant of a skew-symmetric matrix of odd order is zero.
(d) Investigate whether the 3×3 matrix

$$\mathbf{A} = \begin{bmatrix} 0 & 1 & 2 \\ -1 & 0 & -3 \\ -2 & 3 & 0 \end{bmatrix}$$

is skew-symmetric, and compute its determinant.

2.14 Construct elementary matrices that by pre-multiplying the matrix

$$\mathbf{A} = \begin{bmatrix} 1 & 2 & 3 \\ 4 & 5 & 6 \\ 7 & 8 & 9 \end{bmatrix}$$

effect the following row operations:

(a) The addition of twice row 1 to row 2. Let the elementary matrix in this case be $\mathbf{E}_1$.
(b) The multiplication of row 3 by 0.5. Let the elementary matrix in this case be $\mathbf{E}_2$.

(c) The interchange of rows 2 and 3. Let the elementary matrix in this case be $\mathbf{E}_3$.

Compare the results of $\mathbf{E}_3\mathbf{E}_2\mathbf{E}_1\mathbf{A}$, $\mathbf{E}_1\mathbf{E}_2\mathbf{E}_3\mathbf{A}$ and $\mathbf{A}\mathbf{E}_3\mathbf{E}_2\mathbf{E}_1$.

2.15 Let $\mathbf{A} = \mathrm{diag}[a_i]_{n\times n}$ and $\mathbf{B} = \mathrm{diag}[b_i]_{n\times n}$. Prove that $\det(\mathbf{AB}) = \prod_{i=1}^{n} a_i b_i$.

2.16 Compute the inverses of the matrices

$$\mathbf{A} = \begin{bmatrix} 2 & -4 \\ 7 & -5 \end{bmatrix}, \quad \mathbf{B} = \begin{bmatrix} 2 & 0 & 6 \\ -1 & -2 & 0 \\ 5 & 3 & 1 \end{bmatrix}$$

using:

(a) the inverse formula given in Theorem 2.4.3; and
(b) only elementary row operations.

2.17 State and prove Cramer's theorem. Try to find an alternative proof to the one presented above for Theorem 2.5.1.

3 Eigenvalues and eigenvectors

3.1 Introduction

Some of the basic ideas and issues encountered in the previous chapters are often covered in an introductory course in mathematics for economics and finance. The fundamental ideas of eigenvalues and eigenvectors and the associated theorems introduced in this chapter are probably not. Many readers are therefore likely to be encountering these concepts for the first time. Hence this chapter begins by providing definitions and illustrations of eigenvalues and eigenvectors, and explaining how they can be calculated. It goes on to examine some of the uses of these concepts and to establish a number of theorems relating to them that will be useful when we return to the detailed analysis of our various applications.

3.2 Definitions and illustration

Eigenvalues and eigenvectors arise in determining solutions to equations of the form

$$\mathbf{A}\mathbf{x} = \lambda \mathbf{x} \tag{3.1}$$

where $\mathbf{A}$ is an $n \times n$ matrix, $\mathbf{x}$ is a non-zero n-vector and λ is a scalar, and where the solution is for λ and $\mathbf{x}$, given $\mathbf{A}$. We shall call equations like (3.1) **eigenequations**. The scalar λ is called an **eigenvalue** of $\mathbf{A}$, while $\mathbf{x}$ is known as an **eigenvector** of $\mathbf{A}$ associated with λ. Sometimes the value, λ, and the vector, $\mathbf{x}$, are called the **proper**, **characteristic** or **latent** value and vector.

Consider the matrix $\mathbf{A} = \begin{bmatrix} 2 & 0 \\ 8 & -2 \end{bmatrix}$ and the vector $\mathbf{x} = \begin{bmatrix} 1 \\ 2 \end{bmatrix}$. Since

$$\mathbf{A}\mathbf{x} = \begin{bmatrix} 2 & 0 \\ 8 & -2 \end{bmatrix}\begin{bmatrix} 1 \\ 2 \end{bmatrix} = \begin{bmatrix} 2 \\ 4 \end{bmatrix} = 2\mathbf{x} \tag{3.2}$$

$\lambda = 2$ is an eigenvalue of $\mathbf{A}$ and $\mathbf{x}$ is an associated eigenvector.

It is easy to check, by substituting into the eigenequation (3.1), that another eigenvector of $\mathbf{A}$ associated with $\lambda = 2$ is $[-1 \ \ -2]^\top$. Likewise, another eigenvalue of $\mathbf{A}$ is -2, which has associated with it eigenvectors such as $[0 \ 1]^\top$ and $[0 \ -1]^\top$. Thus, for a given λ, we note that there are multiple associated eigenvectors.

For given λ and $\mathbf{x}$, $\mathbf{A}$ may be viewed as the matrix that, by pre-multiplication, changes all of the elements of $\mathbf{x}$ by the same proportion, λ. When $\lambda > 1$, as in the case of the first eigenvalue in our illustration, the elements of $\mathbf{x}$ are increased in absolute value; when $\lambda < 0$, as in the case of the second eigenvalue, the elements change in sign. If $0 < |\lambda| < 1$, then the elements of **x** would be made smaller in absolute value; they would also change sign, if λ was negative.

In this illustration, the numerical eigenvalues and eigenvectors were simply stated. An important question is, if a matrix **A** is given, how can its eigenvalues and eigenvectors be determined? We now develop an answer to this question.

3.3   Computation

3.3.1   Eigenvalues

Consider the eigenequation $\mathbf{Ax} = \lambda\mathbf{x}$ for the $n \times n$ matrix **A**. We may rewrite this as $\mathbf{Ax} = \lambda\mathbf{Ix}$ or $(\mathbf{A} - \lambda\mathbf{I})\mathbf{x} = \mathbf{0}$ or $\mathbf{Bx} = \mathbf{0}$; in other words, as a square homogeneous system of equations. If $\mathbf{B} = \mathbf{A} - \lambda\mathbf{I}$ is non-singular, hence invertible, the solution for **x** is $\mathbf{x} = \mathbf{B}^{-1}\mathbf{0} = \mathbf{0}$, but this trivial result is ruled out by the requirement that an eigenvector is non-zero. For there to be a non-trivial, i.e. non-zero, solution to this system, it must be that $|\mathbf{B}| = |\mathbf{A} - \lambda\mathbf{I}| = 0$. This **determinantal equation** is known as the **characteristic equation** of the matrix **A**. Note that **B** is formed by subtracting λ from each of the principal diagonal elements of **A**. Each term in the expansion of $|\mathbf{B}|$ will contain between 0 and n diagonal elements, so each will yield a polynomial in λ of degree $\leq n$, with one term producing a polynomial of exactly degree n, namely the term $\prod_{i=1}^{n}(a_{ii} - \lambda)$. Thus, collecting the various powers of λ together, evaluation of the left-hand side of the characteristic equation yields the **characteristic polynomial** in λ:

$$|\mathbf{A} - \lambda\mathbf{I}| = k_0\lambda^n + k_1\lambda^{n-1} + k_2\lambda^{n-2} + \cdots + k_n \tag{3.3}$$

where $k_0 = (-1)^n$. In principle, on equating to zero, the characteristic equation may be solved for λ, though solution for large n may be problematical.

Note that, when $\lambda = 0$, equation (3.3) becomes $k_n = |\mathbf{A}|$; also note that the values of λ that satisfy the characteristic polynomial are referred to as its **roots**; and that, by the fundamental theorem!of algebra, allowing for the possibility of pairs of conjugate complex roots and several roots having the same value, the characteristic polynomial has exactly n roots.

We will encounter later conditions that guarantee that the eigenvalues of a matrix are real numbers.

EXAMPLE 3.3.1 To take an example with small n, consider the matrix **A** from the illustration in Section 3.2. In this case,

$$|\mathbf{A} - \lambda\mathbf{I}| = \begin{vmatrix} 2-\lambda & 0 \\ 8 & -2-\lambda \end{vmatrix} = (2-\lambda)(-2-\lambda) - 0 \times 8 = \lambda^2 - 4 \tag{3.4}$$

Equating this characteristic polynomial of degree two to zero gives

$$\lambda^2 - 4 = 0 \tag{3.5}$$

$$(\lambda + 2)(\lambda - 2) = 0 \tag{3.6}$$

and therefore $\lambda = 2$ or $\lambda = -2$, as stated in the illustration. $\diamond$

EXAMPLE 3.3.2 Taking another 2×2 example, let

$$A = \begin{bmatrix} 3 & 2 \\ -1 & 0 \end{bmatrix} \tag{3.7}$$

so that

$$|A - \lambda I| = \begin{vmatrix} 3 - \lambda & 2 \\ -1 & -\lambda \end{vmatrix} = (3 - \lambda)(-\lambda) - 2 \times (-1) = \lambda^2 - 3\lambda + 2 \tag{3.8}$$

Equating to zero yields

$$\lambda^2 - 3\lambda + 2 = 0 \tag{3.9}$$

$$(\lambda - 2)(\lambda - 1) = 0 \tag{3.10}$$

and $\lambda = 1$ or $\lambda = 2$. ◇

EXAMPLE 3.3.3 To indicate how the complexity of the solution increases with n, consider as a third example the case of the 3×3 matrix

$$A = \begin{bmatrix} 0 & 1 & 0 \\ 0 & 0 & 1 \\ 4 & -17 & 8 \end{bmatrix} \tag{3.11}$$

So

$$|A - \lambda I| = \begin{vmatrix} -\lambda & 1 & 0 \\ 0 & -\lambda & 1 \\ 4 & -17 & 8 - \lambda \end{vmatrix} \tag{3.12}$$

In this case, using a co-factor expansion along the first row in order to take advantage of the zero a_{13} element in $|A - \lambda I|$, we have

$$|A - \lambda I| = -\lambda \begin{vmatrix} -\lambda & 1 \\ -17 & 8 - \lambda \end{vmatrix} - \begin{vmatrix} 0 & 1 \\ 4 & 8 - \lambda \end{vmatrix}$$

$$= -\lambda[-\lambda(8 - \lambda) + 17] - (-4)$$

$$= -\lambda^3 + 8\lambda^2 - 17\lambda + 4 \tag{3.13}$$

Therefore, the characteristic equation is the cubic equation in λ:

$$|A - \lambda I| = -\lambda^3 + 8\lambda^2 - 17\lambda + 4 = 0 \tag{3.14}$$

Solution of this equation is not trivial. However, using the standard result that any integer solution of such a polynomial must be a divisor of the constant term, i.e. ± 1, ± 2 or ± 4 in this case, one of the roots, namely $\lambda = 4$, may be found by trial and error. It follows that $\lambda - 4$ must be a factor and, hence, using polynomial long division, that the characteristic equation may be written as

$$(\lambda - 4)(-\lambda^2 + 4\lambda - 1) = 0 \tag{3.15}$$

Therefore, solution of the quadratic

$$-\lambda^2 + 4\lambda - 1 = 0 \tag{3.16}$$

will give the remaining two eigenvalues. Using the standard formula for the roots of a quadratic gives

$$\lambda = \frac{-4 \pm \sqrt{16 - 4}}{-2} \tag{3.17}$$

$$\lambda = 2 + \sqrt{3} \quad \text{and} \quad \lambda = 2 - \sqrt{3} \tag{3.18}$$

$\diamondsuit$

Only very minor changes in the elements of the 3×3 matrix in Example 3.3.1, e.g. replacing $a_{32} = -17$ by $a_{32} = 1$, would lead to very much greater computational complexity, and elements of eigenvectors, as well as eigenvalues, that are complex numbers. To see, more simply, how complex eigenvalues might arise, consider

$$\mathbf{A} = \begin{bmatrix} 0 & 1 \\ -1 & 0 \end{bmatrix} \tag{3.19}$$

The characteristic equation in this case is

$$|\mathbf{A} - \lambda \mathbf{I}| = \begin{vmatrix} -\lambda & 1 \\ -1 & -\lambda \end{vmatrix} = \lambda^2 + 1 = 0 \tag{3.20}$$

the solutions for which are easily found to be $\lambda = \pm\sqrt{-1}$. More generally, complex roots take the form $a \pm bi$, where a and b are real numbers and i denotes the imaginary number $\sqrt{-1}$, as in the next example.[1] The example shows that matrices do not have to be large or have numerically complicated elements to have complex eigenvalues.

EXAMPLE 3.3.4 Let

$$\mathbf{A} = \begin{bmatrix} 1 & -2 \\ 3 & -2 \end{bmatrix} \tag{3.21}$$

Then

$$|\mathbf{A} - \lambda \mathbf{I}| = \begin{vmatrix} 1 - \lambda & -2 \\ 3 & -2 - \lambda \end{vmatrix} = (1 - \lambda)(-2 - \lambda) + 6 = \lambda^2 + \lambda + 4 = 0 \tag{3.22}$$

is the characteristic equation, and the solutions for this are

$$\lambda = \frac{-1 \pm \sqrt{1 - 16}}{2} = \frac{-1 \pm \sqrt{-15}}{2} \tag{3.23}$$

$$\lambda = -\frac{1}{2} + \frac{\sqrt{15}}{2}i \quad \text{and} \quad \lambda = -\frac{1}{2} - \frac{\sqrt{15}}{2}i \tag{3.24}$$

The calculation of the eigenvectors is left as an exercise; see Exercise 3.1. $\diamondsuit$

3.3.2 *Eigenvectors*

Given an eigenvalue of some $n \times n$ matrix, $\mathbf{A}$, the corresponding eigenvectors are those $n \times 1$ matrices, $\mathbf{x} \neq \mathbf{0}$, satisfying $\mathbf{Ax} = \lambda\mathbf{x}$ or $(\mathbf{A} - \lambda\mathbf{I})\mathbf{x} = \mathbf{0}$. Consider the solution of this equation for the second of our previous examples, Example 3.3.1, in which $\mathbf{A} = \begin{bmatrix} 3 & 2 \\ -1 & 0 \end{bmatrix}$ and $\lambda_1 = 1$ and $\lambda_2 = 2$, where we now use subscripts to distinguish the different eigenvalues. We have, for λ_1, using an obvious subscript notation for the individual elements of the corresponding eigenvector, $\mathbf{x}_1$,

$$(\mathbf{A} - \lambda_1\mathbf{I})\mathbf{x}_1 = \begin{bmatrix} 2 & 2 \\ -1 & -1 \end{bmatrix}\begin{bmatrix} x_{11} \\ x_{21} \end{bmatrix} = \begin{bmatrix} 0 \\ 0 \end{bmatrix} \tag{3.25}$$

i.e. using the operation of matrix multiplication,

$$2x_{11} + 2x_{21} = 0 \tag{3.26}$$
$$-x_{11} - x_{21} = 0 \tag{3.27}$$

Note that the second equation is simply a scalar multiple $\left(-\frac{1}{2}\right)$ of the first.[2] Hence either equation yields $x_{11} = -x_{21}$ or

$$\mathbf{x}_1 = \begin{bmatrix} -x_{21} \\ x_{21} \end{bmatrix} = s\begin{bmatrix} -1 \\ 1 \end{bmatrix} \tag{3.28}$$

where s is an arbitrary non-zero scalar. There are therefore infinitely many eigenvectors corresponding to the eigenvalue λ_1.

Similarly, for λ_2, we have

$$(\mathbf{A} - \lambda_2\mathbf{I})\mathbf{x}_2 = \begin{bmatrix} 1 & 2 \\ -1 & -2 \end{bmatrix}\begin{bmatrix} x_{12} \\ x_{22} \end{bmatrix} = \begin{bmatrix} 0 \\ 0 \end{bmatrix} \tag{3.29}$$

or the equations

$$x_{12} + 2x_{22} = 0 \tag{3.30}$$
$$-x_{12} - 2x_{22} = 0 \tag{3.31}$$

Thus the (infinitely many) eigenvectors corresponding to λ_2 are

$$\mathbf{x}_2 = \begin{bmatrix} -2x_{22} \\ x_{22} \end{bmatrix} = t\begin{bmatrix} -2 \\ 1 \end{bmatrix} \tag{3.32}$$

where t is an arbitrary non-zero scalar.

The calculation of the eigenvectors corresponding to the eigenvalues $\lambda_1 = 2$ and $\lambda_2 = -2$, which are given in our illustration in Section 3.2, would be a useful exercise; see Exercise 3.2. Such calculations for the 2×2 case pose no difficulty but, as n increases, things are not so simple, and there arises a need for a more general method of solving systems of homogeneous equations than Cramer's rule, which was introduced and used in Chapter 2. Cramer's rule is not applicable to the calculation of eigenvectors because of the non-singularity of $\mathbf{A} - \lambda\mathbf{I}$.

3.3.3 Normalization

The infinity of eigenvectors corresponding to any one eigenvalue may be something of an inconvenience. To avoid the arbitrariness in the choice of eigenvectors, a unique **x** may be chosen according to some **normalization** rule. The most common form of normalization is to make **x** that eigenvector, the squares of whose elements sum to unity. This rule is not as strange as it may first appear because, as will be seen later, it corresponds to the idea of an eigenvector with unit **length** when the eigenvector is given a geometric interpretation. Thus, a vector with this property is sometimes called a unit vector (see Definition 5.2.12); but for the present we shall concentrate on the algebra of normalization.

To illustrate using a concrete example, let us take the case of the eigenvector $\mathbf{x}_1 = [-s \; s]^\top$ derived in the previous subsection. We want to choose s such that

$$\mathbf{x}_1^\top \mathbf{x}_1 = (-s)^2 + s^2 = 1 \tag{3.33}$$

or $2(s)^2 = 1$, hence $s = \pm\frac{1}{\sqrt{2}}$. Thus, taking the positive square root for s, a normalized eigenvector corresponding to $\lambda_1 = 1$ is $\mathbf{x}_1 = \left[\frac{-1}{\sqrt{2}} \; \frac{1}{\sqrt{2}}\right]^\top$. Note, however, that $-\mathbf{x}_1 = \left[\frac{1}{\sqrt{2}} \; \frac{-1}{\sqrt{2}}\right]^\top$ is also a normalized eigenvector corresponding to $\lambda_1 = 1$. Similarly, though the details of the calculations are left as an exercise, we find that normalized eigenvectors corresponding to $\lambda_2 = 2$ are $\mathbf{x}_2 = \left[\frac{-2}{\sqrt{5}} \; \frac{1}{\sqrt{5}}\right]^\top$ and $-\mathbf{x}_2 = \left[\frac{2}{\sqrt{5}} \; \frac{-1}{\sqrt{5}}\right]^\top$, i.e. in (3.32) we choose $t = \frac{1}{\sqrt{5}}$ and $t = \frac{-1}{\sqrt{5}}$ for $\mathbf{x}_2$ and $-\mathbf{x}_2$, respectively. When there is a repeated eigenvalue, there can be an infinity of associated normalized eigenvectors; for an example of this, see Exercise 3.6.

3.4 Unit eigenvalues

An important situation is that in which one or more eigenvalues take on the values ± 1. For instance, such unit eigenvalues or **unit roots** delineate cases of stability and instability of systems of difference equations, as will be discussed in Section 8.5. In that section, we show that certain eigenvalues must be less than unity in absolute value (or have modulus strictly less than one if they are complex eigenvalues) for such systems to be stable; otherwise the system will be unstable and lack the property of convergence towards a steady state or equilibrium over time.

Unit eigenvalues are also relevant in determining whether certain stochastic processes, such as autoregressive and vector autoregressive processes, as will be discussed in Section 14.4, satisfy conditions for statistical stationarity. The concept of stationarity will be explained in Section 14.4, together with the requirement for eigenvalues with modulus strictly less than one in order to guarantee that stationarity holds. It is an interesting fact, however, that many variables in economics and finance appear to be generated by non-stationary rather than stationary processes, and that these non-stationary processes are often characterized by positive unit roots. Indeed, certain theories, such as the "efficient markets hypothesis", may imply the existence of unit roots; see Section 16.6.

The presence of one or more unit roots in the processes that generate the data used in empirical analyses may pose serious difficulties, but not always. As will be mentioned in Chapter 8, unit roots may allow models to be modified to refer to first- or higher-order differences of variables rather than the raw levels of the variables; and in causal econometric models, they give rise to the possibility of co-integration, a concept of considerable importance. Though not pursued in this book, unit root and co-integration econometrics makes a great deal of use of the material in this chapter and, in the case of co-integrated systems of equations, of the generalization of the eigenvalue problem presented in Section 4.4.3.

3.5 Similar matrices

DEFINITION 3.5.1 The matrices $\mathbf{P}^{-1}\mathbf{AP}$ and $\mathbf{A}$ are said to be **similar** matrices, where $\mathbf{A}$ is square and $\mathbf{P}$ is a conformable non-singular matrix.

Similarity is an equivalence relation; see Exercise 3.9. Matrices in the same equivalence class share lots of properties: determinants, traces, characteristic polynomials and eigenvalues, in particular. It is relatively easy to show this, as will be seen in some of the results to follow in this chapter. Note, however, that the eigenvectors of similar matrices are generally different. Specifically, if $\mathbf{x}$ is an eigenvector of $\mathbf{A}$ corresponding to the eigenvalue λ, then $\mathbf{P}^{-1}\mathbf{x}$ is an eigenvector of $\mathbf{P}^{-1}\mathbf{AP}$ corresponding to the same eigenvalue, since $(\mathbf{P}^{-1}\mathbf{AP})\mathbf{P}^{-1}\mathbf{x} = \mathbf{P}^{-1}\mathbf{Ax} = \mathbf{P}^{-1}\lambda\mathbf{x} = \lambda\mathbf{P}^{-1}\mathbf{x}$.

We shall encounter similar matrices again in Section 6.5.

3.6 Diagonalization

Eigenvalues and eigenvectors relate to the useful concept of **diagonalization** of a matrix. The precise meaning of this concept is contained in the following definition, while the result that immediately follows the definition is basic to a study of the issue.

DEFINITION 3.6.1 A square matrix, $\mathbf{A}$, is **diagonalizable** if there is an invertible (i.e. non-singular) matrix, $\mathbf{P}$, such that $\mathbf{P}^{-1}\mathbf{AP}$ is diagonal; and $\mathbf{P}$ is said to **diagonalize A**.

But when, precisely, does there exist a matrix, $\mathbf{P}$, such that $\mathbf{A}$ is diagonalizable? The answer to this question is given in the following theorem.

THEOREM 3.6.1 *If $\mathbf{A}$ is $n \times n$, then $\mathbf{A}$ is diagonalizable if and only if $\mathbf{A}$ has n linearly independent eigenvectors.*

The idea of linear dependence and linear independence was alluded to in Section 3.3.2. Before we prove the theorem, we must now make this idea more rigorous.

DEFINITION 3.6.2 If $S = \{\mathbf{x}_1, \mathbf{x}_2, \ldots, \mathbf{x}_r\}$ is a set of r n-vectors, and if the equation

$$k_1\mathbf{x}_1 + k_2\mathbf{x}_2 + \cdots + k_r\mathbf{x}_r = \mathbf{0}_{n \times 1} \qquad (3.34)$$

has only the solution $k_1 = 0$, $k_2 = 0$, $\ldots$, $k_r = 0$, then S is called a **linearly independent** set of vectors. If there are other solutions in which some $k_i \neq 0$, then S is called a **linearly dependent** set of vectors.

Writing (3.34) as $\mathbf{Xk} = \mathbf{0}$, where $\mathbf{X} = [\mathbf{x}_1 \quad \mathbf{x}_2 \quad \ldots \quad \mathbf{x}_r]$ and $\mathbf{k} = [k_1 \quad k_2 \quad \ldots \quad k_r]^\top$, we may note the following theorem for the case when $r = n$.

THEOREM 3.6.2 *If $\mathbf{Xk} = \mathbf{0}$, where $\mathbf{X} = [\mathbf{x}_1 \quad \mathbf{x}_2 \quad \ldots \quad \mathbf{x}_n]$, $\mathbf{k} = [k_1 \quad k_2 \quad \ldots \quad k_n]^\top$ and $\mathbf{0}$ is the $n \times 1$ zero vector, then the linear independence of the columns of $\mathbf{X}$ or of the set $S = \{\mathbf{x}_1, \mathbf{x}_2, \ldots, \mathbf{x}_n\}$ implies that $\mathbf{X}$ is non-singular and vice versa.*

The proof of Theorem 3.6.2 is left as Exercise 3.13. The proof of one part of Theorem 3.6.1 now follows.

Proof: ($\Rightarrow$) Let $\mathbf{A}$ be $n \times n$ and diagonalizable. Then there exists a non-singular matrix, $\mathbf{P}$, such that

$$\mathbf{P}^{-1}\mathbf{A}\mathbf{P} = \mathbf{D} = \text{diag}[d_i] \tag{3.35}$$

Therefore, pre-multiplying by $\mathbf{P}$,

$$\mathbf{A}\mathbf{P} = \mathbf{P}\mathbf{D} \tag{3.36}$$

and rewriting the matrix products as partitioned matrices,

$$[\mathbf{A}\mathbf{p}_1 \quad \mathbf{A}\mathbf{p}_2 \quad \ldots \quad \mathbf{A}\mathbf{p}_n] = [d_1\mathbf{p}_1 \quad d_2\mathbf{p}_2 \quad \ldots \quad d_n\mathbf{p}_n] \tag{3.37}$$

where $\mathbf{p}_i$, $i = 1, 2, \ldots, n$, denotes the ith column of $\mathbf{P}$. Thus $\mathbf{A}\mathbf{p}_i = d_i\mathbf{p}_i$ for all i, which is an eigenequation from which we recognize the d_i as eigenvalues of $\mathbf{A}$ and the $\mathbf{p}_i$ as the corresponding eigenvectors of $\mathbf{A}$. Finally, since $\mathbf{P}$ is non-singular, $\mathbf{p}_i \neq \mathbf{0}$ for all i, and the set of $\mathbf{p}_i$ is a linearly independent set by Theorem 3.6.2. $\quad\square$

The second part of the proof ($\Leftarrow$) is left as an exercise (see Exercise 3.14); there is enough in what has been proved already to provide the essential material.

3.6.1 Diagonalization procedure

The result on diagonalization that has just been stated and proved provides us with a procedure for diagonalizing a matrix. First we find n linearly independent eigenvectors of the matrix, $\mathbf{A}$. Next we form the matrix $\mathbf{P}$, having the eigenvectors from the first step as its columns. Finally, we compute $\mathbf{P}^{-1}\mathbf{A}\mathbf{P}$, which will be diagonal with diagonal elements equal to the eigenvalues of $\mathbf{A}$. Note, though, that $\mathbf{P}$ is not unique, since any of its columns can be multiplied by a non-zero scalar without affecting the diagonalization property. Similarly, any two columns of $\mathbf{P}$ can be interchanged without affecting the diagonalization property.

For example, using the matrix $\mathbf{A} = \begin{bmatrix} 3 & 2 \\ -1 & 0 \end{bmatrix}$ from Section 3.3.1, which has eigenvalues $\lambda_1 = 1$ and $\lambda_2 = 2$ with associated eigenvectors $\mathbf{x}_1 = \begin{bmatrix} -1 \\ 1 \end{bmatrix}$ and $\mathbf{x}_2 = \begin{bmatrix} -2 \\ 1 \end{bmatrix}$, respectively, it is straightforward to verify that $\mathbf{x}_1$ and $\mathbf{x}_2$ are linearly independent. Consider the equation

$$k_1\mathbf{x}_1 + k_2\mathbf{x}_2 = [\mathbf{x}_1 \quad \mathbf{x}_2]\begin{bmatrix} k_1 \\ k_2 \end{bmatrix} = \begin{bmatrix} -1 & -2 \\ 1 & 1 \end{bmatrix}\begin{bmatrix} k_1 \\ k_2 \end{bmatrix} = \mathbf{P}\mathbf{k} = \mathbf{0} \tag{3.38}$$

Since $\det(\mathbf{P}) = 1$, $\mathbf{P}$ is non-singular and $\mathbf{P}^{-1}$ exists. Therefore, the solution of (3.38) is $\mathbf{k} = \mathbf{P}^{-1}\mathbf{0} = \mathbf{0}$ and, by Theorem 3.6.2, the eigenvectors $\mathbf{x}_1$ and $\mathbf{x}_2$, the columns of $\mathbf{P}$, are linearly independent.

Now

$$\mathbf{P}^{-1} = \begin{bmatrix} 1 & 2 \\ -1 & -1 \end{bmatrix} \tag{3.39}$$

and

$$\mathbf{P}^{-1}\mathbf{A}\mathbf{P} = \begin{bmatrix} 1 & 2 \\ -1 & -1 \end{bmatrix} \begin{bmatrix} 3 & 2 \\ -1 & 0 \end{bmatrix} \begin{bmatrix} -1 & -2 \\ 1 & 1 \end{bmatrix}$$

$$= \begin{bmatrix} 1 & 2 \\ -2 & -2 \end{bmatrix} \begin{bmatrix} -1 & -2 \\ 1 & 1 \end{bmatrix}$$

$$= \begin{bmatrix} 1 & 0 \\ 0 & 2 \end{bmatrix} = \begin{bmatrix} \lambda_1 & 0 \\ 0 & \lambda_2 \end{bmatrix} = \mathrm{diag}[\lambda_i] \tag{3.40}$$

An important question is *when* is a given matrix diagonalizable, i.e. when does it have n linearly independent eigenvectors, as in the case of matrix $\mathbf{A}$ in this illustration? There are several useful results on this.

THEOREM 3.6.3 *If* $\mathbf{x}_1, \mathbf{x}_2, \ldots, \mathbf{x}_k$ *are eigenvectors of* $\mathbf{A}$ *corresponding to* distinct *eigenvalues* $\lambda_1, \lambda_2, \ldots, \lambda_k$, *then* $\{\mathbf{x}_1, \mathbf{x}_2, \ldots, \mathbf{x}_k\}$ *is a linearly independent set.*

Proof: The proof is by contradiction. Let $\mathbf{A}$ have k distinct eigenvalues and $\mathbf{x}_1, \mathbf{x}_2, \ldots, \mathbf{x}_r$ $(r < k)$ be the largest set of linearly independent eigenvectors. Then

$$c_1\mathbf{x}_1 + c_2\mathbf{x}_2 + \cdots + c_{r+1}\mathbf{x}_{r+1} = \mathbf{0}, \quad \text{not all } c_i = 0 \tag{3.41}$$

Pre-multiplying (3.41) by $\mathbf{A}$, we have

$$c_1\mathbf{A}\mathbf{x}_1 + c_2\mathbf{A}\mathbf{x}_2 + \cdots + c_{r+1}\mathbf{A}\mathbf{x}_{r+1} = \mathbf{0} \tag{3.42}$$

Therefore,

$$c_1\lambda_1\mathbf{x}_1 + c_2\lambda_2\mathbf{x}_2 + \cdots + c_{r+1}\lambda_{r+1}\mathbf{x}_{r+1} = \mathbf{0} \tag{3.43}$$

Multiplying (3.41) by λ_{r+1} gives

$$c_1\lambda_{r+1}\mathbf{x}_1 + c_2\lambda_{r+1}\mathbf{x}_2 + \cdots + c_{r+1}\lambda_{r+1}\mathbf{x}_{r+1} = \mathbf{0} \tag{3.44}$$

and subtracting (3.44) from (3.43) gives

$$c_1(\lambda_1 - \lambda_{r+1})\mathbf{x}_1 + c_2(\lambda_2 - \lambda_{r+1})\mathbf{x}_2 + \cdots + c_r(\lambda_r - \lambda_{r+1})\mathbf{x}_r = \mathbf{0} \tag{3.45}$$

Since $\mathbf{x}_1, \mathbf{x}_2, \ldots, \mathbf{x}_r$ are linearly independent, all of the coefficients in (3.45) are zero, and, since the λ_i are distinct, $c_i = 0$, $i = 1, 2, \ldots, r$. Substitution of these zero values in (3.41) implies that $c_{r+1} = 0$ also, contradicting the assumption that not all $c_i = 0$.   □

The following theorem also follows from Theorem 3.6.3.

THEOREM 3.6.4 *If an* $n \times n$ *matrix* $\mathbf{A}$ *has n distinct eigenvalues, then* $\mathbf{A}$ *is diagonalizable.*

Proof: If $\mathbf{A}$ has eigenvectors $\mathbf{x}_1, \mathbf{x}_2, \ldots, \mathbf{x}_n$ corresponding to distinct eigenvalues $\lambda_1, \lambda_2, \ldots, \lambda_n$, then, by Theorem 3.6.3, $\{\mathbf{x}_1, \mathbf{x}_2, \ldots, \mathbf{x}_n\}$ is linearly independent, and, by

Theorem 3.6.1, $\mathbf{A}$ is diagonalizable, i.e. $\mathbf{P}^{-1}\mathbf{AP} = \text{diag}[\lambda_i]$, where $\mathbf{P}$ is a matrix whose columns are eigenvectors of $\mathbf{A}$. □

DEFINITION 3.6.3 If $\mathbf{A}$ is diagonalizable, then we may define a **square root of the matrix A** as

$$\mathbf{A}^{\frac{1}{2}} = \mathbf{P}\mathbf{D}^{\frac{1}{2}}\mathbf{P}^{-1} \tag{3.46}$$

where $\mathbf{D}^{\frac{1}{2}} = \text{diag}[\sqrt{\lambda_i}]$.

Whatever combination of positive or negative values of $\sqrt{\lambda_i}$ is used for the non-zero eigenvalues, we see that

$$\mathbf{A}^{\frac{1}{2}}\mathbf{A}^{\frac{1}{2}} = \mathbf{P}\mathbf{D}^{\frac{1}{2}}\mathbf{P}^{-1}\mathbf{P}\mathbf{D}^{\frac{1}{2}}\mathbf{P}^{-1} = \mathbf{P}\mathbf{D}\mathbf{P}^{-1} = \mathbf{A} \tag{3.47}$$

Clearly, therefore, a matrix square root is not unique. As in the case of scalars, $-\mathbf{A}^{\frac{1}{2}}$ is also a square root, though there are further possibilities, and these are pursued in Exercise 3.11. In general, $\mathbf{A}^{\frac{1}{2}}$ is possibly complex, and possibly neither symmetric nor invertible. To be invertible, it is necessary and sufficient that $\mathbf{D}^{\frac{1}{2}}$ is invertible and, thus, that none of the eigenvalues is zero.

3.6.2 *Orthogonal diagonalization*

Consider the following example, which involves finding the eigenvalues and eigenvectors of a symmetric matrix. If

$$\mathbf{A} = \begin{bmatrix} 4 & 2 \\ 2 & 1 \end{bmatrix} \tag{3.48}$$

then

$$\begin{aligned} |\mathbf{A} - \lambda\mathbf{I}| &= \begin{vmatrix} 4 - \lambda & 2 \\ 2 & 1 - \lambda \end{vmatrix} \\ &= (4 - \lambda)(1 - \lambda) - 4 = 0 \end{aligned} \tag{3.49}$$

is the characteristic equation. Simplifying gives

$$\lambda^2 - 5\lambda = \lambda(\lambda - 5) = 0 \tag{3.50}$$

Therefore, $\lambda_1 = 5$ and $\lambda_2 = 0$. Notice that these eigenvalues are distinct real numbers. Substituting back, we have, for $\lambda_1 = 5$,

$$(\mathbf{A} - \lambda_1\mathbf{I})\mathbf{x}_1 = \begin{bmatrix} -1 & 2 \\ 2 & -4 \end{bmatrix}\begin{bmatrix} x_{11} \\ x_{21} \end{bmatrix} = \begin{bmatrix} 0 \\ 0 \end{bmatrix} \tag{3.51}$$

from which we derive that $x_{11} = 2x_{21}$. Therefore, $\mathbf{x}_1 = [2x_{21} \quad x_{21}]^\top$ or in normalized form $\left[\frac{2}{\sqrt{5}} \quad \frac{1}{\sqrt{5}}\right]^\top$.

For $\lambda_2 = 0$,

$$(\mathbf{A} - \lambda_2\mathbf{I})\mathbf{x}_2 = \begin{bmatrix} 4 & 2 \\ 2 & 1 \end{bmatrix} \begin{bmatrix} x_{12} \\ x_{22} \end{bmatrix} = \begin{bmatrix} 0 \\ 0 \end{bmatrix} \tag{3.52}$$

from which we have that $x_{22} = -2x_{12}$. Therefore, $\mathbf{x}_2 = [x_{12} \quad -2x_{12}]^\top$ or in normalized form $\left[\frac{1}{\sqrt{5}} \quad \frac{-2}{\sqrt{5}}\right]^\top$.

Now, notice that $\mathbf{x}_1^\top \mathbf{x}_2 = 0$. We call vectors for which this product is zero **orthogonal vectors**.[3] Also notice that, for the normalized eigenvectors, we have $\mathbf{x}_1^\top \mathbf{x}_2 = 0$ and $\mathbf{x}_i^\top \mathbf{x}_i = 1$, $i = 1, 2$. We call such vectors **orthonormal vectors**. It follows that, when these eigenvectors are used as columns to form the matrix $\mathbf{P} = [\mathbf{x}_1 \quad \mathbf{x}_2]$, then $\mathbf{P}^\top\mathbf{P} = \mathbf{I}_2$, i.e. $\mathbf{P}^\top = \mathbf{P}^{-1}$ and $\mathbf{P}$ is the form of matrix that we defined as orthogonal in Section 1.5.9.[4]

DEFINITION 3.6.4 **A** is **orthogonally diagonalizable** if and only if there exists an orthogonal **P** such that $\mathbf{P}^{-1}\mathbf{A}\mathbf{P} = \mathbf{P}^\top\mathbf{A}\mathbf{P}$ is diagonal.

Symmetric matrices with real elements are important in economic and econometric applications, and the next theorems show that the special features that hold for the 2×2 symmetric matrix in our example also hold in the general case as well, namely:

- eigenvalues are real; and
- eigenvectors are orthogonal.

The next three theorems provide formal statements and proofs of these general results.

THEOREM 3.6.5 *The eigenvalues of a real symmetric matrix are real.*

Proof: This proof relies on some of the basic properties of complex numbers; see p. xxi. Note that the **conjugate of a vector** or **matrix** with complex elements is just the vector whose elements are the conjugates of the original elements.

Let **A** be an $n \times n$ real symmetric matrix and let λ be an eigenvalue of **A** with corresponding eigenvector **x**.

We will denote the conjugates of **A**, λ and **x** by $\bar{\mathbf{A}}$, $\bar{\lambda}$ and $\bar{\mathbf{x}}$, respectively. We have assumed that **A** is real or $\bar{\mathbf{A}} = \mathbf{A}$ and we want to show that λ is real or $\bar{\lambda} = \lambda$.

Now consider the transpose of the conjugate of the product

$$\bar{\mathbf{x}}^\top \mathbf{A}\mathbf{x} = \bar{\mathbf{x}}^\top \lambda \mathbf{x} \tag{3.53}$$

Taking the left-hand side of (3.53) first, we have

$$\overline{(\bar{\mathbf{x}}^\top \mathbf{A}\mathbf{x})}^\top = (\mathbf{x}^\top \bar{\mathbf{A}}\bar{\mathbf{x}})^\top = \bar{\mathbf{x}}^\top \bar{\mathbf{A}}^\top \mathbf{x} = \bar{\mathbf{x}}^\top \mathbf{A}\mathbf{x} = \lambda\bar{\mathbf{x}}^\top \mathbf{x} \tag{3.54}$$

where the penultimate step relies on the fact that **A** is real and symmetric, and the final step on the fact that **x** is an eigenvector of **A**.

Now taking the right-hand side of (3.53), we have

$$\overline{(\bar{\mathbf{x}}^\top \lambda \mathbf{x})}^\top = (\mathbf{x}^\top \bar{\lambda} \bar{\mathbf{x}})^\top = \bar{\lambda} \bar{\mathbf{x}}^\top \mathbf{x} \tag{3.55}$$

So we have

$$\lambda \bar{\mathbf{x}}^\top \mathbf{x} = \bar{\lambda} \bar{\mathbf{x}}^\top \mathbf{x} \tag{3.56}$$

Since

$$\bar{\mathbf{x}}^\top \mathbf{x} = \sum_{i=1}^{n} \bar{x}_i x_i = \sum_{i=1}^{n} |x_i|^2 \tag{3.57}$$

which is positive since an eigenvector is non-zero, we can cancel it from both sides to obtain $\lambda = \bar{\lambda}$, which just says that λ is real, as required.   □

THEOREM 3.6.6 *If* **A** *is* $n \times n$, *then the following are equivalent:*[5]

(a) **A** *is orthogonally diagonalizable;*
(b) **A** *has an orthonormal set of eigenvectors; and*
(c) **A** *is symmetric.*

  Proof:

(a)$\Rightarrow$(b): Let **A** be orthogonally diagonalizable. Then there exists **P** such that $\mathbf{P}^{-1}\mathbf{AP} = \mathbf{P}^\top \mathbf{AP} = \text{diag}[\lambda_j]$. But

$$\mathbf{P}^\top \mathbf{AP} = [x_i^\top \mathbf{A}\mathbf{x}_j] = [\mathbf{x}_i^\top \lambda_j \mathbf{x}_j] = [\lambda_j \mathbf{x}_i^\top \mathbf{x}_j] \tag{3.58}$$

where $\mathbf{x}_i$ denotes the ith column of **P**. Therefore,

$$[\lambda_j \mathbf{x}_i^\top \mathbf{x}_j] = \text{diag}[\lambda_j] \tag{3.59}$$

Hence

$$\mathbf{x}_i^\top \mathbf{x}_j = \begin{cases} 0, & i \neq j \\ 1, & i = j \end{cases} \tag{3.60}$$

(a)$\Rightarrow$(c): Let **A** be orthogonally diagonalizable. Then

$$\mathbf{P}^{-1}\mathbf{AP} = \mathbf{P}^\top \mathbf{AP} = \mathbf{D} = \text{diag}[\lambda_j] \tag{3.61}$$

It follows that

$$\mathbf{A} = \mathbf{PDP}^{-1} = \mathbf{PDP}^\top \tag{3.62}$$

Now, using the properties of transposes,

$$\mathbf{A}^\top = (\mathbf{PDP}^\top)^\top = (\mathbf{P}^\top)^\top \mathbf{D}^\top \mathbf{P}^\top = \mathbf{PDP}^\top = \mathbf{A} \tag{3.63}$$

□

The proofs of the remaining equivalences, namely, (b)⇒(a), (b)⇒(c), (c)⇒(a) and (c)⇒(b), are left as exercises; see Exercise 3.15. There is sufficient material in the proofs just provided to aid in these exercises.

THEOREM 3.6.7 *If* **A** *is a real symmetric matrix, then eigenvectors of* **A** *corresponding to distinct eigenvalues are orthogonal.*

Proof: Let $\mathbf{x}_1$ and $\mathbf{x}_2$ be eigenvectors of **A**, and let λ_1 and λ_2 be their corresponding eigenvalues. It follows that

$$\lambda_1 \mathbf{x}_1^\top \mathbf{x}_2 = (\lambda_1 \mathbf{x}_1)^\top \mathbf{x}_2 = (\mathbf{A}\mathbf{x}_1)^\top \mathbf{x}_2 = \mathbf{x}_1^\top \mathbf{A}^\top \mathbf{x}_2 = \mathbf{x}_1^\top \mathbf{A}\mathbf{x}_2 = \mathbf{x}_1^\top (\lambda_2 \mathbf{x}_2) = \lambda_2 \mathbf{x}_1^\top \mathbf{x}_2 \qquad (3.64)$$

From (3.64) we have

$$(\lambda_2 - \lambda_1)\mathbf{x}_1^\top \mathbf{x}_2 = 0 \qquad (3.65)$$

Therefore, if $\lambda_2 \neq \lambda_1$, then $\mathbf{x}_1^\top \mathbf{x}_2 = 0$, so $\mathbf{x}_1$ and $\mathbf{x}_2$ are orthogonal. □

Another general feature that is not apparent from the earlier example is that, if an eigenvalue is repeated, i.e. has **multiplicity** k, say, there will be k orthogonal eigenvectors corresponding to that eigenvalue.

From Theorem 3.6.6, there follows a procedure for orthogonally diagonalizing a symmetric matrix, **A**:

1. Find the eigenvalues of the symmetric matrix, **A**.
2. Derive an orthonormal set of eigenvectors.[6]
3. Form the matrix **P** with columns equal to the vectors from the previous step.
4. Then **P** orthogonally diagonalizes **A**, i.e. $\mathbf{P}^\top \mathbf{A}\mathbf{P} = \text{diag}[\lambda_i]$, and $\mathbf{P}^\top \mathbf{P} = \mathbf{I}$.

The working of this orthogonal diagonalization procedure should be apparent from the partial proof already given for Theorem 3.6.6 and will be further elucidated in Exercise 3.15.

When a matrix, **A**, is orthogonally diagonalizable, the matrix square root introduced in Section 3.6.1 becomes

$$\mathbf{A}^{\frac{1}{2}} = \mathbf{P}\mathbf{D}^{\frac{1}{2}}\mathbf{P}^{-1} = \mathbf{P}\mathbf{D}^{\frac{1}{2}}\mathbf{P}^\top \qquad (3.66)$$

where $\mathbf{D}^{\frac{1}{2}} = \text{diag}[\sqrt{\lambda_i}]$. Therefore, $\mathbf{A}^{\frac{1}{2}}$ is symmetric in this case, though, depending on the values of the eigenvalues, it still may not be invertible.

3.6.3 Some further results

It is instructive to begin this concluding section by considering a 2×2 matrix $\mathbf{A} = \begin{bmatrix} a_{11} & a_{12} \\ a_{21} & a_{22} \end{bmatrix}$. We know from Section 3.3.1 that the eigenvalues of **A** are found from the characteristic equation $\det(\mathbf{A} - \lambda\mathbf{I}) = 0$, which in this case is

$$\begin{vmatrix} a_{11} - \lambda & a_{12} \\ a_{21} & a_{22} - \lambda \end{vmatrix} = (a_{11} - \lambda)(a_{22} - \lambda) - a_{12}a_{21}$$

$$= \lambda^2 - (a_{11} + a_{22})\lambda + a_{11}a_{22} - a_{12}a_{21} = 0 \tag{3.67}$$

Now, according to Viète's formulas[7] on the solution of quadratic equations, the two roots of (3.67), λ_1 and λ_2, satisfy the equations $\lambda_1 + \lambda_2 = a_{11} + a_{22}$ and $\lambda_1\lambda_2 = a_{11}a_{22} - a_{12}a_{21}$, i.e. the sum of the roots equals the coefficient of the linear term in the quadratic, and the product of the roots equals the constant in the quadratic. But $a_{11}a_{22} - a_{12}a_{21} = \det(\mathbf{A})$ and $a_{11} + a_{22} = \operatorname{tr}(\mathbf{A})$, from which we have that $\lambda_1\lambda_2 = \det(\mathbf{A})$ and $\lambda_1 + \lambda_2 = \operatorname{tr}(\mathbf{A})$. Therefore, the product of the eigenvalues of $\mathbf{A}$ is equal to the determinant of $\mathbf{A}$, and the sum of the eigenvalues of $\mathbf{A}$ is equal to the trace of $\mathbf{A}$. These findings are not peculiar to the case of a 2×2 matrix, as we will now show.

Reverting to the more general case, if $\mathbf{A}$ is diagonalizable, then $\mathbf{P}^{-1}\mathbf{A}\mathbf{P} = \operatorname{diag}[\lambda_i] = \mathbf{D}$, say. Therefore, taking determinants, we have

$$|\mathbf{P}^{-1}\mathbf{A}\mathbf{P}| = |\mathbf{D}| \tag{3.68}$$

Using Property 11 of determinants from Section 2.3.2 gives

$$|\mathbf{P}^{-1}||\mathbf{A}||\mathbf{P}| = |\mathbf{D}| \tag{3.69}$$

and

$$|\mathbf{A}| = |\mathbf{D}| = \prod_{i=1}^{n} \lambda_i \tag{3.70}$$

since $|\mathbf{P}^{-1}| = 1/|\mathbf{P}|$ by Corollary 2.3.4 and the determinant of a diagonal matrix is the product of the principal diagonal elements. Thus, we have established the following useful additional theorem.

THEOREM 3.6.8 *The determinant of a matrix, $\mathbf{A}_{n \times n}$, is equal to the product of the eigenvalues of the matrix:* $\det(\mathbf{A}) = \prod_{i=1}^{n} \lambda_i$.

This theorem subsumes the case of a symmetric matrix that is orthogonally diagonalizable. The direct proof for a symmetric matrix, $\mathbf{A}$, is easily obtained by adapting the previous proof and replacing $|\mathbf{P}^{-1}|$ by $|\mathbf{P}^{\top}|$, noting that $|\mathbf{P}^{\top}| = |\mathbf{P}|$, and invoking the following lemma.

LEMMA 3.6.9 *If $\mathbf{P}$ is an orthogonal matrix, then $|\mathbf{P}| = \pm 1$.*

Proof: Given that $\mathbf{P}$ is orthogonal, $\mathbf{P}^{-1} = \mathbf{P}^{\top}$ and, hence, $\mathbf{P}^{\top}\mathbf{P} = \mathbf{I}$.

Taking the determinant of both sides of this last equation, we have $|\mathbf{P}^{\top}\mathbf{P}| = |\mathbf{P}^{\top}||\mathbf{P}| = |\mathbf{P}|^2 = |\mathbf{I}| = 1$.

Therefore, $|\mathbf{P}| = \pm 1$. Recall Exercise 2.12   □

Theorem 3.6.8 also gives rise to the following corollary, whose simple proof is left as Exercise 3.16.

COROLLARY 3.6.10 *A matrix is singular if and only if at least one of its eigenvalues is zero.*

It may be that a square matrix is non-singular, in which case there are no zero eigenvalues, but at least one of the eigenvalues is close to zero. Such a situation is important, as it may signal "**near-singularity**" and have adverse consequences for the computation of the inverse that are not apparent from the value of the determinant. By some measure, we come very close to being able to write one column of the matrix in terms of the other columns. A useful measure of this closeness is provided by the so-called **condition number**, c, of the matrix, which is the positive square root of the ratio of the maximum eigenvalue, λ_{max}, to the minimum eigenvalue, λ_{min}, i.e. $c = \sqrt{\lambda_{max}/\lambda_{min}}$. For a singular matrix, c is indeterminately large. In practice, values of c greater than 20 are considered to be large and, therefore, to imply near-singularity.

The generalization of the result relating to the sum of the eigenvalues and the trace of a diagonalizable matrix, $\mathbf{A}_{n \times n}$, is that $\text{tr}(\mathbf{A}) = \sum_{i=1}^{n} \lambda_i$. It is left as an exercise to show that, given $\mathbf{P}^{-1}\mathbf{A}\mathbf{P} = \text{diag}[\lambda_i]$, this result follows by taking the trace of both sides of this diagonalization equation and using the rule for the traces of matrix products featured in Exercise 1.14; see Exercise 3.18.

Let us explore these further results via a final worked example.

EXAMPLE 3.6.1 Evaluate the eigenvalues and eigenvectors of the matrix

$$\mathbf{A} = \begin{bmatrix} 1 & 2 \\ 0 & 3 \end{bmatrix} \tag{3.71}$$

Hence verify that $|\mathbf{A}| = \prod_{i=1}^{2} \lambda_i$, and find the condition number of $\mathbf{A}$ and a square root of $\mathbf{A}$.
Evaluation of the determinant in the characteristic equation

$$|\mathbf{A} - \lambda\mathbf{I}| = \begin{vmatrix} 1-\lambda & 2 \\ 0 & 3-\lambda \end{vmatrix} = 0 \tag{3.72}$$

gives the immediate factorization $(1-\lambda)(3-\lambda) = 0$ and so $\lambda_1 = 1$ and $\lambda_2 = 3$.
Using a method from Chapter 2, $|\mathbf{A}| = 3$. We note that $\lambda_1\lambda_2 = 1 \times 3 = 3$ also, in accordance with the equality $|\mathbf{A}| = \prod_{i=1}^{2} \lambda_i$.
The condition number of $\mathbf{A}$ is

$$c = \sqrt{\frac{\lambda_{max}}{\lambda_{min}}} = \sqrt{\frac{3}{1}} = \sqrt{3} \tag{3.73}$$

Normalization of the solutions of the 2×2 systems

$$(\mathbf{A} - \lambda_1\mathbf{I})\mathbf{x}_1 = \begin{bmatrix} 0 & 2 \\ 0 & 2 \end{bmatrix}\begin{bmatrix} x_{11} \\ x_{21} \end{bmatrix} = \begin{bmatrix} 0 \\ 0 \end{bmatrix} \tag{3.74}$$

and

$$(\mathbf{A} - \lambda_2\mathbf{I})\mathbf{x}_2 = \begin{bmatrix} -2 & 2 \\ 0 & 0 \end{bmatrix}\begin{bmatrix} x_{21} \\ x_{22} \end{bmatrix} = \begin{bmatrix} 0 \\ 0 \end{bmatrix} \tag{3.75}$$

yields the normalized eigenvectors

$$\mathbf{x}_1 = \begin{bmatrix} 1 \\ 0 \end{bmatrix} \quad \text{and} \quad \mathbf{x}_2 = \begin{bmatrix} \frac{1}{\sqrt{2}} \\ \frac{1}{\sqrt{2}} \end{bmatrix} \tag{3.76}$$

Defining

$$\mathbf{D}^{\frac{1}{2}} = \text{diag}[1, \sqrt{3}] = \begin{bmatrix} 1 & 0 \\ 0 & \sqrt{3} \end{bmatrix} \tag{3.77}$$

and

$$\mathbf{P} = \begin{bmatrix} 1 & \frac{1}{\sqrt{2}} \\ 0 & \frac{1}{\sqrt{2}} \end{bmatrix} \tag{3.78}$$

we have that

$$\mathbf{P}^{-1} = \begin{bmatrix} 1 & -1 \\ 0 & \sqrt{2} \end{bmatrix} \tag{3.79}$$

and that a square root of $\mathbf{A}$ is

$$\mathbf{A}^{\frac{1}{2}} = \mathbf{P}\mathbf{D}^{\frac{1}{2}}\mathbf{P}^{-1}$$

$$= \begin{bmatrix} 1 & \frac{1}{\sqrt{2}} \\ 0 & \frac{1}{\sqrt{2}} \end{bmatrix} \begin{bmatrix} 1 & 0 \\ 0 & \sqrt{3} \end{bmatrix} \begin{bmatrix} 1 & -1 \\ 0 & \sqrt{2} \end{bmatrix}$$

$$= \begin{bmatrix} 1 & \sqrt{3} - 1 \\ 0 & \sqrt{3} \end{bmatrix} \tag{3.80}$$

It is a straightforward matter to check that

$$\mathbf{A}^{\frac{1}{2}}\mathbf{A}^{\frac{1}{2}} = \begin{bmatrix} 1 & \sqrt{3} - 1 \\ 0 & \sqrt{3} \end{bmatrix} \begin{bmatrix} 1 & \sqrt{3} - 1 \\ 0 & \sqrt{3} \end{bmatrix} = \begin{bmatrix} 1 & 2 \\ 0 & 3 \end{bmatrix} = \mathbf{A} \tag{3.81}$$

$\diamond$

EXERCISES

3.1 Find the eigenvectors of the matrix

$$\mathbf{A} = \begin{bmatrix} 1 & -2 \\ 3 & -2 \end{bmatrix}$$

whose (complex) eigenvalues were calculated in Example 3.3.1.

3.2 Calculate the eigenvectors of the matrix

$$\begin{bmatrix} 2 & 0 \\ 8 & -2 \end{bmatrix}$$

which was used in the illustration in Section 3.2.

3.3 Find the eigenvalues and *all* normalized eigenvectors of the following matrices:

(a) $\mathbf{A} = \begin{bmatrix} 3 & 0 \\ 0 & -2 \end{bmatrix}$

(b) $\mathbf{B} = \begin{bmatrix} 1 & 2 \\ 0 & 1 \end{bmatrix}$

3.4 Show that $\mathbf{Q} = \begin{bmatrix} \frac{1}{\sqrt{6}} & \frac{2}{\sqrt{5}} & \frac{1}{\sqrt{30}} \\ \frac{-2}{\sqrt{6}} & \frac{1}{\sqrt{5}} & \frac{-2}{\sqrt{30}} \\ \frac{1}{\sqrt{6}} & 0 & \frac{-5}{\sqrt{30}} \end{bmatrix}$ is orthogonal, i.e. that $\mathbf{Q}^\top = \mathbf{Q}^{-1}$, and find its

eigenvalues.

3.5 Evaluate the eigenvalues and eigenvectors of the following matrices:

(a) $\mathbf{A} = \begin{bmatrix} 3 & 1 & 1 \\ 0 & -2 & 1 \\ 0 & 0 & 2 \end{bmatrix}$

(b) $\mathbf{B} = \begin{bmatrix} 5 & -6 & -6 \\ -1 & 4 & 2 \\ 3 & -6 & -4 \end{bmatrix}$

Normalize the eigenvectors in case (b).

3.6 Given $\mathbf{X} = \begin{bmatrix} 1 & 1 & 1 \\ 1 & 2 & 1 \end{bmatrix}^\top$, compute $\mathbf{A} = \mathbf{I}_3 - \mathbf{X}(\mathbf{X}^\top\mathbf{X})^{-1}\mathbf{X}^\top$. Show that $\mathbf{A}$ is idempotent and determine how many of its columns are linearly independent. Find the eigenvalues and associated eigenvectors of $\mathbf{A}$; normalize the eigenvectors and, hence, obtain the orthogonal matrix that diagonalizes $\mathbf{A}$.

3.7 Consider the matrix $\mathbf{A} = \begin{bmatrix} 3 & 2 \\ -1 & 0 \end{bmatrix}$ that was used in Section 3.6.1. Obtain alternative eigenvectors to those given in Section 3.6.1 and use them to diagonalize $\mathbf{A}$. Compare your result to that given in Section 3.6.1.

3.8 Show that any 3×3 matrix whose elements on each row are consecutive integers has an eigenvalue equal to zero and a corresponding eigenvector $[1 \quad -2 \quad 1]^\top$.

3.9 Show that similarity of matrices is an equivalence relation.

3.10 Find a square root for each of the following matrices:

(a) $\begin{bmatrix} 3 & 2 \\ -1 & 0 \end{bmatrix}$, which was used in Section 3.6.1; and

(b) $\begin{bmatrix} 4 & 2 \\ 2 & 1 \end{bmatrix}$, which was used in Section 3.6.2.

3.11 Determine all of the square roots of the matrix $\mathbf{A} = \begin{bmatrix} 5 & 4 \\ 4 & 5 \end{bmatrix}$, using Definition 3.6.3. How many square roots of this type are there for an $n \times n$ symmetric matrix with distinct real non-zero eigenvalues?

3.12 If $\mathbf{x}$ is an eigenvector of the non-singular matrix $\mathbf{A}$, prove that it is also an eigenvector of $\mathbf{A}^2$ and of $\mathbf{A}^{-1}$. Establish a relationship between the eigenvalues of $\mathbf{A}$ and those of $\mathbf{A}^{-1}$ and $\mathbf{A}^2$.

3.13 Prove that, if $\mathbf{Xk} = \mathbf{0}$, where $\mathbf{X} = [\mathbf{x}_1 \quad \mathbf{x}_2 \quad \ldots \quad \mathbf{x}_n]$, the $\mathbf{x}_i$ $(i = 1, 2, \ldots, n)$ are n-vectors, $\mathbf{k} = [k_1 \quad k_2 \quad \ldots \quad k_n]^\top$ and $\mathbf{0}$ is the $n \times 1$ zero vector, then the linear independence of the columns of $\mathbf{X}$ or of $S = \{\mathbf{x}_1, \mathbf{x}_2, \ldots, \mathbf{x}_n\}$ implies that $\mathbf{X}$ is non-singular and vice versa. Recall Theorem 3.6.2.

3.14 Prove that an $n \times n$ matrix with n linearly independent eigenvectors is diagonalizable (i.e. the second part of Theorem 3.6.1).

3.15 Prove the remaining parts of Theorem 3.6.6, i.e. that (b)$\Rightarrow$(a), (b)$\Rightarrow$(c), (c)$\Rightarrow$(a) and (c)$\Rightarrow$(b).

3.16 Using only the eigenequation, show that a square matrix is singular iff at least one of its eigenvalues is zero.

3.17 Extend the result of Exercise 1.14 to show that

$$\mathrm{tr}\,(\mathbf{ABC}) = \mathrm{tr}\,(\mathbf{BCA}) = \mathrm{tr}\,(\mathbf{CAB})$$

where $\mathrm{tr}(\cdot)$ denotes trace, assuming that the matrices are conformable for multiplication. Hence show that the traces of two similar matrices are equal.

3.18 Prove that the trace of a diagonalizable matrix is equal to the sum of the eigenvalues of the matrix.

3.19 Prove that the eigenvalues of a triangular matrix are equal to the elements on the principal diagonal of the matrix; recall Exercise 2.5.

3.20 Prove that the eigenvalues of an idempotent matrix are equal to 0 or 1.

4 Conic sections, quadratic forms and definite matrices

4.1 Introduction

So far, we have concentrated on linear equations, which represent lines in the plane, planes in three-dimensional space or, as will be seen in Section 7.4.1, hyperplanes in higher dimensions. In this chapter, we consider equations that also include second-order or squared terms, and that represent the simplest types of nonlinear curves and surfaces.

The concepts of matrix quadratic form and definite matrix are of considerable importance in economics and finance, as will be seen in the detailed study of our several applied examples later. Quadratic forms relate importantly to the algebraic representation of conic sections. Also, in Theorem 10.2.5, it will be seen that the definiteness of a matrix is an essential idea in the theory of convex functions. This chapter gives definitions and simple illustrations of the concepts of quadratic form and definiteness before going on to establish a number of theorems relating to them. We will return to quadratic forms in Chapter 14, where the important general problem of maximization or minimization of a quadratic form subject to linear inequality constraints is studied.

4.2 Conic sections

In this section, we consider equations representing **conic sections** in two dimensions. There are a number of equivalent ways of describing and classifying conic sections. We begin with a geometric approach.[1]

Consider the curve traced out in the coordinate plane $\mathbb{R}^2$ by a point $P = (x, y)$, which moves so that its distance from a fixed point (the **focus** S) is always in a constant ratio (the **eccentricity** $\epsilon \geq 0$) to its perpendicular distance from a fixed straight line (the **directrix** L). This curve is called:

- an **ellipse** when $0 < \epsilon < 1$;
- a **parabola** when $\epsilon = 1$;
- a **hyperbola** when $\epsilon > 1$; and
- a **circle** as $\epsilon \to 0$, as we shall see later.

4.2.1 Parabola

Consider first the case of $\epsilon = 1$, i.e. the parabola. The equation of the parabola takes its simplest form when the focus S is a point on the positive x axis, say $(a, 0)$, where $a > 0$, and the directrix is the vertical line with equation $x = -a$. An example of a parabola is shown in Figure 4.1.

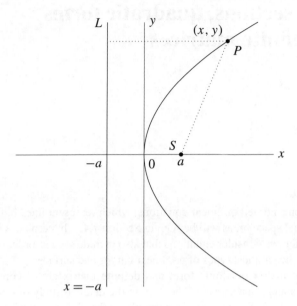

Figure 4.1 Parabola with focus $(a, 0)$ and directrix $x = -a$

By Pythagoras's theorem, the square of the distance of P from the focus S is $(x - a)^2 + y^2$ and the square of the distance of P from the directrix L is $(x + a)^2$. Thus the equation of this parabola is

$$(x - a)^2 + y^2 = (x + a)^2 \tag{4.1}$$

which simplifies to

$$y^2 = 4ax \tag{4.2}$$

Note that, when x is negative, y is imaginary, so the graph of the parabola must lie entirely to the right of the y axis. When $x = 0$, then $y = 0$, so the graph just touches the y axis at the **origin** $(0, 0)$, which is called the **vertex** of the parabola. Finally, for each value of $x > 0$, there are two possible values of y, namely $\pm 2\sqrt{ax}$. Thus the parabola is symmetric about the x axis, which is also called the **axis** of the parabola.

Similarly, $a < 0$ yields a parabola tangential to but entirely to the left of the y axis.

Note that, for any value of the real parameter t, the point $(at^2, 2at)$ lies on the parabola. Expressing the typical point on the parabola in this form often simplifies the solution of related problems.

By a simple change of coordinates, it can be seen that the equation of a parabola with vertex at (α, β), focus at $(\alpha + a, \beta)$ and directrix $x = \alpha - a$ is

$$(y - \beta)^2 = 4a(x - \alpha) \tag{4.3}$$

or

$$x = \alpha + \frac{1}{4a}(y - \beta)^2 \tag{4.4}$$

4.2.2 Ellipse

Consider now the case of $0 < \epsilon < 1$, i.e. the ellipse. The equation of the ellipse takes its simplest form when the focus S is a point on the negative x axis, say $(-a\epsilon, 0)$, where $a > 0$, and the directrix is the vertical line with equation $x = -a/\epsilon$. Such an ellipse is depicted in Figure 4.2.

The square of the distance of P from the focus S is $(x + a\epsilon)^2 + y^2$ and the square of the distance of P from the directrix L is $(x + a/\epsilon)^2$. Thus the equation of this ellipse is

$$(x + a\epsilon)^2 + y^2 = \epsilon^2 \left(x + \frac{a}{\epsilon}\right)^2 \tag{4.5}$$

Gathering up terms in x^2 and terms in y^2 and noting that identical terms in x on each side cancel, this simplifies to

$$x^2(1 - \epsilon^2) + y^2 = a^2(1 - \epsilon^2) \tag{4.6}$$

or

$$\frac{x^2}{a^2} + \frac{y^2}{a^2(1 - \epsilon^2)} = 1 \tag{4.7}$$

If we define b by $b^2 = a^2(1 - \epsilon^2)$ (which we can do because we have assumed that $\epsilon < 1$, which guarantees that b is not imaginary), then the equation of the ellipse becomes

$$\frac{x^2}{a^2} + \frac{y^2}{b^2} = 1 \tag{4.8}$$

Note that by construction $b < a$.

Note also that this equation contains only even powers of both x and y, so that the ellipse must be symmetric about both coordinate axes. From this symmetry, we can deduce the existence of a second focus S' at $(a\epsilon, 0)$ and a second directrix, the line L' with equation $x = a/\epsilon$. When $x = 0$, $y = \pm b$, and when $y = 0$, $x = \pm a$, so the ellipse cuts the coordinate axes in the points $(a, 0)$, $(0, b)$, $(-a, 0)$ and $(0, -b)$, as indicated. The longer (horizontal)

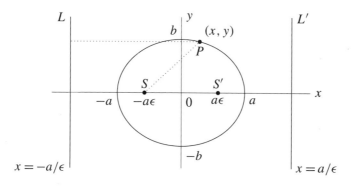

Figure 4.2 Ellipse with foci $(\pm a\epsilon, 0)$ and directrices $x = \pm a/\epsilon$

axis of the ellipse, which is of length $2a$, is called the **major axis**; the shorter (vertical) axis, which is of length $2b$, is called the **minor axis**. As the eccentricity approaches unity, b approaches zero, so the ellipse collapses onto the x axis.

Finally, note that, for any value of the angle ϕ, the point $(a\cos\phi, b\sin\phi)$ lies on the parabola, since $\sin^2\phi + \cos^2\phi = 1$. Expressing the typical point on the ellipse in this form often simplifies the solution of related problems.

4.2.3   Hyperbola

The equation of the hyperbola can be derived in exactly the same way as that of the ellipse (4.7), and we obtain

$$\frac{x^2}{a^2} + \frac{y^2}{a^2(1-\epsilon^2)} = 1 \tag{4.9}$$

In this case, however, $\epsilon > 1$ and $1 - \epsilon^2 < 0$, so to guarantee that b is real we must define it now by

$$b^2 = -a^2(1-\epsilon^2) \tag{4.10}$$

so that the equation of the hyperbola is

$$\frac{x^2}{a^2} - \frac{y^2}{b^2} = 1 \tag{4.11}$$

or

$$\frac{y^2}{b^2} = \frac{x^2}{a^2} - 1 \tag{4.12}$$

This equation still contains only even powers of both x and y, so that the hyperbola must be symmetric about both coordinate axes. Like the ellipse, the hyperbola also has a second focus and a second directrix. The hyperbola cuts the x axis at vertices $x = \pm a$. When $x^2 < a^2$, (4.12) has no real solutions for y, so no part of the hyperbola can lie between the vertices $x = \pm a$. Figure 4.3 shows the general shape of a hyperbola.

Just like the ellipse, as the eccentricity approaches unity, b approaches zero, so the hyperbola collapses onto the x axis.

Now let us consider where the straight line with equation $y = mx + c$ intersects the hyperbola with equation $x^2/a^2 - y^2/b^2 = 1$. Substituting for y yields

$$\frac{x^2}{a^2} - \frac{(mx+c)^2}{b^2} = 1 \tag{4.13}$$

Arranging this as a quadratic in x yields

$$\left(\frac{1}{a^2} - \frac{m^2}{b^2}\right)x^2 - \frac{2mc}{b^2}x - \frac{c^2}{b^2} - 1 = 0 \tag{4.14}$$

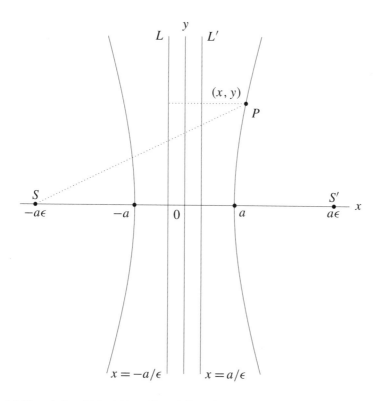

Figure 4.3 Hyperbola with foci $(\pm a\epsilon, 0)$ and directrices $x = \pm a/\epsilon$

Multiplying across by $a^2 b^2/x^2$ produces

$$(b^2 - a^2 m^2) - 2mca^2 \frac{1}{x} - a^2(b^2 + c^2)\frac{1}{x^2} = 0 \tag{4.15}$$

If the constant and linear terms in this equation vanish, then its only solution will be at $1/x = 0$, or as x approaches infinity; in other words, the line will be **asymptotic** to the hyperbola. This will happen when $c = 0$ and $m = \pm b/a$. In other words, the pair of lines passing through the origin with equations $y = \pm(b/a)x$ are the **asymptotes** of the hyperbola.

In the special case where $a = b$, the asymptotes are perpendicular and make angles of 45° with the coordinate axes. Thus, the hyperbola in this case is called a **rectangular hyperbola**. The rectangular hyperbola is more commonly encountered when it is rotated so that the asymptotes are the coordinate axes. This case will be considered again in Section 5.4.8.

By a similar change of coordinates to that employed above for the parabola in (4.4), it can be seen that the equation of a hyperbola with centre at (α, β), vertices at $(\alpha \pm a, \beta)$, foci at $(\alpha \pm a\epsilon, \beta)$ and directrices $x = \alpha \pm a/\epsilon$ is

$$\frac{(y - \beta)^2}{b^2} = \frac{(x - \alpha)^2}{a^2} - 1 \tag{4.16}$$

4.2.4   *Limiting behaviour*

As the eccentricity of an ellipse tends to zero, the geometric interpretation founders: the foci converge and meet at the origin, but the directrices diverge to infinity. However, a and b also come together, so that the equation of the ellipse converges to

$$x^2 + y^2 = a^2 \tag{4.17}$$

which is the familiar equation of a circle of radius a centred at the origin. Thus the circle can be viewed as the conic section with eccentricity zero.

We have seen that both hyperbola and ellipse collapse onto the horizontal axis as ϵ approaches unity; the parabola, which arises when $\epsilon = 1$, is a singularity or discontinuity in this process.

At the other extreme, as the eccentricity of a hyperbola tends to infinity, b also approaches infinity, and the term in y vanishes from the equation, leaving the pair of parallel vertical lines with equation $x^2 = a^2$ or $x = \pm a$.

The shape of a conic section also changes qualitatively as the focus parameter a approaches zero. The parabola collapses onto the positive x axis and the ellipse collapses onto the origin. The hyperbola, however, converges to the pair of lines that are its asymptotes. The details are left as Exercises 4.1 and 4.2.

4.2.5   *General quadratic equation in two variables*

In the coordinate plane, the graph of a quadratic equation in two variables x and y is always a conic section, and all conic sections arise in this way. The equation will be of the form

$$Ax^2 + Bxy + Cy^2 + Dx + Ey + F = 0 \tag{4.18}$$

with A, B and C not all zero. The equation also relates to the concept of matrix quadratic form, introduced in the following section.

4.3   Quadratic forms

DEFINITION 4.3.1 A **quadratic form** in $\mathbf{x}_{n \times 1}$ is a matrix product of the form $\mathbf{x}^\top \mathbf{A} \mathbf{x}$, where $\mathbf{x} \neq \mathbf{0}$.

Clearly $\mathbf{A}$ is of order $n \times n$. Some texts require that $\mathbf{A}$ be symmetric, but this is not essential and sometimes using non-symmetric matrices is relevant. For present purposes, we also assume that $\mathbf{A}$ is a real symmetric matrix. It should also be noted that $\mathbf{x}^\top \mathbf{A} \mathbf{x}$ is a scalar.

EXAMPLE 4.3.1 Suppose $\mathbf{x} = \begin{bmatrix} x_1 \\ x_2 \end{bmatrix}$, $\mathbf{A} = \begin{bmatrix} a_{11} & a_{12} \\ a_{21} & a_{22} \end{bmatrix}$ and $a_{12} = a_{21}$ by symmetry. Then

$$
\begin{aligned}
\mathbf{x}^\top \mathbf{A} \mathbf{x} &= [x_1 \quad x_2] \begin{bmatrix} a_{11} & a_{12} \\ a_{21} & a_{22} \end{bmatrix} \begin{bmatrix} x_1 \\ x_2 \end{bmatrix} \\
&= a_{11} x_1^2 + a_{12} x_1 x_2 + a_{21} x_1 x_2 + a_{22} x_2^2 \\
&= a_{11} x_1^2 + a_{22} x_2^2 + 2 a_{12} x_1 x_2
\end{aligned} \tag{4.19}
$$

$\diamond$

The final version of the 2×2 quadratic form in this example may be recognized as a special case of the left-hand side of (4.18) and, hence, as an expression that appears in the equations of certain conic sections: parabolas, ellipses and hyperbolas, including as a special case the circle.

More generally, for $\mathbf{x}_{n \times 1}$ and $\mathbf{A}_{n \times n}$, we may write

$$\mathbf{x}^\top \mathbf{A} \mathbf{x} = \sum_{i=1}^{n} a_{ii} x_i^2 + 2 \sum_{i=1}^{n-1} \sum_{j=i+1}^{n} a_{ij} x_i x_j \tag{4.20}$$

from which follow two special cases. When $\mathbf{A}$ is diagonal, we have

$$\mathbf{x}^\top \mathbf{A} \mathbf{x} = \sum_{i=1}^{n} a_{ii} x_i^2 \tag{4.21}$$

a weighted sum of squares of the elements of $\mathbf{x}$; and when $\mathbf{A} = \mathbf{I}$, we have

$$\mathbf{x}^\top \mathbf{A} \mathbf{x} = \sum_{i=1}^{n} x_i^2 \tag{4.22}$$

the simple (unweighted) sum of squares of the elements of $\mathbf{x}$.

EXAMPLE 4.3.2 Consider

$$6x_1^2 + 49x_2^2 + 51x_3^2 - 82x_2x_3 + 20x_1x_3 - 4x_1x_2 \tag{4.23}$$

and write it in the form $\mathbf{x}^\top \mathbf{A} \mathbf{x}$. This amounts to finding $\mathbf{A}$.

By inspection, and referring to Example 4.3 and its generalization, we may determine

$$\mathbf{A} = \begin{bmatrix} 6 & -2 & 10 \\ -2 & 49 & -41 \\ 10 & -41 & 51 \end{bmatrix} \tag{4.24}$$

$\diamond$

Exercise 4.3 asks the reader to replicate this exercise using a selection of other quadratic forms.

4.4 Definite matrices

DEFINITION 4.4.1 A square matrix $\mathbf{A}$, and the associated quadratic form $\mathbf{x}^\top \mathbf{A} \mathbf{x}$, are said to be

positive definite	if and only if	$\mathbf{x}^\top \mathbf{A} \mathbf{x} > 0$ for all $\mathbf{x} \neq \mathbf{0}$;
positive semi-definite	if and only if	$\mathbf{x}^\top \mathbf{A} \mathbf{x} \geq 0$ for all $\mathbf{x}$;
negative definite	if and only if	$\mathbf{x}^\top \mathbf{A} \mathbf{x} < 0$ for all $\mathbf{x} \neq \mathbf{0}$; and
negative semi-definite	if and only if	$\mathbf{x}^\top \mathbf{A} \mathbf{x} \leq 0$ for all $\mathbf{x}$.

The cases of semi-definiteness are rather trivial when $\mathbf{x} = \mathbf{0}$ and interest may focus more on the situation in which the condition is satisfied for $\mathbf{x} \neq \mathbf{0}$. When $\mathbf{x}^\top \mathbf{A} \mathbf{x}$ may take both positive and negative values, $\mathbf{A}$ and $\mathbf{x}^\top \mathbf{A} \mathbf{x}$ are said to be **indefinite**.

The following theorem is a useful result, which is easily verified from the above definitions and, hence, whose proof is left as an exercise; see Exercise 4.9.

THEOREM 4.4.1 *A square matrix* **A** *is negative definite if and only if* $-$**A** *is positive definite.*

There are some further useful theorems concerning definite matrices. We state the first of these for positive definite matrices. The proof of the theorem is the first instance in this book of a **proof by contrapositive**. This method exploits the logical equivalence of "P implies Q" and "not Q implies not P", sometimes referred to as the **universal law of sufficiency**. It establishes that "if P, then Q" by showing that "if not Q, then not P".

THEOREM 4.4.2 *If* **A** *is positive definite, then* **A** *is non-singular.*

Proof: Suppose **A** is singular or $\det(\mathbf{A}) = 0$. Then $\mathbf{Ax} = \mathbf{0}$ has non-trivial solutions and so $\mathbf{x}^{\top}\mathbf{Ax} = 0$ for some $\mathbf{x} \neq \mathbf{0}$, and therefore **A** is not positive definite.
Since a singular matrix is not positive definite, the contrapositive is also true.
A positive definite matrix must be non-singular.   □

In the remaining proofs, we let **A** be a symmetric positive definite matrix, the most important case when we return to applications later. However, certain of the theorems may be readily adapted using the result given in Theorem 4.4.1. Several of the theorems listed refer to the rank of a matrix, so we first define this concept, using two approaches.[2] Recall the definition of linear independence (Definition 3.6.2).

DEFINITION 4.4.2

(a) The **column rank** of a matrix, denoted $\rho_c(\mathbf{A})$, is the maximum number of linearly independent columns.
(b) The **row rank** of a matrix, denoted $\rho_r(\mathbf{A})$, is the maximum number of linearly independent rows.
(c) A matrix has **full row (column) rank** if all of its rows (columns) are linearly independent.

THEOREM 4.4.3 *A square matrix is non-singular if and only if it has full column rank if and only if it has full row rank.*

Proof: This proof is left as an exercise; see Exercise 4.10.   □

As symmetric matrices have the same rows and columns, the number of linearly independent rows or columns is the same and this number will be denoted by $\rho(\mathbf{A})$. Theorem 5.4.8 will show that, in fact, column rank equals row rank even when **A** is not symmetric and, indeed, not square.

THEOREM 4.4.4 *Given a symmetric positive definite matrix* **A** *of order* $n \times n$, *let* **B** *be* $n \times s$ ($s \leq n$) *with column rank* $\rho_c(\mathbf{B}) = s$. *Then* $\mathbf{B}^{\top}\mathbf{AB}$ *is symmetric positive definite.*

Proof:

(a) $(\mathbf{B}^{\top}\mathbf{AB})^{\top} = \mathbf{B}^{\top}\mathbf{A}(\mathbf{B}^{\top})^{\top}$ by the rule for transposition of a matrix product; and $\mathbf{B}^{\top}\mathbf{A}(\mathbf{B}^{\top})^{\top} = \mathbf{B}^{\top}\mathbf{AB}$, since $(\mathbf{B}^{\top})^{\top} = \mathbf{B}$.
Therefore $\mathbf{B}^{\top}\mathbf{AB}$ is symmetric.

(b) Consider any $s \times 1$ matrix $\mathbf{y} \neq \mathbf{0}$, and put $\mathbf{x} = \mathbf{By}$. Thus $\mathbf{x}$ is a weighted sum of the columns of $\mathbf{B}$. It follows from the fact that $\mathbf{B}$ has full column rank that $\mathbf{x} \neq \mathbf{0}$.
Therefore $\mathbf{y}^\top (\mathbf{B}^\top \mathbf{A} \mathbf{B}) \mathbf{y} = (\mathbf{B} \mathbf{y})^\top \mathbf{A} (\mathbf{B} \mathbf{y}) = \mathbf{x}^\top \mathbf{A} \mathbf{x} > 0$, since $\mathbf{A}$ is positive definite.
Therefore $\mathbf{B}^\top \mathbf{A} \mathbf{B}$ is positive definite. □

The following corollary is a special case of Theorem 4.4.4 when $s = n$ and $\rho(\mathbf{B}) = n$, i.e. when $\mathbf{B}$ is non-singular.

COROLLARY 4.4.5 *We have that $\mathbf{B}^\top \mathbf{A} \mathbf{B}$ is positive definite for symmetric positive definite $\mathbf{A}$ and non-singular $\mathbf{B}$.*

If $\mathbf{B} = \mathbf{A}^{-1}$, in which case $\mathbf{B}^\top \mathbf{A} \mathbf{B} = (\mathbf{A}^{-1})^\top \mathbf{A} \mathbf{A}^{-1} = (\mathbf{A}^{-1})^\top = \mathbf{A}^{-1}$, since the inverse of a symmetric matrix is also symmetric by Lemma 1.5.1, we also have the following result.

COROLLARY 4.4.6 *The inverse of a symmetric positive definite matrix is positive definite.*

THEOREM 4.4.7 *The identity matrix, $\mathbf{I}_n$, is positive definite.*

Proof: We have $\mathbf{x}^\top \mathbf{I} \mathbf{x} = \mathbf{x}^\top \mathbf{x} = \sum_{i=1}^{n} x_i^2 > 0$ for $\mathbf{x} \neq \mathbf{0}$.
Therefore $\mathbf{I}$ is positive definite by definition. □

THEOREM 4.4.8 *Let $\mathbf{B}$ be $n \times s$ ($s \leq n$) with full column rank $\rho_{\mathrm{c}}(\mathbf{B}) = s$. Then $\mathbf{B}^\top \mathbf{B}$ is symmetric positive definite.*

Proof: $\mathbf{B}^\top \mathbf{B} = \mathbf{B}^\top \mathbf{I} \mathbf{B}$, and from Theorem 4.4.7 $\mathbf{I}$ is positive definite.
Also, from Theorem 4.4.4, $\mathbf{B}^\top \mathbf{I} \mathbf{B}$ is symmetric positive definite.
Therefore $\mathbf{B}^\top \mathbf{B}$ is symmetric positive definite. □

The next result follows from Theorem 4.4.2.

COROLLARY 4.4.9 *If $\mathbf{B}$ has full column rank, then $\mathbf{B}^\top \mathbf{B}$ is non-singular, i.e. $\det(\mathbf{B}^\top \mathbf{B}) \neq 0$, and, hence, $(\mathbf{B}^\top \mathbf{B})^{-1}$ exists.*

We note, however, that, unless $s = n$, the product $\mathbf{B} \mathbf{B}^\top$ is not invertible. The case of $s < n$ is the subject of the following theorem, while the case of $s = n$ is the subject of Exercise 4.13.

THEOREM 4.4.10 *Let $\mathbf{B}$ be an $n \times s$ matrix with row rank $\rho_{\mathrm{r}}(\mathbf{B}) < n$. Then the $n \times n$ matrix $\mathbf{B} \mathbf{B}^\top$ is symmetric positive semi-definite, but not positive definite and not invertible.*

Proof:

(a) $(\mathbf{B} \mathbf{B}^\top)^\top = (\mathbf{B}^\top)^\top \mathbf{B}^\top$ by the rule for transposition of a matrix product; and $(\mathbf{B}^\top)^\top \mathbf{B}^\top = \mathbf{B} \mathbf{B}^\top$, since $(\mathbf{B}^\top)^\top = \mathbf{B}$.
Therefore $\mathbf{B} \mathbf{B}^\top$ is symmetric.
(b) Consider any $n \times 1$ matrix $\mathbf{y} \neq \mathbf{0}$, and put $\mathbf{x} = \mathbf{B}^\top \mathbf{y}$.
Therefore $\mathbf{y}^\top (\mathbf{B} \mathbf{B}^\top) \mathbf{y} = (\mathbf{B}^\top \mathbf{y})^\top (\mathbf{B}^\top \mathbf{y}) = \mathbf{x}^\top \mathbf{x} = \sum_{i=1}^{s} x_i^2 \geq 0$.
Therefore $\mathbf{B} \mathbf{B}^\top$ is positive semi-definite.
However, $\mathbf{x}$ is a weighted sum of the n rows of $\mathbf{B}$. It follows from the fact that $\rho_{\mathrm{r}}(\mathbf{B}) < n$ that there exists some $\mathbf{y}^* \neq \mathbf{0}$ for which $\mathbf{x} = \mathbf{0}$.
But then $\mathbf{y}^{*\top} (\mathbf{B} \mathbf{B}^\top) \mathbf{y}^* = \mathbf{0}^\top \mathbf{0} = 0$, so $\mathbf{B} \mathbf{B}^\top$ is not positive definite. □

An important property of positive definite matrices is the subject of the following theorem.

THEOREM 4.4.11 *A symmetric matrix, $\mathbf{A}_{n \times n}$, is positive definite if and only if all the eigenvalues of $\mathbf{A}$ are positive.*

Proof: Preliminaries. By Theorem 3.6.6, there exists an orthogonal matrix $\mathbf{P}$ that diagonalizes $\mathbf{A}$, i.e. $\mathbf{P}^\top \mathbf{A} \mathbf{P} = \mathbf{D} = \text{diag}[\lambda_i]$, where the λ_i are the eigenvalues of $\mathbf{A}$.

Let $\mathbf{x}_{n \times 1} \neq \mathbf{0}$ and define $\mathbf{y} = \mathbf{P}^\top \mathbf{x} \neq \mathbf{0}$. Then $\mathbf{x} = \mathbf{P} \mathbf{y}$, since $\mathbf{P}$ is orthogonal.

Thus $\mathbf{x}^\top \mathbf{A} \mathbf{x} = (\mathbf{P} \mathbf{y})^\top \mathbf{A} \mathbf{P} \mathbf{y} = \mathbf{y}^\top \mathbf{P}^\top \mathbf{A} \mathbf{P} \mathbf{y} = \mathbf{y}^\top \mathbf{D} \mathbf{y} = \sum_{i=1}^n \lambda_i y_i^2$.

To prove necessity ($\Leftarrow$). Let all $\lambda_i > 0$.

Then $\mathbf{x}^\top \mathbf{A} \mathbf{x} = \sum_{i=1}^n \lambda_i y_i^2 > 0$, and $\mathbf{A}$ is positive definite.

To prove sufficiency ($\Rightarrow$). Let $\mathbf{A}$ be positive definite so that $\mathbf{x}^\top \mathbf{A} \mathbf{x} = \sum_{i=1}^n \lambda_i y_i^2 > 0$. The proof is by contradiction.

Now suppose λ_1, say, is not positive; and choose $\mathbf{x}$ to be a corresponding eigenvector. It follows that $\mathbf{x}^\top \mathbf{A} \mathbf{x} = \lambda_1 \mathbf{x}^\top \mathbf{x} = \lambda_1 \sum_{i=1}^n x_i^2 \leq 0$, which contradicts the assumption that $\mathbf{A}$ is positive definite, i.e. not all λ_i positive implies that $\mathbf{A}$ is not positive definite, which completes the proof. $\square$

COROLLARY 4.4.12 *The matrix $\mathbf{A}$ is positive semi-definite if and only if all the eigenvalues of $\mathbf{A}$ are non-negative.*

Given that $\det(\mathbf{A}) = \prod_{i=1}^n \lambda_i$ (see Theorem 3.6.8), another useful corollary follows from Theorem 4.4.11.

COROLLARY 4.4.13 *We have $\det(\mathbf{A}) > 0$ for positive definite $\mathbf{A}$.*

A similar result to that given in Corollary 4.4.13 relates to the following concept.

DEFINITION 4.4.3 A **principal minor** is a determinant of a submatrix formed by deleting corresponding rows and columns from a square matrix. The **order** of the principal minor is the number of rows (or columns) of the submatrix in question.

Thus, for an $n \times n$ matrix, $\mathbf{A}$, the principal minors of order 1 are a_{ii} ($i = 1, 2, \ldots, n$), obtained by deleting all rows and columns except the ith. The principal minors of order 2 are $\begin{vmatrix} a_{ii} & a_{ij} \\ a_{ji} & a_{jj} \end{vmatrix}$, obtained by deleting all rows and columns except the ith and jth, and so on. It is left as an exercise to determine how many principal minors there are in total; see Exercise 4.14.

THEOREM 4.4.14 *All the principal minors of a symmetric positive definite matrix (in particular, all the entries on the principal diagonal) are positive.*

Proof: The result on positive principal minors follows from Theorem 4.4.4 by using an $n \times n$ identity matrix with columns deleted corresponding to those deleted from $\mathbf{A}$, and putting the result equal to **B**. For example,

$$\mathbf{B} = \begin{bmatrix} 1 & 0 \\ 0 & 1 \\ 0 & 0 \\ \vdots & \vdots \\ 0 & 0 \end{bmatrix}_{n \times 1}$$

when all but the first and second columns of **A** and $\mathbf{I}_n$ are to be deleted.

Then $\mathbf{B}^\top \mathbf{AB} = \begin{bmatrix} a_{11} & a_{12} \\ a_{21} & a_{22} \end{bmatrix}$; and $\mathbf{B}^\top \mathbf{AB}$ is positive definite by Theorem 4.4.4. So from

Corollary 4.4.13 we have that $\begin{vmatrix} a_{11} & a_{12} \\ a_{21} & a_{22} \end{vmatrix} > 0$. Similarly for all the other principal minors. $\square$

Thus we have the important results that symmetric positive definite matrices are non-singular, have positive eigenvalues, have positive determinants and positive principal minors of all orders. Similar results apply in the case of symmetric negative definite matrices, which are also non-singular but have negative eigenvalues, though in this case the signs of the determinant and the principal minors depend on the orders of the matrix and submatrices involved. The definiteness of a symmetric matrix can therefore be determined by checking the signs of its eigenvalues or its principal minors. Semi-definite matrices that are not definite have at least one zero eigenvalue and therefore are singular. Matrices that have eigenvalues with different signs are indefinite.

We have also shown that the inverse of a positive (negative) definite matrix is positive (negative) definite. This fact will be used in connection with a positive definite matrix called a variance–covariance matrix to be introduced in Section 13.6.1. We also note that, if **P** is an invertible $n \times n$ matrix and **A** is any $n \times n$ matrix, then **A** is positive or negative (semi-)definite if and only if $\mathbf{P}^{-1}\mathbf{AP}$ is positive or negative (semi-)definite, respectively; recall the discussion of similar matrices in Section 3.5.

4.4.1 Decomposition of matrices

This section establishes some useful decompositions of symmetric positive definite matrices.

THEOREM 4.4.15 *If $\mathbf{A}_{n \times n}$ is symmetric positive definite, there exists a non-singular matrix, **R**, such that **A** is decomposable as $\mathbf{A} = \mathbf{RR}^\top$.*

Proof: Recalling that **A** is orthogonally diagonalizable, we may write $\mathbf{P}^\top \mathbf{AP} = \mathbf{D} = \text{diag}[\lambda_i]$, where **P** is the orthogonal matrix that diagonalizes **A** and $\lambda_i, i = 1, 2, \ldots, n$, are the eigenvalues of **A**. Since all λ_i are positive, define the diagonal matrix $\mathbf{D}^{\frac{1}{2}} \equiv \text{diag}\left[\sqrt{\lambda_i}\right]$ and let $\mathbf{R} \equiv \mathbf{PD}^{\frac{1}{2}}$. Then

$$\mathbf{RR}^\top = \mathbf{PD}^{\frac{1}{2}}(\mathbf{PD}^{\frac{1}{2}})^\top = \mathbf{PDP}^\top = \mathbf{P}(\mathbf{P}^\top \mathbf{AP})\mathbf{P}^\top = (\mathbf{PP}^\top)\mathbf{A}(\mathbf{PP}^\top) = \mathbf{IAI} = \mathbf{A} \qquad (4.25)$$

(since **P** is orthogonal). $\square$

Like the square root defined earlier, **R** is not unique: $-\mathbf{R}$ has the same property. The columns of **R** are orthonormal eigenvectors of **A** multiplied by the square roots of the corresponding eigenvalues. Re-ordering these columns preserves the property that $\mathbf{A} = \mathbf{R}\mathbf{R}^\top$. More generally, $\mathbf{A} = \mathbf{R}\mathbf{F}(\mathbf{R}\mathbf{F})^\top$, where $\mathbf{F}$ is any conformable orthogonal matrix.

The square root defined in Section 3.6.1 can be written $\mathbf{A}^{\frac{1}{2}} = \mathbf{R}\mathbf{P}^\top = \mathbf{P}\mathbf{D}^{\frac{1}{2}}\mathbf{P}^\top$, or $\mathbf{A}^{\frac{1}{2}} = \mathbf{P}\mathbf{R}^\top = \mathbf{P}\mathbf{D}^{\frac{1}{2}}\mathbf{P}^\top$; see Exercise 4.17. Since the eigenvalues are all positive and, hence, $\mathbf{D}$ and $\mathbf{D}^{\frac{1}{2}}$ are invertible, and the eigenvectors are orthogonal, the square root $\mathbf{A}^{\frac{1}{2}} = \mathbf{P}\mathbf{D}^{\frac{1}{2}}\mathbf{P}^\top$ is both symmetric and invertible when $\mathbf{A}$ is symmetric positive definite. If the elements on the diagonal of $\mathbf{D}^{\frac{1}{2}}$ are the positive square roots of the eigenvalues, then (by Corollary 4.4.5) $\mathbf{A}^{\frac{1}{2}}$ is also positive definite.

The non-uniqueness of the decomposition in Theorem 4.4.15 is remedied by the so-called **triangular factorization** or **triangular decomposition**, which is the subject of the next theorem.

THEOREM 4.4.16 *Let $\mathbf{A}_{n \times n}$ be a symmetric positive definite matrix; then there exists a unique decomposition of $\mathbf{A}$ as $\mathbf{A} = \mathbf{L}\mathbf{D}\mathbf{L}^\top$, where $\mathbf{L}$ is a lower triangular matrix whose diagonal elements are all unity, and $\mathbf{D}$ is a diagonal matrix whose diagonal elements are all positive.*

Proof: We know from Section 2.5.5 that elementary row and column operations may be performed on $\mathbf{A}$ by means of pre- and post-multiplication by suitably defined elementary matrices. Details of the form of the elementary and other matrices referred to in this outline proof are to be found in Hamilton (1994, Section 4.4). Let $\mathbf{E}_1$ be the product of elementary matrices that by means of the multiplication

$$\mathbf{E}_1 \mathbf{A} \mathbf{E}_1^\top = \mathbf{B} \tag{4.26}$$

makes b_{11} non-zero but makes all other entries in the first row and first column of $\mathbf{B}$ zero. Similarly, let $\mathbf{E}_2$ be the product of elementary matrices that by the multiplication

$$\mathbf{E}_2 \mathbf{B} \mathbf{E}_2^\top = \mathbf{C} \tag{4.27}$$

retains the first row and first column of $\mathbf{B}$ but makes c_{22} non-zero and all other entries in the second row and second column of $\mathbf{C}$ zero. Continuing in this manner, we establish that, for symmetric positive definite $\mathbf{A}$, there exist products of elementary matrices $\mathbf{E}_1, \mathbf{E}_2, \ldots, \mathbf{E}_{n-1}$ such that

$$\mathbf{E}_{n-1} \cdots \mathbf{E}_2 \mathbf{E}_1 \mathbf{A} \mathbf{E}_1^\top \mathbf{E}_2^\top \cdots \mathbf{E}_{n-1}^\top = \mathbf{D} \tag{4.28}$$

where $\mathbf{D} = \mathrm{diag}[d_i]$ and $d_i > 0$ for all i. The required $\mathbf{E}_i$ matrices are guaranteed to exist by the positive definiteness of $\mathbf{A}$ and the positive definiteness of $\mathbf{B}$, $\mathbf{C}$, etc., which follows from Theorem 4.4.4 and the fact that the $\mathbf{E}_i$ are non-singular. Moreover, the $\mathbf{E}_i$ are all lower triangular, with non-zero elements below the principal diagonal, ones along the principal diagonal and zeros above the principal diagonal. Thus, by Property 3 of determinants in Section 2.3.2, $\det(\mathbf{E}_i) = 1$ for all i and, hence, $\mathbf{E}_i^{-1}$ exists for all i. We can therefore define the matrix

$$\mathbf{L} = (\mathbf{E}_{n-1} \cdots \mathbf{E}_2 \mathbf{E}_1)^{-1} = \mathbf{E}_1^{-1} \mathbf{E}_2^{-1} \cdots \mathbf{E}_{n-1}^{-1} \tag{4.29}$$

which, because of the lower triangularity of the $\mathbf{E}_i$ and, therefore, the lower triangularity of the $\mathbf{E}_i^{-1}$, is a lower triangular matrix itself. Pre-multiplication of (4.28) by $\mathbf{L}$ and post-multiplication by $\mathbf{L}^\top$ then gives the triangular factorization result that

$$\mathbf{A} = \mathbf{LDL}^\top \tag{4.30}$$

Compared with its detailed derivation, the uniqueness of this factorization is easy to establish, using a proof by contradiction; see Exercise 4.18. □

COROLLARY 4.4.17 *A symmetric positive definite matrix* $\mathbf{A}$ *may be decomposed as* $\mathbf{A} = \mathbf{L}^*\mathbf{L}^{*\top}$, *where* $\mathbf{L}^*$ *is a lower triangular matrix.*

Proof: If $\mathbf{A}$ is a symmetric positive definite matrix, then the triangular factorization gives $\mathbf{A} = \mathbf{LDL}^\top$, where $\mathbf{L}$ is a lower triangular matrix whose diagonal elements are all unity, and $\mathbf{D} = \mathrm{diag}\,[d_i]$ with $d_i > 0$ for all i. Define $\mathbf{D}^{\frac{1}{2}} = \mathrm{diag}\left[\sqrt{d_i}\right]$, such that $\mathbf{D}^{\frac{1}{2}}\mathbf{D}^{\frac{1}{2}} = \mathbf{D}$. Then

$$\mathbf{A} = \mathbf{LDL}^\top = \mathbf{LD}^{\frac{1}{2}}\mathbf{D}^{\frac{1}{2}}\mathbf{L}^\top = \mathbf{LD}^{\frac{1}{2}}(\mathbf{LD}^{\frac{1}{2}})^\top \tag{4.31}$$

or $\mathbf{A} = \mathbf{L}^*\mathbf{L}^{*\top}$, where $\mathbf{L}^* = \mathbf{LD}^{\frac{1}{2}}$. □

The form of the triangular factorization in this corollary is known as the **Cholesky decomposition** or **Cholesky factorization** of $\mathbf{A}$.[3] Like the matrix $\mathbf{L}$ in Theorem 4.4.16, the matrix $\mathbf{L}^*$ is lower triangular, though $\mathbf{L}^*$ has the $\sqrt{d_i}$ along its principal diagonal rather than ones. In fact, $\mathbf{L}^*$ is the matrix that results by multiplying each of the columns of $\mathbf{L}$ by its corresponding value of $\sqrt{d_i}$, i.e.

$$\mathbf{L}^* = \begin{bmatrix} \sqrt{d_1}\mathbf{l}_1 & \sqrt{d_2}\mathbf{l}_2 & \cdots & \sqrt{d_n}\mathbf{l}_n \end{bmatrix} \tag{4.32}$$

where $\mathbf{l}_i$ denotes the ith column of $\mathbf{L}$. As the matrix $\mathbf{L}^{*\top}$ is upper triangular, the Cholesky decomposition is seen to be a particular case of the **LU-decomposition** or **LU-factorization**, which, as its name implies, is a type of decomposition that allows certain matrices to be written as the product of a lower triangular matrix and an upper triangular matrix, though the upper triangular matrix is not generally the transpose of the lower triangular matrix.[4]

4.4.2 Comparing matrices

In Definition 1.3.1, we defined the concept of matrix equality, but we have said nothing so far about matrix inequalities. The concept of definiteness provides a means of comparing certain matrices that are not equal and thereby filling this gap.

Assuming that two matrices, $\mathbf{A}$ and $\mathbf{B}$, are symmetric and of the same dimension, a comparison between them may be based on

$$d \equiv \mathbf{x}^\top \mathbf{A}\mathbf{x} - \mathbf{x}^\top \mathbf{B}\mathbf{x} = \mathbf{x}^\top (\mathbf{A} - \mathbf{B})\mathbf{x} \tag{4.33}$$

Specifically, if $d > 0$ for all $\mathbf{x} \neq \mathbf{0}$, then $\mathbf{A} - \mathbf{B}$ is positive definite by Definition 4.4.1. In this sense, we write $\mathbf{A} > \mathbf{B}$ and say that $\mathbf{A}$ is "larger" than $\mathbf{B}$. Similarly, we may write $\mathbf{A} \geq \mathbf{B}$ if $d \geq 0$ for all $\mathbf{x} \neq \mathbf{0}$; $\mathbf{A} < \mathbf{B}$ if $d < 0$ for all $\mathbf{x} \neq \mathbf{0}$; and $\mathbf{A} \leq \mathbf{B}$ if $d \leq 0$ for all $\mathbf{x} \neq \mathbf{0}$; i.e. $\mathbf{A}$ is "not less than", "less than" and "not greater than" $\mathbf{B}$, if $\mathbf{A} - \mathbf{B}$ is positive semi-definite, negative definite and negative semi-definite, respectively. Note, however, that this categorization is not complete: $\mathbf{A} - \mathbf{B}$ could be indefinite. In this situation, no simple comparison of the two matrices is possible.

It follows from the first of the comparative criteria that, if $\mathbf{A}$ is positive definite and $\mathbf{B}$ is positive semi-definite, then $\mathbf{A} + \mathbf{B} \geq \mathbf{A}$. Another intuitive result is that, if $\mathbf{A}$ and $\mathbf{B}$ are both positive definite and $\mathbf{A} > \mathbf{B}$, then $\mathbf{B}^{-1} > \mathbf{A}^{-1}$. The proof of this, which is left as an exercise, follows from Theorem 4.4.11 and the relationship between the eigenvalues of a matrix and those of its inverse; see Exercise 4.19 and recall Exercise 3.12.

4.4.3   *Generalized eigenvalues and eigenvectors*

Knowledge of definite matrices and matrix square roots allows us to introduce a generalization of the eigenvalue problem discussed in Chapter 3. This problem, which arises, for example, in co-integration analysis in time-series econometrics, is concerned with the solution of the **generalized eigenequation**

$$\mathbf{A}\mathbf{x} = \lambda \mathbf{B}\mathbf{x} \tag{4.34}$$

for the scalar λ and the n-vector $\mathbf{x} \neq \mathbf{0}$, where $\mathbf{A}$ and $\mathbf{B}$ are given $n \times n$ matrices. The standard eigenvalue problem arises when $\mathbf{B} = \mathbf{I}_n$. In the more general context, we refer to λ as a **generalized eigenvalue** and to $\mathbf{x}$ as a **generalized eigenvector** of $\mathbf{A}$, relative to $\mathbf{B}$. Values of λ may, in principle, be obtained from solution of the determinantal equation

$$|\mathbf{A} - \lambda \mathbf{B}| = 0 \tag{4.35}$$

as in the standard case; and for a given λ, the associated vector $\mathbf{x}$ follows from solution of the homogeneous system of linear equations

$$(\mathbf{A} - \lambda \mathbf{B})\mathbf{x} = \mathbf{0} \tag{4.36}$$

Unfortunately, though the solution procedure seems essentially the same as for the standard case discussed in detail earlier, there are complications. For instance, if $\mathbf{B}$ is singular, there may not be n eigenvalues; indeed, there may be no solutions for λ at all.

EXAMPLE 4.4.1 Consider the matrices

$$\mathbf{A} = \begin{bmatrix} -1 & 0 \\ 0 & 1 \end{bmatrix} \quad \text{and} \quad \mathbf{B} = \begin{bmatrix} 2 & 2 \\ 2 & 2 \end{bmatrix} \tag{4.37}$$

Then

$$
\begin{aligned}
|\mathbf{A} - \lambda \mathbf{B}| &= \begin{vmatrix} -1 - 2\lambda & -2\lambda \\ -2\lambda & 1 - 2\lambda \end{vmatrix} \\
&= -(1 + 2\lambda)(1 - 2\lambda) - 4\lambda^2 \\
&= -1
\end{aligned}
\tag{4.38}
$$

and so (4.35) and the generalized eigenvalue problem have no solution.   ◇

We can see what is going on in the example a little more generally by using the 2×2 matrices $\mathbf{A} = [a_{ij}]$ and $\mathbf{B} = [b_{ij}]$, for which we have that

$$
\begin{aligned}
|\mathbf{A} - \lambda\mathbf{B}| &= \begin{vmatrix} a_{11} - \lambda b_{11} & a_{12} - \lambda b_{12} \\ a_{21} - \lambda b_{21} & a_{22} - \lambda b_{22} \end{vmatrix} \\
&= (a_{11} - \lambda b_{11})(a_{22} - \lambda b_{22}) - (a_{12} - \lambda b_{12})(a_{21} - \lambda b_{21}) \\
&= a_{11}a_{22} - a_{12}a_{21} + (a_{21}b_{12} - a_{22}b_{11} - a_{11}b_{22} + a_{12}b_{21})\lambda \\
&\quad + (b_{11}b_{22} - b_{12}b_{21})\lambda^2
\end{aligned}
\tag{4.39}
$$

If $\mathbf{B}$ is singular, then $b_{11}b_{22} - b_{12}b_{21} = |\mathbf{B}| = 0$ and the term in λ^2 vanishes. In this case, if $a_{21}b_{12} - a_{22}b_{11} - a_{11}b_{22} + a_{12}b_{21} \neq 0$, there will be one generalized eigenvalue, but if $a_{21}b_{12} - a_{22}b_{11} - a_{11}b_{22} + a_{12}b_{21} = 0$ and $\det(\mathbf{A}) = a_{11}a_{22} - a_{12}a_{21} \neq 0$, then generalized eigenvalues do not exist.

Imposing a non-singularity condition on $\mathbf{B}$ will eliminate the difficulty illustrated in Example 4.4.3. It will also permit solution for the generalized eigenvalues using the standard procedure on $|\mathbf{B}^{-1}\mathbf{A} - \lambda\mathbf{I}| = 0$, which follows from pre-multiplication of (4.34) by $\mathbf{B}^{-1}$. Equivalently, the solution may also be obtained from $|\mathbf{A}\mathbf{B}^{-1} - \lambda\mathbf{I}| = 0$; see Exercise 4.20.

If $\mathbf{B}$ is symmetric positive definite, then the inverse of a square root of $\mathbf{B}$ may be used so that the solution for λ may be obtained from the standard problem

$$
|\mathbf{B}^{-\frac{1}{2}}\mathbf{A}\mathbf{B}^{-\frac{1}{2}} - \lambda\mathbf{I}| = 0
\tag{4.40}
$$

If $\mathbf{x}$ is an eigenvector of $\mathbf{B}^{-\frac{1}{2}}\mathbf{A}\mathbf{B}^{-\frac{1}{2}}$ corresponding to a solution, λ, of (4.40), then

$$
\mathbf{B}^{-\frac{1}{2}}\mathbf{A}\mathbf{B}^{-\frac{1}{2}}\mathbf{x} = \lambda\mathbf{x} = \lambda\mathbf{B}^{\frac{1}{2}}\mathbf{B}^{-\frac{1}{2}}\mathbf{x}
\tag{4.41}
$$

which implies that

$$
\mathbf{A}(\mathbf{B}^{-\frac{1}{2}}\mathbf{x}) = \lambda\mathbf{B}(\mathbf{B}^{-\frac{1}{2}}\mathbf{x})
\tag{4.42}
$$

Hence $\mathbf{y} = \mathbf{B}^{-\frac{1}{2}}\mathbf{x}$ is the generalized eigenvector of $\mathbf{A}$ relative to $\mathbf{B}$ corresponding to the generalized eigenvalue λ.

Generalized eigenvectors are not, in general, orthogonal, even if both $\mathbf{A}$ and $\mathbf{B}$ are symmetric. However, for pairs of generalized eigenvectors, $\mathbf{y}_i$ and $\mathbf{y}_j$, say, it is the case that, if $\mathbf{A}$ and $\mathbf{B}$ are symmetric and $\mathbf{B}$ is positive definite, then $\mathbf{y}_i^\top \mathbf{B} \mathbf{y}_j = 0$, which follows simply from the definitions $\mathbf{y}_i = \mathbf{B}^{-\frac{1}{2}}\mathbf{x}_i$ and $\mathbf{y}_j = \mathbf{B}^{-\frac{1}{2}}\mathbf{x}_j$ and the results on the eigenvectors in the standard problem.

EXERCISES

4.1 Using equations (4.1) and (4.5), respectively, show that, as the focus parameter a approaches zero, then:

(a) the parabola collapses onto the positive x axis; and
(b) the ellipse collapses onto the origin.

4.2 Using equation (4.9), show that as the focus parameter a tends to zero, a hyperbola converges to the pair of lines $y = \pm(b/a)x$, which otherwise are its asymptotes.

4.3 Write the following quadratic forms in the matrix notation $\mathbf{x}^\top \mathbf{A} \mathbf{x}$:

$$4x_1^2 + 9x_2^2 + 2x_3^2 - 8x_2x_3 + 6x_3x_1 - 6x_1x_2$$

$$x_1^2 + 16x_2^2 + 12x_3^2 + x_4^2$$

$$x_1^2 + x_2^2 + x_3^2 + x_4^2 + x_5^2$$

4.4 Examine the following quadratic forms for positive definiteness:

$$6x_1^2 + 49x_2^2 + 51x_3^2 - 82x_2x_3 + 20x_1x_3 - 4x_1x_2$$

$$4x_1^2 + 9x_2^2 + 2x_3^2 + 6x_1x_2 + 6x_1x_3 + 8x_2x_3$$

4.5 Let $\mathbf{A} = \begin{bmatrix} 1 & 0 & 0 \\ 0 & 2 & 0 \\ 0 & 0 & 3 \end{bmatrix}$ and $\mathbf{B} = \begin{bmatrix} 1 & 2 \\ 0 & 1 \\ 2 & 0 \end{bmatrix}$. Find the values of all the principal minors of $\mathbf{A}$ and, hence, demonstrate that $\mathbf{A}$ is positive definite. Determine the row rank and the column rank of $\mathbf{B}$. Form the product $\mathbf{B}^\top \mathbf{A} \mathbf{B}$ and show that it is positive definite also.

4.6 Examine the definiteness of the matrix $\mathbf{A} = \begin{bmatrix} 0 & 1 \\ -1 & 0 \end{bmatrix}$. Relate your finding to the eigenvalues of the matrix, which were derived in Section 3.3.1.

4.7 Using a decomposition of the type defined in Theorem 4.4.15, find a non-singular matrix, $\mathbf{R}$, such that the matrix $\mathbf{A} = \begin{bmatrix} 4 & 2 \\ 2 & 2 \end{bmatrix}$ may be written as $\mathbf{A} = \mathbf{R}\mathbf{R}^\top$. Compare your result with the Cholesky decomposition $\mathbf{A} = \mathbf{L}^*\mathbf{L}^{*\top}$, where $\mathbf{L}^*$ is a lower triangular matrix.

4.8 Find the generalized eigenvalues, and an associated pair of generalized eigenvectors, of $\mathbf{A}$ relative to $\mathbf{B}$, where $\mathbf{A} = \begin{bmatrix} 1 & 0 \\ 0 & -1 \end{bmatrix}$ and:

(a) $\mathbf{B} = \begin{bmatrix} 2 & 1 \\ 2 & 3 \end{bmatrix}$; and

(b) $\mathbf{B} = \begin{bmatrix} 2 & 1 \\ 1 & 2 \end{bmatrix}$.

4.9 Prove that a square matrix $\mathbf{A}$ is negative definite if and only if $-\mathbf{A}$ is positive definite.

4.10 Prove Theorem 4.4.3.

4.11 Let $\mathbf{A}$ be a symmetric negative definite matrix of order $n \times n$ and let $\mathbf{B}$ be $n \times s$ ($s \leq n$) with rank $\rho(\mathbf{B}) = s$. Prove that $\mathbf{B}^\top \mathbf{A} \mathbf{B}$ is symmetric negative definite. Recall the proof of Theorem 4.4.4.

4.12 Prove that, if $\mathbf{A}$ is negative definite, then $\mathbf{A}$ is non-singular and $\mathbf{A}^{-1}$ is also negative definite.

4.13 Let $\mathbf{B}$ be $n \times s$ with $s \leq n$ and row rank $\rho_r(\mathbf{B}) = r \leq s$. Prove the following:

(a) $\mathbf{B}\mathbf{B}^\top$ is positive semi-definite if $r < s \leq n$; and
(b) $\mathbf{B}\mathbf{B}^\top$ is positive definite if $r = s = n$.

4.14 Let $\mathbf{A}$ be an $n \times n$ matrix. Excluding $\det(\mathbf{A})$, how many principal minors in total does $\mathbf{A}$ have?

4.15 Let $\mathbf{B}$ be a matrix of order $n \times n$ with column rank $\rho_c(\mathbf{B}) = s$ $(s \leq n)$. Show that $\rho_c(\mathbf{B}^\top \mathbf{B}) = s$. Show also that $\rho_c(\mathbf{B}\mathbf{B}^\top) = s$.

4.16 Consider an idempotent matrix $\mathbf{M}$. Show that $\rho_c(\mathbf{M}) = \text{tr}(\mathbf{M})$, i.e. that the rank of $\mathbf{M}$ equals the trace of $\mathbf{M}$.

4.17 Let $\mathbf{A}$ be a symmetric positive definite matrix. Using the definition of $\mathbf{R}$ given in Theorem 4.4.15 and the definition of $\mathbf{A}^{\frac{1}{2}}$ given in Definition 3.6.3, show that $\mathbf{R}\mathbf{R}^\top = \mathbf{A}^{\frac{1}{2}}\mathbf{A}^{\frac{1}{2}}$.

4.18 Let $\mathbf{A}$ be a symmetric positive definite matrix. Prove that the triangular factorization $\mathbf{A} = \mathbf{L}\mathbf{D}\mathbf{L}^\top$, where $\mathbf{L}$ is a lower triangular matrix and $\mathbf{D}$ is a diagonal matrix with positive diagonal elements, is unique, using the method of proof by contradiction (i.e. starting with the assumption of non-uniqueness: $\mathbf{A} = \mathbf{L}_1\mathbf{D}_1\mathbf{L}_1^\top$ and $\mathbf{A} = \mathbf{L}_2\mathbf{D}_2\mathbf{L}_2^\top$, say).

4.19 Let $\mathbf{A}$ and $\mathbf{B}$ be symmetric positive definite matrices of the same order.

(a) Show that, if every eigenvalue of $\mathbf{A}$ is larger than the corresponding eigenvalue of $\mathbf{B}$ when both sets of eigenvalues are ordered from smallest to largest, then $\mathbf{A} - \mathbf{B}$ is positive definite.
(b) Hence show that $\mathbf{B}^{-1} - \mathbf{A}^{-1}$ is positive definite.

4.20 Let $\mathbf{B}$ be a non-singular matrix. Prove that, if $\mathbf{A} - \lambda\mathbf{B}$ is singular, then so are $\mathbf{B}^{-1}\mathbf{A} - \lambda\mathbf{I}$ and $\mathbf{A}\mathbf{B}^{-1} - \lambda\mathbf{I}$. Hence show that $|\mathbf{B}^{-1}\mathbf{A} - \lambda\mathbf{I}| = 0$ and $|\mathbf{A}\mathbf{B}^{-1} - \lambda\mathbf{I}| = 0$ yield the same solutions for λ.

5 Vectors and vector spaces

5.1 Introduction

It is customary to think of **vectors** as entities with magnitudes and directions, and **spaces** as like the two-dimensional space we write in and the three-dimensional space we live in and move around in. Vectors are therefore distinct from scalars, which have magnitudes only. Our aim in this chapter is to develop a collection of results that apply to such vector entities in real n-dimensional space, or simply n-space. Our approach will be both geometric and analytic. The vector geometry will provide fresh insights into what we have already encountered in our algebraic study of $n \times 1$ and $1 \times n$ matrices, while the analysis will echo the matrix algebra itself. However, as we are familiar with one, two and three spatial dimensions, and can visualize more easily in these cases, we begin with vectors in 2-space ($\mathbb{R}^2$) and 3-space ($\mathbb{R}^3$) in order to fix the main ideas intuitively. It will quickly be seen that the vectors in these cases may readily be associated with 2×1 and 3×1 matrices, respectively. However, later generalization is intended not only to take us from 2- and 3-space to n-space and $n \times 1$ matrices, but also to abstract the main properties of vectors in n-space so that they apply as well to kinds of objects other than real row or column matrices.

5.2 Vectors in 2-space and 3-space

5.2.1 Vector geometry

In 2-space, also known as the **plane** or **Euclidean plane** or **Cartesian plane**, a simple geometric approach is to represent vectors by arrows, where the length of the arrow represents the magnitude of the vector and the direction of the arrow, relative to some arbitrary datum in the plane, represents the direction of the vector. To draw a vector in the plane, we must know not only its magnitude and direction, but also its location. Thus three vectors, **v**, **w** and **z**, may be depicted as in Figure 5.1. As with matrices earlier, bold font will be used for vectors to distinguish them from scalars.

The vector **v** starts at initial point A and ends at terminal point B whereas vector **w** runs from initial point C to terminal point D and **z** from E to F. Where the initial and terminal points are to be made explicit, we may write $\mathbf{v} = \overrightarrow{AB}$, $\mathbf{w} = \overrightarrow{CD}$ and $\mathbf{z} = \overrightarrow{EF}$, which denote that **v** is the vector joining initial point A and terminal point B, and similarly for **w** and **z**. With this geometric representation in mind, we may state the following definitions.

DEFINITION 5.2.1 Vectors are called **equivalent** if they have the same magnitude and direction, even if they have different locations.

DEFINITION 5.2.2 Equivalent vectors are said to be **equal** vectors.

Equivalent vectors are regarded as equal because we are only concerned with magnitude and direction. For example, **v** and **w** in Figure 5.1 are equivalent vectors as they have the same magnitude (length) and the same direction. Hence we can write **v** = **w**. By contrast, **v** ≠ **z**; **z** has neither the same magnitude nor the same direction as **v**.

DEFINITION 5.2.3 (ADDITION OF VECTORS). Locating the initial point of one vector at the terminal point of the other, the **sum** of the two vectors is the vector that joins the initial point of the first to the terminal point of the second.

For example, taking vectors $\mathbf{v} = \overrightarrow{AB}$ and $\mathbf{z} = \overrightarrow{EF}$ and locating the initial point of **z** at B, $\mathbf{v} + \mathbf{z} = \overrightarrow{AB} + \overrightarrow{BC^*} = \overrightarrow{AC^*}$. The operation is illustrated in Figure 5.2, which also shows that, by a similar geometric construction, locating the initial point of **z** at A and the initial point of **v** at the terminal point of **z**, $\mathbf{z} + \mathbf{v} = \overrightarrow{AB^*} + \overrightarrow{B^*C^*} = \overrightarrow{AC^*} = \mathbf{v} + \mathbf{z}$; the addition of vectors is commutative. Locating the two separate summations at the same initial point and forming a parallelogram of vectors is a neat way of illustrating commutativity.

DEFINITION 5.2.4 The **zero vector** is the vector of zero magnitude (length).

Denoting the zero vector by **0**, it follows from the definition of vector addition that $\mathbf{0} + \mathbf{v} = \mathbf{v} + \mathbf{0} = \mathbf{v}$ for any vector **v**. The zero vector has no obvious direction.

DEFINITION 5.2.5 The vector having the same magnitude as, but opposite direction to, the specified vector, **v**, is the **negative of the vector** and is denoted by $-\mathbf{v}$; see Figure 5.3.

We define $-\mathbf{0} \equiv \mathbf{0}$.

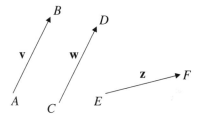

Figure 5.1 Vectors in 2-space

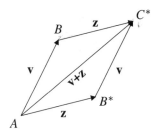

Figure 5.2 Addition of vectors

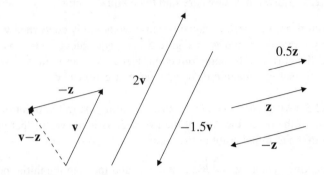

Figure 5.3 Multiplication of vectors by a constant and subtraction of vectors

DEFINITION 5.2.6 (SUBTRACTION OF VECTORS). Given two vectors, $\mathbf{z}$ and $\mathbf{v}$, and using the definitions for the sum and the negative of vectors, we see that $\mathbf{v} - \mathbf{z} \equiv \mathbf{v} + (-\mathbf{z})$; see Figure 5.3.

It follows that $\mathbf{v} - \mathbf{v} = \mathbf{v} + (-\mathbf{v}) = \mathbf{0}$.

DEFINITION 5.2.7 (MULTIPLICATION OF A VECTOR BY A SCALAR). Let $\mathbf{v}$ be a vector and k be a scalar. Then $k\mathbf{v}$ is the vector whose length is $|k|$ times that of $\mathbf{v}$, and whose direction is the same as that of $\mathbf{v}$ if $k > 0$ and opposite to that of $\mathbf{v}$ if $k < 0$. We define $k\mathbf{v} \equiv \mathbf{0}$ if $k = 0$ or $\mathbf{v} = \mathbf{0}$.

The negative of $\mathbf{v}$ is the special case where $k = -1$. Figure 5.3 illustrates a few other cases.

5.2.2   Analytical geometry

Problems involving vectors and vector geometry can often be simplified by the use of some **coordinate system**. Using a rectangular coordinate system in the plane, and locating the initial point of a vector, $\mathbf{v}$, at the origin of the coordinate system, the terminal point of the vector may be represented by two coordinates, v_1 and v_2. Figure 5.4 illustrates this. Indeed, we may write $\mathbf{v} = (v_1, v_2)$, and call the coordinates v_1 and v_2 the components or elements of $\mathbf{v}$.[1] The magnitude of $\mathbf{v}$, which we are representing by the length of the vector, is given by Pythagoras's theorem as $\sqrt{v_1^2 + v_2^2}$, where the positive square root is taken. The direction of $\mathbf{v}$ may be described by the angle it makes with either the horizontal axis or the vertical axis. The axes may also be used to define positive and negative directions. We will have more to say about coordinate systems *per se* later.

If equivalent vectors are located with their initial points at the same origin, then they coincide and have the same elements. Therefore, we say that two vectors $\mathbf{v} = (v_1, v_2)$ and $\mathbf{w} = (w_1, w_2)$ are equal if and only if $v_1 = w_1$ and $v_2 = w_2$. Other operations corresponding to those defined geometrically above are defined using coordinate elements as follows, where we let $\mathbf{v} = (v_1, v_2)$ and $\mathbf{z} = (z_1, z_2)$, and k is a scalar.

DEFINITION 5.2.8   Addition: $\mathbf{v} + \mathbf{z} = (v_1 + z_1, v_2 + z_2)$.

DEFINITION 5.2.9   Multiplication by a scalar: $k\mathbf{v} = (kv_1, kv_2)$.

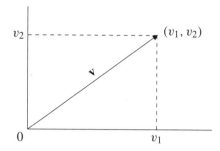

Figure 5.4 Vector coordinates

DEFINITION 5.2.10 Subtraction: $\mathbf{v} - \mathbf{z} = (v_1 - z_1, v_2 - z_2)$.

The proofs of these statements are not provided but each statement may be easily verified by appeal to a coordinate diagram such as Figure 5.5 and simple geometric arguments. Figure 5.5 depicts the cases of vector addition ($\mathbf{v} + \mathbf{z}$) and the multiplication of a vector by a scalar ($k\mathbf{z}$, where $k = 2.5$).

Translations

If a vector's initial point is not at the origin, its components may be obtained by subtracting the coordinates of its initial point from those of its terminal point. Thus, for the vector $\mathbf{v}$ depicted in Figure 5.6, for example, we have

$$\mathbf{v} = \overrightarrow{P_1 P_2} = \overrightarrow{0 P_2} - \overrightarrow{0 P_1}$$
$$= (x_2, y_2) - (x_1, y_1) = (x_2 - x_1, y_2 - y_1) \tag{5.1}$$

This operation is called a **translation**.

Alternatively, we may wish to translate the axes of a coordinate system so that the origin coincides with the initial point (x_1, y_1) of an arbitrary vector. The coordinates of (x_2, y_2) in

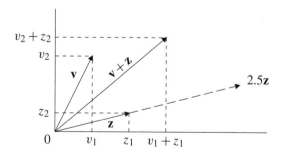

Figure 5.5 Vector addition with coordinates

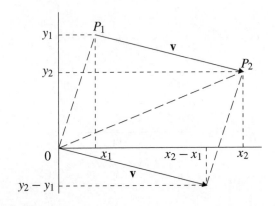

Figure 5.6 A translation

the new system, x' and y', are obtained by means of **translation equations**

$$x' = x_2 - x_1 \quad \text{and} \quad y' = y_2 - y_1 \tag{5.2}$$

This translation, which effects a parallel shift in each of the axes, is illustrated in Figure 5.7.

3-space

Just as pairs of real numbers may be used to represent vectors in 2-space analytically, so vectors in 3-space, and operations on them, may be represented by **triples** of real numbers, (x, y, z). Let these numbers denote the coordinates within a three-dimensional rectangular coordinate system, and let the axes of this system define positive and negative directions. We now have that each pair of axes determines a coordinate plane. It is conventional to view the first two coordinates as giving the location of a vector in a horizontal plane and the third coordinate as giving the vertical distance of the terminal point of the vector (x, y, z) above (if positive) or below (if negative) that plane. An example of a vector in such a coordinate system is depicted in Figure 5.8.

Writing vectors in 3-space as $\mathbf{v} = (v_1, v_2, v_3)$ and $\mathbf{w} = (w_1, w_2, w_3)$, equality, addition, multiplication by a scalar, and subtraction are defined analogously to the same operations in 2-space. However, graphical illustration of the operations is more tricky for 3-space than

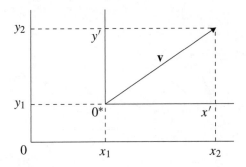

Figure 5.7 Another translation

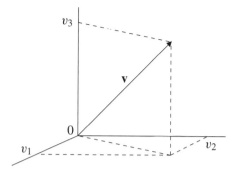

Figure 5.8 A 3-vector

for 2-space. The definitional and graphical details of these operations for 3-space are left as an exercise; see Exercise 5.3.

If a vector is considered to be a one-dimensional matrix, i.e. a single column or a single row of elements, the matrix Definitions 1.3.2, 1.3.3 and 1.3.4 apply. For the time being, we shall adopt this view. Doing this has the further advantage that, except for the distributive rule, the rules of vector algebra for 2- and 3-space correspond to the rules of matrix algebra developed earlier. The distributive rule for matrices involves matrix multiplication. We have not considered the multiplication of vectors in 2- and 3-space; therefore, no meaning yet attaches to it. However, we shall shortly encounter a concept called the "dot product" of vectors, and this will be seen to have an association with matrix multiplication. There are several other new concepts that are useful, and the next section introduces these and their use in the analytical treatment of vectors.

5.2.3 *Further concepts and vector algebra*

There are five concepts presented in this subsection. Associated with some of them are some simple theorems, which will be stated and proved.

DEFINITION 5.2.11 The **norm** (or **Euclidean norm**) of a vector, **v**, is another name for the magnitude or length of **v** and is denoted by $\|\mathbf{v}\|$.

DEFINITION 5.2.12 A **unit vector** is a vector whose norm is unity.

We have already used the unit vector concept when finding normalized eigenvectors in Section 3.3.3, and an expression for the length of a vector in 2-space was given above. This easily generalizes to the case of 3-space as

$$\|\mathbf{v}\| = \sqrt{v_1^2 + v_2^2 + v_3^2} = \sqrt{\sum_{i=1}^{3} v_i^2} \qquad (5.3)$$

where the positive square root is used. This computation may be justified graphically by careful use of Figure 5.8.

DEFINITION 5.2.13 The **distance** between two vectors $\mathbf{v} = \overrightarrow{0P_1}$ and $\mathbf{w} = \overrightarrow{0P_2}$, which we denote as $d(\mathbf{v}, \mathbf{w})$, is defined, for the case of 3-space, as

$$
\begin{aligned}
d(\mathbf{v}, \mathbf{w}) &= \left\| \overrightarrow{P_1 P_2} \right\| = \|(w_1 - v_1, w_2 - v_2, w_3 - v_3)\| \\
&= \sqrt{(w_1 - v_1)^2 + (w_2 - v_2)^2 + (w_3 - v_3)^2} \\
&= \sqrt{\sum_{i=1}^{3} (w_i - v_i)^2} = \|\mathbf{v} - \mathbf{w}\| = \|\mathbf{w} - \mathbf{v}\|
\end{aligned}
\tag{5.4}
$$

Similarly for the distance between two vectors in 2-space, where the required simplification is obvious.

Figure 5.9 illustrates the distance concept in 2-space.

DEFINITION 5.2.14 The **angle** between two vectors, $\mathbf{v}$ and $\mathbf{w}$, is the angle, θ, such that $0 \leq \theta \leq \pi$, where the angle is measured in radians.

Three such angles are illustrated in Figure 5.10.

DEFINITION 5.2.15 The **dot product** of two vectors, $\mathbf{v}$ and $\mathbf{w}$, denoted by $\mathbf{v} \cdot \mathbf{w}$, is defined as

$$
\mathbf{v} \cdot \mathbf{w} = \|\mathbf{v}\| \|\mathbf{w}\| \cos \theta
\tag{5.5}
$$

Note that $\mathbf{v} \cdot \mathbf{w} = 0$ if $\mathbf{v} = \mathbf{0}$ or $\mathbf{w} = \mathbf{0}$ or if $\theta = \pi/2$.

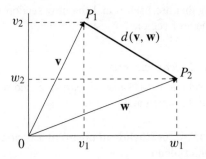

Figure 5.9 Distance between vectors

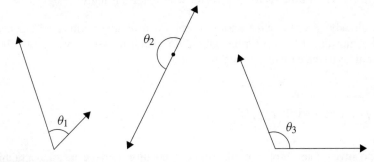

Figure 5.10 Angles between vectors

The definitional formula for a dot product simplifies to yield a more convenient computational formula. The simplification, for 3-space, is given in the following theorem.

THEOREM 5.2.1 *Let* **v** *and* **w** *be vectors in 3-space. Then the dot product may be written as*

$$\mathbf{v} \cdot \mathbf{w} = \sum_{i=1}^{3} v_i w_i \tag{5.6}$$

Proof: Using the cosine rule, we have that

$$(\overrightarrow{P_1 P_2})^2 = \|\mathbf{v} - \mathbf{w}\|^2 = \|\mathbf{v}\|^2 + \|\mathbf{w}\|^2 - 2\|\mathbf{v}\|\|\mathbf{w}\| \cos \theta \tag{5.7}$$

Therefore,

$$\mathbf{v} \cdot \mathbf{w} = \|\mathbf{v}\|\|\mathbf{w}\| \cos \theta = \frac{1}{2}(\|\mathbf{v}\|^2 + \|\mathbf{w}\|^2 - \|\mathbf{v} - \mathbf{w}\|^2)$$

$$= \frac{1}{2}\left(\sum_{i=1}^{3} v_i^2 + \sum_{i=1}^{3} w_i^2 - \sum_{i=1}^{3}(v_i - w_i)^2\right) \tag{5.8}$$

Expanding the final term in the large brackets and simplifying yields the required result. □

If **v** and **w** are treated as 3×1 matrices, the similarity of $\mathbf{v} \cdot \mathbf{w}$ and $\mathbf{v}^\top \mathbf{w}$ is clear.

Exploiting the definition of the dot product of two non-zero vectors, we obtain a useful means of obtaining information on the angle between the two vectors, namely

$$\cos \theta = \frac{\mathbf{v} \cdot \mathbf{w}}{\|\mathbf{v}\|\|\mathbf{w}\|} \tag{5.9}$$

EXAMPLE 5.2.1 Let $\mathbf{v} = (2, 1)$ and $\mathbf{w} = (1, 2)$ be vectors in 2-space. Let us find the distance between these vectors and the angle between them.

First, $d(\mathbf{v}, \mathbf{w}) = \|\mathbf{v} - \mathbf{w}\| = \sqrt{\sum_{i=1}^{2}(v_i - w_i)^2} = \sqrt{2}$.

To find the angle between the vectors we require

$$\mathbf{v} \cdot \mathbf{w} = \sum_{i=1}^{2} v_i w_i = 4, \quad \|\mathbf{v}\| = \sqrt{\sum_{i=1}^{2} v_i^2} = \sqrt{5} \quad \text{and} \quad \|\mathbf{w}\| = \sqrt{\sum_{i=1}^{2} w_i^2} = \sqrt{5} \tag{5.10}$$

Therefore, second, we have from (5.9) that $\cos \theta = \frac{4}{5}$ and, hence, $\theta = \cos^{-1}\left(\frac{4}{5}\right) \approx 0.643$ radians or 36 degrees and 52 minutes. ◇

It is a straightforward matter to verify the findings in Example 5.2.3 geometrically by means of a carefully drawn graph. It would be more difficult to do such geometry if the vectors were vectors in 3-space. Yet the analytical approach remains simple, as the following example demonstrates.

EXAMPLE 5.2.2 Let $\mathbf{v} = (2, -1, 1)$ and $\mathbf{w} = (1, 1, 2)$ be vectors in 3-space. Again, let us find the distance between them and the angle between them. The calculations are as follows:

$$d(\mathbf{v}, \mathbf{w}) = \|\mathbf{v} - \mathbf{w}\| = \sqrt{\sum_{i=1}^{3}(v_i - w_i)^2} = \sqrt{6} \qquad (5.11)$$

$$\mathbf{v} \cdot \mathbf{w} = \sum_{i=1}^{3} v_i w_i = 3 \qquad (5.12)$$

$$\|\mathbf{v}\| = \sqrt{\sum_{i=1}^{3} v_i^2} = \sqrt{6} \quad \text{and} \quad \|\mathbf{w}\| = \sqrt{\sum_{i=1}^{3} w_i^2} = \sqrt{6} \qquad (5.13)$$

Therefore, using (5.9) again, we have that $\cos\theta = \frac{3}{6} = \frac{1}{2}$ and, hence, $\theta = \cos^{-1}\left(\frac{1}{2}\right) = \pi/3$ radians or 60 degrees. $\Diamond$

We may now state the first of the theorems associated with these further concepts. Specifically, the following theorems relate to the dot product, the norm and the angle.

THEOREM 5.2.2 *Let* $\mathbf{v}$ *and* $\mathbf{w}$ *be vectors in 2- or 3-space. Then*

$$\mathbf{v} \cdot \mathbf{v} = \|\mathbf{v}\|^2 \qquad (5.14)$$

and

$$\|\mathbf{v}\| = (\mathbf{v} \cdot \mathbf{v})^{\frac{1}{2}} = \sqrt{\mathbf{v} \cdot \mathbf{v}} \qquad (5.15)$$

Also, for $\mathbf{v}$ *and* $\mathbf{w}$ *non-zero,*

$$\theta \text{ is acute if and only if } \mathbf{v} \cdot \mathbf{w} > 0 \qquad (5.16)$$
$$\theta \text{ is obtuse if and only if } \mathbf{v} \cdot \mathbf{w} < 0 \qquad (5.17)$$
$$\theta \text{ is a right angle if and only if } \mathbf{v} \cdot \mathbf{w} = 0 \qquad (5.18)$$

Proof:

(a) From the definition of dot product we have directly that

$$\mathbf{v} \cdot \mathbf{v} = \|\mathbf{v}\|\|\mathbf{v}\| \cos(0) = \|\mathbf{v}\|^2 \qquad (5.19)$$

(b) From (5.9) we have that $\cos\theta > 0$ if and only if $\mathbf{v} \cdot \mathbf{w} > 0$, since $\|\mathbf{v}\|$ and $\|\mathbf{w}\|$ are positive by definition. Similarly for the remainder of the second part of the theorem. $\square$

In the case of $\mathbf{v} \cdot \mathbf{w} = 0$, i.e. $\theta = \pi/2$ radians or 90 degrees, the vectors $\mathbf{v}$ and $\mathbf{w}$ are perpendicular. We say that perpendicular vectors are **orthogonal**. For example, it is easy to show that, for the vectors $\mathbf{v} = (0, 2)$ and $\mathbf{w} = (6, 0)$ in 2-space, $\mathbf{v} \cdot \mathbf{w} = 0$. Therefore, these vectors are orthogonal. Their perpendicularity may be seen with the aid of a graph, which would show the vectors lying along the axes of the rectangular coordinate system. Similarly for the vectors $\mathbf{v} = (2, 2)$ and $\mathbf{w} = (4, -4)$ in 2-space, $\mathbf{v} \cdot \mathbf{w} = 0$, and again these vectors are orthogonal, although in this case the vectors do not coincide with the axes of the graph. The details of the graphs in both cases are left as an exercise; see Exercise 5.1.

THEOREM 5.2.3 *Let* $\mathbf{u}$, $\mathbf{v}$ *and* $\mathbf{w}$ *be vectors in 2- or 3-space. Then*

$$\mathbf{u} \cdot \mathbf{v} = \mathbf{v} \cdot \mathbf{u} \tag{5.20}$$

$$\mathbf{u} \cdot (\mathbf{v} + \mathbf{w}) = \mathbf{u} \cdot \mathbf{v} + \mathbf{u} \cdot \mathbf{w} \tag{5.21}$$

$$k(\mathbf{u} \cdot \mathbf{v}) = (k\mathbf{u}) \cdot \mathbf{v} = \mathbf{u} \cdot (k\mathbf{v}) \tag{5.22}$$

and

$$\mathbf{v} \cdot \mathbf{v} \begin{cases} > 0 & \text{if } \mathbf{v} \neq \mathbf{0} \\ = 0 & \text{if } \mathbf{v} = \mathbf{0} \end{cases} \tag{5.23}$$

where k is any scalar.

The first two of these results state that the dot product operation is commutative and distributive over vector addition; the third and fourth also have simple interpretations but no special terms to describe them. All four are easy to prove by direct appeal to the definition of dot product or, in the case of the distributive property, to (5.6). The proof of the theorem is therefore left as an exercise; see Exercise 5.12. Note that, if $\mathbf{u}$, $\mathbf{v}$ and $\mathbf{w}$ are treated as $n \times 1$ non-zero matrices, the first of these results echoes $\mathbf{u}^{\top}\mathbf{v} = \mathbf{v}^{\top}\mathbf{u}$, while the remainder echo, in a similar way, other of the results on the multiplication operation for matrices given in Chapter 1.

5.2.4 Projections

A vector $\mathbf{u}$ may be decomposed into a sum of two vector terms, one parallel to a specified non-zero vector $\mathbf{v}$, the other perpendicular to $\mathbf{v}$. By way of illustration, in Figure 5.11 the terminal point of $\mathbf{u}$ is projected perpendicularly onto $\mathbf{v}$ to give the first of these terms, $\mathbf{w}_1$, and the second term is formed as the difference $\mathbf{w}_2 = \mathbf{u} - \mathbf{w}_1$.

Thus $\mathbf{w}_1$ is parallel to $\mathbf{v}$, $\mathbf{w}_2$ is perpendicular (orthogonal) to it and also, by construction, $\mathbf{w}_1 + \mathbf{w}_2 = \mathbf{w}_1 + \mathbf{u} - \mathbf{w}_1 = \mathbf{u}$. The vector $\mathbf{w}_1$ is called the **orthogonal projection of u on v**.

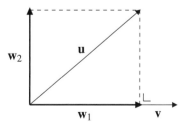

Figure 5.11 Orthogonal projection

It is denoted as $\mathbf{w}_1 = \text{proj}_{\mathbf{v}}\mathbf{u}$. The vector $\mathbf{w}_2$ is the **vector component of u orthogonal to v**. Since $\mathbf{w}_2 = \mathbf{u} - \mathbf{w}_1$, we have that $\mathbf{w}_2 = \mathbf{u} - \text{proj}_{\mathbf{v}}\mathbf{u}$. A convenient method of computing the two elements in this decomposition is provided by the following theorem.

THEOREM 5.2.4 *If* **u** *and* **v** *are vectors in 2- or 3-space and* $\mathbf{v} \neq \mathbf{0}$*, then*

$$\mathbf{w}_1 = \text{proj}_{\mathbf{v}}\mathbf{u} = \frac{\mathbf{u} \cdot \mathbf{v}}{\|\mathbf{v}\|^2}\mathbf{v} \quad and \quad \mathbf{w}_2 = \mathbf{u} - \text{proj}_{\mathbf{v}}\mathbf{u} = \mathbf{u} - \frac{\mathbf{u} \cdot \mathbf{v}}{\|\mathbf{v}\|^2}\mathbf{v} \tag{5.24}$$

Proof: Since $\mathbf{w}_1$ is parallel to $\mathbf{v}$, $\mathbf{w}_1 = k\mathbf{v}$, where k is a scalar to be computed. Therefore $\mathbf{u} = \mathbf{w}_1 + \mathbf{w}_2 = k\mathbf{v} + \mathbf{w}_2$.

Now consider the dot product $\mathbf{u} \cdot \mathbf{v}$. Using (5.21) and (5.22), we have

$$\mathbf{u} \cdot \mathbf{v} = k\mathbf{v} \cdot \mathbf{v} + \mathbf{w}_2 \cdot \mathbf{v} = k\|\mathbf{v}\|^2 \tag{5.25}$$

since $\mathbf{v} \cdot \mathbf{v} = \|\mathbf{v}\|^2$ and $\mathbf{w}_2 \cdot \mathbf{v} = 0$. So $k = (\mathbf{u} \cdot \mathbf{v})/\|\mathbf{v}\|^2$ and, since $\text{proj}_{\mathbf{v}}\mathbf{u} = \mathbf{w}_1 = k\mathbf{v}$,

$$\text{proj}_{\mathbf{v}}\mathbf{u} = \frac{\mathbf{u} \cdot \mathbf{v}}{\|\mathbf{v}\|^2}\mathbf{v} \tag{5.26}$$

$\square$

We may note that the length of the orthogonal projection of **u** on **v** is

$$\|\text{proj}_{\mathbf{v}}\mathbf{u}\| = \left\| \frac{\mathbf{u} \cdot \mathbf{v}}{\|\mathbf{v}\|^2}\mathbf{v} \right\| = \left| \frac{\mathbf{u} \cdot \mathbf{v}}{\|\mathbf{v}\|^2} \right| \|\mathbf{v}\|, \quad \text{since } \frac{\mathbf{u} \cdot \mathbf{v}}{\|\mathbf{v}\|^2} \text{ is a scalar}$$

$$= \frac{|\mathbf{u} \cdot \mathbf{v}|}{\|\mathbf{v}\|^2}\|\mathbf{v}\| = \frac{|\mathbf{u} \cdot \mathbf{v}|}{\|\mathbf{v}\|} \tag{5.27}$$

Also, since $\mathbf{u} \cdot \mathbf{v} = \|\mathbf{u}\|\|\mathbf{v}\|\cos\theta$, $(|\mathbf{u} \cdot \mathbf{v}|)/\|\mathbf{v}\| = \|\mathbf{u}\||\cos\theta|$. Therefore, we have

$$\|\text{proj}_{\mathbf{v}}\mathbf{u}\| = \frac{|\mathbf{u} \cdot \mathbf{v}|}{\|\mathbf{v}\|} = \|\mathbf{u}\||\cos\theta| \tag{5.28}$$

It is left as an exercise to derive an expression for the length of the vector component orthogonal to **v**; see Exercise 5.13. Two numerical examples follow.

EXAMPLE 5.2.3 Let $\mathbf{u} = (1, 2)$ and $\mathbf{v} = (1, -1)$. Let us find $\mathbf{u} \cdot \mathbf{v}$, $\|\mathbf{u}\|$, $\|\mathbf{v}\|^2$, $\text{proj}_{\mathbf{v}}\mathbf{u}$, $\|\text{proj}_{\mathbf{v}}\mathbf{u}\|$, θ (the angle between **u** and **v**), and $d(\mathbf{u}, \mathbf{v})$, and confirm that $\mathbf{w}_2 = \mathbf{u} - \text{proj}_{\mathbf{v}}\mathbf{u}$ is orthogonal to **v**.

Using the appropriate definitions and equations above, we have that $\mathbf{u} \cdot \mathbf{v} = \sum_{i=1}^{2} u_i v_i = 1 + (-2) = -1$, and $\|\mathbf{u}\| = \sqrt{\sum_{i=1}^{2} u_i^2} = \sqrt{1^2 + 2^2} = \sqrt{5}$. Similarly, $\|\mathbf{v}\| = \sqrt{\sum_{i=1}^{2} v_i^2} = \sqrt{1^2 + (-1)^2} = \sqrt{2}$ and $\|\mathbf{v}\|^2 = 2$. Then we find that

$$\text{proj}_{\mathbf{v}}\mathbf{u} = \frac{\mathbf{u} \cdot \mathbf{v}}{\|\mathbf{v}\|^2}\mathbf{v} = \frac{-1}{2}(1, -1) = \left(\frac{-1}{2}, \frac{1}{2} \right) = \mathbf{w}_1 \tag{5.29}$$

But

$$\|\text{proj}_{\mathbf{v}}\mathbf{u}\| = \sqrt{\sum_{i=1}^{2} w_{1i}^2} = \sqrt{\left(\frac{-1}{2}\right)^2 + \left(\frac{1}{2}\right)^2} = \frac{1}{\sqrt{2}} \tag{5.30}$$

calculating the norm directly, and

$$\|\text{proj}_{\mathbf{v}}\mathbf{u}\| = \frac{|\mathbf{u}\cdot\mathbf{v}|}{\|\mathbf{v}\|} = \frac{|-1|}{\sqrt{2}} = \frac{1}{\sqrt{2}} \tag{5.31}$$

using (5.28). So the angle is

$$\theta = \cos^{-1}\left(\frac{\mathbf{u}\cdot\mathbf{v}}{\|\mathbf{u}\|\|\mathbf{v}\|}\right) \quad \text{where} \quad \frac{\mathbf{u}\cdot\mathbf{v}}{\|\mathbf{u}\|\|\mathbf{v}\|} = \frac{-1}{\sqrt{5}\sqrt{2}} = \frac{-1}{\sqrt{10}} \tag{5.32}$$

By Theorem 5.2.2, we know that $\theta > \pi/2$, i.e. θ is obtuse. We find that $\theta = \cos^{-1}\left(\frac{-1}{\sqrt{10}}\right) \approx$ $108°26'6''$, to the nearest second of arc. Finally $d(\mathbf{u}, \mathbf{v}) = \|\mathbf{u} - \mathbf{v}\| = \sqrt{\sum_{i=1}^{2}(u_i - v_i)^2} =$ $\sqrt{(1-1)^2 + (2-(-1))^2} = \sqrt{0+9} = \sqrt{9} = 3$.

Now, $\mathbf{w}_2 = \mathbf{u} - \text{proj}_{\mathbf{v}}\mathbf{u} = (1, 2) - \left(-\frac{1}{2}, \frac{1}{2}\right) = \left(\frac{3}{2}, \frac{3}{2}\right)$. Checking reveals that $\mathbf{w}_2 \cdot \mathbf{v} = 0$. Therefore, $\mathbf{w}_2 \perp \mathbf{v}$ is confirmed, where the symbol $\perp$ signifies that $\mathbf{w}_2$ is orthogonal or perpendicular to $\mathbf{v}$. Similarly, it may be checked that $\mathbf{w}_1 \perp \mathbf{w}_2$. ◇

As the vectors in Example 5.2.4 are in 2-space, it is easy to draw a scale diagram and verify most of the numerical results geometrically. The exception is the result on θ: it is unlikely that this angle can be checked by a protractor, no matter how carefully the scale diagram is drawn. The next example uses vectors from 3-space. It is therefore much less easy to confirm results graphically. However, the arithmetic required by the analytic approach remains straightforward.

EXAMPLE 5.2.4 Let $\mathbf{u} = (2, 1, 1)$ and $\mathbf{v} = (1, -1, 1)$. Again let us find $\mathbf{u}\cdot\mathbf{v}$, $\|\mathbf{u}\|$, $\|\mathbf{v}\|^2$, $\text{proj}_{\mathbf{v}}\mathbf{u}$, $\|\text{proj}_{\mathbf{v}}\mathbf{u}\|$, θ and $d(\mathbf{u}, \mathbf{v})$, and confirm that $\mathbf{w}_2 = \mathbf{u} - \text{proj}_{\mathbf{v}}\mathbf{u}$ is orthogonal to $\mathbf{v}$.

Using the same methods as in the previous example, we have $\mathbf{u}\cdot\mathbf{v} = \sum_{i=1}^{3} u_i v_i = 2 - 1 + 1 = 2$, $\|\mathbf{u}\| = \sqrt{\sum_{i=1}^{3} u_i^2} = \sqrt{2^2 + 1^2 + 1^2} = \sqrt{6}$, $\|\mathbf{v}\| = \sqrt{\sum_{i=1}^{3} v_i^2} = \sqrt{1^2 + (-1)^2 + 1^2} = \sqrt{3}$ and $\|\mathbf{v}\|^2 = 3$. Therefore,

$$\text{proj}_{\mathbf{v}}\mathbf{u} = \frac{\mathbf{u}\cdot\mathbf{v}}{\|\mathbf{v}\|^2}\mathbf{v} = \frac{2}{3}(1, -1, 1) = \left(\frac{2}{3}, \frac{-2}{3}, \frac{2}{3}\right) = \mathbf{w}_1 \tag{5.33}$$

and

$$\|\text{proj}_{\mathbf{v}}\mathbf{u}\| = \sqrt{\sum_{i=1}^{3} w_{1i}^2} = \sqrt{\frac{4}{9} + \frac{4}{9} + \frac{4}{9}} = \frac{2}{\sqrt{3}} \quad \text{or} \quad \|\text{proj}_{\mathbf{v}}\mathbf{u}\| = \frac{|\mathbf{u}\cdot\mathbf{v}|}{\|\mathbf{v}\|} = \frac{2}{\sqrt{3}} \tag{5.34}$$

using the more computationally efficient (5.28).

The angle is

$$\theta = \cos^{-1}\left(\frac{\mathbf{u}\cdot\mathbf{v}}{\|\mathbf{u}\|\,\|\mathbf{v}\|}\right), \quad \text{where} \quad \frac{\mathbf{u}\cdot\mathbf{v}}{\|\mathbf{u}\|\,\|\mathbf{v}\|} = \frac{2}{\sqrt{6}\sqrt{3}} = \frac{2}{\sqrt{18}} = \frac{2}{3\sqrt{2}} = \frac{\sqrt{2}}{3} \qquad (5.35)$$

This time we see by Theorem 5.2.2 that $\theta < \pi/2$, i.e. θ is acute. More precisely, $\theta = \cos^{-1}\left(\frac{\sqrt{2}}{3}\right) \approx 61°52'28''$, to the nearest second of arc. Then $d(\mathbf{u}, \mathbf{v}) = \|\mathbf{u} - \mathbf{v}\| =$ $\sqrt{\sum_{i=1}^{3}(u_i - v_i)^2} = \sqrt{(2-1)^2 + (1-(-1))^2 + (1-1)^2} = \sqrt{1+4+0} = \sqrt{5}$.

Finally, $\mathbf{w}_2 = \mathbf{u} - \text{proj}_{\mathbf{v}}\mathbf{u} = (2, 1, 1) - \left(\frac{2}{3}, -\frac{2}{3}, \frac{2}{3}\right) = \left(\frac{4}{3}, \frac{5}{3}, \frac{1}{3}\right)$.

Checking confirms that $\mathbf{w}_2 \cdot \mathbf{v} = \mathbf{w}_1 \cdot \mathbf{w}_2 = 0$. Therefore, $\mathbf{w}_2 \perp \mathbf{v}$ and $\mathbf{w}_1 \perp \mathbf{w}_2$. $\qquad \diamond$

5.3 *n*-Dimensional Euclidean vector spaces

The previous basic ideas, rooted in the geometry of 2- and 3-space, may be easily generalized to *n*-dimensional spaces and even to **infinite-dimensional spaces**.[2]

To begin our generalization, recall the definition of an ordered *n*-tuple as a sequence of real numbers, $(a_1, a_2, \ldots, a_n)$, say, and the definition of *n*-space as the set of all ordered *n*-tuples, denoted $\mathbb{R}^n$; see p. xx. Therefore, $\mathbb{R}^1 = \mathbb{R}$ is the set of all real numbers; $\mathbb{R}^2$ is the set of all ordered pairs of real numbers, which up to this point we have referred to as 2-space; and $\mathbb{R}^3$ is the set of all ordered triples of real numbers, or 3-space.

By analogy with ordered pairs in 2-space ($\mathbb{R}^2$) and ordered triples in 3-space ($\mathbb{R}^3$), an ordered *n*-tuple can be regarded as the generalization of a "point" or of a "vector". Hence $(6, 2, 1, 3, 4, 5)$, an ordered six-tuple, may be described as a point or vector in $\mathbb{R}^6$ or 6-space. It is not possible to draw diagrams of vectors in $\mathbb{R}^n$ for $n > 3$, of course, but all of the basic definitions given in the previous section carry over to them.

In particular, for vectors $\mathbf{u}$ and $\mathbf{v}$ in $\mathbb{R}^n$, the standard notions of equality ($\mathbf{u} = \mathbf{v}$), addition ($\mathbf{u} + \mathbf{v}$) and multiplication by a scalar ($k\mathbf{v}$) apply. We also define analogously the zero vector ($\mathbf{0}_n$), the vector negative ($-\mathbf{v}$) and the vector difference ($\mathbf{u} - \mathbf{v}$). Moreover, all of the rules of vector algebra that we have encountered above apply to vectors in $\mathbb{R}^n$. These are stated formally in the following theorem.

THEOREM 5.3.1 *Let* $\mathbf{u}$, $\mathbf{v}$, $\mathbf{w}$ *and* $\mathbf{0} \in \mathbb{R}^n$ *and* k *and* $l \in \mathbb{R}$, *then*

$$\mathbf{u} + \mathbf{v} = \mathbf{v} + \mathbf{u} \qquad (5.36)$$

$$(\mathbf{u} + \mathbf{v}) + \mathbf{w} = \mathbf{u} + (\mathbf{v} + \mathbf{w}) \qquad (5.37)$$

$$\mathbf{u} + \mathbf{0} = \mathbf{0} + \mathbf{u} = \mathbf{u} \qquad (5.38)$$

$$\mathbf{u} + (-\mathbf{u}) = \mathbf{0} \qquad (5.39)$$

$$k(l\mathbf{u}) = (kl)\mathbf{u} \qquad (5.40)$$

$$k(\mathbf{u} + \mathbf{v}) = k\mathbf{u} + k\mathbf{v} \qquad (5.41)$$

$$(k + l)\mathbf{u} = k\mathbf{u} + l\mathbf{u} \qquad (5.42)$$

and

$$1\mathbf{u} = \mathbf{u} \qquad (5.43)$$

A proof of this theorem is easy to do for 2- or 3-space, using an analytical or geometric approach. Alternatively, as mentioned in Section 5.2.2, treating vectors as $n \times 1$ matrices allows the proofs of the analogous rules in Chapter 1 to be used.

We also retain the same basic definitions for vector norms (lengths), distances between vectors and dot products of vectors in $\mathbb{R}^n$. Specifically, for $\mathbf{u}$ and $\mathbf{v} \in \mathbb{R}^n$, the norm of $\mathbf{u}$, the distance between $\mathbf{u}$ and $\mathbf{v}$, and the dot product of $\mathbf{u}$ and $\mathbf{v}$ are, respectively,

$$\|\mathbf{u}\| \equiv \sqrt{\sum_{i=1}^{n} u_i^2} \tag{5.44}$$

$$d(\mathbf{u}, \mathbf{v}) \equiv \|\mathbf{u} - \mathbf{v}\| = \sqrt{\sum_{i=1}^{n} (u_i - v_i)^2} \tag{5.45}$$

$$\mathbf{u} \cdot \mathbf{v} \equiv \|\mathbf{u}\| \|\mathbf{v}\| \cos \theta = \sum_{i=1}^{n} u_i v_i \tag{5.46}$$

where, as before, θ denotes the angle between $\mathbf{u}$ and $\mathbf{v}$ and (5.46) follows from a slight modification of the proof of Theorem 5.2.1. It is more difficult to visualize the angle between two vectors once the dimension of a space exceeds three, but we will see in Section 5.4.3 that the cosine rule, and hence the proof of Theorem 5.2.1, are still valid in n dimensions.

It is because so many of the familiar ideas from 2- and 3-space carry over to it that $\mathbb{R}^n$ is frequently referred to as "Euclidean" n-space, and norms, distances and dot products of vectors in $\mathbb{R}^n$ as "Euclidean norms", "Euclidean distances" and "Euclidean dot products". Also, because of the similarity of the vector operations and rules we have encountered in this chapter and those we encountered earlier for matrices, it is not surprising that matrices of appropriate order may be used to denote vectors. Thus for vectors in $\mathbb{R}^n$, $n \times 1$ or $1 \times n$ matrices may be used. In the remainder of this book we will adopt the convention that points in $\mathbb{R}^n$ are represented by $n \times 1$ matrices or n-vectors. With this vertical notation, we have, in particular,

$$\mathbf{u} \cdot \mathbf{v} = \sum_{i=1}^{n} u_i v_i = \mathbf{u}^\top \mathbf{v} \tag{5.47}$$

5.4 General vector spaces

Further generalization of our ideas on vectors and vector spaces is possible by using axioms that abstract the main properties of vectors in $\mathbb{R}^n$, but that apply to other objects as well. Thus what we now speak of as "vectors" will include our original concept of a vector as an ordered n-tuple of real numbers, but many new kinds of vectors. However, our main concern will remain with our original notion of vectors. To effect this further generalization, we state the following definitions.

DEFINITION 5.4.1 A **field** is a set of elements that is closed under two binary operations, which we may call **addition** and **multiplication**, and where addition and multiplication are both commutative and associative and both have well-defined **identities** and **inverses**, and multiplication is distributive over addition.

There are several examples of fields, but the most obvious ones are the set of rational numbers, the set of real numbers and the set of complex numbers.

DEFINITION 5.4.2 Let F be a field and let V be an arbitrary set of objects (which we will call **vectors**) on which the two operations of **addition** and **multiplication by a scalar** are defined, i.e. we have some rule defining $\mathbf{u} + \mathbf{v}$ and some rule defining $k\mathbf{u}$, where $\mathbf{u}, \mathbf{v} \in V$ and $k \in F$. Then V is called a **vector space** if, for all $\mathbf{u}$, $\mathbf{v}$ and $\mathbf{w} \in V$, and all scalars k and $l \in F$, the following axioms are satisfied:

1. $\mathbf{u} + \mathbf{v} \in V$;
2. $\mathbf{u} + \mathbf{v} = \mathbf{v} + \mathbf{u}$;
3. $(\mathbf{u} + \mathbf{v}) + \mathbf{w} = \mathbf{u} + (\mathbf{v} + \mathbf{w})$;
4. there exists an object $\mathbf{0} \in V$, called the zero vector, such that $\mathbf{u} + \mathbf{0} = \mathbf{0} + \mathbf{u} = \mathbf{u}$;
5. there exists an object $-\mathbf{u} \in V$, called the negative of $\mathbf{u}$, such that $\mathbf{u} + (-\mathbf{u}) = \mathbf{0}$;
6. $k\mathbf{u} \in V$;
7. $k(l\mathbf{u}) = (kl)\mathbf{u}$;
8. $k(\mathbf{u} + \mathbf{v}) = k\mathbf{u} + k\mathbf{v}$;
9. $(k + l)\mathbf{u} = k\mathbf{u} + l\mathbf{u}$;
10. $1\mathbf{u} = \mathbf{u}$.

If F is the set of real numbers $\mathbb{R}$, then we call V a **real** or **Euclidean vector space**. The examples in this book will be confined to real vector spaces, but the theory of vector spaces developed here applies equally to vector spaces over other fields.

Most of the axioms in Definition 5.4.2 echo the rules stated in Theorem 5.3.1 and the corresponding rules of matrices from Chapter 1, including the commutative, associative and distributive rules. Two axioms that are not familiar from earlier material are Axioms 1 and 6. These define the concepts of **closure under addition** and **closure under scalar multiplication**, respectively, and they will be important for us when we meet the idea of a subspace presently. Note that neither the objects (vectors) nor the operations on them are defined. The axioms therefore define a diversity of possible spaces. For example, all of the following, which are not exhaustive, constitute vector spaces:

- The set $\{\mathbf{0}\}$. This is the simplest example, for which both vector addition and scalar multiplication are trivial.
- The field F itself is another simple example, for which vector addition is just field addition and scalar multiplication is just field multiplication.
- The set of all points in $\mathbb{R}^n$. As already mentioned, this is the most important vector space for our purposes, with addition and scalar multiplication defined as for n-vectors in Chapter 1.
- The set of all of the points in a straight line (or plane) through the origin, with vector addition and scalar multiplication defined in an obvious way. This case will be the subject of Section 5.4.1.
- The set of all $m \times n$ real matrices, often denoted $\mathbb{R}^{m \times n}$, with vector addition and scalar multiplication defined as for matrices in Chapter 1.
- The set of all polynomials of degree less than or equal to n and with coefficients in F, with vector addition and scalar multiplication defined in an obvious way.
- The set of all random variables on a given sample space; see Section 13.4.

Various results follow from the axioms of Definition 5.4.2 and four of them are contained in the next theorem.

THEOREM 5.4.1 *Let V be a vector space,* **u** *be a vector in V and k be a scalar. Then*

$$0\mathbf{u} = \mathbf{0} \tag{5.48}$$

$$k\mathbf{0} = \mathbf{0} \tag{5.49}$$

$$(-1)\mathbf{u} = -\mathbf{u} \tag{5.50}$$

$$if\, k\mathbf{u} = \mathbf{0}, \quad then\, k = 0\, or\, \mathbf{u} = \mathbf{0} \tag{5.51}$$

The proofs of the results in this theorem are not difficult to establish. By way of example, let us prove the third one, leaving the remaining proofs as exercises; see Exercise 5.14.

Proof: From Axiom 5, we must show that $\mathbf{u} + (-1)\mathbf{u} = \mathbf{0}$.
We have, using Axiom 10, that

$$\mathbf{u} + (-1)\mathbf{u} = 1\mathbf{u} + (-1)\mathbf{u} \tag{5.52}$$

Then, using Axiom 9,

$$1\mathbf{u} + (-1)\mathbf{u} = (1 - 1)\mathbf{u} = 0\mathbf{u} \tag{5.53}$$

Finally, using (5.48) of the theorem,

$$0\mathbf{u} = \mathbf{0} \tag{5.54}$$

$\square$

5.4.1 Subspaces

Certain subsets of a vector space V are themselves vector spaces under the vector addition and scalar multiplication defined on V. Such a subset, W say, is called a **subspace** of V. For example, any straight line through the origin of the plane, such as the line in Figure 5.12, is a subspace of $\mathbb{R}^2$.

In general, the ten axioms of Definition 5.4.2 must be verified for W to show that it is a subspace of V. However, if V is known to be a vector space, certain axioms need not be checked as they are "inherited" from V, namely, Axioms 2, 3, 7, 8, 9 and 10. Thus Axioms 1,

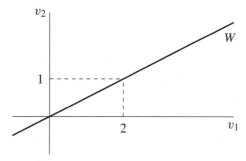

Figure 5.12 Subspace of $\mathbb{R}^2$

4, 5 and 6 only need be checked. Fortunately, the following theorem reduces the list further to Axioms 1 and 6 only.

THEOREM 5.4.2 *Let W be a non-empty subset of a vector space V. Then W is a subspace of V if and only if the following hold:*

(a) *if* $\mathbf{u}, \mathbf{v} \in W$, *then* $\mathbf{u} + \mathbf{v} \in W$; *and*
(b) *if k is a scalar and* $\mathbf{u} \in W$, *then* $k\mathbf{u} \in W$.

In other words, W is a subspace if and only if it is closed under addition and closed under scalar multiplication.

   Proof:

$\Rightarrow$ If W is a subspace of V, all vector space axioms are satisfied; in particular, Axioms 1 and 6, but these are just (a) and (b) of the theorem.
$\Leftarrow$ Conversely, assume that (a) and (b) hold. Axioms 2, 3, 7, 8, 9 and 10 hold for W since they are satisfied for all vectors in V. With regard to Axioms 4 and 5, we have from (b) that, if $\mathbf{u} \in W$, then $k\mathbf{u} \in W$. Let $k = 0$; it follows that $0\mathbf{u} = \mathbf{0} \in W$, which is Axiom 4. Let $k = -1$; it follows that $(-1)\mathbf{u} = -\mathbf{u} \in W$, which is Axiom 5. $\square$

5.4.2   Solution of equation systems

Consider the system of m linear equations in n unknowns:

$$a_{11}x_1 + a_{12}x_2 + \cdots + a_{1n}x_n = b_1$$
$$a_{21}x_1 + a_{22}x_2 + \cdots + a_{2n}x_n = b_2$$

$$\vdots$$

$$a_{m1}x_1 + a_{m2}x_2 + \cdots + a_{mn}x_n = b_m \tag{5.55}$$

or in concise matrix notation $\mathbf{Ax} = \mathbf{b}$, where the dimensions of the matrices are obvious. A vector $\mathbf{s} \in \mathbb{R}^n$ is a **solution!vector** if $x_i = s_i$, $i = 1, 2, \ldots, n$, is a solution of the system. Suppose the system is homogeneous, i.e. $\mathbf{b} = \mathbf{0}$. Then it can be shown that the set of solution vectors is a subspace of $\mathbb{R}^n$, as follows.

Given $\mathbf{Ax} = \mathbf{0}$, let W be the set of solution vectors, and let $\mathbf{s}_1 \in W$ and $\mathbf{s}_2 \in W$. Thus $\mathbf{As}_1 = \mathbf{0}$ and $\mathbf{As}_2 = \mathbf{0}$. Therefore, $\mathbf{A}(\mathbf{s}_1 + \mathbf{s}_2) = \mathbf{As}_1 + \mathbf{As}_2 = \mathbf{0} + \mathbf{0} = \mathbf{0}$ and $\mathbf{A}(k\mathbf{s}_1) = k\mathbf{As}_1 = k\mathbf{0} = \mathbf{0}$. Hence $\mathbf{s}_1 + \mathbf{s}_2$ and $k\mathbf{s}_1$ are also solution vectors. As W is closed under addition and scalar multiplication, it is a subspace of $\mathbb{R}^n$ by Theorem 5.4.2. We call this W the **solution space** of the system $\mathbf{Ax} = \mathbf{0}$, or the **null space** or **kernel** of the matrix $\mathbf{A}$. These concepts will be defined and explained further in Section 6.3.1.

The non-homogeneous equation $\mathbf{Ax} = \mathbf{b}$ may or may not have solutions. The system is **consistent** if and only if the right-hand side is in the so-called column space of $\mathbf{A}$ and there is a solution. Such a solution is called a **particular solution**. A **general solution** is obtained by adding to some particular solution a generic element of the solution space of the homogeneous system $\mathbf{Ax} = \mathbf{0}$. In Chapter 2, the solution of a non-singular square system of linear equations was studied. Now, we can solve any system by describing the solution space. The concept of the column space of a matrix and related ideas are explained in Section 5.4.6, while the issue of the solution of non-square systems and singular square systems of linear equations is dealt with further in Section 6.3.2.

5.4.3 *Linear combinations and spanning*

In the interest of finding the "smallest" subspace of a vector space, V, that contains a specified set of vectors such as $\{\mathbf{v}_1, \mathbf{v}_2, \ldots, \mathbf{v}_r\}$, we state the following definition.

DEFINITION 5.4.3 A **linear combination** of the vectors $\mathbf{v}_1, \mathbf{v}_2, \ldots, \mathbf{v}_r$ is defined as

$$k_1 \mathbf{v}_1 + k_2 \mathbf{v}_2 + \cdots + k_r \mathbf{v}_r = \sum_{i=1}^{r} k_i \mathbf{v}_i \tag{5.56}$$

where r is any positive integer and $k_1, k_2, \ldots, k_r$ are scalars.

EXAMPLE 5.4.1 The vector $\mathbf{w} = (2, 3, 4)$ in $\mathbb{R}^3$ is a linear combination of $\mathbf{v}_1 = (5, 6, 8)$ and $\mathbf{v}_2 = (3, 3, 4)$. It is easy to show that $\mathbf{w} = k_1 \mathbf{v}_1 + k_2 \mathbf{v}_2$, where $k_1 = 1$ and $k_2 = -1$:

$$\begin{aligned}
1(5, 6, 8) + (-1)(3, 3, 4) &= (5, 6, 8) + (-3, -3, -4) \\
&= (5 - 3, 6 - 3, 8 - 4) \\
&= (2, 3, 4) = \mathbf{w}
\end{aligned} \tag{5.57}$$

$\diamond$

On the other hand, $\mathbf{w}$ from Example 5.4.3 is *not* a linear combination of $\mathbf{v}_1 = (5, 6, 8)$ and $\mathbf{v}_3 = (-3, 3, -4)$. It is left as an exercise to verify this and to find a vector $\mathbf{w}^*$ that *is* a linear combination of $\mathbf{v}_1$ and $\mathbf{v}_3$; see Exercise 5.3.

The expression in Definition 5.4.3 should be familiar from the definitions of linear independence and linear dependence given in Definition 3.6.2.

An implication of Theorem 5.4.2 is that a subspace is closed under arbitrary linear combinations of vectors contained within it. The following theorem, whose proof follows by induction from Theorem 5.4.2 and is the subject of Exercise 5.15, formalizes this.

THEOREM 5.4.3 *Let W be a subset of a real vector space V. Then W is a vector subspace of V if and only if $\sum_{i=1}^{r} k_i \mathbf{v}_i \in W$ where r is any positive integer, $k_1, k_2, \ldots, k_r$ are scalars and the vectors $\mathbf{v}_1, \mathbf{v}_2, \ldots, \mathbf{v}_r \in W$.*

DEFINITION 5.4.4 If S is a set of vectors in V and if every vector in V can be expressed as a linear combination of vectors in S, then we say that S **spans** or **generates** V.

When S is finite, say $S = \{\mathbf{v}_1, \mathbf{v}_2, \ldots, \mathbf{v}_r\}$, then we often just say that the vectors $\mathbf{v}_1, \mathbf{v}_2, \ldots, \mathbf{v}_r$ span or generate V.

EXAMPLE 5.4.2 The vectors $\mathbf{i} = (1, 0, 0), \mathbf{j} = (0, 1, 0)$ and $\mathbf{k} = (0, 0, 1)$ span $\mathbb{R}^3$ since every vector (a, b, c) in $\mathbb{R}^3$ can be written as $a\mathbf{i} + b\mathbf{j} + c\mathbf{k}$, a linear combination of $\mathbf{i}, \mathbf{j}$ and $\mathbf{k}$:

$$\begin{aligned}
a\mathbf{i} + b\mathbf{j} + c\mathbf{k} &= a(1, 0, 0) + b(0, 1, 0) + c(0, 0, 1) \\
&= (a, 0, 0) + (0, b, 0) + (0, 0, c) \\
&= (a, b, c)
\end{aligned} \tag{5.58}$$

$\diamond$

The vectors $(2, 0, 0)$, $(0, -5, 0)$ and $(0, 0, 3)$ also span $\mathbb{R}^3$; so do $\mathbf{i}$, $\mathbf{j}$, $\mathbf{k}$ and $\mathbf{l} = (2, 3, -4)$. Verification of these last two statements is left as an exercise; see Exercise 5.4.

A given (finite or infinite) set of vectors may or may not span a vector space V. However, if we group together all vectors that *are* expressible as linear combinations of the vectors in S, then we obtain a subspace of V, called the (linear) space spanned by S and denoted lin S. The following definition formalizes this idea.

DEFINITION 5.4.5 Let S be a set of vectors in the vector space V. Then the **linear space spanned** or **generated** by the **spanning set** or **generating set** S is

$$\text{lin } S = \left\{ \sum_{i=1}^{r} k_i \mathbf{v}_i : \mathbf{v}_i \in S, \ k_i \in \mathbb{R}, \ i = 1, 2, \ldots, r; \ r = 1, 2, \ldots \right\} \tag{5.59}$$

This is the subject of the following theorem.

THEOREM 5.4.4 *If S is a set of vectors in the vector space V and $W \equiv \text{lin } S$, then*

(a) *W is a subspace of V; and*
(b) *W is the smallest subspace of V that contains S, in the sense that every other subspace that contains S must contain W.*

The proof of the first part of this theorem follows from Theorem 5.4.2 by showing that W is closed under addition and scalar multiplication; the proof of the second part is even more straightforward. The details are left as an exercise; see Exercise 5.16.

For example, if $\mathbf{v}_1 \in \mathbb{R}^2$ (or $\mathbb{R}^3$), then lin$\{\mathbf{v}_1\}$ is the set of all scalar multiples of $\mathbf{v}_1$, i.e. a straight line through the origin. More specifically, if $\mathbf{v}_1 = (0, 1)$, then lin$\{\mathbf{v}_1\}$ is one of the axes. If $\mathbf{v}_2 = (2, 1)$, then lin$\{\mathbf{v}_2\}$ is the line shown in Figure 5.12, while lin$\{\mathbf{v}_1, \mathbf{v}_2\} = \mathbb{R}^2$.

Similarly, taking the case of $\mathbb{R}^3$, if $\mathbf{v}_1 = (1, 0, 0)$ and $\mathbf{v}_2 = (0, 1, 0)$, then lin$\{\mathbf{v}_1, \mathbf{v}_2\}$ is the entire horizontal plane in $\mathbb{R}^3$. For arbitrary vectors $\mathbf{v}_3 = (a_3, b_3, c_3)$ and $\mathbf{v}_4 = (a_4, b_4, c_4)$, lin$\{\mathbf{v}_3, \mathbf{v}_4\}$ is a plane through the origin, assuming $\mathbf{v}_3$ and $\mathbf{v}_4$ are linearly independent.

We are now in a position, as noted in Section 5.3, to provide an interpretation of the angle between any two vectors in n-space, say, $\mathbf{u}$ and $\mathbf{v}$. $W \equiv \text{lin}\{\mathbf{u}, \mathbf{v}\}$ is a (two-dimensional) plane through the origin in n-space, as long as $\mathbf{u}$ and $\mathbf{v}$ are linearly independent. The angle between $\mathbf{u}$ and $\mathbf{v}$ can be easily measured in this plane. If $\mathbf{u}$ and $\mathbf{v}$ are linearly dependent, then the angle between them is 0, if they are in the same direction, or π, if they are in opposite directions.

5.4.4 Linear independence

The idea of linear independence, introduced in Definition 3.6.2, is of importance in finding spanning sets of vectors with the smallest number of vectors. Recall that the set of vectors $S = \{\mathbf{v}_1, \mathbf{v}_2, \ldots, \mathbf{v}_r\}$ is said to be linearly independent if and only if $k_1 \mathbf{v}_1 + k_2 \mathbf{v}_2 + \cdots + k_r \mathbf{v}_r = \mathbf{0}$ holds only for $k_i = 0$, for all $i = 1, 2, \ldots, r$. Otherwise, S is a linearly dependent set.

For example, for $\mathbf{v}_1, \mathbf{v}_2 \in \mathbb{R}^3$, where $\mathbf{v}_1 = (2, 3, 4)$ and $\mathbf{v}_2 = (5, 6, 8)$, $k_1 \mathbf{v}_1 + k_2 \mathbf{v}_2 = \mathbf{0}$ only if $k_1 = k_2 = 0$; therefore $S = \{\mathbf{v}_1, \mathbf{v}_2\}$ is a linearly independent set of vectors in $\mathbb{R}^3$. Similarly, it can be shown that if $\mathbf{e}_i = (0, 0, \ldots, 0, 1, 0, \ldots, 0) \in \mathbb{R}^n$ are vectors all of whose elements are zero except for the element in the ith position, which is unity, $S = \{\mathbf{e}_1, \mathbf{e}_2, \ldots, \mathbf{e}_n\}$ is a linearly independent set of vectors in $\mathbb{R}^n$. By contrast, $\mathbf{v}_1 = (2, 3, 4)$, $\mathbf{v}_2 = (5, 6, 8)$ and

$\mathbf{v}_3 = (4, 6, 8)$ constitute a linearly dependent set of vectors in $\mathbb{R}^3$ since, as is easily verified, $k_1\mathbf{v}_1 + k_2\mathbf{v}_2 + k_3\mathbf{v}_3 = \mathbf{0}$ with $k_1 = 1$, $k_2 = 0$ and $k_3 = -1/2$.

The following is a useful theorem on the linear dependence of vectors.

THEOREM 5.4.5 *A (finite) set with two or more vectors is linearly dependent if and only if at least one vector is a linear combination of the others.*

Proof:

$\Leftarrow$ Let $S = \{\mathbf{v}_1, \mathbf{v}_2, \ldots, \mathbf{v}_r\}$ and suppose that $\mathbf{v}_1$ is a linear combination of the others, i.e. $\mathbf{v}_1 = k_2\mathbf{v}_2 + \cdots + k_r\mathbf{v}_r$, where not all $k_i = 0$. Then $\mathbf{v}_1 - k_2\mathbf{v}_2 - \cdots - k_r\mathbf{v}_r = \mathbf{0}$, where not all $k_i = 0$, which implies linear dependence, by definition.

$\Rightarrow$ Now suppose $k_1\mathbf{v}_1 + k_2\mathbf{v}_2 + \cdots + k_r\mathbf{v}_r = \mathbf{0}$ and, say, $k_1 \neq 0$, i.e. assume that the vectors are linearly dependent. Then it follows that

$$\mathbf{v}_1 = -\frac{k_2}{k_1}\mathbf{v}_2 - \frac{k_3}{k_1}\mathbf{v}_3 - \cdots - \frac{k_r}{k_1}\mathbf{v}_r \tag{5.60}$$

$\square$

The geometry of linearly dependent and linearly independent vectors is worthy of note. Take the case of two vectors $\mathbf{v}_1, \mathbf{v}_2 \in \mathbb{R}^2$ and suppose, first, that $\mathbf{v}_1$ and $\mathbf{v}_2$ are linearly dependent, say $\mathbf{v}_1 - \frac{1}{2}\mathbf{v}_2 = \mathbf{0}$ ($k_1 = 1$, $k_2 = -1/2$). From this equation, and in accordance with Theorem 5.4.5, we have that $\mathbf{v}_1 = \frac{1}{2}\mathbf{v}_2$ and $\mathbf{v}_2 = 2\mathbf{v}_1$. Each vector is a scalar multiple of the other and lies in the same line (subspace) through the origin, such that the length of $\mathbf{v}_1$ is half that of $\mathbf{v}_2$ and the length of $\mathbf{v}_2$ is twice that of $\mathbf{v}_1$. Now suppose that $\mathbf{v}_1$ and $\mathbf{v}_2$ are linearly independent, hence $k_1\mathbf{v}_1 + k_2\mathbf{v}_2 = \mathbf{0}$ only for $k_1 = k_2 = 0$. It is now not possible to write one of the vectors as a scalar multiple of the other and so, geometrically, the two vectors cannot lie in the same line through the origin; the angle between them, θ, is such that $0 < \theta < \pi$. These two situations for two vectors in $\mathbb{R}^2$ are depicted in Figure 5.13.

The case of two vectors in $\mathbb{R}^3$ is similar. If two vectors are linearly dependent, they lie in the same line through the origin; if they are linearly independent, there is a non-zero angle between them and they point in different directions. The algebraic and diagrammatic details for this case are left as an exercise; see Exercise 5.6. The situation is somewhat more complicated for three linearly dependent vectors in $\mathbb{R}^3$. In this case, the vectors could all

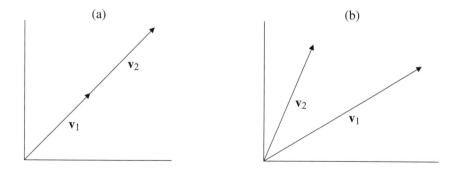

Figure 5.13 Linearly dependent and linearly independent vectors

lie in the same line through the origin but, more generally, they would lie in the same plane through the origin. Three linearly independent vectors in $\mathbb{R}^3$ are not so constrained. However, four or more vectors in $\mathbb{R}^3$, or three or more vectors in $\mathbb{R}^2$, must be linearly dependent. This fact is generalized and stated formally in the next theorem.

THEOREM 5.4.6 *Let* $S = \{\mathbf{v}_1, \mathbf{v}_2, \ldots, \mathbf{v}_r\}$ *be a set of vectors in* $\mathbb{R}^n$. *If* $r > n$, *then* S *is linearly dependent.*

Proof: Let $\mathbf{v}_i = (v_{i1}, v_{i2}, \ldots, v_{in})$, $i = 1, 2, \ldots, r$, and consider the equation

$$k_1\mathbf{v}_1 + k_2\mathbf{v}_2 + \cdots + k_r\mathbf{v}_r = \mathbf{0} \tag{5.61}$$

which is a homogeneous system of n linear equations in the r unknowns $k_1, k_2, \ldots, k_r$. Such a system has non-trivial solutions, as will be proved in Theorem 6.3.5 and its corollary. Therefore, $S = \{\mathbf{v}_1, \mathbf{v}_2, \ldots, \mathbf{v}_r\}$ is a linearly dependent set by Definition 3.6.2. $\square$

5.4.5 Basis and dimension

The following definitions arise from the ideas just discussed and will in their turn be used in the following sections.

DEFINITION 5.4.6 If V is a vector space and $S = \{\mathbf{v}_1, \mathbf{v}_2, \ldots, \mathbf{v}_r\}$ is a finite set of vectors in V, then S is called a **basis** for V if:

(a) S is linearly independent; and
(b) S spans V.

Put another way, S is said to be a basis for V if *all* $\mathbf{v} \in V$ can be expressed as linear combinations of the r linearly independent vectors $\mathbf{v}_1, \mathbf{v}_2, \ldots, \mathbf{v}_r$.

By way of illustration, recall the linearly independent vectors $\mathbf{e}_1, \mathbf{e}_2, \ldots, \mathbf{e}_n \in \mathbb{R}^n$ defined in the previous section, i.e. $\mathbf{e}_1 = (1, 0, \ldots, 0)$, $\mathbf{e}_2 = (0, 1, 0, \ldots, 0)$, $\ldots$, $\mathbf{e}_n = (0, \ldots, 0, 1)$, which form a linearly independent set, S, in $\mathbb{R}^n$. Since this S also spans $\mathbb{R}^n$, i.e. any vector $\mathbf{v} = (v_1, v_2, \ldots, v_n)$ can be written as $\mathbf{v} = v_1\mathbf{e}_1 + v_2\mathbf{e}_2 + \cdots + v_n\mathbf{e}_n$, then S is a basis for $\mathbb{R}^n$. In fact, this basis is called the **standard basis** for $\mathbb{R}^n$. Thus, for instance, $S = \{\mathbf{e}_1, \mathbf{e}_2\}$, where $\mathbf{e}_1 = (1, 0)$ and $\mathbf{e}_2 = (0, 1)$ is the standard basis for $\mathbb{R}^2$.

Note that vector spaces do not have a unique basis: any set of vectors in the space constitutes a basis for the space, as long as it is a linearly independent and spanning set.

EXAMPLE 5.4.3 Defining $\mathbf{v}_1 = (1, 2)$ and $\mathbf{v}_2 = (2, 9)$, the set $S = \{\mathbf{v}_1, \mathbf{v}_2\}$ is a basis for $\mathbb{R}^2$, though not the standard basis. This is because, first, $\mathbf{v}_1$ and $\mathbf{v}_2$ are linearly independent, since

$$k_1\mathbf{v}_1 + k_2\mathbf{v}_2 = \mathbf{0} \tag{5.62}$$

or

$$\mathbf{Ak} \equiv \begin{bmatrix} 1 & 2 \\ 2 & 9 \end{bmatrix} \begin{bmatrix} k_1 \\ k_2 \end{bmatrix} = \begin{bmatrix} k_1 + 2k_2 \\ 2k_1 + 9k_2 \end{bmatrix} = \begin{bmatrix} 0 \\ 0 \end{bmatrix} = \mathbf{0} \tag{5.63}$$

has only the trivial solution $k_1 = k_2 = 0$, which follows from the fact that the matrix $\mathbf{A} = \begin{bmatrix} 1 & 2 \\ 2 & 9 \end{bmatrix}$ is non-singular and, hence, invertible, implying that $\mathbf{k} = \mathbf{A}^{-1}\mathbf{0} = \mathbf{0}$.

Second, an arbitrary vector, say, $\mathbf{b} = (b_1, b_2)$, can be expressed as $\mathbf{b} = k_1\mathbf{v}_1 + k_2\mathbf{v}_2$, a linear combination of the vectors in S. To see this, set

$$
\begin{aligned}
(b_1, b_2) &= k_1(1, 2) + k_2(2, 9) \\
&= (k_1 + 2k_2, 2k_1 + 9k_2)
\end{aligned}
\tag{5.64}
$$

or

$$
\mathbf{b} = \begin{bmatrix} b_1 \\ b_2 \end{bmatrix} = \begin{bmatrix} k_1 + 2k_2 \\ 2k_1 + 9k_2 \end{bmatrix} = \begin{bmatrix} 1 & 2 \\ 2 & 9 \end{bmatrix} \begin{bmatrix} k_1 \\ k_2 \end{bmatrix} = \mathbf{A}\mathbf{k}
\tag{5.65}
$$

For any choice of $\mathbf{b} \neq \mathbf{0}$, this equation has a non-trivial solution for $\mathbf{k} = [k_i]_{2 \times 1}$, namely $\mathbf{k} = \mathbf{A}^{-1}\mathbf{b}$, again since $\mathbf{A}$ is non-singular and, hence, invertible. The k_i can, of course, be obtained by Cramer's rule in this case.

Suppose $\mathbf{b} = (10, -10)$. Then

$$
\begin{aligned}
\mathbf{k} = \mathbf{A}^{-1}\mathbf{b} &= \begin{bmatrix} 1 & 2 \\ 2 & 9 \end{bmatrix}^{-1} \begin{bmatrix} 10 \\ -10 \end{bmatrix} \\
&= \begin{bmatrix} \frac{9}{5} & -\frac{2}{5} \\ -\frac{2}{5} & \frac{1}{5} \end{bmatrix} \begin{bmatrix} 10 \\ -10 \end{bmatrix} \\
&= \begin{bmatrix} 22 \\ -6 \end{bmatrix}
\end{aligned}
\tag{5.66}
$$

by direct matrix calculations using the inverse of $\mathbf{A}$. Alternatively, using Cramer's rule,

$$
k_1 = \frac{|\mathbf{A}_1|}{|\mathbf{A}|} = \frac{110}{5} = 22
\tag{5.67}
$$

and

$$
k_2 = \frac{|\mathbf{A}_2|}{|\mathbf{A}|} = \frac{-30}{5} = -6
\tag{5.68}
$$

where

$$
\mathbf{A}_1 = \begin{bmatrix} 10 & 2 \\ -10 & 9 \end{bmatrix} \quad \text{and} \quad \mathbf{A}_2 = \begin{bmatrix} 1 & 10 \\ 2 & -10 \end{bmatrix}
\tag{5.69}
$$

$\diamondsuit$

Another way of stating the fact illustrated in Example 5.4.5 is that any two **non-collinear** vectors in $\mathbb{R}^2$ form a basis. We may also note the following points.

- If $S = \{\mathbf{v}_1, \mathbf{v}_2, \ldots, \mathbf{v}_r\}$ is a linearly independent set in a vector space V, then S is a basis for the subspace $\mathrm{lin}\{\mathbf{v}_1, \mathbf{v}_2, \ldots, \mathbf{v}_r\}$, since it is linearly independent and, by definition of $\mathrm{lin}\{\mathbf{v}_1, \mathbf{v}_2, \ldots, \mathbf{v}_r\}$, S spans $\mathrm{lin}\{\mathbf{v}_1, \mathbf{v}_2, \ldots, \mathbf{v}_r\}$.

- If $S = \{\mathbf{v}_1, \mathbf{v}_2, \ldots, \mathbf{v}_n\}$ is a basis for V, then every subset of V with more than n vectors is linearly dependent. The proof of this proposition is not difficult and is left as an exercise; see Exercise 5.17.
- Any two, or more, bases for a (finite) vector space V have the same number of vectors. The proof of this follows from the previous assertion.

The next definition relates to the number of vectors in a basis.

DEFINITION 5.4.7 The **dimension** of a (finite) vector space V, denoted $\dim(V)$, is the unique number of vectors in a basis for V. By convention we also define the dimension of the set containing just the zero vector to be zero.

Thus $\dim(\mathbb{R}^n) = n$, and $\dim(\{\mathbf{0}\}) = 0$.

When we know that a vector space, V, has dimension n, then to determine whether $S = \{\mathbf{v}_1, \mathbf{v}_2, \ldots, \mathbf{v}_n\}$ is a basis for V, we only have to check whether S is a linearly independent set or whether S spans V. The remaining condition will hold automatically. We formalize this fact in the following theorem, whose proof is left as an exercise; see Exercise 5.18.

THEOREM 5.4.7 *Let V be an n-dimensional vector space.*

(a) *If $S = \{\mathbf{v}_1, \mathbf{v}_2, \ldots, \mathbf{v}_n\}$ is a set of n linearly independent vectors in V, then S is a basis for V.*

(b) *If $S = \{\mathbf{v}_1, \mathbf{v}_2, \ldots, \mathbf{v}_n\}$ is a set of n vectors that spans V, then S is a basis for V.*

5.4.6   *Row and column space of a matrix, and rank*

Consider the $m \times n$ matrix $\mathbf{A} = \begin{bmatrix} a_{11} & \cdots & a_{1n} \\ \vdots & & \vdots \\ a_{m1} & \cdots & a_{mn} \end{bmatrix}$. This matrix may be written as

$\mathbf{A} = \begin{bmatrix} \mathbf{r}_1 \\ \mathbf{r}_2 \\ \vdots \\ \mathbf{r}_m \end{bmatrix}$, where $\mathbf{r}_i$ denotes the $1 \times n$ matrix that is the ith row of $\mathbf{A}$; or as $\mathbf{A} =$

$[\mathbf{c}_1 \quad \mathbf{c}_2 \quad \cdots \quad \mathbf{c}_n]$, where $\mathbf{c}_j$ is the $m \times 1$ matrix that is the jth column of $\mathbf{A}$. In other words, $\mathbf{A}$ may be partitioned according to its rows or its columns. Further, each row may be thought of as a vector in $\mathbb{R}^n$ and each column as a vector in $\mathbb{R}^m$.

DEFINITION 5.4.8

(a) The **row space** of an $m \times n$ matrix $\mathbf{A}$ is the vector subspace of $\mathbb{R}^n$ spanned by the m rows of $\mathbf{A}$.

(b) The **column space** of an $m \times n$ matrix $\mathbf{A}$ is the vector subspace of $\mathbb{R}^m$ spanned by the n columns of $\mathbf{A}$.

DEFINITION 5.4.9

(a) The **row rank** of a matrix is the dimension of its row space.

(b) The **column rank** of a matrix is the dimension of its column space.[3]

THEOREM 5.4.8 *The row space and the column space of any matrix have the same dimension.*

Proof: The idea of the proof is that performing elementary row operations on a matrix does not change either the row rank or the column rank of the matrix.

By inspection, it is clear that the row rank and the column rank of a matrix in reduced row-echelon form are equal to each other and to the dimension of the identity matrix in the top left corner. In fact, elementary row operations do not change the row space of the matrix. They do change the column space of a matrix, but not the column rank, as we shall now see.

If **A** and **B** are **row-equivalent matrices**, i.e. if they can be formed from each other by elementary row operations, then the equations $\mathbf{Ax} = \mathbf{0}$ and $\mathbf{Bx} = \mathbf{0}$ have the same solution space.

If a subset of columns of **A** is a linearly dependent set, then the solution space contains a vector in which the corresponding entries are non-zero and all other entries are zero.

Similarly, if a subset of columns of **A** is a linearly independent set, then the solution space does not contain a vector in which the corresponding entries are non-zero and all other entries are zero.

The first result implies that the corresponding columns of **B** are also linearly dependent.

The second result implies that the corresponding columns of **B** are also linearly independent.

It follows that the dimension of the column space is the same for both matrices. □

It is for this reason that the dimension of the row and column spaces of **A** is called, simply, the **rank** of **A**, denoted by $\rho(\mathbf{A})$. It follows that $\rho(\mathbf{A}) \leq \min\{m, n\}$.

EXAMPLE 5.4.4 Let $\mathbf{A} = \begin{bmatrix} 3 & 2 & 1 \\ 4 & -2 & 0 \end{bmatrix}$. Here, the rows $\mathbf{r}_1$ and $\mathbf{r}_2$ constitute a set, $S_r = \{\mathbf{r}_1, \mathbf{r}_2\}$, of two vectors in $\mathbb{R}^3$, and since $k_1\mathbf{r}_1 + k_2\mathbf{r}_2 = \mathbf{0}$ has only the solution $k_1 = k_2 = 0$, S_r is a linearly independent set and $\rho(\mathbf{A}) = 2$. The columns of **A** form a set, $S_c = \{\mathbf{c}_1, \mathbf{c}_2, \mathbf{c}_3\}$, of three vectors in $\mathbb{R}^2$ so, by Theorem 5.4.6, they must be linearly dependent. Having established that $\rho(\mathbf{A}) = 2$, we have that any two of the three columns are linearly independent. ◇

In cases such as that illustrated in Example 5.4.6, where the rank of a matrix **A** equals the number of rows in the matrix, we say that **A** has **full row rank**. It is noteworthy that the determinant of at least one of the 2×2 submatrices that may be formed by pairs of columns of **A** is non-zero; in fact, in the case in question, all three such submatrices have non-zero determinants and, hence, are non-singular.

EXAMPLE 5.4.5 Let $\mathbf{A} = \begin{bmatrix} 1 & -2 \\ 2 & -4 \end{bmatrix}$. Here, $S_r = \{\mathbf{r}_1, \mathbf{r}_2\}$ and $S_c = \{\mathbf{c}_1, \mathbf{c}_2\}$ are sets of two vectors in $\mathbb{R}^2$. It is easily verified that $k_1\mathbf{r}_1 + k_2\mathbf{r}_2 = \mathbf{0}$ has non-trivial solutions for k_1 and k_2, and $k_1^*\mathbf{c}_1 + k_2^*\mathbf{c}_2 = \mathbf{0}$ has a non-trivial solution for k_1^* and k_2^*, hence $\rho(\mathbf{A}) < 2$. ◇

In the case in Example 5.4.6, **A** does not have full row, or full column, rank. It is left as an exercise to confirm that, in fact, $\rho(\mathbf{A}) = 1$ in this case. It may also be noted that $|\mathbf{A}| = 0$, signifying that **A** is singular.

EXAMPLE 5.4.6 Let $\mathbf{A} = \begin{bmatrix} 2 & 1 \\ -1 & 0 \end{bmatrix}$. By similar reasoning to that used in the previous examples, it is easily shown that in this case $\rho(\mathbf{A}) = 2$; and we note that as $|\mathbf{A}| \neq 0$, $\mathbf{A}$ is non-singular. $\diamond$

More generally, and combining results from this and previous chapters, if $\mathbf{A}$ is $n \times n$, then $\rho(\mathbf{A}) = n$ implies, among other things, that:

- row vectors of $\mathbf{A}$, $\mathbf{r}_i$, are linearly independent;
- column vectors of $\mathbf{A}$, $\mathbf{c}_j$, are linearly independent;
- $S_r = \{\mathbf{r}_1, \mathbf{r}_2, \ldots, \mathbf{r}_n\}$ is a basis for lin $S_r = \mathbb{R}^n$;
- $S_c = \{\mathbf{c}_1, \mathbf{c}_2, \ldots, \mathbf{c}_n\}$ is a basis for lin $S_c = \mathbb{R}^n$;
- $\det(\mathbf{A}) \neq 0$;
- $\mathbf{A}$ is invertible;
- $\mathbf{A}\mathbf{x} = \mathbf{0}$ has only the trivial solution, $\mathbf{x} = \mathbf{A}^{-1}\mathbf{0} = \mathbf{0}$; and
- $\mathbf{A}\mathbf{x} = \mathbf{b}$ is consistent, i.e. has a unique solution for every $n \times 1$ matrix $\mathbf{b}$, namely $\mathbf{x} = \mathbf{A}^{-1}\mathbf{b}$.

The following theorem concerns consistency for more general systems of equations than that in the immediately preceding point.

THEOREM 5.4.9 *An $m \times n$ system of linear equations $\mathbf{A}\mathbf{x} = \mathbf{b}$ is consistent if and only if $\mathbf{b}$ is in the column space of $\mathbf{A}$.*

Proof: Let $\mathbf{k}$ be a solution of $\mathbf{A}\mathbf{x} = \mathbf{b}$. We may therefore write

$$\mathbf{A}\mathbf{k} = \begin{bmatrix} a_{11} & \cdots & a_{1n} \\ \vdots & & \vdots \\ a_{m1} & \cdots & a_{mn} \end{bmatrix} \begin{bmatrix} k_1 \\ \vdots \\ k_n \end{bmatrix} = \begin{bmatrix} b_1 \\ \vdots \\ b_m \end{bmatrix} = \mathbf{b} \tag{5.70}$$

as

$$k_1 \mathbf{c}_1 + k_2 \mathbf{c}_2 + \cdots + k_n \mathbf{c}_n = \mathbf{b} \tag{5.71}$$

where $\mathbf{c}_1, \mathbf{c}_2, \ldots, \mathbf{c}_n$ are the columns of $\mathbf{A}$. Since the left-hand side of (5.71) is a linear combination of the column vectors, $\mathbf{c}_i$, $i = 1, 2, \ldots, n$, $\mathbf{b}$ is a linear combination of the $\mathbf{c}_i$, hence $\mathbf{b}$ lies in lin$\{\mathbf{c}_1, \mathbf{c}_2, \ldots, \mathbf{c}_n\}$, the column space of $\mathbf{A}$.

The remainder of this proof is straightforward and is left as an exercise; see Exercise 5.19. $\square$

5.4.7 Orthonormal bases

Usually, it is possible to choose a basis for a vector space entirely at one's discretion. However, some kinds of basis are more convenient to work with than others. One such is an **orthogonal** basis; another, which is a special case of an orthogonal basis, is an **orthonormal** basis. As defined above, a pair of vectors is said to be orthogonal if and only if its dot product $\mathbf{u} \cdot \mathbf{v} = 0$ or, alternatively, the angle between the vectors, $\theta = 90°$. A set of vectors is orthogonal if *all* pairs of vectors in the set are orthogonal or, in other words, the vectors are **mutually orthogonal**.

DEFINITION 5.4.10 A set of vectors is **orthonormal** if it is orthogonal and each vector has norm (length) unity.

EXAMPLE 5.4.7 Let $\mathbf{v}_1 = (0, 1, 0)$, $\mathbf{v}_2 = \left(\frac{1}{\sqrt{2}}, 0, \frac{1}{\sqrt{2}}\right)$, $\mathbf{v}_3 = \left(\frac{1}{\sqrt{2}}, 0, \frac{-1}{\sqrt{2}}\right)$. The set $S = \{\mathbf{v}_1, \mathbf{v}_2, \mathbf{v}_3\}$ is orthonormal since

$$\mathbf{v}_1 \cdot \mathbf{v}_2 = \mathbf{v}_1 \cdot \mathbf{v}_3 = \mathbf{v}_2 \cdot \mathbf{v}_3 = 0 \tag{5.72}$$

and

$$\|\mathbf{v}_i\| = 1 \ \forall \ i \tag{5.73}$$

$\diamond$

Normalization

Using $\|k\mathbf{v}\| = |k|\|\mathbf{v}\|$, the vector $\mathbf{w} = (1/\|\mathbf{v}\|)\mathbf{v}$ has norm 1, since

$$\|\mathbf{w}\| = \left\| \frac{1}{\|\mathbf{v}\|} \mathbf{v} \right\| = \left| \frac{1}{\|\mathbf{v}\|} \right| \|\mathbf{v}\| = \frac{1}{\|\mathbf{v}\|} \|\mathbf{v}\| = 1 \tag{5.74}$$

One of the merits of an orthonormal basis is that it is very easy to express a vector in terms of it, as indicated in the following theorem.

THEOREM 5.4.10 *If $S = \{\mathbf{v}_1, \mathbf{v}_2, \ldots, \mathbf{v}_n\}$ is an orthonormal basis for the vector space V, and $\mathbf{u}$ is any vector in V, then $\mathbf{u} = (\mathbf{u} \cdot \mathbf{v}_1)\mathbf{v}_1 + (\mathbf{u} \cdot \mathbf{v}_2)\mathbf{v}_2 + \cdots + (\mathbf{u} \cdot \mathbf{v}_n)\mathbf{v}_n.$*

Proof: Since S is a basis for V, $\mathbf{u} = k_1\mathbf{v}_1 + k_2\mathbf{v}_2 + \cdots + k_n\mathbf{v}_n$.
Consider the dot product $\mathbf{u} \cdot \mathbf{v}_i = (k_1\mathbf{v}_1 + k_2\mathbf{v}_2 + \cdots + k_n\mathbf{v}_n) \cdot \mathbf{v}_i$.
Since S is orthonormal, $k_i\mathbf{v}_i \cdot \mathbf{v}_i = k_i 1$ and $k_j\mathbf{v}_j \cdot \mathbf{v}_i = 0$, $i \neq j$.
Therefore, $\mathbf{u} \cdot \mathbf{v}_i = k_i$. $\square$

The usefulness of this result becomes apparent when a vector is expressed in terms of a non-orthonormal basis, because in this case it is usually necessary to solve a system of equations to obtain the k_i. That orthogonality, hence orthonormality, implies linear independence is confirmed in the next theorem.

THEOREM 5.4.11 *If $S = \{\mathbf{v}_1, \mathbf{v}_2, \ldots, \mathbf{v}_n\}$ is an orthogonal set of non-zero vectors in the vector space V, then S is linearly independent.*

Proof: Consider the equation $k_1\mathbf{v}_1 + k_2\mathbf{v}_2 + \cdots + k_n\mathbf{v}_n = \mathbf{0}$.
For each $\mathbf{v}_i$ in S it follows that $(k_1\mathbf{v}_1 + k_2\mathbf{v}_2 + \cdots + k_n\mathbf{v}_n) \cdot \mathbf{v}_i = \mathbf{0} \cdot \mathbf{v}_i = 0$ or $k_i\mathbf{v}_i \cdot \mathbf{v}_i = 0$, since $k_j\mathbf{v}_j \cdot \mathbf{v}_i = 0$ by orthogonality for $i \neq j$.
Now $\mathbf{v}_i \cdot \mathbf{v}_i = \|\mathbf{v}_i\|^2 > 0$; therefore $k_i = 0$.
This argument applies for all i, therefore S is linearly independent. $\square$

Recalling the vectors $\mathbf{v}_1 = (0, 1, 0)$, $\mathbf{v}_2 = \left(\frac{1}{\sqrt{2}}, 0, \frac{1}{\sqrt{2}}\right)$, and $\mathbf{v}_3 = \left(\frac{1}{\sqrt{2}}, 0, \frac{-1}{\sqrt{2}}\right)$ from Example 5.4.7, in which it was shown that $S = \{\mathbf{v}_1, \mathbf{v}_2, \mathbf{v}_3\}$ is an orthonormal set, then by Theorem 5.4.11 this S is linearly independent. Because $\mathbb{R}^3$ is of dimension three, it follows that $S = \{\mathbf{v}_1, \mathbf{v}_2, \mathbf{v}_3\}$ is an orthonormal basis for $\mathbb{R}^3$. The standard basis for $\mathbb{R}^3$, $S = \{\mathbf{e}_1, \mathbf{e}_2, \mathbf{e}_3\}$, where $\mathbf{e}_1 = (1, 0, 0)$, $\mathbf{e}_2 = (0, 1, 0)$ and $\mathbf{e}_3 = (0, 0, 1)$, is also an orthonormal basis, as may be easily verified; see Exercise 5.21.

By convention, the standard basis is implicitly assumed when referring to vectors.

EXAMPLE 5.4.8 The vector $\mathbf{u} = (1, 2, 3)$, written explicitly in terms of the standard basis, is $\mathbf{u} = k_1\mathbf{e}_1 + k_2\mathbf{e}_2 + k_3\mathbf{e}_3$, where $k_i = \mathbf{u} \cdot \mathbf{e}_i = i$, $i = 1, 2, 3$. The elements of $\mathbf{u}$ are just the scale factors, or coordinates, associated with the corresponding standard basis vectors. In terms of the basis $S = \{\mathbf{v}_1, \mathbf{v}_2, \mathbf{v}_3\}$ from Example 5.4.7, $\mathbf{u} = k_1\mathbf{v}_1 + k_2\mathbf{v}_2 + k_3\mathbf{v}_3$, where the coordinates are $k_1 = \mathbf{u} \cdot \mathbf{v}_1 = 2$, $k_2 = \mathbf{u} \cdot \mathbf{v}_2 = 2\sqrt{2}$ and $k_3 = \mathbf{u} \cdot \mathbf{v}_3 = -\sqrt{2}$, by Theorem 5.4.10.   ◇

Example 5.4.7 shows why the standard basis is generally preferred.

Construction of orthonormal bases: the Gram–Schmidt process

It is often useful to be able to construct orthonormal bases for vector spaces. One well-known method of doing this is the **Gram–Schmidt process**.[4] This process was referred to in note 6 of Chapter 3 in the context of a procedure for orthogonal diagonalization of a symmetric matrix. The details of the process are described in this section, but we begin with a further theorem that is required. The role of the previous two theorems will be clear. It might also be noted that the result involves a generalization of the idea of the projection of one vector, $\mathbf{u}$, onto another, $\mathbf{v}$, defined earlier, i.e.

$$\text{proj}_{\mathbf{v}}\mathbf{u} = \frac{\mathbf{u} \cdot \mathbf{v}}{\|\mathbf{v}\|^2}\mathbf{v} \tag{5.75}$$

although in what follows $\|\mathbf{v}\|^2 = 1$ because of orthonormality. The proof of the following theorem is not given, as it may be obtained as an exercise by adapting the proof of the simpler Theorem 5.2.4; see Exercise 5.22.

THEOREM 5.4.12 *Let V be a vector space and let $S = \{\mathbf{v}_1, \mathbf{v}_2, \ldots, \mathbf{v}_r\}$ be an orthonormal set of vectors in V. If W denotes the subspace spanned by $\mathbf{v}_1, \mathbf{v}_2, \ldots, \mathbf{v}_r$, i.e. $W = \text{lin } S$, then every vector $\mathbf{u}$ in V can be expressed in the form $\mathbf{u} = \mathbf{w}_1 + \mathbf{w}_2$, where $\mathbf{w}_1 \in W$ and $\mathbf{w}_2$ is orthogonal to W by putting*

$$\mathbf{w}_1 = (\mathbf{u} \cdot \mathbf{v}_1)\mathbf{v}_1 + (\mathbf{u} \cdot \mathbf{v}_2)\mathbf{v}_2 + \cdots + (\mathbf{u} \cdot \mathbf{v}_r)\mathbf{v}_r \tag{5.76}$$

and

$$\mathbf{w}_2 = \mathbf{u} - \mathbf{w}_1 \tag{5.77}$$

In keeping with the terminology used for the earlier material on projections onto single vectors, the vector $\mathbf{w}_1$ is called the **orthogonal projection of u onto the subspace** W, denoted $\text{proj}_W\mathbf{u}$, and $\mathbf{w}_2 = \mathbf{u} - \text{proj}_W\mathbf{u}$ is called the component of $\mathbf{u}$ orthogonal to W. The idea is illustrated in Figure 5.14 for a vector $\mathbf{u} \in \mathbb{R}^3$ and a two-dimensional subspace W, spanned by vectors $\mathbf{v}_1$ and $\mathbf{v}_2$.

The idea of orthogonal projection onto a subspace is further illustrated in the following example, which uses the same three vectors in $\mathbb{R}^3$ as were used in the previous example.

EXAMPLE 5.4.9 It has already been shown in the Example 5.4.7 that the vectors $\mathbf{v}_1 = (0, 1, 0)$, $\mathbf{v}_2 = \left(\frac{1}{\sqrt{2}}, 0, \frac{1}{\sqrt{2}}\right)$ and $\mathbf{v}_3 = \left(\frac{1}{\sqrt{2}}, 0, \frac{-1}{\sqrt{2}}\right)$ are orthonormal. By Theorem 5.4.11, we also know that they are linearly independent. Now consider the subset $S = \{\mathbf{v}_2, \mathbf{v}_3\}$, the linear

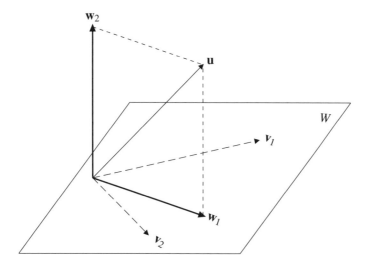

Figure 5.14 Orthogonal projection in $\mathbb{R}^3$

space spanned by S, $W = \mathrm{lin}\, S$, and the additional vector $\mathbf{u} = (1, 1, 1)$ in $\mathbb{R}^3$; and let us find $\mathbf{w}_1$ and $\mathbf{w}_2$. First, the orthogonal projection of $\mathbf{u}$ on W is

$$\mathbf{w}_1 = \mathrm{proj}_W\, \mathbf{u} = (\mathbf{u} \cdot \mathbf{v}_2)\mathbf{v}_2 + (\mathbf{u} \cdot \mathbf{v}_3)\mathbf{v}_3$$

$$= \left(\sum_{i=1}^{3} u_i v_{2i} \right) \left(\frac{1}{\sqrt{2}}, 0, \frac{1}{\sqrt{2}} \right) + \left(\sum_{i=1}^{3} u_i v_{3i} \right) \left(\frac{1}{\sqrt{2}}, 0, \frac{-1}{\sqrt{2}} \right)$$

$$= \sqrt{2} \left(\frac{1}{\sqrt{2}}, 0, \frac{1}{\sqrt{2}} \right) + 0 \left(\frac{1}{\sqrt{2}}, 0, \frac{-1}{\sqrt{2}} \right) = (1, 0, 1) \tag{5.78}$$

The component of $\mathbf{u}$ orthogonal to W is then

$$\mathbf{w}_2 = \mathbf{u} - \mathbf{w}_1 = (1, 1, 1) - (1, 0, 1) = (0, 1, 0) \tag{5.79}$$

$\Diamond$

A useful further exercise, which is left to the reader, is to obtain the result in (5.78) using matrix notation and methods; see Exercise 5.8. This matrix approach will be used in Section 14.2.

The way is now prepared for the main result on the construction of orthonormal bases, which is given in the form of a theorem and proof.

THEOREM 5.4.13 *Every non-zero, finite-dimensional vector space has an orthonormal basis.*

Proof: The proof of this theorem yields the Gram–Schmidt process for constructing orthonormal bases from arbitrary given bases.

Let V be an n-dimensional vector space and let $S = \{\mathbf{u}_1, \mathbf{u}_2, \ldots, \mathbf{u}_n\}$ be any basis for V. An orthonormal basis for V may be obtained as follows:

(a) Put $\mathbf{v}_1 = \mathbf{u}_1 / \|\mathbf{u}_1\|$. Thus $\|\mathbf{v}_1\| = 1$ by the familiar process of normalization.
(b) The vector $\mathbf{v}_2$ is chosen to be the normalized component of $\mathbf{u}_2$ orthogonal to $\mathrm{proj}_{W_1} \mathbf{u}_2$, where W_1 is the subspace spanned by $\mathbf{v}_1$, i.e. $W_1 = \mathrm{lin}\{\mathbf{v}_1\}$. Therefore,

$$\mathbf{v}_2 = \frac{\mathbf{u}_2 - \mathrm{proj}_{W_1} \mathbf{u}_2}{\|\mathbf{u}_2 - \mathrm{proj}_{W_1} \mathbf{u}_2\|} = \frac{\mathbf{u}_2 - (\mathbf{u}_2 \cdot \mathbf{v}_1)\mathbf{v}_1}{\|\mathbf{u}_2 - (\mathbf{u}_2 \cdot \mathbf{v}_1)\mathbf{v}_1\|} \tag{5.80}$$

The denominator in (5.80) cannot be zero, since this would imply that

$$\mathbf{u}_2 = \mathrm{proj}_{W_1} \mathbf{u}_2 = (\mathbf{u}_2 \cdot \mathbf{v}_1)\mathbf{v}_1 = \frac{(\mathbf{u}_2 \cdot \mathbf{v}_1)}{\|\mathbf{u}_1\|}\mathbf{u}_1 = k\mathbf{u}_1 \tag{5.81}$$

i.e. that $\mathbf{u}_2$ is a multiple of $\mathbf{u}_1$, contradicting the linear independence of the basis S.
   The construction of $\mathbf{v}_1$ and $\mathbf{v}_2$ from $\mathbf{u}_1$ and $\mathbf{u}_2$ is illustrated in Figure 5.15.
(c) The vector $\mathbf{v}_3$ is chosen to be the normalized component of $\mathbf{u}_3$ orthogonal to $\mathrm{proj}_{W_2} \mathbf{u}_3$, where W_2 is the subspace spanned by $\mathbf{v}_1$ and $\mathbf{v}_2$, i.e. $W_2 = \mathrm{lin}\{\mathbf{v}_1, \mathbf{v}_2\}$. Therefore,

$$\mathbf{v}_3 = \frac{\mathbf{u}_3 - \mathrm{proj}_{W_2} \mathbf{u}_3}{\|\mathbf{u}_3 - \mathrm{proj}_{W_2} \mathbf{u}_3\|} = \frac{\mathbf{u}_3 - (\mathbf{u}_3 \cdot \mathbf{v}_1)\mathbf{v}_1 - (\mathbf{u}_3 \cdot \mathbf{v}_2)\mathbf{v}_2}{\|\mathbf{u}_3 - (\mathbf{u}_3 \cdot \mathbf{v}_1)\mathbf{v}_1 - (\mathbf{u}_3 \cdot \mathbf{v}_2)\mathbf{v}_2\|} \tag{5.82}$$

Again, the linear independence of S ensures that the denominator in (5.82) is non-zero but the details are left as an exercise. The construction of $\mathbf{v}_3$ from $\mathbf{v}_1$, $\mathbf{v}_2$ and $\mathbf{u}_3$ is illustrated in Figure 5.16.

The process continues like this, with the typical orthonormal basis vector, $\mathbf{v}_i$, being obtained as

$$\mathbf{v}_i = \frac{\mathbf{u}_i - \mathrm{proj}_{W_{i-1}} \mathbf{u}_i}{\|\mathbf{u}_i - \mathrm{proj}_{W_{i-1}} \mathbf{u}_i\|} = \frac{\mathbf{u}_i - (\mathbf{u}_i \cdot \mathbf{v}_1)\mathbf{v}_1 - \cdots - (\mathbf{u}_i \cdot \mathbf{v}_{i-1})\mathbf{v}_{i-1}}{\|\mathbf{u}_i - (\mathbf{u}_i \cdot \mathbf{v}_1)\mathbf{v}_1 - \cdots - (\mathbf{u}_i \cdot \mathbf{v}_{i-1})\mathbf{v}_{i-1}\|} \tag{5.83}$$

though it is not possible to illustrate geometrically the case for $i > 3$. The process terminates when $\mathbf{v}_n$, and thus the orthonormal basis for V has been obtained. $\square$

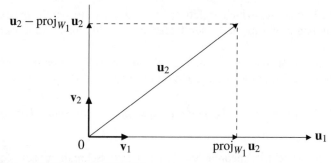

Figure 5.15 Gram–Schmidt process in $\mathbb{R}^2$

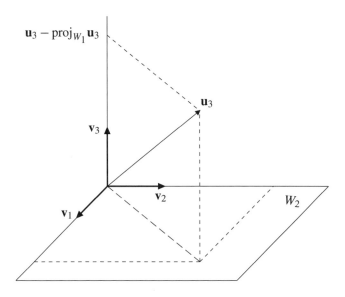

Figure 5.16 Gram–Schmidt process in $\mathbb{R}^3$

EXAMPLE 5.4.10 Consider the basis $S = \{\mathbf{u}_1, \mathbf{u}_2\}$ for $\mathbb{R}^2$, where $\mathbf{u}_1 = (1, 2)$ and $\mathbf{u}_2 = (0, 2)$. It is easy to verify that $\mathbf{u}_1 \cdot \mathbf{u}_2 \neq 0$ and $\|\mathbf{u}_i\| \neq 1$, $i = 1, 2$, and, hence, that S is not an orthonormal basis. Let us use the Gram–Schmidt procedure to construct an orthonormal basis for $\mathbb{R}^2$ based on S.

First, set $\mathbf{v}_1 = \mathbf{u}_1 / \|\mathbf{u}_1\| = \frac{1}{\sqrt{5}}(1, 2) = \left(\frac{1}{\sqrt{5}}, \frac{2}{\sqrt{5}}\right)$. Next, define

$$
\begin{aligned}
\mathbf{v}_2 &= \frac{\mathbf{u}_2 - \mathrm{proj}_{W_1}\mathbf{u}_2}{\|\mathbf{u}_2 - \mathrm{proj}_{W_1}\mathbf{u}_2\|} = \frac{\mathbf{u}_2 - (\mathbf{u}_2 \cdot \mathbf{v}_1)\mathbf{v}_1}{\|\mathbf{u}_2 - (\mathbf{u}_2 \cdot \mathbf{v}_1)\mathbf{v}_1\|} \\[2mm]
&= \frac{(0, 2) - \frac{4}{\sqrt{5}}\left(\frac{1}{\sqrt{5}}, \frac{2}{\sqrt{5}}\right)}{\left\|(0, 2) - \frac{4}{\sqrt{5}}\left(\frac{1}{\sqrt{5}}, \frac{2}{\sqrt{5}}\right)\right\|} = \frac{\left(-\frac{4}{5}, \frac{2}{5}\right)}{\|\left(-\frac{4}{5}, \frac{2}{5}\right)\|} \\[2mm]
&= \frac{\sqrt{5}}{2}\left(-\frac{4}{5}, \frac{2}{5}\right) = \left(\frac{-2}{\sqrt{5}}, \frac{1}{\sqrt{5}}\right)
\end{aligned}
\tag{5.84}
$$

It is straightforward to verify that $\|\mathbf{v}_i\| = 1$, $i = 1, 2$, and $\mathbf{v}_1 \cdot \mathbf{v}_2 = 0$. Thus we have that $S = \{\mathbf{v}_1, \mathbf{v}_2\}$ is an orthonormal basis for $\mathbb{R}^2$. ◇

Example 5.4.7 is very simple. For a somewhat more challenging example, see Exercise 5.9.

5.4.8 *Coordinates and change of basis*

As may now be apparent, there is a close relationship between the idea of a basis and the familiar idea of a coordinate system, based on rectangular or Cartesian coordinates. In such

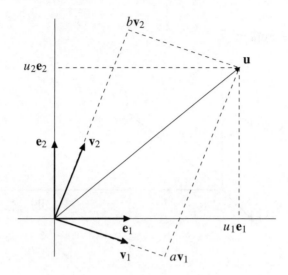

Figure 5.17 Change of basis using rectangular coordinates

a coordinate system, a vector, $\mathbf{u}$, is associated with the numbers (coordinates) obtained by projecting the terminal point of the vector parallel to the direction of each of the axes, as in Figure 5.17. This corresponds to using the standard basis vectors $\mathbf{e}_1$ and $\mathbf{e}_2$, introduced earlier, such that $\mathbf{u} = u_1\mathbf{e}_1 + u_2\mathbf{e}_2$, where u_1 and u_2 are the coordinates. However, we could associate the same vector $\mathbf{u}$ with some other pair of vectors, say, the orthonormal vectors $\mathbf{v}_1$ and $\mathbf{v}_2$, such that $\mathbf{u} = a\mathbf{v}_1 + b\mathbf{v}_2$, as also shown in Figure 5.17. In this case, the numbers a and b are the coordinates relative to the alternative pair of vectors. As any vector in $\mathbb{R}^2$ may be expressed in terms of $\mathbf{v}_1$ and $\mathbf{v}_2$, the set $S = \{\mathbf{v}_1, \mathbf{v}_2\}$ is a basis for $\mathbb{R}^2$.

For the purposes of associating numbers, i.e. coordinates, with points in $\mathbb{R}^2$, given basis vectors $\mathbf{v}_1$ and $\mathbf{v}_2$, it is not essential that the basis be orthonormal. Any basis for $\mathbb{R}^2$ will suffice, as shown in Figure 5.18, where $\mathbf{v}_1$ and $\mathbf{v}_2$ are not orthonormal nor even orthogonal. Nevertheless, we can regard the numbers c and d as legitimate coordinates of the vector $\mathbf{u}$, since $\mathbf{u} = c\mathbf{v}_1 + d\mathbf{v}_2$.

This generalized notion of coordinates is useful and extends to higher-dimensional Euclidean space (and more general vector spaces). Before we examine this, however, we require some preliminary results.

Preliminaries

THEOREM 5.4.14 *If $S = \{\mathbf{v}_1, \mathbf{v}_2, \ldots, \mathbf{v}_n\}$ is a basis for a vector space V, then every vector $\mathbf{u} \in V$ can be expressed uniquely as a linear combination of the $\mathbf{v}_i, i = 1, 2, \ldots, n$, i.e. $\mathbf{u} = c_1\mathbf{v}_1 + c_2\mathbf{v}_2 + \cdots + c_n\mathbf{v}_n$ in exactly one way.*

Proof: Suppose $\mathbf{u} = c_1\mathbf{v}_1 + c_2\mathbf{v}_2 + \cdots + c_n\mathbf{v}_n$ and $\mathbf{u} = k_1\mathbf{v}_1 + k_2\mathbf{v}_2 + \cdots + k_n\mathbf{v}_n$. Subtracting these two equations we have $\mathbf{0} = (c_1 - k_1)\mathbf{v}_1 + (c_2 - k_2)\mathbf{v}_2 + \cdots + (c_n - k_n)\mathbf{v}_n$.

Since $\mathbf{v}_1, \mathbf{v}_2, \ldots, \mathbf{v}_n$ are basis vectors, they are linearly independent and $(c_i - k_i) = 0$ for all i. Therefore $c_i = k_i$ for all i. $\square$

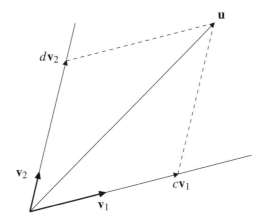

Figure 5.18 Change of basis using non-rectangular coordinates

DEFINITION 5.4.11

(a) If $S = \{\mathbf{v}_1, \mathbf{v}_2, \ldots, \mathbf{v}_n\}$ is a basis for V and $\mathbf{u} = c_1\mathbf{v}_1 + c_2\mathbf{v}_2 + \cdots + c_n\mathbf{v}_n \in V$, then the scalars $c_1, c_2, \ldots, c_n$ are called the **coordinates** of $\mathbf{u}$ relative to, or with respect to, the basis S.
(b) The **coordinate vector** of $\mathbf{u}$ with respect to S is the vector $(\mathbf{u})_S = (c_1, c_2, \ldots, c_n) \in \mathbb{R}^n$.
(c) The **coordinate matrix** of $\mathbf{u}$ with respect to S is the $n \times 1$ matrix $[\mathbf{u}]_S = [c_1 \quad c_2 \quad \ldots \quad c_n]^\top$.

EXAMPLE 5.4.11 Let $\mathbf{v}_1 = (3, 1)$ and $\mathbf{v}_2 = (-1, 2)$. Find the coordinate matrix of $\mathbf{u} = (5, 10)$ relative to $S = \{\mathbf{v}_1, \mathbf{v}_2\}$, which is a basis for $\mathbb{R}^2$.

We have $\mathbf{u} = c_1\mathbf{v}_1 + c_2\mathbf{v}_2 = c_1(3, 1) + c_2(-1, 2) = (5, 10)$. Rewriting this equation, we have the simple system

$$3c_1 - c_2 = 5 \tag{5.85}$$

$$c_1 + 2c_2 = 10 \tag{5.86}$$

the solution of which yields $c_1 = \frac{20}{7}$ and $c_2 = \frac{25}{7}$. Therefore, $[\mathbf{u}]_S = \frac{1}{7}[20 \quad 25]^\top$. Derivation of the solution, which is left as an exercise, is easy by direct manipulation of the two equations or by Cramer's rule or a matrix method. $\diamond$

As a further exercise, the reader might illustrate the solution in Example 5.4.8 graphically; see Exercise 5.10.

Change of basis

THEOREM 5.4.15 *When changing the basis for a vector space, say from* $S = \{\mathbf{v}_1, \mathbf{v}_2, \ldots, \mathbf{v}_n\}$ *to* $S^* = \{\mathbf{v}_1^*, \mathbf{v}_2^*, \ldots, \mathbf{v}_n^*\}$, *the coordinate matrix* $[\mathbf{u}]_S$ *of some vector* $\mathbf{u}$ *is related to the new* coordinate matrix $[\mathbf{u}]_{S*}$ by the following equation:

$$[\mathbf{u}]_S = \mathbf{P}[\mathbf{u}]_{S*}, \quad \text{where } \mathbf{P} = \begin{bmatrix} [\mathbf{v}_1^*]_S & [\mathbf{v}_2^*]_S & \cdots & [\mathbf{v}_n^*]_S \end{bmatrix} \tag{5.87}$$

i.e. the columns of $\mathbf{P}$ *are the coordinate matrices of the new basis vectors relative to the old basis. The matrix* $\mathbf{P}$ *is called the* **transition matrix** *from* S^* *to* S.

Proof: Let $[\mathbf{u}]_{S*} = \begin{bmatrix} k_1 & k_2 & \cdots & k_n \end{bmatrix}^\top$ so that

$$\mathbf{u} = k_1 \mathbf{v}_1^* + k_2 \mathbf{v}_2^* + \cdots + k_n \mathbf{v}_n^* \tag{5.88}$$

Now let $[\mathbf{v}_i^*]_S = [c_{i1} \quad c_{i2} \quad \cdots \quad c_{in}]^\top$, $i = 1, 2, \ldots, n$, i.e.

$$\mathbf{v}_i^* = c_{i1} \mathbf{v}_1 + c_{i2} \mathbf{v}_2 + \cdots + c_{in} \mathbf{v}_n \tag{5.89}$$

Substituting (5.89) in (5.88) we have

$$\mathbf{u} = \sum_{i=1}^{n} k_i (c_{i1} \mathbf{v}_1 + c_{i2} \mathbf{v}_2 + \cdots + c_{in} \mathbf{v}_n) \tag{5.90}$$

Hence

$$[\mathbf{u}]_S = \begin{bmatrix} \sum_i k_i c_{i1} \\ \sum_i k_i c_{i2} \\ \vdots \\ \sum_i k_i c_{in} \end{bmatrix} = \begin{bmatrix} c_{11} & c_{21} & \cdots & c_{n1} \\ c_{12} & c_{22} & \cdots & c_{n2} \\ \vdots & \vdots & & \vdots \\ c_{1n} & c_{2n} & \cdots & c_{nn} \end{bmatrix} \begin{bmatrix} k_1 \\ k_2 \\ \vdots \\ k_n \end{bmatrix} = \mathbf{P}[\mathbf{u}]_{S*} \tag{5.91}$$

□

If this derivation seems involved, a useful exercise would be to do the corresponding derivation for two-dimensional space.

EXAMPLE 5.4.12 For $S = \{\mathbf{v}_1, \mathbf{v}_2\}$, $\mathbf{v}_1 = (1, 0)$ and $\mathbf{v}_2 = (0, 1)$, i.e. the standard basis for $\mathbb{R}^2$, and $S^* = \{\mathbf{v}_1^*, \mathbf{v}_2^*\}$, $\mathbf{v}_1^* = (1, 1)$ and $\mathbf{v}_2^* = (2, 1)$, a non-standard basis for $\mathbb{R}^2$, find the transition matrix $\mathbf{P}$ from S^* to S, find $[\mathbf{u}]_S$ if $[\mathbf{u}]_{S*} = [-3 \quad 5]^\top$, and write $\mathbf{u}$ in terms of each of the bases.

We have by inspection that $\mathbf{v}_1^* = \mathbf{v}_1 + \mathbf{v}_2$ and $[\mathbf{v}_1^*]_S = [1 \quad 1]^\top$. Similarly, $\mathbf{v}_2^* = 2\mathbf{v}_1 + \mathbf{v}_2$ and $[\mathbf{v}_2^*]_S = [2 \quad 1]^\top$. Therefore, $\mathbf{P} = \begin{bmatrix} 1 & 2 \\ 1 & 1 \end{bmatrix}$ and $[\mathbf{u}]_S = \mathbf{P}[\mathbf{u}]_{S*} = \begin{bmatrix} 1 & 2 \\ 1 & 1 \end{bmatrix} \begin{bmatrix} -3 \\ 5 \end{bmatrix} = \begin{bmatrix} 7 \\ 2 \end{bmatrix}$.

In terms of S, $\mathbf{u} = 7\mathbf{v}_1 + 2\mathbf{v}_2 = (7, 2)$. In terms of S^*, $\mathbf{u} = -3\mathbf{v}_1^* + 5\mathbf{v}_2^* = (7, 2)$.   ◇

The derivation of the transition matrix from S to S^* in Example 5.4.8 is left as an exercise; see Exercise 5.11. The result illustrates the following theorem.

THEOREM 5.4.16 *If* **P** *is the transition matrix from a basis* S^* *to a basis* S *for a finite-dimensional vector space* V, *then:*

(a) **P** *is invertible; and*
(b) $\mathbf{P}^{-1}$ *is the transition matrix from* S *to* S^*.

Proof: We establish the result by showing that $\mathbf{PQ} = \mathbf{I}$, where $\mathbf{Q}$ is the transition matrix from S to S^*.

Let $\mathbf{PQ} = [a_{ij}]_{n \times n}$, where n is the dimension of V. From Theorem 5.4.15, we have that

$$[\mathbf{u}]_S = \mathbf{P}[\mathbf{u}]_{S^*} \quad \text{and} \quad [\mathbf{u}]_{S^*} = \mathbf{Q}[\mathbf{u}]_S \tag{5.92}$$

for all $\mathbf{u} \in V$. Substituting for $[\mathbf{u}]_{S^*}$ in the first equation, using the right-hand side of the second, gives

$$[\mathbf{u}]_S = \mathbf{PQ}[\mathbf{u}]_S \tag{5.93}$$

for all $\mathbf{u} \in V$. Putting $\mathbf{u} = \mathbf{e}_1$ in (5.93) gives

$$\begin{bmatrix} 1 \\ 0 \\ \vdots \\ 0 \end{bmatrix} = \begin{bmatrix} a_{11} & a_{12} & \cdots & a_{1n} \\ a_{21} & a_{22} & \cdots & a_{2n} \\ \vdots & \vdots & & \vdots \\ a_{n1} & a_{n2} & \cdots & a_{nn} \end{bmatrix} \begin{bmatrix} 1 \\ 0 \\ \vdots \\ 0 \end{bmatrix} \tag{5.94}$$

or

$$\begin{bmatrix} 1 \\ 0 \\ \vdots \\ 0 \end{bmatrix} = \begin{bmatrix} a_{11} \\ a_{21} \\ \vdots \\ a_{n1} \end{bmatrix} \tag{5.95}$$

Similarly, successive substitutions of $\mathbf{u} = \mathbf{e}_2$, $\mathbf{u} = \mathbf{e}_3$, ..., $\mathbf{u} = \mathbf{e}_n$ in (5.93) yield

$$\begin{bmatrix} 0 \\ 1 \\ 0 \\ \vdots \\ 0 \end{bmatrix} = \begin{bmatrix} a_{12} \\ a_{22} \\ a_{32} \\ \vdots \\ a_{n2} \end{bmatrix}, \quad \cdots, \quad \begin{bmatrix} 0 \\ 0 \\ \vdots \\ 0 \\ 1 \end{bmatrix} = \begin{bmatrix} a_{1n} \\ a_{2n} \\ \vdots \\ a_{(n-1)n} \\ a_{nn} \end{bmatrix} \tag{5.96}$$

Therefore, $\mathbf{PQ} = \mathbf{I}$; hence, $\mathbf{Q} = \mathbf{P}^{-1}$. ☐

Continuing the previous example to further illustrate Theorem 5.4.16, we have $\mathbf{P} = \begin{bmatrix} 1 & 2 \\ 1 & 1 \end{bmatrix}$ and $[\mathbf{u}]_S = [7 \quad 2]^\top$. Inverting **P** we get $\mathbf{P}^{-1} = \begin{bmatrix} -1 & 2 \\ 1 & -1 \end{bmatrix}$, which is precisely the matrix

that results for that part of the example left as an exercise. Now $[\mathbf{u}]_{S*} = \mathbf{P}^{-1}[\mathbf{u}]_S =$
$\begin{bmatrix} -1 & 2 \\ 1 & -1 \end{bmatrix} \begin{bmatrix} 7 \\ 2 \end{bmatrix} = \begin{bmatrix} -3 \\ 5 \end{bmatrix}$, as given in the example.

There follows another interesting theorem on transition matrices.

THEOREM 5.4.17 *If* $\mathbf{P}$ *is the transition matrix from one orthonormal basis to another orthonormal basis for a finite-dimensional vector space* V, *then* $\mathbf{P}^{-1} = \mathbf{P}^\mathsf{T}$, *i.e.* $\mathbf{P}$ *is orthogonal.*[5]

We leave the proof of this theorem as an exercise (see Exercise 5.24) but illustrate it in the next example.

EXAMPLE 5.4.13 Let $S = \{\mathbf{v}_1, \mathbf{v}_2\}$ be the standard orthonormal basis for $\mathbb{R}^2$, as specified fully in the previous example, and let $S^* = \{\mathbf{v}_1^*, \mathbf{v}_2^*\}$, where $\mathbf{v}_1^* = \left(\frac{-1}{\sqrt{2}}, \frac{-1}{\sqrt{2}} \right)$ and $\mathbf{v}_2^* = \left(\frac{1}{\sqrt{2}}, \frac{-1}{\sqrt{2}} \right)$, be another orthonormal basis. Derive the transition matrix from each basis to the other and compare the two.

By inspection we have $\mathbf{v}_1^* = -\frac{1}{\sqrt{2}}\mathbf{v}_1 - \frac{1}{\sqrt{2}}\mathbf{v}_2$ and $\mathbf{v}_2^* = \frac{1}{\sqrt{2}}\mathbf{v}_1 - \frac{1}{\sqrt{2}}\mathbf{v}_2$.

Therefore, $\mathbf{P} = \frac{1}{\sqrt{2}} \begin{bmatrix} -1 & 1 \\ -1 & -1 \end{bmatrix}$.

Similarly, by inspection, $\mathbf{v}_1 = -\frac{1}{\sqrt{2}}\mathbf{v}_1^* + \frac{1}{\sqrt{2}}\mathbf{v}_2^*$ and $\mathbf{v}_2 = -\frac{1}{\sqrt{2}}\mathbf{v}_1^* - \frac{1}{\sqrt{2}}\mathbf{v}_2^*$.

Therefore, $\mathbf{P}^{-1} = \frac{1}{\sqrt{2}} \begin{bmatrix} -1 & -1 \\ 1 & -1 \end{bmatrix}$ and it is immediately clear that $\mathbf{P}^{-1} = \mathbf{P}^\mathsf{T}$. $\diamond$

When a transition matrix is orthogonal, as in Example 5.4.8, we call the change in coordinates an **orthogonal coordinate transformation** in $\mathbb{R}^n$. As $n = 2$ in this example, the orthogonal coordinate transformation is in the plane and it is easy to illustrate this; see Figure 5.19. The illustration includes the vector $\mathbf{u}$ for which $[\mathbf{u}]_S = [1 \quad 1]^\mathsf{T}$. It follows that

$$[\mathbf{u}]_{S*} = \mathbf{P}^{-1}[\mathbf{u}]_S = \frac{1}{\sqrt{2}} \begin{bmatrix} -1 & -1 \\ 1 & -1 \end{bmatrix} \begin{bmatrix} 1 \\ 1 \end{bmatrix} = \begin{bmatrix} -\sqrt{2} \\ 0 \end{bmatrix} \tag{5.97}$$

i.e. $\mathbf{u} = -\sqrt{2}\mathbf{v}_1^* + 0\mathbf{v}_2^* = -\sqrt{2}\left(\frac{-1}{\sqrt{2}}, \frac{-1}{\sqrt{2}} \right) = (1, 1)$.

EXAMPLE 5.4.14 The vector that has coordinates (x, y) with respect to the standard basis for $\mathbb{R}^2$ has coordinates $(x', y') = \left(\frac{1}{2}(x + y), \frac{1}{2}(x - y) \right)$ with respect to the new basis comprising the vectors $(1, 1)$ and $(1, -1)$, since by Theorem 5.4.15 the transition matrix from the standard basis to the new basis is

$$\begin{bmatrix} 1 & 1 \\ 1 & -1 \end{bmatrix}^{-1} = \frac{1}{2} \begin{bmatrix} 1 & 1 \\ 1 & -1 \end{bmatrix} \tag{5.98}$$

Thus, recalling the material on conic sections from Section 4.2, the rectangular hyperbola that has equation $x'^2 - y'^2 = a^2$ with respect to the new basis has equation

$$\left(\frac{x+y}{2} \right)^2 - \left(\frac{x-y}{2} \right)^2 = a^2 \tag{5.99}$$

with respect to the standard basis, in which the coordinate axes are the asymptotes to the hyperbola. Expanding the squares reduces (5.99) to

$$xy = a^2 \qquad (5.100)$$

$\diamond$

The rectangular hyperbola asymptotic to the coordinate axes, which features in this last example, is encountered occasionally in economics; see, for example, Exercise 9.17 and the discussion of total revenue in Section 9.4.

5.4.9 Scalar products

The concept of **scalar product** will be used later in the book. In this section, we define the term and give a few examples.

DEFINITION 5.4.12 Given a real vector space V, a **scalar product** or **inner product** is a **bi-linear** function: $V \times V \to \mathbb{R} : (\mathbf{x}, \mathbf{y}) \mapsto (\mathbf{x} \mid \mathbf{y})$ such that the following properties hold.

(a) Symmetry: $(\mathbf{x} \mid \mathbf{y}) = (\mathbf{y} \mid \mathbf{x})$ for all $\mathbf{x}, \mathbf{y} \in V$.
(b) Linearity: $(a\mathbf{x} \mid \mathbf{y}) = a(\mathbf{x} \mid \mathbf{y})$ for all $\mathbf{x}, \mathbf{y} \in V$ and for all $a \in \mathbb{R}$; also $(\mathbf{x} + \mathbf{y} \mid \mathbf{z}) = (\mathbf{x} \mid \mathbf{z}) + (\mathbf{y} \mid \mathbf{z})$ for all $\mathbf{x}, \mathbf{y}, \mathbf{z} \in V$.
(c) Non-negativity:[6] $(\mathbf{x} \mid \mathbf{x}) \geq 0$ for all $\mathbf{x} \in V$.

A vector space with the additional structure of a scalar product is called a **scalar product space**. Examples of this type of space include the set of real numbers with the standard multiplication as the scalar product and, more generally, any Euclidean space, $\mathbb{R}^n$, with the dot product (see Definition 5.2.15) as the scalar product. Given a positive semi-definite $n \times n$ matrix $\mathbf{A}$, $(\mathbf{x}, \mathbf{y}) \mapsto \mathbf{x}^\top \mathbf{A} \mathbf{y}$ defines a scalar product. If $\mathbf{A} = \mathbf{I}$, we just get the dot product.

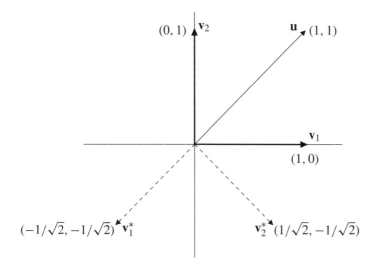

Figure 5.19 Orthogonal transformation

Conversely, any scalar product has a matrix with respect to a given basis $\{\mathbf{u}_1, \mathbf{u}_2, \ldots, \mathbf{u}_n\}$, defined by $a_{ij} = (\mathbf{u}_i \mid \mathbf{u}_j)$. By the non-negativity condition in the definition, $\mathbf{A}$ is positive semi-definite. If $\mathbf{A}$ is positive definite, then the scalar product itself is said to be **positive definite**.

A positive definite symmetric scalar product shares all the familiar properties of the dot product. In particular, two vectors $\mathbf{x}$ and $\mathbf{y}$ are said to be **orthogonal with respect to a scalar product** if $(\mathbf{x} \mid \mathbf{y}) = 0$, and a set of vectors $\{\mathbf{x}_1, \mathbf{x}_2, \ldots, \mathbf{x}_k\}$ is said to be **orthonormal with respect to a scalar product** if $(\mathbf{x}_i \mid \mathbf{x}_j) = \delta_{ij}$ for all i, j, where δ_{ij} is the Kronecker delta introduced in Section 1.5.2. Similarly, the idea of orthogonal projection can be extended to any scalar product space and the notation $\mathbf{x} \perp \mathbf{y}$ can be generalized to denote $(\mathbf{x} \mid \mathbf{y}) = 0$.

Further similarities between a general positive definite symmetric scalar product and the dot product may be seen using a change of basis. We know that for symmetric positive definite $\mathbf{A}$ there exists a non-singular $\mathbf{R}$ such that $\mathbf{A} = \mathbf{R}\mathbf{R}^\top$; and we have seen two different methods of constructing such an $\mathbf{R}$ in Section 4.4.1. If we change basis using $(\mathbf{R}^{-1})^\top$ as transition matrix, then the scalar product of vectors having coordinates $\mathbf{x}$ and $\mathbf{y}$ with respect to the new basis, or $(\mathbf{R}^{-1})^\top \mathbf{x}$ and $(\mathbf{R}^{-1})^\top \mathbf{y}$ with respect to the old basis, is

$$((\mathbf{R}^{-1})^\top \mathbf{x})^\top \mathbf{A} (\mathbf{R}^{-1})^\top \mathbf{y} = \mathbf{x}^\top \mathbf{R}^{-1} \mathbf{R}\mathbf{R}^\top (\mathbf{R}^{-1})^\top \mathbf{y} = \mathbf{x}^\top \mathbf{y} \qquad (5.101)$$

Thus the scalar product is just the dot product for the new coordinate system; therefore the dot product shares all the properties of the scalar product, including symmetry, linearity and non-negativity as defined above. The new basis vectors are orthogonal with respect to the scalar product defined by $\mathbf{A}$.

The following theorem, motivated by the discussion of the construction of orthonormal bases above, makes use of the scalar product concept and introduces the idea of the orthogonal complement of a vector subspace, which will be referred to in the applications discussed in Sections 14.2.2 and 17.4.1.

THEOREM 5.4.18 *If S is a subspace of a real vector space V and $(\cdot \mid \cdot)$ is a scalar product on V, then $S^\perp \equiv \{\mathbf{y} \in V : \mathbf{y} \perp \mathbf{x}$ for all $\mathbf{x} \in S\}$ is a vector subspace of V, called the **orthogonal complement** of S.*

Proof:

(a) $(\mathbf{y} \mid \mathbf{x}) = 0$ for all $\mathbf{x} \in S \Rightarrow (k\mathbf{y} \mid \mathbf{x}) = k(\mathbf{y} \mid \mathbf{x}) = k0 = 0$ for all $k \in \mathbb{R}$.
    Therefore $S^\perp$ is closed under scalar multiplication.
(b) $(\mathbf{y} \mid \mathbf{x}) = (\mathbf{z} \mid \mathbf{x}) = 0$ for all $\mathbf{x} \in S \Rightarrow (\mathbf{y} + \mathbf{z} \mid \mathbf{x}) = (\mathbf{y} \mid \mathbf{x}) + (\mathbf{z} \mid \mathbf{x}) = 0 + 0 = 0$.
    Therefore $S^\perp$ is closed under vector addition. $\qquad\square$

A somewhat more general, though similar, idea to the orthogonal complement of a vector subspace is that of the **direct sum** of vector subspaces. We end this chapter with a definition of this concept, which will be referred to in Sections 7.4.1 and 17.4.1.

DEFINITION 5.4.13 Let W_i, $i = 1, 2, \ldots, n$, be subspaces of a finite vector space, V, such that no W_i contains a non-zero vector belonging to any of the remaining $n - 1$ subspaces, i.e. the intersection $W_1 \cap W_2 \cap \cdots \cap W_n = \{\mathbf{0}\}$ and $W_i - \{\mathbf{0}\}$ are all disjoint. Then the direct sum of the W_i is

$$W_1 \oplus W_2 \oplus \cdots \oplus W_n \equiv \left\{ \sum_{i=1}^{n} \mathbf{w}_i : \mathbf{w}_i \in W_i, \ i = 1, 2, \ldots, n \right\} \tag{5.102}$$

Thus each vector in the direct sum $W_1 \oplus W_2 \oplus \cdots \oplus W_n$ is representable uniquely as $\mathbf{w} = \mathbf{w}_1 + \mathbf{w}_2 + \cdots + \mathbf{w}_n$, where $\mathbf{w}_i \in W_i$.

For example, any two non-collinear lines that pass through the origin of the Euclidean plane are subspaces of $\mathbb{R}^2$. Designating these two subspaces, each of dimension one, by W_1 and W_2, then $W_1 \oplus W_2 = \mathbb{R}^2$. If this plane is thought of as passing through the origin of a three-dimensional Cartesian coordinate system and is designated as subspace V of $\mathbb{R}^3$, then $W_1 \oplus W_2 = V$. Similarly, if W_1, W_2 and W_3 are the subspaces associated with any three linearly independent vectors in $\mathbb{R}^3$, then the direct sum $W_1 \oplus W_2 \oplus W_3$ is the entire three-dimensional space, and any vector, $\mathbf{w} \in \mathbb{R}^3$, can be represented as $\mathbf{w} = \mathbf{w}_1 + \mathbf{w}_2 + \mathbf{w}_3$, where $\mathbf{w}_i \in W_i$, $i = 1, 2, 3$.

We note, without proofs, the following results on direct sums. Though each result is stated for two subspaces, its generalization is straightforward, using proof by induction.

- The direct sum $W_1 \oplus W_2$ of the subspaces W_1 and W_2 of the finite vector space V is also a subspace of V.
- The dimension of the direct sum of vector subspaces is the sum of the dimensions of the constituent subspaces. Hence, if $V = W_1 \oplus W_2$, then $\dim(V) = \dim(W_1) + \dim(W_2)$.
- If a finite vector space V is the direct sum of its subspaces W_1 and W_2, then the union of any basis for W_1 with any basis for W_2 is a basis for V.

Finally, when V is the direct sum of its subspaces W_1 and W_2, we refer to W_1 and W_2 as complementary subspaces of V. Orthogonal complementarity arises as a special case of the direct sum when $W_2 = W_1^{\perp}$, and vice versa; recall Theorem 5.4.18.

EXERCISES

5.1 Illustrate on a graph the vectors $(0, 2)$, $(6, 0)$, $(2, 2)$ and $(4, -4)$ and investigate for each of the six pairs of vectors drawn whether or not they are orthogonal.

5.2 For vectors in 3-space, write out formal definitions of equality, addition, multiplication by a scalar and subtraction and illustrate these definitions graphically.

5.3 Let $\mathbf{w} = (2, 3, 4)$, $\mathbf{v}_1 = (5, 6, 8)$, $\mathbf{v}_2 = (3, 3, 4)$ and $\mathbf{v}_3 = (-3, 3, -4)$.

(a) Find k_1 and k_2 such that $\mathbf{w} = k_1 \mathbf{v}_1 + k_2 \mathbf{v}_2$.
(b) Show that $\mathbf{w}$ is *not* a linear combination of $\mathbf{v}_1$ and $\mathbf{v}_3$.
(c) Find a vector $\mathbf{w}^*$ that *is* a linear combination of $\mathbf{v}_1$ and $\mathbf{v}_3$, and illustrate the construction graphically.

5.4 Verify that each of the following sets of vectors spans $\mathbb{R}^3$:

(a) $S_1 = \{(2, 0, 0), (0, -5, 0), (0, 0, 3)\}$;

(b) $S_2 = \{(2, 3, -4), (2, 0, 0), (0, -5, 0), (0, 0, 3)\}$;
(c) $S_3 = \{(1, 0, 0), (0, \frac{1}{3}, 0), (0, 0, 2), (\frac{1}{3}, 0, \frac{2}{3})\}$; and
(d) $S_4 = \{(0, 1, 0), (2, 3, -4), (1, 0, 0), (0, 0, 1)\}$.

5.5 Show that each of the following sets of vectors does not span $\mathbb{R}^3$:

(a) $S_4 = \{(1, 2, 3), (3, 2, 1)\}$; and
(b) $S_5 = \{(1, 2, 3), (3, -2, -1), (-1, 6, 7)\}$.

5.6 Show that two linearly dependent vectors in Euclidean 3-space lie in the same line through the origin. Also show that, if two vectors in Euclidean 3-space are linearly independent, then they have different directions and a non-zero angle between them. Supplement your demonstration by appropriate diagrams and numerical examples.

5.7 Let $\mathbf{A} = \begin{bmatrix} 1 & -2 \\ 2 & -4 \end{bmatrix}$ and $\mathbf{B} = \begin{bmatrix} 2 & 1 \\ -1 & 0 \end{bmatrix}$. Verify that $\rho(\mathbf{A}) = 1$ and $\rho(\mathbf{B}) = 2$.

5.8 Derive the result in equation (5.78) using matrix notation and methods.

5.9 Consider $\mathbb{R}^3$. Use the Gram–Schmidt process to transform the basis vectors $\mathbf{u}_1 = (1, 1, 1)$, $\mathbf{u}_2 = (1, 2, 0)$ and $\mathbf{u}_3 = (2, 0, 0)$ into an orthonormal basis $\{\mathbf{v}_1, \mathbf{v}_2, \mathbf{v}_3\}$.

5.10 Illustrate the solution of equations (5.85) and (5.86) graphically.

5.11 Find the transition matrix from $S = \{(1, 0), (0, 1)\}$ to $S^* = \{(1, 1), (2, 1)\}$.

5.12 Let $\mathbf{u}$, $\mathbf{v}$ and $\mathbf{w}$ be vectors in 2- or 3-space, and k be a scalar. Prove the following:

(a) $\mathbf{u} \cdot \mathbf{v} = \mathbf{v} \cdot \mathbf{u}$;
(b) $\mathbf{u} \cdot (\mathbf{v} + \mathbf{w}) = \mathbf{u} \cdot \mathbf{v} + \mathbf{u} \cdot \mathbf{w}$;
(c) $k(\mathbf{u} \cdot \mathbf{v}) = (k\mathbf{u}) \cdot \mathbf{v} = \mathbf{u} \cdot (k\mathbf{v})$; and
(d) $\mathbf{v} \cdot \mathbf{v} > 0$ if $\mathbf{v} \neq 0$ and $\mathbf{v} \cdot \mathbf{v} = 0$ if $\mathbf{v} = 0$.

5.13 Define the vector component of $\mathbf{u}$ orthogonal to $\mathbf{v}$, and derive an expression for its length.

5.14 Prove (5.48), (5.49) and (5.51).

5.15 Let W be a subset of a vector space V, and suppose $\mathbf{u}_i \in W$, and $k_i \in \mathbb{R}$, $i = 1, 2, \ldots, n$. Prove that $\sum_{i=1}^{n} k_i \mathbf{u}_i \in W \Leftrightarrow W$ is a subspace of V. (Hint: use proof by induction, introduced in Section 2.4.)

5.16 Let V be a vector space and $\mathbf{v}_1, \mathbf{v}_2, \ldots, \mathbf{v}_r \in V$. Prove Theorem 5.4.4 that:

(a) the set W of all linear combinations of $\mathbf{v}_1, \mathbf{v}_2, \ldots, \mathbf{v}_r$, i.e. $W = \text{lin}\{\mathbf{v}_1, \mathbf{v}_2, \ldots, \mathbf{v}_r\}$, is a subspace of V; and
(b) W is the smallest subspace of V that contains $\mathbf{v}_1, \mathbf{v}_2, \ldots, \mathbf{v}_r$, in the sense that every other subspace that contains $\mathbf{v}_1, \mathbf{v}_2, \ldots, \mathbf{v}_r$ must contain W.

Does your proof work if you start with infinitely many vectors?

5.17 Prove that, if $S = \{\mathbf{v}_1, \mathbf{v}_2, \ldots, \mathbf{v}_n\}$ is a basis for a vector space V, then every subset of V with more than n vectors is linearly dependent.

5.18 Let V be an n-dimensional vector space. Prove the following (Theorem 5.4.7).

(a) If $S = \{\mathbf{v}_1, \mathbf{v}_2, \ldots, \mathbf{v}_n\}$ is a set of n linearly independent vectors in V, then S is a basis for V.

(b) If $S = \{\mathbf{v}_1, \mathbf{v}_2, \ldots, \mathbf{v}_n\}$ is a set of n vectors that spans V, then S is a basis for V.

5.19 Complete the proof of Theorem 5.4.9, which states that an $m \times n$ system of linear equations $\mathbf{Ax} = \mathbf{b}$ is consistent if and only if $\mathbf{b}$ is in the column space of $\mathbf{A}$.

5.20 Let $\mathbf{Q}$ be an $m \times n$ matrix. Prove that $\mathbf{QQ}^\top = \mathbf{I}_m$ implies that $m \leqslant n$ and $\mathbf{Q}^\top\mathbf{Q} = \mathbf{I}_n$ implies $n \leqslant m$. See note 4 of Chapter 3.

5.21 Verify that the standard basis for $\mathbb{R}^3$, $S = \{\mathbf{e}_1, \mathbf{e}_2, \mathbf{e}_3\}$, where $\mathbf{e}_1 = (1, 0, 0)$, $\mathbf{e}_2 = (0, 1, 0)$ and $\mathbf{e}_3 = (0, 0, 1)$, is an orthonormal basis. Recall Theorem 5.4.11.

5.22 Prove Theorem 5.4.12.

5.23 Replicate the proof of Theorem 5.4.15 for two-dimensional space and illustrate graphically.

5.24 Prove that, if $\mathbf{P}$ is the transition matrix from one orthonormal basis to another orthonormal basis, then $\mathbf{P}^{-1} = \mathbf{P}^\top$, i.e. $\mathbf{P}$ is orthogonal. Recall Theorem 5.4.17.

5.25 Show that the set comprising the zero vector and all the eigenvectors of an $n \times n$ matrix $\mathbf{A}$ corresponding to a given (possibly repeated) eigenvalue λ is a vector subspace of $\mathbb{R}^n$.

This subspace is often called the **eigenspace** of $\mathbf{A}$ corresponding to the given λ.

5.26

(a) Is the intersection of two vector subspaces a vector subspace?

(b) Is the union of two vector subspaces a vector subspace?

5.27 Consider the x and y axes in $\mathbb{R}^3$, $S_1 \equiv \mathbb{R} \times \{0\} \times \{0\}$ and $S_2 \equiv \{0\} \times \mathbb{R} \times \{0\}$.

(a) Which of the following is a vector subspace of $\mathbb{R}^3$: S_1, S_2, $S_1 \cap S_2$, $S_1 \cup S_2$?

(b) What are the dimensions of those which are vector subspaces?

(c) What are the dimensions of the vector subspaces $\text{lin } S_1$, $\text{lin } S_2$, $\text{lin}(S_1 \cap S_2)$ and $\text{lin}(S_1 \cup S_2)$?

(d) Is $S_1 \oplus S_2$ well defined? If so, what is its dimension?

6 Linear transformations

6.1 Introduction

In this chapter, we define and develop the idea of linear transformations, and relate the concept to aspects of the material covered in the previous chapters. Linear transformations constitute a special class of vector-valued functions of a vector variable and are of particular interest for economics and finance. The chapter begins with a definition and several illustrations to make the fundamental idea clear. It then goes on to explain a number of related concepts, to address issues concerned with the solution of more general systems of linear equations than the ones we have considered so far (i.e. the square equation systems we encountered in Chapters 1, 2 and 3), to discuss transformations from $\mathbb{R}^n$ to $\mathbb{R}^m$, and finally to consider the topics of matrices of transformations and similarity.

6.2 Definitions and illustrations

6.2.1 Definitions

Readers are assumed to be familiar with the idea of a real-valued function of a single variable, e.g. $g: \mathbb{R} \to \mathbb{R}$; see Definition 0.0.1. We now extend this idea to functions between vector spaces and define some associated notation and terminology.

DEFINITION 6.2.1 A **vector-valued function** is a function whose co-domain is a subset of a Euclidean vector space, say $f: X \to Y$, where $Y \subseteq \mathbb{R}^n$. Such a function has n real-valued **component functions**, usually denoted (using superscripts) $f^1, f^2, \ldots, f^n: X \to \mathbb{R}$.

DEFINITION 6.2.2 A **function of several variables** is a function whose domain is a subset of a (finite-dimensional) vector space. The components of the vector $\mathbf{x}$ are sometimes called the **arguments** of $f(\mathbf{x})$.

DEFINITION 6.2.3 Let $T: U \to V$ be a mapping from the vector space U into the vector space V. Then T is a **linear transformation** if and only if:

(a) $T(\mathbf{u}_1 + \mathbf{u}_2) = T(\mathbf{u}_1) + T(\mathbf{u}_2)$ for all $\mathbf{u}_1, \mathbf{u}_2 \in U$; and
(b) $T(k\mathbf{u}) = kT(\mathbf{u})$ for all $\mathbf{u} \in U$ and for all scalars k (i.e. T is homogeneous of degree one; see Definition 0.0.9).

Definition 6.2.3 implies, by an inductive argument similar to that used in Exercise 5.15 to prove Theorem 5.4.3, that a vector-valued function of several variables is a linear transformation if and only if it preserves arbitrary linear combinations. The generalization is stated in the following theorem.

THEOREM 6.2.1 *Let* $T : U \to V$ *be a mapping from the vector space* U *into the vector space* V. *Then* T *is a linear transformation if and only if*

$$T\left(\sum_{i=1}^{n} k_i \mathbf{u}_i\right) = k_i \sum_{i=1}^{n} T(\mathbf{u}_i) \tag{6.1}$$

for all $\mathbf{u}_i \in U$ *and for all scalars* k_i *and for all positive integers* n

DEFINITION 6.2.4 The mapping $T : U \to V$ such that $T(\mathbf{u}) = \mathbf{0}$ for every $\mathbf{u} \in U$ is called the **zero transformation**.

The zero transformation is a linear transformation. To see this, note that, given $T(\mathbf{u}_1 + \mathbf{u}_2) = \mathbf{0}$, $T(\mathbf{u}_1) = \mathbf{0}$, $T(\mathbf{u}_2) = \mathbf{0}$ and $T(k\mathbf{u}_1) = \mathbf{0}$, it follows easily that $T(\mathbf{u}_1 + \mathbf{u}_2) = T(\mathbf{u}_1) + T(\mathbf{u}_2)$ and $T(k\mathbf{u}_1) = kT(\mathbf{u}_1)$. The dimension of the image $T(\mathbf{u}) = \mathbf{0}$ in this case is equal to the dimension of V, which may be different from the dimension of U.

DEFINITION 6.2.5 The mapping $T : U \to U$ such that $T(\mathbf{u}) = \mathbf{u}$ for every $\mathbf{u} \in U$ is called the **identity transformation**.

The verification that the identity transformation is a linear transformation is left as an exercise; see Exercise 6.6.

DEFINITION 6.2.6 Linear transformations that map a vector space U into itself are referred to as **linear operators**.

The identity transformation is an example of a linear operator.

6.2.2 Illustrations

Mapping from $\mathbb{R}^2$ *to* $\mathbb{R}^3$

Let $T : \mathbb{R}^2 \to \mathbb{R}^3$, where $T(x, y) = (x, y, x + y) \in \mathbb{R}^3$ for $(x, y) \in \mathbb{R}^2$. This function associates, with any given pair of real numbers, a triple of real numbers, where the first number in the triple equals the first number in the original pair, the second number in the triple equals the second number in the original pair, and the third number in the triple is the sum of the two numbers in the original pair. If $\mathbf{u}_1 = (x_1, y_1)$ and $\mathbf{u}_2 = (x_2, y_2)$, then by definition of vector addition $\mathbf{u}_1 + \mathbf{u}_2 = (x_1 + x_2, y_1 + y_2)$. So $T(\mathbf{u}_1 + \mathbf{u}_2) = (x_1 + x_2, y_1 + y_2, x_1 + x_2 + y_1 + y_2)$ by definition of $T(\mathbf{u})$. It is then easy to show, using the property of commutativity of scalar addition and the definition of vector addition, that

$$
\begin{aligned}
T(\mathbf{u}_1 + \mathbf{u}_2) &= (x_1 + x_2, y_1 + y_2, x_1 + x_2 + y_1 + y_2) \\
&= (x_1 + x_2, y_1 + y_2, x_1 + y_1 + x_2 + y_2) \\
&= (x_1, y_1, x_1 + y_1) + (x_2, y_2, x_2 + y_2) \\
&= T(\mathbf{u}_1) + T(\mathbf{u}_2) \tag{6.2}
\end{aligned}
$$

Also, $k\mathbf{u}_1 = (kx_1, ky_1)$ so, using the definition of multiplication of a vector by a scalar,

$$T(k\mathbf{u}_1) = (kx_1, ky_1, kx_1 + ky_1)$$
$$= k(x_1, y_1, x_1 + y_1)$$
$$= kT(\mathbf{u}_1) \tag{6.3}$$

Therefore, T is a linear transformation.

The next illustration relates to the notion of matrix multiplication used extensively in earlier chapters.

Matrix multiplication

Let $\mathbf{A}$ be a fixed $m \times n$ matrix. Using matrix notation for vectors in $\mathbb{R}^m$ and $\mathbb{R}^n$, define the function $T: \mathbb{R}^n \to \mathbb{R}^m$ by $T(\mathbf{x}) = \mathbf{Ax}$, where $\mathbf{x} \in \mathbb{R}^n$ and $T(\mathbf{x}) \in \mathbb{R}^m$. Now, using matrix operations, we have that

$$T(\mathbf{x}_1 + \mathbf{x}_2) = \mathbf{A}(\mathbf{x}_1 + \mathbf{x}_2) = \mathbf{Ax}_1 + \mathbf{Ax}_2 = T(\mathbf{x}_1) + T(\mathbf{x}_2) \tag{6.4}$$

and

$$T(k\mathbf{x}_1) = \mathbf{A}(k\mathbf{x}_1) = k(\mathbf{Ax}_1) = kT(\mathbf{x}_1) \tag{6.5}$$

Therefore, T is a linear transformation.

For obvious reasons, we call such a linear transformation, which involves multiplication by a matrix, a **matrix transformation**.

Linear operator

If U is a vector space, k is any scalar and $T: U \to U$ is defined by $T(\mathbf{u}) = k\mathbf{u}$, T is a linear operator. It is a straightforward exercise to demonstrate this; see Exercise 6.7. Note, too, what T does to $\mathbf{u}$ for different values of k, such as $k < 1$, $k > 1$ and $k < 0$; recall the geometric interpretation of Definition 5.2.7, and note the similarity of $k\mathbf{u}$ to the right-hand side of the eigenequation (3.1).

Orthogonal projection

Let U be a vector space, V be a subspace of U having an orthonormal basis $S = \{\mathbf{v}_1, \mathbf{v}_2, \ldots, \mathbf{v}_r\}$, and $T: U \to V$ be the function that maps every vector $\mathbf{u} \in U$ into its orthogonal projection on V, i.e.

$$T(\mathbf{u}) \equiv \text{proj}_V \mathbf{u} = (\mathbf{u} \cdot \mathbf{v}_1)\mathbf{v}_1 + (\mathbf{u} \cdot \mathbf{v}_2)\mathbf{v}_2 + \cdots + (\mathbf{u} \cdot \mathbf{v}_r)\mathbf{v}_r \tag{6.6}$$

The linearity of the mapping T follows from the properties of the dot product:

$$T(\mathbf{u}_1 + \mathbf{u}_2)$$
$$= ((\mathbf{u}_1 + \mathbf{u}_2) \cdot \mathbf{v}_1)\mathbf{v}_1 + ((\mathbf{u}_1 + \mathbf{u}_2) \cdot \mathbf{v}_2)\mathbf{v}_2 + \cdots + ((\mathbf{u}_1 + \mathbf{u}_2) \cdot \mathbf{v}_r)\mathbf{v}_r$$
$$= (\mathbf{u}_1 \cdot \mathbf{v}_1 + \mathbf{u}_2 \cdot \mathbf{v}_1)\mathbf{v}_1 + (\mathbf{u}_1 \cdot \mathbf{v}_2 + \mathbf{u}_2 \cdot \mathbf{v}_2)\mathbf{v}_2 + \cdots + (\mathbf{u}_1 \cdot \mathbf{v}_r + \mathbf{u}_2 \cdot \mathbf{v}_r)\mathbf{v}_r$$

$$= (\mathbf{u}_1 \cdot \mathbf{v}_1)\mathbf{v}_1 + (\mathbf{u}_2 \cdot \mathbf{v}_1)\mathbf{v}_1 + (\mathbf{u}_1 \cdot \mathbf{v}_2)\mathbf{v}_2 + (\mathbf{u}_2 \cdot \mathbf{v}_2)\mathbf{v}_2 + \cdots + (\mathbf{u}_2 \cdot \mathbf{v}_r)\mathbf{v}_r$$

$$= (\mathbf{u}_1 \cdot \mathbf{v}_1)\mathbf{v}_1 + (\mathbf{u}_1 \cdot \mathbf{v}_2)\mathbf{v}_2 + \cdots + (\mathbf{u}_1 \cdot \mathbf{v}_r)\mathbf{v}_r + (\mathbf{u}_2 \cdot \mathbf{v}_1)\mathbf{v}_1 + \cdots + (\mathbf{u}_2 \cdot \mathbf{v}_r)\mathbf{v}_r$$

$$= T(\mathbf{u}_1) + T(\mathbf{u}_2) \tag{6.7}$$

Similarly for $T(k\mathbf{u}_1) = kT(\mathbf{u}_1)$, which may be easily shown as an exercise; see Exercise 6.8. The illustration and exercise may be taken further by examining a specific example; see Exercise 6.1. The results of these two exercises will verify, and illustrate diagrammatically, the general statements just made.

Other linear transformations

From courses in single-variable or univariate calculus, readers will be familiar with the basic properties of derivatives and integrals, such as the following:

- the derivative of a sum is the sum of the derivatives,

$$(f + g)'(x) = f'(x) + g'(x) \tag{6.8}$$

- the derivative of a scalar multiple of a function is the same scalar multiple of the derivative

$$(kf)'(x) = k(f'(x)) \tag{6.9}$$

- the integral of a sum is the sum of the integrals

$$\int (f(x) + g(x)) \, dx = \int f(x) \, dx + \int g(x) \, dx \tag{6.10}$$

- the integral of a scalar multiple of a function is the same scalar multiple of the integral

$$\int kf(x) \, dx = k \int f(x) \, dx \tag{6.11}$$

If we view the space of all differentiable functions as an infinite-dimensional vector space, then the first two properties merely say that the differentiation operator is a linear transformation on this space. Similarly, if we view the space of all integrable functions as an infinite-dimensional vector space, then the last two properties say that the integration operator is a linear transformation on this space.[1] In Chapter 9, these concepts will be extended from real-valued functions of a single variable to functions of several variables. In Section 13.6, we will encounter another linear transformation on an infinite-dimensional vector space, namely the expectation operator on a space of random variables.

6.3 Properties of linear transformations

In this section, we establish a number of properties of linear transformations and define and illustrate the concepts of the kernel and range of a linear transformation.

THEOREM 6.3.1 *If $T: U \to V$ is a linear transformation, then*

(a) $T(\mathbf{0}) = \mathbf{0}$, *where the two zero vectors are of the relevant dimensions;*

(b) $T(-\mathbf{u}) = -T(\mathbf{u})$ *for all* $\mathbf{u} \in U$; *and*
(c) $T(\mathbf{u}_1 - \mathbf{u}_2) = T(\mathbf{u}_1) - T(\mathbf{u}_2)$ *for all* $\mathbf{u}_1, \mathbf{u}_2 \in U$.

Proof: For any $\mathbf{u} \in U$, we have:

(a) since $0\mathbf{u} = \mathbf{0}$,

$$T(\mathbf{0}) = T(0\mathbf{u}) = 0T(\mathbf{u}) = \mathbf{0} \qquad (6.12)$$

(b) since $-\mathbf{u} = (-1)\mathbf{u}$,

$$T(-\mathbf{u}) = T((-1)\mathbf{u}) = (-1)T(\mathbf{u}) = -T(\mathbf{u}) \qquad (6.13)$$

and for any $\mathbf{u}_1$ and $\mathbf{u}_2 \in U$
(c) since $\mathbf{u}_1 - \mathbf{u}_2 = \mathbf{u}_1 + (-1)\mathbf{u}_2$,

$$T(\mathbf{u}_1 - \mathbf{u}_2) = T(\mathbf{u}_1 + (-1)\mathbf{u}_2) = T(\mathbf{u}_1) + T((-1)\mathbf{u}_2)$$
$$= T(\mathbf{u}_1) + (-1)T(\mathbf{u}_2) = T(\mathbf{u}_1) - T(\mathbf{u}_2) \qquad (6.14)$$

using the previous part. $\qquad\qquad\square$

THEOREM 6.3.2 *A linear transformation* $T: U \to V$ *is completely determined by its values* (*images*) *at a basis.*

Proof: Let $\{\mathbf{u}_1, \mathbf{u}_2, \ldots, \mathbf{u}_n\}$ be a basis for the vector space U, let $T: U \to V$ be a linear transformation, and suppose that the images of the basis vectors, $\mathbf{v}_i = T(\mathbf{u}_i)$, $i = 1, 2, \ldots, n$, are known. Since any $\mathbf{u} \in U$ can be written as

$$\mathbf{u} = k_1\mathbf{u}_1 + k_2\mathbf{u}_2 + \cdots + k_n\mathbf{u}_n \qquad (6.15)$$

we then have that

$$T(\mathbf{u}) = T(k_1\mathbf{u}_1 + k_2\mathbf{u}_2 + \cdots + k_n\mathbf{u}_n)$$
$$= T(k_1\mathbf{u}_1) + T(k_2\mathbf{u}_2) + \cdots + T(k_n\mathbf{u}_n)$$
$$= k_1 T(\mathbf{u}_1) + k_2 T(\mathbf{u}_2) + \cdots + k_n T(\mathbf{u}_n)$$
$$= k_1\mathbf{v}_1 + k_2\mathbf{v}_2 + \cdots + k_n\mathbf{v}_n \qquad (6.16)$$
$$\square$$

6.3.1   *Kernel and range*

We begin with formal definitions, notation and a theorem for the concepts of kernel and range. These are followed by two examples, then the statement of two additional theorems, the second of which has great practical value.

Let $T: U \to V$ be a linear transformation. Then we have the following definition.

DEFINITION 6.3.1 The set of vectors in U that T maps into $\mathbf{0}$ is called the **kernel** (or **null space**) of T, and is denoted ker(T).

This definition is analogous to the definition of solution space for matrix equations given in Section 5.4.2.

We note that $\ker(T) = \{\mathbf{u} \in U : T(\mathbf{u}) = \mathbf{0}\} \subseteq U$, i.e. the kernel of T is a subset of the domain.

DEFINITION 6.3.2 The set of all vectors in V that are images under T of at least one vector in U is called the **range** of T, and is denoted $R(T)$ (or, in the notation used for the range of any function, $T(U)$).

This definition is just Definition 0.0.4 restated in terms of a linear transformation.

We also note that $R(T) = \{\mathbf{v} \in V : \mathbf{v} = T(\mathbf{u}) \text{ for some } \mathbf{u} \in U\} \subseteq V$, i.e. the range of T is a subset of the co-domain. The range is not necessarily the same as the co-domain, as we will see shortly in an example.

We can make somewhat stronger statements than that the kernel and range are subsets of the domain and co-domain, respectively, by dint of the following theorem.

THEOREM 6.3.3 *If* $T : U \to V$ *is a linear transformation, then:*

(a) $\ker(T)$ *is a subspace of* U; *and*
(b) $R(T)$ *is a subspace of* V.

Proof:

(a) We have to show that $\ker(T)$ is closed under addition and scalar multiplication (recall Theorem 5.4.2). So let $\mathbf{u}_1, \mathbf{u}_2 \in \ker(T)$ and k be a scalar. Then $T(\mathbf{u}_1 + \mathbf{u}_2) = T(\mathbf{u}_1) + T(\mathbf{u}_2) = \mathbf{0} + \mathbf{0} = \mathbf{0}$. Therefore $\mathbf{u}_1 + \mathbf{u}_2 \in \ker(T)$. Also $T(k\mathbf{u}_1) = kT(\mathbf{u}_1) = k\mathbf{0} = \mathbf{0}$. Therefore $k\mathbf{u}_1 \in \ker(T)$ and $\ker(T)$ is a subspace of U.
(b) Let $\mathbf{v}_1, \mathbf{v}_2 \in R(T)$. We need to show that $\mathbf{v}_1 + \mathbf{v}_2$ and $k\mathbf{v}_1$ also belong to $R(T)$, i.e. we must find vectors $\mathbf{u}$ and $\mathbf{u}^*$ in U such that $T(\mathbf{u}) = \mathbf{v}_1 + \mathbf{v}_2$ and $T(\mathbf{u}^*) = k\mathbf{v}_1$. Since $\mathbf{v}_1, \mathbf{v}_2 \in R(T)$, there are vectors $\mathbf{u}_1$ and $\mathbf{u}_2 \in U$ such that $T(\mathbf{u}_1) = \mathbf{v}_1$ and $T(\mathbf{u}_2) = \mathbf{v}_2$. So, let $\mathbf{u} = \mathbf{u}_1 + \mathbf{u}_2$ and $\mathbf{u}^* = k\mathbf{u}_1$.
Then $T(\mathbf{u}) = T(\mathbf{u}_1 + \mathbf{u}_2) = T(\mathbf{u}_1) + T(\mathbf{u}_2) = \mathbf{v}_1 + \mathbf{v}_2$ and $T(\mathbf{u}^*) = T(k\mathbf{u}_1) = kT(\mathbf{u}_1) = k\mathbf{v}_1$. $\qquad\square$

DEFINITION 6.3.3

(a) The dimension of $\ker(T)$ is called the **nullity** of T.
(b) The dimension of $R(T)$ is called the **rank** of T.

EXAMPLE 6.3.1 Let $T : U \to V$ be the zero transformation defined earlier. Since T maps every vector in U to $\mathbf{0}$, $\ker(T) = U$. Since $\mathbf{0}$ is the only possible image under T in this case, $R(T) = \{\mathbf{0}\}$. We see from this example that the range is not necessarily the same as the co-domain, i.e. that not all linear transformations are surjective. We note also that in this case the nullity of T is the same as the dimension of the space U, and that the rank of T is zero. $\diamond$

EXAMPLE 6.3.2 Let $T : \mathbb{R}^n \to \mathbb{R}^m$ be the matrix transformation defined by $T(\mathbf{x}) = \mathbf{A}\mathbf{x}$, where $\mathbf{A}$ is $m \times n$ and $\mathbf{x} \in \mathbb{R}^n$. Then $\ker(T)$ comprises all $\mathbf{x} = [x_1 \quad x_2 \quad \cdots \quad x_n]^\top$ that are solution vectors of the homogeneous system of equations $\mathbf{A}\mathbf{x} = \mathbf{0}$, i.e. $\ker(T)$ is the solution space of $\mathbf{A}\mathbf{x} = \mathbf{0}$, while $R(T)$ comprises all vectors $\mathbf{b} = [b_1 \quad b_2 \quad \cdots \quad b_n]^\top$ such that the system $\mathbf{Ax} = \mathbf{b}$ is consistent and therefore has at least one solution. Since, by Theorem 5.4.9, $\mathbf{b}$ must be in the column space of $\mathbf{A}$ for $\mathbf{Ax} = \mathbf{b}$ to be consistent, we have that $R(T)$ is the column space of the matrix $\mathbf{A}$. In this example, therefore, we have that the nullity of T is equal to the dimension of the solution space of $\mathbf{Ax} = \mathbf{0}$ and the rank of T is the dimension of the column space of $\mathbf{A}$, which is the rank of $\mathbf{A}$, $\rho(\mathbf{A})$. $\diamond$

Example 6.3.1 is important for some of the material in the next subsection, as well as for what immediately follows, namely an interesting theorem concerning nullity and rank.

THEOREM 6.3.4 *If $T : U \to V$ is a linear transformation from an n-dimensional vector space U to a vector space V, then*

$$\text{nullity of } T + \text{rank of } T = n \tag{6.17}$$

Proof: The proof of this theorem is beyond the scope of this book, but can be found in, for example, Anton and Rorres (2011, p. 239).   □

We are mainly interested in this theorem for the case in which $U = \mathbb{R}^n$, $V = \mathbb{R}^m$ and $T : U \to V : \mathbf{x} \mapsto \mathbf{Ax}$. For this case, the

$$\text{nullity of } T = n - \text{rank of } T \tag{6.18}$$

or, in view of what was established in Example 6.3.1,

$$\text{nullity of } T = n - \text{rank of } \mathbf{A} \tag{6.19}$$

6.3.2   Solution of general linear equation systems

We state the preceding result more formally in the next theorem, and the result is proved for this special case using matrix results from earlier chapters. Importantly, the proof also provides a method of solution for more general homogeneous systems of linear equations than just the square systems that meet the conditions of Cramer's theorem encountered in Chapter 2.

THEOREM 6.3.5 *If $\mathbf{A}$ is $m \times n$, then the dimension of the solution space of the system of equations $\mathbf{Ax} = \mathbf{0}$ is $n - \rho(\mathbf{A})$.*

Proof: Let $\mathbf{A}$ be an $m \times n$ matrix with rank $\rho(\mathbf{A}) = r$, let $\mathbf{x} \equiv (x_1, x_2, \ldots, x_r, x_{r+1}, \ldots, x_n)$ be an $n \times 1$ vector, and partition $\mathbf{Ax} = \mathbf{0}$ as

$$\begin{bmatrix} \mathbf{A}_{11} & \mathbf{A}_{12} \\ \mathbf{A}_{21} & \mathbf{A}_{22} \end{bmatrix} \begin{bmatrix} \mathbf{x}_1 \\ \mathbf{x}_2 \end{bmatrix} = \begin{bmatrix} \mathbf{0}_r \\ \mathbf{0}_{n-r} \end{bmatrix} \tag{6.20}$$

relabelling variables or equations if necessary, so that $\mathbf{A}_{11}$ is square $r \times r$ and non-singular, $\mathbf{A}_{12}$ is $r \times (n-r)$, $\mathbf{A}_{21}$ is $(m-r) \times r$, $\mathbf{A}_{22}$ is $(m-r) \times (n-r)$, and $\mathbf{x}_1$, $\mathbf{x}_2$, $\mathbf{0}_r$ and $\mathbf{0}_{n-r}$ are conformable partitions of $\mathbf{x}$ and $\mathbf{0}$.

Since the last $m - r$ rows are linearly dependent on the first r rows, and therefore redundant, we consider

$$\mathbf{A}_{11}\mathbf{x}_1 + \mathbf{A}_{12}\mathbf{x}_2 = \mathbf{0}_r \qquad (6.21)$$

The non-singularity of $\mathbf{A}_{11}$ implies that

$$\mathbf{x}_1 = -\mathbf{A}_{11}^{-1}\mathbf{A}_{12}\mathbf{x}_2 \qquad (6.22)$$

where $\mathbf{x}_2$ is an arbitrary $(n - r)$-element vector. This in turn means that all solutions of (6.20) may be written as

$$\mathbf{x} = \begin{bmatrix} \mathbf{x}_1 \\ \mathbf{x}_2 \end{bmatrix} = \begin{bmatrix} -\mathbf{A}_{11}^{-1}\mathbf{A}_{12}\mathbf{x}_2 \\ \mathbf{x}_2 \end{bmatrix}$$
$$= \begin{bmatrix} -\mathbf{A}_{11}^{-1}\mathbf{A}_{12} \\ \mathbf{I}_{n-r} \end{bmatrix} \mathbf{x}_2 \qquad (6.23)$$

Defining

$$\mathbf{B}_{n \times (n-r)} \equiv \begin{bmatrix} -\mathbf{A}_{11}^{-1}\mathbf{A}_{12} \\ \mathbf{I}_{n-r} \end{bmatrix} \qquad (6.24)$$

we may write the solution (6.23) as

$$\mathbf{x} = \mathbf{B}\mathbf{x}_2 = x_{r+1}\mathbf{b}_1 + x_{r+2}\mathbf{b}_2 + \cdots + x_n\mathbf{b}_{n-r} \qquad (6.25)$$

where the $\mathbf{b}_i$ are the columns of $\mathbf{B}$, $i = 1, 2, \ldots, (n - r)$. Thus $\mathbf{b}_1, \mathbf{b}_2, \ldots, \mathbf{b}_{n-r}$ span the solution space. Given the $\mathbf{I}_{n-r}$-component partition of $\mathbf{B}$, it is immediately clear that $\rho(\mathbf{B}) = n - r$ and that the $\mathbf{b}_i$ are linearly independent; $\mathbf{B}$ has full column rank. Therefore, $\mathbf{b}_1, \mathbf{b}_2, \ldots, \mathbf{b}_{n-r}$ form a basis for the solution space and the dimension of the solution space is $n - r = n - \rho(\mathbf{A})$. □

An immediate consequence of this theorem, which was referred to in the proof of Theorem 5.4.6, is stated in the following corollary.

COROLLARY 6.3.6 *A homogeneous system of linear equations with more unknowns than equations has infinitely many solutions.*

EXAMPLE 6.3.3 Find a numerical solution for the x_i, $i = 1, 2, 3$, in the following system of equations:

$$5x_1 - 3x_3 = 0 \qquad (6.26)$$
$$x_1 - 4x_2 - 6x_3 = 0 \qquad (6.27)$$

In matrix notation, these two equations may be written as

$$\mathbf{A}\mathbf{x} = \begin{bmatrix} 5 & 0 & -3 \\ 1 & -4 & -6 \end{bmatrix} \begin{bmatrix} x_1 \\ x_2 \\ x_3 \end{bmatrix} = \begin{bmatrix} 0 \\ 0 \end{bmatrix} \qquad (6.28)$$

The rank of $\mathbf{A}$ is easily determined as $\rho(\mathbf{A}) = r = 2$ so, by Theorem 6.3.5, the dimension of the solution space is $n - \rho(\mathbf{A}) = 3 - 2 = 1$. In light of this, we can write the required partitioned form (6.21) as

$$
\mathbf{A}_{11}\mathbf{x}_1 + \mathbf{A}_{12}\mathbf{x}_2 = \begin{bmatrix} 5 & 0 \\ 1 & -4 \end{bmatrix} \begin{bmatrix} x_1 \\ x_2 \end{bmatrix} + \begin{bmatrix} -3 \\ -6 \end{bmatrix} x_3 = \begin{bmatrix} 0 \\ 0 \end{bmatrix} \tag{6.29}
$$

from which we get the solution

$$
\begin{aligned}
\mathbf{x}_1 = -\mathbf{A}_{11}^{-1}\mathbf{A}_{12}\mathbf{x}_2 &= -\begin{bmatrix} 5 & 0 \\ 1 & -4 \end{bmatrix}^{-1} \begin{bmatrix} -3 \\ -6 \end{bmatrix} x_3 \\
&= -\begin{bmatrix} \frac{1}{5} & 0 \\ \frac{1}{20} & -\frac{1}{4} \end{bmatrix} \begin{bmatrix} -3 \\ -6 \end{bmatrix} x_3 \\
&= \begin{bmatrix} \frac{3}{5} \\ -\frac{27}{20} \end{bmatrix} x_3
\end{aligned} \tag{6.30}
$$

A choice of any value of $x_3 \in \mathbb{R}$ will yield a particular numerical solution, which – recalling (6.25) – will be a scalar multiple, of

$$
\mathbf{b}_1 = \begin{bmatrix} \frac{3}{5} \\ -\frac{27}{20} \\ 1 \end{bmatrix} \tag{6.31}
$$

For example, putting $x_3 = 1$ gives

$$
\mathbf{x} = \begin{bmatrix} \mathbf{x}_1 \\ \mathbf{x}_2 \end{bmatrix} = \begin{bmatrix} \frac{3}{5} \\ -\frac{27}{20} \\ 1 \end{bmatrix} \tag{6.32}
$$

while $x_3 = \frac{5}{3}$ gives

$$
\mathbf{x} = \begin{bmatrix} \mathbf{x}_1 \\ \mathbf{x}_2 \end{bmatrix} = \begin{bmatrix} 1 \\ -\frac{9}{4} \\ \frac{5}{3} \end{bmatrix} \tag{6.33}
$$

It is easy to confirm a particular numerical solution by substitution into the original equations of the system. $\diamond$

Another, slightly more complicated, numerical application of the solution method in Theorem 6.3.5 is the subject of Exercise 6.3.

6.4 Linear transformations from $\mathbb{R}^n$ to $\mathbb{R}^m$

In this section, we will show that every linear transformation from $\mathbb{R}^n$ to $\mathbb{R}^m$ is a matrix transformation. That is, if $T \colon \mathbb{R}^n \to \mathbb{R}^m$ is any linear transformation, then there exists an $m \times n$ matrix $\mathbf{A}$ such that T is multiplication by $\mathbf{A}$.

Let $\mathbf{e}_1, \mathbf{e}_2, \ldots, \mathbf{e}_n$ be the standard basis for $\mathbb{R}^n$, and let $\mathbf{A}$ be the $m \times n$ matrix with columns $T(\mathbf{e}_1), T(\mathbf{e}_2), \ldots, T(\mathbf{e}_n)$. For instance, if $T \colon \mathbb{R}^2 \to \mathbb{R}^2$ is given by

$$T(\mathbf{x}) = T\left(\begin{bmatrix} x_1 \\ x_2 \end{bmatrix}\right) = \begin{bmatrix} x_1 + 2x_2 \\ x_1 - x_2 \end{bmatrix} \tag{6.34}$$

then

$$T(\mathbf{e}_1) = T\left(\begin{bmatrix} 1 \\ 0 \end{bmatrix}\right) = \begin{bmatrix} 1 \\ 1 \end{bmatrix} \tag{6.35}$$

and

$$T(\mathbf{e}_2) = T\left(\begin{bmatrix} 0 \\ 1 \end{bmatrix}\right) = \begin{bmatrix} 2 \\ -1 \end{bmatrix} \tag{6.36}$$

and we have that

$$\mathbf{A} = \begin{bmatrix} 1 & 2 \\ 1 & -1 \end{bmatrix} \tag{6.37}$$

It is easy to verify for this simple case that

$$T(\mathbf{x}) = T\left(\begin{bmatrix} x_1 \\ x_2 \end{bmatrix}\right) = \begin{bmatrix} 1 & 2 \\ 1 & -1 \end{bmatrix}\begin{bmatrix} x_1 \\ x_2 \end{bmatrix} = \mathbf{A}\mathbf{x} \tag{6.38}$$

More generally, we define

$$\mathbf{A} = [\mathbf{a}_1 \quad \mathbf{a}_2 \quad \ldots \quad \mathbf{a}_n] = [T(\mathbf{e}_1) \quad T(\mathbf{e}_2) \quad \ldots \quad T(\mathbf{e}_n)] \tag{6.39}$$

where the $\mathbf{a}_i \equiv T(\mathbf{e}_i)$ denote the columns of $\mathbf{A}$

Now any $\mathbf{x} \in \mathbb{R}^n$, say $\mathbf{x} = [x_1 \quad x_2 \quad \ldots \quad x_n]^{\top}$, can be written as

$$\mathbf{x} = x_1\mathbf{e}_1 + x_2\mathbf{e}_2 + \cdots + x_n\mathbf{e}_n \tag{6.40}$$

Therefore, by the linearity of T, we have

$$T(\mathbf{x}) = x_1 T(\mathbf{e}_1) + x_2 T(\mathbf{e}_2) + \cdots + x_n T(\mathbf{e}_n) \tag{6.41}$$

But we also have

$$\begin{aligned} \mathbf{A}\mathbf{x} &= x_1\mathbf{a}_1 + x_2\mathbf{a}_2 + \cdots + x_n\mathbf{a}_n \\ &= x_1 T(\mathbf{e}_1) + x_2 T(\mathbf{e}_2) + \cdots + x_n T(\mathbf{e}_n) \end{aligned} \tag{6.42}$$

Comparing (6.41) and (6.42) gives $T(\mathbf{x}) = \mathbf{A}\mathbf{x}$, i.e. T is multiplication by $\mathbf{A}$.

DEFINITION 6.4.1 The matrix $\mathbf{A}$ is called the **standard matrix** for T.

The following example illustrates the standard matrix for a linear transformation involving the elements of a 3-vector.

EXAMPLE 6.4.1 Let $T : \mathbb{R}^3 \rightarrow \mathbb{R}^4$ and define T as

$$
T\left(\begin{bmatrix} x_1 \\ x_2 \\ x_3 \end{bmatrix} \right) = \begin{bmatrix} x_1 + x_2 \\ x_1 - x_2 \\ x_3 \\ x_1 \end{bmatrix} \tag{6.43}
$$

The standard matrix for T in this case is

$$
\mathbf{A} = \begin{bmatrix} T(\mathbf{e}_1) & T(\mathbf{e}_2) & T(\mathbf{e}_3) \end{bmatrix} = \begin{bmatrix} 1 & 1 & 0 \\ 1 & -1 & 0 \\ 0 & 0 & 1 \\ 1 & 0 & 0 \end{bmatrix} \tag{6.44}
$$

since

$$
T(\mathbf{e}_1) = T\left(\begin{bmatrix} 1 \\ 0 \\ 0 \end{bmatrix} \right) = \begin{bmatrix} 1+0 \\ 1-0 \\ 0 \\ 1 \end{bmatrix} = \begin{bmatrix} 1 \\ 1 \\ 0 \\ 1 \end{bmatrix} \tag{6.45}
$$

and similarly for $T(\mathbf{e}_2)$ and $T(\mathbf{e}_3)$. $\diamondsuit$

In the case of a matrix transformation, the linear transformation T defined as $T(\mathbf{x}) = \mathbf{A}\mathbf{x}$, the standard matrix for T is simply $\mathbf{A}$. This suggests an interesting way of thinking about matrices, i.e. an arbitrary $m \times n$ matrix, $\mathbf{A}$, may be viewed as the standard matrix for a linear transformation that maps the standard basis for $\mathbb{R}^n$ into the column vectors of $\mathbf{A}$.

6.5 Matrices of linear transformations

With a little ingenuity and care, any linear transformation $T : U \rightarrow V$ between finite-dimensional vector spaces can also be regarded as a matrix transformation. The idea is to choose bases for U and V and to work with the coordinate matrices relative to these bases, rather than with the vectors themselves.

Suppose U is n-dimensional and V is m-dimensional. If we choose bases B and B^* for U and V, then each $\mathbf{u} \in U$ will have a coordinate matrix $[\mathbf{u}]_B \in \mathbb{R}^n$, and the coordinate matrix $[T(\mathbf{u})]_{B^*}$ will be some vector in $\mathbb{R}^m$. Thus in mapping $\mathbf{u}$ into $T(\mathbf{u})$, the linear transformation T generates a mapping from $\mathbb{R}^n$ to $\mathbb{R}^m$. It can be shown that this generated mapping is always a linear transformation. Therefore, it can be carried out using a standard matrix, $\mathbf{A}$, for this transformation, i.e.

$$
\mathbf{A}[\mathbf{u}]_B = [T(\mathbf{u})]_{B^*} \tag{6.46}
$$

If $\mathbf{A}$ can be found, $T(\mathbf{u})$ can be obtained indirectly.

To solve the problem of finding $\mathbf{A}$ to satisfy (6.46), suppose $B = \{\mathbf{u}_1, \mathbf{u}_2, \ldots, \mathbf{u}_n\}$ and $B^* = \{\mathbf{v}_1, \mathbf{v}_2, \ldots, \mathbf{v}_m\}$. Therefore, we want $\mathbf{A}[\mathbf{u}_1]_B = [T(\mathbf{u}_1)]_{B^*}$. But $[\mathbf{u}_1]_B = [1 \quad 0 \quad \ldots \quad 0]^\top$, so that $\mathbf{A}[\mathbf{u}_1]_B = \mathbf{a}_1$, which implies that $[T(\mathbf{u}_1)]_{B^*} = \mathbf{a}_1$, i.e. the first column of $\mathbf{A}$ is the coordinate matrix for the vector $T(\mathbf{u}_1)$ with respect to the basis B^*; and similarly for all other columns of $\mathbf{A}$. So $\mathbf{A}$ is the matrix whose jth column is the coordinate matrix for the vector $T(\mathbf{u}_j)$ with respect to the basis B^*. This unique $\mathbf{A}$ is called the **matrix of T with respect to the bases B and B^***. Symbolically,

$$\mathbf{A} = [[T(\mathbf{u}_1)]_{B^*} \quad [T(\mathbf{u}_2)]_{B^*} \quad \ldots \quad [T(\mathbf{u}_n)]_{B^*}] \tag{6.47}$$

The following two points are noteworthy:

1. If $T: \mathbb{R}^n \to \mathbb{R}^m$ is a linear transformation and if B and B^* are the standard bases for $\mathbb{R}^n$ and $\mathbb{R}^m$, respectively, then the matrix for T with respect to B and B^* is just the standard matrix for T, as previously defined.
2. In the special case where $V = U$ (so that $T: U \to V$ is a linear operator), it is usual to put $B = B^*$ when constructing the matrix of T. In this case we speak of the matrix of T with respect to B.

EXAMPLE 6.5.1 Let $T: \mathbb{R}^2 \to \mathbb{R}^2$ be the linear operator defined by

$$T\left(\begin{bmatrix} x_1 \\ x_2 \end{bmatrix}\right) = \begin{bmatrix} x_1 + x_2 \\ -2x_1 + 4x_2 \end{bmatrix} \tag{6.48}$$

and let us find the matrix of T with respect to the basis $B = \{\mathbf{u}_1, \mathbf{u}_2\}$, where $\mathbf{u}_1 = [1 \quad 1]^\top$ and $\mathbf{u}_2 = [1 \quad 2]^\top$.

From the definition of T, we have

$$T(\mathbf{u}_1) = \begin{bmatrix} 2 \\ 2 \end{bmatrix} = 2\mathbf{u}_1 \quad \text{and} \quad T(\mathbf{u}_2) = \begin{bmatrix} 3 \\ 6 \end{bmatrix} = 3\mathbf{u}_2 \tag{6.49}$$

Therefore,

$$T(\mathbf{u}_1) = 2\mathbf{u}_1 + 0\mathbf{u}_2 \quad \text{and} \quad T(\mathbf{u}_2) = 0\mathbf{u}_1 + 3\mathbf{u}_2 \tag{6.50}$$

Hence

$$[T(\mathbf{u}_1)]_B = \begin{bmatrix} 2 \\ 0 \end{bmatrix} \quad \text{and} \quad [T(\mathbf{u}_2)]_B = \begin{bmatrix} 0 \\ 3 \end{bmatrix} \tag{6.51}$$

and the matrix of T with respect to B is

$$\mathbf{A} = \begin{bmatrix} 2 & 0 \\ 0 & 3 \end{bmatrix} \tag{6.52}$$

This simple diagonal form may be compared with the standard matrix of T, which is

$$\begin{bmatrix} 1 & 1 \\ -2 & 4 \end{bmatrix} \tag{6.53}$$

$\diamond$

The matrix of a linear operator $T: U \to U$ depends on the basis chosen for U. Unfortunately, a simple basis, such as the standard basis, does not usually yield the simplest matrix for T. So consideration is given to changing the basis in order to simplify the matrix for T. In connection with this problem we make use of the following result.

THEOREM 6.5.1 *Let* $T: U \to U$ *be a linear operator on a finite-dimensional vector space* U. *If* **A** *is the matrix of* T *with respect to a basis* B, *and* **A*** *is the matrix of* T *with respect to a basis* B^*, *then* $\mathbf{A}^* = \mathbf{P}^{-1}\mathbf{A}\mathbf{P}$, *where* **P** *is the transition matrix from* B^* *to* B.

Proof: It follows from the conditions of the theorem and the definitions of matrices of T that

$$\mathbf{A}[\mathbf{u}]_B = [T(\mathbf{u})]_B \quad \text{and} \quad \mathbf{A}^*[\mathbf{u}]_{B^*} = [T(\mathbf{u})]_{B^*} \tag{6.54}$$

Also, from what we know of the transition matrix (see Theorem 5.4.15),

$$[\mathbf{u}]_B = \mathbf{P}[\mathbf{u}]_{B^*} \quad \text{and} \quad [T(\mathbf{u})]_{B^*} = \mathbf{P}^{-1}[T(\mathbf{u})]_B \tag{6.55}$$

Therefore,

$$\mathbf{A}[\mathbf{u}]_B = \mathbf{A}\mathbf{P}[\mathbf{u}]_{B^*} = [T(\mathbf{u})]_B \tag{6.56}$$

and

$$\mathbf{P}^{-1}[T(\mathbf{u})]_B = \mathbf{P}^{-1}\mathbf{A}\mathbf{P}[\mathbf{u}]_{B^*} = [T(\mathbf{u})]_{B^*} \tag{6.57}$$

from which it is clear that $\mathbf{A}^* = \mathbf{P}^{-1}\mathbf{A}\mathbf{P}$. $\square$

Note that, if **A** and **A*** are the matrices of a linear transformation with respect to different bases, then they are similar matrices; see Section 3.5.

EXAMPLE 6.5.2   Recall the linear operator used in the previous example, $T: \mathbb{R}^2 \to \mathbb{R}^2$, where

$$T\left(\begin{bmatrix} x_1 \\ x_2 \end{bmatrix}\right) = \begin{bmatrix} x_1 + x_2 \\ -2x_1 + 4x_2 \end{bmatrix} \tag{6.58}$$

Given $B = \{\mathbf{e}_1, \mathbf{e}_2\}$, where $\mathbf{e}_1 = [1 \quad 0]^\top$ and $\mathbf{e}_2 = [0 \quad 1]^\top$, the matrix of T with respect to B, i.e. the standard basis, is

$$\begin{bmatrix} 1 & 1 \\ -2 & 4 \end{bmatrix} \tag{6.59}$$

Using Theorem 6.5.1, let us find the matrix of T with respect to $B^* = \{\mathbf{u}_1, \mathbf{u}_2\}$, where $\mathbf{u}_1 = [1 \quad 1]^\top$ and $\mathbf{u}_2 = [1 \quad 2]^\top$.

First, we obtain the transition matrix **P** from B^* to B. To do this we need the coordinate matrices for the B^* basis vectors relative to the basis B. Now,

$$\mathbf{u}_1 = \mathbf{e}_1 + \mathbf{e}_2 \tag{6.60}$$

$$\mathbf{u}_2 = \mathbf{e}_1 + 2\mathbf{e}_2 \tag{6.61}$$

Therefore,

$$[\mathbf{u}_1]_B = \begin{bmatrix} 1 \\ 1 \end{bmatrix} \quad \text{and} \quad [\mathbf{u}_2]_B = \begin{bmatrix} 1 \\ 2 \end{bmatrix} \tag{6.62}$$

hence

$$\mathbf{P} = \begin{bmatrix} 1 & 1 \\ 1 & 2 \end{bmatrix} \quad \text{and} \quad \mathbf{P}^{-1} = \begin{bmatrix} 2 & -1 \\ -1 & 1 \end{bmatrix} \tag{6.63}$$

Thus by Theorem 6.5.1,

$$
\begin{aligned}
\mathbf{A}^* &= \mathbf{P}^{-1}\mathbf{A}\mathbf{P} \\
&= \begin{bmatrix} 2 & -1 \\ -1 & 1 \end{bmatrix} \begin{bmatrix} 1 & 1 \\ -2 & 4 \end{bmatrix} \begin{bmatrix} 1 & 1 \\ 1 & 2 \end{bmatrix} \\
&= \begin{bmatrix} 2 & 0 \\ 0 & 3 \end{bmatrix}
\end{aligned}
\tag{6.64}
$$

which confirms the result from the direct method used in Example 6.5.   ◇

We see in this last example that the standard basis does not produce the simplest matrix of T. Diagonal matrices are particularly simple and have several desirable properties, as will be recalled from Chapter 1. For instance, the inverse of a diagonal matrix, $\mathbf{D} = [d_i \delta_{ij}] = \text{diag}[d_i]$, is also diagonal: $\mathbf{D}^{-1} = [(1/d_i)\delta_{ij}] = \text{diag}[1/d_i]$, provided that all the d_i are non-zero. Also, any power of $\mathbf{D}$ is diagonal: $\mathbf{D}^k = [d_i \delta_{ij}]^k = [d_i^k \delta_{ij}] = \text{diag}[d_i^k]$.

EXERCISES

6.1 Let $U = \mathbb{R}^3$ and let V be the horizontal (i.e. the xy) plane with basis vectors $\mathbf{v}_1 = (1, 0, 0)$ and $\mathbf{v}_2 = (0, 1, 0)$; and define any two vectors $\mathbf{u}_1 = (x_1, y_1, z_1)$ and $\mathbf{u}_2 = (x_2, y_2, z_2)$. Find $T(\mathbf{u}_1)$, $T(\mathbf{u}_2)$, $T(\mathbf{u}_1 + \mathbf{u}_2)$ and $T(k\mathbf{u}_1)$, using the definition of T in equation (6.6), and draw a sketch to illustrate your results.

6.2 Let $T : \mathbb{R}^5 \to \mathbb{R}^3$ be the multiplication of a 5-vector by

$$\mathbf{A} = \begin{bmatrix} -2 & 3 & 0 & -3 & 4 \\ 5 & -8 & -3 & 2 & -13 \\ 1 & -2 & 3 & -4 & 5 \end{bmatrix}$$

Determine the rank and the nullity of T.

6.3 Recall Theorem 6.3.5 and use equation (6.23) to obtain a numerical solution for the x_i, $i = 1, 2, 3, 4$, in the following system of equations:

$$x_1 - x_2 - 3x_3 - 5x_4 = 0$$
$$2x_1 - 2x_2 - 3x_3 + 4x_4 = 0$$

6.4 Let $T: \mathbb{R}^2 \to \mathbb{R}^3$ be the linear transformation defined by

$$T\left(\begin{bmatrix} x_1 \\ x_2 \end{bmatrix}\right) = \begin{bmatrix} x_1 + x_2 \\ x_1 - x_2 \\ 3x_1 + 2x_2 \end{bmatrix}$$

(a) Find the standard matrix of T.
(b) Find the matrix of T with respect to the basis $B^* = \{\mathbf{u}_1, \mathbf{u}_2\}$ for $\mathbb{R}^2$, where $\mathbf{u}_1 = [-1 \quad 1]^\top$ and $\mathbf{u}_2 = [1 \quad -2]^\top$, and the standard basis for $\mathbb{R}^3$.

6.5 Prove Theorem 6.2.1.

6.6 Verify that the identity transformation $T: U \to U$ such that $T(\mathbf{u}) = \mathbf{u}$ for every $\mathbf{u} \in U$ is a linear transformation.

6.7 Let U be a vector space, k be any scalar and $T: U \to U$ be defined by $T(\mathbf{u}) = k\mathbf{u}$. Demonstrate that T is a linear operator.

6.8 Let U be a vector space, V be a subspace of U having an orthonormal basis $S = \{\mathbf{v}_1, \mathbf{v}_2, \ldots, \mathbf{v}_r\}$, and $T: U \to V$ be a function that maps a vector $\mathbf{u} \in U$ into its orthogonal projection on V, i.e.

$$T(\mathbf{u}) = (\mathbf{u} \cdot \mathbf{v}_1)\mathbf{v}_1 + (\mathbf{u} \cdot \mathbf{v}_2)\mathbf{v}_2 + \cdots + (\mathbf{u} \cdot \mathbf{v}_r)\mathbf{v}_r$$

Show that $T(k\mathbf{u}_1) = kT(\mathbf{u}_1)$, to complete the orthogonal projection illustration in Section 6.2.2.

7 Foundations for vector calculus

7.1 Introduction

This chapter covers selected topics relating to subsets of vector spaces and to functions more general than linear transformations defined on and between vector spaces. It has two simultaneous objectives, namely, to reinforce the reader's grasp of linear algebra and to prepare for its application to multivariate calculus in Chapter 9.

The knowledge of linear algebra acquired in the preceding chapters can now be reinforced by using it first (in this chapter) to generalize further familiar concepts from two- and three-dimensional geometry and later (in Chapter 9) from elementary univariate calculus to a multivariate or vector-based context.

Sections 7.2 and 7.3 introduce the concepts of affine combinations, affine sets, affine hulls and affine functions and their convex equivalents, all of which are natural extensions of similar concepts introduced in the earlier chapters on linear algebra.

Section 7.4 discusses further subsets of n-dimensional spaces, introducing the reader to the properties of various subsets, including multidimensional analogues of familiar objects like circles, spheres, squares and rectangles. Some of these sets will be encountered again later in this chapter, and others in the chapters on economic and financial applications that follow.

Section 7.5 is an elementary introduction to topology, topological spaces and metric spaces. The separating hyperplane theorem is introduced in Section 7.6. Sections 7.7 and 7.8 review some basic material on functions, limits and continuity. Finally, Section 7.9 reviews the univariate fundamental theorem of calculus.

7.2 Affine combinations, sets, hulls and functions

Definition 5.4.3 introduced the reader to the concept of a *linear* combination $\sum_{i=1}^{r} k_i \mathbf{v}_i$ of the vectors $\mathbf{v}_1, \mathbf{v}_2, \ldots, \mathbf{v}_r$, where $k_1, k_2, \ldots, k_r$ are *any* scalars. We now define a special type of linear combination.

DEFINITION 7.2.1 An **affine combination** of the vectors $\mathbf{v}_1, \mathbf{v}_2, \ldots, \mathbf{v}_r$ is defined as

$$k_1 \mathbf{v}_1 + k_2 \mathbf{v}_2 + \cdots + k_r \mathbf{v}_r = \sum_{i=1}^{r} k_i \mathbf{v}_i \tag{7.1}$$

where r is any positive integer, $k_1, k_2, \ldots, k_r$ are scalars *and* $\sum_{i=1}^{r} k_i = 1$.

Note that there is no restriction on the signs of the scalars k_i, provided that they sum to unity. So, for example, all four of $\mathbf{u}$, $\mathbf{v}$, $2\mathbf{u} - \mathbf{v}$ and $(\mathbf{u} + \mathbf{v})/2$ are affine combinations of the vectors $\mathbf{u}$ and $\mathbf{v}$. Every affine combination is a linear combination, but not every linear combination is an affine combination.

In Theorem 5.4.3, it was shown that a subset of a vector space is a *subspace* if and only if it is closed under the taking of linear combinations. We now define a more general class of subsets of vector spaces.

DEFINITION 7.2.2 A subset A of a vector space is an **affine set** if and only if A is closed under the taking of affine combinations; that is, if whenever $k_1, k_2, \ldots, k_r$ are scalars with $\sum_{i=1}^{r} k_i = 1$ and $\mathbf{v}_1, \mathbf{v}_2, \ldots, \mathbf{v}_r \in A$, then $\sum_{i=1}^{r} k_i \mathbf{v}_i \in A$; or, equivalently, if whenever λ is a scalar and $\mathbf{x}$ and $\mathbf{x}'$ are vectors in A, then $\lambda\mathbf{x} + (1 - \lambda)\mathbf{x}' \in A$.

The equivalence between the two characterizations of closure given in the definition follows from the same logic as Theorem 5.4.3. The details are left as an exercise; see Exercise 7.1.

For example, a set containing a single vector is an affine set. The next smallest affine set is a straight line. In fact, in any vector space, the affine set $L = \{\lambda\mathbf{x} + (1 - \lambda)\mathbf{x}' : \lambda \in \mathbb{R}\}$ is the **line from $\mathbf{x}'$**, where $\lambda = 0$, **to $\mathbf{x}$**, where $\lambda = 1$. We say that the parameter λ **parametrizes** the line L. Any affine set containing two distinct points must also contain the entire line connecting those two points.

All subspaces of a vector space are affine sets, but the converse does not hold. For example, one-dimensional subspaces are lines through the origin or zero vector. On the other hand, lines that do not pass through the origin are affine sets, but they are not subspaces.

Just as affine combinations of a set of two points generate a straight line, so affine combinations of a set of three non-collinear points generate a plane, and similarly for higher dimensions.

In fact, if A is an affine set in the vector space V and $\mathbf{a}$ is any vector in A, then the set

$$A - \{\mathbf{a}\} \equiv \{\mathbf{x} - \mathbf{a} : \mathbf{x} \in A\} \tag{7.2}$$

is a vector subspace of V; see Exercise 7.2. We can think of the **dimension** of the affine set as the dimension of this vector subspace. In effect, every affine set is a vector subspace translated away from the origin.

Recall that in Section 5.4.2 it was shown that the solution space of a *homogeneous* system of m linear equations in n unknowns $\mathbf{A}\mathbf{x} = \mathbf{0}$ is a vector subspace of $\mathbb{R}^n$, the null space or kernel of $\mathbf{A}$. The solution set for the corresponding *non-homogeneous* system $\mathbf{A}\mathbf{x} = \mathbf{b}$ is likewise an affine set. If the vector subspace $W \subseteq \mathbb{R}^n$ is the solution space of the homogeneous system and $\mathbf{x}^*$ is a particular solution of the non-homogeneous system, then the affine set

$$W + \{\mathbf{x}^*\} \equiv \{\mathbf{x} + \mathbf{x}^* : \mathbf{x} \in W\} \tag{7.3}$$

is the solution set of the non-homogeneous system.

Definition 5.4.5 introduced the concept of a vector space spanned or generated by a (typically finite) set of vectors, which is just the set of all possible *linear* combinations of the spanning vectors. We now define a similar concept based on *affine* combinations.

DEFINITION 7.2.3 The **affine hull** of the subset S of a vector space, denoted aff(S), is

$$\left\{ \sum_{i=1}^{r} k_i \mathbf{x}_i : \mathbf{x}_i \in S, \; k_i \in \mathbb{R}, \; i = 1, 2, \ldots, r; \; \sum_{i=1}^{r} k_i = 1; r = 1, 2, \ldots \right\} \tag{7.4}$$

in other words the set of all affine combinations of vectors in S.

For example, the affine hull of a set of collinear points is the entire line along which they lie. The affine hull of a set of coplanar points is the entire plane on which they lie.

Definition 6.2.3 introduced the concept of a linear transformation, and Theorem 6.2.1 showed that it is a function that preserves *linear* combinations, and which therefore is homogeneous of degree one. Once again, we can now define a similar concept based on *affine* combinations.

DEFINITION 7.2.4 A vector-valued function of several variables, $f : U \to V$ is said to be an **affine function** if it preserves affine combinations; that is, if

$$f\left(\sum_{i=1}^{r} k_i \mathbf{v}_i \right) = \sum_{i=1}^{r} k_i f(\mathbf{v}_i) \tag{7.5}$$

whenever $k_1, k_2, \ldots, k_r$ are scalars with $\sum_{i=1}^{r} k_i = 1$ and $\mathbf{v}_1, \mathbf{v}_2, \ldots, \mathbf{v}_r$ are vectors in U; or, equivalently, if

$$f(\lambda \mathbf{x} + (1 - \lambda)\mathbf{x}') = \lambda f(\mathbf{x}) + (1 - \lambda) f(\mathbf{x}') \tag{7.6}$$

whenever λ is a scalar and $\mathbf{x}$ and $\mathbf{x}'$ are vectors in U.

The inductive logic used to establish the equivalence between (7.5) and (7.6) should by now be familiar.

THEOREM 7.2.1 *If* $f : \mathbb{R}^n \to \mathbb{R}^m$ *is an affine function, then there exists a unique* $m \times n$ *matrix* $\mathbf{A}$ *and a unique m-dimensional vector* $\mathbf{b}$ *such that*

$$f(\mathbf{x}) = \mathbf{A}\mathbf{x} + \mathbf{b}, \quad \forall \, \mathbf{x} \in \mathbb{R}^n \tag{7.7}$$

Proof: We claim that the function $g : \mathbb{R}^n \to \mathbb{R}^m : \mathbf{x} \mapsto f(\mathbf{x}) - f(\mathbf{0})$ is a linear transformation. To prove this claim, let $k_1, k_2, \ldots, k_r$ be *any* scalars and let $\mathbf{v}_1, \mathbf{v}_2, \ldots, \mathbf{v}_r$ be any vectors in $\mathbb{R}^n$. Then

$$g\left(\sum_{i=1}^{r} k_i \mathbf{v}_i \right) = f\left(\sum_{i=1}^{r} k_i \mathbf{v}_i \right) - f(\mathbf{0})$$

$$= f\left(\sum_{i=1}^{r} k_i \mathbf{v}_i + \left(1 - \sum_{i=1}^{r} k_i \right) \mathbf{0} \right) - f(\mathbf{0}) \tag{7.8}$$

since the additional term inserted in the first argument of f does not change the value of the argument. Since $\sum_{i=1}^{r} k_i + \left(1 - \sum_{i=1}^{r} k_i\right) = 1$, the first argument of f above is an affine combination of elements of $\mathbb{R}^n$, and since f is an affine function (7.8) becomes

$$f\left(\sum_{i=1}^{r} k_i \mathbf{v}_i + \left(1 - \sum_{i=1}^{r} k_i\right)\mathbf{0}\right) - f(\mathbf{0}) = \sum_{i=1}^{r} k_i f(\mathbf{v}_i) + \left(1 - \sum_{i=1}^{r} k_i\right) f(\mathbf{0}) - f(\mathbf{0})$$

$$= \sum_{i=1}^{r} k_i (f(\mathbf{v}_i) - f(\mathbf{0}))$$

$$= \sum_{i=1}^{r} k_i g(\mathbf{v}_i) \tag{7.9}$$

This proves our claim.

We know from Section 6.4 that every linear transformation is a matrix transformation, so that there exists a unique $m \times n$ matrix $\mathbf{A}$ such that $g(\mathbf{x}) = \mathbf{A}\mathbf{x}$, for all $\mathbf{x} \in \mathbb{R}^n$. If we let $f(\mathbf{0}) \equiv \mathbf{b}$, we have

$$f(\mathbf{x}) = g(\mathbf{x}) + f(\mathbf{0}) = \mathbf{A}\mathbf{x} + \mathbf{b}, \quad \forall \mathbf{x} \in \mathbb{R}^n \tag{7.10}$$

as required. $\qquad\qquad\qquad\qquad\qquad\qquad\qquad\qquad\qquad\qquad\qquad\qquad\qquad\qquad\qquad$ $\square$

Thus every affine function is a combination of a linear transformation, represented by the matrix $\mathbf{A}$, and a translation, represented by the vector $\mathbf{b}$. A linear transformation is a special case of an affine function in which there is no translation, or the translation vector $\mathbf{b}$ is the zero vector.

The term "linear function" has historically been used loosely to refer both individually and collectively to linear transformations and affine functions. To avoid confusion, we stick to the term "linear *transformation*" when referring strictly only to the concept defined in Definition 6.2.3. Indeed, the term "linear function" was used loosely as early as the first chapter of this book.

In the language introduced in Section 1.2.2, if $f(\mathbf{x}) = \mathbf{A}\mathbf{x} - \mathbf{b}$, then the system of equations $f(\mathbf{x}) = \mathbf{0}$ is homogeneous if and only if $\mathbf{b} = \mathbf{0}$ if and only if f is a linear function if and only if the function f is homogeneous of degree one. It is non-homogeneous if and only if $\mathbf{b} \neq \mathbf{0}$, in which case f is an affine function. In other words, the distinction between linear homogeneous systems of equations and linear non-homogeneous systems of equations is analogous to the difference between linear transformations and affine functions. The term "linear equation" is used loosely to refer collectively to homogeneous and non-homogeneous systems, for example in the classification of difference equations in Section 8.2.4; the term "affine equation" is rarely if ever seen.

Finally, note that, while a linear transformation is homogeneous of degree one, an affine transformation in general is not homogeneous.

7.3   Convex combinations, sets, hulls and functions

Definitions 5.4.3 and 7.2.1 respectively have introduced the general concept of a *linear* combination and the special case of an *affine* combination. We now further introduce a special type of affine combination.

DEFINITION 7.3.1 For vectors $\mathbf{x}$ and $\mathbf{x}'$ and for $\lambda \in [0, 1]$, $\lambda\mathbf{x} + (1 - \lambda)\mathbf{x}'$ is called a **convex combination** of $\mathbf{x}$ and $\mathbf{x}'$.

More generally, a **convex combination** of the vectors $\mathbf{v}_1, \mathbf{v}_2, \ldots, \mathbf{v}_r$ is defined as

$$k_1\mathbf{v}_1 + k_2\mathbf{v}_2 + \cdots + k_r\mathbf{v}_r = \sum_{i=1}^{r} k_i \mathbf{v}_i \tag{7.11}$$

where r is any positive integer, $k_1, k_2, \ldots, k_r$ are *non-negative* scalars and $\sum_{i=1}^{r} k_i = 1$.

Repeating the example of the previous section, $\mathbf{x}$, $\mathbf{y}$ and $(\mathbf{x}+\mathbf{y})/2$ are convex combinations of the vectors $\mathbf{x}$ and $\mathbf{y}$, but $2\mathbf{x} - \mathbf{y}$ is not, because the coefficient of $\mathbf{y}$ is negative. Every convex combination is an affine combination, but not every affine combination is a convex combination.

We will find that taking affine or convex combinations of two equations or inequalities is often a useful device in proving results; see, for example, the proof of Theorem 10.2.4.

We know that a subspace is a set closed under the taking of linear combinations and an affine set is a set closed under the taking of affine combinations. The next definition is a natural extension of this concept.

DEFINITION 7.3.2 A subset X of a vector space is a **convex set** if and only if every convex combination of elements in X is also in X; that is, if for all $\mathbf{x}, \mathbf{x}' \in X$ and for all $\lambda \in [0, 1]$, $\lambda\mathbf{x} + (1 - \lambda)\mathbf{x}' \in X$; or, equivalently, if whenever $k_1, k_2, \ldots, k_r$ are non-negative scalars with $\sum_{i=1}^{r} k_i = 1$ and $\mathbf{v}_1, \mathbf{v}_2, \ldots, \mathbf{v}_r \in X$, then $\sum_{i=1}^{r} k_i \mathbf{v}_i \in X$.

Once again, the equivalence between the two characterizations of closure given in the definition can be proved using the argument used in the proof of Theorem 5.4.3. The details are left as an exercise; see Exercise 7.3.

For example, a set containing a single vector is a convex set. The next smallest convex set is a line segment. In fact, in any vector space, $L = \{\lambda\mathbf{x} + (1 - \lambda)\mathbf{x}' : \lambda \in [0, 1]\}$ is the **line segment from $\mathbf{x}'$**, where $\lambda = 0$, **to $\mathbf{x}$**, where $\lambda = 1$, in X.

Note that every affine set is a convex set, but that the converse is not true. For example, a line segment is a convex set but is not an affine set.

THEOREM 7.3.1 *A sum of convex sets, such as*

$$X + Y \equiv \{\mathbf{x} + \mathbf{y} : \mathbf{x} \in X, \mathbf{y} \in Y\} \tag{7.12}$$

is also a convex set.

Proof: The proof of this result is left as an exercise; see Exercise 7.4. □

Just as we defined the subspace generated by a set of vectors X to be the set of all linear combinations of vectors in X and the affine hull of X to be the set of all affine combinations of vectors in X, so we can now define the convex hull of X to be the set of all convex combinations of vectors in X.

DEFINITION 7.3.3 If X is a subset of a real vector space, then the **convex hull** of X is

$$\{\lambda\mathbf{x} + (1 - \lambda)\mathbf{x}' : \lambda \in [0, 1], \mathbf{x}, \mathbf{x}' \in X\} \tag{7.13}$$

The next definition is *not* completely analogous to those that have gone before, for it contains an inequality where the reader might have expected to see an equality.

DEFINITION 7.3.4 Let $f: X \to Y$, where X is a convex subset of a real vector space and $Y \subseteq \mathbb{R}$. Then f is a **convex function** if and only if, for all $\mathbf{x} \neq \mathbf{x}' \in X$ and for all $\lambda \in (0, 1)$,

$$f(\lambda \mathbf{x} + (1 - \lambda)\mathbf{x}') \leq \lambda f(\mathbf{x}) + (1 - \lambda)f(\mathbf{x}') \tag{7.14}$$

Note that the definitions of linear transformations and affine functions imply that the domains of such functions must be an entire vector space, but that a convex function need be defined only on a convex subset of a vector space. For example, many commonly encountered convex functions are defined only on the positive orthant of Euclidean space, $\mathbb{R}^n_{++}$. Note also that, while linear transformations and affine functions can be vector-valued functions, convex functions are always real-valued functions.

We will postpone the characterization and properties of convex functions until Section 10.2.1, where they will be treated in much greater depth. First, however, we will present a collection of examples of subsets of n-dimensional spaces, considering in each case whether they are affine sets or convex sets.

7.4   Subsets of n-dimensional spaces

7.4.1   Hyperplanes

Readers will already be familiar with the idea of a vector subspace from Section 5.4.1. A one-dimensional subspace of an n-dimensional vector space is just a line passing through the origin or zero vector, $\mathbf{0}_n$; a two-dimensional subspace is a plane through the origin; more generally, an $(n - 1)$-dimensional subspace of $\mathbb{R}^n$ is a special case of what we call a hyperplane, again passing through the origin. We will now define hyperplanes more formally.

DEFINITION 7.4.1

(a) An **(affine) hyperplane** in $\mathbb{R}^n$ is any set of the form $\{\mathbf{x} \in \mathbb{R}^n : \mathbf{p}^\top \mathbf{x} = c\}$, where $\mathbf{p}$ is a vector in $\mathbb{R}^n$ and c is a scalar.

(b) For any two vectors $\mathbf{x}^*$ and $\mathbf{p} \neq \mathbf{0}$ in $\mathbb{R}^n$, the set $\{\mathbf{x} \in \mathbb{R}^n : \mathbf{p}^\top \mathbf{x} = \mathbf{p}^\top \mathbf{x}^*\}$ is the **affine hyperplane through $\mathbf{x}^*$ with normal $\mathbf{p}$**.

EXAMPLE 7.4.1 The line (or one-dimensional hyperplane) in $\mathbb{R}^2$ with equation $x_1 + x_2 = 1$ cuts the coordinate axes at a 45° angle, passing through both of the standard unit basis vectors, $(1, 0)$ and $(0, 1)$. Any vector of the form $\lambda \mathbf{1}_2 = (\lambda, \lambda)$ $(\lambda \neq 0)$ is normal to this hyperplane.

This line can be described, for example, as the hyperplane through $(1, 0)$ with normal $(1, 1)$; or as the hyperplane through $(0, 1)$ with normal $(2, 2)$; or as the hyperplane through $(0, 1)$ with normal $(-1, -1)$. It is left as an exercise for the reader to confirm that making the appropriate substitutions in each case for $\mathbf{p}$ and $\mathbf{x}^*$ in the second part of Definition 7.4.1 results in the same equation $x_1 + x_2 = 1$; see Exercise 7.8. $\diamond$

As the terminology suggests, an affine hyperplane is an affine set; see Exercise 7.9.

The first part of Definition 7.4.1 is merely an extension of a familiar idea from low dimensions. In two dimensions, an affine hyperplane is just a line with equation like $p_1 x_1 +$

$p_2 x_2 = c$; in three dimensions, it is just a plane with equation like $p_1 x_1 + p_2 x_2 + p_3 x_3 = c$. In n dimensions, it is an $(n-1)$-dimensional subspace if it contains the zero vector; otherwise, it is essentially an $(n-1)$-dimensional subspace translated away from the origin.

The second part of Definition 7.4.1 is illustrated in Figure 7.1. The vector $\mathbf{p}$ is normal to the hyperplane in the sense that, for any two points $\mathbf{x}$ and $\mathbf{x}^*$ in the hyperplane, the vector $\mathbf{x} - \mathbf{x}^*$ is perpendicular or orthogonal to the vector $\mathbf{p}$, since by definition their dot product $\mathbf{p}^\top (\mathbf{x} - \mathbf{x}^*)$ is zero. Any point in the hyperplane can play the role of $\mathbf{x}^*$.

A line is not an affine hyperplane in $\mathbb{R}^3$ but is an affine hyperplane in $\mathbb{R}^2$. An affine hyperplane is defined by a single equation. In $\mathbb{R}^2$, a line is defined by a single (scalar) equation. In $\mathbb{R}^3$, a line is defined by two equations, each specifying a plane, the line being the intersection of the two planes. Similarly, a single plane in $\mathbb{R}^3$ is defined by a single equation and is an affine hyperplane in $\mathbb{R}^3$.

If an affine hyperplane in $\mathbb{R}^n$ passes through the origin (making it an $(n-1)$-dimensional vector subspace), then it is called a **linear hyperplane**. For a linear hyperplane, we must have, in the context of Definition 7.4.1, in the first case, $c = 0$ and, in the second case, $\mathbf{p}^\top \mathbf{x}^* = 0$.

The normal vector $\mathbf{p}$ is unique up to a scalar multiple; in other words, if $\mathbf{p}$ is normal to a given affine hyperplane, then $\lambda \mathbf{p}$ is also normal to that hyperplane for any $\lambda \in \mathbb{R}$ (except, of course, for $\lambda = 0$).

Euclidean space $\mathbb{R}^n$ is the direct sum of any linear hyperplane and the one-dimensional subspace generated by its normal vector. Thus, a linear hyperplane is often described as being of **codimension** one.

In the remainder of this book, we will refer to affine hyperplanes simply as hyperplanes. The conventional terminology here is somewhat unfortunate. The term *hyperplane* on its own is sometimes used to refer to a k-dimensional affine subset of an n-dimensional space, even when $k < n - 1$. The term *affine!hyperplane* is always confined to sets of codimension one, but the lower-dimensional variants are affine sets in the formal sense (and thus also convex sets) and hyperplanes in the informal sense. The $(n-1)$-dimensional affine hyperplane is different from its lower-dimensional analogues in that it is impossible to travel around it

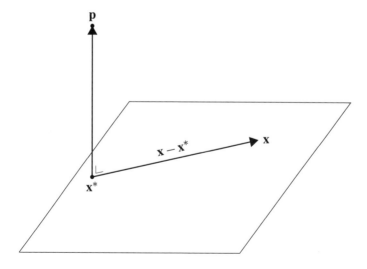

Figure 7.1 The hyperplane through $\mathbf{x}^*$ with normal $\mathbf{p}$

without passing through it. For example, in $\mathbb{R}^3$, it is possible to travel around a line but one can only get to the other side of a plane by travelling through it. This is the fundamental logic behind the separating hyperplane theorem (Theorem 7.6.1).

7.4.2 Simplexes and hyperparallelepipeds

DEFINITION 7.4.2 The **unit simplex** in $\mathbb{R}^n$, also called the **standard simplex** in $\mathbb{R}^n$, is the intersection of the non-negative orthant $\mathbb{R}^n_+$ and the hyperplane through $(1/n)\mathbf{1}_n$ with normal $\mathbf{1}_n$, i.e. the set

$$S^{n-1} \equiv \left\{ \mathbf{p} \in \mathbb{R}^n_+ : \mathbf{p}^\top \mathbf{1}_n = \sum_{i=1}^n p_i = 1 \right\} \tag{7.15}$$

Figures 7.2 and 7.3 show the unit simplexes in $\mathbb{R}^2$ and $\mathbb{R}^3$, respectively. The former is just the intersection of the hyperplane considered in Example 7.4.1 above with the non-negative quadrant.

DEFINITION 7.4.3 More generally, a **simplex** is the convex hull of a finite set of points, which are called the **vertices** of the simplex.

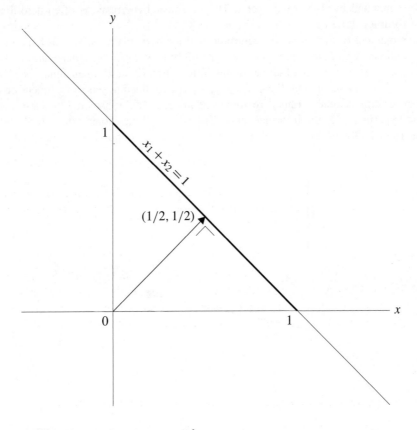

Figure 7.2 Unit or standard simplex in $\mathbb{R}^2$

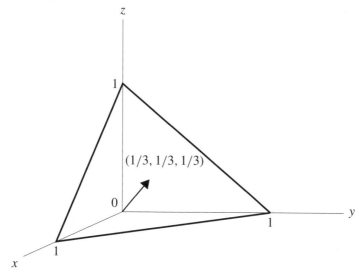

Figure 7.3 Unit or standard simplex in $\mathbb{R}^3$

The n vertices of the unit simplex in $\mathbb{R}^n$ are just the standard basic vectors, $\mathbf{e}_1, \mathbf{e}_2, \ldots, \mathbf{e}_n$.

A simplex with two vertices is just the line segment joining the two points. A simplex with three vertices is a triangle, and so on.

DEFINITION 7.4.4 The m-dimensional **hyperparallelepiped** in $\mathbb{R}^n$ whose **edges** are the vectors $\mathbf{x}_1, \mathbf{x}_2, \ldots, \mathbf{x}_m$ is the set

$$\left\{ \sum_{i=1}^{m} k_i \mathbf{x}_i : k_i \in [0, 1], \ i = 1, 2, \ldots, m \right\} \tag{7.16}$$

In fact, a hyperparallelepiped is just a special type of simplex; see Exercise 7.10. A hyper-parallelepiped is also a convex set; see Exercise 7.11.

When $m = 3$, a hyperparallelepiped is simply called a **parallelepiped**. When $m = 2$, a hyperparallelepiped is just a **parallelogram**. When $m = 1$, it is just a single vector or a line segment.

7.4.3 Hyperspheres

Just as the hyperplane is the higher-dimensional analogue of the line in two dimensions or the plane in three dimensions, so the idea of a circle in two dimensions or of a ball or sphere in three dimensions can be extended to n dimensions. Where ambiguity might arise, we will describe the surface in three dimensions as a **sphere** and the solid volume that it encloses as a **ball**.

DEFINITION 7.4.5

(a) A **hypersphere of radius** r (≥ 0) **in** $\mathbb{R}^n$ **centred at a** is a set of the form

$$\{ \mathbf{x} \in \mathbb{R}^n : (\mathbf{x} - \mathbf{a})^\top (\mathbf{x} - \mathbf{a}) = r^2 \} \tag{7.17}$$

(b) An **open ball of radius** r (≥ 0) **in** $\mathbb{R}^n$ **centred at a** is a set of the form

$$\{\mathbf{x} \in \mathbb{R}^n : (\mathbf{x} - \mathbf{a})^\top (\mathbf{x} - \mathbf{a}) < r^2\} \tag{7.18}$$

(c) A **closed ball of radius** r (≥ 0) **in** $\mathbb{R}^n$ **centred at a** is a set of the form

$$\{\mathbf{x} \in \mathbb{R}^n : (\mathbf{x} - \mathbf{a})^\top (\mathbf{x} - \mathbf{a}) \leq r^2\} \tag{7.19}$$

In other words, an open ball is the set contained (strictly) inside a hypersphere, while a closed ball comprises both the corresponding hypersphere and the set inside it.

Recall that $(\mathbf{x} - \mathbf{a})^\top (\mathbf{x} - \mathbf{a}) = (d(\mathbf{x}, \mathbf{a}))^2 = \|\mathbf{x} - \mathbf{a}\|^2$.

The hypersphere in two dimensions is just the circle with equation $(x_1 - a_1)^2 + (x_2 - a_2)^2 = r^2$ and similarly the hypersphere in three dimensions is just a normal sphere.

Like a hyperplane, a hypersphere is essentially a set of one dimension less than the space inside which it sits. Unlike a hyperplane, a hypersphere cannot be a vector subspace or an affine set. Open balls and closed balls are, however, convex sets.

7.4.4 Hypercubes and hyperrectangles

The closed rectangle in the plane with sides parallel to the coordinate axes and with corners (a_1, a_2), (a_1, b_2), (b_1, a_2) and (b_1, b_2) is just the Cartesian product $[a_1, b_1] \times [a_2, b_2]$. (It is also a simplex whose four vertices are the corners of the rectangle.) The corresponding open rectangle is just the Cartesian product[1] $(a_1, b_1) \times (a_2, b_2)$.

A closed **hyperrectangle** in $\mathbb{R}^n$ is just the natural higher-dimensional analogue of a closed rectangle, namely the Cartesian product of n closed intervals, $\prod_{i=1}^{n} [a_i, b_i]$. If the side lengths $b_i - a_i$ are all equal, then we have a **hypercube**. A closed hypercube or closed hyperrectangle is a simplex with 2^n vertices.[2]

Similarly, an open hyperrectangle in $\mathbb{R}^n$ is just the Cartesian product of n open intervals, $\prod_{i=1}^{n} (a_i, b_i)$. The volume of the hyperrectangle, whether open or closed, is $\prod_{i=1}^{n} (b_i - a_i)$. Hypercubes and hyperrectangles are convex sets, whether they are open or closed.

The notion of an open hypercube or hyperrectangle is sometimes just as useful as that of an open ball; see Exercise 7.15 for the connections between the two. The proof of Theorem 7.8.1 below will involve fitting hyperrectangles inside open balls.

A hyperrectangle with a vertex at the origin is just a special case of a hyperparallelepiped with orthogonal edges. The next theorem shows that calculating the volume of a hyperparallelepiped is slightly more difficult than calculating the volume of a hyperrectangle.

THEOREM 7.4.1 *The volume of the m-dimensional hyperparallelepiped P in $\mathbb{R}^n$ with edges $\mathbf{x}_1, \mathbf{x}_2, \ldots, \mathbf{x}_m$ is $\sqrt{\det(\mathbf{X}^\top \mathbf{X})}$, where $\mathbf{X}$ is the $n \times m$ matrix whose columns are $\mathbf{x}_1, \mathbf{x}_2, \ldots, \mathbf{x}_m$.*

Proof: The proof is by induction on m.

When $m = 1$, the hyperparallelepiped has just a single edge $\mathbf{x}_1 \in \mathbb{R}^n$ and the volume is just the length of this edge, $\|\mathbf{x}_1\| = \sqrt{\mathbf{x}_1^\top \mathbf{x}_1} = \sqrt{\det(\mathbf{x}_1^\top \mathbf{x}_1)}$, since $\mathbf{x}_1^\top \mathbf{x}_1$ is just a non-negative scalar.

When $m > 1$, we can think of the volume of the hyperparallelepiped as being the volume of its base times its height. For example, for $m = 2$, the area of the hyperparallelepiped (in this case, a parallelogram) is the length of one edge times the perpendicular distance to the opposite edge. When $m = 3$, the volume of the parallelepiped is the area of one face times the perpendicular distance to the opposite face, and so on for higher dimensions.

In general, if $S \equiv \mathrm{lin}\{x_1, x_2, \ldots, x_{m-1}\}$, then the volume of the m-dimensional hyperparallelepiped P is the volume of the $(m-1)$-dimensional hyperparallelepiped with edges $x_1, x_2, \ldots, x_{m-1}$ multiplied by the length of the component of x_m orthogonal to S, and this length equals $\|x_m - \mathrm{proj}_S x_m\|$.

Now suppose our hypothesis is true for dimensions up to $m-1$ and let $\mathbf{X}^*$ denote the $n \times m$ matrix obtained by replacing the last column of $\mathbf{X}$ (which is x_m) by $x_m^* \equiv x_m - \mathrm{proj}_S x_m$.

If $\mathrm{proj}_S x_m = \sum_{i=1}^{m-1} \lambda_i x_i$, then $\mathbf{X}^* = \mathbf{XE}$, where

$$
\mathbf{E} \equiv \begin{bmatrix}
1 & 0 & \cdots & 0 & -\lambda_1 \\
0 & 1 & \cdots & 0 & -\lambda_2 \\
\vdots & \vdots & & \vdots & \vdots \\
0 & 0 & \cdots & 1 & -\lambda_{m-1} \\
0 & 0 & \cdots & 0 & 1
\end{bmatrix}
\tag{7.20}
$$

Since $\mathbf{E}$ is upper triangular with ones on the principal diagonal, $\det(\mathbf{E}) = \det(\mathbf{E}^\top) = 1$,

$$
\begin{aligned}
\det((\mathbf{X}^*)^\top \mathbf{X}^*) &= \det((\mathbf{XE})^\top \mathbf{XE}) \\
&= \det(\mathbf{E}^\top \mathbf{X}^\top \mathbf{XE}) \\
&= \det(\mathbf{E}^\top) \det(\mathbf{X}^\top \mathbf{X}) \det(\mathbf{E}) \\
&= \det(\mathbf{X}^\top \mathbf{X})
\end{aligned}
\tag{7.21}
$$

If we partition $\mathbf{X}^*$ as $[\mathbf{X}_- \; x_m^*]$, then

$$
\begin{aligned}
(\mathbf{X}^*)^\top \mathbf{X}^* &= \begin{bmatrix}
\mathbf{X}_-^\top \mathbf{X}_- & \mathbf{X}_-^\top x_m^* \\
(x_m^*)^\top \mathbf{X}_- & (x_m^*)^\top x_m^*
\end{bmatrix} \\
&= \begin{bmatrix}
\mathbf{X}_-^\top \mathbf{X}_- & \mathbf{0}_{m-1} \\
\mathbf{0}_{m-1}^\top & \|x_m^*\|^2
\end{bmatrix}
\end{aligned}
\tag{7.22}
$$

Hence, using a co-factor expansion along the last row (or last column),

$$
\det(\mathbf{X}^\top \mathbf{X}) = \det((\mathbf{X}^*)^\top \mathbf{X}^*) = \det(\mathbf{X}_-^\top \mathbf{X}_-) \times \|x_m^*\|^2
\tag{7.23}
$$

which is just the square of the product of the volume of the base of the hyperparallelepiped (by the inductive hypothesis) times its height, as required. $\qquad\square$

The proofs of the following corollaries are left as exercises; see Exercise 7.13.

COROLLARY 7.4.2 *The volume of a hyperparallelepiped whose edges are linearly dependent is zero.*

COROLLARY 7.4.3 *The volume of an n-dimensional hyperparallelepiped in $\mathbb{R}^n$ whose n edges are the columns of the matrix $\mathbf{X}$ is $|\det(\mathbf{X})|$.*

COROLLARY 7.4.4 *If $m > n$, then the volume of an m-dimensional hyperparallelepiped in $\mathbb{R}^n$ is zero.*

COROLLARY 7.4.5 *If P is a hyperparallelepiped in $\mathbb{R}^n$ with volume $\sqrt{\det(\mathbf{X}^\top \mathbf{X})}$ and $T \colon \mathbb{R}^n \to \mathbb{R}^p$ is a linear transformation with $p \times n$ matrix $\mathbf{A}$, then $T(P)$ is a hyperparallelepiped in $\mathbb{R}^p$ with volume $\sqrt{\det((\mathbf{AX})^\top \mathbf{AX})}$.*

In particular, if $n = m = p$, then, since $\mathbf{A}$ and $\mathbf{X}$ are square matrices, the linear transformation T multiplies the volume of the hyperparallelepiped P by $|\det(\mathbf{A})|$.

7.4.5 Curves and hypersurfaces

We are already familiar with the idea that a line in $\mathbb{R}^2$ and a plane in $\mathbb{R}^3$ are each defined by a single *linear* equation, while in $\mathbb{R}^3$, a line is defined by two independent *linear* equations, each specifying a plane, the line being the intersection of the two (non-parallel) planes.

More generally, a curve in $\mathbb{R}^2$ is defined by a single equation, usually nonlinear. (Recall, for example, the various conic sections discussed in Section 4.2.) Similarly, a surface in $\mathbb{R}^3$ is defined by a single equation, again usually nonlinear. A curve in $\mathbb{R}^3$, however, is generally defined by two independent (and usually nonlinear) equations, each specifying a surface, and the curve being the intersection of the two surfaces. The concept of linear independence, however, does not extend in any natural way to nonlinear equations.

Just as one linear equation in n variables defines an affine hyperplane, so one (nonlinear) equation in n variables defines a **hypersurface**.

More generally, k (independent) equations in n variables define a solution set, which could be loosely described as of dimension $n - k$. If the equations are linear (and thus linearly independent), then, as noted on p. 144 above, the solution set is an affine set of dimension $n - k$.

7.5 Basic topology

Just as linear algebra is the study of vector spaces, so **topology** is the study of topological spaces; see Definition 7.5.4 below. This section begins with a brief review of metric spaces, which are a special type of topological space with additional structure; see Definition 7.5.1 below. We will see that every scalar product space is also a metric space. Not all metric spaces are scalar product spaces and not all topological spaces are vector spaces. All of the topological spaces that will be discussed in detail in this book are metric spaces, and, in fact, real vector spaces. The general definition of a topological space is included only for completeness.

Many of the results from topology needed in this book can be, and will be, proved using familiar properties of the real numbers, especially standard **epsilon–delta arguments**. A thorough knowledge of topology, however, can provide access to simpler, more elegant and more general versions of some of these proofs. Some theorems of critical importance in mathematical economics, in particular the general versions of Brouwer's fixed-point theorem (Theorem 12.5.5) and Kakutani's fixed-point theorem (Theorem 12.5.6), can be proved only using topological concepts. For all these reasons, those intending to study mathematical economics to a level higher than that to which the subject is taken in this book would be well advised to familiarize themselves with one of the standard textbooks on topology, such as Simmons (1963) or Mendelson (1975). Chiang and Wainwright (2005, pp. 64–5) cover the relevant topics in topology from an economic perspective.

The aim of this section is to provide sufficient introduction to topology to motivate the supporting and separating hyperplane theorems in Section 7.6 and the definitions of continuity of functions and correspondences in Section 7.8, but no more.

The first definition generalizes the notion of distance introduced in Definition 5.2.13.

DEFINITION 7.5.1 A **metric space** is a non-empty set X equipped with a **metric**, i.e. a distance function $d: X \times X \to \mathbb{R}_+$ such that:

(a) $d(\mathbf{x}, \mathbf{y}) = 0 \Rightarrow \mathbf{x} = \mathbf{y}$;
(b) $d(\mathbf{x}, \mathbf{y}) = d(\mathbf{y}, \mathbf{x})$ for all $\mathbf{x}, \mathbf{y} \in X$; and
(c) the **triangular inequality** holds, namely,

$$d(\mathbf{x}, \mathbf{z}) + d(\mathbf{z}, \mathbf{y}) \geq d(\mathbf{x}, \mathbf{y}) \quad \forall \, \mathbf{x}, \mathbf{y}, \mathbf{z} \in X \tag{7.24}$$

As noted in Section 5.4.9, any finite-dimensional vector space, say $\mathbb{R}^n$, can be turned into a scalar product space, given a positive definite $n \times n$ matrix, say $\mathbf{A}$, by defining a scalar product

$$(\mathbf{x} \mid \mathbf{y}) \equiv \mathbf{x}^\top \mathbf{A} \mathbf{y} \tag{7.25}$$

Such a scalar product space can in turn be made into a metric space by letting

$$d(\mathbf{x}, \mathbf{y}) \equiv \|\mathbf{x} - \mathbf{y}\| \equiv \sqrt{(\mathbf{x} - \mathbf{y} \mid \mathbf{x} - \mathbf{y})} \tag{7.26}$$

The proof that this d has all the properties of a metric, apart from an outline of the proof that it satisfies the triangular inequality, is left as an exercise; see Exercise 7.16. To show that the triangular inequality is satisfied, note that

$$\begin{aligned}
\|t\mathbf{x} + \mathbf{y}\|^2 &= (t\mathbf{x} + \mathbf{y} \mid t\mathbf{x} + \mathbf{y}) \\
&= t^2 (\mathbf{x} \mid \mathbf{x}) + 2t (\mathbf{x} \mid \mathbf{y}) + (\mathbf{y} \mid \mathbf{y}) \\
&\geq 0 \quad \forall \, t
\end{aligned} \tag{7.27}$$

by positive definiteness of $\mathbf{A}$.

Thus the quadratic expression in t on the right-hand side of (7.27) has no real roots (except possibly a repeated root). Using the standard quadratic equation formula for these roots yields

$$t = \frac{-2(\mathbf{x} \mid \mathbf{y}) \pm \sqrt{4(\mathbf{x} \mid \mathbf{y})^2 - 4(\mathbf{x} \mid \mathbf{x})(\mathbf{y} \mid \mathbf{y})}}{2(\mathbf{x} \mid \mathbf{x})} \tag{7.28}$$

Since there cannot be two distinct real roots, the expression under the square root sign must be non-positive, which (after taking square roots) leads to the **Cauchy–Schwarz inequality**,[3]

$$|(\mathbf{x} \mid \mathbf{y})| \leq \|\mathbf{x}\| \times \|\mathbf{y}\| \tag{7.29}$$

In the case of the dot product, since $|\cos \theta| \leq 1$ for all θ, the Cauchy–Schwarz inequality actually follows immediately from Definition 5.2.15.

Hence,

$$
\begin{aligned}
d(\mathbf{x}, \mathbf{y})^2 &= (\mathbf{x} - \mathbf{y} \mid \mathbf{x} - \mathbf{y}) \\
&= (\mathbf{x} - \mathbf{z} + \mathbf{z} - \mathbf{y} \mid \mathbf{x} - \mathbf{z} + \mathbf{z} - \mathbf{y}) \\
&= \|\mathbf{x} - \mathbf{z}\|^2 + \|\mathbf{z} - \mathbf{y}\|^2 + 2(\mathbf{x} - \mathbf{z} \mid \mathbf{z} - \mathbf{y}) \\
&\leq \|\mathbf{x} - \mathbf{z}\|^2 + \|\mathbf{z} - \mathbf{y}\|^2 + 2\|\mathbf{x} - \mathbf{z}\| \times \|\mathbf{z} - \mathbf{y}\| \\
&= (\|\mathbf{x} - \mathbf{z}\| + \|\mathbf{z} - \mathbf{y}\|)^2 \\
&= (d(\mathbf{x}, \mathbf{z}) + d(\mathbf{z}, \mathbf{y}))^2
\end{aligned}
\tag{7.30}
$$

where the inequality in (7.30) follows from the Cauchy–Schwarz inequality. Taking square roots in (7.30) yields the triangular inequality as required.

It will be seen in Section 13.6.3 that a variance–covariance matrix is an example of a positive definite matrix often used to define a scalar product (covariance) and a metric (standard deviation), and that the cosine of the angle between two vectors is equivalent to a measure of correlation.

The next set of definitions describes subsets of metric spaces that have special properties.

DEFINITION 7.5.2 Let A be a subset of the metric space X.

(a) An **open ball** is a subset of X of the form

$$
B_\epsilon(\mathbf{x}) = \{\mathbf{y} \in X : d(\mathbf{y}, \mathbf{x}) < \epsilon\}
\tag{7.31}
$$

in other words, the set of all points in the metric space less than the distance ϵ away from the centre, $\mathbf{x}$, of the open ball.

(b) A is **open** if and only if, for all $\mathbf{x} \in A$, there exists $\epsilon > 0$ such that $B_\epsilon(\mathbf{x}) \subseteq A$.

(c) The **interior** of A, denoted int A, is defined by

$$
\mathbf{x} \in \text{int } A \iff B_\epsilon(\mathbf{x}) \subseteq A \text{ for some } \epsilon > 0
\tag{7.32}
$$

(d) $\mathbf{x}$ is a **boundary point** of A if, for every $\epsilon > 0$, $B_\epsilon(\mathbf{x})$ contains points both of A and of its complement $X \setminus A$.

The first part of Definition 7.5.2 merely generalizes the relevant part of Definition 7.4.5 from Euclidean space $\mathbb{R}^n$ to an arbitrary metric space. Every point in an open set is in the interior of the set, or an open set does not contain any of its boundary points.

Confining attention to open sets means that we will not have to be concerned with what happens at boundary points when dealing with limits and continuity; on the other hand, confining attention to open sets when using calculus to search for maxima and minima means that we will have to use other techniques to check for boundary solutions to optimization problems.

DEFINITION 7.5.3 A is **bounded** if and only if there exist $\mathbf{x}$, $K > 0$ such that $A \subseteq B_K(\mathbf{x})$.

Note that K generally denotes a large number, whereas ϵ generally denotes a small number.

Metric spaces are special cases of the more general concept of a topological space.

DEFINITION 7.5.4 A **topological space** is a set X together with τ, a collection of subsets of X, satisfying the following axioms:

(a) The empty set and X are in τ.
(b) The union of any collection of sets in τ is also in τ.
(c) The intersection of any finite collection of sets in τ is also in τ.

The collection τ is called a **topology** on X. The elements of X are usually called **points**, though they can be any mathematical objects (including vectors).

It is left as an exercise to confirm that, if X is a metric space and τ is the collection of all open sets in X, then τ is a topology on X; see Exercise 7.17. For this reason, the elements of the collection τ are called the **open sets** of the topology, whether or not X is a metric space.

DEFINITION 7.5.5 Let X be a topological space and $A \subseteq X$. Then A is **closed** if and only if its complement $X \setminus A$ is open.

Note that many subsets of metric spaces, and of topological spaces more generally, are neither open nor closed.

For completeness, we include these last two definitions.

DEFINITION 7.5.6 A **neighbourhood** of the point $\mathbf{x}$ in the topological space X is an open set containing $\mathbf{x}$.

DEFINITION 7.5.7 Let X be a metric space. Then $A \subseteq X$ is **compact** if and only if A is both closed and bounded.

7.6 Supporting and separating hyperplane theorems

These theorems will be required for the proof of the second welfare theorem (Theorem 12.6.2). They can also be used to provide an alternative proof of Jensen's inequality (Theorem 13.10.1). We introduced the concept of a hyperplane in Section 7.4. In this section, we look at it in more detail. Berger (1993, Section 5.2.5) and Rockafellar (1970, Section 11) provide a thorough development of these theorems.

Note that any hyperplane divides $\mathbb{R}^n$ into two closed **half-spaces**,

$$\{\mathbf{x} \in \mathbb{R}^n : \mathbf{p}^\top \mathbf{x} \le \mathbf{p}^\top \mathbf{x}^*\} \tag{7.33}$$

and

$$\{\mathbf{x} \in \mathbb{R}^n : \mathbf{p}^\top \mathbf{x} \ge \mathbf{p}^\top \mathbf{x}^*\} \tag{7.34}$$

The intersection of these two closed half-spaces is the hyperplane itself.

DEFINITION 7.6.1 Let X and Y be subsets of $\mathbb{R}^n$ and let $\mathbf{z}^* \in \mathbb{R}^n$. Then the affine hyperplane through $\mathbf{z}^*$ with normal $\mathbf{p}$ is:

(a) a **supporting hyperplane** to X if $\mathbf{z}^*$ is a boundary point of X and $\mathbf{p}^\top \mathbf{x} \ge \mathbf{p}^\top \mathbf{z}^*$ for all $\mathbf{x} \in X$; and

(b) a **separating hyperplane** for X and Y if $\mathbf{p}^\top\mathbf{x}\geq\mathbf{p}^\top\mathbf{z}^*$ for all $\mathbf{x}\in X$ and $\mathbf{p}^\top\mathbf{y}\leq\mathbf{p}^\top\mathbf{z}^*$ for all $\mathbf{y}\in Y$.

In other words, a set lies entirely in one of the closed half-spaces associated with its supporting hyperplane, while two sets each lie entirely in the respective closed half-spaces associated with their separating hyperplane.

The idea behind the supporting hyperplane theorem is quite intuitive: if we take any boundary point of a *convex* set, then we can find a supporting hyperplane through that point. The supporting hyperplane can be thought of as being tangent to the set that it supports.

THEOREM 7.6.1 (SUPPORTING HYPERPLANE THEOREM). *If Z is a convex subset of $\mathbb{R}^n$ and $\mathbf{z}^*\in Z$, $\mathbf{z}^*\notin\text{int }Z$, then there exists $\mathbf{p}^*\neq\mathbf{0}$ in $\mathbb{R}^n$ such that $\mathbf{p}^{*\top}\mathbf{z}^*\leq\mathbf{p}^{*\top}\mathbf{z}$ for all $\mathbf{z}\in Z$, or Z is contained in one of the closed half-spaces associated with the hyperplane through $\mathbf{z}^*$ with normal $\mathbf{p}^*$.*

Proof: The proof of this theorem is beyond the scope of this book, but can be found in Berger (1993, p. 341). $\qquad\qquad\Box$

THEOREM 7.6.2 (SEPARATING HYPERPLANE THEOREM). *If X and Y are disjoint convex subsets of $\mathbb{R}^n$, then there exists a vector $\mathbf{p}\in\mathbb{R}^n$ such that*

$$\mathbf{p}^\top\mathbf{x}\geq\mathbf{p}^\top\mathbf{y}\quad\forall\,\mathbf{x}\in X,\,\mathbf{y}\in Y \tag{7.35}$$

If we define $c\equiv\sup\{\mathbf{p}^\top\mathbf{y}:\mathbf{y}\in Y\}$, then the hyperplane $\{\mathbf{x}\in\mathbb{R}^n:\mathbf{p}^\top\mathbf{x}=c\}$ separates X and Y.

Proof: The proof of this theorem is beyond the scope of this book, but can be found in Berger (1993, p. 342). $\qquad\qquad\Box$

Some writers call the penultimate theorem (Theorem 7.6.1) the separating hyperplane theorem.

7.7 Visualizing functions of several variables

In Section 6.2, we introduced terminology which can be used to describe real-valued functions, and also correspondences, of n variables. There are at least three very useful ways of visualizing such functions and correspondences, namely:

1. as a graph, which is an n-dimensional surface in $\mathbb{R}^{n+1}$ (see Definition 7.7.1);
2. as a collection of indifference curves, to use economic parlance (see Definition 7.7.2); and
3. as a collection of restrictions to lines in $\mathbb{R}^n$ (see Section 9.5).

DEFINITION 7.7.1 Every function or correspondence $f:X\to Y$, $X\subseteq\mathbb{R}^n$, $Y\subseteq\mathbb{R}$, has a **graph**, defined as

$$G_f=\{(\mathbf{x},y)\in X\times Y:y=f(\mathbf{x})\} \tag{7.36}$$

in the case of a function; and as

$$G_f=\{(\mathbf{x},y)\in X\times Y:y\in f(\mathbf{x})\} \tag{7.37}$$

in the case of a correspondence.

Most readers will probably be familiar with the concept of a graph of a single-valued function of one or two variables, something easily illustrated on a two-dimensional page. Definition 7.7.1 merely extends this idea to n dimensions. The very difficulty of visualizing such graphs in n dimensions suggests the alternative approaches listed above.

The concept of the graph of a correspondence will be important not only in relation to the continuity of correspondences, but also when we deal with fixed points of correspondences in Section 12.5.4.

DEFINITION 7.7.2 Consider the real-valued function $f : X \to \mathbb{R}$.

(a) The **upper contour sets** of f are the sets $\{\mathbf{x} \in X : f(\mathbf{x}) \geq \alpha\}$ $(\alpha \in \mathbb{R})$.
(b) The **level sets** or **indifference curves** of f are the sets $\{\mathbf{x} \in X : f(\mathbf{x}) = \alpha\}$ $(\alpha \in \mathbb{R})$.
(c) The **lower contour sets** of f are the sets $\{\mathbf{x} \in X : f(\mathbf{x}) \leq \alpha\}$ $(\alpha \in \mathbb{R})$.

In Definition 7.7.2, X does not have to be a vector space. If $X = \mathbb{R}^2$, however, a selection of indifference curves can be plotted on a diagram usually called an **indifference map**. One example is the isoquant map encountered in Section 1.2.3. Another example familiar from everyday life is the isobar map seen in many weather forecasts. Numerous further examples will be encountered in later chapters, for example, Figures 9.2, 12.2 and 12.4.

The function $f : X \to \mathbb{R}$ defines an equivalence relation on X by $\mathbf{x} R \mathbf{y} \Leftrightarrow f(\mathbf{x}) = f(\mathbf{y})$. The level sets of f are the equivalence classes for this relation.

Furthermore, the level sets of f are invariant under increasing transformations of f, or changes of scale, even nonlinear transformations. In other words, if $g : \mathbb{R} \to \mathbb{R}$ is a strictly (monotonically) increasing function, then f and $g \circ f$ have the same level sets.

These concepts will be of central importance when we deal with utility functions from Section 12.2 onwards and with expected-utility functions from Section 16.4 onwards.

7.8 Limits and continuity

DEFINITION 7.8.1 The real-valued function $f : X \to Y$ $(X \subseteq \mathbb{R}^n, Y \subseteq \mathbb{R})$ approaches the **limit** y^* as $\mathbf{x} \to \mathbf{x}^*$ if and only if, for all $\epsilon > 0$, there exists $\delta > 0$ such that $\|\mathbf{x} - \mathbf{x}^*\| < \delta \Rightarrow |f(\mathbf{x}) - y^*| < \epsilon$.

In other words, the values of the function at vectors in the domain close to $\mathbf{x}^*$ come arbitrarily close to the number y^*.

This is usually denoted as

$$\lim_{\mathbf{x} \to \mathbf{x}^*} f(\mathbf{x}) = y^* \tag{7.38}$$

DEFINITION 7.8.2 The function $f : X \to Y$ $(X \subseteq \mathbb{R}^n, Y \subseteq \mathbb{R})$ is **continuous at** $\mathbf{x}^*$ if and only if, for all $\epsilon > 0$, there exists $\delta > 0$ such that $\|\mathbf{x} - \mathbf{x}^*\| < \delta \Rightarrow |f(\mathbf{x}) - f(\mathbf{x}^*)| < \epsilon$.

This definition just says that f is continuous at $\mathbf{x}^*$ provided that

$$\lim_{\mathbf{x} \to \mathbf{x}^*} f(\mathbf{x}) = f(\mathbf{x}^*) \tag{7.39}$$

DEFINITION 7.8.3 The real-valued function $f : X \to Y$ $(X \subseteq \mathbb{R}^n, Y \subseteq \mathbb{R})$ is **continuous** if and only if it is continuous at every point of its domain.

We will say that a *vector-valued* or *matrix-valued* function is continuous if and only if each of its *real-valued* component functions is continuous.

Most commonly encountered functions are continuous. An example of a function that is not continuous (or is **discontinuous**) at a single point is

$$f:\mathbb{R} \to \mathbb{R}: x \mapsto \begin{cases} 1/x & \text{if } x \neq 0 \\ 0 & \text{if } x = 0 \end{cases} \tag{7.40}$$

Since $1/0$ is not defined (in fact, it is infinite), $f(0)$ must be defined in some other arbitrary way. Since $\lim_{x \to 0} f(x)$ does not exist, no such arbitrary definition can make f into a continuous function.

An example of a function with a natural singularity that can be made continuous is

$$g:\mathbb{R} \to \mathbb{R}: x \mapsto \begin{cases} \dfrac{\sin x}{x} & \text{if } x \neq 0 \\ 1 & \text{if } x = 0 \end{cases} \tag{7.41}$$

As $x \to 0$, $\sin x/x \to 0/0$, so we need to define $g(0)$ in some other way. In fact, l'Hôpital's rule tells us that

$$\lim_{x \to 0} g(x) = \lim_{x \to 0} \frac{\cos x}{1} = 1 \tag{7.42}$$

Making the value of g explicitly equal to 1 at the singularity makes the function continuous.

A more practical example of a discontinuous function is the function relating the size of university grants to parental income in certain jurisdictions. Sometimes, a full grant is payable if parental income falls at or below a threshold level and nothing is payable if parental income exceeds the threshold level. Thus the function is discontinuous at the threshold level of income. Governments are slowly becoming aware of the inequity of such rules and gradually eliminating them.

We note, without proof, that pointwise sums, differences, products and inverses of continuous functions are also continuous. Similarly, the pointwise reciprocal of a continuous function is continuous provided that the function nowhere takes the value zero; in the latter case, the reciprocal is ill defined.

If, given a continuous function f defined on a vector space and a scalar λ, we define the function λf by $(\lambda f)(\mathbf{x}) \equiv \lambda(f(\mathbf{x}))$, then λf will be a continuous function. Since the sum of two (or more) continuous functions is continuous and the product in this sense of a scalar and a continuous function is continuous, it follows that the set of all continuous functions on a given domain is a vector space, often called a **function space**. This vector space will typically not be of finite dimension.

Note that a quadratic form is continuous. Thus, if a quadratic form is positive at some vector $\mathbf{x}$, then it is positive within some neighbourhood of $\mathbf{x}$. Similarly, if a matrix-valued function has a positive definite value at $\mathbf{x}$, then it has positive definite values within some neighbourhood of $\mathbf{x}$.

DEFINITION 7.8.4 The real-valued function $f: X \to Y$ ($X \subseteq \mathbb{R}^n$, $Y \subseteq \mathbb{R}$) is **uniformly continuous** if and only if, for all $\epsilon > 0$, there exists $\delta > 0$ such that, for all $\mathbf{x}, \mathbf{x}' \in X$, $\|\mathbf{x} - \mathbf{x}'\| < \delta \implies |f(\mathbf{x}) - f(\mathbf{x}')| < \epsilon$.

Continuity itself is a local (or, more precisely, pointwise) property of a function, i.e. a function f is continuous, or not, at a particular point. When we speak of a function being continuous on a subset of its domain, we mean only that it is continuous at each point of that subset. The δ in the definition of continuity at $\mathbf{x}^*$ (Definition 7.8.2) will generally depend both on the ϵ chosen and also on the point $\mathbf{x}^*$.

In contrast, uniform continuity is a global property of f, in the sense that Definition 7.8.4 refers to pairs of points, whereas Definition 7.8.2 refers to individual points. In the definition of uniform continuity, a single value of δ must work for all values of $\mathbf{x}$ and $\mathbf{x}'$.

Many familiar functions defined on $\mathbb{R}$ or $\mathbb{R}_{++}$ are continuous but not uniformly continuous; for example, on $\mathbb{R}_{++}$, $x \mapsto 1/x$; and, on $\mathbb{R}$, $x \mapsto e^x$ and $x \mapsto x^2$.

The trigonometric tangent function, which is undefined at $(1+2n)\pi/2$ for all integers n, is continuous but not uniformly continuous on the open interval $(-\pi/2, \pi/2)$, throughout which it is well defined.

The confirmation of all of these statements is left as an exercise; see Exercise 7.20.

THEOREM 7.8.1 (HEINE–CANTOR THEOREM). *If the function f is continuous on the closed hyperrectangle $X = \prod_{i=1}^{n}[a_i, b_i] \subset \mathbb{R}^n$, then it is uniformly continuous on X.*[4]

Proof: Choose $\epsilon > 0$.

We claim that X can be divided into sub-hyperrectangles such that, for $\mathbf{x}$ and $\mathbf{x}'$ in the same sub-hyperrectangle, $|f(\mathbf{x}) - f(\mathbf{x}')| < \epsilon$. This claim can be proved by contradiction. Suppose not.

Halve each of the intervals $[a_i, b_i]$, so dividing X into 2^n sub-hyperrectangles. By our hypothesis, we can pick a sub-hyperrectangle in which there exist $\mathbf{x}$ and $\mathbf{x}'$ such that $|f(\mathbf{x}) - f(\mathbf{x}')| \geq \epsilon$. Repeat the halving procedure for this sub-hyperrectangle.

As this process continues, the vertices of this sequence of sub-hyperrectangles converge to a common limit, say the vector $\mathbf{x}^* \in \mathbb{R}^n$. Since f is continuous on X, it is continuous at $\mathbf{x}^*$, and thus there exists $\delta > 0$ such that $\|\mathbf{x} - \mathbf{x}^*\| < \delta \Rightarrow |f(\mathbf{x}) - f(\mathbf{x}^*)| < \epsilon/2$. Hence, by the triangular inequality, for $\mathbf{x}, \mathbf{x}' \in B_\delta(\mathbf{x}^*)$, $|f(\mathbf{x}) - f(\mathbf{x}')| < \epsilon$.

The repeated halving procedure will eventually reach a stage at which the sub-hyperrectangle lies entirely within $B_\delta(\mathbf{x}^*)$. This will certainly happen after k halvings if k is large enough that the length of the diagonal of the sub-hyperrectangle is less than δ or

$$\sqrt{\sum_{i=1}^{n} \left(\frac{b_i - a_i}{2^k}\right)^2} < \delta \tag{7.43}$$

This inequality can be solved for the equivalent condition on k, which is

$$k > \frac{\ln\sqrt{\sum_{i=1}^{n}(b_i - a_i)^2} - \ln\delta}{\ln 2} \tag{7.44}$$

We have now established the required contradiction since, by our hypothesis, there exist $\mathbf{x}$ and $\mathbf{x}'$ in this kth sub-hyperrectangle such that $|f(\mathbf{x}) - f(\mathbf{x}')| \geq \epsilon$ but, by definition of δ, for all $\mathbf{x}, \mathbf{x}'$ in the kth sub-hyperrectangle, $|f(\mathbf{x}) - f(\mathbf{x}')| < \epsilon$.

A similar argument can now be used to prove uniform continuity.

Choose $\epsilon > 0$ once again.

Divide X into sub-hyperrectangles such that, for $\mathbf{x}$ and $\mathbf{x}'$ in the same sub-hyperrectangle, $|f(\mathbf{x}) - f(\mathbf{x}')| < \epsilon/2$ and let δ be the length of the longest edge of any of these sub-hyperrectangles.

Any vectors $\mathbf{x}$ and $\mathbf{x}'$ within distance δ of each other must then lie either in the same sub-hyperrectangle or in two sub-hyperrectangles with a common vertex, say $\mathbf{x}^*$. In the former case, we have $|f(\mathbf{x}) - f(\mathbf{x}')| < \epsilon/2$ and thus $|f(\mathbf{x}) - f(\mathbf{x}')| < \epsilon$. In the latter case, we have $|f(\mathbf{x}) - f(\mathbf{x}^*)| < \epsilon/2$ and $|f(\mathbf{x}') - f(\mathbf{x}^*)| < \epsilon/2$ and thus by the triangular inequality $|f(\mathbf{x}) - f(\mathbf{x}')| < \epsilon$.

This completes the proof. □

Assuming uniform continuity will enable us to prove Leibniz's integral rule (Theorem 9.7.4) without calling on Fubini's theorem (Theorem 9.7.1), the proof of which is beyond the scope of this book.

The notion of continuity of a function described above is probably familiar to most readers from earlier courses. Its extension to the notion of continuity of a correspondence, however, while fundamental to consumer theory, general equilibrium theory and much of micro-economics, is probably not. Recall that, especially in economics, the term *correspondence* is used in preference to the oxymoron *multivalued function*; see Definition 0.0.3.

DEFINITION 7.8.5

(a) The correspondence $f: X \to Y$ ($X \subseteq \mathbb{R}^n$, $Y \subseteq \mathbb{R}$) is **upper hemi-continuous**[5] (u.h.c.) at $\mathbf{x}^*$ if and only if, for every open set N *containing* the set $f(\mathbf{x}^*)$, there exists $\delta > 0$ such that $\|\mathbf{x} - \mathbf{x}^*\| < \delta \Rightarrow f(\mathbf{x}) \subseteq N$.

(b) The correspondence $f: X \to Y$ ($X \subseteq \mathbb{R}^n$, $Y \subseteq \mathbb{R}$) is **lower hemi-continuous** (l.h.c.) at $\mathbf{x}^*$ if and only if, for every open set N *intersecting* the set $f(\mathbf{x}^*)$, there exists $\delta > 0$ such that $\|\mathbf{x} - \mathbf{x}^*\| < \delta \Rightarrow f(\mathbf{x})$ intersects N.

(c) The correspondence $f: X \to Y$ ($X \subseteq \mathbb{R}^n$, $Y \subseteq \mathbb{R}$) is **continuous** (at $\mathbf{x}^*$) if and only if it is both upper hemi-continuous and lower hemi-continuous (at $\mathbf{x}^*$).

Upper hemi-continuity basically means that the graph of the correspondence is a closed and connected set, or that the set $f(\mathbf{x})$ does not suddenly become much larger (explode) when there is a small change in $\mathbf{x}$. Similarly, lower hemi-continuity means that the graph does not suddenly implode.

These concepts are illustrated in Figure 7.4, which shows the graph of a correspondence $f: X \to Y$. At x_1, x_3 and x_5, f is both upper and lower hemi-continuous. At x_2, f is upper hemi-continuous but not lower hemi-continuous. At x_4, f is lower hemi-continuous but not upper hemi-continuous.

For single-valued correspondences (i.e. functions), lower hemi-continuity, upper hemi-continuity and continuity are equivalent.

We will encounter the continuity of correspondences again in Theorems 10.5.4 and 12.5.6.[6]

7.9 Fundamental theorem of calculus

We conclude this chapter with an important result from univariate calculus that will be used frequently in what follows, but which does not have a direct multivariate analogue. This theorem sets out the precise rules for cancelling integration and differentiation operations.

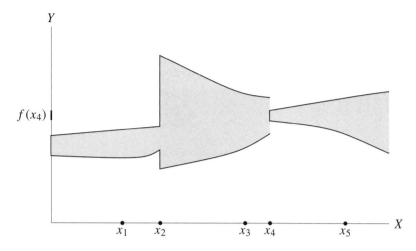

Figure 7.4 Upper and lower hemi-continuity of correspondences

Note that the differentiation in the first part of this theorem is with respect to the upper limit of integration.

THEOREM 7.9.1 (FUNDAMENTAL THEOREM OF CALCULUS). *The differentiation and integration operators are inverse operators in the following senses:*

(a)

$$\frac{d}{db}\int_a^b f(x)\,dx = f(b) \tag{7.45}$$

(b)

$$\int_a^b g'(x)\,dx = g(b) - g(a) \tag{7.46}$$

Proof: The proofs of the two parts of this theorem are illustrated graphically in Figures 7.5 and 7.6 respectively.

(a) The shaded area A in Figure 7.5, which shows the graph of the *function* f, represents the integral in (7.45), since integration can be used to compute the area under a curve. The more lightly shaded area $\Delta A \approx f(b) \times \Delta b$ represents the increase in this area when the upper limit of integration goes from b to $b + \Delta b$. The derivative in (7.45) is

$$\lim_{\Delta b \to 0}\frac{\Delta A}{\Delta b} = \lim_{\Delta b \to 0}\frac{f(b)\Delta b}{\Delta b} = f(b) \tag{7.47}$$

(b) Similarly, the shaded area in Figure 7.6, which shows the graph of the *derivative* g', represents the integral in (7.46).

For a full proof of both parts of this theorem, see Binmore (1982, pp. 126–8).   □

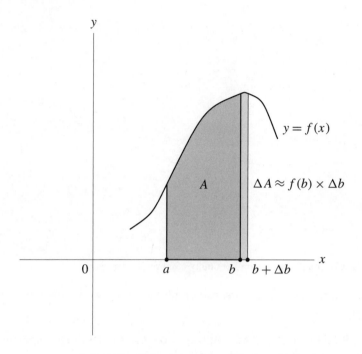

Figure 7.5 Motivation for the fundamental theorem of calculus, part (a)

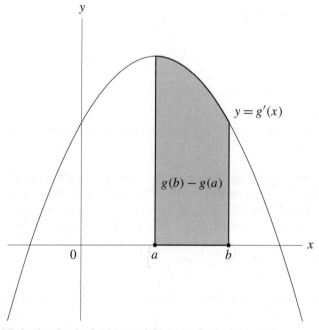

Figure 7.6 Motivation for the fundamental theorem of calculus, part (b)

EXERCISES

7.1 Show that the following statements are equivalent:

(a) if $k_1, k_2, \ldots, k_r$ are scalars with $\sum_{i=1}^{r} k_i = 1$ and $\mathbf{v}_1, \mathbf{v}_2, \ldots, \mathbf{v}_r \in X$, then $\sum_{i=1}^{r} k_i \mathbf{v}_i \in X$; and

(b) if λ is a scalar and $\mathbf{x}$ and $\mathbf{x}'$ are vectors in X, then $\lambda \mathbf{x} + (1 - \lambda)\mathbf{x}' \in X$.

7.2 Let A be an affine set in the vector space X and let $\mathbf{a}$ be any vector in A. Prove that the set

$$A - \{\mathbf{a}\} \equiv \{\mathbf{x} - \mathbf{a} : \mathbf{x} \in A\}$$

is a vector subspace of X.

7.3 Show that the following statements are equivalent:

(a) if $k_1, k_2, \ldots, k_r$ are non-negative scalars with $\sum_{i=1}^{r} k_i = 1$ and $\mathbf{v}_1, \mathbf{v}_2, \ldots, \mathbf{v}_r \in X$, then $\sum_{i=1}^{r} k_i \mathbf{v}_i \in X$; and

(b) if $\lambda \in [0, 1]$ and $\mathbf{x}$ and $\mathbf{x}'$ are vectors in X, then $\lambda \mathbf{x} + (1 - \lambda)\mathbf{x}' \in X$.

7.4 Prove Theorem 7.3.1.

7.5 Let $X_1, X_2, \ldots, X_k$ be convex subsets of $\mathbb{R}^n$. Prove that the intersection of these k sets, $X_1 \cap X_2 \cap \cdots \cap X_k$, is also a convex set.

7.6 Show that the convex hull of any set is itself a convex set.

7.7 Show that the convex hull of the set X is the smallest convex set containing X.

7.8 Sketch and find the equations of each of the following three hyperplanes in $\mathbb{R}^2$:

(a) the hyperplane through $(1, 0)$ with normal $(1, 1)$;
(b) the hyperplane through $(0, 1)$ with normal $(2, 2)$; and
(c) the hyperplane through $(0, 1)$ with normal $(-1, -1)$.

7.9 Show that a hyperplane, as defined in Definition 7.4.1, is both an affine set and a convex set.

7.10 Show that the following are the same:

(a) the hyperparallelepiped with m edges $\mathbf{x}_1, \mathbf{x}_2, \ldots, \mathbf{x}_m$; and
(b) the simplex with 2^m vertices of the form $\sum_{i \in I} \mathbf{x}_i$ where I is any subset of $\{1, 2, \ldots, m\}$ and a sum of vectors over the empty set is defined to be the zero vector.

7.11 Show that a hyperparallelepiped is a convex set.

7.12 Show that the image of a hyperparallelepiped under a linear transformation is also a hyperparallelepiped.

7.13 Prove Corollaries 7.4.2–7.4.5.

7.14 Give an example of a subset of $\mathbb{R}$ that is neither open nor closed.

7.15 What is the largest open hypercube that can be completely enclosed in the open ball of radius ϵ centred at the origin in $\mathbb{R}^n$? What are the coordinates of its vertices? What are the lengths of its edges? What is its volume?

7.16 Prove that the d defined in (7.26) has all the properties of a metric as set out in Definition 7.5.1.

7.17 Let X be a metric space and let τ be the collection of all open sets in X. Show that τ is a topology on X.

7.18 Let $f: X \to Y$ ($X \subseteq \mathbb{R}^n$, $Y \subseteq \mathbb{R}$) and let $\mathbf{x}^* \in X$. Show that the following definitions of continuity of f at $\mathbf{x}^*$ are equivalent:

(a) for all $\epsilon > 0$, $\exists\, \delta > 0$ s.t. $\|\mathbf{x} - \mathbf{x}^*\| < \delta \Rightarrow |f(\mathbf{x}) - f(\mathbf{x}^*)| < \epsilon$;
(b) for each open sphere $B_\epsilon(f(\mathbf{x}^*))$ centred on $f(\mathbf{x}^*)$ there exists an open sphere $B_\delta(\mathbf{x}^*)$ centred on $\mathbf{x}^*$ such that $f(B_\delta(\mathbf{x}^*)) \subseteq B_\epsilon(f(\mathbf{x}^*))$; and
(c) $f^{-1}(B)$ is open (in X) whenever B is open (in Y).

(Hint: see p. xxi for the relevant definitions.)

7.19 Let $f: X \to \mathbb{R}$ and $g: X \to \mathbb{R}$ be continuous functions and let $\lambda \in \mathbb{R}$.
  Define $h: X \to \mathbb{R}$ by $h(x) = f(x) + g(x)$ for all $x \in X$ and $k: X \to \mathbb{R}$ by $k(x) = \lambda f(x)$ for all $x \in X$.
  Show that h and k are continuous functions and, hence, that the set of all continuous functions defined on X is a real vector space.

7.20 Prove that the following functions are not uniformly continuous:

(a) $f: \mathbb{R} \to \mathbb{R}: x \mapsto e^x$;
(b) $g: \mathbb{R}_{++} \to \mathbb{R}_{++}: x \mapsto 1/x$;
(c) $h: \mathbb{R} \to \mathbb{R}: x \mapsto x^2$; and
(d) $k: (-\pi/2, \pi/2) \to \mathbb{R}: x \mapsto \tan x$.

7.21 Consider the set $\mathbb{C}$ of all complex numbers.

(a) Show that $\mathbb{C}$ becomes a real vector space if the operations of addition of complex numbers and multiplication of a complex number by a real number are defined in the natural way.
(b) What is the dimension of this real vector space?
(c) Give two examples of bases for $\mathbb{C}$.
(d) Is multiplication of complex numbers a scalar product on this real vector space?
(e) Is $\mathbb{C}$ a metric space if the distance between two complex numbers is defined to be the modulus of their difference, i.e. $d(y, z) \equiv |y - z|$?
(f) Show that the conjugate operator on $\mathbb{C}$ is a linear operator.
(g) Find the matrix of the conjugate operator with respect to each of the bases that you gave in your answer to part (c) of this exercise.

8 Difference equations

8.1 Introduction

The behaviour of economic and financial variables and systems over time, and in particular how they respond to policy changes or other national or international shocks, is a matter of major concern. Consideration of the evolution of individual variables, such as exchange rates, interest rates, unemployment, national income and other macroeconomic magnitudes, and of the processes of adjustment of systems, such as markets or entire economies, leads us to the subject of dynamics, in which time is made explicit.

The mathematical treatment of dynamics is based principally on the use of difference equations or differential equations – as well as many of the mathematical tools already presented – depending on whether time is modelled as being discrete or continuous. Empirical analysis is limited to a finite number of observations, and thus to discrete time. Theoretical analysis can – and, at an advanced level, does – deal with continuous time. Economic variables are measured and reported only for discrete intervals: for example, monthly (like the Consumer Price Index) or quarterly (like some unemployment and trade series) or annually (like the major macroeconomic variables in the national accounts of many economies). In finance, data are often available (almost) continuously, yet they are still frequently presented for short intervals of time such as tick-by-tick or daily periods. Thus discrete-time models, and hence difference equations, are of more relevance to empirical data analysis, while continuous-time models and differential equations lend themselves more to theoretical analysis.

This chapter provides a grounding in the mathematics of difference equations. Section 8.2 confines itself to definitions, classifications and examples. The next two sections discuss linear difference equations in some detail, Section 8.3 focusing on the first-order case and Section 8.4 on higher orders, but especially the linear, autonomous, second-order difference equation. Section 8.5 examines systems of linear difference equations, starting with general linear systems and then concentrating on the linear, autonomous, first-order system. A few illustrative examples are provided as the material is developed, but the main detailed application is left until Chapter 14.

8.2 Definitions and classifications

The idea of a time series was first mentioned in the econometric example in Section 1.2.1. There, the series in question were stochastic. In this chapter, we consider only non-stochastic time series, leaving stochastic time-series issues aside until Section 14.4. The present section begins with a more formal definition of a (non-stochastic) time series. It then gives definitions of some useful related operators, before explaining the classification of difference equations and their solutions.

8.2.1   Non-stochastic time series

DEFINITION 8.2.1 A non-stochastic **time series** is a collection of values of a variable indexed by the time periods with which each value is associated, and denoted $\{y_t\}$.

The set of all time series of the form $\{y_t : t \in T\}$ is a vector space. The index set, T, is typically the set of natural numbers, $\mathbb{N}$, or the set of integers, $\mathbb{Z}$. Unless stated otherwise, we assume $t = 1, 2, \ldots$.

The use of this subscript notation to denote the time dependence of the values of certain variables will be recalled from Section 1.2.1, and particularly from (1.6) and (1.7).

8.2.2   Lag and difference operators

Let y_t be the value of some variable of interest in time period t. Then prior, or lagged, values of y_t can be handled conveniently by means of the following lag operator.

DEFINITION 8.2.2 The **lag operator**, L, is a linear operator on the vector space of discrete time series defined by

$$Ly_t \equiv y_{t-1} \tag{8.1}$$

To see that L is a linear operator, note that $L(x_t + y_t) = Lx_t + Ly_t$ and $L(cy_t) = cLy_t$, where c is a scalar.

The lag operator satisfies the following further properties:

1. $L^0 y_t = y_t$ for all t or L^0 is the identity operator, which – for reasons that will become apparent in Section 8.4.1 – will be denoted 1.
2. If $y_t = y^*$ for all t, i.e. $\{y_t\}$ is a constant series, then $Ly_t = Ly^* = 1y^* = y^* = y_t$ for all t.
3. $L^i y_t \equiv y_{t-i}$.
4. $(L^i + L^j)y_t = L^i y_t + L^j y_t = y_{t-i} + y_{t-j}$.
5. $(L^i L^j)y_t = L^i(L^j y_t) = L^i y_{t-j} = y_{t-i-j} = L^{i+j} y_t$.
6. $L^{-i} y_t = y_{t+i}$.
7. If $C(x) = c_0 + c_1 x + c_2 x^2 + \cdots + c_p x^p$ is a polynomial of degree p in x, then we may need to refer to the corresponding linear operator $C(L) = \sum_{i=0}^{p} c_i L^i$, a polynomial in the lag operator or **lag polynomial**.[1] For the constant series $\{y_t\}$ with $y_t = y^*$ for all t, $C(L)y_t = c_0 y_t + c_1 y_{t-1} + c_2 y_{t-2} + \cdots + c_p y_{t-p} = (c_0 + c_1 + c_2 + \cdots + c_p)y^* = C(1)y^*$.
8. $(1 + cL + c^2 L^2 + c^3 L^3 + \cdots)y_t = (1 - cL)^{-1} y_t$, also denoted $y_t/(1 - cL)$, for $|c| < 1$.
9. $(1 + (cL)^{-1} + (c^2 L^2)^{-1} + (c^3 L^3)^{-1} + \cdots)y_t = -cLy_t/(1 - cL) = -cy_{t-1}/(1 - cL)$, for $|c| > 1$.

Given the lag operator, the change or difference in the value of y_t between adjacent time periods $t - 1$ and t may be represented by a simple, related operator called the **difference operator**.

DEFINITION 8.2.3 The **first-difference operator**, denoted by Δ, is defined by

$$\Delta y_t \equiv (1 - L)y_t = y_t - Ly_t = y_t - y_{t-1} \tag{8.2}$$

We can generalize the first-difference operator to the **p-period difference operator** $\Delta_p \equiv 1 - L^p$, where p is an integer greater than one, with

$$\Delta_p y_t = (1 - L^p) y_t = y_t - L^p y_t = y_t - y_{t-p} \tag{8.3}$$

which is the change in the variable over p periods. Hence, Δ generates one-period differences (the subscript $p = 1$ being implicit), while Δ_2 yields two-period differences:

$$\Delta_2 y_t = (1 - L^2) y_t = y_t - L^2 y_t = y_t - y_{t-2} \tag{8.4}$$

and so on for higher values of p. For example, the annual inflation rate published in some countries is based on the 12-period (12-month) difference of the monthly Consumer Price Index.

It is important to distinguish between the very different entities Δ_p and Δ^p. For example, compare

$$\Delta^2 y_t = \Delta \Delta y_t = (1 - L)(1 - L) y_t = (1 - 2L + L^2) y_t$$
$$= y_t - 2L y_t + L^2 y_t = y_t - 2y_{t-1} + y_{t-2} \tag{8.5}$$

with (8.4).

8.2.3 Difference equations

A **difference equation** is any equation that involves terms of the form $\Delta_p y_t = y_t - y_{t-p}$, although these p-period changes do not have to appear explicitly. For example, suppose that y_t denotes the money supply for time period t, and the monetary authority has a policy of increasing this each period by a fixed proportion, θ. Then the change in the money supply from one period to the next is given by the simple difference equation

$$\Delta y_t = y_t - y_{t-1} = \theta y_{t-1}, \quad t = 2, 3, \ldots, \theta > 0 \tag{8.6}$$

Adding y_{t-1} to both sides of (8.6) gives

$$y_t = (\theta + 1) y_{t-1} \equiv \phi y_{t-1}, \quad t = 2, 3, \ldots, \phi \equiv \theta + 1 > 1 \tag{8.7}$$

Equations (8.6) and (8.7) are alternative ways of writing this particular difference equation, the first containing Δy_t explicitly and the second containing the first difference implicitly.

As a second illustration, consider from microeconomics a simplified form of the well-known cobweb model of a competitive market:

$$\text{demand:} \quad Q_t = \alpha + \beta P_t \tag{8.8}$$
$$\text{supply:} \quad Q_t = \gamma + \delta P_{t-1} \tag{8.9}$$

where Q_t and P_t are endogenous quantity and price variables for time period t, and α, β, γ and δ are parameters of the demand and supply functions. If demand equals supply, we may equate the right-hand sides of (8.8) and (8.9) to give

$$\alpha + \beta P_t = \gamma + \delta P_{t-1} \tag{8.10}$$

A little rearrangement then gives

$$P_t = \mu + \phi P_{t-1} \tag{8.11}$$

where $\mu = (\gamma - \alpha)/\beta$ and $\phi = \delta/\beta$. This is a difference equation involving price, with the price change being implicit. To make the price change or first difference explicit, we simply subtract P_{t-1} from both sides of (8.11). Thus

$$P_t - P_{t-1} = \mu + (\phi - 1)P_{t-1} \tag{8.12}$$

or

$$\Delta P_t = \mu + \theta P_{t-1} \tag{8.13}$$

where $\theta \equiv \phi - 1$.

8.2.4   Classifications and solutions

Difference equations are conventionally classified according to their functional form, whether or not they depend on time explicitly and, recalling that they are equations containing $\Delta_p y_t = y_t - y_{t-p}$, the highest value of p they involve. A further important distinction concerns whether or not they contain a non-zero constant or intercept term.

A difference equation is said to be **linear** if it is linear in y_{t-i}, $i = 0, 1, \ldots, p$ $(p \geq 1)$; otherwise it is **nonlinear**. If a difference equation does not contain time, t, explicitly as a variable (as distinct from a subscript), then it is described as **autonomous**; otherwise it is said to be **non-autonomous**. The highest value of p that a difference equation involves defines its **order**. Finally, if a difference equation does not contain a constant term (i.e. a term independent of t), it is called **homogeneous**; otherwise, when there is a non-zero constant in the equation, it is said to be **non-homogeneous**.

For example, both (8.7) and (8.11) are linear, autonomous difference equations of order one. However, (8.7) is homogeneous, whereas (8.11) in non-homogeneous by dint of the presence of the constant μ, assuming $\mu \neq 0$, of course. Similarly,

$$y_t = 0.5 y_{t-2} \tag{8.14}$$

and

$$y_t = \phi_0 + \phi_1 y_{t-1} + \phi_2 y_{t-2} \tag{8.15}$$

are both linear, autonomous, second-order difference equations, but (8.14) is homogeneous and (8.15) is non-homogeneous, if $\phi_0 \neq 0$. The equation

$$y_t = \frac{1}{t} + e^t y_{t-1} \tag{8.16}$$

is a linear, non-autonomous, first-order, homogeneous difference equation, and the equations

$$y_t = 2 \log y_{t-1} + t \tag{8.17}$$

and

$$y_t = \phi y_{t-1} y_{t-2} \tag{8.18}$$

are examples of nonlinear, homogeneous difference equations, though (8.17) is non-autonomous of order one, while (8.18) is autonomous of order two.

The ideas on classification of difference equations are simple; nonetheless, a few further examples appear in the exercises; see Exercise 8.1. By contrast, the solution of difference equations is not usually a straightforward matter, with nonlinear difference equations being in general insoluble analytically. Linear difference equations of arbitrary order are soluble. However, their solutions are not unique, unless information is available on certain initial values, in particular the initial value, y_1, in the case of a first-order difference equation.

By way of illustration, consider (8.7), which describes a money supply adjustment process as a linear, autonomous, first-order, homogeneous difference equation. Substituting for y_{t-1}, we have that

$$y_t = \phi y_{t-1} = \phi\phi y_{t-2} = \phi^2 y_{t-2} \tag{8.19}$$

Substituting similarly for y_{t-2} gives

$$y_t = \phi^2 y_{t-2} = \phi^2 \phi y_{t-3} = \phi^3 y_{t-3} \tag{8.20}$$

Continuing this process of substitution for the lagged value on the right-hand side eventually produces

$$y_t = \phi^{t-1} y_1, \quad t = 1, 2, \ldots \tag{8.21}$$

which shows that the solution for y_t depends on the growth factor, ϕ, and the initial value of the money supply, y_1, chosen by the monetary authority. For a given value of ϕ, the infinity of solution possibilities is clear and depends on the initial value, y_1.

We refer to the solution (8.21) as the **general solution** for (8.7), and to y_1 as the **initial condition** of the difference equation. If y_1 is known, then a **particular solution** for the equation arises. Since $\theta > 0$ in (8.6), and therefore $\phi \equiv \theta + 1 > 1$, the money supply increases period by period without bound in this illustration, for any positive value of y_1. If $y_1 = 0$, however, then $y_t = 0$ for all t. This zero value is therefore known as the **steady-state value** or **steady-state solution** for y_t.

Although these various solution concepts have been introduced using a linear, autonomous, homogeneous difference equation of order one, none of them is specific to a particular class of difference equation; they all apply generally. The distinction between the particular solution, expressed in terms of a known initial condition, and the general solution, in which y_1 is free to take on any value, is analogous to the distinction we made for non-homogeneous systems of m linear equations in n unknowns in Section 5.4.2. The steady-state solution, if it exists, is not necessarily zero or equal to y_1; it may be any constant value, y^*, say, as we shall see in the next section.

In this book, we focus on linear difference equations only, as these are the most widely used difference equations in economics and finance. In the sections that follow, we shall discuss the general solution for linear, autonomous, first-order, non-homogeneous difference equations, as well as solutions for linear, non-autonomous, first-order difference equations, and for linear, autonomous, second-order difference equations and systems of linear difference equations. As mentioned earlier, an important application is presented later, in Section 14.4.

8.3 Linear, first-order difference equations

In this section, we consider both linear, autonomous and linear, non-autonomous, first-order difference equations, focusing mainly on their solution and dynamic properties.

8.3.1 *Autonomous case*

Examples of linear, autonomous, first-order difference equations appeared as (8.7) and (8.11) in Section 8.2.3. In a more general notation, such equations may be written as

$$y_t = \phi_0 + \phi_1 y_{t-1}, \quad t = 2, 3, \ldots \tag{8.22}$$

If $\phi_0 = 0$, we have the homogeneous case, and if $\phi_0 \neq 0$, we have the non-homogeneous case. Let us, first, consider the solution of this difference equation.

Solutions

In the manner illustrated in the previous section, continuous substitution for the lagged value on the right-hand side of (8.22) gives

$$y_t = \phi_0 + \phi_0\phi_1 + \phi_0\phi_1^2 + \cdots + \phi_0\phi_1^{t-2} + \phi_1^{t-1}y_1$$

$$= \phi_0 \sum_{i=0}^{t-2} \phi_1^i + \phi_1^{t-1}y_1 \tag{8.23}$$

This result emerges after $t-2$ substitutions, for any $t \geq 2$. The last term in (8.23) is the same as the solution for our earlier illustration, (8.21); the other term arises from the non-homogeneity in the present case. That (8.23) is a solution for y_t may be checked by noting that, if we use it to substitute for y_t and y_{t-1} in (8.22), the difference equation is satisfied. Equivalently, as (8.23) holds for all values of t, we have

$$y_{t-1} = \phi_0 \sum_{i=0}^{t-3} \phi_1^i + \phi_1^{t-2}y_1 \tag{8.24}$$

and so

$$\phi_0 + \phi_1 y_{t-1} = \phi_0 + \phi_1 \left(\phi_0 \sum_{i=0}^{t-3} \phi_1^i + \phi_1^{t-2}y_1 \right)$$

$$= \phi_0 + \phi_0 \sum_{i=1}^{t-2} \phi_1^i + \phi_1^{t-1}y_1$$

$$= \phi_0 \sum_{i=0}^{t-2} \phi_1^i + \phi_1^{t-1}y_1$$

$$= y_t \tag{8.25}$$

which confirms (8.22).

Another way of writing the general solution for y_t results from simplification of the sum in the intercept term in (8.23). Using the formula for the sum of a geometric series on p. xx, for $\phi_1 = 1$, we may rewrite (8.23) as

$$y_t = \phi_0(t-1) + y_1 \tag{8.26}$$

and for $\phi_1 \neq 1$, we may rewrite (8.23) as

$$y_t = \phi_0 \frac{1 - \phi_1^{t-1}}{1 - \phi_1} + \phi_1^{t-1} y_1 \tag{8.27}$$

The solution (8.26) is interesting in that it shows that, when $\phi_1 = 1$, the values of y_t when plotted against t lie along a straight line, whose intercept is $y_1 - \phi_0$ and whose slope is ϕ_0. Such a line is called a **linear time trend**.

The following example makes use of (8.23) and (8.27) to obtain a particular solution when the initial condition y_1 is known.

EXAMPLE 8.3.1 If a saver puts €500 into an account regularly at the beginning of each year and earns interest of 5% per annum, what is the value of the investment after five years, i.e. at the beginning of year six?

This problem can be formulated in terms of the difference equation

$$\begin{aligned} y_t &= \phi_0 + \phi_1 y_{t-1} \\ &= 500 + 1.05 y_{t-1} \end{aligned} \tag{8.28}$$

where y_t denotes the value of the investment immediately after the annual deposit at the beginning of year t. The first deposit is made at the start of the first year, say, $t = 1$, and maturity is at the start of the sixth year, $t = 6$. At the start of the second year, the individual has $1.05 \times €500 = €525$ from the initial investment and adds another €500 to this, and so on. Using (8.23), we have

$$\begin{aligned} y_6 &= \phi_0 \sum_{i=0}^{6-2} \phi_1^i + \phi_1^{6-1} y_1 \\ &= 500 \sum_{i=0}^{4} 1.05^i + 1.05^5 \times 500 \\ &= 500(1 + 1.05 + 1.05^2 + 1.05^3 + 1.05^4) + 1.05^5 \times 500 \end{aligned} \tag{8.29}$$

The required arithmetic in (8.29) gives $y_6 \approx €3400.89$ to two places of decimals or the nearest cent. Using the alternative (8.27), we have

$$\begin{aligned} y_6 &= \phi_0 \frac{1 - \phi_1^{6-1}}{1 - \phi_1} + \phi_1^{6-1} y_1 \\ &= 500 \times \frac{1 - 1.05^5}{1 - 1.05} + 1.05^5 \times 500 \end{aligned} \tag{8.30}$$

which gives the same numerical result, $y_6 \approx €3400.89$. Note that this solution includes the new deposit just made at the start of year six.   ◇

We see from Example 8.3.1 that (8.27) may be preferred to (8.23) in practice as it is computationally more efficient, especially if t is large. We also note that, as in the case of the money supply illustration in the previous section, the value of y_t (savings) would increase continuously, if the time horizon were to be increased beyond five years. This brings us to the matter of the dynamic behaviour of a variable, y_t, that is implicit in any particular difference equation.

Dynamic behaviour

If the process governing the behaviour of a variable, y_t, may be described by a linear, autonomous, first-order difference equation, then it is relatively straightforward to answer questions about the dynamic behaviour of that variable. In general, the value of the variable will change over time but it may be important to know more about the nature of this change. Does the value of the variable increase monotonically over time ($y_t > y_{t-1}$ for all t) – as in the earlier money supply illustration and in Example 8.3.1 about the level of savings in a certain investment plan – or does it decrease monotonically over time ($y_t < y_{t-1}$ for all t)? Or is the time path of the variable oscillatory in nature, and, if so, does it have a tendency to converge to a particular value or to diverge as time passes? All of these questions can be answered by reference to the general solution just presented, using, as appropriate, the ideas on limits referred to in Section 7.8.

The key is the value of the parameter ϕ_1. First, it is clear from (8.27) that, if $\phi_1 = 0$, then, trivially, $y_t = \phi_0$ for all t; y_t stays at the constant value ϕ_0 over time. If $\phi_1 = 1$, y_t increases monotonically (linearly) or decreases monotonically (linearly), depending on whether ϕ_0 is positive or negative, respectively. In fact, as noted above, the dynamic behaviour of y_t in this case is described by a deterministic linear time trend and therefore the variable has no tendency to approach any particular value. Rather, the time path of y_t is divergent, and the divergence is at a constant rate, ϕ_0. The case in which $\phi_1 = 1$ is important for certain economic and financial theories,[2] and for econometric work involving unit roots and co-integration.[3] If $\phi_1 = -1$, the time path of y_t is characterized by continually oscillating behaviour between two values, since

$$y_t = \begin{cases} \phi_0 - y_1 & \text{if } t \text{ is even} \\ y_1 & \text{if } t \text{ is odd} \end{cases} \tag{8.31}$$

The exception is if $y_1 = \phi_0/2$, in which case $y_t = \phi_0/2$ for all t.

Second, when $\phi_1 \neq 1$, rearrangement of (8.27) as

$$y_t = \frac{\phi_0}{1 - \phi_1} + \phi_1^{t-1}\left(y_1 - \frac{\phi_0}{1 - \phi_1}\right) \tag{8.32}$$

indicates that divergent behaviour also arises when $|\phi_1| > 1$, assuming that $y_1 \neq \phi_0/(1 - \phi_1)$. In this case, however, the divergence is not linear, since ϕ_1^{t-1} increases geometrically in absolute terms as $t \to \infty$. Moreover, the nature of the divergence is fundamentally different depending on the signs of $(y_1 - \phi_0/(1 - \phi_1))$ and ϕ_1. For example, if $(y_1 - \phi_0/(1 - \phi_1)) > 0$ and $\phi_1 > 1$, the value of y_t increases monotonically from its initial value, whereas if

$(y_1 - \phi_0/(1 - \phi_1)) > 0$ and $\phi_1 < -1$, the divergence in the value of y_t is oscillatory, i.e. the value changes in a series of alternate upward and downward steps of increasing magnitude. Determination of the other possibilities, when $(y_1 - \phi_0/(1 - \phi_1))$ is negative, is left as an exercise; see Exercise 8.4. The various dynamic possibilities will also be illustrated presently, in Example 8.3.1.

Third, and by contrast with the previous two cases, we conclude from (8.32) that, when $|\phi_1| < 1$ and $y_1 \neq \phi_0/(1 - \phi_1)$, y_t is subject to convergent behaviour, evolving over time towards a fixed value. This convergence may be such that the value of y_t increases monotonically towards a constant, decreases monotonically towards a constant or approaches a fixed value in an oscillatory fashion. Once again, the precise mode of convergence depends on the sign of ϕ_1. If, for example, $(y_1 - \phi_0/(1 - \phi_1)) > 0$ and $0 < \phi_1 < 1$, then ϕ_1^{t-1} remains positive for all finite values of t but tends to zero as $t \to \infty$, and so y_t declines monotonically towards the value $\phi_0/(1 - \phi_1)$. If $(y_1 - \phi_0/(1 - \phi_1)) > 0$ and $-1 < \phi_1 < 0$, then ϕ_1^{t-1} again becomes smaller in magnitude as t gets bigger but it alternates in sign. Thus the convergence in this case is oscillatory, with successive values of y_t being below and then above $\phi_0/(1 - \phi_1)$. In the limit, as $t \to \infty$, we again have that $y_t \to \phi_0/(1 - \phi_1)$. If $y_1 = \phi_0/(1 - \phi_1)$ and $\phi_1 \neq 1$, then $y_t = \phi_0/(1 - \phi_1)$ for all t, so $\phi_0/(1 - \phi_1)$ is the steady-state value or steady-state solution for y_t introduced in the previous section.

In economic and financial theory, the steady-state concept coincides with the idea of an equilibrium; and when $|\phi_1| < 1$, we have the important case of convergence towards an equilibrium. The concept of equilibrium in a dynamic model connotes a situation that, once achieved, will be maintained. It therefore suggests an alternative way of determining the steady-state value of a variable when $|\phi_1| \neq 1$. Let y^* denote this steady-state value of y_t. Then in the steady state, $y_t = y_{t-1} = y^*$ for all t by definition, so that

$$y^* = \phi_0 + \phi_1 y^* \tag{8.33}$$

and therefore

$$y^* = \frac{\phi_0}{1 - \phi_1} \tag{8.34}$$

in agreement with what was inferred from (8.32).

Another approach to establishing the steady-state solution makes use of the lag operator. Thus

$$y_t = \phi_0 + \phi_1 y_{t-1} = \phi_0 + \phi_1 L y_t \tag{8.35}$$

and so we may write

$$\phi(L) y_t = (1 - \phi_1 L) y_t = \phi_0 \tag{8.36}$$

where $\phi(L) \equiv 1 - \phi_1 L$ is a polynomial in the lag operator of degree one. It is conventional to denote the inverse of this polynomial as $\phi^{-1}(L) \equiv 1/(1 - \phi_1 L)$, so that $\phi(L)\phi^{-1}(L)$ is the identity transformation, although the inverse could be written as $(\phi(L))^{-1}$. If the inverse exists, then

$$y_t = \phi^{-1}(L)\phi_0 = \frac{\phi_0}{1 - \phi_1}, \quad \forall t \tag{8.37}$$

as in (8.34).

The existence of $\phi^{-1}(L)$ is guaranteed if $|\phi_1| < 1$, and it takes the form of an infinite polynomial in the lag operator similar to that in Property 8 of the lag operator listed in Section 8.2.2, namely

$$\phi^{-1}(L) = \frac{1}{1 - \phi_1 L} = \sum_{i=0}^{\infty} \phi_1^i L^i \tag{8.38}$$

This result is easily verified by showing that (see Exercise 8.5)

$$(1 - \phi_1 L)(1 + \phi_1 L + \phi_1^2 L^2 + \phi_1^3 L^3 + \cdots) = 1 \tag{8.39}$$

Therefore, assuming $|\phi_1| < 1$, substitution for $\phi^{-1}(L)$ in (8.37) gives

$$\begin{aligned}
y_t &= (1 + \phi_1 L + \phi_1^2 L^2 + \phi_1^3 L^3 + \cdots)\phi_0 \\
&= (1 + \phi_1 + \phi_1^2 + \phi_1^3 + \cdots)\phi_0 \\
&= \frac{\phi_0}{1 - \phi_1} = y^* \tag{8.40}
\end{aligned}$$

since $L^i \phi_0 = \phi_0$ for all i, given that ϕ_0 is a constant. In this approach, the notion of t tending to infinity from some initial value, y_1, is replaced by that of the process generating y_t having started in the infinite past ($t \to -\infty$), which is implicit in the infinite lag polynomial involved in $\phi^{-1}(L)$.

The lag polynomial, $\phi(L)$, also provides another way of stating the convergence condition, which is that the root of $\phi(z) = 1 - \phi_1 z = 0$ must be greater than 1 in absolute value. Since the solution of $\phi(z) = 0$ is $z = 1/\phi_1$, the requirement that $|\phi_1| < 1$ implies that $|z| > 1$, and the equivalence is established.

We may summarize the points made about the dynamics of a variable whose behaviour is governed by a linear, autonomous, first-order difference equation in the following theorem.

THEOREM 8.3.1 *Let $y_t = \phi_0 + \phi_1 y_{t-1}$. Then*

(a) *there exists a steady-state solution $y^* = \phi_0/(1 - \phi_1)$ if and only if $\phi_1 \neq 1$;*
(b) *y_t converges to the steady-state solution y^* for all y_1 if and only if $|\phi_1| < 1$, the convergence being monotonic if and only if $0 < \phi_1 < 1$ and oscillatory if and only if $-1 < \phi_1 < 0$;*
(c) *y_t diverges monotonically from its initial value, y_1, if and only if $\phi_1 \geq 1$, and diverges from its initial value in a series of increasing oscillations if and only if $\phi_1 < -1$;*
(d) *y_t neither converges to, nor diverges from, its initial value if and only if $\phi_1 = -1$; rather, it alternates in value between y_1 and $\phi_0 - y_1$ (provided $y_1 \neq \phi_0/2$); and*
(e) *y_t is constant for all t if $\phi_1 = 0$ or if $y_1 = \phi_0/2$ and $\phi_1 = -1$.*

The next example illustrates some of the dynamic possibilities numerically and graphically, while other possibilities are the subject of Exercise 8.6.

EXAMPLE 8.3.2 Given the difference equation

$$y_t = 1 + \phi_1 y_{t-1} \tag{8.41}$$

let us determine the steady-state and the dynamic behaviour of y_t for the alternative values $\phi_1 = 2, -2, 1/2$ and $-1/2$, and initial value $y_1 = 1$.

First, for $\phi_1 = 2$, we have

$$y^* = \frac{\phi_0}{1 - \phi_1} = \frac{1}{1 - 2} = -1 \qquad (8.42)$$

and because $\phi_1 = 2 > 1$, we anticipate monotonically increasing values of y_t as t increases. Specifically, we have

$$y_2 = 1 + \phi_1 y_1 = 1 + 2 \times 1 = 3 \qquad (8.43)$$
$$y_3 = 1 + \phi_1 y_2 = 1 + 2 \times 3 = 7 \qquad (8.44)$$
$$y_4 = 1 + \phi_1 y_3 = 1 + 2 \times 7 = 15 \qquad (8.45)$$

and so on for values of $t > 4$. Figure 8.1 shows the divergent (monotonically increasing) time path of y_t for $t = 1$ to $t = 8$, inclusive.

Second, for $\phi_1 = -2$, we have

$$y^* = \frac{\phi_0}{1 - \phi_1} = \frac{1}{1 - (-2)} = \frac{1}{3} \qquad (8.46)$$

and because $\phi_1 = -2 < -1$, we anticipate oscillatory increasing absolute values of y_t as t increases. Specifically, we have

$$y_2 = 1 + \phi_1 y_1 = 1 + (-2) \times 1 = -1 \qquad (8.47)$$
$$y_3 = 1 + \phi_1 y_2 = 1 + (-2) \times (-1) = 3 \qquad (8.48)$$
$$y_4 = 1 + \phi_1 y_3 = 1 + (-2) \times 3 = -5 \qquad (8.49)$$
$$y_5 = 1 + \phi_1 y_4 = 1 + (-2) \times (-5) = 11 \qquad (8.50)$$

and so on for values of $t > 5$. Figure 8.2 shows the oscillatory dynamics of y_t for $t = 1$ to $t = 8$, inclusive, for this case.

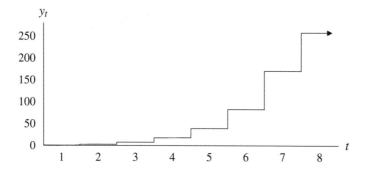

Figure 8.1 Divergent time path of $y_t = 1 + 2y_{t-1}$ from $y_1 = 1$

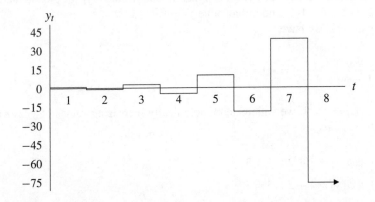

Figure 8.2 Divergent time path of $y_t = 1 - 2y_{t-1}$ from $y_1 = 1$

Third, for $\phi_1 = \frac{1}{2}$, we have

$$y^* = \frac{\phi_0}{1 - \phi_1} = \frac{1}{1 - \frac{1}{2}} = 2 \tag{8.51}$$

and because $0 < \phi_1 < 1$, we anticipate convergence of y_t towards this value of y^* as t increases. Specifically,

$$y_2 = 1 + \phi_1 y_1 = 1 + \frac{1}{2} \times 1 = \frac{3}{2} \tag{8.52}$$

$$y_3 = 1 + \phi_1 y_2 = 1 + \frac{1}{2} \times \frac{3}{2} = \frac{7}{4} \tag{8.53}$$

$$y_4 = 1 + \phi_1 y_3 = 1 + \frac{1}{2} \times \frac{7}{4} = \frac{15}{8} \tag{8.54}$$

and so on for values of $t > 4$. The tendency for y_t to converge to $y^* = 2$ monotonically is apparent in the numbers calculated, but Figure 8.3 plots the path of y_t from its initial value towards its steady-state value for $t = 1$ to $t = 8$, inclusive.

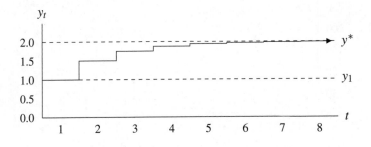

Figure 8.3 Convergent time path of $y_t = 1 + \frac{1}{2} y_{t-1}$ from $y_1 = 1$ to $y^* = 2$

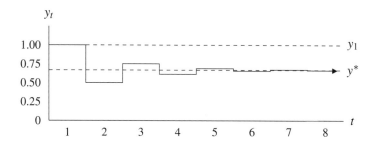

Figure 8.4 Convergent time path of $y_t = 1 - \frac{1}{2}y_{t-1}$ from $y_1 = 1$ to $y^* = \frac{2}{3}$

Finally, for $\phi_1 = -\frac{1}{2}$, we have

$$y^* = \frac{\phi_0}{1 - \phi_1} = \frac{1}{1 - (-\frac{1}{2})} = \frac{2}{3} \tag{8.55}$$

and similar calculations to those carried out for the previous cases lead to $y_2 = \frac{1}{2}$, $y_3 = \frac{3}{4}$, $y_4 = \frac{5}{8}$, $y_5 = \frac{11}{16}$, etc. The anticipated oscillatory convergence of y_t towards $y^* = \frac{2}{3}$ is shown for $t = 1$ to 8, inclusive, in Figure 8.4. ◇

8.3.2 *Non-autonomous case*

The linear, non-autonomous, first-order difference equation may be represented by a generalization of (8.22) in which the intercept and slope, as well as the variable in question, are dependent on time. Using the usual subscript to denote this time dependence, we have

$$y_t = \phi_{0(t-1)} + \phi_{1(t-1)}y_{t-1}, \quad t = 2, 3, \dots \tag{8.56}$$

A specific example of this general equation was given in (8.16). It is clear that (8.56) subsumes a number of special cases. First, the case when $\phi_{0(t-1)}$, but not $\phi_{1(t-1)}$, is constant for all t; second, the case when $\phi_{1(t-1)}$, but not $\phi_{0(t-1)}$, is constant for all t; and third, the autonomous case, in which $\phi_{0(t-1)}$ and $\phi_{1(t-1)}$ are both constant for all t. For the moment, we consider the most general variant.

As in the autonomous case, the general solution for (8.56) may be obtained by a process of continuous substitution for the lagged value of the variable on the right-hand side; and if the initial value, y_1, is known, then a particular solution may be written down. The first few substitutions pose little difficulty:

$$y_2 = \phi_{01} + \phi_{11}y_1 \tag{8.57}$$

$$y_3 = \phi_{02} + \phi_{12}y_2 = \phi_{02} + \phi_{12}(\phi_{01} + \phi_{11}y_1)$$
$$= \phi_{02} + \phi_{12}\phi_{01} + \phi_{12}\phi_{11}y_1 \tag{8.58}$$

$$y_4 = \phi_{03} + \phi_{13}y_3 = \phi_{03} + \phi_{13}(\phi_{02} + \phi_{12}\phi_{01} + \phi_{12}\phi_{11}y_1)$$
$$= \phi_{03} + \phi_{13}\phi_{02} + \phi_{13}\phi_{12}\phi_{01} + \phi_{13}\phi_{12}\phi_{11}y_1 \tag{8.59}$$

and so on, but the algebra becomes tedious as t gets bigger. The result for y_t when t is large can be written as

$$y_t = \phi_{0(t-1)} + \phi_{0(t-2)}\phi_{1(t-1)} + \phi_{0(t-3)}\prod_{i=t-2}^{t-1}\phi_{1i} + \cdots + \phi_{01}\prod_{i=2}^{t-1}\phi_{1i} + y_1\prod_{i=1}^{t-1}\phi_{1i} \quad (8.60)$$

or

$$y_t = \phi_{0(t-1)} + \sum_{j=1}^{t-2}\phi_{0j}\prod_{i=j+1}^{t-1}\phi_{1i} + y_1\prod_{i=1}^{t-1}\phi_{1i} \quad (8.61)$$

The proof that (8.61) is a solution for y_t in the non-autonomous case may be established in similar fashion to what was done earlier for solution (8.23) in the autonomous case. We may either use (8.61) to substitute for y_t and y_{t-1} in (8.56) and show that the difference equation is satisfied, or use (8.61) to rewrite $\phi_{1(t-1)}y_{t-1}$ and add $\phi_{0(t-1)}$ to confirm that y_t results; see Exercise 8.7.

Solutions for the three special cases are easily established using (8.61). In particular, if $\phi_{0(t-1)}$ and $\phi_{1(t-1)}$ are both constant for all t, say, $\phi_{0(t-1)} = \phi_0$ for all t and $\phi_{1(t-1)} = \phi_1$ for all t, then we have

$$y_t = \phi_0 + \sum_{j=1}^{t-2}\phi_{0j}\prod_{i=j+1}^{t-1}\phi_{1i} + y_1\prod_{i=1}^{t-1}\phi_{1i}$$

$$= \phi_0 + \phi_0\phi_1^{t-2} + \phi_0\phi_1^{t-3} + \cdots + \phi_0\phi_1 + y_1\phi_1^{t-1}$$

$$= \phi_0\sum_{i=0}^{t-2}\phi_1^i + \phi_1^{t-1}y_1 \quad (8.62)$$

which is solution (8.23) from the autonomous case. The form that (8.61) takes in the other two special cases is considered in Exercise 8.8.

EXAMPLE 8.3.3 Calculate y_5, given that

$$y_t = \frac{t-1}{10} + \left(\frac{1}{2}\right)^{t-1}y_{t-1} \quad \text{and} \quad y_1 = 16 \quad (8.63)$$

and establish the limiting behaviour of y_t as $t \to \infty$.

In this example, y_t is generated by a linear, non-autonomous, first-order difference equation with $\phi_{0(t-1)} = (t-1)/10$ and $\phi_{1(t-1)} = \left(\frac{1}{2}\right)^{t-1}$; therefore (8.61) gives us directly that

$$y_5 = \phi_{04} + \sum_{j=1}^{5-2}\phi_{0j}\prod_{i=j+1}^{5-1}\phi_{1i} + y_1\prod_{i=1}^{5-1}\phi_{1i}$$

$$= \frac{4}{10} + \frac{3}{10}\left(\frac{1}{2}\right)^4 + \frac{2}{10}\left(\frac{1}{2}\right)^3\left(\frac{1}{2}\right)^4 + \frac{1}{10}\left(\frac{1}{2}\right)^2\left(\frac{1}{2}\right)^3\left(\frac{1}{2}\right)^4$$

$$+ 16 \left(\frac{1}{2}\right)\left(\frac{1}{2}\right)^2 \left(\frac{1}{2}\right)^3 \left(\frac{1}{2}\right)^4$$

$$= \frac{4}{10} + \frac{3}{10}\frac{1}{2^4} + \frac{2}{10}\frac{1}{2^7} + \frac{1}{10}\frac{1}{2^9} + 16\frac{1}{2^{10}} \qquad (8.64)$$

Simple arithmetic then yields the result $y_5 = \frac{2233}{5120} \approx 0.4361$ to four places of decimals. We note that this value is much lower than the initial value $y_1 = 16$. The intermediate values y_2, y_3 and y_4 would be easy to find, either by direct calculations similar to those in (8.64) or by successive substitution in the original difference equation. However, the issue now is the limiting behaviour of y_t, i.e. its tendency as $t \to \infty$. Clearly, $\phi_{0(t-1)} = (t-1)/10 \to \infty$ as $t \to \infty$, while $\phi_{1(t-1)} = \left(\frac{1}{2}\right)^{t-1} \to 0$ as $t \to \infty$. We conclude that y_t increases without bound as t gets large: it becomes indeterminately large in the limit. We note, however, that this increase is not monotonic from the initial value. Further examination of the precise form of the time path of y_t is left to the reader; see Exercise 8.9. ◇

This example illustrates how consideration of the behaviour of the coefficients $\phi_{0(t-1)}$ and $\phi_{1(t-1)}$ as time passes, in conjunction with the solution (8.61), in general, may provide insight into the dynamic behaviour of the variable y_t.

8.4 Linear, autonomous, higher-order difference equations

8.4.1 Linear, autonomous, pth-order difference equations

The linear, first-order difference equation easily generalizes, though the problem of solution becomes more complicated. Concentrating on the autonomous case, we may write the linear difference equation of order p as

$$y_t = \phi_0 + \phi_1 y_{t-1} + \phi_2 y_{t-2} + \cdots + \phi_p y_{t-p} \qquad (8.65)$$

or, using a polynomial in the lag operator,

$$\phi(L)y_t = \phi_0 \qquad (8.66)$$

where

$$\phi(L) = 1 - \phi_1 L - \phi_2 L^2 - \cdots - \phi_p L^p \qquad (8.67)$$

is called the lag polynomial of (8.65).

As a prelude to our consideration of how (8.65) is solved for y_t, we note two important points that arise from the previous section. First, as is evident from (8.32), the general solution for the linear, autonomous, first-order difference equation may be thought of as comprising two components. One is a particular solution, namely the steady-state solution $y^* = \phi_0/(1 - \phi_1)$, and the other is a component of the form $A\phi_1^{t-1}$, where $A = y_1 - \phi_0/(1 - \phi_1)$ depends on the initial value y_1. Second, the component $A\phi_1^{t-1}$ is precisely the form of solution given in (8.21) for an equation – equation (8.7), which describes a money supply adjustment process – in which $\phi_0 = 0$, i.e. a homogeneous difference equation. Thus we may describe the earlier general solution (8.32) as being made up of a particular solution for the linear, autonomous, first-order difference equation and the solution for the homogeneous equation associated with it.

There are few straightforward methods for solving higher-order difference (or differential) equations. Often the simplest method is trial and error. Based on experience with similar and simpler equations, we conjecture a solution containing several arbitrary parameters. Then we substitute the conjectured solution into the original equation and derive any implied restrictions on these arbitrary parameters. In light of this, let us conjecture that the general solution for (8.65) may be written like the first-order case in terms of a particular solution and the general solution of the associated homogeneous equation.

It is straightforward to obtain the particular, steady-state solution. Applying the procedure used previously, but making use of the lag polynomial, we have

$$\phi(L)y^* = \phi(1)y^* = \phi_0 \tag{8.68}$$

and so

$$y^* = \frac{\phi_0}{\phi(1)} = \frac{\phi_0}{1 - \phi_1 - \phi_2 - \cdots - \phi_p} \tag{8.69}$$

assuming, of course, that $\sum_{i=1}^{p} \phi_i \neq 1$.

We see now why it is conventional to denote the identity operator on the vector space of time series by 1 rather than by I. In (8.68), $\phi(L)$ is clearly a linear operator, but the $\phi(1)$ that replaces it can be seen either as a linear operator applied to the constant time series $\{y^*\}$ or as a scalar multiplying the constant value of that time series, $y^* \in \mathbb{R}$. Only the latter interpretation allows us to make the division and substitution in (8.69).

The next step is more tricky. The homogeneous equation associated with (8.65) is $\phi(L)y_t = 0$ or

$$y_t - \phi_1 y_{t-1} - \phi_2 y_{t-2} - \cdots - \phi_p y_{t-p} = 0 \tag{8.70}$$

Suppose a general solution for this takes the form $y_t^h = As^{t-1}$, as for the first-order homogeneous case, where the superscript "h" signifies that the solution is for the associated homogeneous equation, A is an arbitrary, non-zero constant and s is as yet unknown. Then

$$As^{t-1} - \phi_1 As^{t-2} - \phi_2 As^{t-3} - \cdots - \phi_p As^{t-p-1} = 0 \tag{8.71}$$

and, removing the common factor As^{t-p-1},

$$As^{t-p-1}(s^p - \phi_1 s^{p-1} - \phi_2 s^{p-2} - \cdots - \phi_p) = 0 \tag{8.72}$$

Eliminating trivial solutions, i.e. $s = 0$ and $A = 0$, we may find non-trivial solutions by solving the so-called **characteristic equation** of the linear, autonomous, pth-order difference equation,

$$s^p - \phi_1 s^{p-1} - \phi_2 s^{p-2} - \cdots - \phi_p = 0 \tag{8.73}$$

Note the similarities and differences between the lag polynomial (8.67) and the characteristic equation (8.73). In particular, $s = s^* (\neq 0)$ is a solution of the characteristic equation if and only if $L = 1/s^*$ is a root of the lag polynomial; see Exercise 8.10. The close similarity between (8.73) and the characteristic polynomial encountered when the solution of the eigenvalue problem was discussed in Section 3.3.1 should be apparent. Similar remarks to those made in Section 3.3.1 concerning the number and type of roots of the characteristic polynomial therefore also apply to (8.73).

Let us denote the p roots of (8.73), which include the possibility of complex values and values that occur with multiplicity greater than one, by s_j, $j = 1, 2, \ldots, p$. Then $y_t = A_j s_j^{t-1}$ satisfies (8.70) for each $j = 1, 2, \ldots, p$, as does

$$y_t^{\mathrm{h}} = \sum_{j=1}^{p} A_j s_j^{t-1} \tag{8.74}$$

where $A_1, A_2, \ldots, A_p$ are arbitrary constants. Moreover, as certain subsets of these functions of time may be linearly dependent, more general solutions for (8.70) are possible. The most straightforward case is when all of the s_j are different real roots. The other cases, which require obvious modifications of (8.74), are as follows.

1. If a real root s_j has multiplicity $k > 1$, then s_j^{t-1}, $(t-1)s_j^{t-1}, \ldots, (t-1)^{k-1}s_j^{t-1}$ are the linearly independent solutions that enter the linear combination that gives the general solution.
2. If there is a pair of complex roots $s_j = a \pm bi$, and each root has multiplicity one, then $r^{t-1} \cos(t-1)\theta$ and $r^{t-1} \sin(t-1)\theta$ are the individual solutions that enter the general solution, where $r = \sqrt{a^2 + b^2}$ and the angle $\theta = \tan^{-1}(b/a)$.
3. If a pair of complex roots $s_j = a \pm bi$ has multiplicity $m > 1$, then $r^{t-1} \cos(t-1)\theta$, $r^{t-1} \sin(t-1)\theta$, $(t-1)r^{t-1} \cos(t-1)\theta$, $(t-1)r^{t-1} \sin(t-1)\theta, \ldots, (t-1)^{m-1}r^{t-1} \cos(t-1)\theta$ and $(t-1)^{m-1}r^{t-1} \sin(t-1)\theta$ are the individual solutions that enter the general solution, where r and θ are as defined in point 2.

Just as (7.3) shows that solution spaces of matching homogeneous and non-homogeneous systems of linear equations are respectively vector subspaces and affine subsets of $\mathbb{R}^n$, it can be seen that the solution space of a linear, homogeneous, autonomous, pth-order difference equation such as (8.70) is typically a p-dimensional subspace of the vector space of all relevant time series. For example, if the characteristic equation has p distinct real roots, $s_1, s_2, \ldots, s_p \in \mathbb{R}$ and $\{y_t\}$ is to be defined for $t \in \mathbb{Z}$, then the solution space specified by (8.74) is

$$V \equiv \mathrm{lin}\{\{s_1^t : t \in \mathbb{Z}\}, \{s_2^t : t \in \mathbb{Z}\}, \ldots, \{s_p^t : t \in \mathbb{Z}\}\} \tag{8.75}$$

If y^* is the steady-state solution of the associated non-homogeneous equation (8.65), as given by (8.69), then the set

$$V + \{y^*\} \equiv \{\{y_t^{\mathrm{h}} + y^* : t \in \mathbb{Z}\} : \{y_t^{\mathrm{h}} : t \in \mathbb{Z}\} \in V\} \tag{8.76}$$

is the solution set of the non-homogeneous equation and is effectively an affine subset of the vector space of all time series indexed by $\mathbb{Z}$.

In principle, the general solution of the homogeneous equation can be added to the steady-state solution to yield the overall general solution: $y_t = y^* + y_t^{\mathrm{h}}$. However, the additional complexity in dealing with higher-order difference equations is becoming clear. Therefore, rather than validating the above results using (8.70), let us examine in detail the particular case of $p = 2$.

8.4.2 Linear, autonomous, second-order difference equations

The linear, autonomous, second-order difference equation

$$y_t = \phi_0 + \phi_1 y_{t-1} + \phi_2 y_{t-2} \tag{8.77}$$

has the steady-state solution

$$y^* = \frac{\phi_0}{1 - \phi_1 - \phi_2} \tag{8.78}$$

assuming $\phi_1 + \phi_2 \neq 1$, and the associated homogeneous equation

$$y_t - \phi_1 y_{t-1} - \phi_2 y_{t-2} = 0 \tag{8.79}$$

The characteristic equation corresponding to (8.79) is the quadratic equation

$$s^2 - \phi_1 s - \phi_2 = 0 \tag{8.80}$$

and this is easily solved to give

$$s_1 = \frac{\phi_1}{2} + \frac{1}{2}\sqrt{\phi_1^2 + 4\phi_2} \quad \text{and} \quad s_2 = \frac{\phi_1}{2} - \frac{1}{2}\sqrt{\phi_1^2 + 4\phi_2} \tag{8.81}$$

We immediately see three possibilities, depending on the value of $\phi_1^2 + 4\phi_2$.

1. When $\phi_1^2 + 4\phi_2 > 0$, (8.80) has the two different real roots s_1 and s_2 given in (8.81), and the general solution for (8.79) is

$$y_t^h = A_1 s_1^{t-1} + A_2 s_2^{t-1} \tag{8.82}$$

2. When $\phi_1^2 + 4\phi_2 = 0$, (8.80) has one real root, $s_1 = \phi_1/2$, with multiplicity two, so that the general solution for (8.79) is

$$y_t^h = [A_1 + A_2(t-1)]s_1^{t-1} \tag{8.83}$$

3. When $\phi_1^2 + 4\phi_2 < 0$, (8.80) has the pair of conjugate complex roots

$$s_1 = \frac{\phi_1}{2} + \frac{i}{2}\sqrt{-(\phi_1^2 + 4\phi_2)} \equiv a + bi \equiv r(\cos\theta + i\sin\theta) \tag{8.84}$$

$$s_2 = \frac{\phi_1}{2} - \frac{i}{2}\sqrt{-(\phi_1^2 + 4\phi_2)} \equiv a - bi \equiv r(\cos\theta - i\sin\theta) \tag{8.85}$$

where $r \equiv \sqrt{a^2 + b^2} \equiv \sqrt{\frac{1}{4}\phi_1^2 - \frac{1}{4}(\phi_1^2 + 4\phi_2)} = \sqrt{-\phi_2}$, and $\theta \equiv \tan^{-1}(b/a) = \tan^{-1}\left[\sqrt{-(\phi_1^2 + 4\phi_2)}/\phi_1\right]$. In this case, making use of de Moivre's theorem (see p. xx), the general solution for (8.79) is

$$y_t^h = r^{t-1}[A_1 \cos(t-1)\theta + A_2 \sin(t-1)\theta]$$
$$= Ar^{t-1} \cos[(t-1)\theta + \omega] \tag{8.86}$$

where A and ω are arbitrary constants. The derivation of the second equality in (8.86) is the subject of Exercise 8.11.

The case of $\phi_1 + \phi_2 = 1$ is not without interest, for although no steady-state solution exists, it implies that the characteristic equation (8.80) has a unit root. It is easily verified that, with $\phi_1 = 1 - \phi_2$, the left-hand side of $s^2 - \phi_1 s - \phi_2 = 0$ factorizes to give

$$(s + \phi_2)(s - 1) = 0 \tag{8.87}$$

which manifests the unit root directly. Similarly, one of the roots of the lag polynomial equation $\phi(z) = 0$ is unity. The factorization in this case is

$$(1 + \phi_2 z)(1 - z) = 0 \tag{8.88}$$

which allows (8.77) to be rewritten as

$$(1 + \phi_2 L)(1 - L) y_t = \phi_0 \tag{8.89}$$

or

$$(1 + \phi_2 L)\Delta y_t = \phi_0 \tag{8.90}$$

so that the change in y_t is described by the first-order difference equation

$$\Delta y_t = \phi_0 - \phi_2 \Delta y_{t-1} \tag{8.91}$$

Linear independence and validity of solutions

The linear independence of the individual solutions in each of the above three cases is easy to verify. It is sufficient to treat the solution values for $t = 1$ and $t = 2$ as the elements of a pair of 2-vectors and show that these vectors are linearly independent. Recalling Theorem 3.6.2, this boils down to showing that the determinant of a 2×2 matrix is non-zero.

Consider the first possibility, in which s_1 and s_2 are two distinct real roots. Then, noting that $s_1^{t-1} = s_1^0 = 1$ when $t = 1$, and $s_1^{t-1} = s_1^1 = s_1$ when $t = 2$, and similarly for s_2^{t-1}, we examine the determinant

$$\begin{vmatrix} 1 & 1 \\ s_1 & s_2 \end{vmatrix} = s_2 - s_1 \tag{8.92}$$

This is non-zero, since s_1 and s_2 are different real numbers in this first case.

The same approach to the second possibility, where there is a real root $s_1 = \phi_1/2$ with multiplicity two, and corresponding solutions s_1^{t-1} and $(t-1)s_1^{t-1}$, produces the determinant

$$\begin{vmatrix} 1 & 0 \\ s_1 & s_1 \end{vmatrix} = s_1 \tag{8.93}$$

which is immediately seen to be non-zero, unless $\phi_1 = 0$ (which would imply $\phi_2 = 0$ also, since in this case $\phi_1^2 + 4\phi_2 = 0$).

The third possibility gives rise to the determinant

$$\begin{vmatrix} 1 & 0 \\ r\cos\theta & r\sin\theta \end{vmatrix} = r\sin\theta = \sqrt{-\frac{1}{4}(\phi_1^2 + 4\phi_2)} \tag{8.94}$$

by (8.84). Given that $\phi_1^2 + 4\phi_2 < 0$, once again a non-zero value for the relevant determinant emerges, establishing the linear independence of the individual solutions.

Likewise, it is quite straightforward to establish the validity of the three possible general solutions for the homogeneous equation. In each case, use of the solution to substitute for y_t, y_{t-1} and y_{t-2} on the left-hand side of (8.79) will produce the required value of zero. We illustrate this for the second possibility only, leaving as exercises the first and third cases; see Exercise 8.12.

The second possibility is that there is one real root with multiplicity two, for which the suggested general solution of (8.79) is given in (8.83) as $y_t^h = [A_1 + A_2(t - 1)]s_1^{t-1}$, where $s_1 = \phi_1/2$. Thus we have that the left-hand side of (8.79) is

$$y_t - \phi_1 y_{t-1} - \phi_2 y_{t-2}$$
$$= [A_1 + A_2(t-1)]s_1^{t-1} - \phi_1[A_1 + A_2(t-2)]s_1^{t-2} - \phi_2[A_1 + A_2(t-3)]s_1^{t-3}$$
$$= A_1[s_1^{t-1} - \phi_1 s_1^{t-2} - \phi_2 s_1^{t-3}]$$
$$\quad + A_2[(t-1)s_1^{t-1} - \phi_1(t-2)s_1^{t-2} - \phi_2(t-3)s_1^{t-3}] \tag{8.95}$$

The two square-bracketed components on the last line of (8.95) are treated similarly, so here we consider just the second, slightly more complicated, component. Removing the common factor s_1^{t-3} from the second square-bracketed expression, then substituting for s_1 gives

$$A_2 s_1^{t-3}[(t-1)s_1^2 - \phi_1(t-2)s_1 - (t-3)\phi_2]$$

$$= A_2 s_1^{t-3}\left[(t-1)\frac{\phi_1^2}{4} - \phi_1(t-2)\frac{\phi_1}{2} - (t-3)\phi_2\right]$$

$$= A_2 s_1^{t-3}\left[(t-1)\frac{\phi_1^2}{4} - 2(t-2)\frac{\phi_1^2}{4} - (t-3)\phi_2\right] \tag{8.96}$$

Now, since $\phi_1^2 + 4\phi_2 = 0$ in this case, we have $\phi_2 = -\phi_1^2/4$. Therefore, the expression in (8.96) can be written as

$$A_2 s_1^{t-3}\left[(t-1)\frac{\phi_1^2}{4} - 2(t-2)\frac{\phi_1^2}{4} + (t-3)\frac{\phi_1^2}{4}\right]$$

$$= A_2 s_1^{t-3}\frac{\phi_1^2}{4}[t - 1 - 2t + 4 + t - 3]$$

$$= 0 \tag{8.97}$$

Similar algebra establishes that the contents of the first square brackets in (8.95) also sum to zero. Thus we show that $A_2(t-1)s_1^{t-1}$ and $A_1 s_1^{t-1}$ are solutions for (8.79) and, hence, so is $y_t^h = [A_1 + A_2(t-1)]s_1^{t-1}$.

The constants A_i, $i = 1, 2$, in each of the three cases may be uniquely determined to give a particular solution for (8.79), if y_1 and y_2 are known. These two given numbers constitute the initial conditions for the linear, autonomous, second-order difference equation. Consider, for instance, the first possibility, where the general solution for (8.79) is $y_t^h = A_1 s_1^{t-1} + A_2 s_2^{t-1}$. It follows that

$$y_1 = A_1 + A_2 \quad \text{and} \quad y_2 = A_1 s_1 + A_2 s_2 \tag{8.98}$$

where s_1 and s_2 are the two real roots given in (8.81). These two simultaneous equations are readily solved for A_1 and A_2, using one of the methods from Chapter 2, such as Cramer's rule described in Section 2.5.1. The procedure for determining A_1 and A_2 is the same in the other two cases, though the forms of the solutions are different.

Dynamic behaviour

The dynamic behaviour of any variable, y_t, which evolves according to some linear, autonomous, second-order difference equation, may be deduced from the three solution possibilities listed above. In particular, interest often focuses on the possibility of stability, in the sense that the general solution of the associated homogeneous difference equation, y_t^h, converges to the steady-state solution zero as $t \to \infty$ and, hence, the effect of the initial conditions dies out over time. This concept of stability is referred to as **asymptotic stability** and requires, for example, in the first case, where the characteristic equation has two different real roots, s_1 and s_2, that both $A_1 s_1^{t-1} \to 0$ and $A_2 s_2^{t-1} \to 0$ as $t \to \infty$. If this applies, then assuming that a steady-state solution exists, y_t will converge over time to its steady-state value, y^*, for all values of A_1 and A_2.

In the first case, it is clear from (8.82) that $y_t^h \to 0$ if and only if $|s_1| < 1$ and $|s_2| < 1$. Second, when the characteristic equation has a single real root, s_1, with multiplicity two, (8.83) indicates that $y_t^h \to 0$ if and only if $|s_1| < 1$. Third, when the characteristic equation has two complex roots, $s_1 = r(\cos\theta + i\sin\theta)$ and $s_2 = r(\cos\theta - i\sin\theta)$, the solution (8.86) implies that $y_t^h \to 0$ if and only if $r < 1$, where r is the common modulus of the two complex roots. We also note that complex roots imply that the values of y_t oscillate over time; and that when $r < 1$, these oscillations are **damped**.

Since the modulus of a real number is just the absolute value of that number, we may summarize these convergence conditions in the following theorem.

THEOREM 8.4.1 *The linear, autonomous, second-order difference equation $y_t = \phi_0 + \phi_1 y_{t-1} + \phi_2 y_{t-2}$ is asymptotically stable if and only if the roots of the characteristic equation $s^2 - \phi_1 s - \phi_2 = 0$ have modulus strictly less than unity.*

An alternative to the phrase "have modulus strictly less that unity" in Theorem 8.4.1 is "must lie inside the unit circle". This alternative phrase is often encountered in discussions and applications of difference equations. Two conditions equivalent to the condition stated in the theorem are the subject of Exercise 8.14. They are that:

1. $|\phi_1| < 1 - \phi_2$ and $-\phi_2 < 1$; and

2. the roots of the lag polynomial associated with the difference equation, i.e. the roots of $\phi(z) = 1 - \phi_1 z - \phi_2 z^2 = 0$, have modulus strictly greater than unity or lie outside the unit circle.

When a root of the characteristic equation is greater than or equal to unity, i.e. it lies on or outside the unit circle – or, equivalently, a root of the lag polynomial $\phi(L)$ lies on or inside the unit circle – the variable y_t is not asymptotically stable, but is subject to divergent or, at least, non-convergent behaviour over time. The case of unit roots, i.e. roots having modulus equal to one and, therefore, lying on the unit circle, is rather special, as mentioned in Section 3.4 and, later, in Section 14.4.1. We simply note here that, if one of the two roots of the characteristic equation for a linear, autonomous, second-order difference equation for y_t lies on the unit circle, while the other lies within the unit circle, and therefore y_t is unstable, then the first difference Δy_t behaves according to an asymptotically stable linear, autonomous, first-order difference equation; see Exercise 8.15.

The following example illustrates numerically some of the theoretical results just discussed.

EXAMPLE 8.4.1 Find the general solutions for the linear, autonomous, second-order difference equation $y_t - 2.5 y_{t-1} + y_{t-2} = 0$ and comment on the dynamic behaviour of the variable.

This is a difference equation in homogeneous form. The corresponding characteristic equation is

$$s^2 - 2.5s + 1 = 0 \tag{8.99}$$

The solutions for this may be obtained using (8.81), but in fact the right-hand side of (8.99) factorizes to give

$$(s - 2)(s - \tfrac{1}{2}) = 0 \tag{8.100}$$

and, hence, $s_1 = 2$ and $s_2 = \tfrac{1}{2}$ more directly. Recalling (8.82), and substituting these two different real solutions, the required general solution is

$$y_t^h = A_1 s_1^{t-1} + A_2 s_2^{t-1} = A_1 2^{t-1} + A_2 \left(\frac{1}{2} \right)^{t-1} \tag{8.101}$$

Given that the root $s_1 = 2 > 1$, the condition for convergence is not satisfied and we may assert that y_t^h diverges over time. This is also apparent from the fact that $A_1 2^{t-1} \to \infty$ as $t \to \infty$, though $A_2 \left(\frac{1}{2} \right)^{t-1} \to 0$ as $t \to \infty$. The roots of the relevant polynomial in the lag operator, $\phi(L) = 1 - 2.5L + L^2$, yield the same conclusion concerning instability.   ◇

8.4.3 *Generalizations*

The principles concerning solution and dynamic behaviour established in our discussion of linear, autonomous, second-order difference equations generalize straightforwardly to linear, autonomous, pth-order difference equations. Thus, in addition to the general form of solution (8.74), and its various modifications presented above, we may note, first, that the linear independence of the individual solutions of the corresponding pth-order characteristic equation (8.73) requires that a $p \times p$ matrix have a non-zero determinant. For example, when the characteristic equation has p different real roots, s_j, $j = 1, 2, \ldots, p$, it is required that

$$\begin{vmatrix} 1 & 1 & \cdots & 1 \\ s_1 & s_2 & \cdots & s_p \\ \vdots & \vdots & & \vdots \\ s_1^{p-1} & s_2^{p-1} & \cdots & s_p^{p-1} \end{vmatrix} \neq 0 \tag{8.102}$$

Second, asymptotic stability, or convergent dynamic behaviour, requires that all p roots of the characteristic equation lie inside the unit circle, each with modulus strictly less than unity. As with $p = 2$ (see Theorem 8.4.1), it is possible to cast this condition in terms of the coefficients of the characteristic equation. The relevant result is known as Schur's theorem[4] and the interested reader is referred to Sydsæter *et al.* (2008, Section 11.5) for a little more detail on this. Also as before, the stability condition may be expressed in terms of the roots of the lag polynomial. Thus all of the solutions of $\phi(z) = 1 - \phi_1 z - \cdots - \phi_p z^p = 0$ must lie outside the unit circle. A necessary requirement for this is that $\sum_{i=1}^{p} \phi_i < 1$, while a sufficient condition is $\sum_{i=1}^{p} |\phi_i| < 1$. If $\sum_{i=1}^{p} \phi_i = 1$, then at least one root of the lag polynomial is unity. If the characteristic equation (or lag polynomial) has one unit root, it is possible to model the change Δy_t as a linear difference equation of order $p - 1$. More generally, if the characteristic equation (or lag polynomial) has k ($k = 1, 2, \ldots, p$) unit roots, then $\Delta^k y_t$ may be modelled as a linear difference equation of order $p - k$. For further details on the pth-order difference equation, see Hamilton (1994, Section 1.2 and Appendix 1.A).

8.5 Systems of linear difference equations

As was apparent in the two motivational examples in Sections 1.2.2 and 1.2.3, which dealt with a simple Keynesian macroeconomic model and a Leontief input–output model of an economy, respectively, it is often important to take account of the fact that certain variables may be jointly dependent and therefore determined simultaneously. If the relationships among such variables are dynamic, then systems of difference equations may naturally arise. In Part II, two applications involving such systems will be discussed. One concerns a structural dynamic macroeconomic model; the other concerns the vector autoregressive model used widely in econometrics. In this section, we sketch out some of the mathematical background, concentrating on the first-order case and leaving certain details until our later discussion of the applications.

8.5.1 *General systems of linear, autonomous difference equations*

Suppose that concern focuses on m variables. So far, we have dealt with scalar time series $\{y_t\}$, where $y_t \in \mathbb{R}$ for all t. Now we are dealing with vector time series, $\{\mathbf{y}_t\}$, where $\mathbf{y}_t \in \mathbb{R}^m$ for all t. The lag operator can be likewise extended naturally to operate on vector time series with $L\mathbf{y}_t \equiv \mathbf{y}_{t-1}$ and similarly for the difference operator. Then the general system of linear, autonomous, pth-order difference equations can be represented conveniently using matrix notation as

$$\mathbf{y}_t = \mathbf{\Phi}_0 + \mathbf{\Phi}_1 \mathbf{y}_{t-1} + \mathbf{\Phi}_2 \mathbf{y}_{t-2} + \cdots + \mathbf{\Phi}_p \mathbf{y}_{t-p} \tag{8.103}$$

or

$$\Phi(L)\mathbf{y}_t = \Phi_0 \tag{8.104}$$

where $\mathbf{y}_t = [y_{it}]$ is an m-vector of values for the m different variables in time period t, $\Phi_0 = [\phi_{0_i}]$ is an m-vector of constants, the $\Phi_i, i = 1, 2, \ldots, p$, are $m \times m$ matrices of parameters,

$$\Phi(L) = \mathbf{I}_m - \Phi_1 L - \Phi_2 L^2 - \cdots - \Phi_p L^p \tag{8.105}$$

is a matrix polynomial in the lag operator and

$$\Phi_i L^i = [\phi_{i_{jk}}]L^i = [\phi_{i_{jk}} L^i]_{m \times m} \tag{8.106}$$

We note the typical (jkth) term, $\phi_{i_{jk}}$, of the matrix Φ_i, and that post-multiplication of Φ_i by the lag operator is treated like multiplication of the matrix by a scalar.

In (8.103), there is an equation for each of the m variables and, in general, the first p lags of all variables appear in all of these equations. To illustrate, suppose $m = p = 2$. Then

$$\mathbf{y}_t = \begin{bmatrix} y_{1t} \\ y_{2t} \end{bmatrix}, \quad \Phi_0 = \begin{bmatrix} \phi_{0_1} \\ \phi_{0_2} \end{bmatrix}, \quad \Phi_1 = \begin{bmatrix} \phi_{1_{11}} & \phi_{1_{12}} \\ \phi_{1_{21}} & \phi_{1_{22}} \end{bmatrix} \quad \text{and} \quad \Phi_2 = \begin{bmatrix} \phi_{2_{11}} & \phi_{2_{12}} \\ \phi_{2_{21}} & \phi_{2_{22}} \end{bmatrix} \tag{8.107}$$

Substituting these matrices into (8.103), and employing the operations of matrix addition and multiplication, yields the following scalar form of the system:

$$\begin{aligned} y_{1t} &= \phi_{0_1} + \phi_{1_{11}} y_{1(t-1)} + \phi_{1_{12}} y_{2(t-1)} + \phi_{2_{11}} y_{1(t-2)} + \phi_{2_{12}} y_{2(t-2)} \\ y_{2t} &= \phi_{0_2} + \phi_{1_{21}} y_{1(t-1)} + \phi_{1_{22}} y_{2(t-1)} + \phi_{2_{21}} y_{1(t-2)} + \phi_{2_{22}} y_{2(t-2)} \end{aligned} \tag{8.108}$$

In practical applications, some of the $\phi_{i_{jk}}$ coefficients may be zero, so that not all lags of all variables appear in all equations.

8.5.2   Systems of linear, autonomous, first-order difference equations

The m-variable, first-order version of (8.104) is

$$\Phi(L)\mathbf{y}_t = (\mathbf{I}_m - \Phi_1 L)\mathbf{y}_t = \Phi_0 \tag{8.109}$$

or, using (8.103),

$$\mathbf{y}_t = \Phi_0 + \Phi_1 \mathbf{y}_{t-1} \tag{8.110}$$

or, in full scalar form,

$$\begin{aligned} y_{1t} &= \phi_{0_1} + \phi_{1_{11}} y_{1(t-1)} + \phi_{1_{12}} y_{2(t-1)} + \cdots + \phi_{1m} y_{m(t-1)} \\ y_{2t} &= \phi_{0_2} + \phi_{1_{21}} y_{1(t-1)} + \phi_{1_{22}} y_{2(t-1)} + \cdots + \phi_{1_{2m}} y_{m(t-1)} \\ &\vdots \\ y_{mt} &= \phi_{0_m} + \phi_{1_{ml}} y_{1(t-1)} + \phi_{1_{m2}} y_{2(t-1)} + \cdots + \phi_{1_{mm}} y_{m(t-1)} \end{aligned} \tag{8.111}$$

As in the case of single linear, autonomous difference equations, the complete general solution in this case comprises the general solution of the homogeneous form of the system and the system's steady-state solution, i.e. $\mathbf{y}_t = \mathbf{y}_t^h + \mathbf{y}^*$. Also as before, it is easier to derive the steady-state solution, $\mathbf{y}^*$, assuming it exists, than it is to obtain the general homogeneous solution, $\mathbf{y}_t^h$.

If there is a steady-state solution, $\mathbf{y}^*$, then from (8.109), we have that

$$\boldsymbol{\Phi}(L)\mathbf{y}^* = (\mathbf{I} - \boldsymbol{\Phi}_1 L)\mathbf{y}^* = \boldsymbol{\Phi}_0 \qquad (8.112)$$

and so

$$\mathbf{y}^* = \boldsymbol{\Phi}^{-1}(L)\boldsymbol{\Phi}_0 = \boldsymbol{\Phi}^{-1}(1)\boldsymbol{\Phi}_0 = (\mathbf{I} - \boldsymbol{\Phi}_1)^{-1}\boldsymbol{\Phi}_0 \qquad (8.113)$$

where $\boldsymbol{\Phi}^{-1}(L)$ denotes the matrix inverse $[\boldsymbol{\Phi}(L)]^{-1} = (\mathbf{I} - \boldsymbol{\Phi}_1)^{-1}$, and assuming, of course, that the $m \times m$ matrix $\mathbf{I} - \boldsymbol{\Phi}_1$ is non-singular and therefore has an inverse. We will return to the related matters of the existence of a steady-state solution and of the inverse of $\mathbf{I} - \boldsymbol{\Phi}_1$ a little later, when we consider the asymptotic stability of the system. For the moment, we consider the solution of the homogeneous form of the system, in which $\boldsymbol{\Phi}_0 = \mathbf{0}_{m \times 1}$.

By analogy with the single-equation case, in which a solution for the homogeneous equation has the form As^{t-1}, let us postulate that a solution for the homogeneous system $\mathbf{y}_t = \boldsymbol{\Phi}_1 \mathbf{y}_{t-1}$ is of the form $\mathbf{a}s^{t-1}$, where $\mathbf{a}$ denotes an m-vector of constants and s is a scalar. Then, substituting into the homogeneous system of equations, we have that

$$\mathbf{a}s^{t-1} = \boldsymbol{\Phi}_1 \mathbf{a}s^{t-2} \qquad (8.114)$$

Dividing both sides of (8.114) by s^{t-2} gives $\mathbf{a}s = \boldsymbol{\Phi}_1 \mathbf{a}$, and therefore

$$\boldsymbol{\Phi}_1 \mathbf{a} - \mathbf{a}s = (\boldsymbol{\Phi}_1 - s\mathbf{I})\mathbf{a} = \mathbf{0} \qquad (8.115)$$

Recalling the discussion in Chapter 3, (8.115) is recognizable as an eigenequation, in which s is an eigenvalue and $\mathbf{a}$ is an associated eigenvector of the matrix $\boldsymbol{\Phi}_1$. Application of the methods for the computation of eigenvalues and eigenvectors explained in Sections 3.3.1 and 3.3.2 will, therefore, provide the required solutions. An important component in these calculations is the characteristic equation

$$|\boldsymbol{\Phi}_1 - s\mathbf{I}| = 0 \qquad (8.116)$$

which, in the present context, yields m values for s ($s_1, s_2, \ldots, s_m$, say). Substitution of each of these eigenvalues in turn into (8.115) then allows (8.115) to be solved for the associated eigenvectors ($\mathbf{a}_1, \mathbf{a}_2, \ldots, \mathbf{a}_m$, say). It is easy to confirm that $A_1\mathbf{a}_1, A_2\mathbf{a}_2, \ldots, A_m\mathbf{a}_m$, where A_i, $i = 1, 2, \ldots, m$, are arbitrary non-zero constants, are also eigenvectors of $\boldsymbol{\Phi}_1$: when each $A_i\mathbf{a}_i$ is used with its corresponding eigenvalue, s_i, it also satisfies (8.115). Thus the general solution for the homogeneous system is suggested as

$$\mathbf{y}_t^h = \sum_{i=1}^{m} A_i \mathbf{a}_i s_i^{t-1} \qquad (8.117)$$

and the complete general solution as

$$\mathbf{y}_t = \mathbf{y}^* + \mathbf{y}_t^h = (\mathbf{I} - \boldsymbol{\Phi}_1)^{-1}\boldsymbol{\Phi}_0 + \sum_{i=1}^{m} A_i \mathbf{a}_i s_i^{t-1} \tag{8.118}$$

Equation (8.117) is of precisely the same form as (8.74). The modifications that are necessary if there are roots with multiplicity greater than one, or pairs of complex roots, or pairs of complex roots with multiplicity greater than one, are therefore the same as those listed immediately following (8.74).

It is evident from (8.118) that, when all of the s_i have modulus strictly less than unity, then $\mathbf{y}_t^h \to \mathbf{0}$ and $\mathbf{y}_t \to \mathbf{y}^* = (\mathbf{I} - \boldsymbol{\Phi}_1)^{-1}\boldsymbol{\Phi}_0$ as $t \to \infty$. This condition on the eigenvalues is important in guaranteeing convergence of the system of difference equations to its steady state and, as previously, is referred to as the condition for asymptotic stability.

Another approach to establishing the complete general solution, which will allow us to investigate the convergence properties of the system in a different, though equivalent, way, is to employ repeated substitution on the right-hand side of (8.110), in the same way as was done in Section 8.3.1 to obtain (8.23). In the present case, we derive

$$\mathbf{y}_t = \boldsymbol{\Phi}_0 + \boldsymbol{\Phi}_1\boldsymbol{\Phi}_0 + \boldsymbol{\Phi}_1^2\boldsymbol{\Phi}_0 + \cdots + \boldsymbol{\Phi}_1^{t-2}\boldsymbol{\Phi}_0 + \boldsymbol{\Phi}_1^{t-1}\mathbf{y}_1$$
$$= (\mathbf{I} + \boldsymbol{\Phi}_1 + \boldsymbol{\Phi}_1^2 + \cdots + \boldsymbol{\Phi}_1^{t-2})\boldsymbol{\Phi}_0 + \boldsymbol{\Phi}_1^{t-1}\mathbf{y}_1 \tag{8.119}$$

Given the initial conditions, i.e. the value of the vector $\mathbf{y}_1$, if $\boldsymbol{\Phi}_1^{t-1} \to \mathbf{0}$ as $t \to \infty$, then $\boldsymbol{\Phi}_1^{t-1}\mathbf{y}_1 \to \mathbf{0}$ as $t \to \infty$ also. The following theorem establishes that, subject to certain conditions on $\boldsymbol{\Phi}_1$, this happens if and only if all of the eigenvalues of $\boldsymbol{\Phi}_1$ have modulus strictly less than unity, i.e. if and only if the condition for asymptotic stability just noted is satisfied. The proof of the theorem makes use of material on the diagonalization of a matrix covered in Chapter 3.

THEOREM 8.5.1 *Let $\mathbf{A}$ be an $n \times n$ square, diagonalizable matrix (i.e. assume that $\mathbf{A}$ has n linearly independent eigenvectors); then $\mathbf{A}^j \to \mathbf{0}$ as $j \to \infty$ if and only if $|\lambda_i| < 1$ for all i, where λ_i denotes the ith eigenvalue of $\mathbf{A}$.*

Proof: Let $\mathbf{P}$ be the $n \times n$ matrix whose columns are the eigenvectors of $\mathbf{A}$. We know by Theorem 3.6.1 that $\mathbf{P}$ exists and is invertible. Then

$$\mathbf{P}^{-1}\mathbf{A}\mathbf{P} = \boldsymbol{\Lambda} = \text{diag}[\lambda_i] \tag{8.120}$$

and

$$(\mathbf{P}^{-1}\mathbf{A}\mathbf{P})(\mathbf{P}^{-1}\mathbf{A}\mathbf{P}) = \mathbf{P}^{-1}\mathbf{A}^2\mathbf{P} = \boldsymbol{\Lambda}^2 \tag{8.121}$$

Similarly, for all j,

$$\mathbf{P}^{-1}\mathbf{A}^j\mathbf{P} = \boldsymbol{\Lambda}^j = \text{diag}[\lambda_i^j] \tag{8.122}$$

and, hence,

$$\mathbf{A}^j = \mathbf{P}\boldsymbol{\Lambda}^j\mathbf{P}^{-1} \tag{8.123}$$

Since the result that the limit of a product equals the product of the limits holds for matrices as well as for scalars, $\lim_{j\to\infty} \mathbf{A}^j = \mathbf{P}\lim_{j\to\infty}\mathbf{\Lambda}^j\mathbf{P}^{-1}$ and $\lim_{j\to\infty}\mathbf{\Lambda}^j = \mathbf{P}^{-1}\lim_{j\to\infty}\mathbf{A}^j\mathbf{P}$. Thus $\mathbf{\Lambda}^j$ tends to $\mathbf{0}$ as $j\to\infty$ if and only if $\mathbf{A}^j \to \mathbf{0}$. But $\mathbf{\Lambda}^j \to \mathbf{0}$ if and only if $\lambda_i^j \to 0$ for all i. This will be the case if and only if $|\lambda_i| < 1$ for all i. $\qquad\square$

To see what happens to the other term on the right-hand side of (8.119) when $t\to\infty$ and this necessary and sufficient condition for asymptotic stability is satisfied, note, first, the matrix form of the formula on p. xx for the sum of the first $t-1$ terms of a geometric series:

$$(\mathbf{I}+\mathbf{\Phi}_1+\mathbf{\Phi}_1^2+\cdots+\mathbf{\Phi}_1^{t-2})(\mathbf{I}-\mathbf{\Phi}_1) = \mathbf{I}-\mathbf{\Phi}_1^{t-1} \tag{8.124}$$

Multiplying out the left-hand side of (8.124) establishes this result. Second, note that as all the eigenvalues of $\mathbf{\Phi}_1$ have modulus less than one by assumption, then unity is not an eigenvalue of $\mathbf{\Phi}_1$. It follows that $\det(\mathbf{I}-\mathbf{\Phi}_1) \neq 0$ and $\mathbf{I}-\mathbf{\Phi}_1$ is non-singular and, therefore, has an inverse. Post-multiplication of both sides of (8.124) by the inverse of $\mathbf{I}-\mathbf{\Phi}_1$ then yields

$$(\mathbf{I}+\mathbf{\Phi}_1+\mathbf{\Phi}_1^2+\cdots+\mathbf{\Phi}_1^{t-2}) = (\mathbf{I}-\mathbf{\Phi}_1^{t-1})(\mathbf{I}-\mathbf{\Phi}_1)^{-1} \tag{8.125}$$

From (8.125), we see that when all eigenvalues of $\mathbf{\Phi}_1$ have modulus less than unity, and therefore, by Theorem 8.5.1, $\mathbf{\Phi}_1^{t-1}$ tends to zero as $t\to\infty$, the **matrix series expansion** $\mathbf{I}+\mathbf{\Phi}_1+\mathbf{\Phi}_1^2+\cdots+\mathbf{\Phi}_1^{t-2} \to (\mathbf{I}-\mathbf{\Phi}_1)^{-1}$ as $t\to\infty$ and so the first term on the right-hand side of (8.119) tends to the steady-state value $\mathbf{y}^* = (\mathbf{I}-\mathbf{\Phi}_1)^{-1}\mathbf{\Phi}_0$. We conclude that for the existence of $(\mathbf{I}-\mathbf{\Phi}_1)^{-1}$ – as well as for asymptotic stability and convergence to the steady-state solution – all of the eigenvalues of $\mathbf{\Phi}_1$ must have modulus strictly less than one.

The following theorem is useful in confirming that all of the eigenvalues of certain matrices have modulus less than one.

THEOREM 8.5.2 *Let $\mathbf{A} = [a_{ij}]_{m\times m}$ and suppose that all the row sums of the absolute values of the elements in $\mathbf{A}$ are less than unity, i.e. $\sum_{j=1}^{m}|a_{ij}| < 1$ for all $i = 1, 2, \ldots, m$.*
Then all of the eigenvalues of $\mathbf{A}$ have modulus strictly less than unity.

Proof: Let $|\lambda|$ be the maximum absolute eigenvalue of $\mathbf{A}$ and let $|x_i|$ be the maximum absolute value of the elements in an eigenvector, $\mathbf{x}$, associated with λ. Then, from the ith component of the eigenequation $\mathbf{Ax} = \lambda\mathbf{x}$, we have that

$$|\lambda x_i| = |\lambda||x_i| = |\mathbf{a}_i^\top\mathbf{x}| \tag{8.126}$$

where $\mathbf{a}_i^\top$ is the ith row of $\mathbf{A}$. Letting $|\mathbf{a}_i|$ and $|\mathbf{x}|$ denote the vectors whose elements are the absolute values of the corresponding elements of $\mathbf{a}_i$ and of $\mathbf{x}$, respectively, and dividing across by $|x_i|$,

$$|\lambda| = \left|\mathbf{a}_i^\top\left(\frac{1}{|x_i|}\mathbf{x}\right)\right| \leq |\mathbf{a}_i|^\top\left(\frac{1}{|x_i|}|\mathbf{x}|\right) \tag{8.127}$$

where the inequality is a generalization of the triangular inequality.

Replacing each component of $|\mathbf{x}|/|x_i|$ by unity (which is at least as large), we obtain

$$|\mathbf{a}_i|^\top \left(\frac{1}{|x_i|} |\mathbf{x}| \right) \leq |\mathbf{a}_i|^\top \mathbf{1} = \sum_{j=1}^m |a_{ij}| \tag{8.128}$$

It follows that

$$|\lambda| \leq \sum_{j=1}^m |a_{ij}| \tag{8.129}$$

Hence, if $\sum_{j=1}^m |a_{ij}| < 1$ for all i, the result is established.

In fact, we have established the more general result that the maximum absolute eigenvalue of any square matrix $\mathbf{A}$ is less than or equal to the maximum row sum of the absolute values of the elements in $\mathbf{A}$.[5]   □

Since $\mathbf{A}$ and $\mathbf{A}^\top$ have the same eigenvalues, Theorem 8.5.2 could also be written in terms of the column sums of $\mathbf{A}$, which are the row sums of $\mathbf{A}^\top$.

When $\boldsymbol{\Phi}_0$ is set to $\mathbf{0}$, (8.119) yields a solution for the homogeneous equation, namely

$$\mathbf{y}_t^h = \boldsymbol{\Phi}_1^{t-1} \mathbf{y}_1 \tag{8.130}$$

This is entirely consistent with the form of the solution given in (8.118). From (8.117), we have that $\mathbf{y}_1 = \sum_{i=1}^m A_i \mathbf{a}_i s_i^0 = \sum_{i=1}^m A_i \mathbf{a}_i$. Substituting this into (8.130) gives

$$\mathbf{y}_t^h = \boldsymbol{\Phi}_1^{t-1} \sum_{i=1}^m A_i \mathbf{a}_i = \boldsymbol{\Phi}_1^{t-2} \sum_{i=1}^m \boldsymbol{\Phi}_1 A_i \mathbf{a}_i = \boldsymbol{\Phi}_1^{t-2} \sum_{i=1}^m A_i \mathbf{a}_i s_i \tag{8.131}$$

since $A_i \mathbf{a}_i$ is an eigenvector of $\boldsymbol{\Phi}_1$ with eigenvalue s_i. Repeating this procedure yields

$$\mathbf{y}_t^h = \boldsymbol{\Phi}_1^{t-2} \sum_{i=1}^m A_i \mathbf{a}_i s_i = \boldsymbol{\Phi}_1^{t-3} \sum_{i=1}^m \boldsymbol{\Phi}_1 A_i \mathbf{a}_i s_i = \boldsymbol{\Phi}_1^{t-3} \sum_{i=1}^m A_i \mathbf{a}_i s_i^2 \tag{8.132}$$

and, ultimately,

$$\mathbf{y}_t^h = \sum_{i=1}^m \boldsymbol{\Phi}_1 A_i \mathbf{a}_i s_i^{t-2} = \sum_{i=1}^m A_i \mathbf{a}_i s_i^{t-1} \tag{8.133}$$

Hence $\sum_{i=1}^m A_i \mathbf{a}_i s_i^{t-1} = \boldsymbol{\Phi}_1^{t-1} \mathbf{y}_1 = \mathbf{y}_t^h$.

8.5.3 *Scalar approach*

Finally, we note that, when the number of equations in the system, m, is small, the homogeneous solution and the steady-state solution may be obtained using only scalar algebra. The main idea is that, with appropriate substitution, the homogeneous system may be reduced to a single linear, autonomous difference equation of order m, which may be solved using the method described in Section 8.4.2.

Suppose, for instance, that $m = 2$, so that the system may be written as

$$y_{1t} = \delta_1 + \phi_{111} y_{1(t-1)} + \phi_{112} y_{2(t-1)} \tag{8.134}$$

$$y_{2t} = \delta_2 + \phi_{121} y_{1(t-1)} + \phi_{122} y_{2(t-1)} \tag{8.135}$$

If a steady-state solution exists then we have

$$y_1^* = \delta_1 + \phi_{111} y_1^* + \phi_{112} y_2^* \tag{8.136}$$

$$y_2^* = \delta_2 + \phi_{121} y_1^* + \phi_{122} y_2^* \tag{8.137}$$

or

$$(1 - \phi_{111}) y_1^* = \delta_1 + \phi_{112} y_2^* \tag{8.138}$$

$$(1 - \phi_{122}) y_2^* = \delta_2 + \phi_{121} y_1^* \tag{8.139}$$

These two simultaneous equations are easily solved to give

$$y_1^* = \frac{(1 - \phi_{122})\delta_1 + \phi_{112}\delta_2}{(1 - \phi_{111})(1 - \phi_{122}) - \phi_{112}\phi_{121}} \tag{8.140}$$

$$y_2^* = \frac{\phi_{121}\delta_1 + (1 - \phi_{111})\delta_2}{(1 - \phi_{111})(1 - \phi_{122}) - \phi_{112}\phi_{121}} \tag{8.141}$$

The existence of y_1^* and y_2^* depends on the denominator in (8.140) and (8.141) being non-zero. This denominator is just the determinant of the matrix $\mathbf{I} - \boldsymbol{\Phi}_1$ in the two-variable case, which we already know from the earlier general analysis must be non-zero for there to be a steady-state solution. The reader should confirm that the solution is identical to that which emerges from use of the general formula (8.113) when $m = 2$.

To solve the homogeneous form of the system, namely

$$y_{1t} = \phi_{111} y_{1(t-1)} + \phi_{112} y_{2(t-1)} \tag{8.142}$$

$$y_{2t} = \phi_{121} y_{1(t-1)} + \phi_{122} y_{2(t-1)} \tag{8.143}$$

we proceed as follows. First, using (8.143), substitute for $y_{2(t-1)}$ in (8.142) to obtain

$$y_{1t} = \phi_{111} y_{1(t-1)} + \phi_{112}(\phi_{121} y_{1(t-2)} + \phi_{122} y_{2(t-2)}) \tag{8.144}$$

Second, substitute for $y_{2(t-2)}$ in (8.144), using the fact that, from (8.142),

$$y_{2(t-2)} = \frac{y_{1(t-1)} - \phi_{111} y_{1(t-2)}}{\phi_{112}} \tag{8.145}$$

assuming $\phi_{112} \neq 0$. We then have

$$\begin{aligned} y_{1t} &= \phi_{111} y_{1(t-1)} + \phi_{122} y_{1(t-1)} + \phi_{112}\phi_{121} y_{1(t-2)} - \phi_{111}\phi_{122} y_{1(t-2)} \\ &= (\phi_{111} + \phi_{122}) y_{1(t-1)} + (\phi_{112}\phi_{121} - \phi_{111}\phi_{122}) y_{1(t-2)} \end{aligned} \tag{8.146}$$

which is a linear, autonomous, second-order, homogeneous difference equation in one variable that can be solved using the results from Section 8.4.2.

We note, in particular, the characteristic equation associated with (8.146), which, recalling (8.80), is

$$s^2 - (\phi_{1_{11}} + \phi_{1_{22}})s - (\phi_{1_{12}}\phi_{1_{21}} - \phi_{1_{11}}\phi_{1_{22}}) = 0 \tag{8.147}$$

The three possibilities for the two solutions of this type of equation are given in (8.82), (8.83) and (8.86). Taking the first case of two different real roots by way of illustration, the general homogeneous solution for y_{1t} is

$$y_{1t}^h = A_1 s_1^{t-1} + A_2 s_2^{t-1} \tag{8.148}$$

Making use of (8.145), we may substitute this result to find the general homogeneous solution for y_{2t}, which is

$$y_{2t}^h = \frac{A_1 s_1^t + A_2 s_2^t - \phi_{1_{11}}(A_1 s_1^{t-1} + A_2 s_2^{t-1})}{\phi_{1_{12}}} \tag{8.149}$$

or, after slight rearrangement,

$$y_{2t}^h = \frac{(s_1 - \phi_{1_{11}})A_1 s_1^{t-1} + (s_2 - \phi_{1_{11}})A_2 s_2^{t-1}}{\phi_{1_{12}}} \tag{8.150}$$

The full details of the general homogeneous solutions in the other two possible cases, i.e. when there is a real root with multiplicity two, and when there is a pair of conjugate complex roots, are the subject of Exercise 8.16.

The complete general solutions then follow simply as $y_{1t} = y_1^* + y_{1t}^h$, using (8.140) and (8.148) to substitute for the components on the right-hand side, and $y_{2t} = y_2^* + y_{2t}^h$, using (8.141) and (8.150) for the right-hand side substitutions. If the initial values y_{11} and y_{21} are given, then unique numerical values for the constants A_1 and A_2 may be found by a similar method to that described earlier.

EXAMPLE 8.5.1 Using the direct matrix method and the method of substitution, solve the system of difference equations

$$y_{1t} = 1 + y_{1(t-1)} + \tfrac{1}{2}y_{2(t-1)} \tag{8.151}$$

$$y_{2t} = 2 + 4y_{1(t-1)} + 3y_{2(t-1)} \tag{8.152}$$

Find the complete particular solution when $y_{11} = 2$ and $y_{21} = 1$, and comment on the dynamic behaviour of the system over time.

In matrix notation this system can be written as $\mathbf{y}_t = \mathbf{\Phi}_0 + \mathbf{\Phi}_1\mathbf{y}_{t-1}$, where the parameter matrices are

$$\mathbf{\Phi}_0 = \begin{bmatrix} 1 \\ 2 \end{bmatrix} \quad \text{and} \quad \mathbf{\Phi}_1 = \begin{bmatrix} 1 & \tfrac{1}{2} \\ 4 & 3 \end{bmatrix} \tag{8.153}$$

Using the direct method, the steady-state solution, which exists because $\det(\mathbf{I} - \mathbf{\Phi}_1) = -2$ and $\mathbf{I} - \mathbf{\Phi}_1$ is non-singular, is

$$\mathbf{y}^* = (\mathbf{I} - \mathbf{\Phi}_1)^{-1}\mathbf{\Phi}_0 = \begin{bmatrix} 0 & -\frac{1}{2} \\ -4 & -2 \end{bmatrix}^{-1} \begin{bmatrix} 1 \\ 2 \end{bmatrix}$$

$$= -\frac{1}{2}\begin{bmatrix} -2 & \frac{1}{2} \\ 4 & 0 \end{bmatrix}\begin{bmatrix} 1 \\ 2 \end{bmatrix} = \begin{bmatrix} \frac{1}{2} \\ -2 \end{bmatrix} = \begin{bmatrix} y_1^* \\ y_2^* \end{bmatrix} \tag{8.154}$$

The general solution of the homogeneous form of the system requires the eigenvalues and eigenvectors of $\mathbf{\Phi}_1$. The eigenvalues are given by the solution of the characteristic equation $\det(\mathbf{\Phi}_1 - s\mathbf{I}) = 0$ or

$$(1 - s)(3 - s) - 2 = s^2 - 4s + 1 = 0 \tag{8.155}$$

Using the usual method of solution of a quadratic equation, they are found to be

$$s_1 = 2 + \sqrt{3} \quad \text{and} \quad s_2 = 2 - \sqrt{3} \tag{8.156}$$

Substitution of these two different real roots, in turn, into the equation $(\mathbf{\Phi}_1 - s\mathbf{I})\mathbf{a} = \mathbf{0}$, and solution of this equation for $\mathbf{a}$, yields the corresponding eigenvectors

$$\mathbf{a}_1 = \begin{bmatrix} 1 \\ 2 + 2\sqrt{3} \end{bmatrix} \quad \text{and} \quad \mathbf{a}_2 = \begin{bmatrix} 1 \\ 2 - 2\sqrt{3} \end{bmatrix} \tag{8.157}$$

Hence, from (8.117), we have the general homogeneous solution

$$\mathbf{y}_t^h = \sum_{i=1}^{2} A_i \mathbf{a}_i s_i^{t-1} = A_1\begin{bmatrix} 1 \\ 2 + 2\sqrt{3} \end{bmatrix}(2 + \sqrt{3})^{t-1} + A_2\begin{bmatrix} 1 \\ 2 - 2\sqrt{3} \end{bmatrix}(2 - \sqrt{3})^{t-1}$$

$$= \begin{bmatrix} A_1(2 + \sqrt{3})^{t-1} \\ A_1(2 + 2\sqrt{3})(2 + \sqrt{3})^{t-1} \end{bmatrix} + \begin{bmatrix} A_2(2 - \sqrt{3})^{t-1} \\ A_2(2 - 2\sqrt{3})(2 - \sqrt{3})^{t-1} \end{bmatrix} = \begin{bmatrix} y_{1t}^h \\ y_{2t}^h \end{bmatrix} \tag{8.158}$$

and then, from (8.118), the complete general solution is

$$\mathbf{y}_t = \mathbf{y}^* + \mathbf{y}_t^h = \begin{bmatrix} \frac{1}{2} \\ -2 \end{bmatrix} + \begin{bmatrix} A_1(2 + \sqrt{3})^{t-1} \\ A_1(2 + 2\sqrt{3})(2 + \sqrt{3})^{t-1} \end{bmatrix} + \begin{bmatrix} A_2(2 - \sqrt{3})^{t-1} \\ A_2(2 - 2\sqrt{3})(2 - \sqrt{3})^{t-1} \end{bmatrix} \tag{8.159}$$

Turning to the method of substitution, (8.140) and (8.141) produce the particular solutions

$$y_1^* = \frac{(1 - \phi_{1_{22}})\delta_1 + \phi_{1_{12}}\delta_2}{(1 - \phi_{1_{11}})(1 - \phi_{1_{22}}) - \phi_{1_{12}}\phi_{1_{21}}} = \frac{(1 - 3) + (\frac{1}{2} \times 2)}{(1 - 1)(1 - 3) - (\frac{1}{2} \times 4)} = \frac{-1}{-2} = \frac{1}{2} \tag{8.160}$$

and

$$y_2^* = \frac{\phi_{121}\delta_1 + (1 - \phi_{111})\delta_2}{(1 - \phi_{111})(1 - \phi_{122}) - \phi_{112}\phi_{121}} = \frac{4 + [(1-1) \times 2]}{(1-1)(1-3) - (\frac{1}{2} \times 4)} = \frac{4}{-2} = -2$$

(8.161)

in agreement with (8.154), while (8.147) gives the characteristic equation associated with the homogeneous form

$$
\begin{aligned}
&s^2 - (\phi_{111} + \phi_{122})s - (\phi_{112}\phi_{121} - \phi_{111}\phi_{122})\\
&= s^2 - (1+3)s - (\tfrac{1}{2} \times 4 - 1 \times 3)\\
&= s^2 - 4s + 1 = 0
\end{aligned}
$$

(8.162)

which we note is identical to the characteristic equation (8.155) used in the earlier eigen-value calculations. Thus the required roots are as already given in (8.156), and the individual general homogeneous solutions follow by substitution into (8.148) and (8.150) as

$$y_{1t}^h = A_1 s_1^{t-1} + A_2 s_2^{t-1} = A_1 \left(2 + \sqrt{3}\right)^{t-1} + A_2 \left(2 - \sqrt{3}\right)^{t-1}$$

(8.163)

and

$$
\begin{aligned}
y_{2t}^h &= \frac{(s_1 - \phi_{111})A_1 s_1^{t-1} + (s_2 - \phi_{111})A_2 s_2^{t-1}}{\phi_{112}}\\
&= \frac{(2 + \sqrt{3} - 1)A_1(2 + \sqrt{3})^{t-1} + (2 - \sqrt{3} - 1)A_2(2 - \sqrt{3})^{t-1}}{1/2}\\
&= A_1(2 + 2\sqrt{3})(2 + \sqrt{3})^{t-1} + A_2(2 - 2\sqrt{3})(2 - \sqrt{3})^{t-1}
\end{aligned}
$$

(8.164)

These results are identical to those contained in (8.158).

With $\sqrt{3}$ evaluated to five places of decimals, the complete general solution becomes

$$y_{1t} = y_1^* + y_{1t}^h \approx 0.5 + A_1(3.73205)^{t-1} + A_2(0.26795)^{t-1}$$

(8.165)

$$y_{2t} = y_2^* + y_{2t}^h \approx -2 + 5.46410A_1(3.73205)^{t-1} - 1.46410A_2(0.26795)^{t-1}$$

(8.166)

Given the initial conditions $y_{11} = 2$ and $y_{21} = 1$, we have, using (8.158), or (8.163) and (8.164), that

$$y_{11} = A_1 + A_2 = 2$$

(8.167)

$$y_{21} = (2 + 2\sqrt{3})A_1 + (2 - 2\sqrt{3})A_2 = 1$$

(8.168)

Solution of these two simultaneous equations yields $A_1 = 1 - \frac{\sqrt{3}}{4} \approx 0.56699$ and $A_2 = 1 + \frac{\sqrt{3}}{4} \approx 1.43301$. Substitution of these values into (8.165) and (8.166) gives the unique, complete particular solution for the system.

As one of the two eigenvalues of Φ_1 is greater than one in absolute value, namely $s_1 = 2 + \sqrt{3} \approx 3.73205$, this system is not asymptotically stable and will not, therefore, converge towards its steady state from the initial state in which $y_{11} = 2$ and $y_{21} = 1$. Rather, y_{1t} and y_{2t} will both diverge and become ever larger as $t \to \infty$. A plot of the time path of the two variables for the first few time periods is left as an exercise; see Exercise 8.17. ◇

EXERCISES

8.1 Classify the following difference equations:

(a) $y_t = 2y_{t-1} + \sqrt[3]{t}$
(b) $y_t = 1/y_{t-1} + 2y_{t-3}$
(c) $y_t = \phi_0 + \phi_1 y_{t-1} + \phi_2 y_{t-2}^2$
(d) $y_t = \phi_0 + \phi_1 y_{t-1} + \cdots + \phi_p y_{t-p}$
(e) $\Delta_k y_t = 5y_{t-k+1} + t$

8.2 Consider the following first-order difference equations:

(a) $y_t = 2y_{t-1}$
(b) $y_t = 1 + y_{t-1}$
(c) $y_t = 0.1 - 0.5y_{t-1}$
(d) $y_t = 3y_{t-1} - 1$

In each case, (i) write the equation explicitly in terms of the change Δy_t, (ii) derive the general solution for the equation, (iii) find the particular solution when the initial condition is $y_1 = 5$, and (iv) evaluate the particular solution for $t = 10$.

8.3 Suppose you put €500 on deposit at a monthly interest rate of 0.25%. What will be the value of your investment after (a) two years, (b) five years, and (c) ten years? How many months will it take for your investment to exceed €1000 in value?

8.4 Using equation (8.32), determine and illustrate with a sketch the dynamic behaviour of y_t when $(y_1 - \phi_0/(1 - \phi_1))$ is negative and (a) $\phi_1 > 1$ and (b) $\phi_1 < -1$.

8.5 Let $C(L) = 1 - \phi_1 L$ and $D(L) = 1 + \phi_2 L$, where L is the lag operator and $|\phi_i| < 1$, $i = 1, 2$.

(a) Verify that $C^{-1}(L) = (1 + \phi_1 L + \phi_1^2 L^2 + \phi_1^3 L^3 + \cdots)$.
(b) Find the inverse lag polynomial $D^{-1}(L)$.

8.6 Consider the difference equation $y_t = \phi_0 + \phi_1 y_{t-1}$. Assuming $\phi_0 = 2$, examine and draw a sketch of the dynamic behaviour of y_t for (a) $\phi_1 = -1$, (b) $\phi_1 = -\frac{1}{4}$, (c) $\phi_1 = 0$, (d) $\phi_1 = \frac{1}{4}$ and (e) $\phi_1 = 1$. In each case, find the steady-state value of y_t, if a steady-state solution exists. Repeat the exercise assuming $\phi_0 = -2$.

8.7 For the linear, non-autonomous, first-order difference equation $y_t = \phi_{0(t-1)} + \phi_{1(t-1)} y_{t-1}$, with initial value y_1, show that the solution for y_t ($t = 2, 3, \ldots$) is

$$y_t = \phi_{0(t-1)} + \sum_{j=1}^{t-2} \phi_{0j} \prod_{i=j+1}^{t-1} \phi_{1i} + y_1 \prod_{i=1}^{t-1} \phi_{1i}$$

Now, suppose that $\phi_{0(t-1)}=1/(t-1)$, $\phi_{1(t-1)}=2/(t-1)$ and $y_1=1$. Calculate the values of y_2 to y_7, inclusive, plot these values against time, and comment on the steady-state solution, if one exists.

8.8 Using equation (8.61), reproduced in Exercise 8.7, find the solution for y_t $(t=2,3,\ldots)$ when (a) $\phi_{0(t-1)}$, but not $\phi_{1(t-1)}$, is constant for all t, and (b) $\phi_{1(t-1)}$, but not $\phi_{0(t-1)}$, is constant for all t. Now suppose that $\phi_{0(t-1)}=2$ for all t and $\phi_{1(t-1)}=t-1$. Examine the time path of y_t for $t=2$ to $t=5$, inclusive.

8.9 Consider the linear, non-autonomous, first-order difference equation used in Example 8.3.2, namely

$$y_t = \frac{t-1}{10} + \left(\frac{1}{2}\right)^{t-1} y_{t-1}$$

with initial condition $y_1=16$.

(a) Find the values for y_2, y_3, y_4, and y_6, y_7 and y_8 (recall that y_5 is calculated in Example 8.3.2).
(b) How many time periods must elapse before y_t reaches its minimum value?
(c) How many time periods must elapse before y_t exceeds its initial value?
(d) Indicate your answers to parts (b) and (c) on a sketch of the time path of y_t.

8.10 Working to four places of decimals, find the roots of the lag polynomial and the roots of the characteristic equation associated with the difference equation $y_t=0.5y_{t-1}+0.25y_{t-2}$. Comment on the relationship between the two pairs of roots.

8.11 Show that the solution for the homogeneous form of the linear, autonomous, second-order difference equation, given in the first line of equation (8.86) as

$$y_t^h = r^{t-1}[A_1 \cos(t-1)\theta + A_2 \sin(t-1)\theta]$$

may also be written as

$$y_t^h = Ar^{t-1} \cos[(t-1)\theta + \omega]$$

where A and ω are arbitrary constants.

8.12 Verify that $y_t^h = A_1 s_1^{t-1} + A_2 s_2^{t-1}$, where s_1 and s_2 are as defined in equation (8.81), is a solution for the homogeneous equation $y_t - \phi_1 y_{t-1} - \phi_2 y_{t-2}=0$ when $\phi_1^2+4\phi_2>0$; and that $y_t^h = r^{t-1}[A_1 \cos(t-1)\theta + A_2 \sin(t-1)\theta]$ is a solution to the same homogeneous equation when $\phi_1^2+4\phi_2<0$.

8.13 Find the general solution for each of the following difference equations:

(a) $y_t=0.8y_{t-1}+0.2y_{t-2}$; and
(b) $y_t=1+y_{t-1}-0.25y_{t-2}$.

Also find the particular solution in each case, given initial conditions $y_1=0.5$ and $y_2=1$, and compare the dynamic behaviour of y_t in the two situations.

8.14 The stability condition for the linear, autonomous, second-order difference equation is that the roots of the characteristic equation $s^2 - \phi_1 s - \phi_2 = 0$ lie inside the unit circle, i.e. have modulus strictly less than unity; recall Theorem 8.4.1. Show that each of the following is equivalent to this condition.

(a) $|\phi_1| < 1 - \phi_2$ and $-\phi_2 < 1$.
(b) The roots of the lag polynomial associated with the difference equation, i.e. the solutions of $\phi(z) = 1 - \phi_1 z - \phi_2 z^2 = 0$, lie outside the unit circle, i.e. have modulus strictly greater than unity.

8.15 Show that the lag polynomial associated with the difference equation $y_t = 0.75 y_{t-1} + 0.25 y_{t-2}$ has one unit root and one root with modulus greater than unity. Compare the roots of the characteristic equation of the difference equation with the roots of its lag polynomial, and comment. Making use of the factors of the lag polynomial, derive an asymptotically stable difference equation for Δy_t.

8.16 For the two-variable system of first-order difference equations given in equations (8.134) and (8.135), derive the general homogeneous solutions in the cases when the associated characteristic equation has (a) a real root with multiplicity two, and (b) a pair of conjugate complex roots.

8.17 Plot the time paths of y_{1t} and y_{2t}, given the system and particular complete solution in Example 8.5.3.

8.18 Solve the following systems of difference equations, and comment on their dynamic behaviour.

(a) $y_{1t} = 0.5 y_{2(t-1)}$
    $y_{2t} = y_{1(t-1)}$
(b) $y_{1t} = 1 + 0.5 y_{1(t-1)} - 0.5 y_{2(t-1)}$
    $y_{2t} = 2 + 0.5 y_{1(t-1)} + y_{2(t-1)}$
    with initial conditions $y_{11} = y_{21} = 1$
(c) $y_{1t} = 1 + 0.5 y_{1(t-1)} - 0.5 y_{2(t-1)} + 0.2 y_{3(t-1)}$
    $y_{2t} = 0.5 + 0.6 y_{2(t-1)} - 0.4 y_{3(t-1)}$
    $y_{3t} = 1.25 + 0.75 y_{3(t-1)}$
    with initial conditions $y_{11} = 2$, $y_{21} = 2.5$, $y_{31} = 3$

9 Vector calculus

9.1 Introduction

This chapter covers selected topics from multivariate or vector calculus. It has two simultaneous objectives, namely, to reinforce the reader's grasp of topics from earlier chapters, once again including linear algebra, and to prepare for their application to economic and financial problems.

As noted on p. xxii, the reader is assumed to be familiar with both the theory and practice of single-variable differential and integral calculus.

From an economic perspective, we are now in a position to introduce new concepts and techniques that will be essential tools in the later chapters dealing with economic and financial applications. The notion of differentiation is extended in Section 9.2 from a univariate context to a multivariate context. Some examples of differentiation in matrix notation follow. We go on in Section 9.3 to provide matrix generalizations of the chain rule and the product rule. The concept of elasticity, which may be familiar to some readers from introductory economics courses, is introduced in Section 9.4, as is that of a directional derivative in Section 9.5. This allows us to introduce a matrix generalization of Taylor's theorem in Section 9.6.

The chapter continues with an introduction to the use of multiple integrals in Section 9.7, incorporating several standard but important results on changing the order of differentiation and integration. The final section deals with the implicit function theorem.

The results developed in this chapter will be applied to economics and finance throughout the remainder of this book.

9.2 Partial and total derivatives

It is often required to differentiate functions of several variables, vector-valued functions or functions defined by matrix expressions more generally. This section extends familiar concepts from univariate calculus to a multivariate context.

We begin by giving the definition of a partial derivative, which is the rate of change of a function of several variables with respect to one of those variables, holding all the other variables constant. We then introduce the Jacobian matrix, the gradient vector and the Hessian matrix. Gradient vectors and Hessian matrices are used in the theory of optimization in connection with the determination of the maxima and minima of real-valued functions of several variables. They will be encountered again in the next chapter. The section concludes with some examples.

Readers should be aware of subtle differences in notation between the univariate case ($n = 1$) and the multivariate case ($n > 1$). Statements and shorthands that make sense in univariate calculus must generally be modified for multivariate calculus.

9.2.1 *Definitions*

DEFINITION 9.2.1 The jth **first-order partial derivative** of a real-valued function of n variables, say $f : \mathbb{R}^n \to \mathbb{R}$, at a vector $\mathbf{x}$ in its domain is its derivative at that point with respect to the jth variable, treating the other $n - 1$ variables as constants, and is denoted

$$\frac{\partial f}{\partial x_j}(\mathbf{x}) \tag{9.1}$$

DEFINITION 9.2.2 The jkth **second-order partial derivative** of a real-valued function of n variables, say $f : \mathbb{R}^n \to \mathbb{R}$, at a vector $\mathbf{x}$ in its domain is the derivative at that point with respect to the jth variable of its kth first-order partial derivative, treating all variables except the jth as constants, and is denoted

$$\frac{\partial^2 f}{\partial x_j \partial x_k}(\mathbf{x}) \tag{9.2}$$

The partial derivatives of a vector-valued function are the partial derivatives of its real-valued component functions.

DEFINITION 9.2.3 The **(total) derivative** of a real-valued function of n variables, say $f : \mathbb{R}^n \to \mathbb{R}$, at a vector $\mathbf{x}$ in its domain is the n-dimensional *row vector* of its first-order partial derivatives at $\mathbf{x}$, denoted

$$f'(\mathbf{x}) \equiv \left[\frac{\partial f}{\partial x_j}(\mathbf{x}) \right]_{1 \times n} \tag{9.3}$$

The (total) derivative of a vector-valued function with values in $\mathbb{R}^m$ is the $m \times n$ matrix of partial derivatives whose ith row is the total derivative of the ith component function, likewise denoted

$$f'(\mathbf{x}) \equiv \left[\frac{\partial f^i}{\partial x_j}(\mathbf{x}) \right]_{m \times n} \tag{9.4}$$

The total derivative is often referred to also as the **Jacobian** or **Jacobian matrix** of the function.[1]

The total derivative can also be defined using a limiting argument similar to that used to define the derivative of a function of a single variable. The interested reader is referred to Binmore (1982, pp. 203–6) for further details of that alternative approach.

Jacobian matrices are of fundamental importance in most of the results from vector calculus presented later in the book.

DEFINITION 9.2.4 The **gradient** or **gradient vector** of a real-valued function of n variables, say $f: \mathbb{R}^n \to \mathbb{R}$, at a vector $\mathbf{x}$ in its domain is the *column vector* formed by transposing its total derivative or Jacobian, variously denoted

$$\nabla f(\mathbf{x}) \equiv \operatorname{grad} f(\mathbf{x}) \equiv f'(\mathbf{x})^\top \equiv \frac{\partial f}{\partial \mathbf{x}}(\mathbf{x}) \equiv \left[\frac{\partial f}{\partial x_i}(\mathbf{x}) \right]_{n \times 1} \tag{9.5}$$

Note that $f'(\mathbf{x})^\top \equiv [f'(\mathbf{x})]^\top$, i.e. the transpose symbol applies to the vector of partial derivatives, and not to the vector $\mathbf{x}$ at which the partial derivatives are evaluated. For a real-valued function ($n = 1$),

$$f'(x) = \frac{df}{dx}(x) = \frac{\partial f}{\partial x}(x) = \nabla f(x) = \operatorname{grad} f(x) = f'(x)^\top$$

since all of these quantities are scalars (1×1 vectors).

DEFINITION 9.2.5 A real-valued or vector-valued function of several variables $f: X \to Y$ ($X \subseteq \mathbb{R}^n, Y \subseteq \mathbb{R}^m$) is said to be **differentiable at x** if all its (first-order) partial derivatives exist at $\mathbf{x}$.

DEFINITION 9.2.6 The function $f: X \to Y$ ($X \subseteq \mathbb{R}^n, Y \subseteq \mathbb{R}^m$) is said to be:

(a) **differentiable** if and only if it is differentiable at every point of its domain X;
(b) **continuously differentiable** (C^1) if and only if it is differentiable at every point of its domain X and its derivative is a continuous function; and
(c) **twice continuously differentiable** (C^2) if and only if f' exists and is C^1.

DEFINITION 9.2.7 The **Hessian matrix**[2] or simply the **Hessian** of a real-valued function of n variables is the square matrix of its second-order partial derivatives, denoted

$$f''(\mathbf{x}) \equiv \frac{\partial^2 f}{\partial \mathbf{x} \partial \mathbf{x}^\top}(\mathbf{x}) \equiv \left[\frac{\partial^2 f}{\partial x_j \partial x_k}(\mathbf{x}) \right]_{n \times n} \tag{9.6}$$

We will see later (Theorem 9.7.2) that, provided that certain quite general conditions are satisfied, the Hessian matrix is a symmetric matrix.

Note that $\partial f / \partial x_j$ and $\partial^2 f / \partial x_j \partial x_k$ are real-valued functions defined on the same domain as f, $j, k = 1, 2, \ldots, n$; and f' and f'' are matrix-valued functions also defined on the same domain as f.

Note that, if $f: \mathbb{R}^n \to \mathbb{R}$, then, strictly speaking, the second derivative (Hessian) of f is the derivative of the vector-valued function

$$(f')^\top : \mathbb{R}^n \to \mathbb{R}^n : \mathbf{x} \mapsto f'(\mathbf{x})^\top = \nabla f(\mathbf{x}) = \operatorname{grad} f(\mathbf{x}) \tag{9.7}$$

We will see in the next chapter that the positive or negative definiteness or semi-definiteness of the Hessian matrix is of major significance in the theory of optimization.

We conclude this collection of definitions with some notation for maximization and minimization. We will use $\max_{\mathbf{x} \in X} f(\mathbf{x})$ to denote the maximum value that f can take on the set X, and $\min_{\mathbf{x} \in X} f(\mathbf{x})$ to denote the minimum value that f can take on the set X.

We also need a notation for the value(s) of $\mathbf{x}$ at which this maximum or minimum is attained. These will be denoted, respectively, by $\text{argmax}_{\mathbf{x} \in X} f(\mathbf{x})$ and $\text{argmin}_{\mathbf{x} \in X} f(\mathbf{x})$. Thus

$$\max_{\mathbf{x} \in X} f(\mathbf{x}) = f\left(\underset{\mathbf{x} \in X}{\text{argmax}}\, f(\mathbf{x}) \right) \tag{9.8}$$

and

$$\min_{\mathbf{x} \in X} f(\mathbf{x}) = f\left(\underset{\mathbf{x} \in X}{\text{argmin}}\, f(\mathbf{x}) \right) \tag{9.9}$$

9.2.2 Examples

In this section we develop a number of useful results that allow us to handle differentiation using matrix notation. We begin with the differentiation of matrix linear forms and then go on to the differentiation of quadratic forms.

Linear forms

Let $\mathbf{a}$ and $\mathbf{x}$ be $n \times 1$ matrices, and consider

$$\mathbf{a}^\top \mathbf{x} = a_1 x_1 + a_2 x_2 + \cdots + a_n x_n \tag{9.10}$$

The expression in (9.10) can be viewed either as a function of $\mathbf{x}$ with $\mathbf{a}$ treated as a constant vector, say $f(\mathbf{x})$, or as a function of $\mathbf{a}$ with $\mathbf{x}$ treated as a constant vector, say $g(\mathbf{a})$. Using the former interpretation,

$$\frac{\partial(\mathbf{a}^\top \mathbf{x})}{\partial x_1} = a_1, \quad \frac{\partial(\mathbf{a}^\top \mathbf{x})}{\partial x_2} = a_2, \quad \ldots, \quad \text{for all } \mathbf{x} \tag{9.11}$$

Thus the typical first-order partial derivative is $\partial(\mathbf{a}^\top \mathbf{x})/\partial x_i = a_i$, and arranging all n partial derivatives as elements in an $n \times 1$ matrix, we have

$$f'(\mathbf{x})^\top = \frac{\partial(\mathbf{a}^\top \mathbf{x})}{\partial \mathbf{x}} = \left[\frac{\partial(\mathbf{a}^\top \mathbf{x})}{\partial x_j} \right] = \mathbf{a} \tag{9.12}$$

All second-order partial derivatives are zero and may be associated with the elements of a zero matrix. Thus we may write the Hessian matrix as

$$f''(\mathbf{x}) = \frac{\partial^2(\mathbf{a}^\top \mathbf{x})}{\partial \mathbf{x} \partial \mathbf{x}^\top} = \left[\frac{\partial^2(\mathbf{a}^\top \mathbf{x})}{\partial x_j \partial x_k} \right] = \mathbf{0}_{n \times n} \tag{9.13}$$

Using the alternative interpretation in which $\mathbf{a}$ is variable and $\mathbf{x}$ fixed, we have by similar reasoning:

$$g'(\mathbf{a})^\top = \frac{\partial(\mathbf{a}^\top \mathbf{x})}{\partial \mathbf{a}} = \left[\frac{\partial(\mathbf{a}^\top \mathbf{x})}{\partial a_j} \right] = \mathbf{x} \tag{9.14}$$

$$g''(\mathbf{a}) = \frac{\partial^2(\mathbf{a}^\top \mathbf{x})}{\partial \mathbf{a} \partial \mathbf{a}^\top} = \left[\frac{\partial(\mathbf{a}^\top \mathbf{x})}{\partial a_j \partial a_k} \right] = \mathbf{0}_{n \times n} \tag{9.15}$$

The fact that the second-order derivatives of a linear form vanish indicates that such a function has zero **curvature**. The sign and magnitude of the second-order derivatives of a nonlinear function provide information about the nature and magnitude of the curvature of that function. We will return to this interpretation when discussing Taylor's theorem in Section 9.6 and when discussing convex and concave functions in Section 10.2.

Quadratic forms

First, however, we consider the derivatives and curvature of the simplest type of nonlinear function, a quadratic form

$$\mathbf{x}^\top \mathbf{A}\mathbf{x} = a_{11}x_1^2 + a_{22}x_2^2 + \cdots + a_{nn}x_n^2 + 2a_{12}x_1x_2$$
$$+ \cdots + 2a_{1n}x_1x_n + \cdots + 2a_{(n-1)n}x_{n-1}x_n \tag{9.16}$$

where $\mathbf{A}$ is taken to be symmetric of order $n \times n$, so that we can use the fact that $a_{12} = a_{21}$ and so on in the expansion.

Once again, the expression in (9.16) can be viewed either as a function of $\mathbf{x}$ with $\mathbf{A}$ treated as a constant matrix, or as a function of $\mathbf{A}$ with $\mathbf{x}$ treated as a constant vector. Using the former interpretation,

$$\frac{\partial(\mathbf{x}^\top \mathbf{A}\mathbf{x})}{\partial x_1} = 2(a_{11}x_1 + a_{12}x_2 + \cdots + a_{1n}x_n) \tag{9.17}$$

$$\frac{\partial(\mathbf{x}^\top \mathbf{A}\mathbf{x})}{\partial x_2} = 2(a_{21}x_1 + a_{22}x_2 + \cdots + a_{2n}x_n) \tag{9.18}$$

and so on. The typical first-order partial derivative in this case is

$$\frac{\partial(\mathbf{x}^\top \mathbf{A}\mathbf{x})}{\partial x_j} = 2\sum_{k=1}^{n} a_{jk}x_k \tag{9.19}$$

Stacking all of the first-order partial derivatives in an $n \times 1$ matrix, we have

$$\frac{\partial(\mathbf{x}^\top \mathbf{A}\mathbf{x})}{\partial \mathbf{x}} = \left[\frac{\partial(\mathbf{x}^\top \mathbf{A}\mathbf{x})}{\partial x_j} \right] = 2\mathbf{A}\mathbf{x} \tag{9.20}$$

Differentiating each of the n derivatives in $\partial(\mathbf{x}^\top \mathbf{A}\mathbf{x})/\partial \mathbf{x}$ in turn by each of the elements x_k produces n^2 second-order partial derivatives. The details of the differentiation are left as an exercise for the reader; see Exercise 9.3. Putting the resulting second-order derivatives as the elements of an $n \times n$ matrix, we have

$$\frac{\partial^2(\mathbf{x}^\top \mathbf{A}\mathbf{x})}{\partial \mathbf{x}\partial \mathbf{x}^\top} = \left[\frac{\partial^2(\mathbf{x}^\top \mathbf{A}\mathbf{x})}{\partial x_j \partial x_k} \right] = 2\mathbf{A} \tag{9.21}$$

In the case of non-symmetric $\mathbf{A}$, then

$$\frac{\partial(\mathbf{x}^\top \mathbf{A}\mathbf{x})}{\partial \mathbf{x}} = (\mathbf{A} + \mathbf{A}^\top)\mathbf{x} \quad \text{and} \quad \frac{\partial^2(\mathbf{x}^\top \mathbf{A}\mathbf{x})}{\partial \mathbf{x}\partial \mathbf{x}^\top} = (\mathbf{A} + \mathbf{A}^\top) \tag{9.22}$$

Just as we can differentiate a function of a vector variable with respect to that vector variable, so we can differentiate a function of a matrix variable with respect to that matrix variable, although the notation can quickly become cumbersome.

This reasoning leads to the result that

$$\frac{\partial (\mathbf{x}^\top \mathbf{A} \mathbf{x})}{\partial \mathbf{A}} = \left[\frac{\partial (\mathbf{x}^\top \mathbf{A} \mathbf{x})}{\partial a_{ij}}\right] = [x_i x_j]_{n \times n} = \mathbf{x} \mathbf{x}^\top \tag{9.23}$$

In this case, as we are differentiating with respect to a_{ij} while holding the other matrix entries, in particular a_{ji}, constant, the matrix $\mathbf{A}$ is not constrained to be symmetric. Note that the derivative in (9.23) is just the outer product discussed in Exercise 1.9.

9.3   Chain rule and product rule

9.3.1   *Univariate chain rule*

It will be recalled from scalar calculus that, if $z = f(y)$ and $y = g(x)$ are scalars and thus $z = f(g(x)) = f \circ g(x)$, then

$$\frac{dz}{dx} = \frac{dz}{dy}\frac{dy}{dx} \tag{9.24}$$

This is called the **chain rule** of differential calculus. For a proof, see Binmore (1982, p. 99).

EXAMPLE 9.3.1 Let $z = 2y$ and $y = x^3$. Then $dz/dy = 2$ and $dy/dx = 3x^2$, hence

$$\frac{dz}{dx} = \frac{dz}{dy}\frac{dy}{dx} = 2(3x^2) = 6x^2 \tag{9.25}$$

It is easy to check the validity of the chain rule in this simple illustration directly, by substituting in z for y and differentiating z with respect to x:

$$\frac{dz}{dx} = \frac{d(2x^3)}{dx} = 6x^2 \tag{9.26}$$

$\diamond$

9.3.2   *Chain rule for partial derivatives*

Now, if $\mathbf{z} = f(\mathbf{y})$ and $\mathbf{y} = g(\mathbf{x})$ are vectors and thus $\mathbf{z} \equiv h(\mathbf{x}) = f(g(\mathbf{x})) = f \circ g(\mathbf{x})$, then the chain rule generalizes. Specifically, if $\mathbf{z}$ is $p \times 1$, $\mathbf{y}$ is $m \times 1$ and $\mathbf{x}$ is $n \times 1$, then the chain rule for partial derivatives states that, for $i = 1, 2, \ldots, p$ and $j = 1, 2, \ldots, n$,

$$\frac{\partial h^i}{\partial x_j}(\mathbf{x}) = \sum_{k=1}^{m} \frac{\partial f^i}{\partial y_k}(\mathbf{y})\frac{\partial g^k}{\partial x_j}(\mathbf{x})$$

$$= \frac{\partial f^i}{\partial y_1}(\mathbf{y})\frac{\partial g^1}{\partial x_j}(\mathbf{x}) + \frac{\partial f^i}{\partial y_2}(\mathbf{y})\frac{\partial g^2}{\partial x_j}(\mathbf{x}) + \cdots + \frac{\partial f^i}{\partial y_m}(\mathbf{y})\frac{\partial g^m}{\partial x_j}(\mathbf{x}) \tag{9.27}$$

The proof of this result is beyond the scope of this book; for details, the interested reader is referred to Binmore (1982, pp. 213–14).

EXAMPLE 9.3.2 Let $f: \mathbb{R}^2 \to \mathbb{R}^2$ and $g: \mathbb{R} \to \mathbb{R}^2$ be defined by

$$\mathbf{z} = \begin{bmatrix} z_1 \\ z_2 \end{bmatrix} = \begin{bmatrix} y_1 + y_2 \\ y_1 y_2 \end{bmatrix} = f(\mathbf{y}) \quad \text{and} \quad \mathbf{y} = \begin{bmatrix} y_1 \\ y_2 \end{bmatrix} = \begin{bmatrix} x \\ x^2 \end{bmatrix} = g(x) \tag{9.28}$$

with $h \equiv f \circ g$. Then

$$\frac{\partial h^1}{\partial x}(x) = \frac{\partial f^1}{\partial y_1}(\mathbf{y}) \frac{\partial g^1}{\partial x}(x) + \frac{\partial f^1}{\partial y_2}(\mathbf{y}) \frac{\partial g^2}{\partial x}(x) = 1 \times 1 + 1 \times 2x = 1 + 2x \tag{9.29}$$

$$\frac{\partial h^2}{\partial x}(x) = \frac{\partial f^2}{\partial y_1}(\mathbf{y}) \frac{\partial g^1}{\partial x}(x) + \frac{\partial f^2}{\partial y_2}(\mathbf{y}) \frac{\partial g^2}{\partial x}(x) = y_2 \times 1 + y_1 \times 2x = x^2 + 2x^2 = 3x^2 \tag{9.30}$$

The solution is easy to check directly:

$$z_1 = y_1 + y_2 = x + x^2 \quad \therefore \frac{dz_1}{dx} = 1 + 2x \tag{9.31}$$

$$z_2 = y_1 y_2 = x(x^2) = x^3 \quad \therefore \frac{dz_2}{dx} = 3x^2 \tag{9.32}$$

$\diamondsuit$

9.3.3 Chain rule for total derivatives

Stacking the results for partial derivatives (9.27) in matrix form, we have the following theorem.

THEOREM 9.3.1 (CHAIN RULE). *Let $f: \mathbb{R}^m \to \mathbb{R}^p$ and $g: \mathbb{R}^n \to \mathbb{R}^m$ be continuously differentiable functions and let $h: \mathbb{R}^n \to \mathbb{R}^p$ be defined by*

$$h(\mathbf{x}) \equiv f(g(\mathbf{x})) \tag{9.33}$$

Then

$$\underbrace{h'(\mathbf{x})}_{p \times n} = \underbrace{f'(g(\mathbf{x}))}_{p \times m} \underbrace{g'(\mathbf{x})}_{m \times n} \tag{9.34}$$

Proof: This is easily shown using the chain rule for partial derivatives; see Exercise 9.9. $\square$

EXAMPLE 9.3.3 Returning to Example 9.3.2,

$$f'(\mathbf{y}) = \begin{bmatrix} 1 & 1 \\ y_2 & y_1 \end{bmatrix} \quad \text{and} \quad g'(x) = \begin{bmatrix} 1 \\ 2x \end{bmatrix} \tag{9.35}$$

Therefore, by the chain rule for total derivatives,

$$h'(\mathbf{x}) = f'(\mathbf{y})\, g'(x) = \begin{bmatrix} 1 & 1 \\ y_2 & y_1 \end{bmatrix}\begin{bmatrix} 1 \\ 2x \end{bmatrix} = \begin{bmatrix} 1+2x \\ y_2 + 2xy_1 \end{bmatrix}$$

$$= \begin{bmatrix} 1+2x \\ x^2 + 2xx \end{bmatrix} = \begin{bmatrix} 1+2x \\ 3x^2 \end{bmatrix} \tag{9.36}$$

coinciding with the partial derivatives obtained by scalar calculations. ◇

One of the most common applications of the chain rule is the following corollary.

COROLLARY 9.3.2 *Let* $f:\mathbb{R}^{m+n} \to \mathbb{R}^p$ *and* $g:\mathbb{R}^n \to \mathbb{R}^m$ *be continuously differentiable functions, let* $\mathbf{x} \in \mathbb{R}^n$, *and define* $h:\mathbb{R}^n \to \mathbb{R}^p$ *by*

$$h(\mathbf{x}) \equiv f(g(\mathbf{x}), \mathbf{x}) \tag{9.37}$$

Partition the total derivative of f *into two submatrices comprising its first m columns and its last n columns and denoted*

$$\underbrace{f'(\cdot)}_{p\times(m+n)} = \Big[\underbrace{D_g f(\cdot)}_{p\times m} \quad \underbrace{D_{\mathbf{x}} f(\cdot)}_{p\times n} \Big] \tag{9.38}$$

Then the following holds:

$$h'(\mathbf{x}) = D_g f(g(\mathbf{x}), \mathbf{x}) g'(\mathbf{x}) + D_{\mathbf{x}} f(g(\mathbf{x}), \mathbf{x}) \tag{9.39}$$

Proof: The chain rule for partial derivatives can be used to calculate $\partial h^i(\mathbf{x})/\partial x_j$ in terms of partial derivatives of f and g for $i = 1, 2, \ldots, p$ and $j = 1, 2, \ldots, n$:

$$\frac{\partial h^i}{\partial x_j}(\mathbf{x}) = \sum_{k=1}^{m} \frac{\partial f^i}{\partial x_k}(g(\mathbf{x}), \mathbf{x})\frac{\partial g^k}{\partial x_j}(\mathbf{x}) + \sum_{k=m+1}^{m+n} \frac{\partial f^i}{\partial x_k}(g(\mathbf{x}), \mathbf{x})\frac{\partial x^{k-m}}{\partial x_j}(\mathbf{x}) \tag{9.40}$$

Note that

$$\frac{\partial x^l}{\partial x_j}(\mathbf{x}) = \delta_{lj} \equiv \begin{cases} 1 & \text{if } l = j \\ 0 & \text{otherwise} \end{cases} \tag{9.41}$$

which is just the Kronecker delta, defined in Section 1.5.2. Thus all but one of the terms in the second summation in (9.40) vanish, giving

$$\frac{\partial h^i}{\partial x_j}(\mathbf{x}) = \sum_{k=1}^{m} \frac{\partial f^i}{\partial x_k}(g(\mathbf{x}), \mathbf{x})\frac{\partial g^k}{\partial x_j}(\mathbf{x}) + \frac{\partial f^i}{\partial x_{m+j}}(g(\mathbf{x}), \mathbf{x}) \tag{9.42}$$

Stacking these scalar equations in matrix form and factoring yields:

$$
\begin{bmatrix}
\dfrac{\partial h^1}{\partial x_1}(\mathbf{x}) & \cdots & \dfrac{\partial h^1}{\partial x_n}(\mathbf{x}) \\
\vdots & \ddots & \vdots \\
\dfrac{\partial h^p}{\partial x_1}(\mathbf{x}) & \cdots & \dfrac{\partial h^p}{\partial x_n}(\mathbf{x})
\end{bmatrix}
$$

$$
=
\begin{bmatrix}
\dfrac{\partial f^1}{\partial x_1}(g(\mathbf{x}),\mathbf{x}) & \cdots & \dfrac{\partial f^1}{\partial x_m}(g(\mathbf{x}),\mathbf{x}) \\
\vdots & \ddots & \vdots \\
\dfrac{\partial f^p}{\partial x_1}(g(\mathbf{x}),\mathbf{x}) & \cdots & \dfrac{\partial f^p}{\partial x_m}(g(\mathbf{x}),\mathbf{x})
\end{bmatrix}
\begin{bmatrix}
\dfrac{\partial g^1}{\partial x_1}(\mathbf{x}) & \cdots & \dfrac{\partial g^1}{\partial x_n}(\mathbf{x}) \\
\vdots & \ddots & \vdots \\
\dfrac{\partial g^m}{\partial x_1}(\mathbf{x}) & \cdots & \dfrac{\partial g^m}{\partial x_n}(\mathbf{x})
\end{bmatrix}
$$

$$
+
\begin{bmatrix}
\dfrac{\partial f^1}{\partial x_{m+1}}(g(\mathbf{x}),\mathbf{x}) & \cdots & \dfrac{\partial f^1}{\partial x_{m+n}}(g(\mathbf{x}),\mathbf{x}) \\
\vdots & \ddots & \vdots \\
\dfrac{\partial f^p}{\partial x_{m+1}}(g(\mathbf{x}),\mathbf{x}) & \cdots & \dfrac{\partial f^p}{\partial x_{m+n}}(g(\mathbf{x}),\mathbf{x})
\end{bmatrix}
\tag{9.43}
$$

Now, we can use (9.43) to write out the total derivative $h'(\mathbf{x})$ as a product of partitioned matrices:

$$
h'(\mathbf{x}) = D_{\mathbf{g}}f(g(\mathbf{x}),\mathbf{x})g'(\mathbf{x}) + D_{\mathbf{x}}f(g(\mathbf{x}),\mathbf{x})
\tag{9.44}
$$

as required.   □

EXAMPLE 9.3.4 We consider a simple example in which $m = n = p$. Let $f(\mathbf{x},\mathbf{y}) = \mathbf{x} - \mathbf{y}$ and $g(\mathbf{x}) = \mathbf{Gx}$, where $\mathbf{G}$ is an $n \times n$ matrix. Then

$$
h(\mathbf{x}) \equiv f(g(\mathbf{x}),\mathbf{x}) = \mathbf{Gx} - \mathbf{x}
\tag{9.45}
$$

We have $f'(\mathbf{x},\mathbf{y}) = [\mathbf{I} \quad -\mathbf{I}]$, $D_{\mathbf{g}}f = \mathbf{I}$ and $D_{\mathbf{x}}f = -\mathbf{I}$ for all $\mathbf{x},\mathbf{y}$, where all the identity matrices are of dimension n.
   Thus

$$
h'(\mathbf{x}) = D_{\mathbf{g}}fg'(\mathbf{x}) + D_{\mathbf{x}}f = \mathbf{IG} - \mathbf{I} = \mathbf{G} - \mathbf{I}
\tag{9.46}
$$

Of course, this result could have been computed directly by differentiating in (9.45).   ◇

9.3.4   Product rule

The multivariate equivalent of the product rule or **Leibniz's law** comes in two versions.[3]

THEOREM 9.3.3 (PRODUCT RULE FOR VECTOR CALCULUS). *The following two versions hold.*

(a) *Let $f, g: \mathbb{R}^m \to \mathbb{R}^n$, and define $h: \mathbb{R}^m \to \mathbb{R}$ by*

$$\underbrace{h(\mathbf{x})}_{1 \times 1} \equiv \underbrace{f(\mathbf{x})^\top}_{1 \times n} \underbrace{g(\mathbf{x})}_{n \times 1} \tag{9.47}$$

Then

$$\underbrace{h'(\mathbf{x})}_{1 \times m} = \underbrace{g(\mathbf{x})^\top}_{1 \times n} \underbrace{f'(\mathbf{x})}_{n \times m} + \underbrace{f(\mathbf{x})^\top}_{1 \times n} \underbrace{g'(\mathbf{x})}_{n \times m} \tag{9.48}$$

(b) *Let $f: \mathbb{R}^m \to \mathbb{R}$ and $g: \mathbb{R}^m \to \mathbb{R}^n$ and define $h: \mathbb{R}^m \to \mathbb{R}^n$ by*

$$\underbrace{h(\mathbf{x})}_{n \times 1} \equiv \underbrace{f(\mathbf{x})}_{1 \times 1} \underbrace{g(\mathbf{x})}_{n \times 1} \tag{9.49}$$

Then

$$\underbrace{h'(\mathbf{x})}_{n \times m} = \underbrace{g(\mathbf{x})}_{n \times 1} \underbrace{f'(\mathbf{x})}_{1 \times m} + \underbrace{f(\mathbf{x})}_{1 \times 1} \underbrace{g'(\mathbf{x})}_{n \times m} \tag{9.50}$$

Proof: This is easily shown using the product rule from univariate calculus to calculate the relevant partial derivatives and then stacking the results in matrix form. The details are left as exercises; see Exercises 9.10 and 9.11. □

EXAMPLE 9.3.5 Let $f(\mathbf{x}) = (\mathbf{x} \cdot \mathbf{x})^{-\frac{1}{2}} = 1/\|\mathbf{x}\|$, $g(\mathbf{x}) = \mathbf{x}$ and $h(\mathbf{x}) = \mathbf{x}/\|\mathbf{x}\| = f(\mathbf{x})g(\mathbf{x})$. In other words, let $h(\mathbf{x})$ be the unit vector in the direction of $\mathbf{x}$.

Then, using the chain rule,

$$\frac{\partial f}{\partial x_i}(\mathbf{x}) = -\frac{1}{2}\frac{2x_i}{(\sqrt{\mathbf{x} \cdot \mathbf{x}})^3} \tag{9.51}$$

and

$$f'(\mathbf{x}) = -\frac{1}{\|\mathbf{x}\|^3}\mathbf{x}^\top \tag{9.52}$$

Also $g'(\mathbf{x}) = \mathbf{I}$ for all $\mathbf{x}$. So, by the product rule,

$$h'(\mathbf{x}) = -\frac{1}{\|\mathbf{x}\|^3}\mathbf{x}\mathbf{x}^\top + \frac{1}{\|\mathbf{x}\|}\mathbf{I} \tag{9.53}$$

◇

9.4 Elasticities

DEFINITION 9.4.1 Let $f: X \to \mathbb{R}_{++}$ be a positive-valued function defined on $X \subseteq \mathbb{R}^k_{++}$. Then for $i = 1, 2, \ldots, k$ and $\mathbf{x}^* \in X$ the **elasticity** of f with respect to x_i at $\mathbf{x}^*$ is

$$\frac{x_i^*}{f(\mathbf{x}^*)}\frac{\partial f}{\partial x_i}(\mathbf{x}^*) \tag{9.54}$$

Roughly speaking, as can be shown by application of the chain rule (see Exercise 9.12), the elasticity is just $\partial \ln f / \partial \ln x_i$, or the slope of the graph of the function on log–log graph paper. Since only positive numbers have real logarithms, the restriction to positive-valued functions of positive-valued variables in the definition is required.

DEFINITION 9.4.2 A function (of a single variable) is said to be **inelastic** when the absolute value of the elasticity is less than unity; and **elastic** when the absolute value of the elasticity is greater than unity. The borderline case is called a function of **unit elasticity**.

Note that a function can be elastic in some parts of its domain and inelastic in other parts of its domain.

EXAMPLE 9.4.1 One useful application of elasticity, which will be familiar to economics students, is in analysing the behaviour of the total revenue function associated with a particular inverse demand function, $P(Q)$, which gives the maximum price P at which the quantity Q of some commodity could be sold. Using the product rule of scalar differential calculus, we find that the derivative of total revenue $P \times Q$ with respect to the quantity sold, or **marginal revenue**, is

$$\frac{dP(Q)Q}{dQ} = Q\frac{dP}{dQ} + P \tag{9.55}$$

$$= P\left(1 + \frac{1}{\eta}\right) \tag{9.56}$$

where

$$\eta \equiv \frac{P}{Q}\frac{dQ}{dP} \tag{9.57}$$

denotes the elasticity of quantity demanded with respect to price.

From univariate calculus, we know that total revenue is increasing in quantity when this derivative is positive, decreasing when it is negative, and constant, or at a maximum or minimum, when it is zero.

Hence, total revenue is constant or maximized or minimized where the elasticity equals -1; increasing in quantity when elasticity is less than -1 (i.e. demand is elastic); and decreasing in quantity when the elasticity is between 0 and -1 (i.e. demand is inelastic).

Note that the indifference curves of the total revenue function in price–quantity space are the rectangular hyperbolas $PQ = k$ for all $k > 0$. ◇

EXAMPLE 9.4.2 A function of n variables frequently used in economic modelling is the **Cobb–Douglas function**,[4] which takes the general form

$$f(\mathbf{x}) = \prod_{i=1}^{n} x_i^{\alpha_i} \tag{9.58}$$

In most applications, the components x_i and the parameters α_i are all assumed to be positive. Note that f is homogeneous of degree $\sum_{i=1}^{n} \alpha_i$.

Assuming that the x_i are all positive, taking natural logarithms on both sides of (9.58) yields

$$\ln f(\mathbf{x}) = \sum_{i=1}^{n} \alpha_i \ln x_i \tag{9.59}$$

Thus

$$\frac{\partial \ln f}{\partial \ln x_i} = \alpha_i \quad \forall\, i = 1, 2, \ldots, n \tag{9.60}$$

In other words, each α_i is the elasticity of the Cobb–Douglas function with respect to the corresponding variable x_i. $\diamond$

9.5   Directional derivatives and tangent hyperplanes

We now come to the third way of visualizing functions of n variables that was referred to in Section 7.7. This involves visualizing the two-dimensional cross-section, along an arbitrary line in the domain of the function, of the $(n + 1)$-dimensional graph of the function.

DEFINITION 9.5.1 Let X be a real vector space, let $\mathbf{x} \neq \mathbf{x}' \in X$ and let L be the line containing $\mathbf{x}$ and $\mathbf{x}'$.

(a) The **restriction** of the function $f : X \to \mathbb{R}$ to the line L is the function

$$f|_L : \mathbb{R} \to \mathbb{R} : \lambda \mapsto f(\lambda \mathbf{x} + (1 - \lambda)\mathbf{x}') \tag{9.61}$$

(b) If f is a differentiable function, then the **directional derivative** of f at $\mathbf{x}'$ in the direction from $\mathbf{x}'$ to $\mathbf{x}$ is $f|'_L(0)$.

Note that, by the chain rule,

$$f|'_L(\lambda) = f'(\lambda \mathbf{x} + (1 - \lambda)\mathbf{x}')(\mathbf{x} - \mathbf{x}') \tag{9.62}$$

and, hence, the directional derivative reduces to[5]

$$f|'_L(0) = f'(\mathbf{x}')(\mathbf{x} - \mathbf{x}') \tag{9.63}$$

Note also that, returning to first principles,

$$f|'_L(0) = \lim_{\lambda \to 0} \frac{f(\mathbf{x}' + \lambda(\mathbf{x} - \mathbf{x}')) - f(\mathbf{x}')}{\lambda} \tag{9.64}$$

Sometimes it is neater to write $\mathbf{x} - \mathbf{x}' \equiv \mathbf{h}$. Using the chain rule, it can be shown (see Exercise 9.15) that the second derivative of $f|_L$ is

$$f|''_L(\lambda) = \mathbf{h}^\top f''(\mathbf{x}' + \lambda \mathbf{h})\mathbf{h} \tag{9.65}$$

and

$$f|''_L(0) = \mathbf{h}^\top f''(\mathbf{x}')\mathbf{h} \tag{9.66}$$

We endeavour, wherever possible, to stick to the convention that $\mathbf{x}'$ denotes the point at which the derivative is to be evaluated and $\mathbf{x}$ denotes the point in the direction of which it is measured.

We have already implicitly met the concept of restriction to a line in Definition 7.3.4, since (7.14) effectively just says that a function of several variables is convex if its restriction to *every* line segment in its domain is a convex function of one variable, in a sense that can be illustrated in a two-dimensional diagram.

The ith partial derivative of f at $\mathbf{x}$ is the directional derivative of f at $\mathbf{x}$ in the direction from $\mathbf{x}$ to $\mathbf{x} + \mathbf{e}_i$, where $\mathbf{e}_i$ is the ith standard basis vector. In other words, partial derivatives are a special case of directional derivatives or directional derivatives are a generalization of partial derivatives.

EXAMPLE 9.5.1 Consider the restriction to a line of the function

$$f : \mathbb{R}^2 \to \mathbb{R} : (x, y) \mapsto x^2 + y^2 \tag{9.67}$$

The restriction of this function to a line in the xy plane is a parabola. Cross-sections parallel to the xy plane are circles. $\diamond$

DEFINITION 9.5.2 The **tangent hyperplane** at $\mathbf{x}'$ to the graph in $\mathbb{R}^{n+1}$ of a function $f : \mathbb{R}^n \to \mathbb{R}$ is the hyperplane in $\mathbb{R}^{n+1}$ through $(\mathbf{x}', f(\mathbf{x}'))$ with normal $(f'(\mathbf{x}'), -1)$.

The equation of a tangent hyperplane is given by the orthogonality condition satisfied by the typical point $(\mathbf{x}, y)$ on the hyperplane, which is

$$(f'(\mathbf{x}'), -1)^\top ((\mathbf{x}, y) - (\mathbf{x}', f(\mathbf{x}'))) = 0 \tag{9.68}$$

which simplifies to

$$y = f(\mathbf{x}') + f'(\mathbf{x}')(\mathbf{x} - \mathbf{x}') \tag{9.69}$$

Figure 9.1 shows the tangent hyperplane to the graph of a function of one variable, which is just the line tangent to the graph at the point labelled P. Using the coordinates marked on the diagram, the slope of the tangent line can be seen to be

$$\tan \theta = \frac{f'(x')(x - x')}{x - x'} \tag{9.70}$$

which of course simplifies in this one-variable case to $f'(x')$.

Figure 9.1 can be re-interpreted as a cross-section of the graph of a function of n variables and its tangent hyperplane by changing every scalar x to vector $\mathbf{x}$ and likewise for x' and $\mathbf{x}'$. The horizontal axis is labelled L to indicate that it is just an arbitrary line in the domain of the function, parametrized by λ, which equals 0 at $\mathbf{x}'$ and 1 at $\mathbf{x}$. The height of the thick line is $f'(\mathbf{x}')(\mathbf{x} - \mathbf{x}')$, which by (9.61) is just the directional derivative of f at $\mathbf{x}'$ in the direction of $\mathbf{x}$.

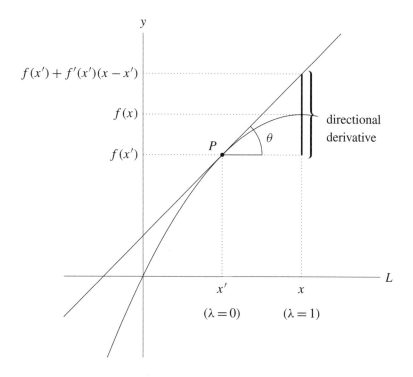

Figure 9.1 Tangent hyperplane to the graph of a function of one variable

The slope of the cross-section of the tangent hyperplane is now (in vector notation)

$$\frac{f'(\mathbf{x}')(\mathbf{x} - \mathbf{x}')}{\|\mathbf{x} - \mathbf{x}'\|} \tag{9.71}$$

The vector terms in the numerator and denominator of (9.71) do *not* cancel, unlike the similar (but quite different) scalar terms in the numerator and denominator of (9.70). Expression (9.71) is the directional derivative of f at $\mathbf{x}'$ in the direction of the unit vector $(\mathbf{x} - \mathbf{x}')/\|\mathbf{x} - \mathbf{x}'\|$. It can be interpreted as the dot product of a column vector of partial derivatives and this unit vector. From Definition 5.2.15, this is $\|f'(\mathbf{x}')^{\top}\|$ multiplied by the cosine of the angle between the two vectors. Thus the slope of the cross-section is zero when the total derivative (transposed) and the unit vector are perpendicular and the slope is at its maximum when they are collinear.

EXAMPLE 9.5.2 Returning to Example 9.5, where $f'(x, y) = [2x \; 2y]$, we illustrate these points in Figure 9.2. We can conclude that the cross-section along the line indicated M through the origin and (x, y) gives the steepest tangent of any cross-section through (x, y), but the cross-section along the line L through (x, y) perpendicular to this has a horizontal tangent. In the indifference map, L is tangent to an indifference curve of f at (x, y) and the value of f on that indifference curve is the minimum value that f takes anywhere along the line L. The family of lines parallel to L are the contour lines of the tangent hyperplane to the graph of f at (x, y). $\diamond$

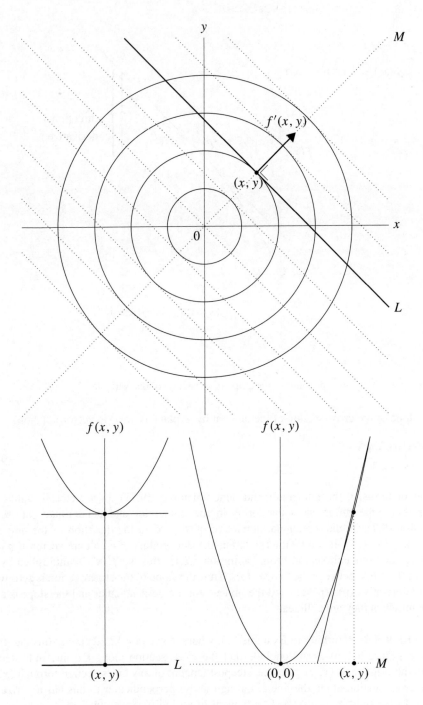

Figure 9.2 Indifference map and two cross-sections of the graph of the function $f \colon \mathbb{R}^2 \to$
$\mathbb{R} \colon (x, y) \mapsto x^2 + y^2$

9.6 Taylor's theorem: deterministic version

Taylor's theorem and its many variants have numerous practical and theoretical uses. These theorems essentially show that most functions can be closely approximated, if not by linear functions, then by low-order polynomial functions. This result alone is the justification behind much of econometrics and other forms of empirical data analysis. In linear algebra terms, the theorem just says that the finite-dimensional subspace of low-order polynomials is in a certain sense representative of an infinite-dimensional function space. Readers may be familiar with the practice of using linear functions to approximate more general functions; Taylor's theorem says that, in cases where a linear approximation is a poor fit, a polynomial of slightly higher order will often give a good fit.[6]

At a more theoretical level, Taylor's theorem will prove useful in proving some results on optimization in the next chapter. More precisely, we will see that Taylor's expansion provides a means of approximating the value of a continuous, differentiable function at some point, using a polynomial expression in which the coefficients involve the value of the function at some other point, together with the values of its low-order derivatives at that other point. The uses of such approximations might not be immediately obvious, but the preceding comments should put them in context.

We will return to Taylor's theorem in Section 13.9, and later practical applications will include the analysis of risk preferences in the theory of choice under uncertainty.

9.6.1 Univariate Taylor's theorem

We begin with the simplest forms of Taylor's theorem, namely, the intermediate value theorem, which is the zeroth-order version (i.e. not involving any derivative), and the mean value theorem, which is the first-order version (i.e. involving only the first derivative). These theorems are illustrated in Figure 9.3.

THEOREM 9.6.1 (INTERMEDIATE VALUE THEOREM). *If $f : \mathbb{R} \to \mathbb{R}$ is continuous on $[a, b]$, and λ lies between $f(a)$ and $f(b)$, then there exists $x^* \in [a, b]$ such that $f(x^*) = \lambda$.*

Proof: This proof is based on Goursat (1959, pp. 144–5).

Let $\phi(x) \equiv f(x) - \lambda$. Since f is continuous, ϕ is also continuous. We must show that there exists $x^* \in [a, b]$ such that $\phi(x^*) = 0$.

If $f(b) = \lambda$ or $f(a) = \lambda$, then there is nothing to prove. So, without loss of generality, suppose that $f(b) > \lambda > f(a)$ and thus that $\phi(b) > 0 > \phi(a)$.

Let $A = \{y \in [a, b] : \phi(y) > 0\}$. Then A is non-empty, because it contains b. We claim that $x^* \equiv \inf A$ is the required point.

Note that, by definition of A and x^*, for $h > 0$, $\phi(x^* - h) \leq 0$. Hence, by continuity of ϕ, $\phi(x^*) \leq 0$.

Now suppose that $\phi(x^*) < 0$, say $\phi(x^*) = -\epsilon$. We will derive a contradiction.

Again by continuity of ϕ, there exists $\delta > 0$ such that

$$|z - x^*| < \delta \implies |\phi(z) - \phi(x^*)| < \epsilon \implies \phi(z) < 0 \implies z \notin A \tag{9.72}$$

The fact that this range of values, including x^* and values on both sides of x^*, are not in A contradicts the fact that $x^* = \inf A$.

So we must have $\phi(x^*) = 0$, as required. $\qquad\square$

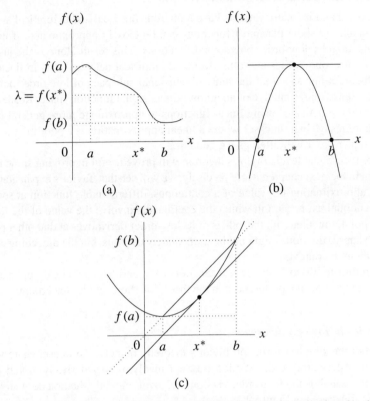

Figure 9.3 (a) Intermediate value theorem, (b) Rolle's theorem and (c) mean value theorem

Before proceeding to the mean value theorem, which is the first-order version of Taylor's theorem, we need one more result, namely Rolle's theorem.[7]

THEOREM 9.6.2 (ROLLE'S THEOREM). *If $f: \mathbb{R} \to \mathbb{R}$ is continuous on $[a, b]$ and differentiable on (a, b) with $f(a) = f(b) = 0$, then there exists $x^* \in (a, b)$ such that $f'(x^*) = 0$.*

Proof: This proof is based on Goursat (1959, pp. 7–8).

Note that Rolle's theorem is trivially true if f is uniformly equal to 0. So we can assume, without loss of generality, that f takes on strictly positive values somewhere on (a, b).

Let $x^* \equiv \mathrm{argmax}_{(a,b)} f(x)$. Then

$$\frac{f(x^* + h) - f(x^*)}{h} \leq 0 \quad \text{for } h > 0 \tag{9.73}$$

and

$$\frac{f(x^* + h) - f(x^*)}{h} \geq 0 \quad \text{for } h < 0 \tag{9.74}$$

Letting $h \to 0$, the former inequality becomes $f'(x^*) \leq 0$ and the latter inequality becomes $f'(x^*) \geq 0$. Thus it must be the case that $f'(x^*) = 0$, as required. $\qquad \square$

THEOREM 9.6.3 (MEAN VALUE THEOREM). *If $f: \mathbb{R} \to \mathbb{R}$ is continuous on $[a, b]$ and differentiable on (a, b), then there exists $x^* \in (a, b)$ such that $f'(x^*) = (f(b) - f(a))/(b - a)$.*

Proof: This proof is based on Goursat (1959, p. 8). We will derive a more general result, of which the mean value theorem is a special case.

Let ϕ be any other function sharing the properties assumed for f.

Define a third function ψ by

$$\psi(y) \equiv Af(y) + B\phi(y) + C \tag{9.75}$$

where A, B and C are any real numbers. The function ψ is also continuous on $[a, b]$ and differentiable on (a, b).

We want to apply Rolle's theorem to ψ, so we chose A, B and C to guarantee that $\psi(a) = \psi(b) = 0$. This gives us the following system of two linear equations in three unknowns:

$$Af(a) + B\phi(a) + C = 0 \tag{9.76}$$
$$Af(b) + B\phi(b) + C = 0 \tag{9.77}$$

or, in matrix form,

$$\begin{bmatrix} f(a) & \phi(a) \\ f(b) & \phi(b) \end{bmatrix} \begin{bmatrix} A \\ B \end{bmatrix} = -\begin{bmatrix} C \\ C \end{bmatrix} \tag{9.78}$$

If we let C equal the determinant of the 2×2 matrix on the left-hand side, we get the neat solution

$$\begin{bmatrix} A \\ B \end{bmatrix} = -\begin{bmatrix} \phi(b) & -\phi(a) \\ -f(b) & f(a) \end{bmatrix} \begin{bmatrix} 1 \\ 1 \end{bmatrix} \tag{9.79}$$

or

$$A = \phi(a) - \phi(b) \tag{9.80}$$
$$B = f(b) - f(a) \tag{9.81}$$
$$C = f(a)\phi(b) - f(b)\phi(a) \tag{9.82}$$

Note that (9.80)–(9.82) solve (9.76) and (9.77) even when $C = 0$.

Now, by Rolle's theorem, there exists $x^* \in (a, b)$ such that $\psi'(x^*) = 0$ or

$$(\phi(a) - \phi(b)) f'(x^*) + (f(b) - f(a))\phi'(x^*) = 0 \tag{9.83}$$

Equation (9.83) is often called the **generalized law of the mean**. If we set $\phi(y) \equiv y$ and rearrange, we obtain

$$f'(x^*) = \frac{f(b) - f(a)}{b - a} \tag{9.84}$$

as required. $\square$

The mean value theorem can be rearranged to give

$$f(b) = f(a) + (b-a)f'(x^*) \tag{9.85}$$

which looks more like the first-order version of Taylor's theorem itself; see below.

Before we give the general statement of the nth-order version of Taylor's theorem, we consider a simple case of a function of a single variable to provide motivation for the theorem, and then an example.

Suppose that $f(x_0)$ and $df(x_0)/dx$ are known and we wish to approximate $f(x)$, where x is close to x_0. Inspired by the mean value theorem, we can see that a rough approximation to $f(x)$ is

$$A = f(x_0) + \frac{df}{dx}(x_0)(x - x_0) \tag{9.86}$$

as shown in Figure 9.4.

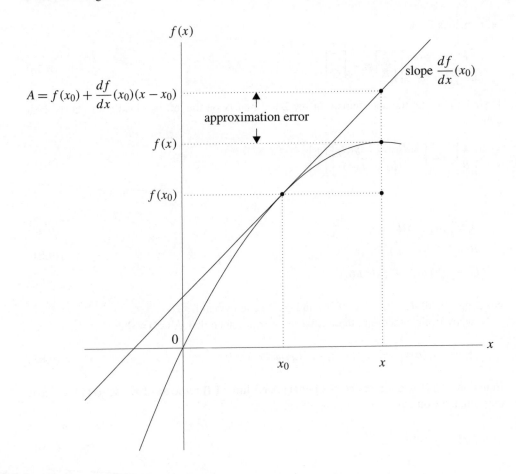

Figure 9.4 Taylor approximation

The value A is an overestimate of $f(x)$ in the case illustrated in the sketch, as it ignores the curvature of the function. Taylor's theorem says that a better approximation is

$$A^* = A + \frac{1}{2!}\frac{d^2 f}{dx^2}(x_0)(x - x_0)^2$$

$$= f(x_0) + \frac{df}{dx}(x_0)(x - x_0) + \frac{1}{2!}\frac{d^2 f}{dx^2}(x_0)(x - x_0)^2 \tag{9.87}$$

In this illustration, $d^2 f(x_0)/dx^2$ is negative, since df/dx is decreasing, and $(x - x_0)^2$ is positive, so $A^* < A$. The $\frac{1}{2!}$ term gives a smaller weight to the additional second-order term than is given to the original first-order term. A still better approximation would be one with further terms in $\frac{1}{3!}[d^3 f(x_0)/dx^3](x - x_0)^3$, etc. As an exercise, the reader should show that, for a quadratic function $f(x) = ax^2 + bx + c$, A^* exactly equals $f(x)$; see Exercise 9.21. The theorem is as follows.

THEOREM 9.6.4 (UNIVARIATE TAYLOR'S THEOREM). *If $f(x)$ is a function with continuous derivatives up to order n on the closed interval $[x_0, x]$, then there exists $x^* \in (x_0, x)$ such that*

$$f(x) = f(x_0) + \frac{df}{dx}(x_0)(x - x_0) + \frac{1}{2!}\frac{d^2 f}{dx^2}(x_0)(x - x_0)^2$$

$$+ \cdots + \frac{1}{(n-1)!}\frac{d^{n-1} f}{dx^{n-1}}(x_0)(x - x_0)^{n-1} + \frac{1}{n!}\frac{d^n f}{dx^n}(x^*)(x - x_0)^n \tag{9.88}$$

Proof: The proof of this theorem can be found, for example, in Binmore (1982, pp. 106–7). □

Note that the first $n - 1$ derivatives at x_0 of the polynomial on the right-hand side of (9.88) are the same as those of f; see Exercise 9.22.

Equation (9.88) is known as **Taylor's expansion**. Note that the first-order term in the expansion is just the directional derivative of f at x_0 in the direction of x.

Taylor's expansion (or more correctly its first few terms) is typically used as an approximation to the value of the underlying function. In this context, the distinction between the derivatives at x_0 in the general terms and the derivative at x^* in the final or **remainder term** is immaterial. Similarly, the distinction between the finite version of Taylor's expansion in (9.88) and the infinite version

$$f(x) = f(x_0) + \sum_{k=1}^{\infty} \frac{1}{k!}\frac{d^k f}{dx^k}(x_0)(x - x_0)^k \tag{9.89}$$

is immaterial in most practical applications.

A special case of Taylor's expansion is when $x_0 = 0$. This produces a form known as **Maclaurin's series**:[8]

$$f(x) = f(0) + \frac{df}{dx}(0)x + \frac{1}{2!}\frac{d^2 f}{dx^2}(0)x^2 + \cdots + \frac{1}{k!}\frac{d^k f}{dx^k}(0)x^k + \cdots \tag{9.90}$$

Maclaurin's series allows us to approximate any function of x using a straightforward polynomial in x, at least for x near zero.

The detailed study of the speed of convergence of Taylor's expansion or Maclaurin's series to the underlying function is beyond the scope of this book. We note, however, that the infinite version of Taylor's expansion in (9.89) does not necessarily converge at all, or to $f(x)$. Functions for which it does are called **analytic**. Binmore (1982, Exercise 15.6(6)) gives an example of a function that is not analytic.

In practice, a **Taylor approximation** or **Maclaurin approximation** usually involves the evaluation of only the first three terms or so of whichever form of the expansion we work with.

EXAMPLE 9.6.1   Consider the simple function $f(x) = (2 + x)^2$. We have, using the Maclaurin form of Taylor's expansion,

$$f(x) = (2+x)^2 \quad \therefore f(0) = 4 \tag{9.91}$$

$$\frac{df}{dx}(x) = 2(2+x) \quad \therefore \frac{df}{dx}(0) = 4 \tag{9.92}$$

$$\frac{d^2 f}{dx^2}(x) = 2(2+x)^0 = 2 \quad \forall x \tag{9.93}$$

$$\frac{d^3 f}{dx^3}(x) = 0 \quad \forall x \tag{9.94}$$

Thus we can write

$$f(x) = f(0) + \frac{df}{dx}(0)x + \frac{1}{2!}\frac{d^2 f}{dx^2}(0)x^2 + \frac{1}{3!}\frac{d^3 f}{dx^3}(0)x^3 + \cdots$$

$$= 4 + 4x + \frac{1}{2}2x^2 + \frac{1}{6}0x^3 + \cdots \tag{9.95}$$

i.e. $(2+x)^2 = 4 + 4x + x^2 + 0 + \cdots \tag{9.96}$

In this simple case, we see that the fourth term and all higher-order terms are zero but that the first three terms of Maclaurin's series actually produce the exact result: $(2 + x)^2 = 4 + 4x + x^2$. $\diamond$

For another interesting practical application of the Maclaurin approximation, see Exercise 15.6.

9.6.2   Multivariate Taylor's theorem

Taylor's theorem is easily generalized to functions of several variables. The multivariate version of Taylor's theorem can be obtained by applying the univariate versions to the restriction to a line of a function of n variables.

THEOREM 9.6.5 (MULTIVARIATE TAYLOR'S THEOREM). *Let $f: X \to \mathbb{R}$ be twice differentiable, where $X \subseteq \mathbb{R}^n$ is open. Then for any $\mathbf{x}, \mathbf{x}' \in X$, there exists $\lambda \in (0, 1)$ such that*

$$f(\mathbf{x}) = f(\mathbf{x}') + f'(\mathbf{x}')(\mathbf{x} - \mathbf{x}') + \tfrac{1}{2}(\mathbf{x} - \mathbf{x}')^\top f''(\mathbf{x}' + \lambda(\mathbf{x} - \mathbf{x}'))(\mathbf{x} - \mathbf{x}') \tag{9.97}$$

Note that the second term on the right-hand side of (9.97) is a linear form in $\mathbf{x} - \mathbf{x}'$ (more precisely, a directional derivative) and that the third term is a quadratic form in $\mathbf{x} - \mathbf{x}'$.

Proof: Let L be the line from $\mathbf{x}'$ to $\mathbf{x}$.

Then the univariate version tells us that there exists $\lambda \in (0, 1)$ such that

$$f|_L(1) = f|_L(0) + f'|_L(0) + \tfrac{1}{2}f''|_L(\lambda) \tag{9.98}$$

Since $f|_L(1) = f(\mathbf{x})$, $f|_L(0) = f(\mathbf{x}')$, $f'|_L(0) = f'(\mathbf{x}')(\mathbf{x} - \mathbf{x}')$ and $f''|_L(\lambda) = (\mathbf{x} - \mathbf{x}')^\top f''(\mathbf{x}' + \lambda(\mathbf{x} - \mathbf{x}'))(\mathbf{x} - \mathbf{x}')$, making the appropriate substitutions gives the multivariate version in the theorem. □

In principle, the multivariate Taylor's theorem can be extended to any order, but the matrix notation becomes cumbersome. In scalar notation, the typical (kth) term of Taylor's expansion of a function of n variables is

$$\frac{1}{k!} \sum_{i_1=1}^{n} \sum_{i_2=1}^{n} \cdots \sum_{i_k=1}^{n} \frac{\partial^k f}{\partial x_{i_1} \partial x_{i_2} \ldots \partial x_{i_k}} (\mathbf{x}') \prod_{j=1}^{k} \left(x_{i_j} - x'_{i_j} \right) \tag{9.99}$$

We will appeal to higher-order versions of the univariate Taylor expansion in the nth derivative test for optima of functions of a single variable in the proof of Theorem 10.3.4.

EXAMPLE 9.6.2 Suppose that $y = f(\mathbf{x}) = x_1^3 + 2x_1 x_2^2$. Let us write down Taylor's expansion of y around the point $\mathbf{x}' = (1, 0)$. Straightforward differentiation yields

$$f'(x_1, x_2) = [3x_1^2 + 2x_2^2 \quad 4x_1 x_2] \quad \text{and} \quad f''(x_1, x_2) = \begin{bmatrix} 6x_1 & 4x_2 \\ 4x_2 & 4x_1 \end{bmatrix} \tag{9.100}$$

(Note again that $f''(x_1, x_2)$ is a symmetric matrix.) Hence,

$$f'(1, 0) = [3 \quad 0] \quad \text{and} \quad f''(1, 0) = \begin{bmatrix} 6 & 0 \\ 0 & 4 \end{bmatrix} \tag{9.101}$$

while

$$f(1, 0) = 1 \quad \text{and} \quad \mathbf{x} - \mathbf{x}' = \begin{bmatrix} x_1 - 1 \\ x_2 \end{bmatrix} \tag{9.102}$$

Substituting in (9.97) with $\lambda = 0$ gives the following approximation:

$$\begin{aligned}
f(\mathbf{x}) &\approx 1 + [3 \quad 0] \begin{bmatrix} (x_1 - 1) \\ x_2 \end{bmatrix} + \frac{1}{2!} [(x_1 - 1) \quad x_2] \begin{bmatrix} 6 & 0 \\ 0 & 4 \end{bmatrix} \begin{bmatrix} x_1 - 1 \\ x_2 \end{bmatrix} \\
&= 1 + 3x_1 - 3 + \tfrac{1}{2}(6x_1^2 - 12x_1 + 6 + 4x_2^2) \\
&= 3x_1^2 + 2x_2^2 - 3x_1 + 1
\end{aligned} \tag{9.103}$$

Thus we approximate a function that is quadratic in x_2 but cubic in x_1 by a function that is quadratic in both variables. Experimentation with values of $\mathbf{x}$ "close to" $\mathbf{x}' = (1, 0)$ will give an idea of the accuracy of the approximation. For example, for $\mathbf{x}'$ itself, the approximation gives the exact result, $f(\mathbf{x}) = 1$. In fact, for $x_1 = 1$, the approximation will always give the exact result, $1 + 2x_2^2$, since both the original function and the approximation are the same quadratic in x_2.

However, for $\mathbf{x} = (1.1, 0.1)$, the approximation produces 1.35 while the exact value is $f(\mathbf{x}) = 1.353$, an approximation error of -0.003.

For $\mathbf{x}$ "further away from" $\mathbf{x}'$ the results are less accurate: for $\mathbf{x} = (0.8, 0.2)$, for instance, the approximation error is 0.024 and for $\mathbf{x} = (0, 0)$, the approximation error is 1.000.   ◇

Applications of Taylor's theorem later in this book include:

- quantifying the size of Jensen's inequality (see (13.86));
- deriving linear and quadratic approximations to the relationship between bond prices and discount rates (see Section 15.4.5); and
- motivating the use of a mean–variance utility function to approximately represent preferences with the expected-utility property (see Section 16.8).

9.7 Multiple integration

9.7.1 *Definitions and notation*

Just as we defined the partial derivative of a function of several variables to be the derivative with respect to one variable, treating the others as constants, so we define the **multiple integral** of a function of several variables (the **integrand**) to be the integral of one variable at a time, treating the other variables as constants at each stage.

Just as not all functions of a single variable are integrable, so too not all functions of several variables are integrable.

The simplest form of double integral, where the limits of integration are constants for both integrals, represents integration over a rectangle. We will now consider the simplest such integral.

EXAMPLE 9.7.1 Multiple integrals are evaluated, by convention, from the inside out.

$$\int_a^b \int_c^d dy\, dx \equiv \int_a^b \left(\int_c^d dy \right) dx$$

$$= \int_a^b (d-c)\, dx$$

$$= (d-c) \int_a^b dx$$

$$= (d-c)(b-a) \tag{9.104}$$

◇

Example 9.7.1 illustrates several important aspects of multiple integration:

- Constants can be taken outside multiple integrals in the same way as with single integrals, for example, as we have done with the factor $d - c$ in the third line of (9.104).

- Readers will be familiar with the interpretation of a single integral as the area under a curve. Similarly, a double integral gives the volume under a surface and other multiple integrals give higher-dimensional analogues.
- In particular, Example 9.7.1 represents the computation of the volume of a solid block of unit height sitting on the rectangle $[a, b] \times [c, d]$, which of course also equals the area of the rectangle. More generally, if we wish to compute the volume $V(A)$ of any set $A \subseteq \mathbb{R}^n$ and if $f: A \rightarrow \mathbb{R}: \mathbf{x} \mapsto 1$ is the constant function taking the value 1 everywhere on A, then we may be able to find the volume using the fact that

$$V(A) = \iint \cdots \int_A f(\mathbf{x}) \, dx_1 \, dx_2 \ldots dx_n \tag{9.105}$$

- The preceding point presumes that the notation $\iint \cdots \int_A$ can be defined unambiguously; this follows from Fubini's theorem, which we consider in the next section.
- A double integral can also represent the volume under a surface over any (non-rectangular) region of the xy plane, for example a circle, as in the concluding example of this section.

EXAMPLE 9.7.2 Consider the function defined by $f(x, y) = \sqrt{1 - x^2 - y^2}$. This function takes on real values only within the unit circle in the xy plane defined as

$$C \equiv \left\{ (x, y) \in \mathbb{R}^2 : -1 \leq x \leq 1, \ -\sqrt{1 - x^2} \leq y \leq \sqrt{1 - x^2} \right\} \tag{9.106}$$

Outside the unit circle, the function f takes on imaginary values. The hemisphere of unit radius centred at the origin has equation $z = f(x, y) = \sqrt{1 - x^2 - y^2}$. The volume enclosed by the hemisphere and the xy plane is given by the double integral

$$V \equiv \int_{-1}^{1} \int_{-\sqrt{1-x^2}}^{\sqrt{1-x^2}} f(x, y) \, dy \, dx$$

$$= \int_{-1}^{1} \int_{-\sqrt{1-x^2}}^{\sqrt{1-x^2}} \sqrt{1 - x^2 - y^2} \, dy \, dx \tag{9.107}$$

To evaluate this integral, we will make some substitutions using the trigonometric identities mentioned on p. xxii. We begin by evaluating the inner integral

$$A \equiv \int_{-\sqrt{1-x^2}}^{\sqrt{1-x^2}} \sqrt{1 - x^2 - y^2} \, dy \tag{9.108}$$

holding x fixed. The single integral A is just the area of the cross-section through the hemisphere at x parallel to the yz plane, which is a semi-circle of radius $\sqrt{1 - x^2}$. To simplify the notation, we will denote this radius $r \equiv \sqrt{1 - x^2}$ and make the substitution $y = r \sin \theta$. Then, since we are holding x, and hence r, constant, $dy = r \cos \theta \, d\theta$. Using the fundamental trigonometric identity from p. xxii, the integrand becomes $\sqrt{1 - x^2 - y^2} = \sqrt{r^2 - y^2} = r \cos \theta$. When $y = r$, $\sin \theta = 1$ and $\theta = \pi/2$, and similarly when

$y = -r, \theta = -\pi/2$. Hence,

$$A = \int_{-r}^{r} \sqrt{r^2 - y^2}\, dy$$

$$= \int_{-\pi/2}^{\pi/2} r\cos\theta\, r\cos\theta\, d\theta$$

$$= r^2 \int_{-\pi/2}^{\pi/2} \cos^2\theta\, d\theta \tag{9.109}$$

Using the double-angle formula, this becomes

$$A = r^2 \int_{-\pi/2}^{\pi/2} \frac{1}{2}(\cos 2\theta + 1)\, d\theta$$

$$= \frac{r^2}{2}\left[\frac{1}{2}\sin 2\theta + \theta\right]_{-\pi/2}^{\pi/2}$$

$$= \frac{r^2}{2}\left[\frac{1}{2}\times 0 + \frac{\pi}{2} - \left(\frac{1}{2}\times 0 + \left(-\frac{\pi}{2}\right)\right)\right]$$

$$= \frac{\pi r^2}{2} \tag{9.110}$$

This reassuringly coincides with the well-known formula for the area of a semi-circle of radius $r = \sqrt{1 - x^2}$.

Substituting this result in (9.107) yields

$$V = \frac{\pi}{2}\int_{-1}^{1}(1 - x^2)\, dx$$

$$= \frac{\pi}{2}\left[x - \frac{x^3}{3}\right]_{-1}^{1}$$

$$= \frac{\pi}{2}\left[1 - \frac{1}{3} - \left(-1 - \frac{(-1)^3}{3}\right)\right]$$

$$= \frac{\pi}{2}\left[1 - \frac{1}{3} + 1 - \frac{1}{3}\right]$$

$$= \frac{2\pi}{3} \tag{9.111}$$

If the volume bounded by a hemisphere of radius 1 is $2\pi/3$, it follows that the volume of a ball of radius 1 is $4\pi/3$. $\diamond$

9.7.2 Fubini's theorem

In defining partial derivatives and multiple integrals, we have been careful to state the order in which differentiation and integration with respect to different variables is to be carried out, although we have hinted that the order in both cases is often immaterial. We will now present a number of theorems that give conditions under which the order of differentiation or integration does not matter (Theorems 9.7.1, 9.7.2 and 9.7.4). We begin with Fubini's theorem, which concerns the order of integration in a multiple integral.[9]

THEOREM 9.7.1 (FUBINI'S THEOREM). *If the integrand is a continuous and bounded function and its absolute value is integrable, then the value of a multiple integral is independent of the order of integration, for example,*

$$\int_a^b \int_c^d f(x, y) \, dy \, dx = \int_c^d \int_a^b f(x, y) \, dx \, dy \tag{9.112}$$

Proof: The full proof of this theorem is beyond the scope of this book. It can be found in Spivak (1965, pp. 58–9). □

Fubini's theorem will be used in this section to prove Young's theorem and Leibniz's integral rule. In fact, all three theorems are effectively equivalent. Fubini's theorem will also be used later to prove Stein's lemma (Theorem 13.7.1).

We can appeal to Fubini's theorem to allow us to denote the double integral of the function f over the rectangle $R \equiv [a, b] \times [c, d]$, or over any specified set, simply as

$$\iint_R f(x, y) \, dy \, dx \tag{9.113}$$

For example, the integral in (9.107) defining the volume of a hemisphere could have been written

$$\iint_C \sqrt{1 - x^2 - y^2} \, dy \, dx \tag{9.114}$$

Provided that the volume of the set $A \subseteq \mathbb{R}^n$ is finite, the expression for the volume of A given in (9.105) is now seen, using Fubini's theorem, to be unambiguous.

EXAMPLE 9.7.3 By Theorem 7.4.1, if P is the hyperparallelepiped in $\mathbb{R}^n$ whose edges are the columns of the matrix $\mathbf{X}$, then

$$\iint \dots \int_P 1 \, dx_1 \, dx_2 \dots dx_n = \sqrt{\det(\mathbf{X}^\top \mathbf{X})} \tag{9.115}$$

◇

We conclude this section with a counter-example to demonstrate what can happen if the assumptions of Fubini's theorem do not hold.

EXAMPLE 9.7.4 Consider the double integral

$$\int_0^1 \int_0^1 \frac{x^2 - y^2}{(x^2 + y^2)^2} \, dy \, dx \tag{9.116}$$

Fubini's theorem tells us that *if the integral of the absolute value of the integrand is finite, then the order of integration does not matter*; if we integrate first with respect to x and then with respect to y, we get the same result as if we integrate first with respect to y and then with respect to x.

Note in this example, however, that reversing the order of integration has the effect of multiplying the value of the integral by -1 because of the antisymmetry of the function being integrated – unless the value of the integral is zero. We will show that the value of the integral in (9.116) is non-zero and then confirm the implication of Fubini's theorem – that the integral of the absolute value of the integrand is infinite.

First, we consider the indefinite form of the inside integral:

$$\int \frac{x^2 - y^2}{(x^2 + y^2)^2}\, dy = \int \frac{x^2 + y^2 - 2y^2}{(x^2 + y^2)^2}\, dy$$

$$= \int \frac{1}{x^2 + y^2}\, dy + \int \frac{-2y^2}{(x^2 + y^2)^2}\, dy$$

$$= \int \frac{1}{x^2 + y^2}\, dy + \int y \left(\frac{d}{dy} \frac{1}{x^2 + y^2} \right) dy$$

$$= \int \frac{1}{x^2 + y^2}\, dy + \left(\frac{y}{x^2 + y^2} - \int \frac{1}{x^2 + y^2}\, dy \right) \quad \text{(by parts)}$$

$$= \frac{y}{x^2 + y^2} + C \tag{9.117}$$

where C is the arbitrary constant of integration. Thus

$$\int_0^1 \frac{x^2 - y^2}{(x^2 + y^2)^2}\, dy = \left[\frac{y}{x^2 + y^2} \right]_{y=0}^1 = \frac{1}{1 + x^2} \tag{9.118}$$

This takes care of the inside integral with respect to y; now we do the outside integral with respect to x.

$$\int_0^1 \frac{1}{1 + x^2}\, dx = [\tan^{-1}(x)]_{x=0}^1$$

$$= \tan^{-1}(1) - \tan^{-1}(0) = \frac{\pi}{4} \tag{9.119}$$

Thus we have

$$\int_0^1 \int_0^1 \frac{x^2 - y^2}{(x^2 + y^2)^2}\, dy\, dx = \frac{\pi}{4} \tag{9.120}$$

and, by antisymmetry,

$$\int_0^1 \int_0^1 \frac{x^2 - y^2}{(x^2 + y^2)^2}\, dx\, dy = -\frac{\pi}{4} \tag{9.121}$$

The prediction of Fubini's theorem does not hold in this case, so the assumption underlying Fubini's theorem must also fail to hold. So let us check whether the integral of the absolute value of the integrand is finite. Note that the integrand is zero when $x = y$, positive when

$x > y$ and negative when $x < y$. So we can divide up the region of integration into two triangular subregions and, using (9.117) again, write

$$
\int_0^1 \int_0^1 \left| \frac{x^2 - y^2}{(x^2 + y^2)^2} \right| dy\, dx = \int_0^1 \left[\int_0^x \frac{x^2 - y^2}{(x^2 + y^2)^2}\, dy + \int_x^1 \frac{y^2 - x^2}{(x^2 + y^2)^2}\, dy \right] dx
$$

$$
= \int_0^1 \left(\left[\frac{y}{x^2 + y^2} \right]_{y=0}^{x} + \left[\frac{y}{x^2 + y^2} \right]_{y=1}^{x} \right) dx
$$

$$
= \int_0^1 \left(\frac{1}{2x} - 0 + \frac{1}{2x} - \frac{1}{1 + x^2} \right) dx
$$

$$
= \int_0^1 \left(\frac{1}{x} - \frac{1}{1 + x^2} \right) dx
$$

$$
= [\ln x]_{x=0}^{1} - [\tan^{-1}(x)]_{x=0}^{1}
$$

$$
= 0 - (-\infty) - (\frac{\pi}{4} - 0)
$$

$$
= \infty \tag{9.122}
$$

This confirms our inference from Fubini's theorem that, since the two iterated integrals differ, the integral of the absolute value must be ∞.   ◇

9.7.3 Young's theorem

Fubini's theorem (Theorem 9.7.1) gave sufficient conditions under which the order of integration in a multiple integral is immaterial. The question naturally arises as to whether we can say the same thing about the order of differentiation in a higher-order mixed partial derivative. Multiple integration of partial derivatives allows us to show, under certain conditions, using Fubini's theorem, that the order of differentiation too is immaterial. This result, variously known as Clairaut's theorem, Schwarz's theorem or Young's theorem, gives conditions under which the symmetry of second derivatives (also called the equality of mixed partials) obtains.[10] We will refer to it by the last-mentioned name, Young's theorem. The reader has probably already noticed that this symmetry has obtained in most of the examples and exercises encountered so far.

THEOREM 9.7.2 (YOUNG'S THEOREM). *If the ijth and jith second-order partial derivatives of a real-valued function of n variables both exist and are continuous at* **x***, then they are equal at* **x***.*

If these conditions are satisfied for all $i, j = 1, 2, \ldots, n$*, then the Hessian matrix of the function is symmetric at* **x***.*

Without loss of generality, we will confine attention to the case of $n = 2$. The proof of Young's theorem, following Spivak (1965, Problem 3-28), is based on the following lemma.

LEMMA 9.7.3 *Let* f *be a function whose second-order partial derivatives are continuous on the rectangle* $[a, b] \times [c, d]$. *Then*

$$\int_c^d \int_a^b \frac{\partial^2 f}{\partial x_1 \partial x_2}(x_1, x_2)\, dx_1\, dx_2 = f(b, d) - f(a, d) - f(b, c) + f(a, c)$$

$$= \int_a^b \int_c^d \frac{\partial^2 f}{\partial x_2 \partial x_1}(x_1, x_2)\, dx_2\, dx_1 \qquad (9.123)$$

This lemma bears a close resemblance to Fubini's theorem but is a completely different proposition, as we do not know (yet) whether the integrands on both sides of (9.123) are equal.

Proof of lemma: The proof will be given for $\partial^2 f/\partial x_1 \partial x_2$. The proof for $\partial^2 f/\partial x_2 \partial x_1$ is left as an exercise; see Exercise 9.25.

Using the fundamental theorem of calculus to evaluate the inner integral yields

$$\int_c^d \int_a^b \frac{\partial^2 f}{\partial x_1 \partial x_2}(x_1, x_2)\, dx_1\, dx_2 = \int_c^d \left(\frac{\partial f}{\partial x_2}(b, x_2) - \frac{\partial f}{\partial x_2}(a, x_2) \right) dx_2 \qquad (9.124)$$

Using the fundamental theorem of calculus once again to evaluate the outer integral,

$$\int_c^d \left(\frac{\partial f}{\partial x_2}(b, x_2) - \frac{\partial f}{\partial x_2}(a, x_2) \right) dx_2 = (f(b, d) - f(a, d)) - (f(b, c) - f(a, c))$$

$$(9.125)$$

Removing the brackets from the right-hand side of (9.125) completes the proof of the lemma. □

Note that the expression $f(b, d) - f(a, d) - f(b, c) + f(a, c)$ is the difference between the increase in the value of the function f along the top of the rectangle (from (a, d) to (b, d)) and the increase along the bottom of the rectangle (from (a, c) to (b, c)). By swapping the two middle terms, the same expression can be seen to equal the difference between the increase in the value of the function f along the right side of the rectangle and the increase along the left side of the rectangle. This simple geometric observation is the basic principle underlying Young's theorem.

The above lemma and proof extend to functions of any number of variables. For n variables, however, the number of terms on the right-hand side of (9.125) increases to 2^n, the number of vertices of a hyperrectangle in n dimensions; see Exercise 9.26.

Proof of Young's theorem: The proof is by contradiction. We begin by assuming that there is some point, (z, t), say, at which the second-order partials differ, so that

$$\frac{\partial^2 f}{\partial x_1 \partial x_2}(z, t) - \frac{\partial^2 f}{\partial x_2 \partial x_1}(z, t) = h \qquad (9.126)$$

for some $h \neq 0$. We can relabel the variables if necessary to ensure that $h > 0$.

By continuity of the second-order partials, there is some neighbourhood of (z, t) throughout which

$$\frac{\partial^2 f}{\partial x_1 \partial x_2}(x_1, x_2) - \frac{\partial^2 f}{\partial x_2 \partial x_1}(x_1, x_2) \geq \frac{h}{2} \tag{9.127}$$

We can now pick a rectangle $[a, b] \times [c, d]$ within this neighbourhood.

The inequality is preserved when we integrate both sides of inequality (9.127) over this rectangle:

$$\int_c^d \int_a^b \left(\frac{\partial^2 f}{\partial x_1 \partial x_2}(x_1, x_2) - \frac{\partial^2 f}{\partial x_2 \partial x_1}(x_1, x_2) \right) dx_1 \, dx_2 \geq \int_c^d \int_a^b \frac{h}{2} \, dx_1 \, dx_2 \tag{9.128}$$

By Fubini's theorem and Lemma 9.7.3, the integral on the left-hand side vanishes. The integrand on the right-hand side of the inequality is constant, so the integral is merely the volume $h \times (b - a) \times (d - c)/2$, which is positive. Thus we have the required contradiction. $\qquad\square$

Just as Lemma 9.7.3 extends straightforwardly to higher dimensions, so too does this proof of Young's theorem. Hence, the order of differentiation is also immaterial in higher-order partial derivatives, again assuming that the relevant partial derivatives exist and are continuous.

9.7.4 Differentiation under the integral sign

In later applications, we will frequently have to differentiate integrals in various ways, and not just with respect to the upper limit of integration, as was the case in the fundamental theorem of calculus (Theorem 7.9.1). We will need more general results. Techniques such as reversing the limits of integration (when we wish to differentiate with respect to the lower limit of integration) and the chain rule (when the limit of integration is a function of the relevant variable) are sometimes sufficient.

More generally, just as we need Fubini's theorem to tell us whether we can shuffle or permute the order of integration and Young's theorem to tell us whether we can permute the order of differentiation, we need to know under what circumstances the order of integration and differentiation can be reversed when both operations occur in the same expression. Before presenting a theorem giving conditions under which reversing the order is legitimate, we begin with a counter-example to show that we cannot perform this procedure indiscriminately and to suggest why the various conditions of the theorem will be required.

EXAMPLE 9.7.5 Let

$$F(\alpha) = \int_0^\infty \frac{\sin \alpha x}{x} \, dx \quad \text{with } \alpha > 0 \tag{9.129}$$

We are interested in evaluating $F'(\alpha)$, i.e. in differentiating this integral with respect to a variable that appears as a constant under the integral sign. Differentiating with respect to α

under the integral sign would yield

$$F'(\alpha) = \int_0^\infty \frac{\cos \alpha x}{x} x \, dx$$

$$= \int_0^\infty \cos \alpha x \, dx$$

$$= \left[\frac{\sin \alpha x}{\alpha} \right]_{x=0}^\infty \tag{9.130}$$

The expression inside the square brackets does not converge as $x \to \infty$ for fixed α; rather, it oscillates between $-1/\alpha$ and $+1/\alpha$.

Now consider an alternative approach to evaluating and then differentiating the integral in (9.129). Let us make the substitution $y = \alpha x$, so that $dy = \alpha \, dx$. Then we obtain, for all α,

$$F(\alpha) = \int_0^\infty \frac{\sin y}{y} \, dy \tag{9.131}$$

In other words, the value of the function F is independent of α and, as with any constant function, $F'(\alpha) = 0$ for all α. In fact, F is related to the **sine integral function**, defined by[11]

$$\mathrm{Si}(z) = \int_0^z \frac{\sin y}{y} \, dy \tag{9.132}$$

It can be shown that the constant value of $F(\alpha)$ is $\pi/2$ for all α; see Exercise 9.27.   ◇

This example is a warning that, when differentiating under the integral sign, we must be sure that both the integral of the original function and the integral of its derivative exist over the relevant range of integration. In the example, the former exists but the latter does not.

9.7.5 Leibniz's integral rule

A full course on integration would present several different theorems concerning differentiation under the integral sign. Here we present only two versions of Leibniz's integral rule for functions of two variables. We provide two alternative proofs in each case, one based on Fubini's theorem (which we have not proved) and a more complete proof assuming uniformly continuous partial derivatives.

THEOREM 9.7.4 (LEIBNIZ'S INTEGRAL RULE WITH FIXED LIMITS OF INTEGRATION). *Suppose that the function* $f : [x_0, x_1] \times [y_0, y_1] \to \mathbb{R}$ *has a continuous partial derivative* $\partial f / \partial x$ *and define* $F : [x_0, x_1] \to \mathbb{R}$ *by*

$$F(x) = \int_{y_0}^{y_1} f(x, y) \, dy \tag{9.133}$$

Then, for $x \in (x_0, x_1)$, *F is differentiable at x and*

$$F'(x) = \int_{y_0}^{y_1} \frac{\partial f}{\partial x}(x, y) \, dy \tag{9.134}$$

provided that either

(a) $\partial f / \partial x$ *is bounded and*

$$\int_{x_0}^{x^*} \int_{y_0}^{y_1} \left| \frac{\partial f}{\partial x}(x, y) \right| dy \, dx$$

exists for all $x^* \in (x_0, x_1)$ *or*

(b) $\partial f / \partial x$ *is uniformly continuous on* $[x_0, x_1]$.

Proof:

(a) The first set of assumptions will allow us to apply Fubini's theorem.
For $x^* \in (x_0, x_1)$, let

$$I(x^*) \equiv \int_{x_0}^{x^*} \int_{y_0}^{y_1} \frac{\partial f}{\partial x}(x, y) \, dy \, dx \tag{9.135}$$

Note that by the first part of the fundamental theorem of calculus

$$I'(x^*) = \int_{y_0}^{y_1} \frac{\partial f}{\partial x}(x^*, y) \, dy \tag{9.136}$$

By Fubini's theorem, given our assumptions, we can interchange the order of integration in (9.135) to obtain

$$I(x^*) = \int_{y_0}^{y_1} \int_{x_0}^{x^*} \frac{\partial f}{\partial x}(x, y) \, dx \, dy \tag{9.137}$$

Using the second part of the fundamental theorem of calculus to evaluate the inner integral,

$$I(x^*) = \int_{y_0}^{y_1} (f(x^*, y) - f(x_0, y)) \, dy$$

$$= \int_{y_0}^{y_1} f(x^*, y) \, dy - \int_{y_0}^{y_1} f(x_0, y) \, dy$$

$$= F(x^*) - F(x_0) \tag{9.138}$$

The choice x^* was arbitrary, so by differentiating (9.138) with respect to x^* we have

$$I'(x^*) = F'(x^*) \quad \forall x^* \in (x_0, x_1) \tag{9.139}$$

Hence, by (9.136),

$$F'(x) = \int_{y_0}^{y_1} \frac{\partial f}{\partial x}(x, y) \, dy \quad \forall x \in (x_0, x_1) \tag{9.140}$$

Thus the rule is proved.

Note also that, while the theorem is expressed in terms of finite limits of integration with respect to y, the proof would still hold if either or both of these limits was infinite.

(b) If $\partial f / \partial x$ is uniformly continuous on $[x_0, x_1]$, then we can prove the result without calling on Fubini's theorem.

By definition, returning to first principles,

$$F'(x) = \lim_{h \to 0} \frac{F(x+h) - F(x)}{h} \tag{9.141}$$

Substituting the definition of F from (9.133) into (9.141) and using the linearity of the integration operator yields

$$
\begin{aligned}
F'(x) &= \lim_{h \to 0} \frac{\int_{y_0}^{y_1} f(x+h, y)\, dy - \int_{y_0}^{y_1} f(x, y)\, dy}{h} \\
&= \lim_{h \to 0} \frac{\int_{y_0}^{y_1} (f(x+h, y) - f(x, y))\, dy}{h} \\
&= \lim_{h \to 0} \int_{y_0}^{y_1} \frac{f(x+h, y) - f(x, y)}{h}\, dy
\end{aligned}
\tag{9.142}
$$

We now claim that

$$\lim_{h \to 0} \int_{y_0}^{y_1} \left(\frac{f(x+h, y) - f(x, y)}{h} - \frac{\partial f}{\partial x}(x, y) \right) dy = 0 \tag{9.143}$$

The mean value theorem (Theorem 9.6.3) tells us that, for each x and h, there exists a number $\theta(x, h) \in [0, 1]$ such that

$$\frac{f(x+h, y) - f(x, y)}{h} = \frac{\partial f}{\partial x}(x + \theta(x, h)h, y) \tag{9.144}$$

So we need to confirm that

$$\lim_{h \to 0} \int_{y_0}^{y_1} \left(\frac{\partial f}{\partial x}(x + \theta(x, h)h, y) - \frac{\partial f}{\partial x}(x, y) \right) dy = 0 \tag{9.145}$$

Now choose $\epsilon > 0$. By uniform continuity of the partial derivative, there exists $\delta > 0$ such that, whenever $|h| < \delta$,

$$\left| \frac{\partial f}{\partial x}(x+h, y) - \frac{\partial f}{\partial x}(x, y) \right| < \frac{\epsilon}{y_1 - y_0} \tag{9.146}$$

and, in particular,

$$\left| \frac{\partial f}{\partial x}(x + \theta(x, h)h, y) - \frac{\partial f}{\partial x}(x, y) \right| < \frac{\epsilon}{y_1 - y_0} \tag{9.147}$$

Since the absolute value of the integrand in (9.145) is less than $\epsilon/(y_1 - y_0)$, the value of the integral is less than

$$(y_1 - y_0) \times \frac{\epsilon}{y_1 - y_0} = \epsilon \tag{9.148}$$

This proves the claim.

Leibniz's integral rule follows by combining (9.142) and (9.143) and again using the linearity of the integration operator.

Like the previous one, this proof would still hold if either or both of the limits of integration with respect to y was infinite.

Note that, when x_0, x_1, y_0 and y_1 are finite, the Heine–Cantor theorem allows us to relax the assumption that $\partial f/\partial x$ is uniformly continuous, since in that case continuity of $\partial f/\partial x$ implies uniform continuity. $\qquad\square$

The first part of the fundamental theorem of calculus and the preceding version of Leibniz's integral rule can be combined into a more general result.

THEOREM 9.7.5 (LEIBNIZ'S INTEGRAL RULE WITH VARIABLE LIMITS OF INTEGRA-TION). *Suppose that the function* $f:[x_0, x_1] \times [y_0, y_1] \to \mathbb{R}$ *has a uniformly continuous partial derivative* $\partial f/\partial x$ *and define*

$$F(x) = \int_{a(x)}^{b(x)} f(x, y)\, dy \tag{9.149}$$

where a and b are continuously differentiable functions defined on (x_0, x_1). Then, for $x \in (x_0, x_1)$, we have

$$F'(x) = b'(x) f(x, b(x)) - a'(x) f(x, a(x)) + \int_{a(x)}^{b(x)} \frac{\partial f}{\partial x}(x, y)\, dy \tag{9.150}$$

Proof: The proof of this result, which also requires the chain rule, is left as an exercise; see Exercise 9.28. $\qquad\square$

9.7.6 Change of variables

In Examples 9.7.1 and 9.7.4 above, we used the familiar change-of-variable technique to simplify single integrals. We now present, without proof, the corresponding result for multiple integrals.

THEOREM 9.7.6 *Let A, $B \subseteq \mathbb{R}^n$, let $f: B \to \mathbb{R}$ be a real-valued function, which it is desired to integrate over B, and let $g: A \to B$ be an invertible vector-valued function of several variables. Then*

$$\iint \cdots \int_B f(\mathbf{y})\, dy_1\, dy_2 \ldots dy_n = \iint \cdots \int_A f(g(\mathbf{x}))|\det(g'(\mathbf{x}))|\, dx_1\, dx_2 \ldots dx_n \tag{9.151}$$

Proof: The full proof of this theorem is beyond the scope of this book. It can be found in Spivak (1965, pp. 67–72). □

We conclude this section by revisiting two familiar examples in the notation of Theorem 9.7.6.

EXAMPLE 9.7.6 Suppose $A = B = \mathbb{R}_+$. Let $f(y) \equiv (\sin \alpha y)/y$ and $g(x) \equiv x/\alpha$ where $\alpha > 0$. Then $|\det(g'(x))| = g'(x) = 1/\alpha$ and (9.151) becomes

$$\int_0^\infty \frac{\sin \alpha y}{y}\, dy = \int_0^\infty \frac{\sin x}{x/\alpha} \frac{1}{\alpha}\, dx = \int_0^\infty \frac{\sin x}{x}\, dx \tag{9.152}$$

This is just the substitution carried out in Example 9.7.4, but presented in slightly different notation. ◇

EXAMPLE 9.7.7 Suppose A is the unit cube $[0, 1]^n$ in $\mathbb{R}^n$. Let g be the linear transformation with matrix $\mathbf{T}$ so that $B \equiv g(A)$ is the hyperparallelepiped in $\mathbb{R}^n$ whose edges are the columns of the matrix $\mathbf{T}$. Let $f(\mathbf{y}) \equiv 1$ for all $\mathbf{y} \in B$. Since $g(\mathbf{x}) = \mathbf{Tx}$, $g'(\mathbf{x}) = \mathbf{T}$ for all $\mathbf{x} \in A$ and (9.151) becomes

$$\begin{aligned}
V(B) &= \iint \cdots \int_B 1\, dy_1\, dy_2 \ldots dy_n \\
&= \iint \cdots \int_A |\det(\mathbf{T})|\, dx_1\, dx_2 \ldots dx_n \\
&= |\det(\mathbf{T})| \iint \cdots \int_A dx_1\, dx_2 \ldots dx_n \\
&= |\det(\mathbf{T})| \prod_{i=1}^n \int_0^1 dx_i \\
&= |\det(\mathbf{T})|
\end{aligned} \tag{9.153}$$

since each integral in the product in the penultimate line evaluates to unity. This is the familiar result from Theorem 7.4.1 and Example 9.7.2 for the case in which $\mathbf{T}$ is square. ◇

We will meet multiple integrals again in Section 13.6 and throughout much of the remainder of this book.

9.8 Implicit function theorem

The implicit function theorem states conditions under which it is possible to solve a system of m equations in n unknowns, where $n > m$, in a continuously differentiable way, which will allow the first m unknowns to be written as functions of the last $n - m$. The theorem merely tells us that a solution exists and how to compute its derivative: it does not tell us how to compute the closed-form solution itself. For a more detailed introduction, the interested reader is referred to Chiang and Wainwright (2005, Section 8.5).

THEOREM 9.8.1 (IMPLICIT FUNCTION THEOREM). *Let* $g: X \to \mathbb{R}^m$, *where* $X \subseteq \mathbb{R}^n$ *and* $m < n$. *Consider the system of m scalar equations in n variables,* $g(\mathbf{x}^*) = \mathbf{0}_m$.

Partition the n-dimensional vector $\mathbf{x}$ *as* $(\mathbf{y}, \mathbf{z})$ *where* $\mathbf{y} = (x_1, x_2, \ldots, x_m)$ *is m-dimensional and* $\mathbf{z} = (x_{m+1}, x_{m+2}, \ldots, x_n)$ *is* $(n - m)$-*dimensional. Similarly, partition the total derivative of g at* $\mathbf{x}^*$ *as*

$$\underbrace{g'(\mathbf{x}^*)}_{(m \times n)} = \begin{bmatrix} \underbrace{D_{\mathbf{y}}g}_{(m \times m)} & \underbrace{D_{\mathbf{z}}g}_{(m \times (n-m))} \end{bmatrix} \tag{9.154}$$

[We aim to solve these equations for the first m variables, $\mathbf{y}$*, which will then be written as functions,* $h(\mathbf{z})$*, of the last* $n - m$ *variables,* $\mathbf{z}$*.]*

Suppose g is continuously differentiable in a neighbourhood of $\mathbf{x}^*$*, and that the* $m \times m$ *matrix*

$$D_{\mathbf{y}}g \equiv \begin{bmatrix} \dfrac{\partial g^1}{\partial x_1}(\mathbf{x}^*) & \cdots & \dfrac{\partial g^1}{\partial x_m}(\mathbf{x}^*) \\ \vdots & \ddots & \vdots \\ \dfrac{\partial g^m}{\partial x_1}(\mathbf{x}^*) & \cdots & \dfrac{\partial g^m}{\partial x_m}(\mathbf{x}^*) \end{bmatrix} \tag{9.155}$$

formed by the first m columns of the total derivative of g at $\mathbf{x}^*$ *is non-singular.*

Then there exist neighbourhoods Y of $\mathbf{y}^*$ *and Z of* $\mathbf{z}^*$ *such that* $Y \times Z \subseteq X$*, and a continuously differentiable function* $h: Z \to Y$ *such that*

(a) $\mathbf{y}^* = h(\mathbf{z}^*)$,
(b) $g(h(\mathbf{z}), \mathbf{z}) = \mathbf{0}$ *for all* $\mathbf{z} \in Z$*, and*
(c) $h'(\mathbf{z}^*) = -(D_{\mathbf{y}}g)^{-1} D_{\mathbf{z}}g$.

Proof: The full proof of this theorem is beyond the scope of this book, but can be found in Spivak (1965, pp. 40–3). However, part (c) of the theorem follows easily from material in Section 9.3. The aim is to derive an expression for the total derivative $h'(\mathbf{z}^*)$ in terms of the partial derivatives of g, using the chain rule.

We know from part (b) that

$$f(\mathbf{z}) \equiv g(h(\mathbf{z}), \mathbf{z}) = \mathbf{0}_m \quad \forall\, \mathbf{z} \in Z \tag{9.156}$$

Thus

$$f'(\mathbf{z}) \equiv \mathbf{0}_{m \times (n-m)} \quad \forall\, \mathbf{z} \in Z \tag{9.157}$$

in particular at $\mathbf{z}^*$. But we know from Corollary 9.3.2 that

$$f'(\mathbf{z}) = D_{\mathbf{y}}g h'(\mathbf{z}) + D_{\mathbf{z}}g \tag{9.158}$$

Hence,

$$D_{\mathbf{y}}g h'(\mathbf{z}) + D_{\mathbf{z}}g = \mathbf{0}_{m \times (n-m)} \tag{9.159}$$

and, since the statement of the theorem requires that $D_{\mathbf{y}}g$ is invertible, we have

$$h'(\mathbf{z}^*) = -(D_{\mathbf{y}}g)^{-1}D_{\mathbf{z}}g \tag{9.160}$$

as required.   □

EXAMPLE 9.8.1   Consider the equation $g(x, y) \equiv x^2 + y^2 - 1 = 0$.
   Here we have $m = 1$ and $n = 2$. Note that $g'(x, y) = [2x \quad 2y]$.
   We have $h(y) = \sqrt{1 - y^2}$ or $h(y) = -\sqrt{1 - y^2}$, each of which describes a single-valued, differentiable function on $(-1, 1)$. At $(x, y) = (0, 1)$, $\partial g/\partial x = 0$ and $h(y)$ is undefined (for $y > 1$) or multivalued (for $y < 1$) in any neighbourhood of $y = 1$.   ◇

EXAMPLE 9.8.2   Now consider the system of m linear equations in n variables, $g(\mathbf{x}) \equiv \mathbf{Bx} = \mathbf{0}$, where $\mathbf{B}$ is an $m \times n$ matrix.
   We have $g'(\mathbf{x}) = \mathbf{B}$ for all $\mathbf{x}$, so the implicit function theorem applies, provided that the equations are linearly independent or, equivalently, that the matrix $\mathbf{B}$ is of full rank m ($m < n$).
   If we partition $\mathbf{x}$ as $(\mathbf{y}, \mathbf{z})$ and partition $\mathbf{B}$ conformably as $[\mathbf{C} \ \mathbf{D}]$, then we can solve for the first m variables $\mathbf{y}$ in terms of the last $n - m$ variables $\mathbf{z}$:

$$\mathbf{y} = -\mathbf{C}^{-1}\mathbf{Dz} \tag{9.161}$$

provided that the first m columns of $\mathbf{C}$ are linearly independent.
   This is consistent with the results set out in Section 5.4.6.   ◇

The next theorem can be derived as a special case of the implicit function theorem.

THEOREM 9.8.2 (INVERSE FUNCTION THEOREM).   *Let* $f: X \to \mathbb{R}^m$, *where* $X \subseteq \mathbb{R}^m$.
Consider the system of m scalar equations in $2m$ variables, $g(\mathbf{y}, \mathbf{z}) \equiv f(\mathbf{y}) - \mathbf{z} = \mathbf{0}_m$.
   The total derivative of g can be partitioned as

$$\underbrace{g'(\mathbf{y}, \mathbf{z})}_{(m \times 2m)} = \Big[\underbrace{f'(\mathbf{y})}_{(m \times m)} \quad \underbrace{-\mathbf{I}_m}_{(m \times m)} \Big] \tag{9.162}$$

[We aim to solve these equations for the first m variables, $\mathbf{y}$, which will then be written as functions, $h(\mathbf{z})$, of the last m variables, $\mathbf{z}$. In this case, we will have $h = f^{-1}$.]
   Suppose f is continuously differentiable in a neighbourhood of $\mathbf{y}^*$, and that the $m \times m$ matrix $f'(\mathbf{y}^*)$ is non-singular.
   Then there exist neighbourhoods Y of $\mathbf{y}^*$ and Z of $\mathbf{z}^*$ such that $Y \subseteq X$, and a continuously differentiable function $h: Z \to Y$ such that

(a) $\mathbf{y}^* = h(\mathbf{z}^*)$,
(b) $f(h(\mathbf{z})) = \mathbf{z}$ *for all* $\mathbf{z} \in Z$, *and*
(c) $h'(\mathbf{z}^*) = (f'(\mathbf{y}^*))^{-1}$.

Proof: The proof is left as an exercise; see Exercise 9.29.   □

The inverse function theorem says that, if the Jacobian matrix $f'(\mathbf{y}^*)$ is non-singular, then f is locally invertible around $\mathbf{y}^*$. The converse is not true.

To see this, just consider the function of one variable defined by $f(x) = x^3$, which is invertible everywhere, with $f^{-1}(y) = \sqrt[3]{y}$. However, $f'(0) = 0$ and f has an **inflexion point** or **point of inflexion** at $x = 0$. So invertible functions can have singular Jacobians.

EXERCISES

9.1 Write down the total derivative and the Hessian matrix associated with the function $f: \mathbb{R}^3_{++} \to \mathbb{R}: (x_1, x_2, x_3) \mapsto x_1^3 + x_1^2 \ln x_2 - 3x_3$. Is $f''(x_1, x_2, x_3)$ semi-definite? If so, why? If not, why not?

9.2 Consider the quadratic form defined by the function

$$f: \mathbb{R}^2 \to \mathbb{R}: (x_1, x_2) \mapsto ax_1^2 + 2bx_1x_2 + cx_2^2$$

where a, b and c are real numbers.

(a) Write $f(x_1, x_2)$ in terms of the vector $\mathbf{x} = \begin{bmatrix} x_1 \\ x_2 \end{bmatrix}$ and a 2×2 real symmetric matrix.

(b) Using scalar methods only, calculate the total derivative $f'(x_1, x_2)$ and the Hessian $f''(x_1, x_2)$ of f at the point (x_1, x_2).

9.3 Show, using scalar differentiation, that

$$\frac{\partial^2 (\mathbf{x}^\top \mathbf{A} \mathbf{x})}{\partial \mathbf{x} \partial \mathbf{x}^\top} = 2\mathbf{A}$$

where $\mathbf{x}$ is $n \times 1$, and $\mathbf{A}$ is symmetric $n \times n$.

9.4 Differentiate the following matrix expressions with respect to $\mathbf{x}$, where λ is a scalar, $\mathbf{1}$ is an $n \times 1$ matrix all of whose elements are unity, $\mathbf{a}$ and $\mathbf{x}$ are both $n \times 1$, and $\mathbf{A}$ is $n \times n$.

(a) $\mathbf{1}^\top \mathbf{x}$
(b) $\lambda \mathbf{a}^\top \mathbf{x}$
(c) $\mathbf{x}^\top \mathbf{x}$
(d) $\mathbf{x}^\top \mathbf{A} \mathbf{x}$
(e) $2\mathbf{A}\mathbf{x}$
(f) $2\lambda + 3\mathbf{x}^\top \mathbf{A} \mathbf{x}$

9.5 If $f(\mathbf{x}) = 10 + 20\mathbf{a}^\top \mathbf{x} + 30\mathbf{x}^\top \mathbf{A} \mathbf{x}$, where $\mathbf{a}$ and $\mathbf{x}$ are $n \times 1$ matrices, and $\mathbf{A}$ is an $n \times n$ matrix, what are the total derivative $f'(\mathbf{x})$ and the Hessian matrix $f''(\mathbf{x})$?

9.6 (This exercise presumes a basic knowledge of some material covered in more detail later in the book; some readers may like to come back to it after studying Chapter 13 and Section 15.2.) An investor divides her wealth in the proportions $\mathbf{a} = (a_1, a_2, \ldots, a_N)$ among a portfolio of N possible investments. The payoffs per unit invested (gross rates of return or simply gross returns) on each asset are given by the (random) vector $\tilde{\mathbf{r}} = (\tilde{r}_1, \tilde{r}_2, \ldots, \tilde{r}_N)$.

(a) Write down (in matrix notation) an expression for the overall gross return on the investor's portfolio.
(b) Assuming that $\tilde{\mathbf{r}}$ has mean $\mathbf{e}$ and variance–covariance matrix $\mathbf{V}$, calculate the mean and variance of this overall gross return.
(c) Calculate the Jacobian and Hessian of the variance of the portfolio gross return (viewed as a function of the portfolio proportions $\mathbf{a}$).
(d) What can you say about the definiteness or semi-definiteness of the Hessian matrix?

9.7 Calculate the first- and second-order partial derivatives of the function

$$f:\mathbb{R}^2 \to \mathbb{R}: (x, y) \mapsto \begin{cases} \dfrac{xy(x^2 - y^2)}{x^2 + y^2} & \text{for } (x, y) \neq (0, 0) \\ 0 & \text{for } (x, y) = (0, 0) \end{cases}$$

Do the second-order mixed partial derivatives exist at $(0, 0)$? Are they continuous? Are they equal?

9.8 A tax is said to be **progressive** if it accounts for an increasing proportion of income as income rises and **regressive** if it accounts for a decreasing proportion of income as income rises.

Let the tax due on an income of Y be denoted by $t = f(Y)$. Show that this tax is progressive wherever f is elastic and regressive wherever f is inelastic.

9.9 Prove Theorem 9.3.1.

9.10 Let $f, g:\mathbb{R}^m \to \mathbb{R}^n$ and define $h:\mathbb{R}^m \to \mathbb{R}$ by

$$\underbrace{h(\mathbf{x})}_{1\times 1} \equiv \underbrace{(f(\mathbf{x}))^\top}_{1\times n} \underbrace{g(\mathbf{x})}_{n\times 1}$$

Show that

$$\underbrace{h'(\mathbf{x})}_{1\times m} = \underbrace{(g(\mathbf{x}))^\top}_{1\times n} \underbrace{f'(\mathbf{x})}_{n\times m} + \underbrace{(f(\mathbf{x}))^\top}_{1\times n} \underbrace{g'(\mathbf{x})}_{n\times m}$$

(This is the first version of the product rule for vector calculus, Theorem 9.3.3.)

9.11 Let $f:\mathbb{R}^m \to \mathbb{R}$ and $g:\mathbb{R}^m \to \mathbb{R}^n$ and define $h:\mathbb{R}^m \to \mathbb{R}^n$ by

$$\underbrace{h(\mathbf{x})}_{n\times 1} \equiv \underbrace{f(\mathbf{x})}_{1\times 1} \underbrace{g(\mathbf{x})}_{n\times 1}$$

Show that

$$\underbrace{h'(\mathbf{x})}_{n\times m} = \underbrace{g(\mathbf{x})}_{n\times 1} \underbrace{f'(\mathbf{x})}_{1\times m} + \underbrace{f(\mathbf{x})}_{1\times 1} \underbrace{g'(\mathbf{x})}_{n\times m}$$

(This is the second version of the product rule for vector calculus, Theorem 9.3.3.)

9.12 Show, using the chain rule, that

$$\frac{x_i^*}{f(\mathbf{x}^*)} \frac{\partial f}{\partial x_i}(\mathbf{x}^*) = \frac{\partial \ln f}{\partial \ln x_i}(\mathbf{x}^*)$$

Show further that

$$\frac{x_i^*}{f(\mathbf{x}^*)} \frac{\partial f}{\partial x_i}(\mathbf{x}^*) = \frac{\partial \log f}{\partial \log x_i}(\mathbf{x}^*)$$

where the logarithms are taken to any base.

9.13 Let $f: \mathbb{R}^l \to \mathbb{R}^m$ and $g: \mathbb{R}^m \to \mathbb{R}^n$ be continuously differentiable functions. Write down in matrix form the chain rule relating the derivatives of the functions f, g and $h \equiv (g \circ f)$.

9.14 Let $f: \mathbb{R}^3 \to \mathbb{R}^2$ and $g: \mathbb{R}^2 \to \mathbb{R}^3$ be defined by

$$f(\mathbf{y}) \equiv \begin{bmatrix} y_1 + y_2 + y_3 \\ y_1 - y_2 y_3 \end{bmatrix}$$

and

$$g(\mathbf{x}) \equiv \begin{bmatrix} x_1 + x_2 \\ x_1 x_2 \\ x_1 - x_2^2 \end{bmatrix}$$

Find the Jacobian matrices $f'(\mathbf{y})$ and $g'(\mathbf{x})$, and, hence, use the chain rule to find $(f \circ g)'(\mathbf{x})$. Write down the gradient vector and the Hessian matrix associated with the real-valued component function f^2.

9.15 Let L be the line from $\mathbf{x}'$ to $\mathbf{x}$ in $\mathbb{R}^n$ and let $f: \mathbb{R}^n \to \mathbb{R}$. Recall the definition of the restriction of f to L, $f|_L: \mathbb{R} \to \mathbb{R}$; see Definition 9.5.1.

Use the chain rule to calculate the second derivative of $f|_L$ at a generic λ and, hence, at $\lambda = 0$.

9.16 Calculate the directional derivative of the Cobb–Douglas function (of two variables) given by

$$f: \mathbb{R}^2 \to \mathbb{R}: (x, y) \mapsto x^\alpha y^{1-\alpha}$$

at the vector (x, y) in the direction of the vector $(\mu x, \mu y)$ (i.e. when the relevant line L passes through the origin).

9.17 Show that, when $\alpha = 0.5$, the indifference curves of the Cobb–Douglas function in the previous exercise are a family of rectangular hyperbolas.

9.18 Calculate the general form of the restriction to a line of a quadratic form (in n variables). What is the general form of the directional derivative of such a function?

9.19 What is the relationship between the directional derivatives of a function f at $\mathbf{x}$ in the direction of $\mathbf{x}'$ and in the direction of $(\mathbf{x}+\mathbf{x}')/2$? What is it in the direction of $\lambda\mathbf{x}+(1-\lambda)\mathbf{x}'$? Illustrate your answers graphically and explain them verbally.

9.20 Show how to interpret the directional derivative at a point in terms of the rescaling of the parametrization of the relevant line L.

9.21 Show that, for a quadratic function $f(x)=ax^2+bx+c$, the approximation A^* given in equation (9.87) exactly equals $f(x)$.

9.22 Show that the first $n-1$ derivatives at x_0 of the polynomial on the right-hand side of (9.88) are the same as those of f.

9.23 Write down the general form of Taylor's expansion for functions of several variables using notation of your choice.

Using the first three terms of Taylor's expansion, approximate the function

$$x_1 x_2^2 + x_2^2 \ln x_3^2$$

around

$$\mathbf{x} = (x_1, x_2, x_3) = (1, 1, 1)$$

9.24 Differentiate the following integrals with respect to x.

(a) $\displaystyle\int_x^y f(t)\,dt$

(b) $\displaystyle\int_0^{x^2} \sin t\,dt$

(c) $\displaystyle\int_{-\infty}^{\infty} (x^2 y + xy^2)\,dy$

(d) $\displaystyle\int_{-\infty}^{\infty} u(wr + x(s-r))\,dr$

(e) $\displaystyle\int_a^b (c + dt + ft^2)\,dt$

(f) $\displaystyle\int_{-\infty}^{x} e^t\,dt$

9.25 Prove that

$$\int_a^b \int_c^d \frac{\partial^2 f}{\partial x_2 \partial x_1}(x_1, x_2)\,dx_2\,dx_1 = f(b,d) - f(a,d) - f(b,c) + f(a,c)$$

(i.e. the second part of Lemma 9.7.3).

9.26 State and prove the equivalent of Lemma 9.7.3 for functions of three variables.

9.27 Show that

$$\int_0^\infty \frac{\sin y}{y} \, dy = \frac{\pi}{2}$$

9.28 Prove Theorem 9.7.5.

9.29 Derive the inverse function theorem as a special case of the implicit function theorem.

10 Convexity and optimization

10.1 Introduction

Much of economics and decision theory reduces to making optimal choices. This requires specifying the decision-maker's objective as a mathematical function, depending on one or more choice variables. The mathematical theory of optimization tells us whether the decision-maker's problem will have a solution or solutions, and how to find a solution if one exists. The objective of this chapter is to provide the reader with all the tools necessary to solve any optimization problem that may be encountered in economics or finance.

The chapter begins with an extended discussion of convexity and concavity, concepts that are important in determining whether a solution to an optimization problem exists or is unique. The next three sections discuss the solution of optimization problems, first when all choice variables are free to vary independently, then when the choice variables are subject to equality constraints, and finally when there are inequality constraints. The chapter concludes with a section on the duality between the maximization of an objective function subject to a constraint and the minimization of the constraint function subject to the objective function taking on a particular value.

Chiang and Wainwright (2005, Chapters 11–13) cover some of the material in this chapter at a more elementary level. More advanced treatments can be found in de la Fuente (2000, Chapter 6) and Takayama (1994).

10.2 Convexity and concavity

10.2.1 Convex and concave functions

We have already encountered the concept of a convex function in Definition 7.3.4. Roughly speaking, a function of n variables is convex if the set above its graph is a convex subset of $\mathbb{R}^{n+1}$ and concave if the set below its graph is a convex subset of $\mathbb{R}^{n+1}$. From this rough description, it should already be clear that the concepts of concave and convex *functions* are broadly analogous. It is important to bear in mind, however, that there is no such thing as a concave *set*. For this reason, this branch of mathematics is usually described as **convexity** and never as concavity. Thus only the word "convexity" appears in the titles of this chapter and of various sections below.

We can approach the characterization of convex and concave functions of several variables from three perspectives. The formal definitions are in terms of the restrictions of the function in question to line segments in its domain. The other characterizations apply only to those convex and concave functions that are appropriately differentiable. The second characterization is based on properties of the first derivative of the function (see Theorem 10.2.4) and the third characterization is based on properties of the second derivative or Hessian (see Theorem 10.2.5).

DEFINITION 10.2.1 Let $f: X \to Y$, where X is a convex subset of a real vector space and $Y \subseteq \mathbb{R}$. Then f is a **concave function** if and only if, for all $\mathbf{x} \neq \mathbf{x}' \in X$ and $\lambda \in (0, 1)$,

$$f(\lambda \mathbf{x} + (1 - \lambda)\mathbf{x}') \geq \lambda f(\mathbf{x}) + (1 - \lambda) f(\mathbf{x}') \tag{10.1}$$

The reader should note a number of important points concerning Definitions 7.3.4 and 10.2.1.

- The function f is convex if and only if $-f$ is concave. Since every convex function is the mirror image of a concave function, and vice versa, every result derived for one has an obvious corollary for the other. In general, we will consider only concave functions in the text, and leave the derivation of the corollaries for convex functions as exercises.
- A function defined on an n-dimensional vector space, V, is convex if and only if the restriction of the function to the line L is convex for every line L in V, and similarly for concave functions (and for strictly convex and strictly concave functions, which will be defined shortly; see Definition 10.2.2).
- Conditions (7.14) and (10.1) could also have been required to hold, equivalently, for all $\mathbf{x}, \mathbf{x}' \in X$ and $\lambda \in [0, 1]$, since they are satisfied as equalities for all f when $\mathbf{x} = \mathbf{x}'$, when $\lambda = 0$ and when $\lambda = 1$.
- Conditions (7.14) and (10.1) are presented, like the equivalent condition in the definition of a linear transformation (Definition 6.2.3), in terms of pairs of vectors. Using familiar logic, we note that these conditions can equivalently be presented in terms of k vectors, as

$$f\left(\sum_{i=1}^{r} k_i \mathbf{v}_i\right) \leq \sum_{i=1}^{r} k_i f(\mathbf{v}_i) \tag{10.2}$$

and

$$f\left(\sum_{i=1}^{r} k_i \mathbf{v}_i\right) \geq \sum_{i=1}^{r} k_i f(\mathbf{v}_i) \tag{10.3}$$

respectively, where in each case $\mathbf{v}_1, \mathbf{v}_2, \ldots, \mathbf{v}_r \in X$ and $k_1, k_2, \ldots, k_r$ are non-negative scalars with $\sum_{i=1}^{r} k_i = 1$.

THEOREM 10.2.1 *Let $f: A \to Y$, where A is an affine subset of a real vector space and $Y \subseteq \mathbb{R}$. Then f is an affine function if and only if f is both a convex function and a concave function.*

Proof: Inequalities (7.14) and (10.1) can hold simultaneously if and only if

$$f(\lambda \mathbf{x} + (1 - \lambda)\mathbf{x}') = \lambda f(\mathbf{x}) + (1 - \lambda) f(\mathbf{x}') \tag{10.4}$$

If f is an affine function, then the equality, and thus both inequalities, hold for all $\lambda \in \mathbb{R}$ and specifically for $\lambda \in [0, 1]$ as required.

If f is a function that is both convex and concave, then both inequalities, and thus the equality also, hold for $\lambda \in [0, 1]$. All that must be shown is that it holds also for $\lambda < 0$ and for $\lambda > 1$.

Consider the case of $\lambda < 0$.

We know that the inequalities (and thus the equality) hold for the vectors $\mathbf{x}$ and $\mathbf{x}'' \equiv \lambda\mathbf{x} + (1 - \lambda)\mathbf{x}' \in A$ and the scalar $\mu \in [0, 1]$, or that

$$f(\mu\mathbf{x} + (1 - \mu)\mathbf{x}'') = \mu f(\mathbf{x}) + (1 - \mu)f(\mathbf{x}'') \tag{10.5}$$

But for $\lambda < 0$, $\lambda/(\lambda - 1) \in [0, 1]$, so setting $\mu = \lambda/(\lambda - 1)$ in (10.5) gives us

$$f\left(\frac{\lambda}{\lambda - 1}\mathbf{x} + \left(1 - \frac{\lambda}{\lambda - 1}\right)\mathbf{x}''\right) = \frac{\lambda}{\lambda - 1}f(\mathbf{x}) + \left(1 - \frac{\lambda}{\lambda - 1}\right)f(\mathbf{x}'') \tag{10.6}$$

and substituting for $\mathbf{x}''$ yields

$$f\left(\frac{\lambda}{\lambda - 1}\mathbf{x} + \left(1 - \frac{\lambda}{\lambda - 1}\right)(\lambda\mathbf{x} + (1 - \lambda)\mathbf{x}')\right)$$
$$= \frac{\lambda}{\lambda - 1}f(\mathbf{x}) + \left(1 - \frac{\lambda}{\lambda - 1}\right)f(\lambda\mathbf{x} + (1 - \lambda)\mathbf{x}') \tag{10.7}$$

Multiplying across by $\lambda - 1$ and collecting terms yields

$$(\lambda - 1)f(\mathbf{x}') = \lambda f(\mathbf{x}) - f(\lambda\mathbf{x} + (1 - \lambda)\mathbf{x}') \tag{10.8}$$

or

$$f(\lambda\mathbf{x} + (1 - \lambda)\mathbf{x}') = \lambda f(\mathbf{x}) + (1 - \lambda)f(\mathbf{x}') \tag{10.9}$$

as required.

The proof for $\lambda > 1$ is left as an exercise; see Exercise 10.1(e).   □

The next theorem addresses the characterization of convexity and concavity in terms of upper and lower contour sets; see Definition 7.7.2.

THEOREM 10.2.2 *The upper contour sets* $\{\mathbf{x} \in X : f(\mathbf{x}) \geq \alpha\}$ *of a concave function are convex.*

Proof: The proof of this theorem is left as an exercise; see Exercise 10.2.   □

This result will be encountered later in the context of consumer theory; see Theorem 12.3.5. Readers who are familiar with the two-good consumer problem from intermediate economics will recognize the implications of Theorem 10.2.2 for the shape of the indifference curves corresponding to a concave utility function.

Note also that concavity of a function is a sufficient but not a necessary condition for convexity of its upper contour sets. For example, any increasing function $f : \mathbb{R} \to \mathbb{R}$ has convex upper contour sets.

THEOREM 10.2.3 *Let $f: X \to \mathbb{R}$ and $g: X \to \mathbb{R}$ be concave functions. The following hold:*

(a) *if $a, b > 0$, then $af + bg$ is concave;*
(b) *if $a < 0$, then af is convex; and*
(c) $\min\{f, g\}$ *is concave.*

Proof: The proofs of the above properties and their obvious corollaries are left as exercises; see Exercise 10.2 again. □

DEFINITION 10.2.2 Again let $f: X \to Y$ where X is a convex subset of a real vector space and $Y \subseteq \mathbb{R}$. Then we have the following:

(a) f is a **strictly convex** function if and only if, for all $\mathbf{x} \neq \mathbf{x}' \in X$ and $\lambda \in (0, 1)$,

$$f(\lambda \mathbf{x} + (1 - \lambda)\mathbf{x}') < \lambda f(\mathbf{x}) + (1 - \lambda) f(\mathbf{x}') \tag{10.10}$$

(b) f is a **strictly concave** function if and only if, for all $\mathbf{x} \neq \mathbf{x}' \in X$ and $\lambda \in (0, 1)$,

$$f(\lambda \mathbf{x} + (1 - \lambda)\mathbf{x}') > \lambda f(\mathbf{x}) + (1 - \lambda) f(\mathbf{x}') \tag{10.11}$$

Note that there is no longer any flexibility as regards allowing $\mathbf{x} = \mathbf{x}'$ or $\lambda = 0$ or $\lambda = 1$ in these definitions.

Much of what has already been said about the relationship between convex functions and concave functions applies equally well to the relationship between strictly convex functions and strictly concave functions.

10.2.2 *Convexity and differentiability*

In this section, we show that, for differentiable functions, the definitions of convex and concave functions above are equivalent to statements about the first derivative or Jacobian (Theorem 10.2.4) and about the second derivative or Hessian of the function (Theorem 10.2.5). As noted above, we could, equivalently, present the discussion in terms of convex functions.

Since the limit of the first difference of a function at $\mathbf{x}$, say, makes no sense if the function and the first difference are not defined in some open neighbourhood of $\mathbf{x}$ (some $B_\epsilon(\mathbf{x})$), we must assume in this section that the domains of functions are not only convex but also open.

THEOREM 10.2.4 (CONCAVITY CRITERION FOR DIFFERENTIABLE FUNCTIONS). *Let $f: X \to \mathbb{R}$ be differentiable, where $X \subseteq \mathbb{R}^n$ is an open, convex set. Then f is (strictly) concave if and only if, for all $\mathbf{x} \neq \mathbf{x}' \in X$,*

$$f(\mathbf{x}) \, (<) \leq f(\mathbf{x}') + f'(\mathbf{x}')(\mathbf{x} - \mathbf{x}') \tag{10.12}$$

Theorem 10.2.4 says that a function is concave if and only if the tangent hyperplane at any point lies completely above the graph of the function, or the tangent hyperplane is a supporting hyperplane for the set lying below the graph in $\mathbb{R}^{n+1}$.

Another interpretation of Theorem 10.2.4 is that a function is concave if and only if, for any two distinct points in the domain, the directional derivative at one point in the direction of the other exceeds the jump in the value of the function between the two points; see Section 9.5 for the definition of a directional derivative.

Proof:

$\Rightarrow$ We first prove that the weak version of inequality (10.12) is necessary for concavity, and then that the strict version is necessary for strict concavity.

Choose $\mathbf{x}, \mathbf{x}' \in X$.

(a) Suppose that f is concave.

Then, for $\lambda \in (0, 1)$, and using (10.1),

$$f(\mathbf{x}' + \lambda(\mathbf{x} - \mathbf{x}')) \geq f(\mathbf{x}') + \lambda(f(\mathbf{x}) - f(\mathbf{x}')) \tag{10.13}$$

Subtract $f(\mathbf{x}')$ from both sides and divide by λ:

$$\frac{f(\mathbf{x}' + \lambda(\mathbf{x} - \mathbf{x}')) - f(\mathbf{x}')}{\lambda} \geq f(\mathbf{x}) - f(\mathbf{x}') \tag{10.14}$$

Now consider the limits of both sides of this inequality as $\lambda \to 0$. The left-hand side tends to $f'(\mathbf{x}')(\mathbf{x} - \mathbf{x}')$ by definition of a directional derivative; see (9.63) and (9.64). The right-hand side is independent of λ and does not change. The result now follows easily for concave functions.

This proof is illustrated in Figure 10.1. The diagram shows a cross-section of $\mathbb{R}^{n+1}$ along the line L from $\mathbf{x}'$ (where $\lambda = 0$) to $\mathbf{x}$ (where $\lambda = 1$). The curve represents a cross-section of the graph of f and the straight line represents a cross-section of the tangent hyperplane at $\mathbf{x}'$, which touches the graph at P. The theorem says that f is concave if and only if the point A lies above the point B on every such graph. Equivalently it says that the directional derivative at $\mathbf{x}'$ in the direction of $\mathbf{x}$ (the height of AC) exceeds the change in the value of the function between $\mathbf{x}'$ and $\mathbf{x}$ (the height of BC). The definition of concavity says that the point D lies above the point E. Condition (10.14) says that the slope of PB is less than or equal to the slope of PD for any D along the arc PB. The idea of the first part of the proof is that, as $\lambda \to 0$, the slope of PD approaches the slope of PA.

However, (10.14) remains a weak inequality even if f is a strictly concave function.

(b) Now suppose that f is strictly concave and $\mathbf{x} \neq \mathbf{x}'$.

Since f is also concave, we can apply the result that we have just proved to $\mathbf{x}'$ and $\mathbf{x}'' \equiv \frac{1}{2}(\mathbf{x} + \mathbf{x}')$ to show that

$$f'(\mathbf{x}')(\mathbf{x}'' - \mathbf{x}') \geq f(\mathbf{x}'') - f(\mathbf{x}') \tag{10.15}$$

Using the definition of strict concavity, or the strict version of inequality (10.13) with $\lambda = 1/2$, gives

$$f(\mathbf{x}'') - f(\mathbf{x}') > \frac{1}{2}(f(\mathbf{x}) - f(\mathbf{x}')) \tag{10.16}$$

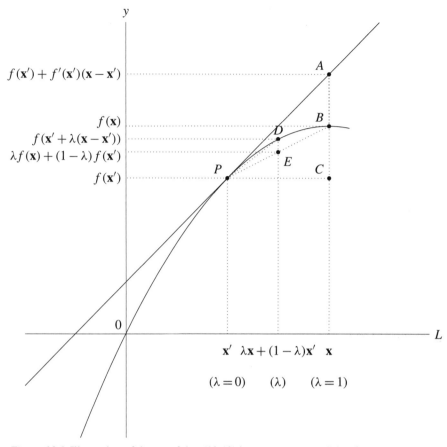

Figure 10.1 Illustration of the proof that (10.12) is a necessary condition for concavity

Combining these two inequalities, multiplying across by 2 and rearranging gives the desired result.

⇐ Conversely, suppose that the derivative satisfies inequality (10.12). We will deal with concavity. To prove the theorem for strict concavity, just replace all the weak inequalities (≥) with strict inequalities (>).

Set $\mathbf{x}'' = \lambda \mathbf{x} + (1 - \lambda)\mathbf{x}'$. Then, applying the hypothesis of the proof in turn to $\mathbf{x}$ and $\mathbf{x}''$ and to $\mathbf{x}'$ and $\mathbf{x}''$ yields

$$f(\mathbf{x}) \leq f(\mathbf{x}'') + f'(\mathbf{x}'')(\mathbf{x} - \mathbf{x}'') \tag{10.17}$$

and

$$f(\mathbf{x}') \leq f(\mathbf{x}'') + f'(\mathbf{x}'')(\mathbf{x}' - \mathbf{x}'') \tag{10.18}$$

Just as we have been taking convex combinations of *vectors*, we can take a convex combination of *inequalities* (10.17) and (10.18), which gives

$$\lambda f(\mathbf{x}) + (1-\lambda) f(\mathbf{x}')$$
$$\leq f(\mathbf{x}'') + f'(\mathbf{x}'')(\lambda(\mathbf{x}-\mathbf{x}'')+(1-\lambda)(\mathbf{x}'-\mathbf{x}''))$$
$$= f(\mathbf{x}'') \tag{10.19}$$

since

$$\lambda(\mathbf{x}-\mathbf{x}'')+(1-\lambda)(\mathbf{x}'-\mathbf{x}'') = \lambda\mathbf{x}+(1-\lambda)\mathbf{x}'-\mathbf{x}'' = \mathbf{0}_n \tag{10.20}$$

Inequality (10.19) is just the definition of concavity as required.

This proof is illustrated in Figure 10.2. Condition (10.17) says that the height of *RS* is less than or equal to the height of *RT*. Condition (10.18) says that the height of *MN* is less than or equal to the height of *MQ*. The convex combination (10.19) says that the height of *UV* is less than or equal to the height of *UW*, which is just the definition of concavity.[1]   □

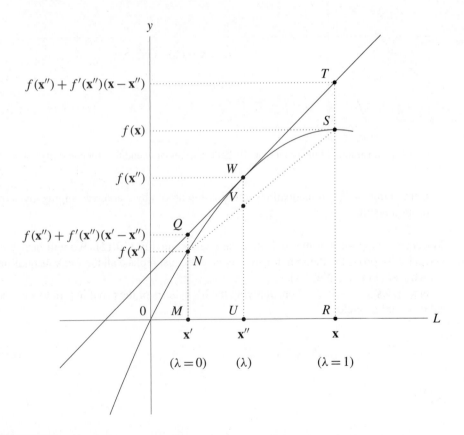

Figure 10.2 Illustration of the proof that (10.12) is a sufficient condition for concavity

THEOREM 10.2.5 (CONCAVITY CRITERION FOR TWICE DIFFERENTIABLE FUNCTIONS).
*Let $f: X \to \mathbb{R}$ be twice continuously differentiable (C^2), where $X \subseteq \mathbb{R}^n$ is open and convex.
Then:*

(a) *f is concave if and only if, for all $\mathbf{x} \in X$, the Hessian matrix $f''(\mathbf{x})$ is negative semi-definite; and*
(b) *If $f''(\mathbf{x})$ is negative definite for all $\mathbf{x} \in X$, then f is strictly concave.*

The fact that the condition in the second part of this theorem is sufficient but not necessary for concavity inspires the search for a counter-example; in other words, for a function that is strictly concave but has a second derivative that is negative semi-definite but not negative definite. The standard counter-example is given by $f(x) = -x^{2n}$ at $x = 0$ for any integer $n > 1$.

Proof: We first use Taylor's theorem (Theorem 9.6.5) to prove the sufficiency of the condition on the Hessian matrices. Then we use the fundamental theorem of calculus (Theorem 7.9.1) and a proof by contrapositive to prove the necessity of this condition in the concave case for $n = 1$. Then we use this result and the chain rule to demonstrate necessity for $n > 1$. Finally, we show how these arguments can be modified to give an alternative proof of sufficiency for functions of one variable.

(a) Suppose first that $f''(\mathbf{x})$ is negative semi-definite for all $\mathbf{x} \in X$. Recall Taylor's theorem. It follows that $f(\mathbf{x}) \leq f(\mathbf{x}') + f'(\mathbf{x}')(\mathbf{x} - \mathbf{x}')$. Theorem 10.2.4 shows that f is then concave.
   A similar proof will work for a negative definite Hessian and a strictly concave function.
(b) To demonstrate necessity, we must consider first functions of a single variable and then functions of several variables.

   (i) First consider a function of a single variable. Instead of trying to show that concavity of f implies a negative semi-definite (i.e. non-positive) second derivative for all $x \in X$, we will prove the contrapositive. In other words, we will show that, if there is any point $x^* \in X$ where the second derivative is positive, then f is locally strictly convex around x^* and so cannot be concave.
   So suppose $f''(x^*) > 0$. Then, since f is twice continuously differentiable, $f''(x) > 0$ for all x in some neighbourhood of x^*, say (a, b). Then f' is an increasing function on (a, b). Consider two points in (a, b), $x < x'$ and let $x'' = \lambda x + (1 - \lambda)x' \in X$, where $\lambda \in (0, 1)$. Using the fundamental theorem of calculus,

$$f(x'') - f(x) = \int_x^{x''} f'(t)\,dt < f'(x'')(x'' - x) \tag{10.21}$$

   and

$$f(x') - f(x'') = \int_{x''}^{x'} f'(t)\,dt > f'(x'')(x' - x'') \tag{10.22}$$

Rearranging each inequality gives

$$f(x) > f(x'') + f'(x'')(x - x'') \qquad (10.23)$$

and

$$f(x') > f(x'') + f'(x'')(x' - x'') \qquad (10.24)$$

which are just the single-variable versions of (10.17) and (10.18). As in the proof of Theorem 10.2.4, a convex combination of these inequalities reduces to

$$f(x'') < \lambda f(x) + (1 - \lambda) f(x') \qquad (10.25)$$

and, hence, f is locally strictly convex on (a, b).

(ii) Now consider a function of several variables. Suppose that f is concave and fix $\mathbf{x} \in X$ and $\mathbf{h} \in \mathbb{R}^n$. (We use an $\mathbf{x}, \mathbf{x} + \mathbf{h}$ argument instead of an $\mathbf{x}, \mathbf{x}'$ argument to tie in with the definition of a negative definite matrix.) Then, at least for sufficiently small λ, $f|_L(\lambda) \equiv f(\mathbf{x} + \lambda \mathbf{h})$ also defines a concave function (of one variable), namely, the restriction of f to the line segment from $\mathbf{x}$ in the direction from $\mathbf{x}$ to $\mathbf{x} + \mathbf{h}$. Thus, using the result we have just proved for functions of one variable, $f|_L$ has non-positive second derivative. But we know from (9.66) that $f|_L''(0) = \mathbf{h}^\top f''(\mathbf{x})\mathbf{h}$, so $f''(\mathbf{x})$ is negative semi-definite.

(c) For functions of one variable, the above arguments can give an alternative proof of sufficiency that does not require Taylor's theorem. The details are left as an exercise; see Exercise 10.5.

In fact, we have the following:

$$\begin{aligned}
f''(x) < 0 \text{ on } (a, b) &\Rightarrow f \text{ locally strictly concave on } (a, b) \\
f''(x) \leq 0 \text{ on } (a, b) &\Rightarrow f \text{ locally concave on } (a, b) \\
f''(x) > 0 \text{ on } (a, b) &\Rightarrow f \text{ locally strictly convex on } (a, b) \\
f''(x) \geq 0 \text{ on } (a, b) &\Rightarrow f \text{ locally convex on } (a, b)
\end{aligned}$$

The same results that we have demonstrated for the interval (a, b) also hold for the entire domain X (which is also just an open interval, as it is an open convex subset of $\mathbb{R}$). $\qquad \square$

Note the implied hierarchy among different classes of functions set out in Table 10.1 (considered again in Exercise 10.7).

We would like to know the extent to which any or all of these properties are qualitative rather than quantitative; in other words, the extent to which they are invariant under changes of scale, or under different types of strictly increasing transformations. In economics, this

Table 10.1 Hierarchy of concave and related functions

negative definite Hessian	$\subset$	strictly concave	$\subset$	concave	$=$	negative semi-definite Hessian	$\subset$	convex upper contour sets *and* consistent directional derivatives

is particularly important, since, when measuring utility or happiness or satisfaction, we will find that ranking alternatives is relatively straightforward, but the choice of a scale on which to measure that ranking is always arbitrary. The following theorem is about as far down this road as we can go for the moment.

THEOREM 10.2.6 *A non-decreasing twice differentiable concave transformation of a twice differentiable concave function (of several variables) is also concave.*

Proof: The details are left as an exercise; see Exercise 10.8. □

Theorem 10.2.6 says that concavity is invariant under non-decreasing *concave* transformations and, hence, under increasing concave transformations. It was noted in the discussion of Definition 7.7.2 that the level sets of a function are invariant under *all* increasing transformations. In the next subsection, another property of real-valued functions, which is closely related to concavity, but which *is* invariant under all increasing transformations, will be introduced.

The second-order condition above is reminiscent of that for optimization and suggests that concave or convex functions will prove useful in developing theories of optimizing behaviour. In fact, there is a wider class of useful functions, leading us now to introduce further definitions.

10.2.3 Quasi-convex and quasi-concave functions

Let $X \subseteq \mathbb{R}^n$ be a convex set and $f: X \to \mathbb{R}$ a real-valued function defined on X.

In order to maintain consistency with earlier notation, for reasons that will become clear in due course, we adopt the convention when labelling vectors $\mathbf{x}$ and $\mathbf{x}'$ that $f(\mathbf{x}') \leq f(\mathbf{x})$.

DEFINITION 10.2.3 Let $C(\alpha) = \{\mathbf{x} \in X : f(\mathbf{x}) \geq \alpha\}$ be an upper contour set of the function $f: X \to \mathbb{R}$. Then f is **quasi-concave** if and only if $C(\alpha)$ is a convex set for all $\alpha \in \mathbb{R}$.

Recall that quasi-concavity, as defined here, was a necessary, but not sufficient, condition for f to be a concave function.

Examples of quasi-concave functions include $f: \mathbb{R}_{++} \to \mathbb{R} : x \mapsto \ln x$ and $f: \mathbb{R} \to \mathbb{R}_{++} : x \mapsto e^x$.

As the first of these examples is *concave*, it is not surprising to find that it is *quasi-concave*; the second of these examples, however, is both *convex* and *quasi-concave*.

THEOREM 10.2.7 *The following statements are equivalent to the definition of quasi-concavity:*

(a) *for all* $\mathbf{x}, \mathbf{x}' \in X$, *and for all* $\lambda \in (0, 1)$, $f(\lambda \mathbf{x} + (1 - \lambda)\mathbf{x}') \geq \min\{f(\mathbf{x}), f(\mathbf{x}')\}$;
(b) *for all* $\mathbf{x}, \mathbf{x}' \in X$ *such that* $f(\mathbf{x}') \leq f(\mathbf{x})$ *and for all* $\lambda \in (0, 1)$, $f(\lambda \mathbf{x} + (1 - \lambda)\mathbf{x}') \geq f(\mathbf{x}')$; *and*
(c) *for all* $\mathbf{x}, \mathbf{x}' \in X$ *such that* $f(\mathbf{x}) - f(\mathbf{x}') \geq 0$, $f'(\mathbf{x}')(\mathbf{x} - \mathbf{x}') \geq 0$ *(provided that X is open and f is differentiable).*

Proof:

(a) We begin by showing the equivalence between the definition and the first condition.

(i) First suppose that the upper contour sets are convex. Let $\mathbf{x}, \mathbf{x}' \in X$ and let $\alpha = \min\{f(\mathbf{x}), f(\mathbf{x}')\}$. Then $\mathbf{x}$ and $\mathbf{x}'$ are in $C(\alpha)$. By the hypothesis of convexity, for any $\lambda \in (0, 1)$, $\lambda\mathbf{x} + (1 - \lambda)\mathbf{x}' \in C(\alpha)$. The desired result now follows.

(ii) Now suppose that condition (a) holds. To show that $C(\alpha)$ is a convex set, we just take $\mathbf{x}, \mathbf{x}' \in C(\alpha)$ and investigate whether $\lambda\mathbf{x} + (1 - \lambda)\mathbf{x}' \in C(\alpha)$. But, by condition (a),

$$f(\lambda\mathbf{x} + (1 - \lambda)\mathbf{x}') \geq \min\{f(\mathbf{x}), f(\mathbf{x}')\} \geq \alpha \tag{10.26}$$

where the final inequality holds because $\mathbf{x}$ and $\mathbf{x}'$ are in $C(\alpha)$.

(b) It is straightforward to show the equivalence of conditions (b) and (a).

(i) First suppose that condition (a) holds. In the case where $f(\mathbf{x}) \geq f(\mathbf{x}')$ or $f(\mathbf{x}') = \min\{f(\mathbf{x}), f(\mathbf{x}')\}$, there is nothing to prove. Otherwise, we can just reverse the labels $\mathbf{x}$ and $\mathbf{x}'$. The statement in condition (b) is true for $\mathbf{x} = \mathbf{x}'$ or $\lambda = 0$ or $\lambda = 1$ even if f is not quasi-concave.

(ii) The proof of the converse is even more straightforward and is left as an exercise; see Exercise 10.9.

(c) Proving that condition (c) is equivalent to quasi-concavity for a differentiable function (by proving that it is equivalent to conditions (b) and (a)) is much the trickiest part of the proof.

(i) Begin by supposing that f satisfies conditions (b) and (a). Proving that condition (c) is necessary for quasi-concavity is the easier part of the proof. Without loss of generality, pick $\mathbf{x}, \mathbf{x}' \in X$ such that $f(\mathbf{x}') \leq f(\mathbf{x})$. By quasi-concavity,

$$f(\lambda\mathbf{x} + (1 - \lambda)\mathbf{x}') \geq f(\mathbf{x}') \quad \forall \lambda \in (0, 1) \tag{10.27}$$

Consider the function defined by

$$f|_L(\lambda) = f(\lambda\mathbf{x} + (1 - \lambda)\mathbf{x}') = f(\mathbf{x}' + \lambda(\mathbf{x} - \mathbf{x}')) \tag{10.28}$$

We want to show that the directional derivative satisfies

$$f|_L'(0) = f'(\mathbf{x}')(\mathbf{x} - \mathbf{x}') \geq 0 \tag{10.29}$$

But

$$f|_L'(0) = \lim_{\lambda \to 0} \frac{f(\mathbf{x}' + \lambda(\mathbf{x} - \mathbf{x}')) - f(\mathbf{x}')}{\lambda} \tag{10.30}$$

Since both the numerator and denominator of the right-hand side are non-negative for small positive values of λ (indeed for any $\lambda < 1$), the derivative must be non-negative as required.

(ii) Now the difficult part – to prove (by contradiction) that condition (c) is a sufficient condition for quasi-concavity.

Suppose that condition (c) is satisfied, but that f is not quasi-concave and therefore does not satisfy condition (a). In other words, there exists $\mathbf{x}, \mathbf{x}' \in X$ and $\lambda^* \in (0, 1)$ such that, letting $\mathbf{x}^* \equiv \lambda^*\mathbf{x} + (1 - \lambda^*)\mathbf{x}'$,

$$f(\mathbf{x}^*) < \min\{f(\mathbf{x}), f(\mathbf{x}')\} \tag{10.31}$$

where, without loss of generality, $f(\mathbf{x}') \leq f(\mathbf{x})$. Condition (c) applied first to $\mathbf{x}$ and $\mathbf{x}^*$ and then to $\mathbf{x}'$ and $\mathbf{x}^*$ tells us that

$$f'(\mathbf{x}^*)(\mathbf{x} - (\lambda^* \mathbf{x} + (1 - \lambda^*)\mathbf{x}')) \geq 0 \qquad (10.32)$$

$$f'(\mathbf{x}^*)(\mathbf{x}' - (\lambda^* \mathbf{x} + (1 - \lambda^*)\mathbf{x}')) \geq 0 \qquad (10.33)$$

or

$$(1 - \lambda^*)f'(\mathbf{x}^*)(\mathbf{x} - \mathbf{x}') \geq 0 \qquad (10.34)$$

$$-\lambda^* f'(\mathbf{x}^*)(\mathbf{x} - \mathbf{x}') \geq 0 \qquad (10.35)$$

Dividing (10.34) by $(1 - \lambda^*)$ and (10.35) by λ^* yields a pair of inequalities that can be satisfied simultaneously only if

$$f'(\mathbf{x}^*)(\mathbf{x} - \mathbf{x}') = 0 \qquad (10.36)$$

In other words, $f|_L'(\lambda^*) = 0$; we already know that

$$f|_L(\lambda^*) < f|_L(0) \leq f|_L(1) \qquad (10.37)$$

Since f is differentiable, it and $f|_L$ are also continuous, and so there exists some interval around λ^*, say $(\lambda^* - \epsilon, \lambda^* + \epsilon)$, on which $f|_L(\lambda) < \min\{f|_L(0), f|_L(1)\}$. By the above argument, $f|_L'(\lambda) = 0$ throughout this interval, and thus $f|_L$ is constant on the interval.

Let $A = \{\lambda \in (0, 1): f|_L(\lambda) = f|_L(\lambda^*)\}$. Since $\lambda^* \in A$, A is non-empty, and, since $f|_L$ is continuous, A is closed. Let $\lambda^{**} = \sup A$. By (10.37) and continuity of $f|_L$, we obtain $\lambda^{**} < 1$.

By applying to λ^{**} the argument previously applied to λ^*, we get an $\epsilon' > 0$ such that $\lambda^{**} + \epsilon' \in A$ and thus $\lambda^* \neq \sup A$, which gives us the required contradiction. $\square$

In words, condition (c) of Theorem 10.2.7 says that, whenever a differentiable quasi-concave function has a higher value at $\mathbf{x}$ than at $\mathbf{x}'$, or the same value at both points, then the directional derivative of f at $\mathbf{x}'$ in the direction of $\mathbf{x}$ is non-negative. It might help to think about this by considering $n = 1$ and separating out the cases $x > x'$ and $x < x'$.

THEOREM 10.2.8 *Let* $f: X \to \mathbb{R}$ *be quasi-concave and* $g: \mathbb{R} \to \mathbb{R}$ *be increasing. Then* $g \circ f$ *is a quasi-concave function.*

Proof: This follows easily from the definition. The details are left as an exercise; see Exercise 10.16. $\square$

We will return to the implications of Theorem 10.2.8 for utility theory in Theorem 12.3.4. It will be seen then that, if preferences can be represented by a quasi-concave utility function, then they can be represented by quasi-concave utility functions only.

10.2.4   *Other variations on the convexity theme*

Let $X \subseteq \mathbb{R}^n$ again be a convex set.

DEFINITION 10.2.4 The function $f: X \to \mathbb{R}$ is **strictly quasi-concave** if and only if, for all $\mathbf{x} \neq \mathbf{x}' \in X$ such that $f(\mathbf{x}) \geq f(\mathbf{x}')$ and for all $\lambda \in (0, 1)$, $f(\lambda\mathbf{x} + (1 - \lambda)\mathbf{x}') > f(\mathbf{x}')$.

DEFINITION 10.2.5 The function $f: X \to \mathbb{R}$ is **(strictly) quasi-convex** if and only if the function $-f$ is (strictly) quasi-concave.

DEFINITION 10.2.6 The function $f: X \to \mathbb{R}$ is **pseudo-concave** if and only if f is differentiable and quasi-concave and

$$f(\mathbf{x}) - f(\mathbf{x}') > 0 \Rightarrow f'(\mathbf{x}')(\mathbf{x} - \mathbf{x}') > 0 \tag{10.38}$$

As usual, the function f is **pseudo-convex** if and only if the function $-f$ is pseudo-concave.

Note that the last definition modifies slightly condition (c) in Theorem 10.2.7, which is equivalent to quasi-concavity for a differentiable function.

Pseudo-concavity will crop up in the second-order conditions for equality-constrained optimization.

EXAMPLE 10.2.1 Consider the interesting case of the affine function

$$f: \mathbb{R}^n \to \mathbb{R}: \mathbf{x} \mapsto M - \mathbf{p}^\top\mathbf{x} \tag{10.39}$$

where $M \in \mathbb{R}$ and $\mathbf{p} \in \mathbb{R}^n$. This function is both concave and convex, but neither strictly concave nor strictly convex. Furthermore,

$$f(\lambda\mathbf{x} + (1 - \lambda)\mathbf{x}') = \lambda f(\mathbf{x}) + (1 - \lambda)f(\mathbf{x}')$$
$$\geq \min\{f(\mathbf{x}), f(\mathbf{x}')\} \tag{10.40}$$

and

$$(-f)(\lambda\mathbf{x} + (1 - \lambda)\mathbf{x}') = \lambda(-f)(\mathbf{x}) + (1 - \lambda)(-f)(\mathbf{x}')$$
$$\geq \min\{(-f)(\mathbf{x}), (-f)(\mathbf{x}')\} \tag{10.41}$$

so f is both quasi-concave and quasi-convex, but not strictly so in either case. The function f is, however, pseudo-concave (and pseudo-convex) since

$$\begin{aligned}
f(\mathbf{x}) > f(\mathbf{x}') \quad &\Leftrightarrow \quad \mathbf{p}^\top\mathbf{x} < \mathbf{p}^\top\mathbf{x}' \\
&\Leftrightarrow \quad \mathbf{p}^\top(\mathbf{x} - \mathbf{x}') < 0 \\
&\Leftrightarrow \quad -f'(\mathbf{x}')(\mathbf{x} - \mathbf{x}') < 0 \\
&\Leftrightarrow \quad f'(\mathbf{x}')(\mathbf{x} - \mathbf{x}') > 0
\end{aligned} \tag{10.42}$$

$\diamondsuit$

10.3 Unconstrained optimization

The background material on convexity in the previous section allows us to proceed to the analysis of the simplest type of optimization problem, in which all the choice variables are free to vary independently, with no constraints.

DEFINITION 10.3.1 Let $X \subseteq \mathbb{R}^n$, $f : X \to \mathbb{R}$. Then we say that:

(a) f has a **(strict) global (or absolute) maximum** at $\mathbf{x}^*$ if and only if, for all $\mathbf{x} \in X$ such that $\mathbf{x} \neq \mathbf{x}^*$, $f(\mathbf{x}) \, (<) \leq f(\mathbf{x}^*)$;
(b) f has a **(strict) local (or relative) maximum** at $\mathbf{x}^*$ if and only if there exists $\epsilon > 0$ such that, for all $\mathbf{x} \in B_\epsilon(\mathbf{x}^*)$ (as introduced in Definition 7.5.2) such that $\mathbf{x} \neq \mathbf{x}^*$, $f(\mathbf{x}) \, (<) \leq f(\mathbf{x}^*)$;
(c) f has a **(strict) global (or absolute) minimum** at $\mathbf{x}^*$ if and only if, for all $\mathbf{x} \in X$ such that $\mathbf{x} \neq \mathbf{x}^*$, $f(\mathbf{x}) \, (>) \geq f(\mathbf{x}^*)$; and
(d) f has a **(strict) local (or relative) minimum** at $\mathbf{x}^*$ if and only if there exists $\epsilon > 0$ such that, for all $\mathbf{x} \in B_\epsilon(\mathbf{x}^*)$ such that $\mathbf{x} \neq \mathbf{x}^*$, $f(\mathbf{x}) \, (>) \geq f(\mathbf{x}^*)$.

THEOREM 10.3.1 *A continuous real-valued function on a compact subset of $\mathbb{R}^n$ attains a global maximum and a global minimum.*

Proof: Intuitively, this theorem just says that a function that cannot drift off towards infinity must have finite bounds. The conditions of the theorem prevent this drift to infinity, since the value of the function must remain finite and well defined on the boundary of its domain. Continuity prevents the function from tending to infinity in the interior of its domain. See Mendelson (1975, p. 161) for a full proof. □

While this is a neat result for functions on compact domains, results in calculus are generally for functions on open domains.

The remainder of this section and the next two sections are each centred around three related theorems:

1. a theorem giving necessary or first-order conditions that must be satisfied by the solution to an optimization problem (Theorems 10.3.2, 10.4.1 and 10.5.1);
2. a theorem giving sufficient or second-order conditions under which a solution to the first-order conditions satisfies the original optimization problem (Theorems 10.3.3, 10.4.2 and 10.5.2); and
3. a theorem giving conditions under which a known solution to an optimization problem is the unique solution (Theorems 10.3.5, 10.4.3 and 10.5.3).

The results are generally presented for maximization problems. However, any minimization problem is easily turned into a maximization problem by reversing the sign of the function to be minimized and maximizing the function thus obtained.

Throughout the present section, we deal with the unconstrained optimization problem

$$\max_{\mathbf{x} \in X} f(\mathbf{x}) \tag{10.43}$$

where $X \subseteq \mathbb{R}^n$ and $f: X \to \mathbb{R}$ is a real-valued function of several variables, i.e. the problem of locating the values of the variables $\mathbf{x}$ at which the function f, called the **objective function** of problem (10.43), takes on its largest value.

EXAMPLE 10.3.1 Before proceeding to the formal theorems, let us consider a simple case of problem (10.43) where $n = 1$ and $f(x) = -x^2$. By graphing this function, it should quickly become apparent that it takes on its maximum value of 0 when $x = 0$. For values of $x < 0$, the function is increasing and its first derivative $f'(x) = -2x > 0$. For values of $x > 0$, the function is decreasing and its first derivative $f'(x) < 0$. At the optimum, $x = 0$, the first derivative $f'(x)$ vanishes. We will see that the same principle holds for the partial derivatives of functions of several variables.

In this example, the second derivative $f''(x) = -2$ and so is negative for all values of x, in particular for $x = 0$. This is because the first derivative $f'(x) = -2x$ is decreasing in x. Another way of expressing the same result is to say that the 1×1 Hessian matrix $f''(x)$ is negative definite. We will see that a similar principle holds for the Hessian matrix of functions of several variables.

Finally, note that the objective function in this example is strictly quasi-concave and has a unique global maximum. ◇

THEOREM 10.3.2 (NECESSARY (FIRST-ORDER) CONDITION FOR UNCONSTRAINED MAXIMA AND MINIMA). *Let $X \subseteq \mathbb{R}^n$ be open and let $f: X \to \mathbb{R}$ be differentiable with a local maximum or minimum at $\mathbf{x}^* \in X$. Then $f'(\mathbf{x}^*) = \mathbf{0}^\top$, or f has a **stationary point** at $\mathbf{x}^*$.*

Proof: Without loss of generality, assume that the function has a local maximum at $\mathbf{x}^*$. Then there exists $\epsilon > 0$ such that, whenever $\|\mathbf{h}\| < \epsilon$,

$$f(\mathbf{x}^* + \mathbf{h}) - f(\mathbf{x}^*) \leq 0 \tag{10.44}$$

It follows that, for $0 < h < \epsilon$,

$$\frac{f(\mathbf{x}^* + h\mathbf{e}_i) - f(\mathbf{x}^*)}{h} \leq 0 \tag{10.45}$$

(where $\mathbf{e}_i$ denotes the ith standard basis vector) and, hence, that

$$\frac{\partial f}{\partial x_i}(\mathbf{x}^*) = \lim_{h \to 0} \frac{f(\mathbf{x}^* + h\mathbf{e}_i) - f(\mathbf{x}^*)}{h} \leq 0 \tag{10.46}$$

Similarly, for $0 > h > -\epsilon$,

$$\frac{f(\mathbf{x}^* + h\mathbf{e}_i) - f(\mathbf{x}^*)}{h} \geq 0 \tag{10.47}$$

and, hence,

$$\frac{\partial f}{\partial x_i}(\mathbf{x}^*) = \lim_{h \to 0} \frac{f(\mathbf{x}^* + h\mathbf{e}_i) - f(\mathbf{x}^*)}{h} \geq 0 \tag{10.48}$$

Combining (10.46) and (10.48) yields the desired result. □

The first-order conditions are useful for identifying local optima only in the interior of the domain of the objective function: Theorem 10.3.2 applies only to functions whose domain X is open. It is important to check also for the possible existence of corner solutions or boundary solutions to optimization problems where the objective function is defined on a domain that is not open.

THEOREM 10.3.3 (SUFFICIENT (SECOND-ORDER) CONDITION FOR UNCONSTRAINED MAXIMA AND MINIMA). *Let $X \subseteq \mathbb{R}^n$ be open and let $f: X \to \mathbb{R}$ be a twice continuously differentiable function with $f'(\mathbf{x}^*) = \mathbf{0}$ and $f''(\mathbf{x}^*)$ negative definite. Then f has a strict local maximum at $\mathbf{x}^*$.*
Similarly for positive definite Hessians and local minima.

Proof: Consider the second-order Taylor expansion (9.97), which we now present in slightly different notation. For any $\mathbf{x} \in X$, there exists $s \in (0, 1)$ such that

$$f(\mathbf{x}) = f(\mathbf{x}^*) + f'(\mathbf{x}^*)(\mathbf{x} - \mathbf{x}^*) + \tfrac{1}{2}(\mathbf{x} - \mathbf{x}^*)^\top f''(\mathbf{x}^* + s(\mathbf{x} - \mathbf{x}^*))(\mathbf{x} - \mathbf{x}^*) \qquad (10.49)$$

or, since the first derivative of f vanishes at $\mathbf{x}^*$,

$$f(\mathbf{x}) = f(\mathbf{x}^*) + \tfrac{1}{2}(\mathbf{x} - \mathbf{x}^*)^\top f''(\mathbf{x}^* + s(\mathbf{x} - \mathbf{x}^*))(\mathbf{x} - \mathbf{x}^*) \qquad (10.50)$$

Since f'' is continuous, $f''(\mathbf{x}^* + s(\mathbf{x} - \mathbf{x}^*))$ will also be negative definite for $\mathbf{x}$ in some open neighbourhood of $\mathbf{x}^*$. Hence, for $\mathbf{x}$ in this neighbourhood, $f(\mathbf{x}) < f(\mathbf{x}^*)$ and f has a strict local maximum at $\mathbf{x}^*$. $\qquad \square$

The weak form of this result does not hold. In other words, semi-definiteness of the Hessian matrix at $\mathbf{x}^*$ is not sufficient to guarantee that f has any sort of maximum at $\mathbf{x}^*$. For example, if $f(x) = x^3$, then the Hessian is negative semi-definite at $x = 0$ but the function does not have a local maximum there; rather, it has an inflexion point.

What about $f(x) = -x^4$? At $x = 0$, the second-order condition is not satisfied, but f has a strict local maximum.

For functions of a single variable, this inspires a more general theorem.

THEOREM 10.3.4 (nTH DERIVATIVE TEST FOR LOCAL MAXIMA AND MINIMA OF A FUNCTION OF A SINGLE VARIABLE). *Let $X \subseteq \mathbb{R}$ be open and let $f: X \to \mathbb{R}$ be a differentiable function with $f'(x^*) = 0$. If in evaluating consecutively the derivatives of f at x^* the first non-zero value encountered is $f^{(n)}(x^*)$ and the nth derivative is continuous, then f has:*

(a) *a local maximum at x^* if n is even and $f^{(n)}(x^*) < 0$;*
(b) *a local minimum at x^* if n is even and $f^{(n)}(x^*) > 0$; and*
(c) *an inflexion point at x^* if n is odd.*

Proof: The nth-order Taylor expansion of f around x^* reduces to

$$f(x) = f(x^*) + \frac{1}{n!} f^{(n)}(x^{**})(x - x^*)^n \qquad (10.51)$$

for some x^{**} between x and x^*, since all the intervening terms are zero.

If n is even, then the last factor takes the sign of the nth derivative at x^{**}, which by continuity is the same as its sign at x^*. The first two results follow easily.

If n is odd, then $f(x)$ is greater than $f(x^*)$ for x on one side of x^* but less for x on the other side, so there is neither a maximum nor a minimum at x^*. In other words, x^* is an inflexion point. □

For a function f of more than one variable, as we have already seen:

- a stationary point where the Hessian matrix is positive or negative definite is a minimum or maximum respectively; and
- a stationary point where the Hessian matrix is semi-definite may be an inflexion point; but
- a stationary point $\mathbf{x}^*$ where the Hessian matrix $\mathbf{A} \equiv f''(\mathbf{x}^*)$ is indefinite is known as a **saddle point**; in this case, there exist vectors $\mathbf{h}$ and $\mathbf{k}$ such that $\mathbf{h}^\top \mathbf{A} \mathbf{h} < 0$ and $\mathbf{k}^\top \mathbf{A} \mathbf{k} > 0$.

We know from Section 9.5 that the first and second derivatives of the restriction of f to the line from $\mathbf{x}^*$ to $\mathbf{x}^* + \mathbf{h}$ equal $f'(\mathbf{x}^*)\mathbf{h}$ and $\mathbf{h}^\top \mathbf{A} \mathbf{h}$, respectively. Thus the first derivative of this restriction is zero and the second derivative is negative at $\mathbf{x}^*$. Similarly, the directional derivative at $\mathbf{x}^*$ in the direction of $\mathbf{x}^* + \mathbf{k}$ is also zero, but the second derivative is positive in this case. Thus, the function appears to achieve a local maximum looking in the direction of $\mathbf{h}$ and a local minimum looking in the direction of $\mathbf{k}$. Hence the surface looks locally like a horse saddle or a mountain pass at $\mathbf{x}^*$, whence the phrase "saddle point".

EXAMPLE 10.3.2 The simplest function with a saddle point is the quadratic form $f: \mathbb{R}^2 \to \mathbb{R}: \mathbf{x} \mapsto \mathbf{x}^\top \mathbf{A} \mathbf{x}$, where

$$\mathbf{A} \equiv \begin{bmatrix} 0 & \frac{1}{2} \\ \frac{1}{2} & 0 \end{bmatrix} \tag{10.52}$$

In this case $f(\mathbf{x}) = x_1 x_2$, $f'(\mathbf{x}) = [x_2 \ x_1]$ and $f''(\mathbf{x}) = 2\mathbf{A}$. There is one stationary point, at the origin. The Hessian matrix is not semi-definite since $\mathbf{h}^\top \mathbf{A} \mathbf{h} = h_1 h_2$ is positive when h_1 and h_2 have the same sign and negative when they have opposite signs.

The contour map for this function is symmetric about the origin and about both coordinate axes, consisting of four sets of rectangular hyperbolas, asymptotic in all cases to the coordinate axes. ◇

For another example of a function with a saddle point, see Exercise 10.18.

THEOREM 10.3.5 (UNIQUENESS CONDITIONS FOR UNCONSTRAINED MAXIMIZATION). *If*

(a) $\mathbf{x}^*$ *solves problem (10.43) and*
(b) *f is strictly quasi-concave (presupposing that X is a convex set),*

then $\mathbf{x}^*$ is the unique (global) maximum.

Proof: The proof is by contradiction. Suppose $\mathbf{x}^*$ is not unique; in other words, assume that there exists $\mathbf{x} \neq \mathbf{x}^*$ such that $f(\mathbf{x}) = f(\mathbf{x}^*)$.

Then, for any $\alpha \in (0, 1)$, by strict quasi-concavity,

$$f(\alpha\mathbf{x} + (1 - \alpha)\mathbf{x}^*) > \min\{f(\mathbf{x}), f(\mathbf{x}^*)\} = f(\mathbf{x}^*) \tag{10.53}$$

so f does not have a maximum at either $\mathbf{x}$ or $\mathbf{x}^*$.
This is a contradiction, so the maximum must be unique. $\qquad\qquad\qquad\square$

The following are tempting, but not quite true, corollaries of Theorem 10.3.3:

- Every stationary point of a twice continuously differentiable strictly concave function is a strict global maximum.
- Every stationary point of a twice continuously differentiable strictly convex function is a strict global minimum.
- So there can be at most one stationary point of any such function.

These conclusions are valid for functions whose Hessian matrices are, respectively, negative definite everywhere or positive definite everywhere. We know from Theorem 10.2.5 that all such functions are, respectively, strictly concave or strictly convex, but that the converse is not true.

For a function whose Hessian matrix is positive definite everywhere or negative definite everywhere, the argument in the proof of Theorem 10.3.3 can be applied for $\mathbf{x} \in X$ and not just for $\mathbf{x} \in B_\epsilon(\mathbf{x}^*)$. If there are points at which the Hessian is merely semi-definite, then the proof breaks down.

Note that some strictly concave or strictly convex functions will have no stationary points, for example,

$$f : \mathbb{R} \to \mathbb{R} : x \mapsto e^x \tag{10.54}$$

10.4 Equality-constrained optimization

10.4.1 Lagrange multiplier theorems

In economics, optimization problems are usually subject to resource constraints. For example, an individual deciding on a basket of goods to consume is invariably subject to a budget constraint involving the quantities to be consumed, as well as the prices of each good and the amount available to spend. The choice variables are no longer free to vary independently. Roughly speaking, the solution is to reduce the number of choice variables by the number of constraints, find the optimal values for this reduced set of choice variables, and then use the constraints to calculate the optimal values for the remaining choice variables.

Throughout this section, we deal with the equality-constrained optimization problem

$$\max_{\mathbf{x} \in X} f(\mathbf{x}) \quad \text{s.t.} \quad g(\mathbf{x}) = \mathbf{0}_m \tag{10.55}$$

where $X \subseteq \mathbb{R}^n$, $f : X \to \mathbb{R}$ is a real-valued function of several variables, which is the objective function of problem (10.55), and $g : X \to \mathbb{R}^m$ is a vector-valued function of several variables, called the **constraint function** of problem (10.55); or, equivalently, $g^j : X \to \mathbb{R}$ are real-valued functions for $j = 1, 2, \ldots, m$. In other words, there are m scalar constraint equations, or, simply, **constraints**, represented by a single vector constraint equation:

$$\begin{bmatrix} g^1(\mathbf{x}) \\ g^2(\mathbf{x}) \\ \vdots \\ g^m(\mathbf{x}) \end{bmatrix} = \begin{bmatrix} 0 \\ 0 \\ \vdots \\ 0 \end{bmatrix} \tag{10.56}$$

The set of vectors in X satisfying all m constraints

$$\{\mathbf{x} \in X : g(\mathbf{x}) = \mathbf{0}_m\} \tag{10.57}$$

is called the **feasible set** or **constraint set** for problem (10.55).

We will introduce and motivate the **Lagrange multiplier method**,[2] which applies to such constrained optimization problems with equality constraints. We will assume, where appropriate, that the objective function f and the m constraint functions $g^1, g^2, \ldots, g^m$ are all once or twice continuously differentiable.

The entire discussion here is again presented in terms of maximization, but can equally be presented in terms of minimization by reversing the sign of the objective function. Similarly, note that the signs of the constraint function(s) can be reversed without altering the underlying problem. We will see, however, that this also reverses the signs of the corresponding **Lagrange multipliers**. The significance of this effect will be seen from the formal results, which are presented here in terms of the usual three theorems, for necessity, sufficiency and uniqueness.

THEOREM 10.4.1 (FIRST-ORDER (NECESSARY) CONDITIONS FOR OPTIMIZATION WITH EQUALITY CONSTRAINTS). *Consider problem (10.55) or the corresponding minimization problem. If*

(a) $\mathbf{x}^*$ *solves this problem (which implies that $g(\mathbf{x}^*) = \mathbf{0}$);*
(b) *f and g are continuously differentiable; and*
(c) *the $m \times n$ matrix*

$$g'(\mathbf{x}^*) = \begin{bmatrix} \dfrac{\partial g^1}{\partial x_1}(\mathbf{x}^*) & \cdots & \dfrac{\partial g^1}{\partial x_n}(\mathbf{x}^*) \\ \vdots & \ddots & \vdots \\ \dfrac{\partial g^m}{\partial x_1}(\mathbf{x}^*) & \cdots & \dfrac{\partial g^m}{\partial x_n}(\mathbf{x}^*) \end{bmatrix} \tag{10.58}$$

is of full rank m (i.e. there are no redundant constraints, both in the sense that there are fewer constraints than variables and in the sense that the constraints that are present are "independent"),[3]

then there exists a vector of Lagrange multipliers $\boldsymbol{\lambda}^* \in \mathbb{R}^m$ such that $f'(\mathbf{x}^*) + \boldsymbol{\lambda}^{*\top} g'(\mathbf{x}^*) = \mathbf{0}^\top$ (i.e. in $\mathbb{R}^n$, $f'(\mathbf{x}^*)$ is in the m-dimensional subspace generated by the m vectors $g^{1'}(\mathbf{x}^*), g^{2'}(\mathbf{x}^*), \ldots, g^{m'}(\mathbf{x}^*)$).

It is conventional to use the letter λ both to parametrize convex combinations and as a Lagrange multiplier. To avoid confusion, in this section we switch to the letter α for the former usage.

Proof: The idea is to solve $g(\mathbf{x}^*) = \mathbf{0}$ for m variables as a function of the other $n - m$ variables and to substitute the solution into the objective function to give an unconstrained problem with $n - m$ variables.

For this proof, we need the implicit function theorem (Theorem 9.8.1). Using this theorem, we must find the m weights $\lambda_1, \lambda_2, \ldots, \lambda_m$ to prove that $f'(\mathbf{x}^*)$ is a linear combination of $g^{1'}(\mathbf{x}^*), g^{2'}(\mathbf{x}^*), \ldots, g^{m'}(\mathbf{x}^*)$.

Without loss of generality, we assume that the first m columns of $g'(\mathbf{x}^*)$ are linearly independent (if not, then we merely relabel the variables accordingly).

Now we can partition the vector $\mathbf{x}^*$ as $(\mathbf{y}^*, \mathbf{z}^*)$, where $\mathbf{y}^* \in \mathbb{R}^m$ and $\mathbf{z}^* \in \mathbb{R}^{n-m}$, and, using the notation of the implicit function theorem, find a neighbourhood Z of $\mathbf{z}^*$ and a function h defined on Z such that

$$g(h(\mathbf{z}), \mathbf{z}) = \mathbf{0} \quad \forall \, \mathbf{z} \in Z \tag{10.59}$$

and also

$$h'(\mathbf{z}^*) = -(D_\mathbf{y} g)^{-1} D_\mathbf{z} g \tag{10.60}$$

Now define a new objective function $F : Z \to \mathbb{R}$ by

$$F(\mathbf{z}) \equiv f(h(\mathbf{z}), \mathbf{z}) \tag{10.61}$$

Since $\mathbf{x}^*$ solves the constrained problem $\max_{\mathbf{x} \in X} f(\mathbf{x})$ subject to $g(\mathbf{x}) = \mathbf{0}$, it follows (see Exercise 10.19) that $\mathbf{z}^*$ solves the unconstrained problem $\max_{\mathbf{z} \in Z} F(\mathbf{z})$.

Hence, $\mathbf{z}^*$ satisfies the first-order conditions for unconstrained maximization of F, namely,

$$F'(\mathbf{z}^*) = \mathbf{0}^\top \tag{10.62}$$

Applying Corollary 9.3.2 yields an equation that can be written in shorthand as

$$F'(\mathbf{z}^*) = D_\mathbf{y} f h'(\mathbf{z}^*) + D_\mathbf{z} f = \mathbf{0}^\top \tag{10.63}$$

where $f'(\mathbf{x}^*) \equiv [D_\mathbf{y} f \quad D_\mathbf{z} f]$. Substituting for $h'(\mathbf{z}^*)$ gives

$$D_\mathbf{y} f (D_\mathbf{y} g)^{-1} D_\mathbf{z} g = D_\mathbf{z} f \tag{10.64}$$

We can also partition $f'(\mathbf{x}^*)$ by inserting an identity matrix in the form $(D_\mathbf{y} g)^{-1} D_\mathbf{y} g$ as

$$[D_\mathbf{y} f (D_\mathbf{y} g)^{-1} D_\mathbf{y} g \quad D_\mathbf{z} f] \tag{10.65}$$

Substituting for the second submatrix yields

$$\begin{aligned} f'(\mathbf{x}^*) &= [D_\mathbf{y} f (D_\mathbf{y} g)^{-1} D_\mathbf{y} g \quad D_\mathbf{y} f (D_\mathbf{y} g)^{-1} D_\mathbf{z} g] \\ &= D_\mathbf{y} f (D_\mathbf{y} g)^{-1} [D_\mathbf{y} g \quad D_\mathbf{z} g] \\ &= -\lambda^\top g'(\mathbf{x}^*) \end{aligned} \tag{10.66}$$

where we define

$$\lambda \equiv -D_{\mathbf{y}} f (D_{\mathbf{y}} g)^{-1} \tag{10.67}$$

$\square$

THEOREM 10.4.2 (SECOND-ORDER (SUFFICIENT OR CONCAVITY) CONDITIONS FOR MAXIMIZATION WITH EQUALITY CONSTRAINTS). *If*

(a) *f and g are differentiable;*
(b) $f'(\mathbf{x}^*) + \lambda^{*\top} g'(\mathbf{x}^*) = \mathbf{0}^\top$ *(i.e. the first-order conditions are satisfied at* $\mathbf{x}^*$*);*
(c) $\lambda_j^* \geq 0$ *for* $j = 1, 2, \ldots, m$;
(d) *f is pseudo-concave; and*
(e) g^j *is quasi-concave for* $j = 1, 2, \ldots, m$,

then $\mathbf{x}^*$ *solves the constrained maximization problem.*

Note that non-positive Lagrange multipliers and quasi-convex constraint functions can take the place of non-negative Lagrange multipliers and quasi-concave constraint functions to give an alternative set of second-order conditions for optimization problems with equality constraints.

Proof: Suppose that the second-order conditions are satisfied, but that $\mathbf{x}^*$ is not a constrained maximum. We will derive a contradiction.

Since $\mathbf{x}^*$ is not a maximum, there exists $\mathbf{x} \neq \mathbf{x}^*$ such that $g(\mathbf{x}) = \mathbf{0}$ but $f(\mathbf{x}) > f(\mathbf{x}^*)$.

By pseudo-concavity, $f(\mathbf{x}) - f(\mathbf{x}^*) > 0$ implies that $f'(\mathbf{x}^*)(\mathbf{x} - \mathbf{x}^*) > 0$.

Since the constraints are satisfied at both $\mathbf{x}$ and $\mathbf{x}^*$, we have $g(\mathbf{x}^*) = g(\mathbf{x}) = \mathbf{0}$.

By quasi-concavity of the constraint functions (see the last part of Theorem 10.2.7), $g^j(\mathbf{x}) - g^j(\mathbf{x}^*) = 0$ implies that $g^{j'}(\mathbf{x}^*)(\mathbf{x} - \mathbf{x}^*) \geq 0$.

By assumption, all the Lagrange multipliers are non-negative, so

$$f'(\mathbf{x}^*)(\mathbf{x} - \mathbf{x}^*) + \lambda^{*\top} g'(\mathbf{x}^*)(\mathbf{x} - \mathbf{x}^*) > 0 \tag{10.68}$$

Rearranging yields

$$(f'(\mathbf{x}^*) + \lambda^{*\top} g'(\mathbf{x}^*))(\mathbf{x} - \mathbf{x}^*) > 0 \tag{10.69}$$

But the first-order condition guarantees that the left-hand side of this inequality is zero (not positive), which is the required contradiction. $\square$

Various slightly different second-order conditions could be proposed – for example, if f was quasi-concave, g^j pseudo-concave and at least one λ_j^* strictly positive, then (10.69) would again provide a contradiction.

THEOREM 10.4.3 (UNIQUENESS CONDITION FOR EQUALITY-CONSTRAINED MAXIMIZATION). *If*

(a) $\mathbf{x}^*$ *is a solution;*
(b) *f is strictly quasi-concave; and*
(c) g^j *is an affine function for* $j = 1, 2, \ldots, m$,

then $\mathbf{x}^*$ *is the unique (global) maximum.*

Proof: The uniqueness result is also proved by contradiction. Note that it does not require any differentiability assumption.

(a) We first show that the feasible set is an affine set (and, hence, a convex set).

Suppose that $\mathbf{x} \neq \mathbf{x}'$ are two distinct vectors satisfying the constraints and $\alpha \in \mathbb{R}$. Consider the affine combination of these two vectors, $\mathbf{x}_\alpha \equiv \alpha \mathbf{x} + (1 - \alpha)\mathbf{x}'$. Since each g^j is affine and $g^j(\mathbf{x}') = g^j(\mathbf{x}) = 0$, we have

$$g^j(\mathbf{x}_\alpha) = \alpha g^j(\mathbf{x}) + (1 - \alpha)g^j(\mathbf{x}') = 0 \tag{10.70}$$

In other words, $\mathbf{x}_\alpha$ also satisfies the constraints, as required.

(b) To complete the proof, we find the required contradiction.

Now suppose that $\mathbf{x} \neq \mathbf{x}'$ are two distinct vectors solving the optimization problem and $\alpha \in (0, 1)$. Consider the *convex* combination of these two vectors, $\mathbf{x}_\alpha \equiv \alpha \mathbf{x} + (1 - \alpha)\mathbf{x}'$. Since f is strictly quasi-concave and $f(\mathbf{x}') = f(\mathbf{x})$, it must be the case that $f(\mathbf{x}_\alpha) > f(\mathbf{x})$ and $f(\mathbf{x}_\alpha) > f(\mathbf{x}')$.

But, by the first part of the proof, $\mathbf{x}_\alpha$ satisfies the constraints, so neither $\mathbf{x}$ nor $\mathbf{x}'$ is a solution, and there can be only one solution, as required. □

The construction of the obvious corollaries for minimization problems is left as an exercise; see Exercise 10.20.

10.4.2 Solution methodology

The first n first-order or **Lagrangian conditions** say that the total derivative (or gradient) of f at $\mathbf{x}$ is a linear combination of the total derivatives (or gradients) of the constraint functions at $\mathbf{x}$.

Consider a picture with $n = 2$ and $m = 1$ as shown in Figure 10.3. Since the directional derivative along a tangent to a level set or indifference curve is zero at the point of tangency, $\mathbf{x}$ (the function is at a maximum or minimum along the tangent), or $f'(\mathbf{x})(\mathbf{x}' - \mathbf{x}) = 0$, the gradient vector, $f'(\mathbf{x})^{\top}$, must be perpendicular to the direction of the tangent, $\mathbf{x}' - \mathbf{x}$.

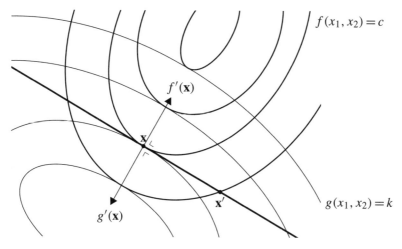

Figure 10.3 Constrained optimization with two variables and one constraint

At the optimum, the level sets of f and g have a common tangent, so $f'(\mathbf{x})$ and $g'(\mathbf{x})$ are collinear, or $f'(\mathbf{x}) = -\lambda g'(\mathbf{x})$. It can also be seen with a little thought that, for the solution to be a local constrained maximum, λ must be positive if g is quasi-concave or negative if g is quasi-convex (in either case, the constraint curve is the boundary of a convex set).

If $\mathbf{x}^*$ is a solution to problem (10.55), then there exist Lagrange multipliers,[4] $\boldsymbol{\lambda} \equiv (\lambda_1, \lambda_2, \ldots, \lambda_m)$, such that

$$f'(\mathbf{x}^*) + \boldsymbol{\lambda}^\top g'(\mathbf{x}^*) = \mathbf{0}_n^\top \tag{10.71}$$

Thus, to find the constrained optimum, we proceed as if optimizing the *Lagrangian*:

$$\mathcal{L}(\mathbf{x}, \boldsymbol{\lambda}) \equiv f(\mathbf{x}) + \boldsymbol{\lambda}^\top g(\mathbf{x}) \tag{10.72}$$

Note that:

- $\mathcal{L} = f$ whenever $g = \mathbf{0}$; and
- $g = \mathbf{0}$ where $\mathcal{L}$ is optimized with respect to $\boldsymbol{\lambda}$.

Roughly speaking, this is why the constrained optimum of f corresponds to the optimum of $\mathcal{L}$.

The Lagrange multiplier method involves the following steps:

1. Introduce the m Lagrange multipliers, $\boldsymbol{\lambda} \equiv (\lambda_1, \lambda_2, \ldots, \lambda_m)$.
2. Define the Lagrangian $\mathcal{L} : X \times \mathbb{R}^m \to \mathbb{R}$, where

$$X \times \mathbb{R}^m \equiv \{(\mathbf{x}, \boldsymbol{\lambda}) : \mathbf{x} \in X, \boldsymbol{\lambda} \in \mathbb{R}^m\} \tag{10.73}$$

by

$$\mathcal{L}(\mathbf{x}, \boldsymbol{\lambda}) \equiv f(\mathbf{x}) + \boldsymbol{\lambda}^\top g(\mathbf{x}) \tag{10.74}$$

3. Find the stationary points of the Lagrangian, i.e. set $\mathcal{L}'(\mathbf{x}, \boldsymbol{\lambda}) = \mathbf{0}^\top$. Since the Lagrangian is a function of $n + m$ variables, this gives $n + m$ first-order conditions. The first n are

$$f'(\mathbf{x}) + \boldsymbol{\lambda}^\top g'(\mathbf{x}) = \mathbf{0}^\top \tag{10.75}$$

or

$$\frac{\partial f}{\partial x_i}(\mathbf{x}) + \sum_{j=1}^{m} \lambda_j \frac{\partial g^j}{\partial x_i}(\mathbf{x}) = 0, \quad i = 1, 2, \ldots, n \tag{10.76}$$

The last m are just the original constraints,

$$g(\mathbf{x}) = \mathbf{0} \tag{10.77}$$

or

$$g^j(\mathbf{x}) = 0, \quad j = 1, 2, \ldots, m \tag{10.78}$$

4. Now we need to solve the first-order conditions, which are usually a highly nonlinear system of simultaneous equations. A method that often works is to:

   (a) solve the first n first-order conditions to obtain an initial solution for **x** in terms of λ; then
   (b) substitute this solution into the constraint equations to obtain a system of m equations in the m Lagrange multipliers only; then
   (c) solve for λ; and
   (d) substitute for λ in the initial solution to obtain a closed-form solution for **x**.

5. Finally, the second-order and uniqueness conditions must be checked.

EXAMPLE 10.4.1 One of the most commonly encountered examples of an equality-constrained optimization problem is the maximization of a Cobb–Douglas utility function subject to a budget constraint. Suppose that a consumer wishes to maximize the Cobb–Douglas utility function given by

$$u(x, y) = x^{0.5} y^{0.5} \tag{10.79}$$

where x and y are the quantities of each of two goods consumed.

We will assume that the consumer always spends all his money income, M per period, on the two goods, which he can buy at prices p_x and p_y, respectively (both constant for the consumer).

The consumer must then

$$\max_{x,y} x^{0.5} y^{0.5} \tag{10.80}$$

subject to the equality budget constraint $p_x x + p_y y = M$. The Lagrangian is

$$\mathcal{L}(x, y, \lambda) \equiv x^{0.5} y^{0.5} + \lambda (M - p_x x - p_y y) \tag{10.81}$$

The first-order conditions are

$$\frac{\partial \mathcal{L}}{\partial x}(x, y, \lambda) = 0.5 x^{0.5-1} y^{0.5} - \lambda p_x = 0 \tag{10.82}$$

$$\frac{\partial \mathcal{L}}{\partial y}(x, y, \lambda) = 0.5 x^{0.5} y^{0.5-1} - \lambda p_y = 0 \tag{10.83}$$

and the budget constraint. Adding x times (10.82) to y times (10.83) yields

$$(0.5 + 0.5) x^{0.5} y^{0.5} - \lambda (p_x x + p_y y) = 0 \tag{10.84}$$

So

$$\lambda = \frac{x^{0.5} y^{0.5}}{p_x x + p_y y} = \frac{x^{0.5} y^{0.5}}{M} \tag{10.85}$$

Substituting this in the first-order conditions gives, after rearrangement,

$$x^*(p_x, p_y, M) = 0.5 \frac{M}{p_x} \tag{10.86}$$

$$y^*(p_x, p_y, M) = 0.5 \frac{M}{p_y} \tag{10.87}$$

Confirmation that the second-order conditions are satisfied and further variations on this problem are considered in Exercises 10.21 and 10.22. $\diamond$

Another commonly encountered example of an equality-constrained optimization problem is the maximization of a quadratic form $\mathbf{x}^\top \mathbf{A}\mathbf{x}$ subject to linear constraints $g(\mathbf{x}) = \mathbf{G}\mathbf{x} = \boldsymbol{\alpha}$. This problem will be solved in full in Section 14.3.

10.4.3   Envelope theorems

We conclude this section with an investigation of how the optimal value of the objective function depends on any exogenous variables that may appear in the objective function, a topic often described as **comparative statics**.

THEOREM 10.4.4 (ENVELOPE THEOREM FOR MAXIMIZATION). *Consider the modified constrained maximization problem*

$$\max_{\mathbf{x}} f(\mathbf{x}, \boldsymbol{\alpha}) \quad \text{subject to } g(\mathbf{x}, \boldsymbol{\alpha}) = 0 \tag{10.88}$$

where $\mathbf{x} \in \mathbb{R}^n$, $\boldsymbol{\alpha} \in \mathbb{R}^q$, $f: \mathbb{R}^{n+q} \to \mathbb{R}$ *and* $g: \mathbb{R}^{n+q} \to \mathbb{R}^m$ *(i.e. as usual f is the real-valued objective function and g is a vector of m real-valued constraint functions, but either or both can depend on exogenous or **control** variables $\boldsymbol{\alpha}$ as well as on the endogenous or **choice** variables $\mathbf{x}$).*

Suppose that the standard conditions for application of the Lagrange multiplier theorems (Theorems 10.4.1, 10.4.2 and 10.4.3) are satisfied.

*Let $\mathbf{x}^*(\boldsymbol{\alpha})$ denote the optimal choice of $\mathbf{x}$ for given $\boldsymbol{\alpha}$ ($\mathbf{x}^*: \mathbb{R}^q \to \mathbb{R}^n$ is called the **optimal response function**) and let $M(\boldsymbol{\alpha})$ denote the maximum value attainable by f for given $\boldsymbol{\alpha}$ ($M: \mathbb{R}^q \to \mathbb{R}$ is called the **envelope function**).[5]*

Then the partial derivative of M with respect to α_i is just the partial derivative of the relevant Lagrangian, $f + \boldsymbol{\lambda}^\top g$, with respect to α_i, evaluated at the optimal value of $\mathbf{x}$. The dependence of the vector of Lagrange multipliers, $\boldsymbol{\lambda}$, on the vector $\boldsymbol{\alpha}$ should be ignored in calculating the last-mentioned partial derivative.

Proof: The envelope theorem can be proved in the following steps.

(a) Write down the identity relating the functions M, f and $\mathbf{x}^*$:

$$M(\boldsymbol{\alpha}) \equiv f(\mathbf{x}^*(\boldsymbol{\alpha}), \boldsymbol{\alpha}) \tag{10.89}$$

(b) Use Corollary 9.3.2 to derive an expression for the partial derivatives $\partial M / \partial \alpha_i$ of M in terms of the partial derivatives of f and $\mathbf{x}^*$:

$$M'(\boldsymbol{\alpha}) = D_\mathbf{x} f(\mathbf{x}^*(\boldsymbol{\alpha}), \boldsymbol{\alpha})\mathbf{x}^{*\prime}(\boldsymbol{\alpha}) + D_{\boldsymbol{\alpha}} f(\mathbf{x}^*(\boldsymbol{\alpha}), \boldsymbol{\alpha}) \tag{10.90}$$

(c) The first-order (necessary) conditions for constrained optimization say that

$$D_{\mathbf{x}} f(\mathbf{x}^*(\boldsymbol{\alpha}), \boldsymbol{\alpha}) = -\lambda(\boldsymbol{\alpha})^\top D_{\mathbf{x}} g(\mathbf{x}^*(\boldsymbol{\alpha}), \boldsymbol{\alpha}) \tag{10.91}$$

and allow us to eliminate the $\partial f / \partial x_i$ terms from (10.90).

(d) Apply Corollary 9.3.2 again to the identity $g(\mathbf{x}^*, \boldsymbol{\alpha}) = \mathbf{0}_m$ to obtain

$$D_{\mathbf{x}} g(\mathbf{x}^*(\boldsymbol{\alpha}), \boldsymbol{\alpha}) \mathbf{x}^{*\prime}(\boldsymbol{\alpha}) + D_{\boldsymbol{\alpha}} g(\mathbf{x}^*(\boldsymbol{\alpha}), \boldsymbol{\alpha}) = \mathbf{0}_{m \times q} \tag{10.92}$$

Finally, use this result to eliminate the $\partial g / \partial x_i$ terms from (10.91).

Combining all these results gives

$$M'(\boldsymbol{\alpha}) = D_{\boldsymbol{\alpha}} f(\mathbf{x}^*(\boldsymbol{\alpha}), \boldsymbol{\alpha}) + \lambda(\boldsymbol{\alpha})^\top D_{\boldsymbol{\alpha}} g(\mathbf{x}^*(\boldsymbol{\alpha}), \boldsymbol{\alpha}) \tag{10.93}$$

which is the required result. $\qquad\square$

Inspection of the Lagrangian (10.81) for the utility-maximization problem above shows that the Lagrange multiplier itself in that problem equals the rate of change of optimal utility with respect to income, M. This result holds whatever the functional form of the utility function.

THEOREM 10.4.5 (ENVELOPE THEOREM FOR MINIMIZATION). *Consider the modified constrained minimization problem*

$$\min_{\mathbf{x}} f(\mathbf{x}, \boldsymbol{\alpha}) \quad \text{subject to } g(\mathbf{x}, \boldsymbol{\alpha}) = \mathbf{0} \tag{10.94}$$

where $\mathbf{x} \in \mathbb{R}^n$, $\boldsymbol{\alpha} \in \mathbb{R}^q$, $f: \mathbb{R}^{n+q} \to \mathbb{R}$ and $g: \mathbb{R}^{n+q} \to \mathbb{R}^m$.

Suppose that the standard conditions for application of the Lagrange multiplier theorems (Theorems 10.4.1, 10.4.2 and 10.4.3) are satisfied.

Let $\mathbf{x}^*(\boldsymbol{\alpha})$ denote the optimal choice of $\mathbf{x}$ for given $\boldsymbol{\alpha}$ and let $m(\boldsymbol{\alpha})$ denote the minimum value attainable by f for given $\boldsymbol{\alpha}$ ($m: \mathbb{R}^q \to \mathbb{R}$ is again called the envelope function).

Then the partial derivative of m with respect to α_i is just the partial derivative of the relevant Lagrangian, $f - \lambda^\top g$, with respect to α_i, evaluated at the optimal value of $\mathbf{x}$. The dependence of the vector of Lagrange multipliers, λ, on the vector $\boldsymbol{\alpha}$ should be ignored in calculating the last-mentioned partial derivative.

Proof: Let us view the problem as one of maximizing $-f$ rather than minimizing f. The envelope functions of the two problems are related by $m = -M$. The first version of the envelope theorem tells us that

$$\frac{\partial M}{\partial \alpha_i} = \frac{\partial (-f)}{\partial \alpha_i} + \sum_{j=i}^{m} \lambda_j \frac{\partial g^j}{\partial \alpha_i} \tag{10.95}$$

Multiplying across by -1 gives

$$\frac{\partial m}{\partial \alpha_i} = \frac{\partial f}{\partial \alpha_i} - \sum_{j=i}^{m} \lambda_j \frac{\partial g^j}{\partial \alpha_i} \tag{10.96}$$

$\qquad\square$

We will encounter envelope functions again throughout the remainder of this chapter. They will also appear later in the book as, for example, indirect utility functions (Section 12.4.4), expenditure functions (Section 12.4.5), representative agents' utility functions (Section 12.6.6) and portfolio frontiers (Section 17.4).

In applications in economics, the most frequently encountered applications of equality-constrained optimization will make sufficient assumptions to guarantee that:

1. $\mathbf{x}^*$ satisfies the first-order conditions with each $\lambda_i \geq 0$,
2. the Hessian $f''(\mathbf{x}^*)$ is a negative definite matrix, and
3. g is an affine function,

so that $\mathbf{x}^*$ is the unique optimal solution to the equality-constrained optimization problem.

10.5   Inequality-constrained optimization

10.5.1   Kuhn–Tucker theorems

In the preceding sections, it was assumed that the various constraints on the choice variables were required to hold simultaneously and exactly. In practice, many constraints take an inequality form. For example, the amount spent can be less than the amount earned, but must not be more. On occasions there will be a number of inequality constraints, some of which will prove **binding** (or **active**) and some **non-binding** (or **inactive**) at the optimum. In this section, we extend the preceding analysis to cover these situations.

Throughout the section, we deal with the inequality-constrained optimization problem

$$\max_{\mathbf{x} \in X} f(\mathbf{x}) \quad \text{s.t. } g^i(\mathbf{x}) \geq 0, \ i = 1, 2, \ldots, m \tag{10.97}$$

where once again $X \subseteq \mathbb{R}^n$, $f: X \to \mathbb{R}$ is a real-valued function of several variables (the objective function of problem (10.97)) and $g: X \to \mathbb{R}^m$ is a vector-valued function of several variables (the constraint function of problem (10.97)). Before presenting general results for problem (10.97), we will look at two special cases.

The first special case is that of $m = n = 1$ and $g(x) = x$, i.e. the maximization of a function of one variable subject to a non-negativity constraint:

$$\max_{x} f(x) \quad \text{s.t. } x \geq 0 \tag{10.98}$$

The first-order conditions in this case can be expressed as

$$f'(x^*) \leq 0 \tag{10.99}$$

$$f'(x^*) = 0 \quad \text{if } x^* > 0 \tag{10.100}$$

The second special case is that in which the constraint functions are given by

$$g(\mathbf{x}, \boldsymbol{\alpha}) = \boldsymbol{\alpha} - h(\mathbf{x}) \tag{10.101}$$

assuming f to be quasi-concave as usual and h^i quasi-convex or (equivalently) g^i quasi-concave. We continue to denote the envelope function by $M(\boldsymbol{\alpha})$ for such inequality-constrained problems.

Figure 10.4 illustrates the solution to this problem for various values of α in the case where $n = 2$ and $m = 1$. This figure provides some graphical motivation concerning the interpretation of Lagrange multipliers. In fact, we will soon begin to refer to the analogue of the Lagrange multipliers in the inequality-constrained problem as the **Kuhn–Tucker multipliers**.[6]

The Lagrangian for the equality-constrained version of the problem is

$$\mathcal{L}(\mathbf{x}, \boldsymbol{\lambda}) = f(\mathbf{x}) + \boldsymbol{\lambda}^\top (\boldsymbol{\alpha} - h(\mathbf{x})) \tag{10.102}$$

Thus, using the envelope theorem (Theorem 10.4.4), it is easily seen that the rate of change of the envelope function for the equality-constrained problem, $M(\boldsymbol{\alpha})$, with respect to the "level" of the ith underlying constraint function h^i, is

$$\frac{\partial M}{\partial \alpha_i} = \frac{\partial \mathcal{L}}{\partial \alpha_i} = \lambda_i \tag{10.103}$$

In the upper part of Figure 10.4, the circles represent indifference curves of the objective function f and the straight lines represent indifference curves of the constraint function h. The ordinary lines represent binding constraints (where the Lagrange or Kuhn–Tucker multiplier λ is positive); the thick line represents the just-binding constraint (where $\lambda = 0$); and the dotted lines represent non-binding constraints (where $\lambda < 0$). Tangency points such as that marked $\mathbf{x}^*(\alpha_1)$ represent solutions to both the equality-constrained and inequality-constrained optimization problems

For $\alpha > \alpha_5$, the inequality constraint in Figure 10.4 is non-binding; at $\alpha = \alpha_5$, it is just binding. In other words, $\mathbf{x}^*(\alpha_5)$ solves the inequality-constrained problem (10.97) for any $\alpha \geq \alpha_5$.

The lower part of Figure 10.4 shows the values of α and of the envelope functions corresponding to the five binding constraint lines in the upper picture. For $\alpha > \alpha_5$, the constraint is non-binding, so the envelope functions for the equality-constrained and inequality-constrained problems differ. For the equality-constrained problem, the envelope function reaches the unconstrained maximum value $f(\mathbf{x}^*(\alpha_5))$ at α_5, but then turns downwards, as represented by the dotted curve. For the inequality-constrained problem, the envelope function equals the unconstrained maximum value $f(\mathbf{x}^*(\alpha_5))$ whenever $\alpha \geq \alpha_5$, as represented by the horizontal line. At any point, the slope of the envelope function for the inequality-constrained problem equals the Kuhn–Tucker multiplier for the corresponding value of α; and similarly the slope of the envelope function for the equality-constrained problem equals the corresponding Lagrange multiplier.

We can now summarize how the nature of the inequality constraint

$$h^i(\mathbf{x}) \leq \alpha_i \tag{10.104}$$

(or $g^i(\mathbf{x}, \boldsymbol{\alpha}) \geq 0$) changes as α_i increases (as illustrated in Figure 10.4, so that the relationship between α_i and λ_i is negative).

- For values of α_i such that the Lagrange multiplier $\lambda_i = 0$, this constraint is just binding.
- For values of α_i such that the Lagrange multiplier $\lambda_i > 0$, this constraint is strictly binding.
- For values of α_i such that the Lagrange multiplier $\lambda_i < 0$, this constraint is non-binding.

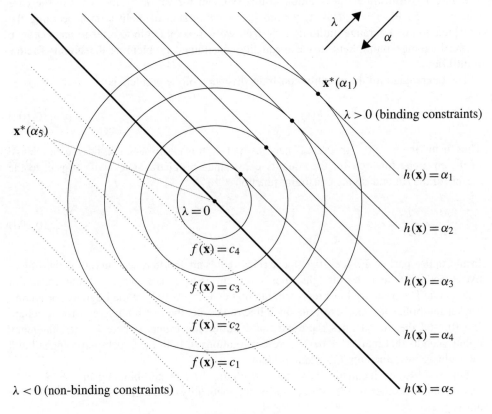

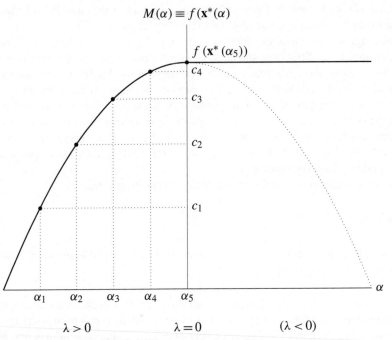

Figure 10.4 Interpretation of Lagrange and Kuhn–Tucker multipliers

Thus, more generally, we can say the following:

1. Wherever the Lagrange multiplier $\lambda_i = 0$, the envelope function for the equality-constrained problem is at its maximum, i.e. the objective function is at its unconstrained maximum, and the ith inequality constraint is just binding.
2. Wherever the Lagrange or Kuhn–Tucker multiplier $\lambda_i > 0$, the envelope function for the equality-constrained problem is increasing and the ith inequality constraint is strictly binding.
3. Wherever the Lagrange multiplier $\lambda_i < 0$, the envelope function for the equality-constrained problem is decreasing, but the corresponding Kuhn–Tucker multiplier is zero, the envelope function for the inequality-constrained problem is constant at its maximum value and the ith inequality constraint is non-binding.

Thus we will find that one of the *necessary* conditions below is that the Kuhn–Tucker multipliers be non-negative. (For equality-constrained optimization, the signs were important only when dealing with second-order conditions.)

Note that in situations such as Figure 10.4

$$\frac{\partial^2 M}{\partial \alpha_i^2} = \frac{\partial \lambda_i}{\partial \alpha_i} < 0 \tag{10.105}$$

so that the envelope function for the equality-constrained problem is strictly concave in each parameter. The envelope function for the inequality-constrained problem is strictly concave in each parameter over the range of values for which the constraint is strictly binding, but constant over the range of values that allow the unconstrained optimum to be achieved.

The various sign conditions that we have looked at are summarized in Table 10.2.

We are now in a position to sum up the above discussion formally in a theorem.

THEOREM 10.5.1 (NECESSARY (FIRST-ORDER) CONDITIONS FOR OPTIMIZATION WITH INEQUALITY CONSTRAINTS). *If*

(a) $\mathbf{x}^*$ *solves problem (10.97), with*

$$g^i(\mathbf{x}^*) = 0, \quad i = 1, 2, \ldots, b \tag{10.106}$$

and

$$g^i(\mathbf{x}^*) > 0, \quad i = b+1, b+2, \ldots, m \tag{10.107}$$

Table 10.2 Sign conditions for inequality-constrained optimization

Type of constraint	Derivative of objective function in (10.98)	Constraint function in (10.97)	Kuhn–Tucker multiplier in (10.97)	Lagrange multiplier
Binding/active	$f'(x^*) \leq 0$	$g(\mathbf{x}^*) = 0$	$\lambda \geq 0$	$\lambda \geq 0$
Just binding	$f'(x^*) = 0$	$g(\mathbf{x}^*) = 0$	$\lambda = 0$	$\lambda = 0$
Non-binding/inactive	$f'(x^*) = 0$	$g(\mathbf{x}^*) > 0$	$\lambda = 0$	$\lambda < 0$

(in other words, the first b constraints are binding at $\mathbf{x}^*$ *and the last* $m - b$ *are non-binding at* $\mathbf{x}^*$, *renumbering the constraints if necessary to achieve this);*

(b) *f and g are continuously differentiable; and*

(c) *the* $b \times n$ *submatrix of* $g'(\mathbf{x}^*)$,

$$
\begin{bmatrix}
\dfrac{\partial g^1}{\partial x_1}(\mathbf{x}^*) & \cdots & \dfrac{\partial g^1}{\partial x_n}(\mathbf{x}^*) \\
\vdots & \ddots & \vdots \\
\dfrac{\partial g^b}{\partial x_1}(\mathbf{x}^*) & \cdots & \dfrac{\partial g^b}{\partial x_n}(\mathbf{x}^*)
\end{bmatrix}
\tag{10.108}
$$

is of full rank b (i.e. there are no redundant binding constraints, both in the sense that there are fewer binding constraints than variables and in the sense that the constraints that are binding are "independent"),

then there exist Kuhn–Tucker multipliers $\lambda \in \mathbb{R}^m$ *such that* $f'(\mathbf{x}^*) + \lambda^\top g'(\mathbf{x}^*) = \mathbf{0}$, *with* $\lambda_i \geq 0$ *for* $i = 1, 2, \ldots, m$ *and* $g^i(\mathbf{x}^*) = 0$ *if* $\lambda_i > 0$.

Proof: The proof is similar to that of Theorem 10.4.1 for the equality-constrained case. It can be broken into seven steps.

(a) Suppose $\mathbf{x}^*$ solves problem (10.97).

We begin by restricting attention to a neighbourhood $B_\epsilon(\mathbf{x}^*)$ within which the constraints that are non-binding at $\mathbf{x}^*$ remain non-binding, i.e.

$$
g^i(\mathbf{x}) > 0 \quad \forall \mathbf{x} \in B_\epsilon(\mathbf{x}^*), \ i = b+1, b+2, \ldots, m \tag{10.109}
$$

Such a neighbourhood exists, since the constraint functions are continuous. Since $\mathbf{x}^*$ solves problem (10.97) by assumption, it also solves the following problem:

$$
\max_{\mathbf{x} \in B_\epsilon(\mathbf{x}^*)} f(\mathbf{x}) \quad \text{s.t. } g^i(\mathbf{x}) \geq 0, \ i = 1, 2, \ldots, b \tag{10.110}
$$

In other words, since the constraints that are non-binding at $\mathbf{x}^*$ remain non-binding for all $\mathbf{x} \in B_\epsilon(\mathbf{x}^*)$ by construction, we can ignore them if we confine our search for a maximum to this neighbourhood. We will return to the non-binding constraints in the very last step of this proof, but until then g will be taken to refer to the vector of b binding constraint functions only and λ to the vector of b Kuhn–Tucker multipliers corresponding to these binding constraints.

(b) We now introduce slack variables $\mathbf{s} \equiv (s_1, s_2, \ldots, s_b)$, one corresponding to each binding constraint, and consider the following equality-constrained maximization problem:

$$
\max_{\mathbf{x} \in B_\epsilon(\mathbf{x}^*), \, \mathbf{s} \in \mathbb{R}_+^b} f(\mathbf{x}) \quad \text{s.t. } G(\mathbf{x}, \mathbf{s}) = \mathbf{0}_b \tag{10.111}
$$

where $G: X \times \mathbb{R}^b \to \mathbb{R}^b$ is defined by $G^i(\mathbf{x}, \mathbf{s}) \equiv g^i(\mathbf{x}) - s_i$, $i = 1, 2, \ldots, b$. Since $\mathbf{x}^*$ solves problem (10.110) and all b constraints in that problem are binding at $\mathbf{x}^*$, it can be seen that $(\mathbf{x}^*, \mathbf{0}_b)$ solves this new problem. For consistency of notation, we define $\mathbf{s}^* \equiv \mathbf{0}_b$.

(c) We proceed with problem (10.111) as in the Lagrange case. In other words, we use the implicit function theorem to solve the system of b equations in $n + b$ unknowns,

$$G(\mathbf{x}, \mathbf{s}) = \mathbf{0}_b \tag{10.112}$$

restricted to $B_\epsilon(\mathbf{x}^*) \times \mathbb{R}^b$, for the first b variables in terms of the last n. To do this, we partition the vector of choice and slack variables in three ways:

$$(\mathbf{x}, \mathbf{s}) \equiv (\mathbf{y}, \mathbf{z}, \mathbf{s}) \tag{10.113}$$

where $\mathbf{y} \in \mathbb{R}^b$ and $\mathbf{z} \in \mathbb{R}^{n-b}$, and correspondingly partition the matrix of partial derivatives evaluated at the optimum:

$$\begin{aligned}
G'(\mathbf{y}^*, \mathbf{z}^*, \mathbf{s}^*) &= G'(\mathbf{x}^*, \mathbf{s}^*) \\
&= [D_\mathbf{y} G \quad D_{\mathbf{z},\mathbf{s}} G] \\
&= [D_\mathbf{y} G \quad D_\mathbf{z} G \quad D_\mathbf{s} G] \\
&= [D_\mathbf{y} g \quad D_\mathbf{z} g \quad -\mathbf{I}_b]
\end{aligned} \tag{10.114}$$

The rank condition allows us to apply the implicit function theorem and to find neighbourhoods $Z \subseteq \mathbb{R}^n$ and $Y \subseteq \mathbb{R}^b$ of $(\mathbf{z}^*, \mathbf{s}^*)$ and $\mathbf{y}^*$, respectively, such that $Y \times Z \subseteq B_\epsilon(\mathbf{x}^*) \times \mathbb{R}^b$, and a function $h: Z \to Y$ such that $\mathbf{y} = h(\mathbf{z}, \mathbf{s})$ is a solution to $G(\mathbf{y}, \mathbf{z}, \mathbf{s}) = \mathbf{0}$ with

$$h'(\mathbf{z}^*, \mathbf{s}^*) = -(D_\mathbf{y} g)^{-1} D_{\mathbf{z},\mathbf{s}} G \tag{10.115}$$

Equation (10.115) can in turn be partitioned to yield

$$D_\mathbf{z} h = -(D_\mathbf{y} g)^{-1} D_\mathbf{z} G = -(D_\mathbf{y} g)^{-1} D_\mathbf{z} g \tag{10.116}$$

and

$$D_\mathbf{s} h = -(D_\mathbf{y} g)^{-1} D_\mathbf{s} G = (D_\mathbf{y} g)^{-1} \mathbf{I}_b = (D_\mathbf{y} g)^{-1} \tag{10.117}$$

(d) This solution can be substituted into the original objective function f to create a new objective function F defined by

$$F(\mathbf{z}, \mathbf{s}) \equiv f(h(\mathbf{z}, \mathbf{s}), \mathbf{z}) \tag{10.118}$$

and another new maximization problem, where there are only (implicit) non-negativity constraints:

$$\max_{\mathbf{z} \in B_\epsilon(\mathbf{z}^*), \, \mathbf{s} \in \mathbb{R}^b_+} F(\mathbf{z}, \mathbf{s}) \tag{10.119}$$

It should be clear that $\mathbf{z}^*, \mathbf{0}_b$ solves problem (10.119). The first-order conditions for problem (10.119) are just that the partial derivatives of F with respect to the remaining $n - b$

choice variables equal zero (according to the first-order conditions for unconstrained optimization), while the partial derivatives of F with respect to the b slack variables must be less than or equal to zero.

(e) The Kuhn–Tucker multipliers can now be found exactly as in the Lagrange case. We know that

$$D_z F = D_y f\, D_z h + D_z f \mathbf{I}_{n-b} = \mathbf{0}_{n-b}^\top \qquad (10.120)$$

Substituting for $D_z h$ from (10.116) gives

$$D_y f (D_y g)^{-1} D_z g = D_z f \qquad (10.121)$$

We can also partition $f'(\mathbf{x}^*)$ as

$$[D_y f (D_y g)^{-1} D_y g \quad D_z f] \qquad (10.122)$$

Substituting for the second submatrix yields

$$\begin{aligned} f'(\mathbf{x}^*) &= [D_y f (D_y g)^{-1} D_y g \quad D_y f (D_y g)^{-1} D_z g] \\ &= D_y f (D_y g)^{-1}[D_y g \quad D_z g] \\ &\equiv -\boldsymbol{\lambda}^\top g'(\mathbf{x}^*) \end{aligned} \qquad (10.123)$$

where we define the Kuhn–Tucker multipliers corresponding to the binding constraints, $\boldsymbol{\lambda}$, by

$$\boldsymbol{\lambda} \equiv -D_y f (D_y g)^{-1} \qquad (10.124)$$

(f) Next, we calculate the partial derivatives of F with respect to the slack variables and show that they can be less than or equal to zero if and only if the Kuhn–Tucker multipliers corresponding to the binding constraints are greater than or equal to zero. This can be seen by differentiating both sides of (10.118) with respect to $\mathbf{s}$ to obtain

$$\begin{aligned} D_s F &= D_y f\, D_s h + D_z f \mathbf{0}_{(n-b)\times b} \\ &= D_y f (D_y g)^{-1} \\ &= -\boldsymbol{\lambda} \end{aligned} \qquad (10.125)$$

where we have used (10.117) and (10.124).

(g) Finally just set the Kuhn–Tucker multipliers corresponding to the non-binding constraints equal to zero. $\qquad \square$

THEOREM 10.5.2 (SECOND-ORDER (SUFFICIENT OR CONCAVITY) CONDITIONS FOR OPTIMIZATION WITH INEQUALITY CONSTRAINTS). *If*

(a) *f and g are differentiable;*
(b) *there exists $\boldsymbol{\lambda} \in \mathbb{R}^m$ such that $f'(\mathbf{x}^*) + \boldsymbol{\lambda}^\top g'(\mathbf{x}^*) = \mathbf{0}$, with $\lambda_i \geq 0$ for $i = 1, 2, \ldots, m$ and $g^i(\mathbf{x}^*) = 0$ if $\lambda_i > 0$ (i.e. the first-order conditions are satisfied at $\mathbf{x}^*$);*

(c) *f is pseudo-concave; and*

(d) $g^i(\mathbf{x}^*) = 0$ *for* $i = 1, 2, \ldots, b$, $g^i(\mathbf{x}^*) > 0$ *for* $i = b+1, b+2, \ldots, m$ *and* g^j *is quasi-concave for* $j = 1, 2, \ldots, b$ *(i.e. the binding constraint functions are quasi-concave)*,

then $\mathbf{x}^*$ *solves the constrained maximization problem.*

Proof: The proof just requires the first-order conditions to be reduced to

$$f'(\mathbf{x}^*) + \sum_{i=1}^{b} \lambda_i g^{i\prime}(\mathbf{x}^*) = \mathbf{0} \tag{10.126}$$

from where it is virtually identical to that for the Lagrange case, and so it is left as an exercise; see Exercise 10.23. □

THEOREM 10.5.3 (UNIQUENESS CONDITION FOR INEQUALITY-CONSTRAINED OPTIMIZATION). *If*

(a) $\mathbf{x}^*$ *is a solution;*

(b) *f is strictly quasi-concave; and*

(c) g^j *is a quasi-concave function for* $j = 1, 2, \ldots, m$,

then $\mathbf{x}^*$ *is the unique (global) optimal solution.*

Proof: The proof is again similar to that for the Lagrange case and is left as an exercise; see Exercise 10.23. The point to note this time is that the feasible set with equality constraints is convex if the constraint functions are affine, whereas the feasible set with inequality constraints is convex in the more general situation in which the constraint functions are quasi-concave. This is because the feasible set (where all the inequality constraints are satisfied simultaneously) is the intersection of m upper contour sets of quasi-concave functions, or the intersection of m convex sets. □

10.5.2 Theorem of the maximum

The last important result on optimization, the theorem of the maximum, is closely related to the envelope theorem. This theorem gives sufficient conditions for the optimal response $\mathbf{x}^*(\boldsymbol{\alpha})$ to change continuously with the parameters $\boldsymbol{\alpha}$, something that greatly facilitates further analysis of economic behaviour in particular. Before proceeding to the statement of the theorem, the reader may want to review Definition 7.8.5.

THEOREM 10.5.4 (THEOREM OF THE MAXIMUM). *Consider the modified inequality-constrained optimization problem:*

$$\max_{\mathbf{x}} f(\mathbf{x}, \boldsymbol{\alpha}) \quad \text{subject to } g^i(\mathbf{x}, \boldsymbol{\alpha}) \geq 0, \ i = 1, 2, \ldots, m \tag{10.127}$$

where $\mathbf{x} \in \mathbb{R}^n$, $\boldsymbol{\alpha} \in \mathbb{R}^q$, $f : \mathbb{R}^{n+q} \to \mathbb{R}$ *and* $g : \mathbb{R}^{n+q} \to \mathbb{R}^m$.

Let $\mathbf{x}^*(\boldsymbol{\alpha})$ *denote the optimal choice of* $\mathbf{x}$ *for given* $\boldsymbol{\alpha}$ $(\mathbf{x}^* : \mathbb{R}^q \to \mathbb{R}^n)$ *and let* $M(\boldsymbol{\alpha})$ *denote the maximum value attainable by* f *for given* $\boldsymbol{\alpha}$ $(M : \mathbb{R}^q \to \mathbb{R})$.

If

(a) *f is continuous; and*
(b) *the constraint set (feasible set)*

$$G(\boldsymbol{\alpha}) \equiv \{\mathbf{x} \in \mathbb{R}^n : g^i(\mathbf{x}, \boldsymbol{\alpha}) \geq 0, \ i = 1, 2, \ldots, m\} \tag{10.128}$$

is a non-empty, compact-valued,[7] continuous correspondence of $\boldsymbol{\alpha}$,

then

(a) *M is a continuous (single-valued) function; and*
(b) *$\mathbf{x}^*$ is a non-empty, compact-valued, upper hemi-continuous correspondence, and, hence, is continuous if it is a (single-valued) function (e.g. if the uniqueness conditions of Theorem 10.5.3 are satisfied).*

Proof: The proof of this theorem is beyond the scope of this book, but can be found in de la Fuente (2000, pp. 301–3). The original version of the theorem of the maximum is credited to Berge (1959), whose original proof is available in an English translation in Berge (1997, Chapter VI, Section 3). Note, however, that the existence of the envelope function M is guaranteed by Theorem 10.3.1.[8] □

Theorem 10.5.4 will be used in consumer theory to prove such critical results as the continuity of demand functions derived from the maximization of continuous utility functions.

10.5.3   Examples

The following are two frequently encountered examples illustrating the use of the Kuhn–Tucker theorems in economics. (The calculations are left as exercises; see Exercises 10.21 and 10.25.)

1. Maximizing a Cobb–Douglas utility function subject to a budget constraint and non-negativity constraints. The applications of this problem in later chapters will include choice under certainty, choice under uncertainty with logarithmic utility where the parameters are re-interpreted as probabilities, the extension to Stone–Geary preferences (see Exercise 12.3), and inter-temporal choice with logarithmic utility, where the parameters are re-interpreted as time discount factors.
2. The canonical quadratic programming problem, which merits a separate treatment in Section 14.3, as it has so many applications throughout economics, econometrics and finance.

Further exercises consider the duals of each of the afore-mentioned problems, and it is to the question of duality that we now turn.

10.6   Duality

Let $X \subseteq \mathbb{R}^n$ be a convex set and let $f, g : X \to \mathbb{R}$ be, respectively, pseudo-concave and pseudo-convex functions. Consider the envelope functions defined by the **dual** families of inequality-constrained problems:

$$M(\alpha) \equiv \max_{\mathbf{x} \in X} f(\mathbf{x}) \quad \text{s.t. } g(\mathbf{x}) \le \alpha \tag{10.129}$$

and

$$N(\beta) \equiv \min_{\mathbf{x} \in X} g(\mathbf{x}) \quad \text{s.t. } f(\mathbf{x}) \ge \beta \tag{10.130}$$

Suppose that each of these problems has a unique solution, say $\mathbf{x}^*(\alpha)$ and $\mathbf{x}^\dagger(\beta)$, respectively, and that the constraints bind at these points. Let the corresponding Kuhn–Tucker multipliers be denoted $\lambda(\alpha)$ and $\mu(\beta)$, respectively.

The first-order conditions for the two families of problems are, respectively,

$$f'(\mathbf{x}) - \lambda(\alpha)g'(\mathbf{x}) = \mathbf{0}^\top \tag{10.131}$$

and

$$-g'(\mathbf{x}) + \mu(\beta)f'(\mathbf{x}) = \mathbf{0}^\top \tag{10.132}$$

Thus if $\mathbf{x}^*$ and $\lambda^* \ne 0$ solve (10.131), then $\mathbf{x}^*$ and $\mu^* \equiv 1/\lambda^*$ solve (10.132).

Furthermore, for the $\mathbf{x}^*$ that solves the original problem, problem (10.129), to also solve problem (10.130), it must also satisfy the constraint, or $f(\mathbf{x}^*) = \beta$. We know, however, that $f(\mathbf{x}^*) = M(\alpha)$. This allows us to conclude that $M(\alpha) = \beta$. Similarly, $N(\beta) = \alpha$.

Combining these equations leads to the conclusion that

$$\alpha = N(M(\alpha)) \tag{10.133}$$

and

$$\beta = M(N(\beta)) \tag{10.134}$$

We also have

$$\mathbf{x}^*(\alpha) = \mathbf{x}^\dagger(M(\alpha)) \tag{10.135}$$

and

$$\mathbf{x}^\dagger(\beta) = \mathbf{x}^*(N(\beta)) \tag{10.136}$$

In other words, the envelope functions for the two dual problems are inverse functions (over any range where the Kuhn–Tucker multipliers are non-zero, i.e. where the constraints are binding). Thus, either α or β, or indeed λ or μ, can be used to parametrize either family of problems.

We will see many examples of these principles in the applications in the next part of the book. In particular, duality will be covered in more detail in the context of its applications to consumer theory in Section 12.4.8.

EXERCISES

10.1 Suppose A is an affine set and

$$f(\lambda \mathbf{x} + (1-\lambda)\mathbf{x}') = \lambda f(\mathbf{x}) + (1-\lambda)f(\mathbf{x}')$$

for all $\lambda \in [0, 1]$ and for all $\mathbf{x} \in A$. Prove that this equality also holds for $\lambda < 0$ and for $\lambda > 1$.

10.2 Let X be a convex set and let $f^1 \colon X \to \mathbb{R}$, $f^2 \colon X \to \mathbb{R}, \ldots, f^m \colon X \to \mathbb{R}$ be concave functions. Prove that:

(a) if $a, b > 0$, then $af^1 + bf^2$ is concave;
(b) if $k_1 > 0, k_2 > 0, \ldots, k_m > 0$, then $\sum_{i=1}^m k_i f^i$ is concave;
(c) if $a < 0$, then af^1 is convex;
(d) $\min\{f^1, f^2, \ldots, f^m\}$ is concave; and
(e) the set $\{\mathbf{x} \in X \colon f^1(\mathbf{x}) \geq \alpha\}$ is convex (i.e. Theorem 10.2.2).

10.3 Write down the properties corresponding to those in Exercise 10.2 that would hold if $f^1 \colon X \to \mathbb{R}$, $f^2 \colon X \to \mathbb{R}, \ldots, f^m \colon X \to \mathbb{R}$ were *convex* functions.

10.4 Let $X, Y \subseteq \mathbb{R}$ be convex sets. Show that:

(a) the inverse of a strictly increasing concave function $f \colon X \to Y$ is convex; and
(b) the inverse of a strictly decreasing concave function $g \colon X \to Y$ is concave.

Deduce the corresponding properties of strictly increasing and strictly decreasing convex functions and give an example for each of the four categories.

10.5 Prove, without using Taylor's theorem, that if $f \colon \mathbb{R} \to \mathbb{R}$ with $f''(x) \leq 0$ for all $x \in \mathbb{R}$, then f is a concave function.

10.6 Sketch the graph of a differentiable function $f \colon \mathbb{R} \to \mathbb{R}$ in each of the following categories:

(a) concave, but not strictly quasi-concave;
(b) strictly concave;
(c) concave and strictly quasi-concave, but not strictly concave;
(d) pseudo-concave and strictly quasi-concave, but not concave;
(e) strictly quasi-concave, but not pseudo-concave;
(f) quasi-concave, but neither pseudo-concave nor strictly quasi-concave; and
(g) pseudo-concave, but neither concave nor strictly quasi-concave.

Write out a definition of each function in the form

$$x \mapsto \begin{cases} f_1(x) & \text{if } \ldots \\ f_2(x) & \text{if } \ldots \\ \ldots & \text{if } \ldots \end{cases}$$

(Hint: use functions that are piecewise linear/affine and/or piecewise quadratic.)

10.7 Draw a Venn diagram to illustrate the hierarchy among different classes of functions set out in Table 10.2.2, and add to it the other classes of functions introduced later in the chapter.

10.8 Prove that a twice differentiable non-decreasing concave transformation of a twice differentiable concave function (of several variables) is concave (Theorem 10.2.6).

10.9 Suppose that, for all $\mathbf{x}, \mathbf{x}' \in X$ such that $f(\mathbf{x}') \le f(\mathbf{x})$ and for all $\lambda \in (0, 1)$,

$$f(\lambda \mathbf{x} + (1 - \lambda)\mathbf{x}') \ge f(\mathbf{x}')$$

Prove that, for all $\mathbf{x}, \mathbf{x}' \in X$, and for all $\lambda \in (0, 1)$,

$$f(\lambda \mathbf{x} + (1 - \lambda)\mathbf{x}') \ge \min\{f(\mathbf{x}), f(\mathbf{x}')\}$$

10.10 Show that $f: \mathbb{R} \to \mathbb{R}: x \mapsto -x^{2n}$ defines a strictly concave function (even though the second-order sufficient condition for concavity is not satisfied at $x = 0$).

10.11 Consider the quadratic form defined by the function

$$f: \mathbb{R}^2 \to \mathbb{R}: (x_1, x_2) \mapsto x_1^2 + 2bx_1x_2 + cx_2^2$$

where b and c are real numbers (i.e. the special case of Exercise 9.2 in which $a = 1$). Let

$$\mathbf{A} \equiv \begin{bmatrix} 1 & b \\ b & c \end{bmatrix}$$

(a) For what values of b and c (if any) is the matrix $\mathbf{A}$:

   (i) positive definite;
   (ii) positive semi-definite;
   (iii) negative definite; or
   (iv) negative semi-definite?

   Graph the relevant values in the bc plane in each case.
(b) For what values of b and c (if any) is the function f:

   (i) concave; or
   (ii) convex?

(c) Find the equations of two lines in the x_1x_2 plane along which the value of f is zero when $b = 3$ and $c = 5$, and graph these lines.
(d) Show that there exists a line in the x_1x_2 plane such that the restriction of f to that line is a concave function when $b = 3$ and $c = 5$.

10.12 Consider the general quadratic form:

$$f: \mathbb{R}^n \to \mathbb{R}: \mathbf{x} \mapsto \mathbf{x}^\top \mathbf{A} \mathbf{x}$$

where $\mathbf{A}$ is any $n \times n$ real symmetric matrix.

(a) Calculate the total derivative $f'(\mathbf{x})$ and the Hessian matrix $f''(\mathbf{x})$ of f at the point $\mathbf{x}$.
(b) Give sufficient conditions on $\mathbf{A}$ for f to be: (i) a concave function; (ii) a strictly concave function; (iii) a convex function; and (iv) a strictly convex function. In each case, state whether your conditions are also necessary.

10.13 Consider the Cobb–Douglas function:

$$f:\mathbb{R}_{++}^2 \to \mathbb{R}_{++}:(x, y) \mapsto x^\alpha y^{1-\alpha}$$

where $\mathbb{R}_{++} \equiv (0, \infty)$.

(a) Calculate the total derivative $f'(x_0, y_0)$ and the Hessian matrix $f''(x_0, y_0)$ of f at the point (x_0, y_0).
(b) Calculate the determinant, eigenvalues and eigenvectors of the matrix $f''(x_0, y_0)$.
(c) For what values of α is f:

  (i) a concave function; or
  (ii) a strictly concave function?

10.14 Using the previous exercise, or otherwise, show that

$$h:\mathbb{R}_{++}^2 \to \mathbb{R}_{++}:(x, y) \mapsto \alpha \ln x + (1 - \alpha) \ln y$$

is a concave function for $0 < \alpha < 1$. Is this function strictly concave?

10.15 Find the values of a for which the functions $f:\mathbb{R} \to \mathbb{R}:x \mapsto e^{ax}$ and $g:\mathbb{R} \to \mathbb{R}:x \mapsto -e^{ax}$ are

(a) concave;
(b) convex;
(c) quasi-concave; and
(e) quasi-convex.

10.16 Let $f:X \to \mathbb{R}$ be quasi-concave and $g:\mathbb{R} \to \mathbb{R}$ be increasing. Prove that $g \circ f$ is a quasi-concave function (Theorem 10.2.8).

10.17 Consider the **constant elasticity of substitution** (CES) function:

$$f:\mathbb{R}_{++}^2 \to \mathbb{R}_{++}:(x, y) \mapsto (\alpha x^\rho + \beta y^\rho)^{1/\rho}$$

What are the most general sufficient conditions that you can find on the values of the parameters α, β and ρ for which this function is increasing in both variables and concave?
  For what additional values of the parameters is the function quasi-concave?
  Be careful to consider the limiting behaviour of the function as $\rho \to 0$, 1, $+\infty$ and $-\infty$.
(The origin of the name of this function will be familiar to readers who have done courses in production theory; see Varian (1992, pp. 19–21) for more details.)

10.18 Find the stationary points of the function defined by

$$f:\mathbb{R}^2 \to \mathbb{R}: (x, y) \mapsto x^3 - 3x^2 - 4y^2$$

and determine for each one whether it is a maximum, a minimum, an inflexion point or a saddle point.

Hence, sketch the indifference map of this function.

10.19 Suppose $\mathbf{x}^*$ solves problem (10.55). Partition the vector $\mathbf{x}^*$ as $(\mathbf{y}^*, \mathbf{z}^*)$, where $\mathbf{y}^* \in \mathbb{R}^m$ and $\mathbf{z}^* \in \mathbb{R}^{n-m}$. Define a new objective function $F: Z \to \mathbb{R}$ by

$$F(\mathbf{z}) \equiv f(h(\mathbf{z}), \mathbf{z})$$

where h is a function (the existence of which is guaranteed by the implicit function theorem) such that

$$g(h(\mathbf{z}), \mathbf{z}) = \mathbf{0} \quad \forall \mathbf{z} \in Z$$

Show, using a proof by contradiction argument, that $\mathbf{z}^*$ solves the unconstrained problem $\max_{\mathbf{z} \in Z} F(\mathbf{z})$.

10.20 State and prove the equivalents of Theorems 10.4.1, 10.4.2 and 10.4.3 for minimization subject to equality constraints.

10.21 Consider the Cobb–Douglas function:

$$f:\mathbb{R}^2_{++} \to \mathbb{R}_{++}: (x, y) \mapsto x^\alpha y^{1-\alpha}$$

where $\mathbb{R}_{++} \equiv (0, \infty)$. Find the maximum value that f can take subject to the following constraints:

(a) $px + qy = M$
(b) $px + qy \le M$

where p, q and M are constants.

10.22 A consumer's utility function is given by

$$u(x, y) = x^\alpha y^\beta$$

where x and y are the quantities of each of two goods consumed.

(a) Assuming that the consumer always spends all his money income, M per period, derive his utility-maximizing consumption of the two goods as functions of their prices p_x and p_y (both constant for the consumer) and his income.
(b) Confirm that the second-order conditions for equality-constrained maximization are satisfied.

(c) Solve for the value of the Lagrange multiplier in the previous part, in terms of p_x, p_y and M, and show that it equals the value of $\partial u^*/\partial M$, where u^* is the optimal utility, expressed as a function of p_x, p_y and M.

(d) What would the consumer's utility-maximizing demands be if his utility function was given by the following?

$$v(x, y) = \alpha \ln x + \beta \ln y$$

10.23 Prove Theorems 10.5.2 and 10.5.3.

10.24 Let the objective function $F: \mathbb{R}^{p+q} \to \mathbb{R}$ and the constraint function $G: \mathbb{R}^{p+q} \to \mathbb{R}^r$ satisfy the assumptions of the Lagrange multiplier theorems, and define

$$M: \mathbb{R}^q \to \mathbb{R}$$

$$: \mathbf{a} \mapsto \max_{\{\mathbf{x} \in \mathbb{R}^p: G^i(\mathbf{x}, \mathbf{a}) = 0, \ i = 1, 2, ..., r\}} F(\mathbf{x}, \mathbf{a})$$

State and prove in full the relation between the partial derivative $\partial M/\partial a_j$, the partial derivatives of F and G, and the Lagrange multipliers, $\lambda_1, \lambda_2, \ldots, \lambda_r$ (i.e. the envelope theorem).

10.25 Consider the following canonical quadratic programming problem.

Find the vector $\mathbf{x} \in \mathbb{R}^n$ that maximizes the value of the quadratic form $\mathbf{x}^\top \mathbf{A} \mathbf{x}$ subject to the m linear inequality constraints $\mathbf{g}^{i\top} \mathbf{x} \geq \alpha_i$, where $\mathbf{A}$ is an $n \times n$ negative definite matrix, $m < n$ and $\mathbf{g}^i \in \mathbb{R}^n$ for $i = 1, 2, \ldots, m$.

Show that the objective function can always be rewritten as a quadratic form in a symmetric (negative definite) matrix and, hence, solve the problem.

(This problem will be considered in greater detail in Chapter 14.)

10.26 Derive and graph the envelope functions for the following pairs of dual inequality-constrained optimization problems:

(a) (i) $\min_{\mathbf{x}} \mathbf{x}^\top \mathbf{A} \mathbf{x}$ subject to $\mathbf{e}^\top \mathbf{x} \geq \mu$ and
   (ii) $\max_{\mathbf{x}} \mathbf{e}^\top \mathbf{x}$ subject to $\mathbf{x}^\top \mathbf{A} \mathbf{x} \leq \sigma^2$,
   where $\mathbf{e}, \mathbf{x} \in \mathbb{R}^n$, $\mathbf{e} \neq \mathbf{0}$, $\mathbf{A} \in \mathbb{R}^{n \times n}$ is positive definite and $\mu, \sigma \in \mathbb{R}$;

(b) (i) $\max_{x_1, x_2, \ldots, x_n} \sum_{i=1}^n \alpha_i \ln x_i$ subject to $\sum_{i=1}^n p_i x_i \leq M$ and
   (ii) $\min_{x_1, x_2, \ldots, x_n} \sum_{i=1}^n p_i x_i$ subject to $\sum_{i=1}^n \alpha_i \ln x_i \geq u$,
   where $\alpha_i, x_i \in \mathbb{R}_{++}$ for $i = 1, 2, \ldots, n$, $M \in \mathbb{R}_{++}$ and $u \in \mathbb{R}$.

How would your solutions change, respectively, if:

(a) $\mathbf{A}$ was not positive definite; or
(b) $\alpha_i \leq 0$ for some i?

Part II
APPLICATIONS

Introduction

The first part of this book has presented a large (though by no means exhaustive) body of mathematics that is of relevance for both theoretical and empirical work in economics and finance. A number of examples, particularly those in Chapter 1 and Chapter 10, were used to illustrate the ways in which mathematics is useful. The examples showed that there are benefits from adopting certain mathematical notation; they indicated how various problems may be formulated mathematically; and they motivated the need for methods for the solution of these problems and, hence, for a thorough understanding of certain aspects of mathematics. In Part II of the book, armed with this body of mathematical knowledge, we return to the matter of applications and investigate – with considerable mathematical rigour – a selection of issues from macroeconomics, microeconomics, econometrics and finance.

In Chapter 11, which draws mainly on matrix algebra, dynamic linear macroeconomic models and input–output models of an economy are discussed. These models constitute generalizations of two of the simple models introduced in Chapter 1; their analysis involves the application of several matrix theorems and leads to results that have important economic interpretations.

The focus switches to microeconomics in Chapter 12, where consumer theory, general equilibrium theory and welfare economics are examined. The main mathematical tools used are those from vector calculus, convexity and optimization, though these in turn draw on a good deal of linear algebra. Among the topics covered are utility, demand and expenditure functions and their properties; fixed-point theorems and the existence of equilibrium; and welfare theorems, complete markets and the representative agent approach.

Chapter 13 introduces some basic probability and statistical theory, mainly for later use in dealing with applications in finance and econometrics, but it also features certain immediate applications to such matters as lotteries, options, spread betting, pari-mutuel and exchange betting. Important results for later use include, in particular, a discussion of vector spaces of random variables, the stochastic version of Taylor's theorem, and Jensen's inequality. The econometric applications are presented in Chapter 14. Drawing on some statistics – and material on matrices, vector geometry and difference equations – this chapter deals with the generic quadratic programming problem, the algebra and geometry of ordinary least squares, restricted least squares, and univariate and multivariate autoregressive processes.

Decision-making over time, but under the assumption of certainty, is the subject of Chapter 15. This chapter builds on the material on single-period choice under certainty in Chapter 12. It extends the idea of equilibrium to multi-period general equilibrium and, using a largely arithmetic approach, examines the measurement of rates of return, theories of the term structure of interest rates, and the duration, volatility and convexity of bonds.

Further substantive discussions of financial applications are contained in Chapters 16 and 17. Chapter 16 deals with topics involving single-period choice under uncertainty, such as pricing state-contingent claims, complete markets, expected utility, risk aversion, arbitrage, risk neutrality and the mean–variance paradigm. The discussion makes considerable use of the mathematics of vector spaces, and convexity and concavity, as well as of some statistical methods and results. Finally, Chapter 17 gives a detailed account of important topics in portfolio theory, including the derivation of the mean–variance portfolio frontier, market equilibrium and the capital asset pricing model. The treatment again makes use of vector space ideas and statistics, and the technique of quadratic programming is also employed.

11 Macroeconomic applications

11.1 Introduction

The macroeconomic applications considered in this chapter have already been introduced in Chapter 1. Section 1.2.2 presented a simple macroeconomic model of a closed economy

$$C = f(Y) = \alpha_1 + \alpha_2 Y \tag{11.1}$$

$$I = g(Y, R) = \beta_1 + \beta_2 Y + \beta_3 R \tag{11.2}$$

$$Y = C + I + G \tag{11.3}$$

where C, I and Y are the endogenous variables, consumption, investment and national income, respectively, and R and G are the exogenous variables, rate of interest and government expenditure, respectively. The problem in this case was to solve this model for the endogenous variables in terms of the exogenous variables. Thus we motivated our study of the solution of systems of linear equations.

Section 1.2.3 introduced the Leontief input–output system and posed another kind of solution problem, namely, that of solving the system

$$f_i + a_{i1}x_1 + a_{i2}x_2 + \cdots + a_{in}x_n = x_i, \quad i = 1, 2, \ldots, n \tag{11.4}$$

for the total industrial outputs, x_i, where the f_i denote the final consumer demands for the outputs, and the $a_{ij}x_j$ denote the intermediate demands by industries. The a_{ij} are the technological input–output coefficients, which indicate the demand for input i per unit of output produced by industry j.

In this chapter, we consider these applications in more detail. Moreover, in the case of the first, we shall also generalize the example in certain ways, drawing on the material on difference equations in Chapter 8. Both of the problems posed will be addressed in general terms, their solutions will be obtained and a variety of associated mathematical facts and economic interpretations established.

11.2 Dynamic linear macroeconomic models

This section develops the example in Section 1.2.2 by considering, more generally, aspects of the analysis of macroeconomic models. Macroeconomic modelling has been of importance for many years, not only for purposes of testing economic theory but also for policy simulation and, especially, forecasting. The use of macroeconomic models dates from the early work of Tinbergen (1937, 1939) and has become widespread. Applications include the simple Klein and Goldberger (1955) model, the very large Brookings model of the US economy, the Treasury model and the London Business School model in the UK, and the

Central Bank, the Department of Finance and the Economic and Social Research Institute models in Ireland.

Most macroeconomic models have been basically Keynesian, with representations of the goods, labour and money markets of an economy. The goods market typically comprises consumption, investment, export (X) and import (M) functions, as well as a national income identity $(Y = C + I + G + X - M)$. The labour market might include a production function, labour supply function and a marginal productivity condition, while the money market is based essentially on an IS function (arising from an investment and saving equilibrium), an LM function (arising from a liquidity preference and money supply equilibrium) and an equilibrium condition. Like these, and the model in Section 1.2.2, the type of model we are about to consider may be thought of as Keynesian in nature.

The earlier example led us to define a static macroeconomic model as

$$\mathbf{Ax} = \mathbf{Bz} \tag{11.5}$$

where $\mathbf{A}$ is a square (say, $m \times m$) matrix of structural parameters associated with m endogenous variables, and $\mathbf{B}$ is an $m \times n$ matrix of structural parameters associated with n exogenous variables in the system. Equation (11.5) is called the **structural form** of the system. The fact that $\mathbf{A}$ is square means that there are as many equations in (11.5) as there are endogenous variables. Such a system is said to be **complete**. As mentioned previously, our concern is with the solubility of systems like these, where the solution is for the endogenous variables, $\mathbf{x}$, in terms of the exogenous variables, $\mathbf{z}$. From our discussion in Chapter 2, we know that a solution will exist if $\mathbf{A}$ is non-singular.

However, we wish to generalize somewhat by considering a complete dynamic, rather than just a static, system. To illustrate simply, suppose we formulate the following model of a closed economy:

$$C_t = f(Y_t) = \alpha_1 + \alpha_2 Y_t \tag{11.6}$$

$$I_t = g(Y_{t-1}, R_t) = \beta_1 + \beta_2 Y_{t-1} + \beta_3 R_t \tag{11.7}$$

$$Y_t = C_t + I_t + G_t \tag{11.8}$$

where the variables are as defined previously, with subscripts added to indicate time. Thus Y_{t-1} denotes a one-period lagged value of national income, whereas all other variables take their current value. Although otherwise very similar to the previous model, this difference is a most important one. The use of a lagged value of an endogenous variable as an explanatory variable in an equation introduces a dynamic component into the model, which has significant behavioural consequences, as well as consequences for the solution of the system. In what follows we distinguish between lagged values of endogenous variables and exogenous variables, even though the lagged values are given for the current period, having been determined by the operation of the system in the past. The term **predetermined variables** subsumes both lagged endogenous and exogenous variables.

It is easy to see that we may rewrite equations (11.6) to (11.8) as

$$\begin{bmatrix} 1 & 0 & -\alpha_2 \\ 0 & 1 & 0 \\ -1 & -1 & 1 \end{bmatrix} \begin{bmatrix} C_t \\ I_t \\ Y_t \end{bmatrix} + \begin{bmatrix} 0 & 0 & 0 \\ 0 & 0 & -\beta_2 \\ 0 & 0 & 0 \end{bmatrix} \begin{bmatrix} C_{t-1} \\ I_{t-1} \\ Y_{t-1} \end{bmatrix}$$

$$= \begin{bmatrix} \alpha_1 & 0 & 0 \\ \beta_1 & \beta_3 & 0 \\ 0 & 0 & 1 \end{bmatrix} \begin{bmatrix} 1 \\ R_t \\ G_t \end{bmatrix} \tag{11.9}$$

or, more compactly, as

$$\mathbf{A}_0 \mathbf{x}_t + \mathbf{A}_1 \mathbf{x}_{t-1} = \mathbf{B} \mathbf{z}_t \qquad (11.10)$$

Hence, if $\mathbf{A}_0$ is non-singular, we have for our solution that

$$\mathbf{x}_t = -\mathbf{A}_0^{-1} \mathbf{A}_1 \mathbf{x}_{t-1} + \mathbf{A}_0^{-1} \mathbf{B} \mathbf{z}_t$$
$$\equiv \mathbf{\Pi}_1 \mathbf{x}_{t-1} + \mathbf{\Pi}_0 \mathbf{z}_t \qquad (11.11)$$

This is the **reduced form** of the dynamic system and the matrices $\mathbf{\Pi}_0$ and $\mathbf{\Pi}_1$ contain the reduced-form parameters as functions of the structural parameters.

The matrix $\mathbf{\Pi}_0$ contains the coefficients that measure the immediate change in endogenous variables in response to a change in an exogenous variable. For example, $\pi_{0_{33}}$ (the third element in the third row of $\mathbf{\Pi}_0$) is $\partial Y / \partial G$ and $\pi_{0_{12}}$ is $\partial C / \partial R$. Such coefficients are called **impact multipliers**, and $\mathbf{\Pi}_0$ is, therefore, called the **impact multiplier matrix**.

If a change in an exogenous variable occurs at time t and is maintained, we can trace out the effects on endogenous variables in subsequent periods, making time explicit, as follows. For period t,

$$\mathbf{x}_t = \mathbf{\Pi}_1 \mathbf{x}_{t-1} + \mathbf{\Pi}_0 \mathbf{z}_t \qquad (11.12)$$

where the immediate effect is given via the impact multipliers in $\mathbf{\Pi}_0$. Equation (11.12) is a non-homogeneous system of linear autonomous first-order difference equations in $\mathbf{x}_t$. For period $t + 1$:

$$\mathbf{x}_{t+1} = \mathbf{\Pi}_1 \mathbf{x}_t + \mathbf{\Pi}_0 \mathbf{z}_{t+1}$$
$$= \mathbf{\Pi}_1 (\mathbf{\Pi}_1 \mathbf{x}_{t-1} + \mathbf{\Pi}_0 \mathbf{z}_t) + \mathbf{\Pi}_0 \mathbf{z}_{t+1}$$
$$= \mathbf{\Pi}_1^2 \mathbf{x}_{t-1} + \mathbf{\Pi}_1 \mathbf{\Pi}_0 \mathbf{z}_t + \mathbf{\Pi}_0 \mathbf{z}_{t+1} \qquad (11.13)$$

So the effect after one period is given via

$$\mathbf{\Pi}_1 \mathbf{\Pi}_0 + \mathbf{\Pi}_0 = (\mathbf{\Pi}_1 + \mathbf{I}) \mathbf{\Pi}_0 \qquad (11.14)$$

For period $t + 2$,

$$\mathbf{x}_{t+2} = \mathbf{\Pi}_1 \mathbf{x}_{t+1} + \mathbf{\Pi}_0 \mathbf{z}_{t+2}$$
$$= \mathbf{\Pi}_1 (\mathbf{\Pi}_1^2 \mathbf{x}_{t-1} + \mathbf{\Pi}_1 \mathbf{\Pi}_0 \mathbf{z}_t + \mathbf{\Pi}_0 \mathbf{z}_{t+1}) + \mathbf{\Pi}_0 \mathbf{z}_{t+2}$$
$$= \mathbf{\Pi}_1^3 \mathbf{x}_{t-1} + \mathbf{\Pi}_1^2 \mathbf{\Pi}_0 \mathbf{z}_t + \mathbf{\Pi}_1 \mathbf{\Pi}_0 \mathbf{z}_{t+1} + \mathbf{\Pi}_0 \mathbf{z}_{t+2} \qquad (11.15)$$

The effect after two periods is given via

$$\mathbf{\Pi}_1^2 \mathbf{\Pi}_0 + \mathbf{\Pi}_1 \mathbf{\Pi}_0 + \mathbf{\Pi}_0 = (\mathbf{\Pi}_1^2 + \mathbf{\Pi}_1 + \mathbf{I}) \mathbf{\Pi}_0 \qquad (11.16)$$

and so on, such that the effect on the endogenous variables after j periods is determined via

$$(\mathbf{\Pi}_1^j + \cdots + \mathbf{\Pi}_1^2 + \mathbf{\Pi}_1 + \mathbf{I}) \mathbf{\Pi}_0 = \left(\sum_{i=0}^{j} \mathbf{\Pi}_1^i \right) \mathbf{\Pi}_0 \equiv \mathbf{D}_j \qquad (11.17)$$

These various effects for different $j > 0$ are known as **dynamic multipliers**. We may call the $\mathbf{D}_j$ the **dynamic multiplier matrix** of order j; its elements measure the changes in endogenous variables in response to a maintained change in an exogenous variable over j periods. For example, d_{j33} is $\partial Y_{t+j}/\partial G_t$.

If the matrix $\mathbf{\Pi}_1$ is such that $\mathbf{\Pi}_1^j \to \mathbf{0}$ as $j \to \infty$, then equilibrium will be approached, the total long-run effects on the endogenous variables being given by the elements of

$$\left(\sum_{i=0}^{\infty} \mathbf{\Pi}_1^i \right) \mathbf{\Pi}_0 = (\mathbf{I} - \mathbf{\Pi}_1)^{-1} \mathbf{\Pi}_0 \equiv \mathbf{E} \tag{11.18}$$

the matrix of **equilibrium multipliers**. The matrix $\mathbf{E}$ is the limit of $\mathbf{D}_j$ as $j \to \infty$. One of its components is seen to be the matrix generalization of the result on the sum of a geometric progression

$$\sum_{i=0}^{\infty} \mathbf{\Pi}_1^i = (\mathbf{I} - \mathbf{\Pi}_1)^{-1} \tag{11.19}$$

The convergence condition required for this inverse to exist was stated and proved in Section 8.5.2 on systems of difference equations and their dynamic properties. For the moment, if we suppose the condition is satisfied, we may derive the result in the following alternative way.

If the system settles down to a steady-state equilibrium with $\mathbf{x}_t = \mathbf{x}^*$ and $\mathbf{z}_t = \mathbf{z}^*$ for all t, then

$$\mathbf{x}^* = \mathbf{\Pi}_1 \mathbf{x}^* + \mathbf{\Pi}_0 \mathbf{z}^* \tag{11.20}$$

It follows immediately that

$$\mathbf{x}^* - \mathbf{\Pi}_1 \mathbf{x}^* = \mathbf{\Pi}_0 \mathbf{z}^* \tag{11.21}$$

$$(\mathbf{I} - \mathbf{\Pi}_1) \mathbf{x}^* = \mathbf{\Pi}_0 \mathbf{z}^* \tag{11.22}$$

$$\mathbf{x}^* = (\mathbf{I} - \mathbf{\Pi}_1)^{-1} \mathbf{\Pi}_0 \mathbf{z}^* \tag{11.23}$$

assuming $(\mathbf{I} - \mathbf{\Pi}_1)$ is non-singular. We see the long-run or equilibrium multiplier effect immediately as $\partial \mathbf{x}^*/\partial \mathbf{z}^* = (\mathbf{I} - \mathbf{\Pi}_1)^{-1} \mathbf{\Pi}_0 \equiv \mathbf{E}$. Each element of $\mathbf{E}$ indicates the change in the equilibrium level of an endogenous variable in response to a maintained change in an exogenous variable.

Recalling Theorem 8.5.1, the condition on $\mathbf{\Pi}_1$ for $\mathbf{\Pi}_1^j \to \mathbf{0}$ as $j \to \infty$ is that all the eigenvalues of $\mathbf{\Pi}_1$ are less than unity in absolute value (or have modulus less than unity for complex eigenvalues). Moreover, the sign and size of these eigenvalues determine the speed of the approach to equilibrium, and whether the approach is monotonic or oscillatory.

If this convergence condition holds, another form of the macroeconomic model called the **final form** may be derived. By repeated substitution for the lagged endogenous variable on the right-hand side of (11.11), as was done in (8.119), the model may be written as

$$\mathbf{x}_t = \mathbf{\Pi}_0 \mathbf{z}_t + \mathbf{\Pi}_1 \mathbf{\Pi}_0 \mathbf{z}_{t-1} + \mathbf{\Pi}_1^2 \mathbf{\Pi}_0 \mathbf{z}_{t-2} + \cdots = \sum_{j=0}^{\infty} \mathbf{\Pi}_1^j \mathbf{\Pi}_0 \mathbf{z}_{t-j} \tag{11.24}$$

Thus the final form gives the current value of each endogenous variable as a function of the current and past values of the exogenous variables only, in contrast to the reduced form, which has lagged endogenous and exogenous variables on the right-hand side. The elements of the matrices $\Pi_1^j \Pi_0$ ($j \geq 1$) associated with the individual variables in the final form give the effect on the current value of endogenous variables of an unsustained change in an exogenous variable j periods in the past. The $\Pi_1^j \Pi_0$ are referred to as **interim multiplier matrices**. From (11.17), it can be seen that the dynamic multiplier $\mathbf{D}_j$ is the sum of the impact multiplier Π_0 and the interim multipliers $\Pi_1^i \Pi_0$, $i = 1, 2, \ldots, j$.

EXAMPLE 11.2.1 Recall the simple macroeconomic model of a closed economy defined by (11.6), (11.7) and (11.8). Suppose the parameters of this system are known, as follows:

$$C_t = 5 + 0.6 Y_t \tag{11.25}$$

$$I_t = 3 + 0.4 Y_{t-1} - 0.2 R_t \tag{11.26}$$

$$Y_t = C_t + I_t + G_t \tag{11.27}$$

Then for the structural form $\mathbf{A}_0 \mathbf{x}_t + \mathbf{A}_1 \mathbf{x}_{t-1} = \mathbf{B} \mathbf{z}_t$ in (11.10), we have

$$\begin{bmatrix} 1 & 0 & -0.6 \\ 0 & 1 & 0 \\ -1 & -1 & 1 \end{bmatrix} \begin{bmatrix} C_t \\ I_t \\ Y_t \end{bmatrix} + \begin{bmatrix} 0 & 0 & 0 \\ 0 & 0 & -0.4 \\ 0 & 0 & 0 \end{bmatrix} \begin{bmatrix} C_{t-1} \\ I_{t-1} \\ Y_{t-1} \end{bmatrix} = \begin{bmatrix} 5 & 0 & 0 \\ 3 & -0.2 & 0 \\ 0 & 0 & 1 \end{bmatrix} \begin{bmatrix} 1 \\ R_t \\ G_t \end{bmatrix}$$

Since

$$\mathbf{A}_0^{-1} = \frac{1}{0.4} \begin{bmatrix} 1 & 0.6 & 0.6 \\ 0 & 0.4 & 0 \\ 1 & 1 & 1 \end{bmatrix} \tag{11.28}$$

solving for the reduced form $\mathbf{x}_t = \Pi_1 \mathbf{x}_{t-1} + \Pi_0 \mathbf{z}_t$ in (11.11), we obtain the matrix of impact multipliers

$$\Pi_0 = \mathbf{A}_0^{-1} \mathbf{B} = \begin{bmatrix} 17 & -0.3 & 1.5 \\ 3 & -0.2 & 0 \\ 20 & -0.5 & 2.5 \end{bmatrix} \tag{11.29}$$

and the matrix

$$\Pi_1 = -\mathbf{A}_0^{-1} \mathbf{A}_1 = \begin{bmatrix} 0 & 0 & 0.6 \\ 0 & 0 & 0.4 \\ 0 & 0 & 1 \end{bmatrix} \tag{11.30}$$

from which the interim and dynamic multipliers may be computed for different values of j. For example, from (11.29), we have that

$$\pi_{0_{33}} = \frac{\partial Y}{\partial G} = 2.5 \tag{11.31}$$

which gives the overall impact, allowing for all interactions in the current period, of a change in government expenditure on national income. Similarly, if we denote the interim multipliers $\mathbf{\Pi}_1^j \mathbf{\Pi}_0$ by $\mathbf{N}_j$, $j = 1, 2, \ldots$, we have

$$n_{1_{33}} = 2.5 \tag{11.32}$$

$$n_{1_{33}} + \pi_{0_{33}} = d_{1_{33}} = 5 \tag{11.33}$$

where (11.32) gives the one-period interim multiplier for government expenditure on national income, and (11.33) gives the one-period dynamic multiplier for government expenditure on national income. It is left as an exercise to compute interim and dynamic multipliers for longer periods; see Exercise 11.2.

However, in this illustration, the equilibrium multipliers do not exist. Insight into this fact is provided by the results that $\mathbf{I} - \mathbf{\Pi}_1$ is singular and $\mathbf{\Pi}_1^2 = \mathbf{\Pi}_1$, both of which may be checked as a final exercise with this application; see Exercise 11.3. Thus the matrix $\mathbf{\Pi}_1$ is idempotent and does not converge to $\mathbf{0}_{3 \times 3}$ as it is raised to higher powers, $j \to \infty$. As $\mathbf{\Pi}_1$ is an upper triangular matrix, its eigenvalues are its diagonal elements, $\lambda_1 = 0$, $\lambda_2 = 0$ and $\lambda_3 = 1$, which is a familiar property of idempotent matrices; recall Exercise 3.20. $\diamond$

11.3 Input–output analysis

As mentioned in Chapter 1, input–output analysis provides an alternative way of describing and analysing an economy to that provided by Keynesian macroeconomic models such as the one examined in the previous section. The background to the input–output approach and the definition of our notation have been given in Chapter 1, while the problem posed by the approach has been restated via (11.4) in the introduction to the present chapter. In matrix form, (11.4) is

$$\mathbf{f} + \mathbf{Ax} = \mathbf{x} \tag{11.34}$$

where

$$\mathbf{f} = [f_i]_{n \times 1}, \quad \mathbf{A} = [a_{ij}]_{n \times n} \quad \text{and} \quad \mathbf{x} = [x_i]_{n \times 1} \tag{11.35}$$

We require the solution for $\mathbf{x}$, the n-vector of total industry outputs, in terms of $\mathbf{f}$, the n-vector of final consumer demands, and $\mathbf{A}$, the matrix of input–output coefficients.

We have noted previously that $f_i \geq 0$ for all i and $a_{ij} \geq 0$ for all i, j. It is sensible to add that $x_i \geq 0$ for all i. Moreover, to ensure a non-negative net output from all industries – net, that is, of industries' demands for their own outputs – we require that

$$x_i - a_{ii}x_i \geq 0 \quad \forall i \tag{11.36}$$

Therefore

$$(1 - a_{ii})x_i \geq 0 \quad \text{and} \quad (1 - a_{ii}) \geq 0 \tag{11.37}$$

which implies that $0 \leq a_{ii} \leq 1$ for all i.

The input–output coefficients in any column of $\mathbf{A}$, say the jth, indicate the amount of every input per unit of output of the corresponding industry, j. They therefore describe the linear production process for industry j.

In approaching the solution of the input–output system, we may, first, use basic matrix operations to rewrite (11.34) as

$$\mathbf{f} = \mathbf{x} - \mathbf{A}\mathbf{x} = \mathbf{I}\mathbf{x} - \mathbf{A}\mathbf{x} = (\mathbf{I} - \mathbf{A})\mathbf{x} \tag{11.38}$$

The typical term of the vector $(\mathbf{I} - \mathbf{A})\mathbf{x}$ in (11.38), namely,

$$x_i - a_{ii}x_i - a_{i1}x_1 - \cdots - a_{i(i-1)}x_{i-1} - a_{i(i+1)}x_{i+1} - \cdots - a_{in}x_n \tag{11.39}$$

is of interest. It denotes the amount of good i left over for consumption by final consumers after all industries, including the ith itself, have taken their quantities of the good for production purposes. Thus for $f_i > 0$, so that some of good i is available to final consumers, the expression in (11.39) must be positive.

The solution we seek is straightforward. If $\mathbf{I} - \mathbf{A}$ is non-singular, then

$$\begin{aligned} \mathbf{x} &= (\mathbf{I} - \mathbf{A})^{-1}\mathbf{f} \\ &\equiv \mathbf{B}\mathbf{f} \equiv f_1\mathbf{b}_1 + f_2\mathbf{b}_2 + \cdots + f_n\mathbf{b}_n \end{aligned} \tag{11.40}$$

where the $\mathbf{b}_i$ are the columns of $(\mathbf{I} - \mathbf{A})^{-1}$. We observe that the solution vector is in the column space of $(\mathbf{I} - \mathbf{A})^{-1}$, i.e. $\mathbf{x} \in \text{lin}\{\mathbf{b}_1, \mathbf{b}_2, \ldots, \mathbf{b}_n\} = \mathbb{R}^n$. We may also note that columns of $(\mathbf{I} - \mathbf{A})^{-1}$ have a useful interpretation as vectors of 'multipliers' with respect to the final demands. For instance,

$$\mathbf{b}_j = \frac{\partial \mathbf{x}}{\partial f_j} = \left[\frac{\partial x_i}{\partial f_j}\right]_{n \times 1} \tag{11.41}$$

which contains the overall effects on the total output from each industry of a unit change in the final consumer demand for good j, allowing for all inter-industry intermediate demands. However, we must look at this solution more carefully and, in particular, examine what is required for the non-singularity of $\mathbf{I} - \mathbf{A}$.

The solution (11.40) may be written in another useful form. By recursively substituting for $\mathbf{x}$ in (11.34), we obtain

$$\begin{aligned} \mathbf{f} + \mathbf{A}\mathbf{x} &= \mathbf{f} + \mathbf{A}(\mathbf{f} + \mathbf{A}\mathbf{x}) = \mathbf{f} + \mathbf{A}\mathbf{f} + \mathbf{A}^2\mathbf{x} \\ &= \mathbf{f} + \mathbf{A}\mathbf{f} + \mathbf{A}^2(\mathbf{f} + \mathbf{A}\mathbf{x}) = \mathbf{f} + \mathbf{A}\mathbf{f} + \mathbf{A}^2\mathbf{f} + \mathbf{A}^3\mathbf{x} \end{aligned} \tag{11.42}$$

Repeating this substitution operation n times, we get

$$\begin{aligned} \mathbf{x} &= \mathbf{f} + \mathbf{A}\mathbf{f} + \mathbf{A}^2\mathbf{f} + \mathbf{A}^3\mathbf{f} + \cdots + \mathbf{A}^n\mathbf{f} + \mathbf{A}^{n+1}\mathbf{x} \\ &= (\mathbf{I} + \mathbf{A} + \mathbf{A}^2 + \mathbf{A}^3 + \cdots + \mathbf{A}^n)\mathbf{f} + \mathbf{A}^{n+1}\mathbf{x} \end{aligned} \tag{11.43}$$

Thus if $\mathbf{A}^{n+1} \to \mathbf{0}$ as $n \to \infty$, we have that

$$\mathbf{x} = (\mathbf{I} + \mathbf{A} + \mathbf{A}^2 + \mathbf{A}^3 + \cdots)\mathbf{f} \tag{11.44}$$

Comparing this solution with the original result in (11.40), we conclude that

$$\mathbf{x} = (\mathbf{I} - \mathbf{A})^{-1}\mathbf{f} = (\mathbf{I} + \mathbf{A} + \mathbf{A}^2 + \mathbf{A}^3 + \cdots)\mathbf{f} \qquad (11.45)$$

and, hence, that

$$(\mathbf{I} - \mathbf{A})^{-1} = (\mathbf{I} + \mathbf{A} + \mathbf{A}^2 + \mathbf{A}^3 + \cdots) = \sum_{i=0}^{\infty} \mathbf{A}^i \qquad (11.46)$$

This is just the matrix series expansion of the inverse $(\mathbf{I} - \mathbf{A})^{-1}$ that we first encountered in Section 8.5.2. The crucial requirement for this to exist is that $\mathbf{A}^i \to \mathbf{0}$ as $i \to \infty$. But this is essentially the same condition we encountered in the previous application involving the dynamic linear model and the matrix $\mathbf{\Pi}_1$; see (11.19) and Theorem 8.5.1. It is therefore left as an exercise to verify that a necessary and sufficient condition for the non-singularity of $\mathbf{I} - \mathbf{A}$, and, hence, for the input–output problem to have a meaningful solution, is that the eigenvalues of $\mathbf{A}$ all have modulus less than one; see Exercise 11.4.[1]

EXAMPLE 11.3.1 Consider the following two-industry case, where the matrix of input–output coefficients is

$$\mathbf{A} = \begin{bmatrix} 0.2 & 0.3 \\ 0.4 & 0.1 \end{bmatrix} \qquad (11.47)$$

Here, for example, industry 1 uses 0.2 of a unit of its own output in producing one unit of its own output; and it uses 0.4 of a unit of industry 2's output to produce one unit of its own output. Note the linear nature of the production process, production function and isoquant map implied by these two numbers. A similar interpretation applies to the numbers in the second column of $\mathbf{A}$.

It follows that

$$\mathbf{I} - \mathbf{A} = \begin{bmatrix} 0.8 & -0.3 \\ -0.4 & 0.9 \end{bmatrix} \qquad (11.48)$$

from which, using (11.39) and assuming, for example, that each industry is producing the same number of units of output, we conclude that 0.5 of a unit of industry 1's output is available for final consumers for every unit of output 1 produced, allowing for industrial demands for industry 1's output. The corresponding figure for industry 2's output is also 0.5.

Further, we have that $\det(\mathbf{I} - \mathbf{A}) = 0.6$,

$$\mathrm{adj}(\mathbf{I} - \mathbf{A}) = \begin{bmatrix} 0.9 & 0.3 \\ 0.4 & 0.8 \end{bmatrix} \qquad (11.49)$$

and

$$(\mathbf{I} - \mathbf{A})^{-1} = \begin{bmatrix} \frac{3}{2} & \frac{1}{2} \\ \frac{2}{3} & \frac{4}{3} \end{bmatrix} \qquad (11.50)$$

This matrix gives the output multipliers. Thus, for example, if final consumer demand for good 1 were to increase by one unit, then total output from industry 1 would need to increase by $\partial x_1 / \partial f_1 = 1\frac{1}{2}$ units to meet that demand and all associated intermediate industrial demands. Similarly, we have that

$$\frac{\partial x_1}{\partial f_2} = \frac{1}{2}, \quad \frac{\partial x_2}{\partial f_1} = \frac{2}{3}, \quad \frac{\partial x_2}{\partial f_2} = \frac{4}{3} \tag{11.51}$$

The matrix in (11.50) also allows us to obtain the solution vector, $\mathbf{x}$, for any given final demands. For example, suppose it is required to provide $f_1 = 10$ and $f_2 = 20$ units of the two goods for consumers, then the two industries would need to produce the amounts given by the elements in

$$\mathbf{x} = (\mathbf{I} - \mathbf{A})^{-1} \mathbf{f} = \begin{bmatrix} \frac{3}{2} & \frac{1}{2} \\ \frac{2}{3} & \frac{4}{3} \end{bmatrix} \begin{bmatrix} 10 \\ 20 \end{bmatrix} = \begin{bmatrix} 25 \\ \frac{100}{3} \end{bmatrix} \tag{11.52}$$

Of the 25 units of output produced by industry 1, $a_{11}x_1 = 0.2 \times 25 = 5$ would be consumed by industry 1 and $a_{12}x_2 = 0.3 \times \frac{100}{3} = 10$ by industry 2, making a total intermediate industrial demand of 15 and the balance of 10 units of output for final consumers, as required. The corresponding decomposition for the $\frac{100}{3}$ units of output produced by industry 2 is obtained in similar fashion.

Having demonstrated the non-singularity of $\mathbf{I} - \mathbf{A}$ and the existence of a solution in this case, it follows that the eigenvalue condition on $\mathbf{A}$ must be satisfied. However, as an exercise, it might be checked that the characteristic equation

$$|\mathbf{A} - \lambda\mathbf{I}| = 0 \tag{11.53}$$

yields the quadratic equation

$$\lambda^2 - 0.3\lambda - 0.1 = 0 \tag{11.54}$$

from which the eigenvalues are determined as

$$\lambda_1 = -0.2 \quad \text{and} \quad \lambda_2 = 0.5 \tag{11.55}$$

both of which are less than unity in absolute value; see Exercise 11.5.

Finally, the convergence implied by these eigenvalues and the non-singularity of $\mathbf{I} - \mathbf{A}$ could be explored directly by computing powers of $\mathbf{A}$. For instance,

$$\mathbf{A}^2 = \begin{bmatrix} 0.16 & 0.09 \\ 0.12 & 0.13 \end{bmatrix} \tag{11.56}$$

and

$$\mathbf{A}^3 = \begin{bmatrix} 0.068 & 0.057 \\ 0.076 & 0.049 \end{bmatrix} \tag{11.57}$$

The rate of convergence of the terms of the matrix series expansion of $(\mathbf{I} - \mathbf{A})^{-1}$ to the zero matrix quickly becomes apparent. $\diamondsuit$

EXERCISES

11.1 Distinguish between the structural form, the reduced form and the final form of a dynamic linear system of equations. Derive the final form of the system from the specific reduced form given in equation (11.11).

11.2 Consider the system given by equations (11.6), (11.7) and (11.8).

(a) Derive the two-period and the three-period dynamic multipliers for national income, Y, with respect to government expenditure, G.
(b) Derive the one-period and the two-period dynamic multipliers for consumption, C, with respect to the rate of interest, R.
(c) Find a condition on the structural parameters such that equilibrium multipliers for this system exist.

11.3 Show that equilibrium multipliers do not exist for the macroeconomic model specified in Example 11.2.

11.4 Let A be a matrix of input–output coefficients. Verify theoretically that, for the non-singularity of $I - A$, and hence for the solubility of the input–output problem, the eigenvalues of A must all be less than one in absolute value, assuming real eigenvalues.

11.5 Consider the input–output matrix given in equation (11.47),

$$A = \begin{bmatrix} 0.2 & 0.3 \\ 0.4 & 0.1 \end{bmatrix}$$

(a) Sketch the isoquants implied by the input–output matrix for both industry 1 and industry 2, i.e. the indifference maps of the functions g_i, where $g_i(x_1, x_2)$ is the maximum amount of good i that can be produced from x_1 units of good 1 and x_2 units of good 2, $i = 1, 2$.
(b) Find the amounts of good 2 used by industry 1 and industry 2, when the final consumer demands for the goods are $f_1 = 15$ and $f_2 = 30$, respectively.
(c) Determine by how much each industry output would need to change if:

 (i) f_1 were to increase from 15 to 20; and
 (ii) f_2 were to decrease from 30 to 25.

(d) Calculate the eigenvalues of the matrix A.

12 Single-period choice under certainty

12.1 Introduction

Economic decisions can always be reduced to optimization problems. The economic reality is that these decisions are invariably subject to resource constraints. This chapter will consider in detail the most basic example, the expenditure decision of an individual consumer, assuming a single time horizon and no risk or uncertainty. Later chapters will relax these assumptions by allowing for multi-period decision-making and decision-making in the face of uncertainty. Decision-making of those kinds is in the realm of finance or financial economics, a particular subfield of microeconomics. The objective function in the consumer's problem will be called a utility function and will often be quasi-concave or concave. We begin by discussing the properties of the utility function using an axiomatic approach to consumer choice.

12.2 Definitions

12.2.1 Economies

There are two possible types of **economy** that we could analyse:

- a **pure exchange economy**, in which households are endowed directly with goods, but there are no firms, there is no production, and economic activity consists solely of pure exchanges of an initial aggregate **endowment**; and
- a **production economy**, in which households are further indirectly endowed with, and can trade, shares in the profit or loss of firms, which can use part of the initial aggregate endowment (including an endowment of labour) as inputs to production processes whose outputs are also available for trade and consumption.

Economies of these types comprise:

- H **households** or **agents** or **consumers** or (later) **investors** or, merely, **individuals**, indexed by the subscript h;
- N **goods** or **commodities**, indexed by the superscript n; and
- (in the case of a production economy only) F **firms**, indexed by f.

This chapter concentrates on the theory of optimal consumer choice and of equilibrium in a pure exchange economy. The theory of optimal production decisions and of equilibrium in an economy with production is mathematically similar; see Varian (1992) or Takayama (1994) for a good mathematical treatment.

Goods can be distinguished from each other (and consequently can differ in value and in cost of production) in many ways:

1. obviously, by intrinsic physical characteristics, e.g. apples or oranges;
2. by the time at which they are consumed, e.g. Christmas decorations or gifts delivered before Christmas day or the same decorations or gifts delivered after Christmas day; an Easter egg delivered before Easter Sunday or an Easter egg delivered after Easter Sunday;[1] and
3. by the state of the world in which they are consumed, e.g. the service provided by an umbrella on a wet day or the service that it provides on a dry day.

This chapter concentrates on the first distinction; the other two are addressed in later chapters.

12.2.2   Prices

In order to exchange goods, consumers must agree on a **price**, and on the units in which that price is to be expressed. In a pure exchange or **barter** system, **relative prices** will be expressed in units of one good per unit of another, e.g. 2.5 apples per orange. In an economy with **money** or **currency**, the **absolute prices** can be expressed in terms of units of currency per unit, e.g. €2.40 per pineapple.

In order to reduce the problem from one of finding absolute prices to one of finding relative prices in a pure exchange economy, we merely choose one commodity or bundle of commodities to be a **numeraire** and express all prices in terms of the price of the numeraire.

There is no money, as such, in the pure exchange economy. In terms of the standard definition of money as something that is both a **medium of exchange** and a **unit of account**:

- the pure exchange economy has no specialized medium of exchange, but
- the numeraire commodity serves as a unit of account.

12.2.3   Consumers and utility functions

The important characteristics of consumer h are that (s)he is faced with the choice of a **consumption vector** or **consumption plan** or **consumption bundle**, $\mathbf{x}_h = (x_h^1, x_h^2, \ldots, x_h^N)$, from a (closed, convex) **consumption set**, $\mathcal{X}_h$. Typically, $\mathcal{X}_h = \mathbb{R}_+^N$. More generally, consumer h's consumption set might require a certain subsistence consumption of some commodities, such as water, and rule out points of $\mathbb{R}_+^N$ not meeting this requirement. The consumer's endowments are denoted $\mathbf{e}_h \in \mathcal{X}_h$ and can be traded. Labour is a good with which most consumers are endowed.

In a production economy, the shareholdings of consumers in firms are denoted $\mathbf{c}_h \in \mathbb{R}^F$: negative shareholdings (short-selling; see Definition 13.3.1) may be allowed.

A consumer's **net demand** or **excess demand** or vector of desired trades is denoted by $\mathbf{z}_h \equiv \mathbf{x}_h - \mathbf{e}_h \in \mathbb{R}^N$.

Each consumer is assumed to have a **(weak) preference relation** or **preference ordering**, which is a binary relation on the consumption set $\mathcal{X}_h$; see Definitions 0.0.10 and 0.0.11 for general information on binary relations and Varian (1992, Chapter 7) for their application in consumer theory.

Since each consumer will have different preferences, we should really denote consumer h's preference relation by $\succeq_h$, but the subscript will be omitted for the time being while we consider a single consumer. Similarly, we will assume for the time being that each consumer chooses from the same consumption set, $\mathcal{X}$, although this is not essential.

An **indifference relation**, $\sim$, and a **strict preference relation**, $\succ$, can be derived from every preference relation:

1. $x \succ y$ means $x \succeq y$ but not $y \succeq x$, while
2. $x \sim y$ means $x \succeq y$ and $y \succeq x$.

The **utility function** $u: \mathcal{X} \to \mathbb{R}$ **represents** the preference relation $\succeq$ if

$$u(\mathbf{x}) \geq u(\mathbf{y}) \iff \mathbf{x} \succeq \mathbf{y} \tag{12.1}$$

If $f: \mathbb{R} \to \mathbb{R}$ is a monotonically increasing function and u represents the preference relation $\succeq$, then $f \circ u$ also represents $\succeq$, since

$$f(u(\mathbf{x})) \geq f(u(\mathbf{y})) \iff u(\mathbf{x}) \geq u(\mathbf{y}) \iff \mathbf{x} \succeq \mathbf{y} \tag{12.2}$$

Furthermore, u and $f \circ u$ have the same level sets or, as they are always called in consumer theory, indifference curves; see Definition 7.7.2.

If $\mathcal{X}$ is a countable set, then there exists a utility function representing any preference relation on $\mathcal{X}$. To prove this, just write out the consumption plans in $\mathcal{X}$ in order of preference, and assign numbers to them, assigning the same number to any two or more consumption plans between which the consumer is indifferent.

If $\mathcal{X}$ is an uncountable set, then there may not exist a utility function representing every preference relation on $\mathcal{X}$.

Sections 12.3 and 12.4 analyse the optimal behaviour of an individual consumer under conditions of certainty or perfect foresight. Sections 12.5 and 12.6 look at interactions among consumers.

12.3 Axioms

We now consider six axioms that are frequently assumed to be satisfied by preference relations when considering **consumer choice under certainty**. Section 16.4.3 will consider further axioms that are often added to simplify the analysis of **consumer choice under uncertainty**. After the definition of each axiom, we will give a brief rationale for its use.

AXIOM 1 (COMPLETENESS). *A (weak) preference relation is complete.*

Completeness means that the consumer is never agnostic.

AXIOM 2 (REFLEXIVITY). *A (weak) preference relation is reflexive.*

Reflexivity means that each bundle is at least as good as itself.

AXIOM 3 (TRANSITIVITY). *A (weak) preference relation is transitive.*

Transitivity means that preferences are rational and consistent.

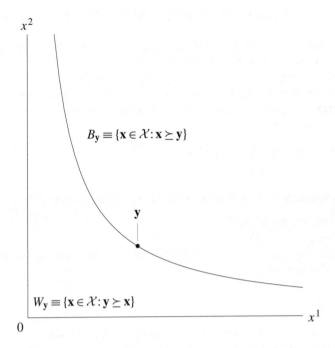

Figure 12.1 Continuous preferences when $N = 2$

These first three axioms are three of the four properties of binary relations that were introduced in Definition 0.0.11. Note that the remaining property, namely, symmetry, would not be a very sensible axiom for a preference relation, since it would imply indifference between all possible consumption bundles on the part of the consumer.

AXIOM 4 (CONTINUITY). *The preference relation $\succeq$ is **continuous**, i.e. for all consumption vectors $\mathbf{y} \in \mathcal{X}$, the sets $B_{\mathbf{y}} \equiv \{\mathbf{x} \in \mathcal{X} : \mathbf{x} \succeq \mathbf{y}\}$ and $W_{\mathbf{y}} = \{\mathbf{x} \in \mathcal{X} : \mathbf{y} \succeq \mathbf{x}\}$ (containing, respectively, the consumption vectors as good as or better than $\mathbf{y}$ and those as good as or worse than $\mathbf{y}$) are closed sets.*

Figure 12.1 shows an indifferences curve of a continuous preference relation for $N = 2$. The sets $W_{\mathbf{y}}$ and $B_{\mathbf{y}}$ both contain the indifference curve, which forms their shared boundary; $W_{\mathbf{y}}$ also contains the coordinate axes. We will see shortly that $B_{\mathbf{y}}$ and $W_{\mathbf{y}}$ are just the upper contour sets and lower contour sets, respectively, of utility functions, if such exist.

Figure 12.2 illustrates **lexicographic preferences**, which violate the continuity axiom. A consumer with such preference prefers more of commodity 1 regardless of the quantities of other commodities, more of commodity 2 if faced with a choice between two consumption vectors having the same amount of commodity 1, and so on.

In Figure 12.2, the consumption vector $\mathbf{y}$ is the only consumption vector that is in both the lower contour set $W_{\mathbf{y}}$ and the upper contour set $B_{\mathbf{y}}$. However, $B_\epsilon(\mathbf{y})$, the open ball of radius ϵ around $\mathbf{y}$, never lies completely in $W_{\mathbf{y}}$ for any ϵ. Thus, lower contour sets are not open, and hence upper contour sets are not closed.

Theorems on the existence of continuous utility functions have been proved by Debreu (1959, pp. 55–9) using Axioms 1–4 only (see also Debreu (1964)) and also by Varian (1992, p. 97), whose proof is simpler by virtue of adding an additional axiom.

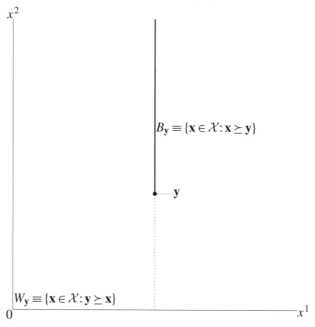

x^2

$B_{\mathbf{y}} \equiv \{\mathbf{x} \in \mathcal{X} : \mathbf{x} \succeq \mathbf{y}\}$

y

$W_{\mathbf{y}} \equiv \{\mathbf{x} \in \mathcal{X} : \mathbf{y} \succeq \mathbf{x}\}$

0

x^1

Figure 12.2 Lexicographic preferences when $N = 2$

THEOREM 12.3.1 (DEBREU: EXISTENCE OF UTILITY FUNCTIONS). *If*

(a) *the consumption set $\mathcal{X}$ is a closed and convex set, and*
(b) *$\succeq$ is a complete, reflexive, transitive and continuous preference relation on $\mathcal{X}$,*

then there exists a continuous utility function $u \colon \mathcal{X} \to \mathbb{R}$ representing $\succeq$.

Proof: For the proof of this theorem, see Debreu (1959, pp. 56–9).   □

AXIOM 5 (GREED). *Greed is incorporated into consumer behaviour by assuming either:*

(a) ***strong monotonicity*** *– if $\mathcal{X} = \mathbb{R}^N_+$, then $\succeq$ is said to be **strongly monotonic** if and only if, whenever $x^n \geq y^n$ for all n but $\mathbf{x} \neq \mathbf{y}$, $\mathbf{x} \succ \mathbf{y}$ (where $\mathbf{x} = (x^1, x^2, \ldots, x^N)$, etc.);*

or the weaker, more general:

(b) ***local non-satiation*** *– for all $\mathbf{x} \in \mathcal{X}$ and $\epsilon > 0$, there exists $\mathbf{x}' \in B_\epsilon(\mathbf{x})$ such that $\mathbf{x}' \succ \mathbf{x}$.*

THEOREM 12.3.2 *If preferences are strongly monotonic, then they exhibit local non-satiation.*

Proof: This should be obvious from the definitions above.   □

The strong monotonicity axiom is a much stronger restriction on preferences than local non-satiation; however, it greatly simplifies the proof of existence of utility functions.

Note that both versions of the greed axiom rule out the existence of **bliss points**, i.e. consumption vectors where a utility function (assuming that it exists) attains a local maximum.

We will prove the existence, but not the continuity, part of the following weaker theorem.

THEOREM 12.3.3 (VARIAN: EXISTENCE OF UTILITY FUNCTIONS). *If*

(a) $\mathcal{X} = \mathbb{R}_+^N$, *and*
(b) $\succeq$ *is a complete, reflexive, transitive, continuous and strongly monotonic preference relation on* $\mathcal{X}$,

then there exists a continuous utility function $u\colon \mathcal{X} \to \mathbb{R}$ *representing* $\succeq$.

Proof: We will prove existence only (Varian 1992, p. 97); the continuity proof is beyond the scope of this book.

Pick a benchmark consumption vector, e.g. $\mathbf{1} \equiv (1, 1, \ldots, 1)$.

The idea is that the utility of $\mathbf{x}$ is the multiple of the benchmark consumption vector to which $\mathbf{x}$ is equally preferred.

By strong monotonicity, the sets $\{t \in \mathbb{R}_+ : t\mathbf{1} \succeq \mathbf{x}\}$ and $\{t \in \mathbb{R}_+ : \mathbf{x} \succeq t\mathbf{1}\}$ are both non-empty.

By continuity of preferences, both are closed (each is the intersection of a ray through the origin and a closed set); and by completeness, they cover $\mathbb{R}_+$.

By connectedness of $\mathbb{R}$, they intersect in at least one point, $u(\mathbf{x})$, say, and $\mathbf{x} \sim u(\mathbf{x})\mathbf{1}$.

Now

$$\mathbf{x} \succeq \mathbf{y} \quad \Leftrightarrow \quad u(\mathbf{x})\mathbf{1} \succeq u(\mathbf{y})\mathbf{1}$$
$$\Leftrightarrow \quad u(\mathbf{x}) \geq u(\mathbf{y}) \tag{12.3}$$

where the first equivalence follows from transitivity of preferences and the second from strong monotonicity.

The assumption that preferences are reflexive is not used in establishing the existence of the utility function, so it can be inferred that it is required to establish continuity. $\square$

Some commonly encountered examples of preference relations are those which can be represented by the following utility functions, details of which can be found in the various exercises referred to. In each case, $\mathbf{x} = (x^1, x^2, \ldots, x^N)$ denotes the consumption vector.

- Two-good linear–quadratic preferences (see Exercise 12.2):

$$u(x^1, x^2) \equiv x^2 + \alpha x^1 + \beta (x^1)^2 \quad (\beta > 0) \tag{12.4}$$

- Cobb–Douglas preferences (see equation (9.58)):

$$u(\mathbf{x}) \equiv \prod_{n=1}^{N} (x^n)^{\beta^n} \quad (0 < \beta^n, \ n = 1, 2, \ldots, N) \tag{12.5}$$

This is really only well defined on $\mathbb{R}_{++}^N$.

- **Stone–Geary**[2] **preferences** (see Exercise 12.3):

$$u(\mathbf{x}) = \sum_{n=1}^{N} \beta^n \ln(x^n - \alpha^n) \tag{12.6}$$

where the α^n and β^n are positive constants with $\sum_{n=1}^{N} \beta^n = 1$. Cobb–Douglas preferences are a special case of Stone–Geary preferences with $\boldsymbol{\alpha} = \mathbf{0}_N$.

- Constant elasticity of substitution (CES) preferences (see Exercise 10.17):

$$u(\mathbf{x}) = \left(\sum_{n=1}^{N} \alpha^n (x^n)^\rho \right)^{1/\rho} \tag{12.7}$$

- Leontief preferences (see Exercise 12.4):

$$u(\mathbf{x}) = \min_{n=1}^{N} \{\beta^n x^n\} \tag{12.8}$$

(Note the similarity to the Leontief production function mentioned in Section 1.2.3.)

We have already considered lexicographic preferences, and shown that they cannot be represented by a real-valued utility function.

Where preferences can be represented by a utility function, the words "preference" and "utility" become interchangeable, so that, for example, we can refer to an individual either as having "linear–quadratic preferences" or as having "linear–quadratic utility".

The rule that the consumer will follow is to choose the most preferred bundle from the set of affordable alternatives (the **budget set**), in other words the bundle at which the utility function is maximized subject to the **budget constraint**, if one exists. We know that an optimal choice will exist if the utility function is continuous and the budget set is closed and bounded.

If the utility function is differentiable, we can go further and use calculus to find the maximum. So we usually assume differentiability.

If u is a concave utility function and f is an increasing function, then $f \circ u$, which also represents the same preferences, is not necessarily a concave function (unless f itself is a convex function). In other words, concavity of a utility function is a property of the particular representation and not of the underlying preferences. Notwithstanding this, convexity of preferences is important, as indicated by the use of one or other of the following axioms, each of which relates to the preference relation itself and not to the particular utility function chosen to represent it.

AXIOM 6 (CONVEXITY). *There are two versions of this axiom, which guarantees the existence of optimal consumer strategies:*

(a) ***convexity** – the preference relation $\succeq$ is **convex** if and only if*

$$\mathbf{x} \succeq \mathbf{y} \quad \Rightarrow \quad \lambda \mathbf{x} + (1 - \lambda)\mathbf{y} \succeq \mathbf{y} \quad \forall \lambda \in (0, 1) \tag{12.9}$$

(b) **strict convexity** – *the preference relation* $\succeq$ *is* **strictly convex** *if and only if*

$$\mathbf{x} \succeq \mathbf{y} \quad \Rightarrow \quad \lambda \mathbf{x} + (1-\lambda)\mathbf{y} \succ \mathbf{y} \quad \forall \lambda \in (0, 1) \tag{12.10}$$

Note that a preference relation that is strictly convex is also convex. The difference between the two versions of the convexity axiom basically amounts to ruling out linear segments in indifference curves in the strict case.

With strictly convex preferences, it will be seen that the consumer's problem generally has a unique solution, allowing (single-valued) demand functions to be derived. With convex preferences, there may be multiple solutions to the consumer's problem, and we may have to deal with (multivalued) demand correspondences.

THEOREM 12.3.4 *The preference relation* $\succeq$ *is (strictly) convex if and only if every utility function representing* $\succeq$ *is a (strictly) quasi-concave function.*

Proof: In either case, both statements are equivalent to saying that

$$u(\mathbf{x}) \geq u(\mathbf{y}) \quad \Rightarrow \quad u(\lambda \mathbf{x} + (1-\lambda)\mathbf{y}) \geq (>) u(\mathbf{y}) \quad \forall \lambda \in (0, 1) \tag{12.11}$$

$\square$

THEOREM 12.3.5 *All upper contour sets of a utility function representing convex preferences are convex sets (and thus indifference curves are convex to the origin for strongly monotonic, convex preferences).*

Proof: This result follows immediately from Theorem 10.2.2. $\square$

It can be seen that convexity of preferences is a generalization of the two-good assumption of a diminishing **marginal rate of substitution**, which will be familiar to some readers from intermediate economics courses.

Even the above axioms are not sufficient to guarantee tractability of the consumer's problem. There are two further desirable properties of utility functions that must sometimes be assumed and that are difficult to state in terms of the underlying preference relation.

1. The first of these properties comprises a pair of conditions known as the **Inada conditions**.[3] They are

$$\lim_{x_i \to 0} \frac{\partial u}{\partial x_i} = \infty \tag{12.12}$$

and

$$\lim_{x_i \to \infty} \frac{\partial u}{\partial x_i} = 0 \tag{12.13}$$

The Inada conditions will be required to rule out corner solutions in the consumer's problem. Intuitively, they just say that indifference curves may be asymptotic to the axes but never reach them.

2. Pseudo-concavity of utility functions is also desirable as it permits the use of Kuhn–Tucker second-order conditions and the development of the duality properties of the consumer's problem.

Note that pseudo-concavity is not invariant under increasing transformations. (Consider, for example, the pseudo-concave function x and the non-pseudo-concave function x^3 each defined on $\mathbb{R}$.)

12.4 The consumer's problem and its dual

12.4.1 Perfect competition and the Walrasian auctioneer

Assume for the time being that trade takes place in a perfectly competitive environment. This means that all consumers face the same prices for all units of a particular good. Neither nonlinear pricing (for example, in the form of discounts on purchases of large quantities) nor price discrimination is allowed. As a practical justification, it is often proposed that there are large numbers of buyers and sellers, making collusion and price-setting difficult and making it easy for any consumer to switch from one counter-party to another offering a more favourable price.

It is sometimes proposed that a (mythical, Walrasian[4]) auctioneer calls out prices and asks consumers how much of each good they wish to purchase at those prices until a full demand function or schedule is revealed. In the search for equilibrium, it can be assumed that this process continues until such time as all N markets clear simultaneously.

12.4.2 Solution

A consumer with consumption set $\mathcal{X}_h$, endowment vector $\mathbf{e}_h \in \mathcal{X}_h$, shareholdings in firms $\mathbf{c}_h \in \mathbb{R}^F$ and preference ordering $\succeq_h$ represented by continuous utility function u_h, who desires to trade his endowment at prices $\mathbf{p} \in \mathbb{R}^N_+$, faces an inequality-constrained optimization problem:

$$\max_{\mathbf{x} \in \mathcal{X}_h} u_h(\mathbf{x}) \quad \text{s.t.} \quad \mathbf{p}^\top \mathbf{x} \le \mathbf{p}^\top \mathbf{e}_h + \mathbf{c}_h^\top \mathbf{\Pi}(\mathbf{p}) \equiv M_h \tag{12.14}$$

where $\mathbf{\Pi}(\mathbf{p})$ is the vector of the F firms' maximized profits when prices are $\mathbf{p}$.

Constraining $\mathbf{x}$ to lie in the consumption set normally just means imposing non-negativity constraints on the problem. These constraints are rarely binding in the examples we will consider, since the Inada conditions usually hold.

Constraining $\mathbf{x}$ to satisfy the inequality

$$\mathbf{p}^\top \mathbf{x} \le \mathbf{p}^\top \mathbf{e}_h + \mathbf{c}_h^\top \mathbf{\Pi}(\mathbf{p}) \tag{12.15}$$

known as the budget constraint, is more important. The budget constraint will almost invariably be binding. The set of consumption vectors satisfying the budget constraint is sometimes called the budget set. It is the intersection of $\mathcal{X}_h$ with the half-space bounded by the hyperplane (sometimes called the **budget hyperplane**) with equation

$$\mathbf{p}^\top \mathbf{x} = \mathbf{p}^\top \mathbf{e}_h + \mathbf{c}_h^\top \mathbf{\Pi}(\mathbf{p}) \tag{12.16}$$

From a mathematical point of view, the source of the household's income is irrelevant, and in particular the distinction between pure exchange and production economy is irrelevant. Thus, income can be represented by M_h in either case, as it is independent of the choice of variables $\mathbf{x}$.

We have the following theorems concerning the consumer's problem.

THEOREM 12.4.1 (EXISTENCE OF SOLUTION). *If the consumption set is $\mathbb{R}_+^N$, then the consumer's problem has at least one solution for all price vectors $\mathbf{p} \in \mathbb{R}_{++}^N$ and income levels $M_h \in \mathbb{R}_+$, and, hence, there exists a well-defined demand correspondence*

$$\mathbf{x}_h^*: \mathbb{R}_{++}^N \times \mathbb{R}_{++} \to \mathbb{R}_+^N : (\mathbf{p}, M_h) \mapsto \mathbf{x}_h^*(\mathbf{p}, M_h) \tag{12.17}$$

Proof: These are the circumstances in which the budget set is closed and bounded. Since the utility function is continuous, we know by Theorem 10.3.1 that it attains a maximum on the budget set. The maximum may not be unique, so we can say only that there is a demand correspondence; further conditions would be required to guarantee the existence of a single-valued demand function.   □

In the case of demand, we will generally omit the * denoting the optimal response function and simply write $\mathbf{x}_h(\mathbf{p}, M_h)$.

Having established the existence of solutions to the consumer's problem, we can now proceed to analyse the corresponding optimal response correspondences – or functions, if we can show that they are single-valued.

THEOREM 12.4.2 (SINGLE-VALUEDNESS OF DEMAND FUNCTIONS).

(a) *If the underlying preference relation is **strictly convex** and can be represented by the utility function u_h, then the consumer's problem, problem (12.14), has a **unique** solution. The corresponding optimal response function, usually denoted just $\mathbf{x}_h$, is called a **Marshallian demand function**.*[5]

(b) *If the underlying preference relation is **convex** and can be represented by the utility function u_h, then problem (12.14) can have **multiple** solutions. The corresponding optimal response correspondence is called a **Marshallian demand correspondence**.*

The truth of these theorems should become clear from the following discussion. We will consider in turn second-order conditions, first-order conditions and uniqueness conditions for the consumer's problem.

1. Since the constraint functions are linear in the choice variables $\mathbf{x}$, the Kuhn–Tucker theorem on second-order conditions (Theorem 10.5.2) can be applied, provided that the utility function u_h is pseudo-concave.
2. In this case, the first-order conditions identify a maximum. The Lagrangian, using multipliers λ for the budget constraint and $\mu \in \mathbb{R}^N$ for the non-negativity constraints, is

$$u_h(\mathbf{x}) + \lambda(M_h - \mathbf{p}^\top \mathbf{x}) + \mu^\top \mathbf{x} \tag{12.18}$$

The first-order conditions are given by the (N-dimensional) vector equation

$$u_h'(\mathbf{x}) + \lambda(-\mathbf{p}^\top) + \mu^\top = \mathbf{0}_N^\top \tag{12.19}$$

and the sign condition $\lambda \geq 0$, with $\lambda > 0$ if the budget constraint is binding. We also have $\mu = \mathbf{0}_N$ unless one of the non-negativity constraints is binding. The Inada conditions (see p. 306) would rule out this possibility.

Now for each $\mathbf{p} \in \mathbb{R}_{++}^N$ (ruling out **bads**, i.e. goods with negative prices,[6] and even – see below – free goods), $\mathbf{c}_h \in \mathcal{X}_h$ and $\mathbf{c}_h \in \mathbb{R}^F$, or, for each $\mathbf{p}$ and M_h combination, there is a corresponding solution to the consumer's utility-maximization problem, denoted $\mathbf{x}_h(\mathbf{p}, \mathbf{e}_h, \mathbf{c}_h)$ or $\mathbf{x}_h(\mathbf{p}, M_h)$. The latter function (correspondence) $\mathbf{x}_h$ is the Marshallian demand function (correspondence).

3. If the utility function u_h is also strictly quasi-concave (i.e. preferences are strictly convex), then the conditions of the Kuhn–Tucker theorem on uniqueness (Theorem 10.5.3) are satisfied. In this case, the consumer's problem has a unique solution for given prices and income, so that the optimal response correspondence is a single-valued demand function. On the other hand, the weak form of the convexity axiom would permit a multivalued demand correspondence.

12.4.3   *Properties of Marshallian demand*

Noteworthy properties of Marshallian demand are listed below. The first three hold regardless of the greed axioms, but the remaining propositions are a consequence of greed.

1. If preferences are strictly convex, then Marshallian demand is a single-valued function.
2. The demand $\mathbf{x}_h(\mathbf{p}, M_h)$ is independent of the representation u_h of the underlying preference relation $\succeq_h$ that is used in the statement of the consumer's problem.
3. The demand function $\mathbf{x}_h$ is homogeneous of degree zero in $\mathbf{p}$ and M_h. In other words, if all prices and income are multiplied by $\alpha > 0$, then demand does not change:

$$\mathbf{x}_h(\alpha\mathbf{p}, \alpha M_h) = \mathbf{x}_h(\mathbf{p}, M_h) \tag{12.20}$$

4. Marshallian demand functions are continuous. This follows from the theorem of the maximum (Theorem 10.5.4) and the discussion at the end of Section 10.5.2. It follows that small changes in prices or income will lead to small changes in quantities demanded.
5. If preferences exhibit local non-satiation, then the budget constraint is binding. This is because no consumption vector in the interior of the budget set can maximize utility, as some nearby consumption vector will always be both preferred and affordable. At the optimum, on the budget hyperplane, the nearby consumption vector that is preferred will not be affordable. This allows the duality analysis which follows in Section 12.4.5.
6. In the case of strongly monotonic preferences, if $\mathbf{p}$ includes a zero price ($p^n = 0$ for some n), then $\mathbf{x}_h(\mathbf{p}, M_h)$ may not be well defined. This is because the consumer will seek to acquire and consume an infinite amount of the free good, thereby increasing utility without bound. For this reason, it is neater to define Marshallian demand only on the open positive orthant in $\mathbb{R}^N$, namely, $\mathbb{R}^N_{++}$.
7. The components of $\mathbf{x}_h(\mathbf{p}, M_h)$ may either increase or decrease in income M_h. Goods are said to be **normal goods** over the range of income where Marshallian demand is increasing in income and **inferior goods** over the range of income where Marshallian demand is decreasing in income; see Section 17.3.2 for a discussion of normal and inferior goods in financial markets.

12.4.4   *Properties of indirect utility*

The envelope function corresponding to the consumer's problem is called the **indirect utility function** and is denoted by:

$$v_h(\mathbf{p}, M) \equiv u_h(\mathbf{x}_h(\mathbf{p}, M)) \tag{12.21}$$

The following are interesting properties of the indirect utility function:

1. By the theorem of the maximum (Theorem 10.5.4), the indirect utility function is continuous for positive prices and income.
2. The indirect utility function is non-increasing in $\mathbf{p}$ and non-decreasing in M.
3. The indirect utility function is quasi-convex in prices. To see this, let $B(\mathbf{p})$ denote the budget set when prices are $\mathbf{p}$ and let $\mathbf{p}_\lambda \equiv \lambda \mathbf{p} + (1-\lambda)\mathbf{p}'$.

   CLAIM. *We claim that* $B(\mathbf{p}_\lambda) \subseteq (B(\mathbf{p}) \cup B(\mathbf{p}'))$.

   Proof: Suppose this was not the case, i.e. for some $\mathbf{x}$, $\mathbf{p}_\lambda^\top \mathbf{x} \leq M$ but $\mathbf{p}^\top \mathbf{x} > M$ and $\mathbf{p}'^\top \mathbf{x} > M$. Then taking a convex combination of the last two inequalities yields

$$\lambda \mathbf{p}^\top \mathbf{x} + (1-\lambda)\mathbf{p}'^\top \mathbf{x} > M \tag{12.22}$$

   which contradicts the first inequality.

   It follows that the maximum value of $u_h(\mathbf{x})$ on the subset $B(\mathbf{p}_\lambda)$ is less than or equal to its maximum value on the superset $B(\mathbf{p}) \cup B(\mathbf{p}')$.

   In terms of the indirect utility function, this says that

$$v_h(\mathbf{p}_\lambda, M) \leq \max\{v_h(\mathbf{p}, M), v_h(\mathbf{p}', M)\} \tag{12.23}$$

   or that v_h is quasi-convex.   □

4. The indirect utility function $v_h(\mathbf{p}, M)$ is homogeneous of degree zero in $\mathbf{p}$ and M, or

$$v_h(\lambda \mathbf{p}, \lambda M) = v_h(\mathbf{p}, M) \tag{12.24}$$

12.4.5 *The dual problem*

Consider also the (dual) expenditure-minimization problem:

$$\min_{\mathbf{x}} \mathbf{p}^\top \mathbf{x} \quad \text{s.t. } u_h(\mathbf{x}) \geq \bar{u} \tag{12.25}$$

where $\bar{u}$ is some desired level of utility. In other words, what happens if expenditure is minimized subject to a certain level of utility, $\bar{u}$, being attained?

The solution (optimal response function) is called the **Hicksian**[7] or **compensated demand function** (or **correspondence**) and is usually denoted $\mathbf{h}_h(\mathbf{p}, \bar{u})$.

If the local non-satiation axiom holds, then the constraints are binding in both the utility-maximization and expenditure-minimization problems, and we have a number of duality relations. In particular, there will be a one-to-one correspondence between income M and utility $\bar{u}$ for a given price vector $\mathbf{p}$.

The envelope function corresponding to the dual problem is called the **expenditure function**:

$$e_h(\mathbf{p}, \bar{u}) \equiv \mathbf{p}^\top \mathbf{h}_h(\mathbf{p}, \bar{u}) \tag{12.26}$$

The expenditure function and the indirect utility function will then act as a pair of inverse envelope functions mapping utility levels to income levels and vice versa, respectively.

The following duality relations (or fundamental identities, as Varian (1992, p. 106) calls them) will prove extremely useful later on:

$$e(\mathbf{p}, v(\mathbf{p}, M)) = M \qquad (12.27)$$

$$v(\mathbf{p}, e(\mathbf{p}, \bar{u})) = \bar{u} \qquad (12.28)$$

$$\mathbf{x}(\mathbf{p}, M) = \mathbf{h}(\mathbf{p}, v(\mathbf{p}, M)) \qquad (12.29)$$

$$\mathbf{h}(\mathbf{p}, \bar{u}) = \mathbf{x}(\mathbf{p}, e(\mathbf{p}, \bar{u})) \qquad (12.30)$$

These are just (10.133)–(10.136) adapted to the notation of the consumer's problem.

The consumer's problem, its dual problem and the associated optimal response functions and envelope functions are summarized in Table 12.1.

12.4.6 Properties of Hicksian demands

The following are interesting properties of the Hicksian demand function:

1. The Hicksian demand is specific to a particular representation of the underlying preferences.
2. Hicksian demands are homogeneous of degree zero in prices:

$$\mathbf{h}_h(\alpha \mathbf{p}, \bar{u}) = \mathbf{h}_h(\mathbf{p}, \bar{u}) \qquad (12.31)$$

3. As in the Marshallian approach, if preferences are strictly convex, then any solution to the expenditure-minimization problem is unique and the Hicksian demands are well-defined single-valued functions.

   It is worth going back to the uniqueness proof on p. 309 with this added interpretation. If two different consumption vectors minimize expenditure, then they both cost the same amount, and any convex combination of the two also costs the same amount. But by strict convexity, a convex combination yields higher utility, and nearby there must, by continuity, be a cheaper consumption vector still yielding utility $\bar{u}$.

   If preferences are not strictly convex, then Hicksian demands may be correspondences rather than functions.

4. By the theorem of the maximum (Theorem 10.5.4), Hicksian demands are continuous.

Table 12.1 The consumer's problem and its dual

Problem	Utility maximization	Expenditure minimization
Objective function	$u_h(\mathbf{x})$	$\mathbf{p}^\top \mathbf{x}$
Constraint	$\mathbf{p}^\top \mathbf{x} \le \mathbf{p}^\top \mathbf{e}_h + \mathbf{c}_h^\top \mathbf{\Pi}(\mathbf{p}) \equiv M_h$	$u_h(\mathbf{x}) \ge \bar{u}$
Optimal response function	Marshallian demand: $\mathbf{x}_h(\mathbf{p}, M)$	Hicksian demand: $\mathbf{h}_h(\mathbf{p}, \bar{u})$
Envelope function	Indirect utility: $v_h(\mathbf{p}, M) \equiv u_h(\mathbf{x}_h(\mathbf{p}, M))$	Expenditure: $e_h(\mathbf{p}, \bar{u}) \equiv \mathbf{p}^\top \mathbf{h}_h(\mathbf{p}, \bar{u})$

12.4.7 *Properties of the expenditure function*

The following are interesting properties of the expenditure function:

1. By the theorem of the maximum (Theorem 10.5.4), the expenditure function is continuous.
2. The expenditure function itself is non-decreasing in prices, since raising the price of one good while holding the prices of all other goods constant cannot reduce the minimum cost of attaining a fixed utility level.

   Raising the price of a good that is not demanded might leave expenditure unchanged, so we cannot say that the expenditure function is strictly increasing in prices in all cases. A counter-example is provided by the linear–quadratic utility function, or any utility function for which demand hits zero at high levels of own price.
3. The expenditure function is concave in prices. To see this, we just fix two price vectors $\mathbf{p}$ and $\mathbf{p}'$ and consider the value of the expenditure function at the convex combination $\mathbf{p}_\lambda \equiv \lambda \mathbf{p} + (1 - \lambda)\mathbf{p}'$:

$$
\begin{aligned}
e(\mathbf{p}_\lambda, \bar{u}) &= (\mathbf{p}_\lambda)^\top \mathbf{h}(\mathbf{p}_\lambda, \bar{u}) \\
&= \lambda \mathbf{p}^\top \mathbf{h}(\mathbf{p}_\lambda, \bar{u}) + (1 - \lambda)(\mathbf{p}')^\top \mathbf{h}(\mathbf{p}_\lambda, \bar{u}) \\
&\geq \lambda \mathbf{p}^\top \mathbf{h}(\mathbf{p}, \bar{u}) + (1 - \lambda)(\mathbf{p}')^\top \mathbf{h}(\mathbf{p}', \bar{u}) \\
&= \lambda e(\mathbf{p}, \bar{u}) + (1 - \lambda)e(\mathbf{p}', \bar{u})
\end{aligned}
\tag{12.32}
$$

   where the inequality follows because the cost of a sub-optimal bundle for the given prices must be no less than the cost of the optimal (expenditure-minimizing) consumption vector for those prices.
4. The expenditure function is homogeneous of degree one in prices:

$$
e_h(\alpha \mathbf{p}, \bar{u}) = \alpha e_h(\mathbf{p}, \bar{u})
\tag{12.33}
$$

Sometimes we meet two other related functions:

- The **money metric utility function**

$$
m_h(\mathbf{p}, \mathbf{x}) \equiv e_h(\mathbf{p}, u_h(\mathbf{x}))
\tag{12.34}
$$

  is the (least) cost at prices $\mathbf{p}$ of being as well off as with the consumption vector $\mathbf{x}$.
- The **money metric indirect utility function**

$$
\mu_h(\mathbf{p}; \mathbf{q}, M) \equiv e_h(\mathbf{p}, v_h(\mathbf{q}, M))
\tag{12.35}
$$

  is the (least) cost at prices $\mathbf{p}$ of being as well off as if prices were $\mathbf{q}$ and income was M.

12.4.8 *Further results in consumer theory*

In this section, we present four important theorems on demand functions and the corresponding envelope functions. **Shephard's lemma** will allow us to recover Hicksian demands from the expenditure function.[8] Similarly, **Roy's identity** will allow us to recover Marshallian demands from the indirect utility function.[9] The **Slutsky symmetry condition** and the **Slutsky equation** provide further insights into the properties of consumer demand.[10]

THEOREM 12.4.3 (SHEPHARD'S LEMMA). *The partial derivatives of the expenditure function with respect to prices are the corresponding Hicksian demand functions, i.e.*

$$\frac{\partial e_h}{\partial p^n}(\mathbf{p}, \bar{u}) = h_h^n(\mathbf{p}, \bar{u}) \tag{12.36}$$

Proof: By differentiating the expenditure function with respect to the price of good n and applying the envelope theorem (Theorem 10.4.4), we obtain

$$\frac{\partial e_h}{\partial p^n}(\mathbf{p}, \bar{u}) = \frac{\partial}{\partial p^n}(\mathbf{p}^\top \mathbf{x} + \lambda(u_h(\mathbf{x}) - \bar{u})) \tag{12.37}$$

$$= x^n \tag{12.38}$$

which, when evaluated at the optimum, is just $h_h^n(\mathbf{p}, \bar{u})$, as required.

(To apply the envelope theorem, we should be dealing with an equality-constrained optimization problem; however, if we assume local non-satiation, we know that the budget constraint or utility constraint will always be binding, and so the inequality-constrained expenditure-minimization problem is essentially an equality-constrained problem.) ☐

THEOREM 12.4.4 (ROY'S IDENTITY). *Marshallian demands may be recovered from the indirect utility function using*

$$x^n(\mathbf{p}, M) = -\frac{\partial v(\mathbf{p}, M)/\partial p^n}{\partial v(\mathbf{p}, M)/\partial M} \tag{12.39}$$

Proof: For a proof of Roy's identity, see Varian (1992, pp. 106–7).

It is obtained by differentiating the duality relation (12.28)

$$v(\mathbf{p}, e(\mathbf{p}, \bar{u})) = \bar{u} \tag{12.40}$$

with respect to p^n, using the chain rule, which implies that

$$\frac{\partial v}{\partial p^n}(\mathbf{p}, e(\mathbf{p}, \bar{u})) + \frac{\partial v}{\partial M}(\mathbf{p}, e(\mathbf{p}, \bar{u}))\frac{\partial e}{\partial p^n}(\mathbf{p}, \bar{u}) = 0 \tag{12.41}$$

and using Shephard's lemma gives

$$\frac{\partial v}{\partial p^n}(\mathbf{p}, e(\mathbf{p}, \bar{u})) + \frac{\partial v}{\partial M}(\mathbf{p}, e(\mathbf{p}, \bar{u}))\mathbf{h}^n(\mathbf{p}, \bar{u}) = 0 \tag{12.42}$$

Hence,

$$\mathbf{h}^n(\mathbf{p}, \bar{u}) = -\frac{\partial v(\mathbf{p}, e(\mathbf{p}, \bar{u}))/\partial p^n}{\partial v(\mathbf{p}, e(\mathbf{p}, \bar{u}))/\partial M} \tag{12.43}$$

and expressing this last equation in terms of the relevant level of income M rather than the corresponding value of utility $\bar{u}$:

$$x^n(\mathbf{p}, M) = -\frac{\partial v(\mathbf{p}, M)/\partial p^n}{\partial v(\mathbf{p}, M)/\partial M} \tag{12.44}$$

$\square$

THEOREM 12.4.5 (SLUTSKY SYMMETRY CONDITION). *All cross-price substitution effects are symmetric:*

$$\frac{\partial h_h^n}{\partial p^m} = \frac{\partial h_h^m}{\partial p^n} \tag{12.45}$$

Proof: From Shephard's lemma, we can easily derive these conditions, assuming that the expenditure function is twice continuously differentiable. Under this assumption, by Young's theorem (Theorem 9.7.2) we have that

$$\frac{\partial^2 e_h}{\partial p^m \partial p^n} = \frac{\partial^2 e_h}{\partial p^n \partial p^m} \tag{12.46}$$

Since $h_h^m = \partial e_h/\partial p^m$ and $h_h^n = \partial e_h/\partial p^n$, the result follows. $\square$

Unlike the other theorems in this section, the next result has no special name.

THEOREM 12.4.6 *Since the expenditure function is concave in prices (see p. 312), the corresponding Hessian matrix is negative semi-definite (by Theorem 10.2.5). In particular, its diagonal entries are non-positive, or*

$$\frac{\partial^2 e_h}{\partial (p^n)^2} \leq 0, \quad n = 1, 2, \ldots, N \tag{12.47}$$

Using Shephard's lemma, it follows that

$$\frac{\partial h_h^n}{\partial p^n} \leq 0, \quad n = 1, 2, \ldots, N \tag{12.48}$$

In other words, Hicksian demand functions, unlike Marshallian demand functions, are uniformly decreasing in own price. Another way of saying this is that own-price substitution effects are always negative.

THEOREM 12.4.7 (SLUTSKY EQUATION). *The total effect $\partial x^m(\mathbf{p}, M)/\partial p^n$ of a price change on (Marshallian) demand can be decomposed as follows into a substitution effect $\partial h^m(\mathbf{p}, \bar{u})/\partial p^n$ and an income effect $-[\partial x^m(\mathbf{p}, M)/\partial M]\, h^n(\mathbf{p}, \bar{u})$:*

$$\frac{\partial x^m}{\partial p^n}(\mathbf{p}, M) = \frac{\partial h^m}{\partial p^n}(\mathbf{p}, \bar{u}) - \frac{\partial x^m}{\partial M}(\mathbf{p}, M)\, h^n(\mathbf{p}, \bar{u}) \tag{12.49}$$

where $\bar{u} \equiv v(\mathbf{p}, M)$.

Before proving this, let us consider the signs of the various terms in the Slutsky equation and look at what it means in a two-good example, as illustrated in Figure 12.3. Figure 12.3 shows one indifference curve of the utility function, $u_h(x^1, x^2) = \bar{u}$, say, and three budget lines in x^1x^2 space. A reduction in the price of the first good from p^1 to $p^{1'}$, keeping income fixed at M, say, causes the budget line to swing outwards from L_1 to L_2. The dotted budget line L_3 represents the expenditure required to maintain the original utility level $\bar{u}$ at the new relative prices.

Point A represents the optimal solution to the utility-maximization problem at the original prices. Point B represents the optimal solution to the expenditure-minimization problem at the new prices with the original utility level. The optimal solution to the utility-maximization problem at the new prices could lie anywhere along the line segment CF. Three possible cases can be distinguished. By Theorem 12.4.6, we know that own-price substitution effects, corresponding to the move along the indifference curve from A to B, are non-positive in all three cases. The total effect of the price change is to move the solution from A to the optimal point along CF and the income effect of the price change is to move the solution from B to the optimal point along CF.

1. If the optimal point lies along the line segment CD, then the income effect more than offsets the substitution effect, and the total effect of a *reduction* in the price of good 1 is to *reduce* demand for good 1. Thus, this graphical interpretation of the Slutsky equation suggests the possible existence of **Giffen goods**, i.e. goods whose Marshallian demand functions are *locally* increasing in own price.[11] However, none of the standard

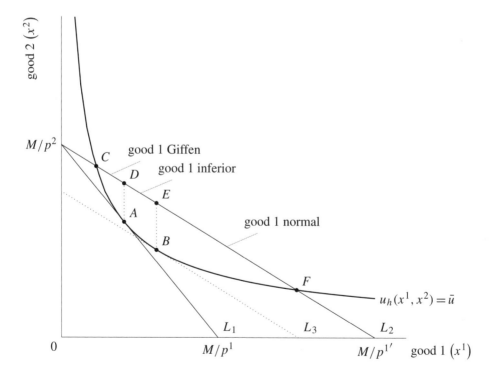

Figure 12.3 Signs in the Slutsky equation

microeconomics textbooks gives an example of a well-behaved utility function that can result in Giffen goods.

2. If the optimal point lies along the line segment *DE*, then the income effect partially offsets the substitution effect, but the total effect of a *reduction* in the price of good 1 is to *increase* demand for good 1. As noted above, goods (including Giffen goods) with this property are said to be inferior goods.

3. Finally, if the optimal point lies along the line segment *EF*, then the income effect reinforces the substitution effect, and good 1 is *locally* a normal good.

Note that goods that are locally normal, inferior or Giffen at one price vector may have different properties at another price vector.

We now return to the algebraic proof of the Slutsky equation for the *N*-good case.

Proof of Slutsky equation: Differentiating the right-hand sides of the *m*th component of the fundamental identity (12.30)

$$\mathbf{h}(\mathbf{p}, \bar{u}) = \mathbf{x}(\mathbf{p}, e(\mathbf{p}, \bar{u})) \tag{12.50}$$

with respect to p^n, using the chain rule, yields

$$\frac{\partial h^m}{\partial p^n}(\mathbf{p}, \bar{u}) = \frac{\partial x^m}{\partial p^n}(\mathbf{p}, e(\mathbf{p}, \bar{u})) + \frac{\partial x^m}{\partial M}(\mathbf{p}, e(\mathbf{p}, \bar{u}))\frac{\partial e}{\partial p^n}(\mathbf{p}, \bar{u}) \tag{12.51}$$

or

$$\frac{\partial x^m}{\partial p^n}(\mathbf{p}, e(\mathbf{p}, \bar{u})) = \frac{\partial h^m}{\partial p^n}(\mathbf{p}, \bar{u}) - \frac{\partial x^m}{\partial M}(\mathbf{p}, e(\mathbf{p}, \bar{u}))\frac{\partial e}{\partial p^n}(\mathbf{p}, \bar{u}) \tag{12.52}$$

To complete the proof, substitute from Shephard's lemma and use the fact that $M \equiv e(\mathbf{p}, \bar{u})$ (since $\bar{u} \equiv v(\mathbf{p}, M)$) to obtain (12.49) as required. □

12.5   General equilibrium theory

12.5.1   *Definitions*

Our analysis of general equilibrium will be confined for the time being to a pure exchange economy.

DEFINITION 12.5.1 An **allocation** is a collection of consumption vectors for each individual in an economy,

$$\mathbf{X} \equiv [\mathbf{x}_1\ \mathbf{x}_2\ \cdots\ \mathbf{x}_h\ \cdots\ \mathbf{x}_H] \tag{12.53}$$

Since we have specified that there are *N* commodities, **X** is an $N \times H$ matrix.

DEFINITION 12.5.2 An allocation is **feasible**, given endowments $\mathbf{e}_1, \mathbf{e}_2, \ldots, \mathbf{e}_H$, provided that

$$\sum_{h=1}^{H} x_h^n \le \sum_{h=1}^{H} e_h^n \quad \forall n \tag{12.54}$$

In other words, a feasible allocation does not require the availability of more than the aggregate endowment of the nth good, or an allocation is feasible, given an aggregate endowment, if it can be achieved by redistributing that aggregate endowment.

DEFINITION 12.5.3 An allocation is **market-clearing**, given endowments $\mathbf{e}_1, \mathbf{e}_2, \ldots, \mathbf{e}_H$, provided that

$$\sum_{h=1}^{H} x_h^n = \sum_{h=1}^{H} e_h^n \quad \forall\, n \tag{12.55}$$

DEFINITION 12.5.4 A **competitive equilibrium** (often called a **Walrasian equilibrium** or just an **equilibrium**) is a price–allocation pair $(\mathbf{p}, \mathbf{X})$ with the following properties, given endowments $\mathbf{e}_1, \mathbf{e}_2, \ldots, \mathbf{e}_H$:

(a) $p^n \geq 0$ for all $n = 1, 2, \ldots, N$ with $p^n > 0$ for some n (in other words, all prices are non-negative but not all goods are free);
(b) for all $h = 1, 2, \ldots, H$, the hth column of $\mathbf{X}$, $\mathbf{x}_h$, maximizes individual h's utility given the prices $\mathbf{p}$; and
(c) $\mathbf{X}$ is a feasible allocation.

Note that we do not require markets to clear exactly in equilibrium, but allow for the possibility that one or more goods may be in excess supply. Equilibrium is the situation in which no good is in excess demand.

Note also that, if $(\mathbf{p}, \mathbf{X})$ is a competitive equilibrium, then $(\lambda \mathbf{p}, \mathbf{X})$ is also a competitive equilibrium for any positive scalar λ. Thus, without loss of generality, we can confine our search for equilibrium prices to the unit simplex, S^{N-1}; see Definition 7.4.2.

If $\mathbf{q}$ is any equilibrium price vector, then the price vector

$$\mathbf{p} \equiv \frac{1}{\mathbf{q}^\top \mathbf{1}} \mathbf{q} \tag{12.56}$$

will also be an equilibrium price vector (with the same equilibrium allocation $\mathbf{X}$). Since by definition not all goods are free and thus $\mathbf{q}^\top \mathbf{1} > 0$, there is no danger of division by zero.

As should be clear from the preceding discussion of the consumer's problem, absolute equilibrium prices will certainly not be uniquely determined. Even relative equilibrium prices may not be uniquely determined, as will be seen below.

While it might be intuitively more appealing to pick a particular good (e.g. gold) as the numeraire, it is mathematically more convenient to use the consumption vector comprising one unit of every commodity, or $\mathbf{1}$. This is effectively what is achieved by restricting the search for equilibrium prices to the unit simplex.

To simplify the notation, given endowments in this pure exchange economy, we will denote individual h's demand at prices $\mathbf{p}$ by $\mathbf{x}_h(\mathbf{p})$ rather than $\mathbf{x}_h(\mathbf{p}, \mathbf{p}^\top \mathbf{e}_h)$. Then we can define **aggregate demand** as $\mathbf{x}(\mathbf{p}) \equiv \sum_{h=1}^{H} \mathbf{x}_h(\mathbf{p})$. Likewise, the **aggregate endowment** vector can be defined as $\mathbf{e} \equiv \sum_{h=1}^{H} \mathbf{e}_h$ and the **aggregate excess demand** vector as $\mathbf{z}(\mathbf{p}) \equiv \sum_{h=1}^{H} \mathbf{z}_h(\mathbf{p})$.

12.5.2   Walras's law

THEOREM 12.5.1 (WALRAS'S LAW). *The value of aggregate excess demand is non-positive for all price vectors, i.e.*

$$\mathbf{p}^\top \mathbf{z}(\mathbf{p}) \le 0 \quad \forall\, \mathbf{p} \in S^{N-1} \tag{12.57}$$

Proof: Walras's law is essentially an aggregate budget constraint.
Individual h's budget constraint guarantees that

$$\mathbf{p}^\top \mathbf{x}_h(\mathbf{p}) \le \mathbf{p}^\top \mathbf{e}_h \quad \forall\, h, \mathbf{p} \tag{12.58}$$

Summing these budget constraints over all individuals yields

$$\sum_{h=1}^{H} \mathbf{p}^\top \mathbf{x}_h(\mathbf{p}) \le \sum_{h=1}^{H} \mathbf{p}^\top \mathbf{e}_h \quad \forall\, \mathbf{p} \tag{12.59}$$

or

$$\mathbf{p}^\top \mathbf{x}(\mathbf{p}) \le \mathbf{p}^\top \mathbf{e} \quad \forall\, \mathbf{p} \tag{12.60}$$

or, in terms of excess demand,

$$\mathbf{p}^\top \mathbf{z}(\mathbf{p}) \le 0 \quad \forall\, \mathbf{p} \tag{12.61}$$

$\square$

There are several useful corollaries to Walras's law, the first of which should be obvious.

COROLLARY 12.5.2 *If individual preferences exhibit local non-satiation, then individual budget constraints and, hence, Walras's law hold as equalities, or*

$$\mathbf{p}^\top \mathbf{z}(\mathbf{p}) = 0 \quad \forall\, \mathbf{p} \tag{12.62}$$

COROLLARY 12.5.3 *If individual preferences exhibit local non-satiation and $N-1$ markets clear at prices $\mathbf{p}$, then either the Nth market also clears or the Nth good is free.*

Proof: It is assumed that

$$x^n(\mathbf{p}) = e^n, \quad n = 1, 2, \ldots, N-1 \tag{12.63}$$

Multiplying each of these equations by p^n and summing yields

$$\sum_{n=1}^{N-1} p^n x^n(\mathbf{p}) = \sum_{n=1}^{N-1} p^n e^n \tag{12.64}$$

Rearranging Walras's law (12.62) yields

$$\mathbf{p}^\top \mathbf{x}(\mathbf{p}) = \mathbf{p}^\top \mathbf{e} \tag{12.65}$$

Finally, subtracting (12.64) from (12.65) yields

$$p^N x^N(\mathbf{p}) = p^N e^N \tag{12.66}$$

from which it follows that either $x^N(\mathbf{p}) = e^N$ (the Nth market clears) or $p^N = 0$ (the Nth good is free). ☐

COROLLARY 12.5.4 *If individual preferences exhibit local non-satiation and* $\mathbf{p}$ *is an equilibrium price vector, then*

(a) *where prices are positive the corresponding markets clear (i.e.* $p^n > 0 \Rightarrow z^n(\mathbf{p}) = 0$*); and*
(b) *goods in excess supply are free (i.e.* $z^n(\mathbf{p}) < 0 \Rightarrow p^n = 0$*).*

Proof: By definition of equilibrium prices, $p^n \geq 0$ for all n and, since the equilibrium allocation must be feasible, $z^n(\mathbf{p}) \leq 0$ for all n. Hence, the product $p^n z^n(\mathbf{p}) \leq 0$ for all n. By local non-satiation, we have the equality version of Walras's law:

$$\sum_{n=1}^{N} p^n z^n(\mathbf{p}) = 0 \tag{12.67}$$

The only way that such a sum of non-positive terms can equal zero is if each individual term is equal to zero, or

$$p^n z^n(\mathbf{p}) = 0 \quad \forall\, n \tag{12.68}$$

For any good for which $p^n > 0$, it must be the case that $z^n(\mathbf{p}) = 0$; in other words, markets clear exactly for goods whose prices are positive. Similarly, for any good for which $z^n(\mathbf{p}) < 0$, it must be the case that $p^n = 0$; in other words, goods in excess supply must be free. ☐

Nowhere in this section have we assumed that individual or aggregate demand is single-valued.

In summary, the two versions of Walras's law state the following:

• Aggregate excess demand always satisfies

$$\mathbf{p}^\top \mathbf{z}(\mathbf{p}) \leq 0 \tag{12.69}$$

• If all individual preferences satisfy the local non-satiation axiom, then aggregate excess demand satisfies

$$\mathbf{p}^\top \mathbf{z}(\mathbf{p}) = 0 \tag{12.70}$$

12.5.3 *Equation counting*

For systems of linear equations, the technique of equation counting (proving that the number of independent equations equals the number of unknowns to be solved for) is often sufficient to prove the existence of a unique solution. In the search for a competitive equilibrium, however, the equations are generally nonlinear, and having the same number of equations and unknowns alone is neither a necessary nor a sufficient condition for the existence of a solution. Given this caveat, it is still reassuring to note that we have:

(a) $N \times H + N - 1$ unknowns, comprising

    (i) $N \times H$ allocations x_h^n, and
    (ii) $N - 1$ relative prices (assuming the existence of a numeraire with a price of unity); and

(b) $N \times H + N - 1$ equations, comprising

    (i) $N \times H$ utility-maximization first-order conditions, and
    (ii) $N - 1$ independent market-clearing equations (the Nth being redundant by Corollary 12.5.3 above).

12.5.4 Fixed-point theorems

If single-valued demand functions exist, then Brouwer's fixed-point theorem can be used to demonstrate the existence of equilibrium.[12] In the case of multivalued demand correspondences, Kakutani's fixed-point theorem can be used.[13]

THEOREM 12.5.5 (BROUWER'S FIXED-POINT THEOREM). *If X is a non-empty, closed, bounded, convex subset of $\mathbb{R}^n$ and $f: X \to X$ is a continuous function mapping X into itself, then f has a fixed point, i.e. there exists $\mathbf{x}^* \in X$ such that $f(\mathbf{x}^*) = \mathbf{x}^*$.*

Proof: The full proof of this theorem is beyond the scope of this book. In fact, it is generally proved as a corollary to Kakutani's fixed-point theorem (Theorem 12.5.6).

When $X = S^1$, the unit simplex in $\mathbb{R}^2$, the proof is a straightforward consequence of the intermediate value theorem (Theorem 9.6.1). The unit simplex S^1 is just a closed line segment and can be identified with the closed unit interval $[0, 1]$.

Define a new continuous function $g: [0, 1] \to [-1, 1]: x \mapsto f(x) - x$. Then

$$g(0) = f(0) - 0 \geq 0 \tag{12.71}$$

and

$$g(1) = f(1) - 1 \leq 0 \tag{12.72}$$

Applying the intermediate value theorem with $\lambda = 0$, and noting that λ then lies between $g(0)$ and $g(1)$, tells us that there exists $x^* \in [0, 1]$ such that $g(x^*) = 0$ or, equivalently, $f(x^*) = x^*$ as required.   □

To see that X must be closed, consider the continuous function $f: (0, 1) \to (0, 1): x \mapsto x^2$. Since $x > x^2$ for all $x \in (0, 1)$, this function has no fixed point. Extending the domain to the closed interval $[0, 1]$, however, gives a function with two fixed points, at 0 and 1.

To see that X must be bounded, consider the continuous function $f: \mathbb{R} \to \mathbb{R}: x \mapsto x + 1$, which has no fixed point.

To see that X must be convex, consider the continuous function $f: \left[0, \frac{1}{3}\right] \cup \left[\frac{2}{3}, 1\right] \to \left[0, \frac{1}{3}\right] \cup \left[\frac{2}{3}, 1\right]: x \mapsto 1 - x$. This function has no fixed point. Extending the domain to the convex interval $[0, 1]$, however, gives a function with a fixed point, at $\frac{1}{2}$.

To see that X must be continuous, consider the function

$$f\colon [0,1] \to [0,1]\colon x \mapsto \begin{cases} \frac{2}{3} & \text{if } x < \frac{1}{2} \\ \frac{1}{3} & \text{if } x \geq \frac{1}{2} \end{cases} \tag{12.73}$$

This function has no fixed point.

The reader is advised to graph each of these functions as an exercise; see Exercise 12.5.

THEOREM 12.5.6 (KAKUTANI'S FIXED-POINT THEOREM). *If X is a non-empty, closed, bounded, convex subset of $\mathbb{R}^n$ and $f\colon X \to X$ is a convex-valued correspondence mapping X into itself that has a closed graph, then f has a **fixed point**, i.e. there exists $\mathbf{x}^* \in X$ such that $\mathbf{x}^* \in f(\mathbf{x}^*)$.*

Proof: Once again, the full proof of this theorem is beyond the scope of this book; see Hildenbrand and Kirman (1988, p. 277) for more detail and Berge (1997, pp. 174–6) for a full proof.   □

12.5.5 Existence of equilibrium

THEOREM 12.5.7 (EXISTENCE OF EQUILIBRIUM IN A PURE EXCHANGE ECONOMY). *If*

(a) *the aggregate excess demand function $\mathbf{z}$ is a continuous (single-valued) function (for which a sufficient condition is strict convexity of preferences), and*
(b) *Walras's law holds as an equality or $\mathbf{p}^\top \mathbf{z}(\mathbf{p}) = 0$ for all $\mathbf{p}$ (for which a sufficient condition is that preferences exhibit local non-satiation),*

then there exists $\mathbf{p}^* \in S^{N-1}$ such that $z^n(\mathbf{p}^*) \leq 0$ for all $n = 1, 2, \ldots, N$, i.e. there exists an equilibrium price vector.

Proof: This proof is based on Varian (1992, pp. 321–2).

Define a vector-valued function $\mathbf{f}\colon S^{N-1} \to S^{N-1}$ by

$$\mathbf{p} \mapsto \frac{1}{\mathbf{1}_N^\top(\mathbf{p} + \max\{\mathbf{0}_N, \mathbf{z}(\mathbf{p})\})}(\mathbf{p} + \max\{\mathbf{0}_N, \mathbf{z}(\mathbf{p})\}) \tag{12.74}$$

where $\max\{\mathbf{x}, \mathbf{y}\}$ denotes the component-by-component maximum of two vectors.

It should be clear from this definition that $\mathbf{1}_N^\top \mathbf{f}(\mathbf{p}) = 1$ and that $f^n(\mathbf{p}) \geq 0$ for all $n = 1, 2, \ldots, N$ so that $\mathbf{f}(\mathbf{p}) \in S^{N-1}$ as stated.

Furthermore, $\mathbf{f}$ is a continuous function since by assumption $\mathbf{z}$ is a continuous function, and sums, ratios and maxima of continuous functions are continuous – unless the denominator vanishes, which cannot happen in this case as it is positive on S^{N-1} by construction.

Thus all the conditions of Brouwer's fixed-point theorem are satisfied and $\mathbf{f}$ has a fixed point, say, $\mathbf{p}^*$, with

$$\mathbf{p}^* = \frac{1}{\mathbf{1}_N^\top(\mathbf{p}^* + \max\{\mathbf{0}_N, \mathbf{z}(\mathbf{p}^*)\})}(\mathbf{p}^* + \max\{\mathbf{0}_N, \mathbf{z}(\mathbf{p}^*)\}) \tag{12.75}$$

Cross-multiplying and using the fact that $\mathbf{1}_N^\top \mathbf{p}^* = 1$ (since $\mathbf{p}^* \in S^{N-1}$) yields

$$(1 + \mathbf{1}_N^\top \max\{\mathbf{0}_N, \mathbf{z}(\mathbf{p}^*)\})\mathbf{p}^* = \mathbf{p}^* + \max\{\mathbf{0}_N, \mathbf{z}(\mathbf{p}^*)\} \tag{12.76}$$

Cancelling a $\mathbf{p}^*$ from each side of the equation gives

$$(\mathbf{1}_N^\top \max\{\mathbf{0}_N, \mathbf{z}(\mathbf{p}^*)\})\mathbf{p}^* = \max\{\mathbf{0}_N, \mathbf{z}(\mathbf{p}^*)\} \tag{12.77}$$

Taking the dot product of each side with $\mathbf{z}(\mathbf{p}^*)$ gives

$$(\mathbf{1}_N^\top \max\{\mathbf{0}_N, \mathbf{z}(\mathbf{p}^*)\})\mathbf{p}^{*\top}\mathbf{z}(\mathbf{p}^*) = \max\{\mathbf{0}_N, \mathbf{z}(\mathbf{p}^*)\}^\top \mathbf{z}(\mathbf{p}^*) \tag{12.78}$$

By assumption, Walras's law holds in its equality form, so the left-hand side of this equation is zero. The right-hand side is a sum of N terms. The nth term in this sum is zero if $z^n(\mathbf{p}^*) \leq 0$ and strictly positive otherwise. But a sum of such non-negative terms can equal zero only if every single term is zero, or $z^n(\mathbf{p}^*) \leq 0$ for all $n = 1, 2, \ldots, N$. In other words, all markets clear at prices $\mathbf{p}^*$, and so $\mathbf{p}^*$ is an equilibrium price vector as required. $\square$

The preceding proof carries over unaltered to a production economy, where $\mathbf{z}$ denotes the excess demand function for such an economy.

12.5.6   No-arbitrage principle

DEFINITION 12.5.5 An **arbitrage opportunity** means the opportunity to acquire a consumption vector or its constituents, directly or indirectly, at one price, and to sell the same consumption vector or its constituents, directly or indirectly, at a higher price.

THEOREM 12.5.8 (NO-ARBITRAGE PRINCIPLE). *Arbitrage opportunities do not exist in equilibrium in an economy in which at least one agent has preferences that exhibit local non-satiation.*[14]

Proof: If an arbitrage opportunity exists, then any individual whose preferences exhibit local non-satiation will seek to exploit it on an infinite scale, thereby increasing wealth without bound and removing the budget constraint. Since local non-satiation rules out bliss points, the individual's utility too can be increased without bound. Thus that individual's Marshallian demand for the components of the arbitrage opportunity is not well defined, i.e. not finite.

If even one individual in an economy has preferences that exhibit local non-satiation, it follows that prices that permit arbitrage opportunities will not allow markets to clear.

Conversely, if such an economy is in equilibrium, then markets must clear, demand for all goods must be finite, and either there are no arbitrage opportunities or all individuals have attained bliss points, which would be impossible if even one individual's preferences exhibited local non-satiation. $\square$

The most useful applications of the no-arbitrage principle are probably those in the financial markets. We will come across several such applications later in this book. Many of these applications occur in a multi-period context, for example, in defining the term structure of interest rates; see Section 15.4. The most powerful application of the no-arbitrage principle in finance is in the derivation of option-pricing formulas, since options can be shown to be identical to various combinations of the underlying security and the risk-free security; see Exercise 16.2. Indeed, it is the no-arbitrage principle itself that allows us to refer to "*the* risk-free security", since it rules out the possibility of several different risk-free securities existing with different risk-free rates of return in a particular currency.

The simple rule for working out how to exploit arbitrage opportunities is "buy low, sell high". With interest rates and currencies, for example, this may be a non-trivial calculation.

EXAMPLE 12.5.1 A common application of the no-arbitrage principle is the theory of **covered interest rate parity**.[15]

Suppose that an individual has access to one-period risk-free investments in both EUR and GBP at interest rates, respectively, of i_{EUR} and i_{GBP}. Let S_t denote the current spot GBP/EUR exchange rate and let F_t denote the current one-period forward GBP/EUR exchange rate.

A risk-free payoff in euro can be engineered in either of two ways:

(a) invest the principal, say 1 EUR, in the risk-free euro investment, for a return of $1 + i_{EUR}$; or

(b) convert the principal to S_t GBP, invest the proceeds in the risk-free sterling investment for a GBP return of $S_t(1 + i_{GBP})$ and sell this GBP return forward for a EUR return of $(S_t/F_t)(1 + i_{GBP})$.

An arbitrage opportunity would exist if these payoffs were different. Thus, in equilibrium, the forward rate must be given by

$$F_t = S_t \frac{1 + i_{GBP}}{1 + i_{EUR}} \tag{12.79}$$

An identical result can be obtained by starting with a principal of 1 GBP.

In other words, apparent gains in the money market when interest rates differ across currencies will be eroded by losses in the foreign exchange market, or, in other words, a currency with a higher interest rate will depreciate against one with a lower interest rate, at least insofar as the comparison between the current forward and spot rates is concerned. ◇

Covered interest rate parity, however, says nothing about the relationship between the current one-period forward exchange rate F_t and the unknown future spot exchange rate $\tilde{S}_{t+1}$; the latter relationship will be discussed later, in particular in Sections 13.10.2 and 16.7.

It will be seen in Section 16.6 that there are strong parallels between

- the no-arbitrage principle,
- the risk-neutral world and
- the efficient markets hypothesis.

12.6 Welfare theorems

12.6.1 Edgeworth box

Let us consider now the simplest possible pure exchange economy, with $N = H = 2$, i.e. two goods and two consumers. Market-clearing allocations in this economy can be represented by points in Figure 12.4, a rectangular diagram with dimensions $(e_1^1 + e_2^1) \times (e_1^2 + e_2^2)$. The coordinates of each point with respect to the lower left corner (or origin) of the so-called **Edgeworth box** (sometimes also referred to as the Edgeworth–Bowley box)[16] represent the consumption vector of consumer 1, and the coordinates with respect to the upper right corner (origin) represent the consumption vector of consumer 2. The (convex) indifference curves of each consumer are also shown; see Exercise 12.9 for details of the calculations involved. At a typical point in the Edgeworth box, such as A, there is a lens-shaped region (shaded) throughout which both consumers would attain higher utility than at A. At A itself, a range of trades are available to the consumers that would leave both strictly better off. At a point such as B, however, the indifference curve of consumer 1 is tangential to the indifference curve of consumer 2; any movement away from this point will leave at least one consumer strictly worse off.

It is easily shown that an equilibrium allocation can only occur at a tangency point such as B, with the ratio of the equilibrium prices equal to the absolute value of the common slope of the two indifference curves. The set of all such points is known as the **contract curve**. The existence proof above shows that, for any initial allocation in the Edgeworth box, there exists

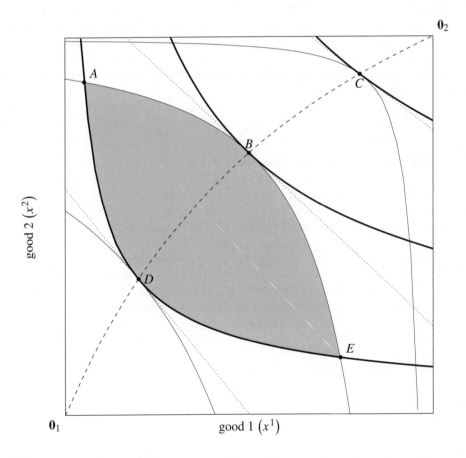

Figure 12.4 Edgeworth box

at least one budget line through that point that is somewhere perpendicular to the contract curve (i.e. tangential to the two indifference curves that meet at that point).

The set of allocations along the contract curve has special properties, which we will now consider in the general case of N goods and H households.

12.6.2 Pareto optimality

We have already alluded to the fact that, at points on the contract curve, and only at such points, it is not possible to make one consumer better off, in terms of utility, without making the other worse off. This is the motivation behind the following more general definition.

DEFINITION 12.6.1 A feasible allocation $\mathbf{X} = [\mathbf{x}_1 \ \mathbf{x}_2 \ \cdots \ \mathbf{x}_H]$ is **Pareto**[17] **optimal**, given initial endowments $\mathbf{e}_1, \mathbf{e}_2, \ldots, \mathbf{e}_H$, if there does not exist any **feasible** way of re-allocating the same initial aggregate endowment, $\sum_{h=1}^{H} \mathbf{e}_h$, which makes one individual better off without making any other worse off.

DEFINITION 12.6.2 The allocation $\mathbf{X} = [\mathbf{x}_1 \ \mathbf{x}_2 \ \cdots \ \mathbf{x}_H]$ is **Pareto dominated by** the allocation $\mathbf{X}' = [\mathbf{x}_1' \ \mathbf{x}_2' \ \cdots \ \mathbf{x}_H']$ if $\sum_{h=1}^{H} \mathbf{x}_h = \sum_{h=1}^{H} \mathbf{x}_h'$, with $\mathbf{x}_h' \succeq_h \mathbf{x}_h$ for all h and $\mathbf{x}_h' \succ_h \mathbf{x}_h$ for at least one h.

For example, in Figure 12.4, the point A is Pareto dominated by any point in the interior of the shaded region. A move from A to any such point would be **Pareto efficient** or **allocatively efficient**. A move to such a point that is also on the contract curve would be Pareto optimal.

The reader should take care to distinguish between various notions of efficiency discussed in this book, including computational efficiency (e.g. Section 2.4), statistical efficiency (Section 14.3), informational efficiency (e.g. Section 13.10.2), mean–variance efficiency (Sections 17.4 and 17.5) and Pareto or allocative efficiency. The notions of allocative efficiency and informational efficiency are both of fundamental importance in economics and finance.

12.6.3 First welfare theorem

THEOREM 12.6.1 (FIRST WELFARE THEOREM). *If the pair $(\mathbf{p}, \mathbf{X})$ is an equilibrium (for given preferences, $\succeq_h$, which exhibit local non-satiation, and given endowments, $\mathbf{e}_h$, $h = 1, 2, \ldots, H$), then $\mathbf{X}$ is a Pareto optimal allocation.*

Proof: The proof, based on Varian (1992, Section 17.6), is by contradiction. Suppose that $\mathbf{X}$ is an equilibrium allocation that is Pareto dominated by a feasible allocation $\mathbf{X}'$. Then each individual h either is strictly better off under $\mathbf{X}'$ or is indifferent between $\mathbf{X}$ and $\mathbf{X}'$.

(a) If individual h is strictly better off under $\mathbf{X}'$, or $\mathbf{x}_h' \succ_h \mathbf{x}_h$, then it follows that individual h cannot afford $\mathbf{x}_h'$ at the equilibrium prices $\mathbf{p}$ or

$$\mathbf{p}^\top \mathbf{x}_h' > \mathbf{p}^\top \mathbf{x}_h = \mathbf{p}^\top \mathbf{e}_h \tag{12.80}$$

The latter equality is just the budget constraint, which is binding since we have assumed local non-satiation.

(b) Similarly, if individual h is indifferent between $\mathbf{X}$ and $\mathbf{X}'$, or $\mathbf{x}'_h \sim_h \mathbf{x}_h$, then it follows that

$$\mathbf{p}^\top \mathbf{x}'_h \geq \mathbf{p}^\top \mathbf{x}_h = \mathbf{p}^\top \mathbf{e}_h \qquad (12.81)$$

since, if $\mathbf{x}'_h$ was to cost strictly less than $\mathbf{x}_h$, then, by local non-satiation, some nearby consumption vector (near enough to $\mathbf{x}'_h$ to also cost less than $\mathbf{x}_h$) would be strictly preferred to $\mathbf{x}_h$. Thus $\mathbf{x}_h$ would not maximize utility given the budget constraint.

Summing (12.80) and (12.81) over all households (at least one of which must fall in the former category) yields

$$\mathbf{p}^\top \sum_{h=1}^{H} \mathbf{x}'_h > \mathbf{p}^\top \sum_{h=1}^{H} \mathbf{x}_h = \mathbf{p}^\top \sum_{h=1}^{H} \mathbf{e}_h \qquad (12.82)$$

where the equality is essentially Walras's law.

But since $\mathbf{X}'$ is feasible we must have, for each good n,

$$\sum_{h=1}^{H} x'^n_h \leq \sum_{h=1}^{H} e^n_h \qquad (12.83)$$

and, hence, multiplying by prices (which are non-negative) and summing over all goods,

$$\mathbf{p}^\top \sum_{h=1}^{H} \mathbf{x}'_h \leq \mathbf{p}^\top \sum_{h=1}^{H} \mathbf{e}_h \qquad (12.84)$$

But the weak inequality in (12.84) contradicts the strict inequality in (12.82), so no such Pareto dominant allocation $\mathbf{X}'_h$ can exist. □

12.6.4　Second welfare theorem

The first welfare theorem showed that every competitive equilibrium allocation is Pareto optimal. In this section, we introduce the second welfare theorem, which shows that every Pareto optimal allocation is, in a sense to be made clear, a competitive equilibrium allocation. We make slightly stronger assumptions than are essential for the proof of this theorem. This allows us to give an easier proof.

THEOREM 12.6.2 (SECOND WELFARE THEOREM). *If all individuals have preferences that are convex, continuous and strongly monotonic, and if X^* is a Pareto optimal allocation such that all households are allocated positive amounts of all goods ($x^{*n}_h > 0$ for all $n = 1, 2, \ldots, N$, $h = 1, 2, \ldots, H$), then a re-allocation of the initial aggregate endowment can yield an equilibrium where the allocation is X^*.*

Proof: There are four main steps in the proof, which is based on Varian (1992, Section 17.7):

(a) First we construct a set of utility-enhancing **endowment perturbations**, so that we can use the supporting hyperplane theorem (Theorem 7.6.1) to find prices at which no such endowment perturbation is affordable.

The supporting hyperplane theorem will essentially be applied to the upper contour sets of quasi-concave utility functions, which are convex sets. We will interpret the supporting hyperplane as a budget hyperplane, and the normal vector as a price vector, so that at those prices nothing giving higher utility than that given by the relevant boundary point is affordable.

We need to use the fact that a sum of convex sets, such as

$$X + Y \equiv \{\mathbf{x} + \mathbf{y} : \mathbf{x} \in X, \ \mathbf{y} \in Y\} \tag{12.85}$$

is also a convex set; see Theorem 7.3.1.

Given the implicit aggregate initial endowment $\mathbf{x}^* = \sum_{h=1}^{H} \mathbf{x}_h^*$, we interpret any vector of the form $\mathbf{z} = \sum_{h=1}^{H} \mathbf{x}_h - \mathbf{x}^*$ as an endowment perturbation. Now consider the set of all ways of changing the aggregate endowment without making anyone worse off:

$$Z \equiv \left\{ \mathbf{z} \in \mathbb{R}^N : \exists\, x_h^n \geq 0 \ \forall\, n, h \ \text{ s.t. } u_h(\mathbf{x}_h) \geq u_h(\mathbf{x}_h^*) \text{ and } \mathbf{z} = \sum_{h=1}^{H} \mathbf{x}_h - \mathbf{x}^* \right\} \tag{12.86}$$

The set Z is a sum of convex sets

$$Z = \sum_{h=1}^{H} X_h - \{\mathbf{x}^*\} \tag{12.87}$$

where

$$X_h \equiv \{\mathbf{x}_h : u_h(\mathbf{x}_h) \geq u_h(\mathbf{x}_h^*)\}, \quad h = 1, 2, \ldots, H \tag{12.88}$$

which are convex since u_h is a quasi-concave utility function representing convex preferences.

(b) Next, we need to show that the zero vector is in the set Z, but not in the interior of Z.

To show that $\mathbf{0} \in Z$, we just set $\mathbf{x}_h = \mathbf{x}_h^*$ and observe that $\mathbf{0} = \sum_{h=1}^{H} \mathbf{x}_h^* - \mathbf{x}^*$.

The zero vector is not, however, in the interior of Z, since then Z would contain some vector, say $\mathbf{z}^*$, in which all components were strictly negative. In other words, we could take away some of the aggregate endowment of every good without making anyone worse off than under the allocation $\mathbf{X}^*$. But by then giving $-\mathbf{z}^*$ back to one individual, he or she could be made better off without making anyone else worse off, contradicting Pareto optimality, using the assumption that preferences are strongly monotonic.

So, applying the supporting hyperplane theorem with $\mathbf{z}^* = \mathbf{0}$, we have a price vector $\mathbf{p}^*$ such that $0 = \mathbf{p}^{*\top} \mathbf{0} \leq \mathbf{p}^{*\top} \mathbf{z}$ for all $\mathbf{z} \in Z$. Since preferences are strongly monotonic, the set Z must contain all the standard unit basis vectors $((1, 0, 0, \ldots, 0), (0, 1, 0, \ldots, 0),$ etc.). This fact can be used to show that all components of $\mathbf{p}^*$ are non-negative, which is essential if it is to be interpreted as an equilibrium price vector.

(c) Next, we specify one way of redistributing the initial endowment in order that the desired prices and allocation emerge as a competitive equilibrium. All we need to do is to value endowments at the equilibrium prices, and to redistribute the aggregate endowment of each good to consumers in proportion to their share in aggregate wealth computed in this way.

(d) Finally, we confirm that utility is maximized by the given Pareto optimal allocation, $\mathbf{X}^*$, at these prices. As usual, the proof is by contradiction: the details are left as an exercise; see Exercise 12.10.   □

12.6.5   *Complete and perfectly competitive markets*

The first welfare theorem tells us that competitive equilibrium allocations are Pareto optimal if markets are **complete**, i.e. if there exists a market corresponding to every good appearing as a variable in the utility function. If there are missing markets, i.e. markets are **incomplete**, then competitive trading may not lead to a Pareto optimal allocation.

We can use the Edgeworth box diagram (Figure 12.4) to illustrate the simplest possible version of this principle: a two-good world requires only one market, in which the first good can be exchanged for the second. Exchange allows the two households to move from an initial endowment point somewhere off the contract curve to a competitive equilibrium on the contract curve. However, if exchange is not possible, then the two households must remain at the initial endowment point, which in general will not be Pareto efficient. (It is possible, though unlikely, that the initial endowment point may by pure coincidence fall on the contract curve.)

If we think of goods as being distinguished by physical characteristics only, then the implicit assumption of complete markets is quite plausible. When we move on in later chapters to consider choice under uncertainty, in which goods can also be distinguished by the state of the world in which they are consumed, completeness of markets may be the exception rather than the rule.

Like the assumption of complete markets, the assumption of perfect competition is implicitly built into everything that we have done so far. Much of modern microeconomics and industrial economics deals with situations where competition is imperfect and where one or more households or firms enjoy market power. Such situations will not be considered in this book, but we will return later to the discussion of complete and incomplete markets.

12.6.6   *The representative agent approach*

Consider a pure exchange economy with an aggregate endowment of N goods represented by the vector $\mathbf{e} \in \mathbb{R}_+^N$ and with H agents with strictly convex preferences represented by the utility functions u_h, $h = 1, 2, \ldots, H$, which are assumed to exhibit local non-satiation. Recall that strictly convex preferences guarantee that the consumer's problem has a unique solution and that local non-satiation guarantees that the budget constraint is strictly binding at the unique solution.

Within this framework, we can define a **representative agent** as follows

DEFINITION 12.6.3 The utility derived by representative agent θ from an aggregate endowment of $\mathbf{e}$ is defined to be $M(\theta, \mathbf{e})$, where M is the envelope function for the problem

$$\max_{\{x_h^n : h=1,2,\ldots,H; \; n=1,2,\ldots,N\}} \sum_{h=1}^{H} \theta_h u_h(\mathbf{x}_h) \quad \text{s.t.} \sum_{h=1}^{H} \mathbf{x}_h = \mathbf{e} \tag{12.89}$$

for some parameter vector of positive weightings[18] $\theta \in \mathbb{R}_{++}^H$.

Each value of θ gives us a different representative agent.

We will use the standard notation for the corresponding optimal response function, namely $\mathbf{X}^*(\boldsymbol{\theta}, \mathbf{e})$. Representative agents differ in the weightings θ_h assigned to the different individuals' utility functions in problem (12.89). Note that the envelope function M is homogeneous of degree one and the optimal response function homogeneous of degree zero in the positive weightings $\boldsymbol{\theta}$. Thus, without loss of generality, we need consider only weights such that $\sum_{h=1}^{H} \theta_h = 1$.

Note that $\mathbf{X}^*(\boldsymbol{\theta}, \mathbf{e})$ will always be a Pareto optimal allocation, since otherwise it would be possible to increase the value of the objective function in problem (12.89) by redistributing the initial endowment $\mathbf{e}$ in such a way as to increase the value of some $u_h(\mathbf{x}_h)$ without reducing the value of any other $u_h(\mathbf{x}_h)$ or of the weighted sum $\sum_{h=1}^{H} \theta_h u_h(\mathbf{x}_h)$. We will now analyse in more detail the correspondence between the set of Pareto optimal allocations and the set of solutions to problem (12.89).

The first-order conditions for problem (12.89) with respect to the x_h^n, obtained by differentiating the Lagrangian with respect to x_h^n, are

$$\theta_h \frac{\partial u_h}{\partial x_h^n}(\mathbf{x}_h) - \phi^n = 0, \quad h = 1, 2, \ldots, H; \; n = 1, 2, \ldots, N \tag{12.90}$$

where $\phi^1, \phi^2, \ldots, \phi^N$ denote non-negative Kuhn–Tucker multipliers. Our aim now is to show that these $N \times H$ equations can be solved for the $N \times H$ variables x_h^n (in terms of the multipliers $\phi^1, \phi^2, \ldots, \phi^N$).

Let us consider at this stage the utility-maximization problem faced by individual h, which is

$$\max_{\{x_h^n : n = 1, 2, \ldots, N\}} u_h(\mathbf{x}_h) \quad \text{s.t.} \quad \sum_{n=1}^{N} p^n x_h^n \leq \sum_{n=1}^{N} p^n e_h^n \tag{12.91}$$

As already noted, problem (12.91) will have a unique solution for each agent, with the further property that, by Theorem 12.5.7, there exist equilibrium prices $\mathbf{p}^* \in \mathbb{R}_+^N$ for which markets clear.

The first-order conditions for problem (12.91) with respect to the x_h^n, obtained by differentiating individual h's Lagrangian with respect to x_h^n, are

$$\frac{\partial u_h}{\partial x_h^n}(\mathbf{x}_h) - \lambda_h p^n = 0, \quad h = 1, 2, \ldots, H; \; n = 1, 2, \ldots, N \tag{12.92}$$

where λ_h is the non-negative Kuhn–Tucker multiplier for individual h's utility-maximization problem.

As we have assumed local non-satiation, the constraints will be binding in both problem (12.89) and problem (12.91). Hence, by the arguments in Section 10.5.1, the λ_h will be strictly positive, so we can divide each individual's set of first-order conditions by the corresponding multiplier to obtain

$$\frac{1}{\lambda_h} \frac{\partial u_h}{\partial x_h^n}(\mathbf{x}_h) - p^n = 0, \quad h = 1, 2, \ldots, H; \; n = 1, 2, \ldots, N \tag{12.93}$$

Note now the equivalence or duality between the systems of equations (12.90) and (12.93). If we choose the weightings in the representative agent's utility function to equal the reciprocals of the corresponding equilibrium Kuhn–Tucker multipliers ($\theta_h = 1/\lambda_h$), then $\mathbf{X}^*(\boldsymbol{\theta}, \mathbf{e})$ will equal the corresponding equilibrium allocation and furthermore the Kuhn–Tucker multipliers for the representative agent's problem will equal the corresponding equilibrium prices ($\phi^n = p^{n*}$).

Conversely, if we set prices equal to the Kuhn–Tucker multipliers from the representative agent's problem and redistribute the initial endowments appropriately, then markets will clear, and the Kuhn–Tucker multipliers from the individuals' problems will equal the reciprocals of their weightings in the representative agent's utility function.

In other words, given the aggregate endowment $\mathbf{e}$, there is a direct correspondence between vectors of weightings $\boldsymbol{\theta}$ and equilibrium distributions $\mathbf{X}^*$ of the initial allocation.

Given an equilibrium allocation, $\mathbf{X}^*$, we can find the corresponding market-clearing prices, $\mathbf{p}^*$, and compute the corresponding individual Kuhn–Tucker multipliers, λ_h. If we then assign the weightings in the representative agent's utility function in inverse proportion to the Kuhn–Tucker multipliers, the allocation given by the representative agent's optimal response function will equal the equilibrium allocation.

Conversely, given a vector of weightings $\boldsymbol{\theta}$, we can use the optimal response function to find an allocation $\mathbf{X}^*(\boldsymbol{\theta}, \mathbf{e})$. If we redistribute the initial endowment accordingly and set prices equal to the Kuhn–Tucker multipliers from the representative agent's problem ($p^n = \phi^n$), then individuals will not wish to trade and the market will be in equilibrium, with individual Kuhn–Tucker multipliers equal to the reciprocals of the corresponding weightings.

12.6.7   Summary of characterizations of Pareto optimal allocations

We have now seen that there are four alternative equivalent characterizations of Pareto optimal allocations, which are summarized in the following theorem.

THEOREM 12.6.3 *Each of the following is an equivalent description of the set of allocations that are Pareto optimal for given fixed initial aggregate endowments $\sum_{h=1}^{H} \mathbf{e}_h$:*

(a) *by definition, feasible allocations such that no other feasible allocation strictly increases at least one individual's utility without decreasing the utility of any other individual;*
(b) *by the welfare theorems, equilibrium allocations for all possible distributions of the fixed initial aggregate endowment;*
(c) *in two dimensions, allocations lying on the contract curve in the Edgeworth box; and*
(d) *allocations given by the optimal response function for problem (12.89), as faced by representative agents, for all possible weighting vectors $\boldsymbol{\theta} \in S^{H-1}$.*

Note that corresponding to each Pareto optimal and equilibrium allocation there is at least:

1. one equilibrium price vector, which is also the vector of Kuhn–Tucker multipliers for problem (12.89); and
2. one vector of positive weights defining the relevant representative agent, which is also the vector of the reciprocals of the equilibrium Kuhn–Tucker multipliers for the individual utility-maximization problems.

The weight θ_h assigned to individual h roughly speaking determines the individual's share of the aggregate endowment in the associated equilibrium.

EXERCISES

12.1 Consider an individual who consumes two goods. Let x represent the quantity of the first good consumed and y the quantity of the second good consumed. Suppose that the individual's preferences over all possible combinations (x, y) can be represented by the Cobb–Douglas utility function

$$U(x, y) = x^\alpha y^{1-\alpha}$$

where $0 < \alpha < 1$. Let M denote the individual's wealth, and p and q the prices of the two goods.

(a) Write down the equality-constrained optimization problem that the utility-maximizing individual must solve.
(b) Solve the problem formulated in (a).
(c) Sketch a selection of the individual's indifference curves and his budget constraint and indicate your solution on the diagram.
(d) What happens if the utility function is

   (i) $V(x, y) = xy$
   (ii) $W(x, y) = \ln x + \ln y$
   (iii) $Z(x, y) = x^\beta + y^\gamma$ where β and γ are positive real numbers?

12.2 In introductory economics, it is often suggested that the demand function for a typical good is linear in its own price (given income and the prices of other goods). In other words, for the two-good case, it is suggested that the demand function for good 1, say, is given by

$$x_1(p_1, p_2, M) = a(p_2, M) + b(p_2, M)p_1$$

By working backwards from the first-order conditions for the consumer's problem, find a two-good utility function that will give such a demand function for good 1, and find the corresponding demand function for good 2.

Which of the standard axioms are satisfied by the preference relation underlying your utility function?

12.3 A consumer has a Stone–Geary utility function defined by

$$u(\mathbf{x}) = \sum_{j=1}^n \beta_j \ln(x_j - \alpha_j)$$

where $\mathbf{x} = (x_1, x_2, \ldots, x_n)$ is her consumption vector and the α_j and β_j are positive constants with $\sum_{j=1}^n \beta_j = 1$.

Prove that maximization of utility subject to the budget constraint yields the demand functions

$$x_i(\mathbf{p}, M) = \alpha_i + \frac{\beta_i}{p_i}\left(M - \sum_{j=1}^{n} \alpha_j p_j\right)$$

for $i = 1, 2, \ldots, n$, where $\mathbf{p} = (p_1, p_2, \ldots, p_n)$ is the price vector and M is income.

These demand functions describe what is known as the **linear expenditure system**. Explain the rationale behind this name.

12.4 For the case of $N = 2$, sketch the indifference curves corresponding to the Leontief preferences given by equation (12.8).

Compare your answer to the isoquant map corresponding to the Leontief production function sketched for Exercise 11.5.

12.5 Draw graphs of all the functions cited on p. 320 in Section 12.5.4 as counter-examples to Brouwer's fixed-point theorem.

12.6 Consider a pure exchange economy in which there are three consumers, endowed with quantities of three goods given by the 3×3 matrix $\mathbf{E}$, where e^i_j, the element in row i and column j of $\mathbf{E}$, denotes consumer j's endowment of good i. Each consumer has Cobb–Douglas preferences, and consumer j's preferences ($j = 1, 2, 3$) are represented by the utility function

$$u_j : \mathbb{R}^3_{++} \to \mathbb{R} : (x_1, x_2, x_3) \mapsto a^1_j \ln x_1 + a^2_j \ln x_2 + a^3_j \ln x_3$$

where (for notational simplicity) $a^1_j + a^2_j + a^3_j = 1$.

(a) What proportion of the value of consumer j's initial endowment is accounted for by the value of his consumption of good i?
(b) Write down the individual demand functions, the aggregate demand function, the aggregate *excess* demand function and the market-clearing equations for this economy.
(c) Calculate the rank of the system of market-clearing equations. (Hint: use Walras's law.)
(d) Calculate the equilibrium prices (using good 1 as numeraire), incomes and allocations when initial endowments are given by

$$\mathbf{E} = \begin{bmatrix} 3 & 0 & 0 \\ 0 & 3 & 0 \\ 0 & 0 & 3 \end{bmatrix}$$

and the parameters of the utility functions are given by

$$\mathbf{A} \equiv \begin{bmatrix} a^1_1 & a^1_2 & a^1_3 \\ a^2_1 & a^2_2 & a^2_3 \\ a^3_1 & a^3_2 & a^3_3 \end{bmatrix} = \frac{1}{4}\begin{bmatrix} 1 & 1 & 2 \\ 1 & 2 & 1 \\ 2 & 1 & 1 \end{bmatrix}$$

(where each column of the matrix again represents a consumer and each row represents a good).

(e) Repeat the previous part of this question for initial endowments given by

$$E = \begin{bmatrix} 1 & 1 & 2 \\ 1 & 2 & 3 \\ 4 & 0 & 0 \end{bmatrix}$$

and utility functions parametrized by

$$A \equiv \begin{bmatrix} a_1^1 & a_2^1 & a_3^1 \\ a_1^2 & a_2^2 & a_3^2 \\ a_1^3 & a_2^3 & a_3^3 \end{bmatrix} = \frac{1}{3} \begin{bmatrix} 1 & 1 & 1 \\ 1 & 1 & 1 \\ 1 & 1 & 1 \end{bmatrix}$$

12.7 Under what circumstances are the equilibrium allocations in perfectly competitive markets Pareto optimal? Using a diagram, show how your answer applies in the context of a two-good, two-consumer economy.

12.8 Sketch the Edgeworth box and contract curve for an economy with two goods and two consumers whose preferences are identical. (Hint: use a symmetry argument.)

12.9 Consider a two-good $(1, 2)$, two-consumer (A, B) world in which both consumers have Cobb–Douglas preferences.

(a) Using appropriate notation, write out the individual demand functions, aggregate demand function and aggregate excess demand function for this economy.
(b) Explain why there is only one (independent) market-clearing equation, derive it and solve it for the equilibrium price ratio p_1/p_2 in terms of the exogenous preference parameters and the individual endowments.
(c) Find the condition(s) under which no trade will take place in equilibrium.
(d) Hence, or otherwise, write down the equation of the contract curve in this economy.
(e) Show that the contract curve is a (segment of a) rectangular hyperbola passing through both origins in the Edgeworth box. (Hint: this will be easier to see if you choose your units of measurement so that the aggregate endowment of each good is 1 unit.)

12.10 Confirm that the Pareto efficient allocation X^* in the statement of the second welfare theorem (Theorem 12.6.2) simultaneously maximizes each individual's utility, given the price vector p^* and the initial endowments found in the proof, even in the case in which one component of p^* is zero.

12.11 Find the first-order conditions for the problem of redistributing a given initial endowment to maximize one individual's utility (say, that of the first individual) subject to the constraint that all other individuals' utilities are unchanged:

$$\max_{\{x_h : h=1,2,\ldots,H\}} u_1(x_1) \quad \text{s.t.} \quad u_h(x_h) \geq u_h(x_h^*), \quad h = 2, 3, \ldots, H$$

and the feasibility constraint

$$\sum_h x_h^n \leq \sum_h x_h^{*n} \quad \forall\, n$$

13 Probability theory

13.1 Introduction

Like Chapter 9 on vector calculus, this chapter aims both to reinforce the reader's understanding of linear algebra, by applying it in this case to probability theory, and to prepare for
the financial applications in the final part of the book. While many readers will be familiar
with the basic concepts, such as descriptive statistics, probability, random variables, probability distributions and single-variable regression analysis, the presentation here is somewhat
more formal than that generally used in elementary courses. At the same time, it is sufficiently self-contained that even a reader with a limited prior knowledge of probability will
be able to pick up the concepts that will be essential for the later applications.

Section 13.2 introduces formal definitions in a purely scalar or univariate context.
To emphasize that most of financial theory and practice can be applied to trading in any quantity of which the final value is uncertain, we then present a number of simple applications
within this formal framework in Section 13.2.2.

The next two sections show different ways in which sets of random variables can be given
a vector space structure.

A discussion of expectations and moments follows in Section 13.6. Section 13.7 presents
important properties of the multivariate normal distribution. Then we consider the twin problems of estimation and forecasting in Section 13.8. The chapter continues with a random or
stochastic version of Taylor's theorem in Section 13.9.

Section 13.10 discusses Jensen's inequality, a simple and purely mathematical result that
follows easily from the material on convex and concave functions in Section 10.2 and the
material on expectations and moments in Section 13.6. While readers may find some of the
consequences of Jensen's inequality counter-intuitive, the result has profound implications
in several areas of financial economics, some of which are set out after the statement and
proof of the theorem. These implications are frequently poorly understood by practitioners
with limited mathematical training.

For a more thorough treatment of probability theory, the reader is referred to the classic
textbook treatments, such as those of Mood, Graybill and Boes (1974) or Hogg and Craig
(1978).

13.2 Sample spaces and random variables

13.2.1 Definitions

When we consider a consumer's choice under conditions of uncertainty, the objective will be
to calculate the consumer's optimal consumption plan, by specifying optimal consumption

for each possible **state of nature** or **state of the world** that might materialize. The optimal consumption plan is then a random or stochastic variable with values in the relevant consumption set or budget set.

We will now review more formally the associated concepts from basic probability theory, namely, probability spaces and random variables.

Let Ω denote the set of all possible states of the world, called the **sample space**. For example, Ω has two elements, heads and tails, if the states of the world are outcomes associated with tossing a coin, and six elements if the states of the world are outcomes of throwing a die.

A collection of states of the world or subset of the sample space, $A \subseteq \Omega$, is called an **event**. For example, an event might correspond to getting heads up when a coin is tossed or to throwing an odd number on a die. In simple cases such as these, the coin and the die would be said to be **fair** if all states of the world are equally likely. In such cases, the "classical" probability of an event is the number of states of the world in the event divided by the total number of states of the world in the sample space. Thus the probability of getting heads up when a fair coin is tossed is $\frac{1}{2} = 0.5$; and the probability of throwing an odd number when a fair die is rolled is $\frac{3}{6} = 0.5$.

Let $\mathcal{A}$ be a collection of events in, or subsets of, Ω. The function $P: \mathcal{A} \to [0, 1]$ is a **probability function** if

1. (a) $\Omega \in \mathcal{A}$,
   (b) $A \in \mathcal{A} \Rightarrow \Omega \setminus A \in \mathcal{A}$ and
   (c) $A_i \in \mathcal{A}$ for $i = 1, 2, \ldots \Rightarrow \bigcup_{i=1}^{\infty} A_i \in \mathcal{A}$
   (i.e. $\mathcal{A}$ is a **sigma-algebra** of events),
   and
2. (a) $P(\Omega) = 1$,
   (b) $P(\Omega \setminus A) = 1 - P(A)$ for all $A \in \mathcal{A}$ and
   (c) $P\left(\bigcup_{i=1}^{\infty} A_i\right) = \sum_{i=1}^{\infty} P(A_i)$, when $A_1, A_2, \ldots$ are pairwise disjoint events in $\mathcal{A}$.

The triple $(\Omega, \mathcal{A}, P)$ is then called a **probability space**.

It can be shown that the second condition on the function P is technically redundant, as it can be derived from the first and third conditions; see Exercise 13.2.

While the properties in the definition of a sigma-algebra may not be completely intuitive, comparison with the more intuitive properties of a probability function will make it clear why they are required.

Probabilities are often expressed as percentages rather than fractions or decimals between 0 and 1. For example, probabilities of 1/5, 0.2 and 20% all mean the same thing.

Note that the theory of choice under certainty (considered in Chapters 12 and 15) is just the special case of choice under uncertainty (considered in Chapter 16) in which the set Ω has only one element.

Suppose we are given a probability space $(\Omega, \mathcal{A}, P)$. The real-valued function $\tilde{x}: \Omega \to \mathbb{R}$ is called a **random variable** (sometimes abbreviated as "rv") if, for all $x \in \mathbb{R}$, $\{\omega \in \Omega : \tilde{x}(\omega) \leq x\} \in \mathcal{A}$, i.e. a function is a random variable if we know the probability that the value of the function is less than or equal to any given real number. Another way of expressing this is to say that the function $\tilde{x}$ is **measurable**.

We generally use the notation $\Pr[\tilde{x} \leq x]$ as a shorthand means of denoting the probability $P(\{\omega \in \Omega : \tilde{x}(\omega) \leq x\})$. The $\Pr[\cdot]$ notation will also be used occasionally in other similar contexts.

The convention of using a tilde over a letter to denote a random variable is common in financial economics; in other fields, capital letters may be reserved for random variables. In either case, small letters usually denote particular real numbers, i.e. particular values or **realizations** of the random variable.

The function $F_{\tilde{x}}: \mathbb{R} \to [0, 1]: x \mapsto \Pr[\tilde{x} \leq x]$ is known as the **cumulative distribution function** (abbreviated as "cdf") of the random variable $\tilde{x}$. Every random variable has a cdf and by definition the cdf is non-decreasing. The cdf completely describes the **distribution** or **probability distribution** of the random variable. Other functions introduced below to describe certain types of random variables or their distributions may not exist for other types of random variables.

A **stochastic process** is a sequence or collection of random variables indexed by time (or, in other contexts, such as geography, by space), e.g. $\{\tilde{x}_t : t \in T\}$ or just $\{\tilde{x}_t\}$ if the times are clear from the context. The realization of a stochastic process indexed by time is therefore a time series; see Definition 8.2.1.

13.2.2   Some common probability distributions

Random variables can be discrete, continuous or mixed. For example, any random variable defined on a finite sample space is **discrete**, taking on each of its possible values with a discrete probability between 0 and 1 (inclusive).

A discrete random variable on a finite sample space containing, say, N states of nature, denoted $\tilde{x}$, say, can also be represented as an $N \times 1$ vector, say $\mathbf{x} = (x_1, x_2, \dots, x_N)$, where x_i denotes the realization of the random variable $\tilde{x}$ in state of nature i.

The **probability function** of a discrete random variable $\tilde{x}$ is the mapping $f_{\tilde{x}}: \mathbb{R} \to [0, 1]$ that associates with each number x the probability that the random variable $\tilde{x}$ takes the value x. The mapping $f_{\tilde{x}}$ will be zero except at, at most, a countably infinite number of points in $\mathbb{R}$. For example, if $\tilde{x}$ is the number of successes in n independent **Bernoulli trials**,[1] where the probability of success for an individual trial is q, then

$$f_{\tilde{x}}(x) = {}^{n}C_x q^x (1-q)^{n-x} \tag{13.1}$$

In this case, $\tilde{x}$ is said to have a **binomial distribution** with **parameters** n and q, denoted $\tilde{x} \sim B(n, q)$. For example, $\tilde{x}$ might be the number of sixes obtained in three throws of an **unbiased** die. In that case, $\tilde{x} \sim B\left(3, \frac{1}{6}\right)$. For a **biased** die, $\tilde{x}$ would still have a binomial distribution, but the second parameter might differ from $\frac{1}{6}$.

Another important discrete probability distribution is the **Poisson distribution**,[2] for which

$$f_{\tilde{x}}(x) = \frac{e^{-\lambda} \lambda^x}{x!}, \quad x = 0, 1, 2, \dots \tag{13.2}$$

where λ is a parameter of the distribution. When the number of trials n is large and the probability of success is relatively small (say, $nq \leq 7$), the binomial probabilities are close to the Poisson probabilities with $\lambda = nq$.

Another simple example of a discrete random variable is the **trivial** random variable that takes on a single value with probability one. We will often assume in applications that random variables are **non-trivial**, in other words that not all the probability mass is concentrated at a single point. However, if the sample space is, say, the interval $[0, 1]$, then a random variable that is, say, **uniformly distributed**, i.e. equally likely to take on any value in the interval, is **continuous**. The probability that such a random variable takes on a value in the sub-interval $(a, b]$ is clearly $b - a$. However, the probability that it takes on the value b exactly is undefined, i.e. unquantifiable or infinitesimally small but not equal to zero.

A practical example of a **mixed** random variable is the liability of an insurer on a particular policy or group of policies. With large positive probability, there will be no claim against the insurer and the liability will be zero. However, if there is a claim, then the insurer's liability could be any positive amount up to the maximum stated in the policy. If we restrict liabilities to whole numbers of euros and cents, for example, the random variable becomes discrete again.

Readers who have taken courses in probability and statistics will be familiar with many other forms of probability distribution, both discrete and continuous. Distribution theory, however, is beyond the scope of the present work.

The **probability density function** (abbreviated "pdf") of a random variable with differentiable cdf is the derivative of the cdf, also denoted $f_{\tilde{x}} : \mathbb{R} \to \mathbb{R}_+$ and defined by

$$f_{\tilde{x}}(x) \equiv \frac{d F_{\tilde{x}}}{dx}(x) \tag{13.3}$$

Since the cdf is non-decreasing, the pdf is non-negative. Also

$$\int_{-\infty}^{\infty} f_{\tilde{x}}(x) \, dx = \lim_{x \to \infty} F_{\tilde{x}}(x) = \Pr[\tilde{x} \leq \infty] = 1 \tag{13.4}$$

Conversely, by the fundamental theorem of calculus (Theorem 7.9.1), the cdf is the integral of the pdf:

$$F_{\tilde{x}}(x^*) = \int_{-\infty}^{x^*} f_{\tilde{x}}(x) \, dx \tag{13.5}$$

Any non-negative function f such that $\int_{-\infty}^{\infty} f(x) \, dx = 1$ is a pdf.

Indeed, in some cases, there is no closed-form solution for the cdf and it can be written in the form (13.5) only. One such example is the **normal** or **Gaussian** distribution, which has pdf given by

$$f_{\tilde{x}}(x) = \frac{1}{\sqrt{2\pi}\sigma} e^{-\frac{1}{2}[(x-\mu)/\sigma]^2} \tag{13.6}$$

where $\mu \in \mathbb{R}$ and $\sigma \in \mathbb{R}_{++}$ are parameters of the distribution. This is denoted $\tilde{x} \sim N(\mu, \sigma^2)$. The pdf of the **standard normal** distribution, i.e. the special case of $\mu = 0$ and $\sigma = 1$, is usually written

$$\phi(x) = \frac{1}{\sqrt{2\pi}} e^{-\frac{1}{2}x^2} \tag{13.7}$$

If $\ln \tilde{x}$ has a normal distribution (i.e. is **normally distributed**), then $\tilde{x}$ is said to have a **lognormal distribution** or to be **lognormally distributed**.

Just as the (probability) distribution of a random variable is completely specified by its cdf, so it is also completely specified by its pdf, if it exists, since the cdf can be recovered from the pdf by integration.

Two different random variables can have the same distribution. For example, if $\tilde{x}$ is the number showing on the top face of an unbiased die and $\tilde{y} \equiv 7 - \tilde{x}$, then $\tilde{x}$ and $\tilde{y}$ both take the values 1, 2, 3, 4, 5 and 6, with probability one-sixth each. From a mathematical point of view, two random variables are equivalent if they have the same distribution. All of the concepts and properties introduced in the remainder of this chapter are properties of the distribution, rather than being specific to the underlying random variable.

Students taking more advanced courses in probability theory will encounter several other functions that, like the cdf or pdf, can be used to completely specify a probability distribution.

13.3   Applications

The sections that follow introduce a selection of **financial instruments**, variously (and interchangeably) called **assets**, **securities**, **contracts** or **investments**, which provide payoffs at one or more future dates. The final values of these payoffs are typically uncertain at the time when the securities are bought and sold, so the underlying instruments are described as **risky** and the payoffs can be represented by random variables.

This chapter considers briefly the probability distributions of these security payoffs. Chapters 16 and 17 will consider in more detail the economic valuation of such risky securities.

Just as a scalar is a special or trivial case of a random variable that takes on a single value with probability 1, we will also encounter securities, generally called **bonds** or **risk-free securities**, whose future payoffs are known with certainty in advance; see Definition 15.4.1.

DEFINITION 13.3.1 Short-selling a security means owning a negative quantity of it.

For example, borrowing is equivalent to short-selling a risk-free security.

In practice short-selling means promising (credibly) to pay someone the same cash flows as would be paid by a security that one does not own, always being prepared, if required, to pay the current market price of the security to end the arrangement.

13.3.1 Lotteries

DEFINITION 13.3.2 A **lottery ticket** is an investment whose payoff is a discrete random variable, with possible values or realizations $x_1, x_2, x_3, \ldots$ occurring with respective probabilities $\pi_1, \pi_2, \pi_3, \ldots$, say.

We will use the notation

$$\pi_1 x_1 \oplus \pi_2 x_2 \oplus \pi_3 x_3 \oplus \cdots \tag{13.8}$$

for such a **lottery**.

Similar notation will be used for **compound lotteries** (or **mixtures** of random variables), where the payoffs themselves are further lotteries. For example,

$$p \in \tilde{x} \oplus q \in \tilde{y} \tag{13.9}$$

is shorthand for a lottery that pays $\in \tilde{x}$ with probability p and $\in \tilde{y}$ with probability $q \equiv 1 - p$, where the realizations of the random variable $\tilde{x}$ or $\tilde{y}$, as appropriate, may be determined simultaneously with, or at a later time than, the choice between $\tilde{x}$ and $\tilde{y}$.

Table 13.1 Probability of winning (π) and size of jackpot (J) for a Lotto-style draw with n balls

n	π	J (€)
6	1	0.50
7	1/7	3.50
8	1/28	14.00
12	1/924	462.00
36	1/1947792	973 896.00
39	1/3262623	1 631 311.50
42	1/5245786	2 622 893.00

As with ordinary multiplication, the $\times$ symbol will be used in numerical examples. For example,

$$0.25 \times €100 \oplus 0.75 \times €0 \tag{13.10}$$

denotes a lottery giving a one-in-four chance of receiving €100 and a 75% chance of receiving nothing.

A familiar type of lottery is based on predicting the numbers on six numbered balls drawn from a drum containing n numbered balls. We will consider an example that is similar to the **Lotto** game run by national or state lottery organizations in many jurisdictions, but with a single jackpot prize only. The structure of the lottery is as follows:

• There are

$$^{n}C_6 = \frac{n!}{6!\,(n-6)!} \tag{13.11}$$

equally likely outcomes, each occurring with probability $\pi \equiv 1/{}^{n}C_6$.
• The lottery player chooses a ticket specifying exactly one of these outcomes.
• With probability $1 - \pi$ the lottery player gets nothing.
• With probability π (s)he gets a jackpot prize of, say, $€J = €\,{}^{n}C_6/2$.

Table 13.1 shows the probability of winning and the value of the jackpot for different numbers of balls.

The reader is encouraged to consider how much (s)he would be willing to pay for a ticket in each lottery in Table 13.1 and to keep a record of these amounts until we return to this topic in Section 16.2. The Lotto game is considered further in Exercise 13.4.

13.3.2 Spread betting

Spread betting or **index betting** on random variables (for example, on the number of goals scored in a soccer match or on the number of horses that finish the Grand National horse race) means buying or selling a security whose value is deemed to equal the value realized by the random variable.

For example, suppose that one buys the number of goals scored in a soccer match between Arsenal and Chelsea at a price of 2.3 (goals) for €10 per goal. If the final score is Arsenal 3, Chelsea 1, then the total number of goals scored is 4.0 and the spread bet bought for €23 has a final value of €40, leaving the buyer with a profit of €17, and the seller with a matching loss of €17.

Similarly, suppose that one sells the number of finishers in the Grand National at a price of 15.5 (finishers) for €20 per horse and that 18 horses complete the course. The spread bet sold for €310 (=15.5 × 20) must be settled for €360 (=18 × 20), resulting in a loss of €50 for the seller.

The term *spread* betting arises from the fact that organizations offering spread betting to the public typically offer an **ask price** at which the public can buy and a (lower) **bid price** at which the public can sell; the difference between the ask price and the bid prices is called the **bid–ask spread** or just the **spread**. This terminology was borrowed from markets for commodities and financial securities.

The term *index* betting arises from the fact that the random variable traded is often an **index**, which takes the value, say, 50 if the chosen contestant wins the contest, with values of, say, 30 for second place, 20 for third place and 10 for fourth place. Most of the firms offering this type of betting have names containing either the word "spread" or the word "index".

Futures and **forward** trading are more traditional forms of spread betting; indeed, they were the original inspiration for spread betting on sporting events. The underlying random variable in these cases is usually the value on a future **settlement date** or **maturity date** of something like a commodity price or interest rate or exchange rate. The difference between futures and forward trading relates essentially to differences in trading conventions. Futures contracts are actively traded on an exchange right up to maturity and usually have a settlement date at the end of a month or quarter. Forward contracts are usually one-off deals, with a maturity date maybe 30 or 90 days after the date of execution.

13.3.3   *Options*

Like spread bets, **options** can also be traded on any random variable.

DEFINITION 13.3.3   The following types of option are defined:

(a) A **European call option** with **exercise price** K *on* a random variable $\tilde{x}$ is an option to buy the random variable for K *on* a fixed **expiry date** in the future.
(b) An **American call option** on a stochastic process is an option to buy *on or before* the expiry date.
(c) A **put option** is an option to sell.

The profit earned by exercising a European call option is $\tilde{x} - K$. The option will obviously be exercised only if this profit is positive. Thus the final value of the option at expiry is $\max\{0, \tilde{x} - K\}$.

In practice, options are most often traded on the future price of a stock market index, an individual stock, or some other financial security.

13.3.4   *State-contingent claims*

DEFINITION 13.3.4   A **state-contingent claim** or **Arrow–Debreu security** is a random variable that takes the value 1 in one particular state of nature and the value 0 in all other states. The price of a state-contingent claim is often simply called a **state price**.

The concept that is now called an Arrow–Debreu security was introduced separately by Arrow (1953) and Debreu (1959, Section 7.3).[3]

When the sample space is finite, the Arrow–Debreu security that takes the value 1 in state i, say $\tilde{e}_i$, can, equivalently, be represented by the vector of its possible payoffs or realizations, which is just the ith standard basis vector, $\mathbf{e}_i = (0, \ldots, 0, 1, 0, \ldots, 0)$.

DEFINITION 13.3.5 A **complex security** is a random variable or lottery that can take on arbitrary values, in other words a **portfolio** of state-contingent claims. A typical complex security in a world with M states of nature can be represented either by a random variable, $\tilde{y}$, say, or by the corresponding column vector of its possible payoffs or realizations, $\mathbf{y} \in \mathbb{R}^M$, say, where y_i is the payoff of the complex security in state i.

The set of all possible complex securities on a given finite sample space containing M states of nature is an M-dimensional vector space, and the M possible Arrow–Debreu securities constitute the standard basis for this vector space. We will return to this point in Section 13.4.

Section 16.3 will consider in depth the pricing of state-contingent claims.

13.3.5 Odds and betting

The **odds** against the occurrence of an outcome whose probability of occurring is p are conventionally expressed as the ratio $\frac{1-p}{p}$, often called the **odds ratio**, and also sometimes denoted $(1-p){:}p$ or $(1-p)/p$. For example, if the probability of throwing a six on a die is one-sixth, the corresponding **fractional odds** against this outcome are $5/1 = \frac{5/6}{1/6}$. A successful €100 **bet** on throwing a six at the **fair** odds of 5/1 produces a profit of €500 for the **backer**, yielding a total payout of €600 including the original €100 **stake**. An unsuccessful bet produces a profit of €100 for the **layer**. The distinction between backer and layer is usually clear when there are many possible outcomes or winners, but becomes blurred when there are only two outcomes, e.g. in a sport such as baseball or tennis where all matches are played to a conclusion.

Conventionally, where betting is legal, the layers and stakeholders are **bookmakers**, who are strictly licensed and regulated, while any (adult) member of the general public can be a customer or backer or **punter**. Betting takes place around the world on a wide variety of sporting contests as well as on political elections and on random experiments such as throwing a die or a Lotto draw. We will avoid the temptation to make ambiguous use of the collective word "event", which has already been given a rigorous mathematical meaning, and will instead use the term **contest** to include any subject on which bookmakers **make a book**.

In certain jurisdictions, particularly in continental Europe, the convention in betting markets is to use **decimal odds**, $\frac{1}{p}$, written in decimal rather than fractional notation. In this notation, the odds *against* drawing an ace or a picture card from a standard deck of 52 cards are $3.25 = \frac{52}{16}$.

Having a bet against a particular event occurring at decimal odds of x is mathematically equivalent to purchasing the corresponding state-contingent claim for a price of $1/x$. Whether or not these two transactions are legally equivalent in all jurisdictions is less certain. The odds ratio is often described by practitioners as the "price" at which the bet is struck; others may view the implied probability as the price. These two "prices" are reciprocals. Each is the relative price of one item in terms of another. The difference is equivalent to the difference between expressing the price of oranges in a barter economy as 2.5 apples per orange or as 0.4 oranges per apple. In odds-based markets, the rule for exploiting arbitrage opportunities is reversed: "back high, lay low".

In practice, the fair odds against (or, equivalently, the true probability of) an outcome are not always as easily computed as in the case of tossing a die. For example, there is no obvious probability model to describe the possible outcomes of a horse race or soccer match. Even when the fair odds are easily computed, such as in a casino, bets may be placed at odds that are more than fair or **favourable** to one party and less than fair or **unfavourable** to the other.

In real-world betting markets, the probabilities implicit in the odds against all possible outcomes do not always sum to unity. In theory, adding up the probabilities for all the possible outcomes of a contest should give a total probability of 1.0 or, equivalently, a **total percentage** of 100%. If the sum of probabilities is less than one, then backing each contestant in proportion to its quoted probability will lead to a certain, riskless profit, or an arbitrage opportunity. In practice, the sum of the quoted or market probabilities is usually (much) greater than 1.0, and the excess over 1.0 (the **over-round**) represents the gross profit of the bookmaker and arises from the regulatory asymmetry between backer and layer.

Fixed-odds betting as defined here is common in many jurisdictions. Under this system, the odds are agreed or fixed between the parties and the potential payout is known at the time at which the bet is struck, and the only uncertainty concerns the binary outcome of the bet.

13.3.6   *Pari-mutuel betting*

Another common betting system is **pari-mutuel** or **pool** or **tote** betting. Under this system, the odds at which a bet is settled are determined when betting ends, which is normally the time at which the relevant contest starts. All money bet into a pari-mutuel pool (less a fixed percentage deduction for taxes and expenses) is shared *pro rata* among those who place successful bets. The potential payout on a pari-mutuel bet thus depends on everyone else's bets and is not known with certainty until betting has ceased.

The pool operator generally quotes prices as **dividends** per unit invested. While the dividend is actually just the decimal odds at which the bet will be settled, it is usually prefixed with the relevant (but redundant) currency symbol. Improvements in communications technology have in recent times allowed pari-mutuel pools to be operated across multiple currency areas, rendering the currency symbol even more irrelevant.

In North America, the convention is to express pari-mutuel dividends in dollars as twice the implicit decimal odds: e.g. a dividend of $3.60 represents decimal odds of 1.8 or fractional odds of 4/5. This anomaly arose because the minimum stake accepted is $2. A dividend of 1.8 also tells us that roughly five-ninths of the net pool (after deductions) was bet on the winning outcome (slightly more or less depending on whether the declared dividend has been rounded up or down to the nearest 10 cents according to the relevant local rounding convention).

The commonest and most popular form of pool betting worldwide is probably the Lotto game mentioned above. However, Lotto operators generally use a slightly different version of the pari-mutuel system, reserving smaller fractions of the pool for consolation prizes for those who miss the jackpot by one or two numbers, etc. (usually called Match 5 and Match 4 prizes in a six-number Lotto).

In many jurisdictions, pari-mutuel betting is the only legal form of betting on horse racing. Pari-mutuel pools are usually run not just for straight win betting, but for **exotic** bets, such as the **exacta** (name the first two finishers in the correct order) or the **trifecta** (name the first three finishers in the correct order). Thus we will usually refer to the subjects of pari-mutuel bets as "outcomes" rather than "contestants".

Pari-mutuel betting evolved in the early twentieth century as it allowed the stakeholder – often a government-owned monopoly – to avoid bearing any risk. Under the pari-mutuel system, the stakeholder remains the layer, while members of the general public are backers.

It will sometimes be the case, particularly when there are large numbers of potential outcomes, as in Lotto and exotic horse race bets, that there is no bet in the pool on the winning outcome. The convention in such cases is normally to **roll over** the money in the pool to the next similar contest.

13.3.7 *Exchange betting*

With the advent of the Internet, another type of betting in which the stakeholder bears no risk has evolved. This is **exchange betting**, which allows members of the public to make offers, either to back or to lay any outcome, at odds of their choice. Offers arriving on the exchange via the Internet and other channels are queued and the offer at the front of the queue is **matched** when an equal and opposite offer arrives. For example, one person's offer to back Brazil for €100 at decimal odds of 1.9 to win the World Cup Final might be matched with two separate offers to lay Brazil at 1.9 for €60 and €40. However, an offer to back Brazil at 2.4 or an offer to lay Brazil at 1.6 might expire unmatched at the start of play. Thus, while the customer no longer faces the **dividend uncertainty** of pari-mutuel betting, he faces some **matching uncertainty**.

The betting exchange operation is funded by a commission deducted from customers' winnings. This new paradigm has removed the asymmetry between backers and layers that is inherent in both the fixed-odds and pari-mutuel paradigms. Some sporting bodies have expressed concern that allowing ordinary, unlicensed market participants to be layers has increased opportunities to profit from corruption in sport. Exchange operators have put forward the counter-argument that the transparency of the exchange model makes it easier to identify, trace and police possible corruption, and thus acts as a disincentive to such corruption.

Most betting exchanges operate in an odds framework, but there are some that operate in the equivalent state-contingent claims framework. The latter are not very different from stock exchanges and similar conventional financial exchanges.

Exchange betting is not suited to markets with large numbers of potential outcomes, as transactions on individual unlikely outcomes would be few and far between. The pari-mutuel system remains the obvious one for exotic betting, where the normal transaction involves a combination bet on a number of individually unlikely outcomes, e.g. a combination trifecta on five horses comprising equal bets on $5 \times 4 \times 3 = 60$ possible outcomes.

On events of any significance, offers to back and offers to lay just one **tick** apart can be observed simultaneously on betting exchanges. In other words, the bid–ask spread is minimal: backers face a total percentage just above 100% and layers face a total percentage just below 100%. The mid-market odds generally provide a very good "estimate" of the true odds against any outcome.

13.4 Vector spaces of random variables

We will now show that any *finite* set of random variables defined on a given sample space, Ω, say, spans a *finite-dimensional* real vector space. We begin with the set of random variables $\{\tilde{x}_1, \tilde{x}_2, \ldots, \tilde{x}_n\}$, say, where $\tilde{x}_i \colon \Omega \to \mathbb{R}$, $i = 1, 2, \ldots, n$. To define a real vector space in accordance with Definition 5.4.2, we must specify the two operations of addition of random variables and multiplication of a random variable by a scalar. This can be done in the following rather obvious way:

$$(\tilde{x} + \tilde{y}) : \Omega \to \mathbb{R} : \omega \mapsto \tilde{x}(\omega) + \tilde{y}(\omega)$$
$$\lambda \tilde{x} : \Omega \to \mathbb{R} : \omega \mapsto \lambda \times (\tilde{x}(\omega)) \tag{13.12}$$

This yields a vector space, which we will denote X. The zero vector in this vector space is the random variable that equals the scalar zero with probability 1. Any set of n random variables on Ω generates such a vector space; it will be of dimension n provided that the spanning vectors are linearly independent; in other words, provided that there is no linear combination of the spanning random variables that equals zero with probability 1. We will assume for the purposes of the remainder of this section that the spanning vectors are indeed linearly independent.

In terms of the spanning set of random variables, the typical vector in X can be written in the form $a_1 \tilde{x}_1 + a_2 \tilde{x}_2 + \cdots + a_n \tilde{x}_n \equiv \mathbf{a}^\top \tilde{\mathbf{x}}$, where $\tilde{\mathbf{x}} \equiv (\tilde{x}_1, \tilde{x}_2, \ldots, \tilde{x}_n)$.

In terms of vectors, as with any vector space, the typical element of this space can be represented by the n-tuple of its coordinates with respect to the chosen basis (in this case, since we have assumed linear independence, the generating set), i.e. as $\mathbf{a} = (a_1, a_2, \ldots, a_n)$.

It is easy to go from the coordinate vector $\mathbf{a} \in \mathbb{R}^n$ to the random variable $\mathbf{a}^\top \tilde{\mathbf{x}}$, but not so easy to find the coordinates of the typical random variable $\tilde{y} \in X$ with respect to the basis of spanning random variables. Given our assumption that the spanning vectors are linearly independent, the method that follows can be used.[4]

Pick n points from the sample space Ω, say $\omega_1, \omega_2, \ldots, \omega_n$. Define an $n \times n$ transition matrix $\mathbf{P}$ by

$$p_{ji} \equiv \tilde{x}_i(\omega_j), \quad i, j = 1, 2, \ldots, n \tag{13.13}$$

and define the vector $\mathbf{y} \in \mathbb{R}^n$ by

$$y_j \equiv \tilde{y}(\omega_j), \quad j = 1, 2, \ldots, n \tag{13.14}$$

Since we have assumed that $\tilde{y} \in X$, we know that there exists $\mathbf{a} \in \mathbb{R}^n$ such that

$$\sum_{i=1}^n a_i \tilde{x}_i(\omega) = \tilde{y}(\omega) \quad \forall \, \omega \in \Omega \tag{13.15}$$

In particular, $\mathbf{a}$ must satisfy

$$\sum_{i=1}^n a_i \tilde{x}_i(\omega_j) = \tilde{y}(\omega_j), \quad j = 1, 2, \ldots, n \tag{13.16}$$

or

$$\sum_{i=1}^n a_i p_{ji} = y_j, \quad j = 1, 2, \ldots, n \tag{13.17}$$

or, in matrix notation,

$$\mathbf{Pa} = \mathbf{y} \tag{13.18}$$

If $\mathbf{P}$ was singular, then there would exist $\mathbf{a}^* \in \mathbb{R}^n$ such that

$$\mathbf{Pa}^* = \mathbf{0} \tag{13.19}$$

or

$$\sum_{i=1}^{n} p_{ji} a_i^* = \sum_{i=1}^{n} \tilde{x}_i(\omega_j) a_i^* = 0, \quad j = 1, 2, \ldots, n \tag{13.20}$$

If (13.20) held for all $\omega \in \Omega$ and not just for the n sample points chosen, then our assumption that the spanning random variables are linearly independent would be violated. So, if (13.20) holds for the n sample points originally chosen, we can find a different set of n sample points for which the transition matrix $\mathbf{P}$ defined by (13.13) will be invertible, and so the coefficients of $\tilde{y}$ are given by $\mathbf{a} = \mathbf{P}^{-1} \mathbf{y}$ (where the vector $\mathbf{y}$ is also redefined in terms of the new $\omega_1, \omega_2, \ldots, \omega_n$).

Note that the elements of X here are both vectors (since they belong to an n-dimensional real vector space) and random variables (since they are measurable functions defined on the sample space Ω).[5] An example of such a vector space is the portfolio space, which will be introduced in Section 17.2. The set of *all* random variables on a given sample space likewise constitutes a vector space, but one that will generally be *infinite-dimensional*.[6] When the underlying sample space is finite, it is *finite-dimensional*: for example, if the sample space consists of the six possible orientations of a die, then the set of random variables defined on the sample space is a six-dimensional vector space. We have already encountered one such vector space in Section 13.3.4 – the set of all possible Arrow–Debreu securities.

13.5 Random vectors

A **random vector** is a vector of random variables. The convenient abbreviation rv is sometimes used to denote a random vector; this will be particularly helpful when stating results that apply to both random variables and random vectors.

An n-dimensional random vector can also be thought of as a measurable, vector-valued function on the sample space Ω, i.e. a function $\tilde{\mathbf{x}} \colon \Omega \to \mathbb{R}^n$, each of whose component functions $\tilde{x}_1 \colon \Omega \to \mathbb{R}$, $\tilde{x}_2 \colon \Omega \to \mathbb{R}, \ldots, \tilde{x}_n \colon \Omega \to \mathbb{R}$ is measurable. The realizations of an n-dimensional random vector lie in the n-dimensional Euclidean vector space $\mathbb{R}^n$. Note the convention of using a combination of a tilde and a boldface letter to denote a quantity that is both random and a vector.

A stochastic process indexed by a time subscript running from 1 to T and whose elements are n-dimensional random vectors is equivalent to an nT-dimensional random vector.

Note that the elements of the vector space X introduced in Section 13.4 are random vectors only in the sense that every scalar is a one-dimensional vector, i.e. the realizations of the elements of X lie not in $\mathbb{R}^n$ but in $\mathbb{R}$.

The **(joint) distribution** and **(joint) cumulative distribution function** of a random vector or stochastic process are the natural extensions of the one-dimensional concept. In other words, the distribution of an n-dimensional random vector $\tilde{\mathbf{x}}$ can be specified by the cdf:

$$F_{\tilde{\mathbf{x}}}(\mathbf{x}) \equiv \Pr[\tilde{x}_1 \leq x_1, \tilde{x}_2 \leq x_2, \ldots, \tilde{x}_n \leq x_n] \tag{13.21}$$

Two random variables $\tilde{x}$ and $\tilde{y}$ are said to be **(statistically) independent** if their joint cdf equals the product of their **marginal** cdfs, i.e. if

$$F_{\tilde{x},\tilde{y}}(x, y) = F_{\tilde{x}}(x) F_{\tilde{y}}(y) \quad \forall\, x, y \in \mathbb{R} \tag{13.22}$$

We say that the distribution of the n-dimensional random vector $\tilde{\mathbf{x}}$ is **continuous** if there exists a non-negative function $f_{\tilde{\mathbf{x}}} \colon \mathbb{R}^n \to \mathbb{R}_+$, called its probability density function, such that

$$F_{\tilde{\mathbf{x}}}(\mathbf{x}^*) = \int_{-\infty}^{x_1^*} \int_{-\infty}^{x_2^*} \ldots \int_{-\infty}^{x_n^*} f_{\tilde{\mathbf{x}}}(x_1, x_2, \ldots, x_n)\, dx_1\, dx_2 \ldots dx_n \tag{13.23}$$

Note that, as $F_{\tilde{\mathbf{x}}}(\mathbf{x}^*) \leq 1$ for all $\mathbf{x}^*$ and $f_{\tilde{\mathbf{x}}}(\mathbf{x}) \geq 0$ for all $\mathbf{x}$, the conditions of Fubini's theorem are automatically satisfied by the multiple integral in (13.23).

Similarly to (13.3), the pdf of a random vector can be calculated from its cdf by repeated application of the fundamental theorem of calculus:

$$\frac{\partial^n F_{\tilde{\mathbf{x}}}(\mathbf{x})}{\partial x_1 \partial x_2 \ldots \partial x_n} = f_{\tilde{\mathbf{x}}}(\mathbf{x}) \tag{13.24}$$

The **conditional distribution** of one or more components of a random vector, given particular realizations of its other components, is defined in the natural way. In the case of discrete random variables, $\tilde{x}_1$ and $\tilde{x}_2$, the probability that $\tilde{x}_1$ takes the value x_1 given that $\tilde{x}_2 = x_2$ is

$$\Pr[\tilde{x}_1 = x_1 \mid \tilde{x}_2 = x_2] = \frac{\Pr[\tilde{x}_1 = x_1 \text{ and } \tilde{x}_2 = x_2]}{\Pr[\tilde{x}_2 = x_2]} \tag{13.25}$$

This defines the conditional distribution of $\tilde{x}_1$ given x_2.

Similarly, in the case of continuous random variables, the **conditional pdf** of the random variable $\tilde{x}_1$ given that $\tilde{x}_2 = x_2$ is defined by

$$f_{\tilde{x}_1 \mid \tilde{x}_2 = x_2}(x_1) = \frac{f_{(\tilde{x}_1, \tilde{x}_2)}(x_1, x_2)}{f_{\tilde{x}_2}(x_2)} \tag{13.26}$$

In both cases, it must be assumed that $\tilde{x}_2$ takes the value x_2 with non-zero probability, or that $\Pr[\tilde{x}_2 = x_2] \neq 0$ and $f_{\tilde{x}_2}(x_2) \neq 0$, respectively.

As with sets of random variables, any finite set of m random vectors in $\mathbb{R}^n$ defined on a given sample space spans a vector space of up to m dimensions, depending on the linear independence or otherwise of the spanning set. Note that in this case m can be greater than n, i.e. the dimension of the vector space of random vectors can be greater than the dimension of the Euclidean space in which realizations of the random vectors lie.

13.6 Expectations and moments

13.6.1 Definitions

The **expectation**, also known as the **mean** or **average**, of a discrete random variable, $\tilde{x}$, with possible values $x_1, x_2, x_3, \ldots$, is given by

$$E[\tilde{x}] \equiv \sum_{i=1}^{\infty} x_i \Pr[\tilde{x} = x_i] \tag{13.27}$$

For a continuous random variable, the summation is replaced by an integral:

$$E[\tilde{x}] \equiv \int_{-\infty}^{\infty} x \, dF_{\tilde{x}}(x) = \int_{-\infty}^{\infty} x f_{\tilde{x}}(x) \, dx \tag{13.28}$$

For a random vector, or a function of a random vector, the single integral in (13.28) is replaced by a multiple integral.

Many random variables do not have finite expectations; see, for example, Exercise 16.11. For the most part, however, we will deal in this book only with expectations that are finite.

In economics and finance, this objective notion of expectation is sometimes referred to as **mathematical expectation** or, less frequently, as **statistical expectation**, in order to distinguish it from the more subjective, but related, notions of expectation that play an important role in economic behaviour. The latter type of expectation will be discussed, for example, in connection with the pure expectations hypothesis introduced on p. 364. There is also an extensive literature in economics on the theory of **rational expectations**, which is essentially the proposition that subjective expectations equal mathematical expectations.

Recall that it was mentioned in Section 6.2.2 that expectation is a linear operator on the set of random variables defined on a given sample space. This is because for scalars $k_1, k_2, \ldots, k_r$ and random variables $\tilde{x}_1, \tilde{x}_2, \ldots, \tilde{x}_r$ it can be shown that $E\left[\sum_{i=1}^{r} k_i \tilde{x}_i\right] = \sum_{i=1}^{r} k_i E[\tilde{x}_i]$; see Exercise 13.5.

When dealing with stochastic processes, the expectation operator is often written with a time subscript, e.g. $E_t[\tilde{x}_{t+s}]$ denotes the expectation of the value of the process s periods in the future given the information available now, at period t.

The kth **moment about the mean** or kth **central moment** of the random variable $\tilde{x}$ is

$$m^k(\tilde{x}) \equiv E[(\tilde{x} - E[\tilde{x}])^k] \tag{13.29}$$

In particular, the first moment about the mean is always zero and the second, third and fourth moments about the mean are called, respectively, the **variance**, **skewness** and **kurtosis** of the random variable, denoted as follows:

$$m^1[\tilde{x}] = E[(\tilde{x} - E[\tilde{x}])^1] \equiv 0 \tag{13.30}$$

$$m^2[\tilde{x}] = E[(\tilde{x} - E[\tilde{x}])^2] \equiv \text{Var}[\tilde{x}] \tag{13.31}$$

$$m^3[\tilde{x}] = E[(\tilde{x} - E[\tilde{x}])^3] \equiv \text{Skew}[\tilde{x}] \tag{13.32}$$

$$m^4[\tilde{x}] = E[(\tilde{x} - E[\tilde{x}])^4] \equiv \text{Kurt}[\tilde{x}] \tag{13.33}$$

Note that even-numbered moments about the mean must be non-negative, as they are based on even powers of the deviation of the random variable from its mean. An even-numbered moment can be zero only if the underlying random variable has a trivial distribution. In fact, any even-numbered moment is zero if and only if all even-numbered moments are zero if and only if the distribution is trivial. In particular, $\text{Var}[\tilde{x}] > 0$ if and only if $\tilde{x}$ is non-trivial; see Example 13.10.1.

Note also that odd-numbered moments about the mean will be zero if the distribution is symmetric around the mean.

As for expectations, it is sometimes the case that higher-order moments do not exist (i.e. are not finite).

The **standard deviation** of a random variable is the square root of its variance.

The **skewness coefficient** of a random variable is the third moment about the mean divided by the standard deviation cubed. The **kurtosis coefficient** of a random variable is the fourth moment about the mean divided by the variance squared. Both of these are dimensionless quantities, i.e. they are invariant under changes of the units of measurement of the underlying variable.

The **covariance** between two random variables $\tilde{x}$ and $\tilde{y}$ is given by

$$\text{Cov}[\tilde{x}, \tilde{y}] \equiv E[(\tilde{x} - E[\tilde{x}])(\tilde{y} - E[\tilde{y}])] = E[\tilde{x}(\tilde{y} - E[\tilde{y}])]$$

$$= E[(\tilde{x} - E[\tilde{x}])\tilde{y}] = E[\tilde{x}\tilde{y}] - E[\tilde{x}]E[\tilde{y}] \tag{13.34}$$

Proving the equivalence between all these expressions is left as an exercise; see Exercise 13.6. We also have that $\text{Cov}[\tilde{x}, \tilde{y}] = \text{Cov}[\tilde{y}, \tilde{x}]$ and the covariance between a random variable and itself is just its variance,

The **correlation** between two non-trivial random variables $\tilde{x}$ and $\tilde{y}$ is given by

$$\text{Corr}[\tilde{x}, \tilde{y}] \equiv \frac{\text{Cov}[\tilde{x}, \tilde{y}]}{\sqrt{\text{Var}[\tilde{x}]\,\text{Var}[\tilde{y}]}} = \text{Corr}[\tilde{y}, \tilde{x}] \tag{13.35}$$

The correlation between a random variable and itself is always unity.

Two random variables are said to be **uncorrelated** if their correlation (or, equivalently, their covariance) is zero.[7]

It can easily be shown that two random variables are uncorrelated if and only if the expectation of their product equals the product of their expectations; see Exercise 13.8.

The expectation of a random vector is just the vector of the expectations of the component random variables.

The **conditional expectation** of one or more components of a random vector, given particular realizations of its other components, is just the expectation calculated from the conditional distribution. For example, for continuous random variables $\tilde{x}_1$ and $\tilde{x}_2$,

$$E[\tilde{x}_1 \mid \tilde{x}_2 = x_2] = \int_{-\infty}^{\infty} x_1 f_{\tilde{x}_1 \mid \tilde{x}_2 = x_2}(x_1)\, dx_1 \tag{13.36}$$

Conditional expectations are also of interest when dealing with stochastic processes. The expectation $E_t[\tilde{x}_{t+s}]$ could also be expressed, at least in a univariate context, as $E[\tilde{x}_{t+s} \mid \tilde{x}_t = x_t]$.

The variance of an n-dimensional random vector, $\tilde{\mathbf{x}}$, say, is the $n \times n$ square matrix of the covariances between the component random variables (provided that they exist), usually called the **variance–covariance matrix**, and denoted $\text{Var}[\tilde{\mathbf{x}}]$. If it exists, a variance–covariance matrix is real, symmetric and positive semi-definite; see Exercise 13.9.

The **correlation matrix** of a random vector is, similarly, defined to be the square matrix of the correlations between the component random variables. It too is always real, symmetric and positive semi-definite. Furthermore, the entries on the leading diagonal of a correlation matrix are all unity.

Note that the variance–covariance matrix is the expectation of the outer product of the deviation of the random vector from its mean with itself, i.e.

$$E[(\tilde{\mathbf{x}} - E[\tilde{\mathbf{x}}])(\tilde{\mathbf{x}} - E[\tilde{\mathbf{x}}])^\top] \tag{13.37}$$

see Exercise 1.9.

Similarly, the covariance between the n-dimensional random vector $\tilde{\mathbf{x}}$ and the p-dimensional random vector $\tilde{\mathbf{y}}$ is the $n \times p$ matrix

$$\text{Cov}[\tilde{\mathbf{x}}, \tilde{\mathbf{y}}] = E[(\tilde{\mathbf{x}} - E[\tilde{\mathbf{x}}])(\tilde{\mathbf{y}} - E[\tilde{\mathbf{y}}])^\top] \tag{13.38}$$

The covariance operator $\text{Cov}[\cdot, \cdot]$ and the correlation operator $\text{Corr}[\cdot, \cdot]$ are bi-linear functions. They are symmetric if the two arguments are of the same dimension; see Section 13.6.3. If the arguments are of different dimensions, reversing the arguments transposes the matrix of covariances or correlations.

Note also that, for an n-dimensional random vector $\tilde{\mathbf{x}}$, using the results of Exercise 13.7:

$$\text{Var}\left[\sum_{i=1}^n \tilde{x}_i\right] = \text{Var}[\mathbf{1}^\top \tilde{\mathbf{x}}] = \text{Cov}[\mathbf{1}^\top \tilde{\mathbf{x}}, \mathbf{1}^\top \tilde{\mathbf{x}}] = \mathbf{1}^\top \text{Var}[\tilde{\mathbf{x}}]\mathbf{1}$$

$$= \sum_{i=1}^n \text{Var}[\tilde{x}_i] + 2\sum_{i=1}^{n-1}\sum_{j=i+1}^n \text{Cov}[\tilde{x}_i, \tilde{x}_j] \tag{13.39}$$

Given any two random variables $\tilde{x}$ and $\tilde{y}$, we can define a third random variable $\tilde{\epsilon}$ by

$$\tilde{\epsilon} \equiv \tilde{y} - \alpha - \beta\tilde{x} \tag{13.40}$$

To specify the **disturbance term** $\tilde{\epsilon}$ completely, we can either specify the scalar constants α and β explicitly or fix them implicitly by imposing (two) conditions on $\tilde{\epsilon}$. We do the latter by insisting that

- $\tilde{\epsilon}$ and $\tilde{x}$ are uncorrelated and
- $E[\tilde{\epsilon}] = 0$.

It follows after a little calculation that

$$\beta = \frac{\text{Cov}[\tilde{x}, \tilde{y}]}{\text{Var}[\tilde{x}]} \tag{13.41}$$

and

$$\alpha = E[\tilde{y}] - \beta E[\tilde{x}] \tag{13.42}$$

Similar results can be derived for random vectors $\tilde{\mathbf{x}}$ and $\tilde{\mathbf{y}}$; see Exercise 13.11.

Note that the conditional expectation, $E[\tilde{y} \mid \tilde{x} = x]$ is not equal to $\alpha + \beta x$, as one might expect, unless $E[\tilde{\epsilon} \mid \tilde{x} = x] = 0$ for all x. This would require the stronger assumption of statistical independence rather than the mere lack of correlation assumed here in order to calculate the expression in (13.41) for β.[8] The implications of this stronger assumption will be considered further when we deal with regression in Section 14.2.

The notion of the **beta** of $\tilde{y}$ with respect to $\tilde{x}$ as given in (13.41) will recur frequently.

13.6.2   *Differentiation and expectation*

We will frequently want to find the derivative of an expectation with respect to some parameter that appears within the expectation, for example when maximizing expected utility in Chapter 16. For discrete random variables, the expectation operator is just a summation operator. As it is well known that the derivative of a sum is the sum of the derivatives, there is no problem about passing the differentiation operator through the expectation operator.

For continuously distributed random variables, the expectation operator is really an integration operator, so the rules for differentiation of integrals in Section 9.7 apply, in particular Leibniz's integral rule. Provided that the appropriate regularity conditions are satisfied, the differentiation operator can again be passed through the expectation operator.

EXAMPLE 13.6.1 Suppose we wish to find the vector $\mathbf{a}$ that minimizes the variance of the random variable $\mathbf{a}^\top \tilde{\mathbf{r}}$. The variance is $\mathbf{a}^\top \mathbf{V} \mathbf{a}$, where $\mathbf{V} \equiv \mathrm{Var}[\tilde{\mathbf{r}}]$ is the positive semi-definite variance–covariance matrix of $\tilde{\mathbf{r}}$. It is clear that the variance is minimized by setting $\mathbf{a} = \mathbf{0}$, but it is worthwhile deriving this result from first principles.

The objective function is

$$E[\mathbf{a}^\top (\tilde{\mathbf{r}} - E[\tilde{\mathbf{r}}])(\tilde{\mathbf{r}} - E[\tilde{\mathbf{r}}])^\top \mathbf{a}] \tag{13.43}$$

Differentiating through the expectation operator, the first-order condition is

$$E[2\mathbf{a}^\top (\tilde{\mathbf{r}} - E[\tilde{\mathbf{r}}])(\tilde{\mathbf{r}} - E[\tilde{\mathbf{r}}])^\top] = \mathbf{0}^\top \tag{13.44}$$

This simplifies to $\mathbf{a}^\top \mathbf{V} = \mathbf{0}^\top$ – the same first-order condition that would have resulted if the objective function had been simplified to $\mathbf{a}^\top \mathbf{V} \mathbf{a}$.

Thus the variance is zero for any vector $\mathbf{a}$ in the null space of the symmetric matrix $\mathbf{V}$. If $\mathbf{V}$ is invertible (positive definite), then there is a unique solution $\mathbf{a} = \mathbf{0}$. If $\mathbf{V}$ is singular (positive semi-definite but not positive definite), then any linear combination of the components of $\tilde{\mathbf{r}}$ with coefficient vector in the null space of $\mathbf{V}$ has zero variance. $\diamond$

This example could easily have been solved without explicitly passing the differentiation operator through the expectation; for an application where there is no such workaround, see Section 17.3.1.

13.6.3   *Covariance as a scalar product*

Consider again the n-dimensional vector space X of random variables introduced in Section 13.4, which was generated by the random variables $\tilde{x}_1, \tilde{x}_2, \ldots, \tilde{x}_n$. We will again let the vector $\mathbf{a} = (a_1, a_2, \ldots, a_n)$ denote the random variable $a_1 \tilde{x}_1 + a_2 \tilde{x}_2 + \cdots + a_n \tilde{x}_n$.

Let $\tilde{\mathbf{x}} = (\tilde{x}_1, \tilde{x}_2, \ldots, \tilde{x}_n)$ be the random vector whose components are the spanning random variables and assume that the variance–covariance matrix of this random vector, to be denoted **V**, exists. (Note that the random vector $\tilde{\mathbf{x}}$ is not itself an element of the vector space of random variables X.)

Recall that **V**, like any variance–covariance matrix, is positive semi-definite (see Exercise 13.9) and that the covariance between the random variables represented by the vectors $\mathbf{a}_1$ and $\mathbf{a}_2$ is $\mathbf{a}_1^\top \mathbf{V} \mathbf{a}_2$ (see Exercise 13.10).

As mentioned in Section 7.5, the variance–covariance matrix **V** of any random vector $\tilde{\mathbf{x}}$ can therefore be used to define a (symmetric, positive semi-definite) scalar product (i.e. covariance) and a metric (i.e. standard deviation) on the vector space of random variables generated by the components of $\tilde{\mathbf{x}}$.

It follows from Section 5.4.9 that any finite-dimensional vector space generated by a set of *non-trivial* random variables has an orthonormal basis, i.e. a basis of uncorrelated random variables, each with unit variance; see Exercise 13.13.

13.7 Multivariate normal distribution

A distribution that is commonly encountered is the **multivariate normal** (MVN). The n-dimensional random vector $\tilde{\mathbf{x}}$ is said to have a multivariate normal distribution with parameters $\boldsymbol{\mu} \in \mathbb{R}^n$ and $\boldsymbol{\Sigma} \in \mathbb{R}^{n \times n}$ if its pdf is

$$f_{\tilde{\mathbf{x}}}(\mathbf{x}) = \frac{1}{\sqrt{(2\pi)^n \det \boldsymbol{\Sigma}}} e^{-\frac{1}{2}(\mathbf{x}-\boldsymbol{\mu})^\top \boldsymbol{\Sigma}^{-1}(\mathbf{x}-\boldsymbol{\mu})} \tag{13.45}$$

where $\boldsymbol{\Sigma}$ is assumed to be symmetric positive definite. This is often written $\tilde{\mathbf{x}} \sim \text{MVN}(\boldsymbol{\mu}, \boldsymbol{\Sigma})$. The one-dimensional multivariate normal distribution is just the familiar normal distribution. The two-dimensional multivariate normal distribution is usually referred to as the **bivariate normal** distribution. Exercises 13.15 and 13.16 give some useful properties of the multivariate normal distribution. We will prove here some less well-known and slightly less straightforward properties of the normal and bivariate normal distributions, which have come to share the name Stein's lemma.[9] Assuming that asset returns are multivariate normal allows some elegant results to be derived from Stein's lemma; see Section 17.5.5.

THEOREM 13.7.1 (STEIN'S LEMMA). *Let $g \colon \mathbb{R} \to \mathbb{R}$ be differentiable.*

(a) *If $\tilde{x} \sim N(0, 1)$ and $E[|g'(\tilde{x})|] < \infty$, then*

$$E[g'(\tilde{x})] = E[g(\tilde{x})\tilde{x}] \tag{13.46}$$

(b) *If $\tilde{x} \sim N(\mu, \sigma^2)$ and $E[|g'(\tilde{x})|] < \infty$, then*

$$E[g'(\tilde{x})] = E\left[g(\tilde{x})\frac{\tilde{x}-\mu}{\sigma^2}\right] \tag{13.47}$$

(c) *If $\tilde{x}$ and $\tilde{y}$ are bivariate normally distributed and $E[|g'(\tilde{x})|] < \infty$, then*

$$\text{Cov}[g(\tilde{x}), \tilde{y}] = \text{Cov}[\tilde{x}, \tilde{y}]E[g'(\tilde{x})] \tag{13.48}$$

Proof:

(a) The proof of this original version of Stein's lemma appears in Stein (1981, p. 1136).

Recall from (13.7) that the standard normal pdf is

$$\phi(x) = \frac{1}{\sqrt{2\pi}} e^{-\frac{1}{2}x^2} \tag{13.49}$$

Note that, as $x \to \pm\infty$, $\phi(x) \to 0$. Hence, using the fundamental theorem of calculus, we can write

$$\phi(x) = \int_{-\infty}^{x} \phi'(z)\,dz = -\int_{x}^{\infty} \phi'(z)\,dz \tag{13.50}$$

Differentiating (13.49) using the chain rule yields

$$\phi'(x) = -x\phi(x) \tag{13.51}$$

so that (13.50) becomes

$$\phi(x) = -\int_{-\infty}^{x} z\phi(z)\,dz = \int_{x}^{\infty} z\phi(z)\,dz \tag{13.52}$$

Hence, we can write

$$
\begin{aligned}
E[g'(\tilde{x})] &= \int_{0}^{\infty} g'(x)\phi(x)\,dx + \int_{-\infty}^{0} g'(x)\phi(x)\,dx \\
&= \int_{0}^{\infty} g'(x)\left(\int_{x}^{\infty} z\phi(z)\,dz\right) dx - \int_{-\infty}^{0} g'(x)\left(\int_{-\infty}^{x} z\phi(z)\,dz\right) dx \\
&= \int_{0}^{\infty}\int_{0}^{z} g'(x)z\phi(z)\,dx\,dz - \int_{-\infty}^{0}\int_{z}^{0} g'(x)z\phi(z)\,dx\,dz
\end{aligned} \tag{13.53}
$$

where we have used Fubini's theorem (since we are assuming that $E[|g'(\tilde{x})|] < \infty$) to reverse the order of integration in the two integrals in the last step, which are respectively over the octants in the xz plane

$$\{(x, z) \in \mathbb{R}^2 : x \geq 0,\ z \geq x\} = \{(x, z) \in \mathbb{R}^2 : z \geq 0,\ 0 \leq x \leq z\} \tag{13.54}$$

and

$$\{(x, z) \in \mathbb{R}^2 : x \leq 0,\ z \leq x\} = \{(x, z) \in \mathbb{R}^2 : z \leq 0,\ 0 \geq x \geq z\} \tag{13.55}$$

Using the fundamental theorem of calculus to evaluate the inner integrals with respect to x, (13.53) becomes

$$
\begin{aligned}
E[g'(\tilde{x})] &= \int_{0}^{\infty} (g(z) - g(0))z\phi(z)\,dz - \int_{-\infty}^{0} (g(0) - g(z))z\phi(z)\,dz \\
&= \int_{-\infty}^{\infty} (g(z) - g(0))z\phi(z)\,dz \\
&= E[g(\tilde{x})\tilde{x}] - g(0)E[\tilde{x}] \\
&= E[g(\tilde{x})\tilde{x}]
\end{aligned} \tag{13.56}
$$

as required.

(b) For a general normal random variable, $\tilde{x} \sim N(\mu, \sigma^2)$, we can apply the previous result to $\tilde{z} \equiv (\tilde{x} - \mu)/\sigma$ and $h(\tilde{z}) \equiv g(\mu + \sigma \tilde{z}) = g(\tilde{x})$ to obtain

$$E[h'(\tilde{z})] = E[h(\tilde{z})\tilde{z}] \tag{13.57}$$

But

$$h'(z) = \sigma g'(\mu + \sigma z) \tag{13.58}$$

or

$$h'(\tilde{z}) = \sigma g'(\tilde{x}) \tag{13.59}$$

Making the appropriate substitutions in (13.57),

$$\sigma E[g'(\tilde{x})] = E\left[g(\tilde{x})\frac{\tilde{x} - \mu}{\sigma}\right] \tag{13.60}$$

or

$$E[g'(\tilde{x})] = E\left[g(\tilde{x})\frac{\tilde{x} - \mu}{\sigma^2}\right] \tag{13.61}$$

as required.

(c) The marginal distribution of $\tilde{x}$ is univariate normal; see Exercise 13.16.

(i) First assume that $\tilde{x} \sim N(0, 1)$.
As in (13.40), we write

$$\tilde{y} \equiv \alpha + \beta \tilde{x} + \tilde{\epsilon} \tag{13.62}$$

and fix α and β so that $\tilde{\epsilon}$ has mean zero and is uncorrelated with $\tilde{x}$. As we have assumed that $\text{Var}[\tilde{x}] = 1$, we have $\beta = \text{Cov}[\tilde{x}, \tilde{y}]$. It follows that

$$\begin{aligned}
\text{Cov}[g(\tilde{x}), \tilde{y}] &= \text{Cov}[g(\tilde{x}), \alpha + \beta \tilde{x} + \tilde{\epsilon}] \\
&= \beta \text{Cov}[g(\tilde{x}), \tilde{x}]
\end{aligned} \tag{13.63}$$

because $\tilde{\epsilon}$ is independent of $\tilde{x}$ (since uncorrelated bivariate normal random variables are independent; see Exercise 13.16) and, hence, also independent of $g(\tilde{x})$. Using (13.34) and the univariate version of Stein's lemma, this becomes

$$\begin{aligned}
\text{Cov}[g(\tilde{x}), \tilde{y}] &= \beta E[g(\tilde{x})\tilde{x}] \\
&= \text{Cov}[\tilde{x}, \tilde{y}]E[g'(\tilde{x})]
\end{aligned} \tag{13.64}$$

as required.

(ii) For a general normal random variable, $\tilde{x} \sim N(\mu, \sigma^2)$, we can again apply the previous result to $\tilde{z} \equiv (\tilde{x} - \mu)/\sigma$ and $h(\tilde{z}) \equiv g(\mu + \sigma \tilde{z}) = g(\tilde{x})$ to obtain

$$\text{Cov}[h(\tilde{z}), \tilde{y}] = \text{Cov}[\tilde{z}, \tilde{y}]E[h'(\tilde{z})] \tag{13.65}$$

Making the appropriate substitutions and again using (13.59) yields

$$\mathrm{Cov}[g(\tilde{x}), \tilde{y}] = \mathrm{Cov}\left[\frac{\tilde{x} - \mu}{\sigma}, \tilde{y}\right] E[\sigma g'(\tilde{x})] \tag{13.66}$$

The two σs cancel and the μ in the covariance vanishes, so that (13.66) reduces to (13.48) as in the standard normal case. $\qquad\qquad\square$

13.8   Estimation and forecasting

In simple examples, such as throwing an unbiased die, the values of the parameters of a probability distribution are intuitively obvious. In most practical applications, however, especially in economic and financial applications, the parameters of the distribution are unobservable and must be estimated. We have already considered estimation of a single-equation econometric model in Section 1.2.1. We also noted in Section 13.3.7 that good estimates of odds can be made by observing highly liquid betting exchange markets.

The most straightforward estimation problems are based on **samples** of many **independent and identically distributed** (iid) observations on a random variable, or **random samples**. For example, we might wish to model the distribution of the number of strokes taken by a golfer to play an 18-hole round of golf, say $\tilde{x}$. In particular, suppose that we wish to estimate the mean or expected number of shots per round, say $\mu \equiv E[\tilde{x}]$. If we have observed the values of the golfer's scores for n past rounds, $\tilde{x}_1 = x_1, \tilde{x}_2 = x_2, \ldots, \tilde{x}_n = x_n$, then an obvious **estimator** of the **population mean** μ is the **sample mean**, denoted

$$\overline{X} \equiv \frac{1}{n}\sum_{i=1}^{n} \tilde{x}_i \tag{13.67}$$

The observed value of the sample mean, denoted

$$\bar{x} \equiv \frac{1}{n}\sum_{i=1}^{n} x_i \tag{13.68}$$

is then called an **estimate**. Similarly, the sample equivalent is an obvious estimator of the population variance, covariance, or other moment.

In theory, any random variable can be used as an estimator of any parameter. To be useful, however, an estimator of a parameter should be **unbiased** in the sense that the expectation of the estimator equals the parameter being estimated. In our example, it was assumed that the sample observations were independent and identically distributed, so that by linearity of the expectation operator

$$E[\overline{X}] = \frac{1}{n}\sum_{i=1}^{n} E[\tilde{x}_i] = \frac{1}{n}\sum_{i=1}^{n} \mu = \frac{1}{n}n\mu = \mu \tag{13.69}$$

Thus the sample mean is an unbiased estimator of the population mean.

It is conventional to denote an estimator of a parameter of a probability distribution by placing a ˆ over the symbol denoting the parameter. For example, one might use $\hat{\mu}$ instead of $\overline{X}$ to denote the estimator of the population mean above.

Forecasting is similar to estimation, except that the quantity to be predicted is not an unobservable population parameter but a future observable value of a random variable or stochastic process, for example, the value of the main European Central Bank interest rate 12 months hence. Many economics graduates will learn that the public perception of their discipline is not so much as being the analysis of rational economic decision-making, but as forecasting of future values of economic variables such as interest rates or asset prices.

The forecasting problem is one to which both statistics and economics can make valuable contributions. For example, prices from futures or forward markets can be used to forecast future prices in spot markets. Alternatively, a time-series model of spot prices can be used to produce forecasts.

Just as any random variable can be proposed as an estimator of a particular parameter, so, in forecasting, any random variable can be proposed as a **predictor** of a value to be observed in the future. The observed value of the predictor is then called a **forecast**. As before, to be useful, a predictor of a random variable should be **unbiased** in the sense that the expectation of the predictor equals the expectation of the random variable being forecasted.

The distinction between estimation and forecasting can become blurred in areas such as sports betting. From a purely statistical point of view, the probabilities that each contestant will win a contest are unobservable parameters to be estimated. From a trading point of view, however, the market odds (or, equivalently, the market probabilities) at the start of the event are observable random variables to be forecasted.

13.9 Taylor's theorem: stochastic version

We will frequently apply the univariate Taylor expansion to a function of a random variable expanded about the mean of the random variable. We will illustrate this procedure using the infinite version of Taylor's expansion (9.87)

$$f(\tilde{x}) = f(E[\tilde{x}]) + \sum_{n=1}^{\infty} \frac{1}{n!} f^{(n)}(E[\tilde{x}])(\tilde{x} - E[\tilde{x}])^n \tag{13.70}$$

Taking expectations on both sides of (13.70) yields

$$E[f(\tilde{x})] = f(E[\tilde{x}]) + \sum_{n=2}^{\infty} \frac{1}{n!} f^{(n)}(E[\tilde{x}])m^n(\tilde{x}) \tag{13.71}$$

The fact that

$$m^1(\tilde{x}) = E[(\tilde{x} - E[\tilde{x}])^1] \equiv 0 \tag{13.72}$$

allows us to start the summation in (13.71) at $n = 2$ rather than $n = 1$. Indeed, we can rewrite (13.71) as

$$E[f(\tilde{x})] = f(E[\tilde{x}]) + \frac{1}{2} f''(E[\tilde{x}])\text{Var}[\tilde{x}] + \frac{1}{6} f'''(E[\tilde{x}])\text{Skew}[\tilde{x}]$$

$$+ \frac{1}{24} f''''(E[\tilde{x}])\text{Kurt}[\tilde{x}] + \sum_{n=5}^{\infty} \frac{1}{n!} f^{(n)}(E[\tilde{x}])m^n(\tilde{x}) \tag{13.73}$$

A similar procedure can be used to obtain a finite-order version of this result. For example, taking the expectation of the second-order Taylor expansion of f around $E[\tilde{x}]$ yields, roughly speaking,

$$E[f(\tilde{x})] = f(E[\tilde{x}]) + \tfrac{1}{2} f''(x^*) \mathrm{Var}[\tilde{x}] \tag{13.74}$$

for "some" x^*. However, this supposes that x^* is fixed, whereas in fact it varies with the value taken on by $\tilde{x}$, and is itself a random variable, correlated with $\tilde{x}$. In other words, an extra degree of approximation is involved in this stochastic version of Taylor's theorem, compared to the deterministic version.

These results are valid also for random vectors, although the notation becomes cumbersome beyond the variance term. The second-order Taylor approximation of $f(\tilde{\mathbf{x}})$ about $E[\tilde{\mathbf{x}}]$ yields

$$E[f(\tilde{\mathbf{x}})] = f(E[\tilde{\mathbf{x}}]) + \tfrac{1}{2} E\big[(\tilde{\mathbf{x}} - E[\tilde{\mathbf{x}}])^\top f''(E[\tilde{\mathbf{x}}])(\tilde{\mathbf{x}} - E[\tilde{\mathbf{x}}])\big] + \cdots \tag{13.75}$$

Since matrix multiplication is not commutative, the second-order term no longer reduces to a multiple of the variance, as it did in the one-dimensional case. However, if concavity or convexity of f allows us to determine that the Hessian matrix f'' is definite or semi-definite, then we will also be able to sign the second-order term in the Taylor approximation.

Most of the applications of Taylor's theorem listed at the end of Section 9.6 use this version of the theorem.

13.10   Jensen's inequality

This section introduces **Jensen's inequality**.[10] This result combines the theory of convexity and concavity in Section 10.2 with the ideas of random variables and vectors and their expectations from the earlier sections of the present chapter.

The first subsection gives the formal statement of the theorem and provides a rigorous proof for continuously differentiable functions alongside some more intuitive motivation and examples. The second subsection presents a selection of more practical applications from financial economics.

Jensen's inequality will be encountered again in a more theoretical context when we study risk aversion in Section 16.5.

13.10.1   *Statement and proof*

THEOREM 13.10.1 (JENSEN'S INEQUALITY). *Let $\tilde{\mathbf{x}}$ be a **non-trivial** random vector with finite expectation taking on values in some convex set $X \subseteq \mathbb{R}^n$.*

(a) *The expected value of a (strictly) concave function of the random vector $\tilde{\mathbf{x}}$ is (strictly) less than the same function of the expected value of the random vector, or*

$$E[u(\tilde{\mathbf{x}})] \,(<)\leq u(E[\tilde{\mathbf{x}}]) \quad \text{when } u\colon X \to \mathbb{R} \text{ is (strictly) concave and } E[u(\tilde{\mathbf{x}})] \text{ exists}$$
$$\tag{13.76}$$

(b) *Similarly, the expected value of a (strictly) convex function of the random vector $\tilde{\mathbf{x}}$ is (strictly) greater than the same function of the expected value of the random variable, or*

$$E[v(\tilde{\mathbf{x}})] \,(>)\geq v(E[\tilde{\mathbf{x}}]) \quad \text{when } v\colon X \to \mathbb{R} \text{ is (strictly) convex and } E[v(\tilde{\mathbf{x}})] \text{ exists}$$
$$\tag{13.77}$$

Proof: Without loss of generality, consider the concave case. We will supply two proofs, the first assuming that $\tilde{\mathbf{x}}$ has a discrete distribution with a finite number of possible values, and the second assuming that u is continuously differentiable. For a fully general proof (based on the separating and supporting hyperplane theorems) that makes no assumptions about either the distribution of $\tilde{\mathbf{x}}$ or the continuity or differentiability of the function u, see Berger (1993, pp. 343–4).

We will consider three ways of motivating this result, starting, respectively, from the definition of a concave function, from the first-order condition for identifying concave functions and from the second-order condition for identifying concave functions.

(a) One can re-interpret the inequality defining a concave function,

$$u(\lambda \mathbf{x} + (1 - \lambda)\mathbf{x}') \geq \lambda u(\mathbf{x}) + (1 - \lambda)u(\mathbf{x}') \tag{13.78}$$

in terms of a discrete random vector $\tilde{\mathbf{x}} \in X$ taking on the vector value $\mathbf{x} \in X$ with probability π and the vector value $\mathbf{x}' \in X$ with probability $1 - \pi$, $\pi \in (0, 1)$. Inequality (13.78) then becomes

$$u(\pi \mathbf{x} + (1 - \pi)\mathbf{x}') \geq \pi u(\mathbf{x}) + (1 - \pi)u(\mathbf{x}') \tag{13.79}$$

which just says that

$$u(E[\tilde{\mathbf{x}}]) \geq E[u(\tilde{\mathbf{x}})] \tag{13.80}$$

Figure 13.1 illustrates this argument for one-dimensional $\tilde{x}$.

An inductive argument can be used to extend the result to all discrete random variables with a finite number of possible values; in fact, all this amounts to is re-interpreting the scalars k_i in the defining inequality (10.3) as probabilities π_i.

This inductive argument runs into problems if the number of possible values is either countably or uncountably infinite.

(b) A rigorous proof of Jensen's inequality for continuously differentiable functions starts from the first-order condition for concavity

$$u(\mathbf{x}) \leq u(\mathbf{x}') + u'(\mathbf{x}')(\mathbf{x} - \mathbf{x}') \tag{13.81}$$

Replace $\mathbf{x}'$ with the expected value of the random vector $E[\tilde{\mathbf{x}}]$ and $\mathbf{x}$ with $\tilde{\mathbf{x}}$, a generic value of the random vector different from its expected value, to obtain

$$u(\tilde{\mathbf{x}}) \leq u(E[\tilde{\mathbf{x}}]) + u'(E[\tilde{\mathbf{x}}])(\tilde{\mathbf{x}} - E[\tilde{\mathbf{x}}]) \tag{13.82}$$

Using a similar approach to that used with Taylor's expansion in (13.71), take expectations on both sides of this inequality. The first term on the right-hand side is non-random and the second term again vanishes, yielding

$$E[u(\tilde{\mathbf{x}})] \leq u(E[\tilde{\mathbf{x}}]) \tag{13.83}$$

This completes the proof.

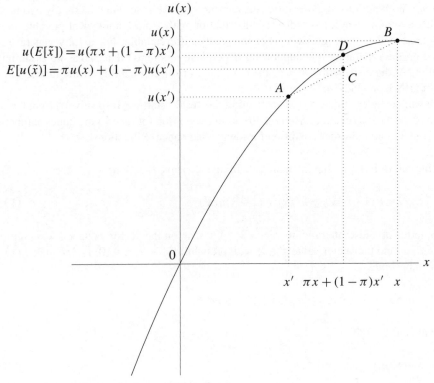

Figure 13.1 Motivation for Jensen's inequality

(c) One can also appeal to the second-order condition for concavity and the second-order Taylor approximation based on (13.75)

$$E[u(\tilde{\mathbf{x}})] \approx u(E[\tilde{\mathbf{x}}]) + \tfrac{1}{2} E\big[(\tilde{\mathbf{x}} - E[\tilde{\mathbf{x}}])^{\top} u''(E[\tilde{\mathbf{x}}])(\tilde{\mathbf{x}} - E[\tilde{\mathbf{x}}])\big] \qquad (13.84)$$

If u is concave, then the Hessian matrix $u''(E[\tilde{\mathbf{x}}])$ will be negative semi-definite and the quadratic form in the second term on the right-hand side will be non-positive for all values of $\tilde{\mathbf{x}}$, so that the expectation will also be non-positive. Thus,

$$E[u(\tilde{\mathbf{x}})] \leq u(E[\tilde{\mathbf{x}}]) \qquad (13.85)$$

approximately.

   This argument, however, cannot be used to prove the strict version of Jensen's inequality for strictly concave and strictly convex functions.

   The arguments for convex functions, and in the first two cases for strictly concave functions and strictly convex functions, are almost identical and are left as exercises; see Exercise 13.27. □

   Note that, when $\tilde{\mathbf{x}}$ has a *trivial* distribution, $\tilde{\mathbf{x}} = E[\tilde{\mathbf{x}}]$ with probability 1 and $f(\tilde{\mathbf{x}}) = f(E[\tilde{\mathbf{x}}]) = E[f(\tilde{\mathbf{x}})]$, whether f is convex or concave or neither. This result for trivial distributions is sometimes referred to as **Jensen's equality**.

To get a feel for the extent to which $E[u(\tilde{x})]$ differs from $u(E[\tilde{x}])$ in the scalar case, we can again use the following second-order Taylor approximation based on (13.73):

$$E[u(\tilde{x})] \approx u(E[\tilde{x}]) + \tfrac{1}{2}u''(E[\tilde{x}])\operatorname{Var}[\tilde{x}] \tag{13.86}$$

This shows that the magnitude of the difference is larger the larger is the curvature of u (as measured by the second derivative at the mean of $\tilde{x}$) and the larger is the variance of $\tilde{x}$.

EXAMPLE 13.10.1 Since $x \mapsto 1/x$ defines a strictly convex function on $\mathbb{R}_{++}$, for any non-trivial positive random variable $\tilde{x}$,

$$E\left[\frac{1}{\tilde{x}}\right] > \frac{1}{E[\tilde{x}]} \tag{13.87}$$

$\diamond$

EXAMPLE 13.10.2 Since $x \mapsto \ln x$ defines a strictly concave function on $\mathbb{R}_{++}$, for any non-trivial positive random variable $\tilde{x}$,

$$E[\ln \tilde{x}] < \ln E[\tilde{x}] \tag{13.88}$$

From this it follows by taking the exponential of each side that

$$E[\tilde{x}] > e^{E[\ln \tilde{x}]} \tag{13.89}$$

Taking reciprocals of both sides gives

$$\frac{1}{E[\tilde{x}]} < e^{-E[\ln \tilde{x}]} \tag{13.90}$$

or, putting $\tilde{y} = 1/\tilde{x}$,

$$\frac{1}{E[1/\tilde{y}]} < e^{E[\ln \tilde{y}]} \tag{13.91}$$

Finally, combining (13.89) and (13.91) yields

$$\frac{1}{E[1/\tilde{y}]} < e^{E[\ln \tilde{y}]} < E[\tilde{y}] \tag{13.92}$$

$\diamond$

EXAMPLE 13.10.3 Since $x \mapsto x^2$ defines a strictly convex function on $\mathbb{R}$, for any non-trivial positive random variable $\tilde{x}$,

$$E[\tilde{x}^2] > (E[\tilde{x}])^2 \tag{13.93}$$

Recall from Exercise 13.21 that $\operatorname{Var}[\tilde{x}] = E[\tilde{x}^2] - (E[\tilde{x}])^2$.

Combining these two results, it can be seen that Jensen's inequality implies that $\operatorname{Var}[\tilde{x}]$ is positive for non-trivial $\tilde{x}$. This result can also be derived directly from the definition of variance in (13.31). $\diamond$

We have presented Jensen's inequality in terms of random variables and their expectations. It applies equally to random samples and the corresponding sample means. As random samples are naturally finite, this is merely another re-interpretation of the defining inequalities (10.2) and (10.3), in the former case as

$$f(\bar{\mathbf{x}}) = f\left(\sum_{i=1}^{n} \frac{1}{n}\mathbf{x}_i\right) \le \sum_{i=1}^{n} \frac{1}{n}f(\mathbf{x}_i) \tag{13.94}$$

where f is convex and n denotes the sample size.

Jensen's inequality has a whole host of classical inequalities as special cases. The first example, which Jensen cites in his original paper, is that the geometric mean of a set of non-negative numbers is less than their arithmetic mean, a result that he credits to Cauchy; see Exercise 13.28. Another special case considered by Jensen (1906, p. 181) is the familiar Cauchy–Schwarz inequality (7.29). The next example is based on Jensen's derivation of that inequality.

EXAMPLE 13.10.4   Just as we wrote a version of Jensen's inequality for an equally weighted sample mean in (13.94), so we can write an equivalent version for any weighted average. For any positive real numbers (weights) $a_1, a_2, \ldots, a_n$, real numbers $b_1, b_2, \ldots, b_n$ and convex function v, Jensen's inequality, interpreted in terms of the weighted average of the b_i with weights a_i, implies that

$$v\left(\frac{\sum_{i=1}^{n} a_i b_i}{\sum_{i=1}^{n} a_i}\right) \le \frac{\sum_{i=1}^{n} a_i v(b_i)}{\sum_{i=1}^{n} a_i} \tag{13.95}$$

If $x_1, x_2, \ldots, x_n$ are any real numbers and $y_1, y_2, \ldots, y_n$ are any non-zero real numbers,[11] then we can set $v(x) \equiv x^2$, $a_i = y_i^2$ and $b_i = x_i/y_i$ $(i = 1, 2, \ldots, n)$ in (13.95) to obtain

$$\left(\frac{\sum_{i=1}^{n} y_i^2(x_i/y_i)}{\sum_{i=1}^{n} y_i^2}\right)^2 \le \frac{\sum_{i=1}^{n} y_i^2(x_i/y_i)^2}{\sum_{i=1}^{n} y_i^2} \tag{13.96}$$

Multiplying across by $\left(\sum_{i=1}^{n} y_i^2\right)^2$ and cancelling terms, this reduces to

$$\left(\sum_{i=1}^{n} x_i y_i\right)^2 \le \sum_{i=1}^{n} x_i^2 \sum_{i=1}^{n} y_i^2 \tag{13.97}$$

which (after taking square roots) becomes the Cauchy–Schwarz inequality (7.29), written in scalar rather than the original vector notation.   ◇

13.10.2   Applications in financial economics

Pari-mutuel betting

Intuition about the operation of markets often suggests hypotheses such as the following:

- The proportions bet on the outcomes in a pari-mutuel pool are unbiased estimators of the respective unobservable true probabilities of the outcomes occurring.

- The pari-mutuel odds, adjusted for deductions, against each of a set of outcomes are unbiased estimators of the respective unobservable true odds against each.

Such hypotheses are often presented as part of an argument asserting the **informational efficiency** of markets.

We will now show, using Jensen's inequality, that these two hypotheses cannot be true simultaneously.

If $\tilde{p}_i$ is the proportion of the pool bet on the ith contestant, π_i denotes the true probability that the ith contestant wins and τ is the proportion of the pool retained by the operator for expenses and taxes, then $\tilde{O}_i \equiv (1 - \tau)/\tilde{p}_i$ will be the pari-mutuel decimal odds and $1/\pi_i$ the true decimal odds against contestant i. The proportion $\tilde{p}_i$ is a random variable that cannot be observed by investors until the close of betting.

The first hypothesis above is that $E[\tilde{p}_i] = \pi_i$, while the second is that $E[\tilde{O}_i] = (1 - \tau)/\pi_i$. Substitution of one hypothesis into the other yields

$$E[\tilde{O}_i] = \frac{(1 - \tau)}{\pi_i} = \frac{(1 - \tau)}{E[\tilde{p}_i]} \tag{13.98}$$

and taking expectations in the definition of $\tilde{O}_i$ yields

$$E[\tilde{O}_i] = (1 - \tau)E\left[\frac{1}{\tilde{p}_i}\right] \tag{13.99}$$

This would imply that

$$\frac{1}{E[\tilde{p}_i]} = E\left[\frac{1}{\tilde{p}_i}\right] \tag{13.100}$$

which is a direct contradiction of Jensen's inequality (unless $\tilde{p}_i$ is known with certainty).

Thus at least one, if not both, of the above hypotheses must be rejected. This is not all bad news, however, for we can deduce a principle that might lead to more profitable pari-mutuel betting. If the proportions bet accurately reflect the true probabilities, i.e. $E[\tilde{p}_i] = \pi_i$, then Jensen's inequality implies that the expected pari-mutuel odds, adjusted for deductions, on average exceed the true odds, i.e.

$$\begin{aligned} E[\tilde{O}_i] &= (1 - \tau)E\left[\frac{1}{\tilde{p}_i}\right] \\ &> (1 - \tau)\frac{1}{E[\tilde{p}_i]} \\ &= \frac{(1 - \tau)}{\pi_i} \end{aligned} \tag{13.101}$$

Thus, the expected return on a 1 unit bet, $\pi_i E[\tilde{O}_i]$, will be greater than $(1 - \tau)$. Profitable betting, however, requires

$$\pi_i E[\tilde{O}_i] > 1 \tag{13.102}$$

In practice, however, the deductions will generally be far larger than the Jensen effect, so that

$$(1 - \tau) < \pi_i E[\tilde{O}_i] < 1 \tag{13.103}$$

In fact, it is easily shown that the expected pari-mutuel odds, again adjusted for deductions, cannot equal the true odds for all outcomes, i.e. that the second hypothesis above is untenable.

Again working with decimal odds, suppose that

$$E[\tilde{O}_i] = \frac{(1 - \tau)}{\pi_i} \tag{13.104}$$

i.e.

$$E\left[\frac{(1 - \tau)}{\tilde{p}_i}\right] = \frac{(1 - \tau)}{\pi_i} \tag{13.105}$$

or

$$E\left[\frac{1}{\tilde{p}_i}\right] = \frac{1}{\pi_i} \tag{13.106}$$

By Jensen's inequality, it follows that

$$\frac{1}{E[\tilde{p}_i]} < \frac{1}{\pi_i} \tag{13.107}$$

Inverting both sides gives

$$E[\tilde{p}_i] > \pi_i \tag{13.108}$$

Summing over outcomes gives

$$E\left[\sum_i \tilde{p}_i\right] > \sum_i \pi_i \tag{13.109}$$

which reduces to $1 > 1$ since both the proportions in the pool and the true probabilities of the outcomes must by definition sum to unity. Hence, we have proved by contradiction that the initial hypothesis is false.

Siegel's paradox

Another nice, but more perplexing, application of Jensen's inequality in finance is **Siegel's paradox**.[12] Before presenting a formal statement and discussion of this paradox, let us consider an example illustrating the relationship between spot rates and forward rates, in particular currency exchange rates.

EXAMPLE 13.10.5 Suppose that it is believed that 30 days hence £1 will be worth either €1.25 or €1.60 with equal probability of one-half, i.e. that the EUR/GBP spot exchange rate will be either 1.25 or 1.60. Then the expected value of the future EUR/GBP spot rate is 1.425. Suppose this is the same as today's EUR/GBP 30-day forward rate.

What is today's GBP/EUR 30-day forward rate? An arbitrage opportunity would exist if it differed from the reciprocal of the EUR/GBP rate, i.e. $1/1.425 \approx 0.701\,754\,386$.

What is the expected value of the GBP/EUR spot exchange rate 30 days hence? Similarly, an arbitrage opportunity would exist in either or both states of the world unless the rate was equal to the reciprocal of the relevant EUR/GBP rate in each state, i.e. unless it was $1/1.25 = 0.8$ with probability one-half and $1/1.60 = 0.625$ with probability one-half. Thus its expected value is 0.7125, over a penny greater than the corresponding forward rate. ◇

This example shows that the apparently reasonable hypothesis that the expected values of all unknown future spot rates equal the corresponding known forward rates is mathematically impossible.

THEOREM 13.10.2 (SIEGEL'S PARADOX). *Current forward prices cannot all equal expected future spot prices without perfect foresight.*

Proof: Siegel's paradox was originally stated in terms of currency exchange rates and is most easily understood if expressed in those terms. Let F_t be the current one-period-ahead forward exchange rate for, say, GBP/EUR, as in Example 13.10.5, and let $\tilde{S}_{t+1}$ be the corresponding unknown future spot GBP/EUR exchange rate. If the forward rate is an unbiased forecast of the future spot rate or

$$E_t[\tilde{S}_{t+1}] = F_t \tag{13.110}$$

then Jensen's inequality tells us that

$$\frac{1}{F_t} = \frac{1}{E_t[\tilde{S}_{t+1}]} < E_t\left[\frac{1}{\tilde{S}_{t+1}}\right] \tag{13.111}$$

except in the degenerate case where $\tilde{S}_{t+1}$ is known with certainty at time t (perfect foresight or Jensen's equality).

For simplicity, we have assumed here that interest rates are zero in both currencies. A full analysis would account for interest earned in both currencies between t and $t + 1$, but the ultimate result would be unchanged.

The reciprocal $1/\tilde{S}_{t+1}$ is the future EUR/GBP spot exchange rate unless there is to be an arbitrage opportunity at time $t + 1$. Likewise, the reciprocal $1/F_t$ is the current forward EUR/GBP exchange rate unless there is an arbitrage opportunity at time t.

Thus our initial hypothesis (expressed in terms of GBP/EUR rates) is untenable when rewritten in terms of EUR/GBP rates. □

Like the previous pari-mutuel example, Siegel's paradox is a warning that all intuitively appealing hypotheses about expectations should be treated with caution.

Siegel's paradox has remained unresolved for over three decades. Chu (2005) provides a summary of the related literature and a proposed resolution. The above discussion of the applications of Jensen's inequality has used the unconditional expectation operator $E[\cdot]$ without any discussion of the information available at the time that expectations are formed. Chu (2005) points out that the discussion should be based around the joint distribution of the unknown future spot exchange rates, say $\tilde{S}_{t+1}$ (GBP/EUR) and $\tilde{S}^*_{t+1}$ (EUR/GBP). The no-arbitrage principle requires that the identity $\tilde{S}_{t+1}\tilde{S}^*_{t+1} = 1$ holds with probability 1. In other words, all the probability mass of the joint probability distribution of $(\tilde{S}_{t+1}, \tilde{S}^*_{t+1})$ is concentrated along the rectangular hyperbola with equation $SS^* = 1$ in the SS^* plane.

It follows that the distribution of $\tilde{S}_{t+1}$ conditional on $\tilde{S}^*_{t+1}$ is trivial and that Jensen's inequality becomes Jensen's equality for this conditional distribution. Chu (2005) argues that this observation resolves the paradox; an alternative resolution will be presented in Section 16.7 below.

Pure expectations hypothesis

Another similar hypothesis is the **pure expectations hypothesis** relating to the term structure of interest rates, which states that forward simple interest rates are unbiased forecasts of future spot simple interest rates. If f_t denotes the current one-period forward simple interest rate and $\tilde{\imath}_{t+1}$ is the corresponding unknown future spot interest rate, then the pure expectations hypothesis is that

$$E_t[\tilde{\imath}_{t+1}] = f_t \tag{13.112}$$

See Section 15.4.4 for a more detailed discussion. In this case, the corollary refuted by Siegel's paradox concerns the relationship between spot and forward pure discount bond prices. In the absence of arbitrage opportunities, a one-period pure discount bond must trade at $1/(1+f_t)$ in the forward market and at $1/(1+\tilde{\imath}_{t+1})$ in the future spot market. Deriving the combined implications of Jensen's inequality and the pure expectations hypothesis for the relationship between these bond prices is left as an exercise; see Exercise 15.14.

While it would be completely irrational to argue that a hypothesis that cannot hold for the euro might hold for the pound or the dollar, bond prices and interest rates are sufficiently different that the pure expectations hypothesis cannot be condemned out of hand solely because of Siegel's paradox.

A more plausible and internally consistent alternative to the simple expectations hypothesis is the **logarithmic expectations hypothesis**, which can be written as follows for the currency exchange rate example above:

$$E_t[\ln \tilde{S}_{t+1}] = \ln F_t \tag{13.113}$$

Multiplying both sides of this equation by -1 gives

$$E_t[-\ln \tilde{S}_{t+1}] = -\ln F_t \tag{13.114}$$

or

$$E_t\left[\ln \frac{1}{\tilde{S}_{t+1}}\right] = \ln \frac{1}{F_t} \tag{13.115}$$

which is exactly the same hypothesis expressed in terms of the reciprocal exchange rate.

Using (13.92), it follows that this hypothesis gives forward rates lying between the two values implied by the original versions of the simple expectations hypothesis:

$$E_t[\tilde{S}_{t+1}] > e^{E_t[\ln \tilde{S}_{t+1}]} > \frac{1}{E_t[1/\tilde{S}_{t+1}]} \tag{13.116}$$

The logarithmic expectations hypothesis also suggests an alternative to the pure expectations hypothesis for interest rates, using continuously compounded rates of return. The details are again left as an exercise; see Exercise 15.15.

Efficient markets hypothesis

Both Siegel's paradox and the proposed solution apply equally well to any theory that uses current prices as a predictor of future values. Another such theory that is enormously popular is the **efficient markets hypothesis** (EMH) of Fama (1970). In its general form, this hypothesis argues that current prices **fully reflect** all available information about (expected) future values, or that markets are **informationally efficient**. Attempts to make the words *fully reflect* in any way mathematically rigorous quickly run into problems. Indeed, all three applications above can be looked on as simple examples of the efficient markets hypothesis. The EMH will be considered in more detail in Section 16.6.

EXERCISES

13.1 Let $\mathcal{A}$ be a sigma-algebra of subsets of the sample space Ω.

(a) Suppose $A_1, A_2 \in \mathcal{A}$. Show that $A_1 \cap A_2 \in \mathcal{A}$.
(b) Show that

$$\{\omega \in \Omega : \tilde{x}_1(\omega) \le x_1, \tilde{x}_2(\omega) \le x_2, \ldots, \tilde{x}_n(\omega) \le x_n\} \in \mathcal{A} \quad \forall \, (x_1, x_2, \ldots, x_n) \in \mathbb{R}^n$$

if and only if

$$\{\omega \in \Omega : \tilde{x}_i(\omega) \le x\} \in \mathcal{A} \quad \forall \, x \in \mathbb{R}, \ i = 1, 2, \ldots, n$$

(This shows that the two natural definitions of a random vector are equivalent.)

13.2 Let $\mathcal{A}$ be a sigma-algebra of events in Ω and $P : \mathcal{A} \to [0, 1]$. Suppose that

(a) $P(\Omega) = 1$ and
(b) $P\left(\bigcup_{i=1}^{\infty} A_i\right) = \sum_{i=1}^{\infty} P(A_i)$ when $A_1, A_2, \ldots$ are pairwise disjoint events in $\mathcal{A}$.

Show that $P(\Omega \setminus A) = 1 - P(A)$ for all $A \in \mathcal{A}$.

13.3 A bookmaker wishes to find the odds (or, equivalently, the probabilities) that he should quote for a contest with n candidates in order to maximize his expected profit, given the following beliefs:

(a) The probability that the ith candidate will win is p_i.
(b) The fraction z of all bets will be placed (on the winner) by "insider traders", i.e. bettors who already know the result (e.g. the Booker Prize winner, an Oscar winner or any contest where the winner is determined in private before it is announced publicly).
(c) The fraction $p_i(1-z)$ of all bets will be placed on the ith candidate by bettors who do not know the result.

The bookmaker's licence to bet is contingent on the sum of the probabilities implicit in the odds that he quotes being less than β, and the odds for all candidates must, of course, be positive.

   By formulating his problem as a Kuhn–Tucker inequality-constrained optimization problem, find the bookmaker's optimal strategy (i.e. his optimal probability quotes), the resulting expected profit, and the value of β (for the given z and $\mathbf{p} = (p_1, p_2, \ldots, p_n)$) at which it would no longer be worth his while to make a book.

13.4 Section 13.3.1 considered a fixed-odds version of the "6-from-n" Lotto game. Now consider the pari-mutuel version. Suppose that N people buy Lotto tickets at a price of €1 each and that all those who pick the correct numbers share a prize fund of €$N/2$.

(a) Is the expected value of a ticket still 50 cent? Explain.
(b) Should a value-seeking Lotto player bet when there is a rollover? Explain.

13.5 Show that, for scalars $k_1, k_2, \ldots, k_r$ and random variables $\tilde{x}_1, \tilde{x}_2, \ldots, \tilde{x}_r$,

$$E\left[\sum_{i=1}^{r} k_i \tilde{x}_i\right] = \sum_{i=1}^{r} k_i E[\tilde{x}_i]$$

13.6 Prove the equivalence of all the expressions in equation (13.34).

13.7 Let $\tilde{x}$ be a random vector with mean $\boldsymbol{\mu}$ and variance–covariance matrix $\boldsymbol{\Sigma}$; and let $\mathbf{A}$ and $\mathbf{b}$ be a conformable fixed matrix and vector, respectively. Find expressions for the mean and the variance–covariance matrix of $\tilde{y} \equiv \mathbf{A}\tilde{x} + \mathbf{b}$.

13.8 Show that two random variables are uncorrelated if and only if the expectation of their product equals the product of their expectations, assuming that all expectations exist.

13.9 Prove that all variance–covariance matrices of non-trivial random vectors are real, symmetric and positive semi-definite.

13.10 Show that the covariance between the random variables $\mathbf{a}_1^\top \tilde{r}$ and $\mathbf{a}_2^\top \tilde{r}$ is $\mathbf{a}_1^\top \mathrm{Var}[\tilde{r}]\mathbf{a}_2$.

13.11 Find matrix expressions equivalent to equations (13.41) and (13.42) for $\boldsymbol{\alpha} \in \mathbb{R}^m$ and $\mathbf{B} \in \mathbb{R}^{m \times n}$ satisfying

$$\tilde{\epsilon} \equiv \tilde{y} - \boldsymbol{\alpha} - \mathbf{B}\tilde{x}$$

where $\tilde{x}$ is an n-dimensional random vector and $\tilde{y}$ is an m-dimensional random vector.

13.12 Suppose $\tilde{x}$ is a random variable; let $f: \mathbb{R}^n \to \mathbb{R}^m$ be differentiable and let $\tilde{y} = f(\tilde{x})$. Using Theorem 9.7.6, explain the relationships between (a) $f_{\tilde{x}}$ and $f_{\tilde{y}}$ and (b) $E[\tilde{x}]$ and $E[\tilde{y}]$.

13.13 Suppose $\mathrm{Var}[\tilde{x}] = \Sigma$ is positive definite and let $\Sigma^{\frac{1}{2}}$ be a symmetric positive definite square root of Σ (which we know exists from Section 4.4.1). Define a new random vector $\tilde{y}$ by

$$\tilde{y} \equiv (\Sigma^{\frac{1}{2}})^{-1}\tilde{x}$$

(a) Show that $\mathrm{Var}[\tilde{y}] = \mathbf{I}$, i.e. that the components of $\tilde{y}$ are uncorrelated random variables with unit variance.
(b) Find an expression for the pdf $f_{\tilde{y}}$ in terms of $f_{\tilde{x}}$. (Hint: see Section 9.7.6.)

13.14 Let $\tilde{x} \sim N(\mu, \sigma^2)$. Recall the normal pdf $f_{\tilde{x}}$ given by equation (13.6).

(a) Evaluate the integral $\int_{-\infty}^{\infty} f_{\tilde{x}}(x)\,dx$.
Does your answer depend on the value of the parameter μ?
(b) Use Leibniz's integral rule and the chain rule to differentiate $\int_{-\infty}^{\infty} f_{\tilde{x}}(x)\,dx$ with respect to μ.
(c) Hence show that $E[\tilde{x} - \mu] = 0$.
(d) Calculate $E[e^{\tilde{x}}]$.

13.15 Suppose $\tilde{x} \sim MVN(\mu, \Sigma)$.

(a) Using the properties of the univariate normal distribution, calculate the mean vector and variance–covariance matrix of $\tilde{x}$ for the case where $\Sigma = \mathbf{I}$.
(b) Hence, calculate the mean vector and variance–covariance matrix of $\tilde{x}$ for any symmetric positive definite Σ. (Hint: use your answers to Exercise 13.7 and use a positive definite symmetric square root of the matrix Σ to simplify the multiple integrals involved; see Section 4.4.1.)

13.16 Suppose that the n-dimensional random vector $\tilde{x}$ has a multivariate normal distribution with mean μ and variance–covariance matrix Σ.

(a) If $\mathbf{A}$ and $\mathbf{b}$ are a conformable fixed matrix and vector, respectively, show that $\tilde{y} \equiv \mathbf{A}\tilde{x} + \mathbf{b}$ also has a multivariate normal distribution.
(b) Prove that $\tilde{x}_1, \tilde{x}_2, \ldots, \tilde{x}_n$ are mutually independent if and only if $\mathrm{Cov}[\tilde{x}_i, \tilde{x}_j] = 0$ for all $i \neq j$.

Now partition $\tilde{x}$ as $(\tilde{x}_1, \tilde{x}_2)$, where $\tilde{x}_1$ is a k-dimensional subvector of $\tilde{x}$ and $\tilde{x}_2$ denotes the complementary $(n-k)$-dimensional subvector.

(c) Show that $\tilde{x}_1$ also has a multivariate normal distribution. (This result is particularly important for the case of $k = 1$.)
(d) Show that the conditional distribution of $\tilde{x}_2$ given $\tilde{x}_1$ is multivariate normal.

13.17 Suppose that a pari-mutuel betting operator retains the fraction τ of the gross pool to cover taxes and expenses. Calculate the total percentage implicit in the resulting pari-mutuel odds.

13.18 Recall that the trifecta bet requires the first three finishers in a race to be nominated in the correct order. A popular method of betting into a trifecta pool is to pick one or more **bankers** and combine them with a (usually larger) number of other horses. Suppose you want to bet on all permutations including one horse from group A (containing a horses) and two horses from group B (containing b horses).

If group A is smaller than group B, this is considered a **banker trifecta**; if group B is smaller than group A, it is considered a **double-banker trifecta**.

Suppose first that groups A and B have no horse in common.

(a) How many (unordered) combinations does your portfolio of bets comprise?
(b) How many (ordered) permutations does your portfolio of bets comprise?
(c) In how many of these bets does the horse nominated to win come from group A?

Now suppose that there is an overlap of c horses between group A and group B and that you want each different permutation included only once in your portfolio.

(d) How many different (unordered) combinations does your portfolio of bets now comprise?
(e) How many (ordered) permutations does your portfolio of bets comprise?

13.19 A pari-mutuel operator offers a "Pick 4" pool, where punters are required to nominate the winners of four consecutive races. The operator retains 25% of the stakes and pays out the remainder *pro rata* to punters who correctly select all four winners. If no punter selects all four winners, the operator retains all stakes and adds them as a rollover to a Pick 4 pool on a future date.

Suppose that each race is a **handicap** with ten runners. In a handicap, the weights carried by each horse are adjusted, by placing a lead-cloth under the saddle, to equalize all horses' chances of winning. Like economics, handicapping is an inexact science, but you may assume for the purposes of this question that the handicapper has achieved his objective.

(a) How many different possible outcomes are available for the Pick 4 punter to select from? (You may ignore the possibility that a race may end in a dead-heat.)
(b) What is the probability of each of these outcomes?
(c) Can you determine the expected value of a €1 Pick 4 ticket at the time that the bet is placed?
(d) Suppose that the operator has sold €20 000 worth of tickets covering 5000 of the possible outcomes.

  (i) What will be the operator's total payout at the end of the day if one of these 5000 outcomes occurs?
  (ii) What is the operator's expected total payout to punters at the end of the day?
  (iii) What is the operator's expected liability to today's punters per €1 ticket sold?

(e) Can you now determine the expected value of a €1 Pick 4 ticket at the time that the bet is placed?

(f) If you hold a €1 Pick 4 ticket and one extra €1 ticket is sold on a combination that has *already* been covered, will the expected value of your ticket rise, fall or stay the same?

(g) If you hold a €1 Pick 4 ticket and one extra €1 ticket is sold on a combination that has *not previously* been covered, will the expected value of your ticket rise, fall or stay the same?

(h) Suppose the operator had guaranteed to pay out at least €25 000 if one or more winning tickets had been sold. How would this change your answers to the preceding parts?

13.20 Suppose that you have the opportunity to bet into a pari-mutuel pool into which there is a rollover of R and from which the proportion τ will be deducted to cover taxes, expenses, etc.

(a) If the total pool (rollover plus bets by others) is P and you believe that the probability that the pool will be won is π, how much should you bet in order to maximize your expected profit? You may assume that your bet will not affect the probability that the pool will be won. You may also assume that the expected return on all bets is the same, i.e. that neither you nor the other participants has any advantage or extra skill in picking outcomes with higher expected returns.

(b) How would your answer change if you were eligible for a rebate or commission of the proportion c of the amount bet?

(c) How much would you bet if the rollover was €1 000 000, another €1 500 000 had been bet into the pool, the takeout rate was 29%, you were eligible for a 5% rebate on your bet, and you believed there would certainly be at least one winning bet?

(d) Find the relationship between the amount bet P and the probability of a winning bet π that would just dissuade you from betting.

13.21 Show that, for any random variable $\tilde{x}$, $\mathrm{Var}[\tilde{x}] = E[\tilde{x}^2] - (E[\tilde{x}])^2$ (provided that all these quantities exist).

13.22 If $\tilde{u}_i$, $i = 1, 2, \ldots, n$, are random variables with $E[\tilde{u}_i] = 0$ for all i, $E[\tilde{u}_i^2] = \sigma^2$ for all i, and $E[\tilde{u}_i \tilde{u}_j] = 0$ for all i, j ($i \neq j$), show that $E[\tilde{\mathbf{u}}^\top \mathbf{A} \tilde{\mathbf{u}}] = \sigma^2 \mathrm{tr}(\mathbf{A})$, where $\tilde{\mathbf{u}} = [\tilde{u}_i]$ is $n \times 1$ and $\mathbf{A}$ is a conformable square matrix.

13.23 Prove that, if two random variables are statistically independent, then:

(a) they are uncorrelated; and
(b) all conditional expectations equal marginal expectations.

13.24 Prove that

$$\int_{-\infty}^{\infty} E[\tilde{x}_1 \mid \tilde{x}_2 = x_2] f_{\tilde{x}_2}(x_2) \, dx_2 = E[\tilde{x}_1]$$

for continuous random variables $\tilde{x}_1$ and $\tilde{x}_2$.

13.25 Calculate the mean and variance of the Poisson distribution; recall equation (13.2).

13.26 Show that a sum of random variables each having a Poisson distribution also has a Poisson distribution.

13.27 Prove Jensen's inequality for continuously differentiable convex functions and for continuously differentiable strictly convex functions.

13.28 Recall that the **geometric mean** of the numbers $x_1, x_2, \ldots, x_n$ is the nth root of their product $\sqrt[n]{\prod_{i=1}^{n} x_i}$.

Show that the geometric mean of a set of non-equal non-negative numbers is less than their arithmetic mean.

13.29 Consider a pari-mutuel pool operated on a contest with N possible outcomes. Suppose that the probability of outcome i is p_i and that you have just bet 1 unit on this outcome. Suppose also that the total stakes bet by other participants on each outcome i are unobservable, but are drawn from independent Poisson distributions with parameters λp_i.

(a) Calculate the expected total amount bet on outcome i.
(b) Calculate the expected dividend on outcome i, conditional on a known value X for the gross pool.
(c) Explain why the product of your two answers is different from the (expected) net pool.

13.30 Consider a 6-from-42 Lotto game with $^{42}C_6 = 5\,245\,786$ possible outcomes and a known jackpot pool of J, including a rollover of R.

(a) Calculate the expected number of jackpot-winning tickets if $N + 1$ tickets have been sold.
(b) Calculate the expected payout to a player who has bought one ticket and whose numbers have just been drawn (i.e. the expected value of the ticket before the National Lottery computer has indicated whether or not there are other winning tickets among the other N tickets sold).

You may assume that players select numbers at random and uniformly (uniform selection). You may also assume that 17.5 cent is added to the jackpot pool for each ticket sold (i.e. that J does not include any subsidy from a reserve fund to bring the jackpot pool up to a guaranteed minimum value). Finally, note that each ticket played can win only one prize in any one draw and that the highest prize will be paid in each case.

13.31 A spread betting firm trades on the product of the winning distances in eight races run at the Breeders' Cup. It offers to sell the product for a^8. Calculate the expected percentage return to a buyer who believes that the winning distances are iid random variables $a + \tilde{\epsilon}_i$, $i = 1, 2, \ldots, 8$, where the mean of $\tilde{\epsilon}_i$ is unknown. Noting that the winning distance cannot be negative, show that buying the product is a good bet even when the seller's price is unbiased, in the sense that the mean of $E[\tilde{\epsilon}_i]$ is drawn from a distribution which itself has mean 0.

13.32 Suppose $\tilde{x}$ and $\tilde{y}$ are independent random variables with finite variances. Show that

$$\text{Var}[\tilde{x}\,\tilde{y}] = \text{Var}[\tilde{x}]\text{Var}[\tilde{y}] + E[\tilde{x}]^2\text{Var}[\tilde{y}] + E[\tilde{y}]^2\text{Var}[\tilde{x}]$$

14 Quadratic programming and econometric applications

14.1 Introduction

At the start of Chapter 1 we considered three examples to illustrate the role of matrices and linear algebra in economics, and to provide the motivation for the mathematical material in the chapters that followed in Part I. The first of these topics involved a single-equation demand relationship

$$\tilde{Q}_t = \alpha + \beta P_t + \gamma Y_t + \tilde{u}_t \tag{14.1}$$

where $\tilde{Q}_t$ (quantity demanded), P_t (price) and Y_t (income) are observable for time periods $t = 1, 2, \ldots, T$, and $\tilde{u}_t$ is an unobservable random disturbance. A typical econometric problem was posed, namely, the estimation of the structural parameters of this population regression equation, α, β and γ, using a sample of time-series data.

In this chapter, we begin by returning to a more general form of this problem that involves estimation of not three unknown parameters, but an arbitrary number, k, of them. There are several approaches to estimation in econometrics: the methods of moments, least squares and maximum likelihood. In Section 14.2, we adopt the method of ordinary least squares, using some of the mathematics relating to matrices and optimization developed earlier to explore the algebra and geometry associated with this approach. We adopt a slightly different approach and notation here compared with Section 13.2 in order to emphasize the fact that we are dealing mainly with data, i.e. realizations of random variables, rather than the random variables themselves.

In Section 14.3, the theory of optimization is applied to maximization or minimization of a quadratic form subject to linear inequality constraints. This is an important subject as all constrained optimization is *locally* like maximizing a quadratic form with linear constraints, as will be discussed in Section 14.3. That section also contains a number of applications of this more general problem, which include an important statistical result associated with least squares regression: the Gauss–Markov theorem. Finally, in Section 14.4, some of the material on difference equations from Chapter 8 is extended to stochastic difference equations and a discussion of vector autoregressive models, which are widely used in empirical work in macroeconomics.

14.2 Algebra and geometry of ordinary least squares

Let the general single-equation economic relationship be denoted by the **population regression equation**

$$\tilde{Y}_t = \beta_1 + \beta_2 X_{t2} + \beta_3 X_{t3} + \cdots + \beta_k X_{tk} + \tilde{u}_t \tag{14.2}$$

where $\tilde{Y}_t$ is the value of the dependent variable, X_{tj}, $j = 1, 2, \ldots, k$, is the jth **explanatory variable**, and $\tilde{u}_t$ is an unknown and unobservable random disturbance in time period $t = 1, 2, \ldots, T$. The β_j, $j = 1, 2, \ldots, k$, are the unknown and unobservable parameters of the relationship, for which numerical estimates are required. The random variable $\tilde{u}_t$ accounts for factors ranging from errors in the measurement of $\tilde{Y}_t$ to the influences of variables missing from the model specification. This model, known as the **linear regression model**, is the basis of classical econometrics. It may be written compactly in a standard matrix notation as

$$\tilde{\mathbf{y}} = \mathbf{X}\boldsymbol{\beta} + \tilde{\mathbf{u}} \tag{14.3}$$

where

$$\tilde{\mathbf{y}} = \begin{bmatrix} \tilde{Y}_1 \\ \tilde{Y}_2 \\ \vdots \\ \tilde{Y}_T \end{bmatrix}_{T \times 1} , \quad \mathbf{X} = \begin{bmatrix} 1 & X_{12} & \cdots & X_{1k} \\ 1 & X_{22} & \cdots & X_{2k} \\ \vdots & \vdots & & \vdots \\ 1 & X_{T2} & \cdots & X_{Tk} \end{bmatrix}_{T \times k} , \quad \tilde{\mathbf{u}} = \begin{bmatrix} \tilde{u}_1 \\ \tilde{u}_2 \\ \vdots \\ \tilde{u}_T \end{bmatrix}_{T \times 1} \tag{14.4}$$

and $\boldsymbol{\beta} = \begin{bmatrix} \beta_1 & \beta_2 & \cdots & \beta_k \end{bmatrix}^\top$.

Note the orders of the matrices and the conformability of the matrices for the matrix multiplication, addition and equality operations. We will make an assumption about the rank of $\mathbf{X}$, for reasons that will become clear later, namely, that $\rho(\mathbf{X}) = k < T$.

14.2.1   Algebra of ordinary least squares

Use of the **ordinary least squares** (OLS) technique is a standard approach to the problem of estimation of $\boldsymbol{\beta}$, which chooses an estimate, $\hat{\boldsymbol{\beta}}$, so as to obtain the best-fitting line or plane (or hyperplane in more than three dimensions) in the sense of minimizing the sum of the squared deviations (or residuals) around the line or plane. The two-dimensional or bivariate situation is familiar from introductory econometrics. For a population regression function $\tilde{Y}_t = \beta_1 + \beta_2 X_t + \tilde{u}_t$, and an observed random sample of realizations of $\tilde{Y}_t$, denoted $Y_1, Y_2, \ldots, Y_T$, and corresponding values $X_1, X_2, \ldots, X_T$ of the explanatory variable, scatter plots with the sample conditional expectation function $\hat{Y}_t = \hat{\beta}_1 + \hat{\beta}_2 X_t$ superimposed are frequently used. In this case, the empirical counterparts of the unobservable random disturbances are the residuals $e_t = Y_t - \hat{Y}_t = Y_t - \hat{\beta}_1 - \hat{\beta}_2 X_t$, and OLS finds the vector $\hat{\boldsymbol{\beta}}^* = \begin{bmatrix} \hat{\beta}_1^* & \hat{\beta}_2^* \end{bmatrix}^\top$ such that $\sum_{t=1}^{T} e_t^2$ is minimized.

Similarly, our general problem is to choose $\hat{\boldsymbol{\beta}}^*$ to minimize

$$\sum_{t=1}^{T} e_t^2 = \sum_{t=1}^{T} (Y_t - \hat{\beta}_1 - \hat{\beta}_2 X_{t2} - \hat{\beta}_3 X_{t3} - \cdots - \hat{\beta}_k X_{tk})^2 \tag{14.5}$$

or in matrix notation

$$\mathbf{e}^\top \mathbf{e} = (\mathbf{y} - \hat{\mathbf{y}})^\top (\mathbf{y} - \hat{\mathbf{y}}) = (\mathbf{y} - \mathbf{X}\hat{\boldsymbol{\beta}})^\top (\mathbf{y} - \mathbf{X}\hat{\boldsymbol{\beta}}) \tag{14.6}$$

with respect to choice of $\hat{\boldsymbol{\beta}}$.

It is noteworthy that $\mathbf{e}^\top\mathbf{e} = \mathbf{e}^\top\mathbf{I}\mathbf{e}$ is a quadratic form in $\mathbf{e}$, which is positive definite since the identity matrix, $\mathbf{I}$, is positive definite, i.e. $\mathbf{e}^\top\mathbf{e} > 0$ for all $\mathbf{e} \neq \mathbf{0}$. It may also be noted that $\mathbf{e}^\top\mathbf{e} = \|\mathbf{e}\|^2$, $\mathbf{e} \in \mathbb{R}^T$, so our problem may also be thought of as finding $\hat{\boldsymbol{\beta}}$ such that the length of the vector of residuals, $\mathbf{e}$, is minimized in Euclidean T-space. To begin, we concentrate on the algebra of the OLS solution.

From (14.6), and using the rules of matrix transposition and the distributive law, we have that

$$
\begin{aligned}
\mathbf{e}^\top\mathbf{e} &= (\mathbf{y}^\top - \hat{\boldsymbol{\beta}}^\top\mathbf{X}^\top)(\mathbf{y} - \mathbf{X}\hat{\boldsymbol{\beta}}) \\
&= \mathbf{y}^\top\mathbf{y} - \hat{\boldsymbol{\beta}}^\top\mathbf{X}^\top\mathbf{y} - \mathbf{y}^\top\mathbf{X}\hat{\boldsymbol{\beta}} + \hat{\boldsymbol{\beta}}^\top\mathbf{X}^\top\mathbf{X}\hat{\boldsymbol{\beta}} \\
&= \mathbf{y}^\top\mathbf{y} - 2\hat{\boldsymbol{\beta}}^\top\mathbf{X}^\top\mathbf{y} + \hat{\boldsymbol{\beta}}^\top\mathbf{X}^\top\mathbf{X}\hat{\boldsymbol{\beta}}
\end{aligned}
\tag{14.7}
$$

Note that, like $\mathbf{e}^\top\mathbf{e}$, all terms on the right-hand side of (14.7) are scalars. The middle two terms in the middle expression are equal since one is the transpose of the other. Note, too, that, expressed as a function of $\hat{\boldsymbol{\beta}}$, $\mathbf{e}^\top\mathbf{e}$ includes a linear form in $\hat{\boldsymbol{\beta}}$ and a quadratic form in $\hat{\boldsymbol{\beta}}$. We can therefore differentiate with respect to $\hat{\boldsymbol{\beta}}$ using the results from Chapter 9. We get

$$
\frac{\partial \mathbf{e}^\top\mathbf{e}}{\partial \hat{\boldsymbol{\beta}}} = -2\mathbf{X}^\top\mathbf{y} + 2\mathbf{X}^\top\mathbf{X}\hat{\boldsymbol{\beta}}
\tag{14.8}
$$

For a maximum or minimum, $\partial \mathbf{e}^\top\mathbf{e}/\partial\hat{\boldsymbol{\beta}} = \mathbf{0}$. Therefore, our first-order condition is

$$
-2\mathbf{X}^\top\mathbf{y} + 2\mathbf{X}^\top\mathbf{X}\hat{\boldsymbol{\beta}}^* = \mathbf{0}
\tag{14.9}
$$

which, dividing by 2 and rearranging, gives

$$
\mathbf{X}^\top\mathbf{X}\hat{\boldsymbol{\beta}}^* = \mathbf{X}^\top\mathbf{y}
\tag{14.10}
$$

The asterisk is used here to denote the value of $\hat{\boldsymbol{\beta}}$ that satisfies the first-order condition, i.e. the optimal OLS solution.[1]

The equations in (14.10) are called the OLS **normal equations**. They constitute a square system of linear simultaneous equations in the k unknown elements of $\hat{\boldsymbol{\beta}}^*$. Recalling Cramer's theorem (Theorem 2.5.1), these equations are uniquely soluble for $\hat{\boldsymbol{\beta}}^*$ if the $k \times k$ matrix $\mathbf{X}^\top\mathbf{X}$ is invertible. We now see the reason for the earlier assumption that $\rho(\mathbf{X}) = k < T$. For, with $\mathbf{X}$ of full rank, Theorem 4.4.8 applies to $\mathbf{X}^\top\mathbf{X}$. Hence we can assert that $\mathbf{X}^\top\mathbf{X}$ is positive definite and therefore non-singular; the inverse of $\mathbf{X}^\top\mathbf{X}$ exists.

The unique solutions for the individual elements of $\hat{\boldsymbol{\beta}}^*$ could be obtained via Cramer's rule as

$$
\hat{\beta}_i^* = \frac{|(\mathbf{X}^\top\mathbf{X})_i|}{|\mathbf{X}^\top\mathbf{X}|}, \quad i = 1, 2, \ldots, k
\tag{14.11}
$$

where, following the convention used in Section 2.5.1, $(\mathbf{X}^\top\mathbf{X})_i$ denotes the matrix formed by replacing the ith column of $\mathbf{X}^\top\mathbf{X}$ by the $k \times 1$ vector $\mathbf{X}^\top\mathbf{y}$. However, we are more interested in using the general solution

$$\hat{\boldsymbol{\beta}}^* = (\mathbf{X}^\top\mathbf{X})^{-1}\mathbf{X}^\top\mathbf{y} \tag{14.12}$$

which follows by pre-multiplying both sides of (14.10) by $(\mathbf{X}^\top\mathbf{X})^{-1}$. If $\rho(\mathbf{X}) < k$ so that $\mathbf{X}^\top\mathbf{X}$ is not positive definite, then $\mathbf{X}^\top\mathbf{X}$ would be singular, $(\mathbf{X}^\top\mathbf{X})^{-1}$ would not exist and therefore $\hat{\boldsymbol{\beta}}^*$ could not be computed. This is the situation known in econometrics as **perfect multicollinearity**, i.e. linear dependence of the columns of $\mathbf{X}$.

Before the implications of our main result (14.12) are explored, let us examine the second-order condition for this problem. Differentiating (14.8) with respect to $\hat{\boldsymbol{\beta}}$ yields the $k \times k$ Hessian matrix of second-order partial derivatives

$$\frac{\partial^2 \mathbf{e}^\top\mathbf{e}}{\partial\hat{\boldsymbol{\beta}}\partial\hat{\boldsymbol{\beta}}^\top} = 2\mathbf{X}^\top\mathbf{X} \tag{14.13}$$

Given the positive definiteness of $\mathbf{X}^\top\mathbf{X}$ that follows from our full rank assumption for $\mathbf{X}$, this Hessian is positive definite and the second-order condition for a minimum is satisfied. The solution for $\hat{\boldsymbol{\beta}}^*$ given in (14.12) does indeed minimize $\mathbf{e}^\top\mathbf{e} = \sum_{t=1}^{T} e_t^2$.

It is of interest to use the general result in the context of the two-variable regression model $\tilde{Y}_t = \beta_1 + \beta_2 X_t + \tilde{u}_t$. It is easy to show that in this case we have

$$\hat{\boldsymbol{\beta}}^* = (\mathbf{X}^\top\mathbf{X})^{-1}\mathbf{X}^\top\mathbf{y} = \begin{bmatrix} T & \sum_{t=1}^{T} X_t \\ \sum_{t=1}^{T} X_t & \sum_{t=1}^{T} X_t^2 \end{bmatrix}^{-1} \begin{bmatrix} \sum_{t=1}^{T} Y_t \\ \sum_{t=1}^{T} X_t Y_t \end{bmatrix} \tag{14.14}$$

It is left as an exercise to show that evaluation of the right-hand side of this equation generates the well-known scalar expressions for the OLS estimates $\hat{\beta}_1^*$ and $\hat{\beta}_2^*$; see, for example, Gujarati (2003, Chapter 3) and Exercise 14.1. It is also of interest for us to explore the algebra of the general OLS solution a little further.

Using the OLS solution (14.12) we may write

$$\mathbf{e} = \mathbf{y} - \hat{\mathbf{y}} = \mathbf{y} - \mathbf{X}\hat{\boldsymbol{\beta}}^* = \mathbf{y} - \mathbf{X}(\mathbf{X}^\top\mathbf{X})^{-1}\mathbf{X}^\top\mathbf{y} \tag{14.15}$$

or, more compactly,

$$\mathbf{e} = \mathbf{M}\mathbf{y} \tag{14.16}$$

where $\mathbf{M} = \mathbf{I} - \mathbf{X}(\mathbf{X}^\top\mathbf{X})^{-1}\mathbf{X}^\top$. Substituting $\mathbf{y} = \mathbf{X}\boldsymbol{\beta} + \tilde{\mathbf{u}}$ in this equation yields

$$\mathbf{e} = \mathbf{M}\tilde{\mathbf{u}} \tag{14.17}$$

since $\mathbf{M}\mathbf{X} = \mathbf{0}$. Therefore, the OLS vector of residuals is both a linear function of $\mathbf{y}$ and the same linear function of the unknown vector $\tilde{\mathbf{u}}$.

Applying the theorems for transposes from Section 1.5.7, it is easy to demonstrate that $\mathbf{M}^\top = \mathbf{M}$. By direct multiplication, it is also straightforward to show that $\mathbf{M}^2 = \mathbf{M}$. Thus $\mathbf{M}$ is symmetric and idempotent.

We have some similar results for the $\hat{\mathbf{y}}$ vector, namely,

$$\hat{\mathbf{y}} = \mathbf{y} - \mathbf{e} = \mathbf{y} - \mathbf{My} = (\mathbf{I} - \mathbf{M})\mathbf{y} = \mathbf{Ny} \tag{14.18}$$

another linear function of $\mathbf{y}$, where $\mathbf{N} = \mathbf{X}(\mathbf{X}^\top\mathbf{X})^{-1}\mathbf{X}^\top = \mathbf{I} - \mathbf{M}$ is symmetric idempotent. However, in this case $\mathbf{NX} = \mathbf{X}$ and $\hat{\mathbf{y}}$ is not the same linear function of $\tilde{\mathbf{u}}$; rather, on substitution for $\mathbf{y}$, using (14.3), we get

$$\hat{\mathbf{y}} = \mathbf{X}\boldsymbol{\beta} + \mathbf{N}\tilde{\mathbf{u}} \tag{14.19}$$

Using the fact that $\mathbf{MX} = \mathbf{0}$, we observe that

$$\mathbf{MN} = \mathbf{0} \quad \text{and} \quad \mathbf{X}^\top\mathbf{e} = \mathbf{0} \tag{14.20}$$

The first of these properties means that $\mathbf{M}$ and $\mathbf{N}$ are "orthogonal" to each other; the second implies that

$$\mathbf{X}^\top\mathbf{y} = \mathbf{X}^\top\hat{\mathbf{y}} \tag{14.21}$$

$$\hat{\mathbf{y}}^\top\mathbf{e} = \hat{\boldsymbol{\beta}}^{*\top}\mathbf{X}^\top\mathbf{e} = 0 \tag{14.22}$$

so that $\hat{\mathbf{y}}$ and $\mathbf{e}$ are orthogonal, and if there is an intercept in the regression equation, so that the first row of $\mathbf{X}^\top$ consists entirely of ones, then

$$\mathbf{1}^\top\mathbf{y} = \sum_{t=1}^T Y_t = \sum_{t=1}^T \widehat{Y}_t = \mathbf{1}^\top\hat{\mathbf{y}} \tag{14.23}$$

and

$$\mathbf{1}^\top\mathbf{e} = \sum_{t=1}^T e_t = 0 \tag{14.24}$$

so that the mean of the Y_t values and the mean of the $\widehat{Y}_t$ values are the same, and the mean of the e_t values is zero. All of these results, which have important roles in econometrics, follow mathematically from the OLS methodology.

14.2.2 Geometry of ordinary least squares

We have already noted that OLS may be thought of as minimizing the length of the vector $\mathbf{e}$ in $\mathbb{R}^n$. In this section, we elaborate on this fact and provide some further geometric insight into the OLS technique.

First note that, if we partition $\mathbf{X}$ by its columns, $\mathbf{x}_i$, $i = 1, 2, \ldots, k$, and partition $\hat{\boldsymbol{\beta}}^*$ conformably by its individual elements, we may write

$$\hat{\mathbf{y}} = \mathbf{X}\hat{\boldsymbol{\beta}}^* = \hat{\beta}_1^*\mathbf{x}_1 + \hat{\beta}_2^*\mathbf{x}_2 + \cdots + \hat{\beta}_k^*\mathbf{x}_k \tag{14.25}$$

and note that $\hat{\mathbf{y}} \in \text{lin}\{\mathbf{x}_1, \mathbf{x}_2, \ldots, \mathbf{x}_k\}$, i.e. $\hat{\mathbf{y}}$ belongs to the linear space spanned by the columns of $\mathbf{X}$, and thus by (14.20) $\mathbf{e}$ belongs to its orthogonal complement. Moreover, $\hat{\mathbf{y}}$ will be shown to be the orthogonal projection of $\mathbf{y}$ onto this column space. To see this most easily, let us consider the simplest possible regression

$$\tilde{Y}_t = \beta X_t + \tilde{u}_t, \quad t = 1, 2, \ldots, T \tag{14.26}$$

For this case, we have

$$\hat{\beta}^* = (\mathbf{X}^\top \mathbf{X})^{-1} \mathbf{X}^\top \mathbf{y}$$
$$= \frac{1}{\mathbf{x}_1^\top \mathbf{x}_1} \mathbf{x}_1^\top \mathbf{y} \tag{14.27}$$

since $\mathbf{X}$ comprises just a single column of observations $\mathbf{x}_1$. Therefore,

$$\hat{\mathbf{y}} = \hat{\beta}^* \mathbf{x}_1 = \frac{\mathbf{x}_1^\top \mathbf{y}}{\mathbf{x}_1^\top \mathbf{x}_1} \mathbf{x}_1$$
$$= \frac{\mathbf{x}_1^\top \mathbf{y}}{\|\mathbf{x}_1\|^2} \mathbf{x}_1 = \text{proj}_{\mathbf{x}_1} \mathbf{y} \tag{14.28}$$

using Theorem 5.2.4; i.e. $\hat{\mathbf{y}}$ is the orthogonal projection of $\mathbf{y}$ on $\mathbf{x}_1$. It follows that $\mathbf{e} = \mathbf{y} - \hat{\mathbf{y}}$ is the component of $\mathbf{y}$ orthogonal to $\mathbf{x}_1$.

The result that

$$\hat{\mathbf{y}} = \mathbf{X}\hat{\beta}^* = \mathbf{N}\mathbf{y} \tag{14.29}$$

is the generalization of (14.28), and $\mathbf{N}$ is known as the OLS projection matrix. The matrix $\mathbf{N}$ projects $\mathbf{y}$ orthogonally onto the column space of $\mathbf{X}$. Thus $\mathbf{e} = \mathbf{M}\mathbf{y} = (\mathbf{I} - \mathbf{N})\mathbf{y}$ is orthogonal to $\hat{\mathbf{y}}$ by construction and the length of $\mathbf{e}$ is minimized. Any other choice of coordinates, $\tilde{\beta}$, say, resulting in the linear combination $\tilde{\mathbf{y}}$ would lead to a residual vector $\mathbf{y} - \tilde{\mathbf{y}}$ with greater length than $\mathbf{e}$. The situation is depicted in Figure 14.1 for the simple regression (14.26) and $T = 3$. In Figure 14.1, the optimal value of $\hat{\beta}^*$ is one-half, which has the shortest residual vector associated with it.

To ensure understanding of these geometric ideas, a useful exercise for the reader would be to produce a corresponding diagram to Figure 14.1 for the regression equation

$$\tilde{Y}_t = \beta_1 X_{t1} + \beta_2 X_{t2} + \tilde{u}_t \tag{14.30}$$

and $T = 3$ observations on $\tilde{Y}_t$, X_{t1} and X_{t2}; see Exercise 14.3.

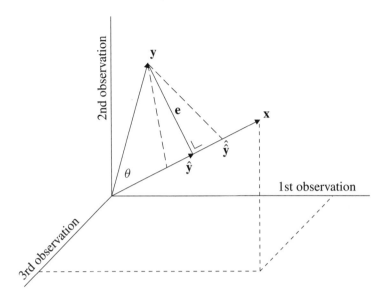

Figure 14.1 Geometry of ordinary least squares regression in $\mathbb{R}^3$

Finally, consider the angle, θ, between $\mathbf{y}$ and $\hat{\mathbf{y}}$. From the definition of the dot product, we have that

$$
\cos\theta = \frac{\mathbf{y} \cdot \hat{\mathbf{y}}}{\|\mathbf{y}\|\,\|\hat{\mathbf{y}}\|} = \frac{(\hat{\mathbf{y}}+\mathbf{e}) \cdot \hat{\mathbf{y}}}{\|\mathbf{y}\|\,\|\hat{\mathbf{y}}\|}
$$

$$
= \frac{(\hat{\mathbf{y}}+\mathbf{e})^\top \hat{\mathbf{y}}}{(\mathbf{y}^\top \mathbf{y})^{\frac{1}{2}}(\hat{\mathbf{y}}^\top \hat{\mathbf{y}})^{\frac{1}{2}}} = \frac{\hat{\mathbf{y}}^\top \hat{\mathbf{y}}}{(\mathbf{y}^\top \mathbf{y})^{\frac{1}{2}}(\hat{\mathbf{y}}^\top \hat{\mathbf{y}})^{\frac{1}{2}}}
$$

$$
= \frac{(\hat{\mathbf{y}}^\top \hat{\mathbf{y}})^{\frac{1}{2}}}{(\mathbf{y}^\top \mathbf{y})^{\frac{1}{2}}} \tag{14.31}
$$

The square of this ratio is similar to the ratio of the so-called **explained sum of squares** to the **total sum of squares** known in regression analysis as the **coefficient of determination**, and denoted as R^2. Similarly, therefore, $\cos^2\theta$ may be interpreted as a measure of goodness of fit of the estimated regression to the data. Given that $0 \le \cos^2\theta \le 1$, we have improving fit as $\cos^2\theta \to 1$. Geometrically, as may be seen from Figure 14.1, this corresponds to $\theta \to 0$ and thus $\hat{\mathbf{y}} \to \mathbf{y}$ and, hence, $\|\mathbf{e}\| \to 0$.

The following section considers a more general optimization problem, of which OLS regression is a special case.

14.3 Canonical quadratic programming problem

14.3.1 Canonical solution

A problem that arises frequently in economics, econometrics and finance is to find the vector $\mathbf{x} \in \mathbb{R}^n$ that maximizes the value of the quadratic form $\mathbf{x}^\top \mathbf{A}\mathbf{x}$ subject to the m linear inequality constraints $\mathbf{g}^{i\top}\mathbf{x} \geq \alpha_i$, where $\mathbf{A} \in \mathbb{R}^{n \times n}$ is negative definite and $\mathbf{g}^i \in \mathbb{R}^n$ for $i = 1, 2, \ldots, m$ $(m \leq n)$, or the equivalent problem with equality constraints.

The objective function can always be rewritten as a quadratic form in a symmetric (negative definite) matrix, since as $\mathbf{x}^\top \mathbf{A} \mathbf{x}$ is a scalar,

$$
\begin{aligned}
\mathbf{x}^\top \mathbf{A} \mathbf{x} &= (\mathbf{x}^\top \mathbf{A} \mathbf{x})^\top \\
&= \mathbf{x}^\top \mathbf{A}^\top \mathbf{x} \\
&= \tfrac{1}{2}(\mathbf{x}^\top \mathbf{A} \mathbf{x} + \mathbf{x}^\top \mathbf{A}^\top \mathbf{x}) \\
&= \mathbf{x}^\top \left(\tfrac{1}{2}(\mathbf{A} + \mathbf{A}^\top)\right)\mathbf{x}
\end{aligned}
\tag{14.32}
$$

and, as is easily demonstrated, $\tfrac{1}{2}(\mathbf{A} + \mathbf{A}^\top)$ is always symmetric; see Exercise 14.6. In what follows, we will assume that $\mathbf{A}$ itself is symmetric.

Let $\mathbf{G}$ be the $m \times n$ matrix whose ith row is $\mathbf{g}^i$, and let $\boldsymbol{\alpha} = [\alpha_i]_{m \times 1}$. The matrix $\mathbf{G}$ must have full rank if we are to apply the Kuhn–Tucker conditions.

The Lagrangian is

$$
\mathbf{x}^\top \mathbf{A} \mathbf{x} + \boldsymbol{\lambda}^\top (\mathbf{G}\mathbf{x} - \boldsymbol{\alpha})
\tag{14.33}
$$

The first-order condition is

$$
2\mathbf{x}^{*\top}\mathbf{A} + \boldsymbol{\lambda}^\top \mathbf{G} = \mathbf{0}_{1 \times n}
\tag{14.34}
$$

or, transposing and pre-multiplying across by $\tfrac{1}{2}\mathbf{A}^{-1}$,

$$
\mathbf{x}^* = -\tfrac{1}{2}\mathbf{A}^{-1}\mathbf{G}^\top \boldsymbol{\lambda}
\tag{14.35}
$$

where the asterisk denotes the optimal solution. Assuming for the time being that the constraints are binding, we have that

$$
\mathbf{G}\mathbf{x}^* = \boldsymbol{\alpha} = -\tfrac{1}{2}\mathbf{G}\mathbf{A}^{-1}\mathbf{G}^\top \boldsymbol{\lambda}
\tag{14.36}
$$

We need to be able to invert $\mathbf{G}\mathbf{A}^{-1}\mathbf{G}^\top$ in order to solve for the Lagrange multipliers $\boldsymbol{\lambda}$. Since $\mathbf{G}$ itself has full rank and $\mathbf{A}$ and thus $\mathbf{A}^{-1}$ are negative definite, it follows by Theorem 4.4.4 that $\mathbf{G}\mathbf{A}^{-1}\mathbf{G}^\top$ is also negative definite, and thus invertible, so we have that

$$
\boldsymbol{\lambda} = -2(\mathbf{G}\mathbf{A}^{-1}\mathbf{G}^\top)^{-1}\boldsymbol{\alpha}
\tag{14.37}
$$

The sign conditions tell us that every component of $\boldsymbol{\lambda}$ must be non-negative. If they are not, then we must examine the problem carefully to see which subset of the constraints is binding and start again using only this subset of constraints.[2]

We can now find the optimal $\mathbf{x}$ by substituting for $\boldsymbol{\lambda}$ in (14.35) from (14.37). Provided that all the constraints are binding, the solution is

$$
\mathbf{x}^* = \mathbf{A}^{-1}\mathbf{G}^\top (\mathbf{G}\mathbf{A}^{-1}\mathbf{G}^\top)^{-1}\boldsymbol{\alpha}
\tag{14.38}
$$

and the envelope function is given by

$$
\begin{aligned}
\mathbf{x}^{*\top}\mathbf{A}\mathbf{x}^* &= \boldsymbol{\alpha}^\top (\mathbf{G}\mathbf{A}^{-1}\mathbf{G}^\top)^{-1}\mathbf{G}\mathbf{A}^{-1}\mathbf{A}\mathbf{A}^{-1}\mathbf{G}^\top (\mathbf{G}\mathbf{A}^{-1}\mathbf{G}^\top)^{-1}\boldsymbol{\alpha} \\
&= \boldsymbol{\alpha}^\top (\mathbf{G}\mathbf{A}^{-1}\mathbf{G}^\top)^{-1}\boldsymbol{\alpha} \\
&= -\tfrac{1}{2}\boldsymbol{\alpha}^\top\boldsymbol{\lambda}
\end{aligned}
\tag{14.39}
$$

The applications of this problem include ordinary least squares, generalized least squares and restricted least squares regression, the mean–variance portfolio choice problem in finance, principal components and factor analysis. By way of illustration, there follows a discussion of the application to an important statistical result associated with ordinary least squares, namely, the Gauss–Markov theorem, and to restricted least squares regression. The mean–variance portfolio choice problem in finance is deferred to Section 17.4.

14.3.2 Gauss–Markov theorem

Consider again the regression equation

$$
\tilde{y} = \mathbf{x}^\top \boldsymbol{\beta} + \tilde{\epsilon}
\tag{14.40}
$$

Here, $\tilde{y}$ and $\tilde{\epsilon}$ are random variables, $\tilde{\epsilon}$ being a zero-mean disturbance term, and $\mathbf{x}$ is a fixed k-vector. Given an unobserved random sample of T values for $\tilde{y}$, $\tilde{\mathbf{y}} \equiv (\tilde{y}_1, \tilde{y}_2, \ldots, \tilde{y}_T)$, corresponding to values $\mathbf{x}_1, \mathbf{x}_2, \ldots, \mathbf{x}_T$, respectively, of $\mathbf{x}$, the vector of disturbance terms, $\tilde{\boldsymbol{\epsilon}}$, with variance–covariance matrix $\mathbf{V}$, say, may be written as $\tilde{\boldsymbol{\epsilon}} = \tilde{\mathbf{y}} - \mathbf{X}\boldsymbol{\beta}$, where $\mathbf{X}$ is the matrix with rows $\mathbf{x}_1^\top, \mathbf{x}_2^\top, \ldots, \mathbf{x}_T^\top$.

Any expression of the form $\sum_{t=1}^{T} \alpha_t \tilde{y}_t$, where $\alpha_1, \alpha_2, \ldots, \alpha_T$ are non-random scalars, is said to be a **linear estimator**. Recall from Section 13.8 the meaning of the term "unbiased estimator". We also require the statistical notion of the **best estimator**, meaning the estimator with minimum variance in a given class of estimators. This idea relates to **statistical efficiency**, and the best estimator is the statistically most efficient in a given class. Using these concepts, we can now state the theorem of interest.[3]

THEOREM 14.3.1 (GAUSS–MARKOV THEOREM). *Assume that $k < T$ and that the $T \times k$ matrix $\mathbf{X}$ with rows $\mathbf{x}_1^\top, \mathbf{x}_2^\top, \ldots, \mathbf{x}_T^\top$ is of full rank. Then the **best linear unbiased estimators** (BLUE) of the parameters β_i, $i = 1, 2, \ldots, k$, are given by*

$$
\hat{\boldsymbol{\beta}}_{GLS} \equiv (\mathbf{X}^\top\mathbf{V}^{-1}\mathbf{X})^{-1}\mathbf{X}^\top\mathbf{V}^{-1}\tilde{\mathbf{y}}
\tag{14.41}
$$

Proof: Rather than proving the precise result directly, it is convenient and easier to consider the more general problem of finding the BLUE of $\mu \equiv \mathbf{c}^\top \boldsymbol{\beta}$ where $\mathbf{c} \in \mathbb{R}^k$. Setting $\mathbf{c}$ equal to the appropriate standard unit basis vector solves the problem posed.

Stack the values, as in (14.3), in the form

$$
\tilde{\mathbf{y}} = \mathbf{X}\boldsymbol{\beta} + \tilde{\boldsymbol{\epsilon}}
\tag{14.42}
$$

A linear estimator of μ will be of the form $\mathbf{a}^\top\tilde{\mathbf{y}}$, where $\mathbf{a} \in \mathbb{R}^T$. Its expected value is

$$
E[\mathbf{a}^\top\tilde{\mathbf{y}}] = \mathbf{a}^\top\mathbf{X}\boldsymbol{\beta}
\tag{14.43}
$$

so our estimator is unbiased (for all values of the true parameter vector $\boldsymbol{\beta}$) if and only if

$$\mathbf{a}^{\top}\mathbf{X}=\mathbf{c}^{\top} \tag{14.44}$$

The variance of $\mathbf{a}^{\top}\tilde{\mathbf{y}}$ is $\mathbf{a}^{\top}\mathbf{V}\mathbf{a}$, which must now be minimized subject to the linear constraint (14.44).

This is the canonical problem above with T and k replacing the dimensions n and m, respectively, and $-\mathbf{V}$, $\mathbf{a}$, $\mathbf{X}^{\top}$ and $\mathbf{c}$ taking the places of $\mathbf{A}$, $\mathbf{x}$, $\mathbf{G}$ and $\boldsymbol{\alpha}$, respectively. So the solution is

$$\mathbf{a}^{*}=\mathbf{V}^{-1}\mathbf{X}(\mathbf{X}^{\top}\mathbf{V}^{-1}\mathbf{X})^{-1}\mathbf{c} \tag{14.45}$$

Thus the BLUE of μ is

$$\mathbf{a}^{*\top}\tilde{\mathbf{y}}=\tilde{\mathbf{y}}^{\top}\mathbf{a}^{*}=\tilde{\mathbf{y}}^{\top}\mathbf{V}^{-1}\mathbf{X}(\mathbf{X}^{\top}\mathbf{V}^{-1}\mathbf{X})^{-1}\mathbf{c} \tag{14.46}$$

and by successive substitution for $\mathbf{c}$ using the standard basis vectors, and stacking results in a column vector, the BLUE of $\boldsymbol{\beta}$ is

$$(\mathbf{X}^{\top}\mathbf{V}^{-1}\mathbf{X})^{-1}\mathbf{X}^{\top}\mathbf{V}^{-1}\tilde{\mathbf{y}} \tag{14.47}$$

$\square$

The result in (14.41) is known as the **generalized least squares** (GLS) estimator; hence the subscript. If the disturbance terms are iid, then $\mathbf{V}=\sigma^{2}\mathbf{I}_{T}$ for some σ and the model is just the ordinary least squares regression model, and (14.41) reduces to the OLS estimator (14.12). If $\mathbf{V}$ is diagonal but the $\tilde{y}_{i}$ values are heteroskedastic, i.e. have different variances, then the estimator is the **weighted least squares** (WLS) estimator. For more general $\mathbf{V}$, we are dealing with the full GLS regression model.

An alternative approach to the derivation of the GLS estimator is to formulate the Lagrangian and solve the first-order conditions for the appropriate constrained optimization problem from first principles; see Exercise 14.8.

The particular estimators for the β_{i}, the variance of the minimum-variance estimator, etc. are easily derived; see Johnston and DiNardo (1997, pp. 89–90) and Exercise 14.9.

14.3.3   *Restricted least squares estimation*

Occasionally, theory may suggest linear restrictions on the parameters of regression models. Suppose, for example, that it suggests that $\beta_{1}+\beta_{3}=1$ in a model with k parameters. Such a restriction can be represented by matrices as

$$\mathbf{R}\boldsymbol{\beta}=\mathbf{r}\quad\text{or}\quad\mathbf{R}\boldsymbol{\beta}-\mathbf{r}=\mathbf{0} \tag{14.48}$$

where $\mathbf{R}=[1\quad 0\quad 1\quad 0\quad\ldots\quad 0]_{1\times k}$, $\mathbf{r}=[1]_{1\times 1}$, $\mathbf{0}=[0]_{1\times 1}$ and $\boldsymbol{\beta}$ is the k-vector of parameters defined in Section 14.2. If theory also suggests that $\beta_{2}=\beta_{4}$, then the two restrictions may be represented by (14.48), where now

$$\mathbf{R}=\begin{bmatrix}1 & 0 & 1 & 0 & 0 & \ldots & 0\\ 0 & 1 & 0 & -1 & 0 & \ldots & 0\end{bmatrix}_{2\times k},\quad \mathbf{r}=\begin{bmatrix}1\\ 0\end{bmatrix}_{2\times 1} \tag{14.49}$$

and $\mathbf{0}$ is a 2×1 zero matrix.

More generally, g linear restrictions on the k parameters may be written as (14.48), where $\mathbf{R}$ is $g \times k$ and $\mathbf{r}$ and $\mathbf{0}$ are $g \times 1$. We assume that $g < k$ and that the g restrictions are linearly independent so that $\rho(\mathbf{R}) = g$. If such restrictions constitute valid additional information, then in the interest of statistical efficiency they should be used in the estimation of $\boldsymbol{\beta}$. Given a sample of data, as in Section 14.2, the estimation problem then becomes that of minimizing the sum of squared residuals $\mathbf{e}^\top \mathbf{e}$, subject to the constraint $\mathbf{R}\hat{\boldsymbol{\beta}} - \mathbf{r} = \mathbf{0}$. Thus we have another form of the quadratic programming problem, but with both linear and quadratic terms in the objective function, for which the Lagrangian is

$$\mathcal{L}(\hat{\boldsymbol{\beta}}, \boldsymbol{\lambda}) = \mathbf{e}^\top \mathbf{e} + \boldsymbol{\lambda}^\top (\mathbf{R}\hat{\boldsymbol{\beta}} - \mathbf{r}) \tag{14.50}$$

where $\boldsymbol{\lambda}$ is a $g \times 1$ matrix of Lagrange multipliers, each element of which is associated with one of the g restrictions.

Substituting for $\mathbf{e}^\top \mathbf{e}$ using (14.6), then differentiating with respect to $\hat{\boldsymbol{\beta}}$ and $\boldsymbol{\lambda}$ and putting the results equal to zero, we have the first-order conditions for a constrained optimum

$$\frac{\partial \mathcal{L}}{\partial \hat{\boldsymbol{\beta}}} = -2\mathbf{X}^\top \mathbf{y} + 2\mathbf{X}^\top \mathbf{X}\hat{\boldsymbol{\beta}}_R^* + \mathbf{R}^\top \hat{\boldsymbol{\lambda}} = \mathbf{0} \tag{14.51}$$

$$\frac{\partial \mathcal{L}}{\partial \boldsymbol{\lambda}} = \mathbf{R}\hat{\boldsymbol{\beta}}_R^* - \mathbf{r} = \mathbf{0} \tag{14.52}$$

where the subscript "R" is used to distinguish the restricted least squares (RLS) estimator from the OLS estimator, $\hat{\boldsymbol{\beta}}^*$. The first two terms in the middle expression for $\partial \mathcal{L}/\partial \hat{\boldsymbol{\beta}}$ are precisely those derived in (14.8). Solution of the first-order equations (14.51) and (14.52) is rather more difficult than for the normal equations (14.10) encountered in the unconstrained OLS problem, however. We consider two approaches to this solution. The first is a two-step procedure, which finds the solution for $\hat{\boldsymbol{\lambda}}$, then uses this to obtain the solution for $\hat{\boldsymbol{\beta}}_R^*$. The second solves for $\hat{\boldsymbol{\beta}}_R^*$ and $\hat{\boldsymbol{\lambda}}$ simultaneously and makes use of the partitioned inverse introduced in Section 1.5.14.

Solution: approach 1

Pre-multiplying (14.51) by $\frac{1}{2}\mathbf{R}(\mathbf{X}^\top \mathbf{X})^{-1}$, assuming $(\mathbf{X}^\top \mathbf{X})^{-1}$ exists, as in Section 14.2, we obtain

$$-\mathbf{R}(\mathbf{X}^\top \mathbf{X})^{-1}\mathbf{X}^\top \mathbf{y} + \mathbf{R}\hat{\boldsymbol{\beta}}_R^* + \frac{1}{2}\mathbf{R}(\mathbf{X}^\top \mathbf{X})^{-1}\mathbf{R}^\top \hat{\boldsymbol{\lambda}} = \mathbf{0} \tag{14.53}$$

Since $(\mathbf{X}^\top \mathbf{X})^{-1}\mathbf{X}^\top \mathbf{y}$ is the OLS estimator, $\hat{\boldsymbol{\beta}}^*$, and $\mathbf{R}\hat{\boldsymbol{\beta}}_R^* = \mathbf{r}$, because the RLS estimator, $\hat{\boldsymbol{\beta}}_R^*$, satisfies the restrictions (see (14.52)), we may rewrite (14.53) as

$$\mathbf{R}(\mathbf{X}^\top \mathbf{X})^{-1}\mathbf{R}^\top \hat{\boldsymbol{\lambda}} = 2(\mathbf{R}\hat{\boldsymbol{\beta}}^* - \mathbf{r}) \tag{14.54}$$

from which it follows, since $\mathbf{R}$ is of full rank and Theorem 4.4.4 again applies, that

$$\hat{\boldsymbol{\lambda}} = 2[\mathbf{R}(\mathbf{X}^\top \mathbf{X})^{-1}\mathbf{R}^\top]^{-1}(\mathbf{R}\hat{\boldsymbol{\beta}}^* - \mathbf{r}) \tag{14.55}$$

Now, using (14.55) to substitute for $\hat{\lambda}$ in (14.51), we have

$$-2\mathbf{X}^\top\mathbf{y} + 2\mathbf{X}^\top\mathbf{X}\hat{\boldsymbol{\beta}}_R^* + 2\mathbf{R}^\top[\mathbf{R}(\mathbf{X}^\top\mathbf{X})^{-1}\mathbf{R}^\top]^{-1}(\mathbf{R}\hat{\boldsymbol{\beta}}^* - \mathbf{r}) = 0 \tag{14.56}$$

and finally, after pre-multiplying this equation by $\frac{1}{2}(\mathbf{X}^\top\mathbf{X})^{-1}$, we can write the solution for $\hat{\boldsymbol{\beta}}_R^*$ as

$$\hat{\boldsymbol{\beta}}_R^* = \hat{\boldsymbol{\beta}}^* - (\mathbf{X}^\top\mathbf{X})^{-1}\mathbf{R}^\top[\mathbf{R}(\mathbf{X}^\top\mathbf{X})^{-1}\mathbf{R}^\top]^{-1}(\mathbf{R}\hat{\boldsymbol{\beta}}^* - \mathbf{r}) \tag{14.57}$$

Solution: approach 2

For the alternative approach, let us minimize $\frac{1}{2}\mathbf{e}^\top\mathbf{e}$ to simplify the calculations. The first-order conditions become (14.51) and (14.52) without the 2s and can be written as a system of equations in the form

$$\begin{bmatrix} \mathbf{X}^\top\mathbf{X} & \mathbf{R}^\top \\ \mathbf{R} & 0 \end{bmatrix}\begin{bmatrix} \hat{\boldsymbol{\beta}}_R^* \\ \hat{\lambda} \end{bmatrix} = \begin{bmatrix} \mathbf{X}^\top\mathbf{y} \\ \mathbf{r} \end{bmatrix} \tag{14.58}$$

using partitioned matrices. The assumptions we have made about $\mathbf{X}$ and $\mathbf{R}$ ensure that the square matrix in (14.58) is invertible (see Exercise 14.10), hence

$$\begin{bmatrix} \hat{\boldsymbol{\beta}}_R^* \\ \hat{\lambda} \end{bmatrix} = \begin{bmatrix} \mathbf{X}^\top\mathbf{X} & \mathbf{R}^\top \\ \mathbf{R} & 0 \end{bmatrix}^{-1}\begin{bmatrix} \mathbf{X}^\top\mathbf{y} \\ \mathbf{r} \end{bmatrix} \tag{14.59}$$

As indicated in Section 1.5.14, there are several forms for the partitioned inverse. The appropriate form in this instance is the one given in Exercise 1.23, which the reader is asked to derive in Exercise 14.10. This allows us to write

$$\begin{bmatrix} \hat{\boldsymbol{\beta}}_R^* \\ \hat{\lambda} \end{bmatrix} = \begin{bmatrix} (\mathbf{X}^\top\mathbf{X})^{-1} - (\mathbf{X}^\top\mathbf{X})^{-1}\mathbf{R}^\top\mathbf{A}\mathbf{R}(\mathbf{X}^\top\mathbf{X})^{-1} & (\mathbf{X}^\top\mathbf{X})^{-1}\mathbf{R}^\top\mathbf{A} \\ \mathbf{A}\mathbf{R}(\mathbf{X}^\top\mathbf{X})^{-1} & -\mathbf{A} \end{bmatrix}\begin{bmatrix} \mathbf{X}^\top\mathbf{y} \\ \mathbf{r} \end{bmatrix} \tag{14.60}$$

where $\mathbf{A} = [\mathbf{R}(\mathbf{X}^\top\mathbf{X})^{-1}\mathbf{R}^\top]^{-1}$. It follows from this that

$$\hat{\boldsymbol{\beta}}_R^* = (\mathbf{X}^\top\mathbf{X})^{-1}\mathbf{X}^\top\mathbf{y} - (\mathbf{X}^\top\mathbf{X})^{-1}\mathbf{R}^\top\mathbf{A}\mathbf{R}(\mathbf{X}^\top\mathbf{X})^{-1}\mathbf{X}^\top\mathbf{y} + (\mathbf{X}^\top\mathbf{X})^{-1}\mathbf{R}^\top\mathbf{A}\mathbf{r}$$
$$= \hat{\boldsymbol{\beta}}^* - (\mathbf{X}^\top\mathbf{X})^{-1}\mathbf{R}^\top\mathbf{A}(\mathbf{R}\hat{\boldsymbol{\beta}}^* - \mathbf{r}) \tag{14.61}$$

which is identical to (14.57) when substitution for $\mathbf{A}$ is undertaken.

It is easy to see from (14.57) or (14.61) that, if the ordinary least squares estimates happen to satisfy the theoretical restrictions, so that $\mathbf{R}\hat{\boldsymbol{\beta}}^* - \mathbf{r} = 0$, then $\hat{\boldsymbol{\beta}}_R^* = \hat{\boldsymbol{\beta}}^*$.

14.4 Stochastic difference equations

In this section, we introduce an extension of the linear, autonomous difference equation that is widely used in statistical time-series analysis and econometrics. The extension simply involves adding a random variable on the right-hand side of the equation, so that the process generating a variable is no longer deterministic, but rather is stochastic. We might think of the deterministic component of the modified difference equation as describing the average behaviour of a variable over time, and the random component as accounting for other non-systematic factors that influence the precise value of the variable in a given time period.

We begin by establishing some basic facts about a class of single stochastic difference equations, then move on to consider aspects of a type of system of stochastic difference equations that has been widely used in empirical macroeconomics, namely, the vector autoregressive model.[4]

14.4.1 Single-equation autoregressive processes

The concept of a stochastic process was defined in Section 13.2.1 as a sequence of random variables indexed by time. Before we introduce our first specific type of stochastic process in this section, we define three variants of an important general property of stochastic processes – namely, the property of **stationarity** – and some related concepts.

DEFINITION 14.4.1 A stochastic process, $\{\tilde{x}_t\}$, is said to be **strictly stationary** if the joint probability distributions (or densities) of $\tilde{x}_1, \tilde{x}_2, \ldots, \tilde{x}_T$ and $\tilde{x}_{1+k}, \tilde{x}_{2+k}, \ldots, \tilde{x}_{T+k}$ are the same for $k = \pm 1, \pm 2, \ldots$.

DEFINITION 14.4.2 A stochastic process, $\{\tilde{z}_t\}$, is said to be **weakly** (or **covariance**) **stationary** if the first and second moments of $\tilde{z}_t$ exist and

$$E[\tilde{z}_t] = \mu \quad \forall \, t \tag{14.62}$$

$$\text{Var}[\tilde{z}_t] = \sigma^2 \quad \forall \, t \tag{14.63}$$

$$\text{Cov}[\tilde{z}_t, \tilde{z}_{t-j}] = \text{Cov}[\tilde{z}_{t+k}, \tilde{z}_{t+k-j}] = \gamma_j \quad \forall \, t, k, j \tag{14.64}$$

where the $\text{Cov}[\tilde{z}_t, \tilde{z}_{t-j}]$ are the **autocovariances** for all t, j, with $\gamma_0 = \sigma^2 \geq 0$.

Thus, if a stochastic process $\{\tilde{z}_t\}$ is weakly stationary, the means, variances and auto-covariances are time invariant, and the autocovariances depend only on the lag length j. It follows that the **autocorrelations**, defined as

$$\text{Corr}[\tilde{z}_t, \tilde{z}_{t-j}] \equiv \frac{\text{Cov}[\tilde{z}_t, \tilde{z}_{t-j}]}{\sqrt{\text{Var}[\tilde{z}_t]\text{Var}[\tilde{z}_{t-j}]}} = \frac{\gamma_j}{\sigma^2} = \frac{\gamma_j}{\gamma_0} \equiv \rho_j \tag{14.65}$$

also depend only on j. Since $\gamma_j = \text{Cov}[\tilde{z}_t, \tilde{z}_{t-j}] = \text{Cov}[\tilde{z}_{t-j}, \tilde{z}_t]$ and, by weak stationarity, $\text{Cov}[\tilde{z}_{t-j}, \tilde{z}_t] = \text{Cov}[\tilde{z}_{t-j+j}, \tilde{z}_{t+j}] = \text{Cov}[\tilde{z}_t, \tilde{z}_{t-(-j)}] = \gamma_{-j}$, it also follows that $\rho_j = \rho_{-j}$.

DEFINITION 14.4.3 A stochastic process, $\{\tilde{w}_t\}$, is said to be **asymptotically stationary** or **integrated to order zero**, denoted $\tilde{w}_t \sim I(0)$, if, as $t \to \infty$, $E[\tilde{w}_t] \to \mu$, $\text{Var}[\tilde{w}_t] \to \sigma^2$, $\text{Cov}[\tilde{w}_t, \tilde{w}_{t-j}] \to \gamma_j$, which depends only on $|j|$, and $\text{Corr}[\tilde{w}_t, \tilde{w}_{t-j}] \to \rho_j$, which also depends only on $|j|$, where μ, σ^2, γ_j and ρ_j for all j are constants.

Asymptotic stationarity is a weaker concept than both strict and weak stationarity: stationary processes are $I(0)$ but $I(0)$ processes are not necessarily stationary. In this section, only the concepts of weak stationarity and asymptotic stationarity are used. As the difference will usually be clear from the context, the adjectives "weak" and "asymptotic" will often be omitted.

We are now ready to define a stochastic linear, autonomous, difference equation of order one called the first-order autoregressive process, and to examine some of its properties; brief generalization will follow.

DEFINITION 14.4.4 The stochastic process $\{\tilde{y}_t\}$ is said to be a **first-order autoregressive process**, denoted AR(1), if

$$\tilde{y}_t = \phi_0 + \phi_1 \tilde{y}_{t-1} + \tilde{\varepsilon}_t \quad \forall t \tag{14.66}$$

where ϕ_0 and ϕ_1 are constants, and $\tilde{\varepsilon}_t$ is a random variable assumed to be independently and identically distributed with mean $E[\tilde{\varepsilon}_t] = 0$ and constant variance $\text{Var}[\tilde{\varepsilon}_t] = \sigma^2$ for all t.[5]

If ϕ_1 is such that $-1 < \phi_1 < 1$, then the AR(1) process is weakly stationary. This condition on ϕ_1 in (14.66) will be recognized as the same as that required to guarantee the asymptotic stability of the linear, autonomous, first-order difference equation discussed in Section 8.3. However, as $\tilde{y}_t$ is a random variable (as indicated by the tilde), because it is a function of the random variable $\tilde{\varepsilon}_t$, the statistical concepts of weak stationarity or asymptotic stationarity usually replace stability as a prime concern in the study of stochastic difference equations, though the three concepts are closely related.

Despite the shift in focus, the asymptotic stability conditions from Section 8.3 underpin the conditions for weak stationarity and asymptotic stationarity. To see this, consider the AR(1) process as having started in time period 1. Employing the technique of repeated substitution for the right-hand side of (14.66), we obtain

$$\tilde{y}_t = \phi_0(1 + \phi_1 + \cdots + \phi_1^{t-2}) + \phi_1^{t-1} y_1 + \tilde{\varepsilon}_t + \phi_1 \tilde{\varepsilon}_{t-1} + \cdots + \phi_1^{t-2} \tilde{\varepsilon}_2$$

$$= \phi_0 \sum_{i=0}^{t-2} \phi_1^i + \phi_1^{t-1} y_1 + \sum_{i=0}^{t-2} \phi_1^i \tilde{\varepsilon}_{t-i} \tag{14.67}$$

where y_1 is a fixed starting value (initial condition) for the process. This is a modified form of (8.23), in which a weighted sum of random variables (the $\tilde{\varepsilon}_{t-i}$) also appears. Applying the properties of expectations, variances and covariances given in Section 13.6.1 to (14.67), it is easy to verify that, conditional on $\tilde{y}_1 = y_1$,

$$E[\tilde{y}_t] = \phi_0(1 + \phi_1 + \cdots + \phi_1^{t-2}) + \phi_1^{t-1} y_1, \quad t \geq 2 \tag{14.68}$$

$$\text{Var}[\tilde{y}_t] = \sigma^2(1 + \phi_1^2 + \cdots + \phi_1^{2(t-2)}), \quad t \geq 2 \tag{14.69}$$

$$\text{Cov}[\tilde{y}_t, \tilde{y}_{t-j}] = \phi_1^j \, \text{Var}[\tilde{y}_{t-j}], \quad 2 \leq j \leq t - 1 \tag{14.70}$$

Since all of these moments depend on time, the AR(1) process is non-stationary; but since $|\phi_1| < 1$ by assumption, as $t \to \infty$ we have that

$$E[\tilde{y}_t] \to \frac{\phi_0}{1 - \phi_1} \tag{14.71}$$

$$\text{Var}[\tilde{y}_t] \to \frac{\sigma^2}{1 - \phi_1^2} \tag{14.72}$$

$$\text{Cov}[\tilde{y}_t, \tilde{y}_{t-j}] \to \frac{\phi_1^j \sigma^2}{1 - \phi_1^2} \tag{14.73}$$

In other words, all of the moments in (14.68), (14.69) and (14.70) become time invariant, and independent of the starting value y_1, as $t \to \infty$, making it clear why $|\phi_1| < 1$ is a sufficient condition for asymptotic stationarity in the AR(1) case. Using (14.70), the autocorrelations are

$$\text{Corr}[\tilde{y}_t, \tilde{y}_{t-j}] = \frac{\text{Cov}[\tilde{y}_t, \tilde{y}_{t-j}]}{\sqrt{\text{Var}[\tilde{y}_t]\text{Var}[\tilde{y}_{t-j}]}} = \phi_1^j \sqrt{\frac{\text{Var}[\tilde{y}_{t-j}]}{\text{Var}[\tilde{y}_t]}} \tag{14.74}$$

which also depend on time; but as $t \to \infty$ we see that, for given j, $\text{Corr}[\tilde{y}_t, \tilde{y}_{t-j}] \to \phi_1^j = \rho_j$, which is time invariant.[6] Thus, in (14.67), $\tilde{y}_t \sim I(0)$ when $|\phi_1| < 1$: the AR(1) process is asymptotically stationary.

If $|\phi_1| \geq 1$, (14.68), (14.69) and (14.70) indicate that the moments of the AR(1) process diverge with t and, therefore, that the process is non-stationary. The case of $|\phi_1| > 1$ is not considered to be of much interest in economics and finance, but that of $|\phi_1| = 1$, and particularly $\phi_1 = 1$, is of great importance, as mentioned in Section 3.4. Such a value implies a **unit root** in the lag polynomial associated with the process. Moreover, a positive unit root also implies that $\Delta \tilde{y}_t = (1 - L)\tilde{y}_t \sim I(0)$ so that first-differencing induces asymptotic stationarity. The case of a negative unit root is considered in Exercise 14.13.

A particular case of a unit root process that has been used to model variables (such as foreign exchange rates and stock prices) that are thought to be determined by efficient markets[7] is the so-called **random walk**. This is simply an AR(1) process in which $\phi_0 = 0$ and $\phi_1 = 1$:

$$\tilde{y}_t = \tilde{y}_{t-1} + \tilde{\varepsilon}_t \tag{14.75}$$

The lag polynomial associated with this process is $\phi(L) = 1 - L$; the unit root – and, hence, the non-stationarity of the random walk – is obvious. It follows that $\Delta \tilde{y}_t = \tilde{\varepsilon}_t$, and so $\Delta \tilde{y}_t$ is iid and so is stationary in all three senses. The practical implication is that if (14.75) is an adequate description of how a particular variable behaves, then the change in that variable is impossible to forecast: the optimal predictor of $\tilde{y}_t$, given information available at time $t - 1$, is the conditional expectation $E[\tilde{y}_t \mid \Omega_{t-1}] = \tilde{y}_{t-1}$, where Ω_{t-1} denotes the available information at $t - 1$.

The existence of unit roots does not mean that all forecasting exercises are as limited as in the case of the random walk model. Suppose, for example, that a variable is generated as

$$\tilde{y}_t = \phi_0 + 0.75\tilde{y}_{t-1} + 0.25\tilde{y}_{t-2} + \tilde{\varepsilon}_t \tag{14.76}$$

This is an example of an AR(2) process, in which a two-period lag of the variable of interest appears in the equation. The lag polynomial associated with (14.76) is

$$\phi(L) = 1 - 0.75L - 0.25L^2 \tag{14.77}$$

This polynomial factorizes as $(1 + 0.25L)(1 - L)$ and so, recalling (8.91), the generating process can be rewritten as

$$\Delta \tilde{y}_t = \phi_0 - 0.25 \Delta \tilde{y}_{t-1} + \tilde{\varepsilon}_t \tag{14.78}$$

In this case, therefore, knowledge of the lagged change Δy_{t-1} is relevant to forecasting, the optimal predictor of $\Delta \tilde{y}_t$ being $E[\Delta \tilde{y}_t \mid \Omega_{t-1}] = \phi_0 - 0.25 \Delta \tilde{y}_{t-1}$, from which a forecast of $\Delta \tilde{y}_t$ can be obtained, namely $\phi_0 - 0.25 \Delta y_{t-1}$, where Δy_{t-1} would be known. In practice, of course, an estimate of ϕ_0 would also be required. A forecast of $\tilde{y}_t$ follows simply by adding the forecast change to the known value y_{t-1}.

The AR(1) process generalizes further to the AR(p) process

$$\tilde{y}_t = \phi_0 + \phi_1 \tilde{y}_{t-1} + \phi_2 \tilde{y}_{t-2} + \cdots + \phi_p \tilde{y}_{t-p} + \tilde{\varepsilon}_t \tag{14.79}$$

or, using the lag operator,

$$\phi(L) \tilde{y}_t = \phi_0 + \tilde{\varepsilon}_t \tag{14.80}$$

where $\phi(L)$ is the polynomial defined in (8.67). In the immediately preceding illustration in (14.76), the value of p is 2.

In the AR(1) case, the stationarity condition is $|\phi_1| < 1$. Another way of stating this is that the root of $\phi(z) = 1 - \phi_1 z = 0$ must be greater than 1, since the solution of $\phi(z) = 0$ is $z = 1/\phi_1$ and the requirement that $|\phi_1| < 1$ implies that $|z| > 1$. This is precisely the idea first introduced in Section 8.3.1. The idea also applies in the AR(p) case, for which (using the phraseology that allows for the possibility of complex solutions), to ensure stationarity, the roots of the lag polynomial equation $\phi(z) = 1 - \phi_1 z - \cdots - \phi_p z^p = 0$ must all lie **outside the unit circle**. This is the general stationarity condition for autoregressive processes. A necessary condition for the roots of the lag polynomial to lie outside the unit circle is that $\sum_{i=1}^{p} \phi_i < 1$; a sufficient condition is that $\sum_{i=1}^{p} |\phi_i| < 1$; see Section 8.4.3.

14.4.2   *Vector autoregressive processes*

The **vector autoregressive process** (VAR process), introduced by Sims (1980), is particularly popular in macroeconomics, where it is used for testing for causality, the analysis of economic shocks, forecasting and forecast error variance decomposition. In this section, following a brief description of the VAR process and some of its properties, we shall illustrate just one of these applications: the analysis of shocks.

The VAR(p) process is a straightforward generalization of the univariate AR(p) process to a vector variable $\tilde{\mathbf{y}}_t$:

$$\tilde{\mathbf{y}}_t = \mathbf{\Phi}_0 + \mathbf{\Phi}_1 \tilde{\mathbf{y}}_{t-1} + \mathbf{\Phi}_2 \tilde{\mathbf{y}}_{t-2} + \cdots + \mathbf{\Phi}_p \tilde{\mathbf{y}}_{t-p} + \tilde{\boldsymbol{\varepsilon}}_t \tag{14.81}$$

where $\tilde{\mathbf{y}}_t$, $\mathbf{\Phi}_0$ and the $\mathbf{\Phi}_i$ and their dimensions are all as defined in Section 8.5.1, and $\tilde{\boldsymbol{\varepsilon}}_t$ is an m-vector of independently and identically distributed stochastic disturbances with mean vector $E[\tilde{\boldsymbol{\varepsilon}}_t] = \mathbf{0}$ and variance–covariance matrix $\text{Var}[\tilde{\boldsymbol{\varepsilon}}_t] = \mathbf{\Sigma}_{m \times m}$. This is a system in which all lags of all variables appear in all equations.

More compactly, we may represent the VAR(p) process as

$$\mathbf{\Phi}(L) \tilde{\mathbf{y}}_t = \mathbf{\Phi}_0 + \tilde{\boldsymbol{\varepsilon}}_t \tag{14.82}$$

where $\mathbf{\Phi}(L)$ is as defined in (8.105).

If Σ and all of the Φ_i are diagonal matrices, then the model reduces to a set of m univariate AR(p) processes. If Σ is not diagonal, but all of the Φ_i are diagonal, then the model may be interpreted as a version of the **seemingly unrelated regression** model used in econometrics. For example, suppose $m = 2$ and $p = 1$; then the unrestricted VAR(1) model may be written as

$$\tilde{\mathbf{y}}_t = \Phi_0 + \Phi_1 L \tilde{\mathbf{y}}_t + \tilde{\boldsymbol{\varepsilon}}_t \tag{14.83}$$

or

$$\begin{bmatrix} \tilde{y}_{1t} \\ \tilde{y}_{2t} \end{bmatrix} = \begin{bmatrix} \phi_{0_1} \\ \phi_{0_2} \end{bmatrix} + \begin{bmatrix} \phi_{1_{11}} L & \phi_{1_{12}} L \\ \phi_{1_{21}} L & \phi_{1_{22}} L \end{bmatrix} \begin{bmatrix} \tilde{y}_{1t} \\ \tilde{y}_{2t} \end{bmatrix} + \begin{bmatrix} \tilde{\varepsilon}_{1t} \\ \tilde{\varepsilon}_{2t} \end{bmatrix} \tag{14.84}$$

or

$$
\begin{aligned}
\tilde{y}_{1t} &= \phi_{0_1} + \phi_{1_{11}} L \tilde{y}_{1t} + \phi_{1_{12}} L \tilde{y}_{2t} + \tilde{\varepsilon}_{1t} \\
&= \phi_{0_1} + \phi_{1_{11}} \tilde{y}_{1(t-1)} + \phi_{1_{12}} \tilde{y}_{2(t-1)} + \tilde{\varepsilon}_{1t} \tag{14.85} \\
\tilde{y}_{2t} &= \phi_{0_2} + \phi_{1_{21}} L \tilde{y}_{1t} + \phi_{1_{22}} L \tilde{y}_{2t} + \tilde{\varepsilon}_{2t} \\
&= \phi_{0_2} + \phi_{1_{21}} \tilde{y}_{1(t-1)} + \phi_{1_{22}} \tilde{y}_{2(t-1)} + \tilde{\varepsilon}_{2t} \tag{14.86}
\end{aligned}
$$

for all t, where

$$\tilde{\boldsymbol{\varepsilon}}_t = \begin{bmatrix} \tilde{\varepsilon}_{1t} \\ \tilde{\varepsilon}_{2t} \end{bmatrix} \sim \text{iid} \left(\begin{bmatrix} 0 \\ 0 \end{bmatrix}, \begin{bmatrix} \sigma_{11} & \sigma_{12} \\ \sigma_{21} & \sigma_{22} \end{bmatrix} \right) = \text{iid}(\mathbf{0}, \Sigma) \quad \forall\, t \tag{14.87}$$

using the statistical shorthand iid($\mathbf{0}, \Sigma$) for the vector of independently and identically distributed random variables with the given mean and variance–covariance matrix. The σ_{ij}, $i = 1, 2$, $j = 1, 2$, denote variances when $i = j$ and contemporaneous covariances when $i \neq j$, with $\sigma_{12} = \sigma_{21}$, of course.

The more restricted form is

$$\begin{bmatrix} \tilde{y}_{1t} \\ \tilde{y}_{2t} \end{bmatrix} = \begin{bmatrix} \phi_{0_1} \\ \phi_{0_2} \end{bmatrix} + \begin{bmatrix} \phi_{1_{11}} L & 0 \\ 0 & \phi_{1_{22}} L \end{bmatrix} \begin{bmatrix} \tilde{y}_{1t} \\ \tilde{y}_{2t} \end{bmatrix} + \begin{bmatrix} \tilde{\varepsilon}_{1t} \\ \tilde{\varepsilon}_{2t} \end{bmatrix} \tag{14.88}$$

or

$$\tilde{y}_{1t} = \phi_{0_1} + \phi_{1_{11}} \tilde{y}_{1(t-1)} + \tilde{\varepsilon}_{1t} \tag{14.89}$$

$$\tilde{y}_{2t} = \phi_{0_2} + \phi_{1_{22}} \tilde{y}_{2(t-1)} + \tilde{\varepsilon}_{2t} \tag{14.90}$$

where

$$\begin{bmatrix} \tilde{\varepsilon}_{1t} \\ \tilde{\varepsilon}_{2t} \end{bmatrix} \sim \text{iid} \left(\begin{bmatrix} 0 \\ 0 \end{bmatrix}, \begin{bmatrix} \sigma_{11} & 0 \\ 0 & \sigma_{22} \end{bmatrix} \right) \tag{14.91}$$

The difference is clear: in the case of each variable in the more general two-variable VAR(1) process, not only the lag of that variable but also the lag of the other variable has a determining role. This is a key characteristic of a VAR process.

Stationarity of VAR processes

To explore the conditions under which a VAR process is stationary, we begin by considering the restricted VAR(1) process specified in (14.88), where $m = 2$. In this case, the matrix polynomial $\mathbf{\Phi}(L) = \mathbf{I}_2 - \mathbf{\Phi}_1 L$ is diagonal and the determinant

$$\det(\mathbf{\Phi}(L)) = \det(\mathbf{I}_2 - \mathbf{\Phi}_1 L) = (1 - \phi_{1_{11}} L)(1 - \phi_{1_{22}} L) \tag{14.92}$$

is the product of the lag polynomials for the individual elements of $\tilde{\mathbf{y}}_t$, $\tilde{y}_{1t}$ and $\tilde{y}_{2t}$. From Section 14.4.1, we know that the roots of these polynomials determine the stationarity or otherwise of the individual AR(1) processes, and that for stationarity the roots have to lie outside the unit circle. We note that the solutions of the determinantal equation

$$\det(\mathbf{I}_2 - \mathbf{\Phi}_1 L) = 0 \tag{14.93}$$

provide the information we require: roots greater than unity in absolute value imply $|\phi_{1_{11}}| < 1$ and $|\phi_{1_{22}}| < 1$.[8]
The same principle applies in the more general case when $\phi_{1_{12}} \neq 0$ or $\phi_{1_{21}} \neq 0$. Indeed, it applies also in the completely general case. Thus for stationarity of a VAR(p) process, we require that the equation

$$\det(\mathbf{\Phi}(L)) = \det(\mathbf{I}_m - \mathbf{\Phi}_1 L - \cdots - \mathbf{\Phi}_p L^p) = 0 \tag{14.94}$$

yields roots all of which are outside the unit circle. It is important to note that, in the general case, the highest power of L in $\det(\mathbf{\Phi}(L))$ is mp and all mp roots must satisfy the condition. The determinantal equation for a bivariate VAR(2) process, i.e. for $m = 2$ and $p = 2$, is the subject of Exercise 14.14(b).

Assuming stationarity, it is easy to derive the mean of the VAR(p) process but rather more difficult to find the variance–covariance matrix and the autocovariance matrices. Let us look briefly at the mean vector and variance–covariance matrix.

Mean of a stationary VAR process

We may take the expectation of both sides of (14.82) and use the fact that $E[\tilde{\mathbf{y}}_t]$ is constant for stationary $\tilde{\mathbf{y}}_t$, and the properties of the lag operator from Section 8.2.2, to obtain

$$\mathbf{\Phi}(L)E[\tilde{\mathbf{y}}_t] = \mathbf{\Phi}(1)E[\tilde{\mathbf{y}}_t] = \mathbf{\Phi}_0 + E[\tilde{\boldsymbol{\varepsilon}}_t] \tag{14.95}$$

Since $E[\tilde{\boldsymbol{\varepsilon}}_t] = \mathbf{0}$, we have that

$$E[\tilde{\mathbf{y}}_t] = \mathbf{\Phi}^{-1}(1)\mathbf{\Phi}_0 = \boldsymbol{\mu} \tag{14.96}$$

the existence of the matrix inverse $\mathbf{\Phi}^{-1}(1)$ being guaranteed by stationarity. To see this, consider the VAR(1) case, in which $\mathbf{\Phi}(1) = \mathbf{I}_m - \mathbf{\Phi}_1$, which is singular if and only if unity is an eigenvalue of $\mathbf{\Phi}_1$. This cannot be the case if the VAR(1) process is stationary, so $\mathbf{\Phi}(1)$ is invertible for stationary VAR(1) processes. A similar argument applies in the VAR(p) case.

Variance of a stationary VAR process

Given the additional complexity involved in the general case, let us examine the variance of a stationary VAR(1) process. As $\tilde{\boldsymbol{\varepsilon}}_t$ is iid, it is independent of $\tilde{\mathbf{y}}_{t-1}$ and so (see Exercise 13.7)

$$\text{Var}[\tilde{\mathbf{y}}_t] = \text{Var}[\boldsymbol{\Phi}_1 \tilde{\mathbf{y}}_{t-1}] + \text{Var}[\tilde{\boldsymbol{\varepsilon}}_t]$$
$$= \boldsymbol{\Phi}_1 \text{Var}[\tilde{\mathbf{y}}_{t-1}] \boldsymbol{\Phi}_1^\top + \text{Var}[\tilde{\boldsymbol{\varepsilon}}_t] \tag{14.97}$$

As $\text{Var}[\tilde{\mathbf{y}}_t] = \text{Var}[\tilde{\mathbf{y}}_{t-1}] = \mathbf{V}$, say, under stationarity we have that

$$\mathbf{V} = \boldsymbol{\Phi}_1 \mathbf{V} \boldsymbol{\Phi}_1^\top + \boldsymbol{\Sigma} \tag{14.98}$$

and $\mathbf{V}$ follows as the solution of this equation. We note that the diagonal elements of $\mathbf{V}$ are the variances of the individual elements of $\tilde{\mathbf{y}}_t$, and that the off-diagonal elements of $\mathbf{V}$ are their contemporaneous covariances.

The non-contemporaneous covariances and autocovariances for a VAR(1) process are derived in a similar way.[9]

Impulse response analysis

Any stationary VAR(p) process, $\boldsymbol{\Phi}(L)\tilde{\mathbf{y}}_t = \boldsymbol{\Phi}_0 + \tilde{\boldsymbol{\varepsilon}}_t$, can be solved for $\tilde{\mathbf{y}}_t$ to yield

$$\tilde{\mathbf{y}}_t = \boldsymbol{\Phi}^{-1}(L)\boldsymbol{\Phi}_0 + \boldsymbol{\Phi}^{-1}(L)\tilde{\boldsymbol{\varepsilon}}_t$$
$$= \boldsymbol{\Phi}^{-1}(1)\boldsymbol{\Phi}_0 + \sum_{i=0}^{\infty} \boldsymbol{\pi}_i \tilde{\boldsymbol{\varepsilon}}_{t-i} \tag{14.99}$$

where the $\boldsymbol{\pi}_i$ are $m \times m$ coefficient matrices, $\boldsymbol{\pi}_0 = \mathbf{I}_m$ and $\boldsymbol{\Phi}^{-1}(L) = \boldsymbol{\pi}(L) = \sum_{i=0}^{\infty} \boldsymbol{\pi}_i L^i$, which exists when the VAR(p) process satisfies the stationarity condition. The interpretation of $\boldsymbol{\pi}_i$ as the multiplier matrix

$$\frac{\partial \tilde{\mathbf{y}}_t}{\partial \tilde{\boldsymbol{\varepsilon}}_{t-i}} = \frac{\partial \tilde{\mathbf{y}}_{t+i}}{\partial \tilde{\boldsymbol{\varepsilon}}_t} = \boldsymbol{\pi}_i, \; i = 0, 1, 2, \ldots \tag{14.100}$$

is noteworthy because it allows identification of the consequences of a unit change in a given variable's disturbance at time t for the value of that, or another, variable at time $t + i$. For instance, the jkth element of $\boldsymbol{\pi}_i$ measures the effect on variable j at time $t + i$ of a unit change in the kth variable's disturbance at time t, holding all other disturbances at all dates constant. A plot of the jkth element of $\boldsymbol{\pi}_i$, $\partial \tilde{y}_{j_{t+i}} / \partial \tilde{\varepsilon}_{k_t}$, as a function of the lag i is known as an **impulse response function**, and this is useful for analysing the dynamic effects of shocks within a VAR system. The matrix $\boldsymbol{\pi}_0$ contains the impact multipliers. The dynamic multiplier matrices, which give the accumulated responses to a shock over s periods, are $\sum_{i=0}^{s} \boldsymbol{\pi}_i$, $s = 1, 2, \ldots$. The long-run or equilibrium multiplier matrix, which gives the total accumulated effects for all future time periods, is $\sum_{i=0}^{\infty} \boldsymbol{\pi}_i$. This is exactly the same terminology that we used when dealing with the equivalent non-stochastic dynamic linear macroeconomic model in Section 11.2

The assumption that other disturbances remain constant in the face of a change in one particular disturbance is problematical, since $\boldsymbol{\Sigma} = E[\tilde{\boldsymbol{\varepsilon}}_t \tilde{\boldsymbol{\varepsilon}}_t^\top]$ is not generally a diagonal matrix.

The structure of $\boldsymbol{\Sigma}$ implies that a shock in one disturbance is likely to be accompanied by changes in other disturbances in the same period. A way round this problem arises from the fact that, because $\boldsymbol{\Sigma}$ is symmetric positive definite and therefore diagonalizable, it can be expressed as $\boldsymbol{\Sigma} = \boldsymbol{\Sigma}^{\frac{1}{2}}\boldsymbol{\Sigma}^{\frac{1}{2}}$, where $\boldsymbol{\Sigma}^{\frac{1}{2}}$ is a square root of $\boldsymbol{\Sigma}$ constructed in the manner described in Section 3.6.1.

Specifically, let $\tilde{\mathbf{u}}_t = \boldsymbol{\Sigma}^{-\frac{1}{2}}\tilde{\boldsymbol{\varepsilon}}_t$, where $\boldsymbol{\Sigma}^{-\frac{1}{2}}$ is the inverse of $\boldsymbol{\Sigma}^{\frac{1}{2}}$, which exists since $\boldsymbol{\Sigma}$ is symmetric positive definite (see Section 4.4.1); then the second term on the right-hand side of (14.99) may be written as

$$\sum_{i=0}^{\infty} \boldsymbol{\pi}_i \tilde{\boldsymbol{\varepsilon}}_{t-i} = \sum_{i=0}^{\infty} \boldsymbol{\pi}_i \boldsymbol{\Sigma}^{\frac{1}{2}}\boldsymbol{\Sigma}^{-\frac{1}{2}}\tilde{\boldsymbol{\varepsilon}}_{t-i} = \sum_{i=0}^{\infty} \boldsymbol{\pi}_i^* \tilde{\mathbf{u}}_{t-i} \tag{14.101}$$

where $\boldsymbol{\pi}_i^* = \boldsymbol{\pi}_i \boldsymbol{\Sigma}^{\frac{1}{2}}$ and $\mathrm{Var}[\tilde{\mathbf{u}}_t] = E[\tilde{\mathbf{u}}_t \tilde{\mathbf{u}}_t^{\top}] = \mathbf{I}$. This last fact is easily established and is left as an exercise; see Exercise 14.15. So the $\tilde{u}_{jt}$ are uncorrelated and have unit variances. A similar orthogonalization procedure is employed later, in Section 17.4.1, and is the subject of Exercise 17.11, while use of the alternative decomposition given in Theorem 4.4.15 is covered in Exercise 14.15.

However, a more usual method of constructing orthogonal (i.e. uncorrelated) disturbances in econometrics exploits the triangular factorization presented in Theorem 4.4.16. As $\boldsymbol{\Sigma}$ is symmetric positive definite, we may write $\boldsymbol{\Sigma} = \mathbf{L}\mathbf{D}\mathbf{L}^{\top}$, where $\mathbf{L}$ is a lower triangular matrix and $\mathbf{D}$ is a diagonal matrix whose diagonal elements are all positive. Now let $\tilde{\mathbf{u}}_t^* = \mathbf{L}^{-1}\tilde{\boldsymbol{\varepsilon}}_t$; then the second term on the right-hand side of (14.99) may be written as

$$\sum_{i=0}^{\infty} \boldsymbol{\pi}_i \tilde{\boldsymbol{\varepsilon}}_{t-i} = \sum_{i=0}^{\infty} \boldsymbol{\pi}_i \mathbf{L}\mathbf{L}^{-1}\tilde{\boldsymbol{\varepsilon}}_{t-i} = \sum_{i=0}^{\infty} \boldsymbol{\pi}_i^{**} \tilde{\mathbf{u}}_{t-i}^* \tag{14.102}$$

where $\boldsymbol{\pi}_i^{**} = \boldsymbol{\pi}_i \mathbf{L}$ and $\mathrm{Var}[\tilde{\mathbf{u}}_t^*] = E[\tilde{\mathbf{u}}_t^* \tilde{\mathbf{u}}_t^{*\top}] = \mathbf{D}$ (again, this variance result is easy to establish and is left as part of Exercise 14.15). So, like the previous $\tilde{u}_{jt}$, the $\tilde{u}_{jt}^*$ are uncorrelated, but, unlike the $\tilde{u}_{jt}$, the variances of the $\tilde{u}_{jt}^*$, given by the diagonal elements of $\mathbf{D}$, are not unity, in general.

A plot of the jkth element of $\boldsymbol{\pi}_i^{**}$, $\partial \tilde{y}_{jt+i}/\partial \tilde{u}_{kt}^*$, as a function of i is called an **orthogonalized impulse response function** (as would be a plot of the jkth element of $\boldsymbol{\pi}_i^*$) and this avoids the objection levelled at the original impulse response function. Notice, however, that $\boldsymbol{\pi}_0^{**} = \boldsymbol{\pi}_0 \mathbf{L} = \mathbf{I}\mathbf{L} = \mathbf{L}$ is lower triangular, which implies that the ordering of variables in $\tilde{\mathbf{y}}_t$ in a recursive fashion is important. A change in $\tilde{u}_k^*$ has an impact on $\tilde{y}_j$ via the jkth element of $\boldsymbol{\pi}_0^{**}$ only if $j \geq k$. Thus $\tilde{y}_{1t}$ is affected only by $\tilde{u}_{1t}^*$; $\tilde{y}_{2t}$ is affected by $\tilde{u}_{1t}^*$ and $\tilde{u}_{2t}^*$; $\tilde{y}_{3t}$ is affected by $\tilde{u}_{1t}^*$, $\tilde{u}_{2t}^*$ and $\tilde{u}_{3t}^*$; and so on. It is difficult to be sure about this recursivity in practice.

EXERCISES

14.1 Derive and evaluate the right-hand side of equation (14.14) to obtain individual scalar expressions for the ordinary least squares estimators $\hat{\beta}_1^*$ and $\hat{\beta}_2^*$ in the simple linear regression model $\tilde{Y}_t = \beta_1 + \beta_2 X + \tilde{u}_t$.

14.2 Estimate the parameters in each of the linear regression models $\tilde{Y}_t = \beta_0 t + \tilde{u}_t$ and $\tilde{Y}_t = \beta_1 + \beta_2 t + \tilde{u}_t$, using the ordinary least squares (OLS) estimator (14.12) and the data in the following table. Confirm your estimates using Cramer's rule. In each case, also obtain the quantities $\sum_{t=1}^{10} e_t$, $\sum_{t=1}^{10} e_t^2$, $\sum_{t=1}^{10} \widehat{Y}_t$, $\sum_{t=1}^{10} t e_t$ and $\sum_{t=1}^{10} \widehat{Y}_t e_t$, where $\widehat{Y}_t$ and e_t denote the fitted value of Y_t and the corresponding OLS residual, respectively; and comment on your comparative findings. Finally, obtain a measure of the goodness of fit to the data for each of the OLS lines, using the appropriate cosine formula.

Table 14.1 Sample data for regressions

t	1	2	3	4	5	6	7	8	9	10
Y_t	8	9	8	11	10	12	9	10	13	14

14.3 Suppose that you have just three observations on each of the variables in the linear regression $\tilde{Y}_t = \beta_1 X_{t1} + \beta_2 X_{t2} + \tilde{u}_t$, i.e. $t = 1, 2, 3$.

(a) Recalling that such an equation may be expressed in matrix notation as $\tilde{y} = X\beta + \tilde{u}$, define the matrices $\tilde{y}$, X, β and $\tilde{u}$.

(b) Draw diagrams to illustrate the fact that ordinary least squares estimation of β_1 and β_2 is equivalent to projecting the vector $y = (Y_1, Y_2, Y_3) \in \mathbb{R}^3$ into the two-dimensional subspace spanned by $x_1 = (X_{11}, X_{21}, X_{31})$ and $x_2 = (X_{12}, X_{22}, X_{32})$, and that the angle between y and its OLS fitted value $\tilde{y}$ constitutes a measure of goodness of fit.

(c) Deduce approximate values for $\hat{\beta}_1$ and $\hat{\beta}_2$, and for the goodness-of-fit measure $\cos^2 \theta$, from your diagram.

14.4 Consider again the regression model $\tilde{Y}_t = \beta_1 X_{t1} + \beta_2 X_{t2} + \tilde{u}_t$, and the following data matrices:

$$y = \begin{bmatrix} 2 \\ 1 \\ 3 \\ 4 \\ 3 \end{bmatrix}, \quad X = \begin{bmatrix} 1 & 3 \\ 1 & 1 \\ 1 & 2 \\ 1 & 4 \\ 1 & 5 \end{bmatrix}$$

(a) Evaluate the OLS estimates $\hat{\beta}_1^*$ and $\hat{\beta}_2^*$ using Cramer's rule. Recall equation (14.11).

(b) Confirm the previous results by evaluating $\hat{\beta}^* = (X^\top X)^{-1} X^\top y$ using matrix methods.

(c) Evaluate the projection matrix $N = X(X^\top X)^{-1} X^\top$; hence, obtain the vector of fitted values $\hat{y}$ and the vector of OLS residuals e.

(d) Confirm numerically that $MX = (I - N)X = 0$; that $e \perp \hat{y}$ and e is orthogonal to each of the columns of X; and that the arithmetic mean of the residuals, $\frac{1}{5}\sum_{t=1}^{5} e_t$, is zero.

14.5 Consider the general linear regression model

$$\tilde{y} = X\beta + \tilde{u}$$

as defined in Section 14.2, and define $\mathbf{y} = \hat{\mathbf{y}} + \mathbf{e}$ and $\hat{\mathbf{y}} = \mathbf{X}\hat{\boldsymbol{\beta}}^*$, where $\hat{\boldsymbol{\beta}}^* = (\mathbf{X}^\top\mathbf{X})^{-1}\mathbf{X}^\top\mathbf{y}$ is an ordinary least squares estimate of $\boldsymbol{\beta}$. Show that $\mathbf{X}^\top\mathbf{e} = \mathbf{0}$, $\mathbf{X}^\top\mathbf{y} = \mathbf{X}^\top\hat{\mathbf{y}}$ and $\hat{\mathbf{y}}^\top\mathbf{e} = 0$.

14.6 Let $\mathbf{A}$ be a square matrix. Prove that $\mathbf{A} + \mathbf{A}^\top$ is symmetric.

14.7 Given T points in a $(k+1)$-dimensional space, $(\mathbf{x}_1, y_1), (\mathbf{x}_2, y_2), \ldots, (\mathbf{x}_T, y_T)$, it is desired to find the hyperplane of best fit, i.e. to find $\boldsymbol{\beta} \equiv (\beta_1, \beta_2, \ldots, \beta_k)$ such that the sum of the squared vertical distances (i.e. the distances parallel to the y axis) between the given points and the hyperplane with equation $y = \mathbf{x}^\top\boldsymbol{\beta}$ is as small as possible. By writing the objective function as a quadratic form, formulate this as an unconstrained optimization problem and find the solution using matrix methods.

14.8 Prove the Gauss–Markov theorem from first principles by formulating the Lagrangian and solving the first-order conditions for the appropriate constrained optimization problem.

14.9 Derive the variance–covariance matrix of the generalized least squares estimator given in equation (14.41). Hence, write down the variance–covariance matrix of the ordinary least squares estimator.

14.10 Prove that the square matrix in equation (14.58) is invertible, and derive its partitioned inverse used in equation (14.60); recall the results on partitioned inverses from Section 1.5.14.

14.11 Examine the nature of the restricted least squares estimator, equation (14.57), when the number of linearly independent restrictions, g, is equal to the number of parameters to be estimated, k.

14.12 Derive the variance–autocovariance matrix and, hence, the autocorrelation matrix of the weakly stationary AR(1) process $\tilde{y}_t = \phi_0 + \phi_1\tilde{y}_{t-1} + \tilde{\varepsilon}_t$ $(-1 < \phi_1 < 1)$, given T consecutive values of $\tilde{y}_t$. Comment on the form of the two matrices and establish some of their properties.

14.13 Show that the autoregressive process

$$\tilde{y}_t = 0.5 - 0.5\tilde{y}_{t-1} + 0.5\tilde{y}_{t-2} + \tilde{\varepsilon}_t$$

has a negative unit root. Find a transformation of y_t that is integrated to order zero, and establish the nature of the process that determines the behaviour of the transformed variable.

14.14 Formulate and solve the following determinantal equations and, hence, derive stationarity conditions on the individual parameters of the bivariate VAR process in each case.

(a) $\det[\boldsymbol{\Phi}(L)] = \det[\mathbf{I}_2 - \boldsymbol{\Phi}_1 L] = 0$, where

$$\boldsymbol{\Phi}_1 = \begin{bmatrix} \phi_{111} & \phi_{112} \\ \phi_{121} & \phi_{122} \end{bmatrix}$$

(b) $\det[\boldsymbol{\Phi}(L)] = \det[\mathbf{I}_2 - \boldsymbol{\Phi}_1 L - \boldsymbol{\Phi}_2 L^2] = 0$, where

$$\boldsymbol{\Phi}_1 = \begin{bmatrix} \phi_{1_{11}} & \phi_{1_{12}} \\ \phi_{1_{21}} & \phi_{1_{22}} \end{bmatrix} \quad \text{and} \quad \boldsymbol{\Phi}_2 = \begin{bmatrix} \phi_{2_{11}} & 0 \\ 0 & \phi_{2_{22}} \end{bmatrix}$$

14.15 Let $\boldsymbol{\Sigma}$ be an $m \times m$ symmetric positive definite matrix and $\tilde{\boldsymbol{\varepsilon}}_t \sim \mathrm{iid}(\mathbf{0}, \boldsymbol{\Sigma})$.

(a) Using a square root of $\boldsymbol{\Sigma}$, $\boldsymbol{\Sigma}^{\frac{1}{2}}$, say, define $\tilde{\mathbf{u}}_t = \boldsymbol{\Sigma}^{-\frac{1}{2}} \tilde{\boldsymbol{\varepsilon}}_t$ and prove that $\mathrm{Var}[\tilde{\mathbf{u}}_t] = \mathbf{I}_m$.

(b) Using the triangular factorization $\boldsymbol{\Sigma} = \mathbf{LDL}^\top$ from Theorem 4.4.16, where $\mathbf{L}$ is a lower triangular matrix and $\mathbf{D}$ is a diagonal matrix, define $\tilde{\mathbf{u}}_t^* = \mathbf{L}^{-1} \tilde{\boldsymbol{\varepsilon}}_t$ and prove that $\mathrm{Var}[\tilde{\mathbf{u}}_t^*] = \mathbf{D}$.

(c) Using the decomposition $\boldsymbol{\Sigma} = \mathbf{RR}^\top$ from Theorem 4.4.15, find a third orthogonalized disturbance, $\tilde{\mathbf{u}}_t^{**}$, and derive its mean, $E[\tilde{\mathbf{u}}_t^{**}]$, and its variance, $V[\tilde{\mathbf{u}}_t^{**}]$.

15 Multi-period choice under certainty

15.1 Introduction

In Section 12.2, it was pointed out that the objects of choice can be differentiated not only by their physical characteristics, but also both by the time at which they are consumed and by the state of nature in which they are consumed. These distinctions were suppressed in the intervening sections but are considered again in this chapter and in Chapter 16, respectively.

Discrete-time multi-period investment problems, such as are considered in this chapter, serve as a stepping stone from the single-period case in Chapter 12 to the continuous-time case, which the reader will encounter in more advanced texts. This chapter will assume that there is only one physical good, but that it can be traded and consumed at many different points in time.

The chapter begins with a survey of various concepts connected with the measurement of rates of return, such as interest rates and growth rates. Section 15.3 discusses the utility-maximization problem in this context. The main point to be taken from this section is that equilibrium interest rates (both spot rates and forward rates) can be derived in a straightforward manner from equilibrium prices. Section 15.4 then discusses various common interest rate concepts and the relationships between them under the general heading of the "Term structure of interest rates". The no-arbitrage principle of Section 12.5.6 is central to this analysis.

15.2 Measuring rates of return

Rates of return, or, equivalently, rates of growth, can be calculated for any quantity that varies over time, whether a stock or a flow.[1] A **flow** is any quantity whose dimension is measured in units per time period, e.g. euro per month; a **stock** is any quantity measured in units without a time dimension. Note that dimension in this context has a different meaning from that encountered in previous chapters.

A typical quantity whose rate of growth is of general interest is the value of an investment (a stock, possibly in both senses of that word), but other quantities whose growth rates concern economists include the Consumer Price Index (also a stock) and the level of national income, as measured, for example, by Gross National Product (a flow).

Note that national income is a flow, the level and growth rate of which are relatively easily defined, valued and measured; consequently, national income is relatively easy to tax. National wealth, on the other hand, is a stock, the level and growth rate of which are relatively difficult to define, value or measure; consequently, national wealth is relatively difficult to tax. Even finance ministers have been known to confuse these two very different concepts.

Similar comments apply to the distinction between individual wealth and individual income. We have not stressed this distinction up to now, since in a single-period world there is little difference between income and wealth. The distinction is more important in multi-period modelling.

In a single-period context, the **net rate of return** on an investment or **principal** of W_0, which yields a **payoff** of W_1, is the ratio of the profit earned, $W_1 - W_0$, to the amount invested, usually expressed as a percentage

$$100 \times \left(\frac{W_1}{W_0} - 1 \right) \% \tag{15.1}$$

Note that W_1 and W_0 are measured in units of currency (e.g. euro) but that the ratio is dimensionless. The net rate of return is equivalent to the **simple interest** earned on the investment. The **gross rate of return** is the ratio of the total payoff to the amount invested, i.e. the payoff per unit invested

$$\frac{W_1}{W_0} \tag{15.2}$$

Again, there is a choice of scale on which to measure gross rates of return: as a percentage of the initial investment or as a simple ratio. Gross returns, like net returns, are dimensionless.

The word **return** on its own is sometimes loosely used interchangeably for the **payoff**, for the **profit** and for the **rate of return**. We will use it only for the last of these, also using **gross return** as a shorthand for gross rate of return and **net return** as a shorthand for net rate of return.

In a multi-period context, the rate of return must have a **time** dimension. This is equivalent to working with **compound interest**. For example, 2% per annum is very different from 2% per month. Furthermore, the method or frequency of compounding must be specified in order to avoid ambiguity. The only exception to this is when the growth rate is zero; growth of 0% per month compounded weekly has exactly the same effect as growth of 0% per year compounded daily. In other words, rates of interest, growth, inflation, etc., are not properly defined unless we state the time interval to which they apply and the method of compounding to be used.

We have already seen how compound interest works in Example 8.3.1 and Exercise 8.3. Table 15.1 illustrates what happens to €100 invested at 10% per annum as we change the frequency of compounding. The final calculation in the table uses the fact (see p. xxiii) that

$$\lim_{n \to \infty} \left(1 + \frac{r}{n} \right)^n = e^r \tag{15.3}$$

There are five related calculations which readers should be familiar with and which will be considered in the subsections that follow.

15.2.1 Discrete compounding

If a quantity with initial value P_0 grows for t periods at a (net) rate of r per period, compounded n times per period, to reach a final value of P_t, then

$$P_t = \left(1 + \frac{r}{n} \right)^{nt} P_0 \tag{15.4}$$

Table 15.1 The effect of an interest rate of 10% per annum at different frequencies of compounding

Compounded	Principal (€)		Payoff (€)
Annually	100	→	$110 = 110.00$
Semi-annually	100	→	$100 \times (1.05)^2 = 110.25$
Quarterly	100	→	$100 \times (1.025)^4 = 110.381\ldots$
Monthly	100	→	$100 \times \left(1 + \dfrac{0.10}{12}\right)^{12} = 110.471\ldots$
Weekly	100	→	$100 \times \left(1 + \dfrac{0.10}{52}\right)^{52} = 110.506\ldots$
Daily	100	→	$100 \times \left(1 + \dfrac{0.10}{365}\right)^{365} = 110.515\ldots$
Continuously	100	→	$100 \times e^{0.10} = 110.517\ldots$

This equation can be solved for any of five quantities, given the other four:

- present value, P_0;
- final value, P_t;
- implicit rate of return, r;
- time, t; or
- frequency of compounding n.

15.2.2 Continuous compounding

Similarly, if a quantity with initial value P_0 grows for t periods at a (net) rate of r per period, compounded continuously, to reach a final value of P_t, then

$$P_t = e^{rt} P_0 \tag{15.5}$$

This equation can be solved for any of the four quantities P_0, P_t, r or t, given the other three.

Note also that the exponential function is convex and is its own derivative; that $y = 1 + r$ is the tangent to $y = e^r$ at $r = 0$, $y = 1$ and, hence, that $e^r > 1 + r$ for all $r \neq 0$. In other words, given an initial value, continuous compounding yields a higher terminal value than discrete compounding for all interest rates, positive and negative, with equality for a zero interest rate only. Similarly, given a final value, continuous discounting yields a strictly lower present value than does discrete discounting, again with the exception of equality for a zero rate of growth.

Finally, note that, if r_d is the discretely compounded rate of return and r_c the continuously compounded rate of return per period on an investment, then, combining (15.4) – with $n = 1$ – and (15.5),

$$1 + r_d = e^{r_c} \tag{15.6}$$

15.2.3 Aggregating and averaging returns

It will now be demonstrated that discretely compounded rates aggregate neatly across portfolios, while continuously compounded rates aggregate neatly across time.

Aggregating and averaging across portfolios

Simple rates of return are additive across portfolios, so we use them in one-period cross-sectional studies, in particular in portfolio theory in Chapter 17.

Consider a *portfolio* of N risky single-period investments comprising the euro amount b_i invested in the ith asset, which has a random gross rate of return over the single period of $\tilde{r}_i$, $i = 1, 2, \ldots, N$.

The total payoff on the portfolio is $\tilde{r}_\mathbf{b} \equiv \sum_{i=1}^{N} b_i \tilde{r}_i = \mathbf{b}^\top \tilde{\mathbf{r}}$, where $\mathbf{b} \equiv (b_1, b_2, \ldots, b_N)$ and $\tilde{\mathbf{r}} \equiv (\tilde{r}_1, \tilde{r}_2, \ldots, \tilde{r}_N)$; recall Exercise 9.6.

If $\tilde{\rho}_i$ denotes the continuously compounded net measure of the same underlying rate of return on asset i, then the portfolio payoff can be expressed as $\sum_{i=1}^{N} e^{\tilde{\rho}_i} b_i$. There is no obvious simple way of manipulating this expression in matrix notation, as there was for the discretely compounded version.

Aggregating and averaging across time

Continuously compounded rates of return, on the other hand, are additive across time, so we use them in multi-period single-variable studies, in particular when considering the term structure of interest rates in Section 15.4.

Consider a *single asset* whose value is changing at a time-varying rate over time. In a *discrete-time framework*, we can write

$$P_{t+1} = e^{r_t} P_t \tag{15.7}$$

where r_t represents the continuously compounded rate of return earned in the single period between time t and time $t + 1$. Hence, the final value of the asset at time T is

$$P_T = \prod_{t=1}^{T} e^{r_{t-1}} P_0 \tag{15.8}$$

Taking logarithms yields

$$\ln P_T = \sum_{t=1}^{T} r_{t-1} + \ln P_0 \tag{15.9}$$

This allows us to compute the average continuously compounded rate of return per period over the T periods

$$\bar{r} \equiv \frac{\sum_{t=1}^{T} r_{t-1}}{T} = \frac{\ln P_T - \ln P_0}{T} \tag{15.10}$$

Note that the average of the single-period growth rates can be computed from just two observations, the initial value P_0 and the final value P_T.

In a *continuous-time framework*, we can write

$$P_t = e^{rt} P_0 \tag{15.11}$$

$$P_{t+\Delta t} = e^{r \Delta t} P_t \tag{15.12}$$

Taking logarithms and rearranging gives

$$r = \frac{\ln P_{t+\Delta t} - \ln P_t}{\Delta t} \tag{15.13}$$

Taking limits as $\Delta t \to 0$, the instantaneous growth rate can be written as

$$r(s) = \frac{d \ln P_s}{ds} \tag{15.14}$$

Integrating yields

$$\int_0^t r(s)\,ds = \int_0^t d \ln P_s = \ln P_t - \ln P_0 \tag{15.15}$$

Taking the exponential of each side and rearranging gives

$$P_t = P_0\, e^{\int_0^t r(s)\,ds} \tag{15.16}$$

Thus the average rate of return over t periods of a quantity that grows at the instantaneous time-varying rate $r(s)$ at time s is

$$\frac{1}{t} \int_0^t r(s)\,ds \tag{15.17}$$

It is left as an exercise for the reader to explore the relative intractability of time-varying discrete rates of return; see Exercise 15.3.

Discrete rates of return are bounded below by -100%; continuously compounded rates of return are unbounded. The former therefore cannot be exactly normally distributed; the latter can. If continuously compounded rates of return are normally distributed, then it can be seen from (15.6) that the corresponding gross discrete rates of return are lognormally distributed.

15.2.4 Net present value

Net present value (NPV) is widely used to value streams of certain or uncertain future cash flows given various assumptions about rates of return. We consider three cases.

1. The **(net) present value** at time 0 of a known stream of cash flows,

$$P_0, P_1, \ldots, P_T \tag{15.18}$$

given a *constant* interest rate or **discount rate** r, is

$$\begin{aligned} \text{NPV}(r) &\equiv \frac{P_0}{(1+r)^0} + \frac{P_1}{(1+r)^1} + \frac{P_2}{(1+r)^2} + \cdots + \frac{P_T}{(1+r)^T} \\ &= P_0 + \frac{P_1}{(1+r)^1} + \frac{P_2}{(1+r)^2} + \cdots + \frac{P_T}{(1+r)^T} \end{aligned} \tag{15.19}$$

The coefficients $1/(1+r)^t$, $t = 1, 2, \ldots, T$, are known as **discount factors**.

2. If there is a sequence of known **time-varying** (really **maturity-varying**) discount rates, $\{r_t\}_{t=1}^{T}$, then

$$\text{NPV}_r \equiv P_0 + \frac{P_1}{(1+r_1)^1} + \frac{P_2}{(1+r_2)^2} + \cdots + \frac{P_T}{(1+r_T)^T} \tag{15.20}$$

3. If the relevant future payoffs and/or discount rates are not known with certainty at time 0, then they can be represented by random variables, resulting in an NPV that is also a random variable:

$$\widetilde{\text{NPV}} \equiv P_0 + \frac{\tilde{P}_1}{(1+\tilde{r}_1)^1} + \frac{\tilde{P}_2}{(1+\tilde{r}_2)^2} + \cdots + \frac{\tilde{P}_T}{(1+\tilde{r}_T)^T} \tag{15.21}$$

Equation (15.21) lies behind the commonly used **discounted cash flow** (DCF) approach to the valuation of projects, companies and financial assets. Note the implications of Jensen's inequality for (15.21). The computation of present values based on replacing the uncertain future discount factors in this formula with point estimates derived from expected future interest rates is commonplace, but the bias that this introduces is poorly understood. Even if cash flows and discount rates are independent, this underestimates present values. In the more likely scenario that cash flows and discount rates are interdependent, we cannot unambiguously sign the bias introduced, but it is unlikely that it vanishes.

15.2.5 *Internal rate of return*

The **internal rate of return** (IRR) of the stream of cash flows,

$$P_0, P_1, \ldots, P_T \tag{15.22}$$

is the solution of the polynomial equation of degree T obtained by setting the NPV, based on a constant discount rate of r, equal to zero:

$$P_0 + \frac{P_1}{(1+r)^1} + \frac{P_2}{(1+r)^2} + \cdots + \frac{P_T}{(1+r)^T} = 0 \tag{15.23}$$

In general, the polynomial defining the IRR has T (complex) roots. Conditions have been derived under which there is only one meaningful real root of this polynomial equation, in other words only one root corresponding to a positive IRR.[2]

If a financial institution issues a loan, e.g. a mortgage, at a fixed interest rate, to be repaid in a fixed number of equal instalments, the problem of calculating the size of the repayments can be viewed as either an NPV calculation or an IRR calculation. The repayment is set to equate the NPV of the repayments at the relevant interest rate to the amount borrowed. Equivalently, the repayment is set so that the IRR on the stream of cash flows comprising the sum borrowed (negative, from the perspective of the financial institution) and the sequence of repayments (positive) equals the interest rate associated with the loan. The precise details of the calculation depend on the method of compounding specified (e.g. continuous or discrete); see Exercise 15.5.

We will meet the concept of IRR again in Definition 15.4.5.

15.3 Multi-period general equilibrium

Consider a world in which there is only one physical good. (It may help to think of the good as "money".) Suppose that an individual in this world knows with certainty that he will live for a further T periods. (A world with uncertainty will be considered in later chapters). The individual's endowment comprises an "income" in each period of m_t units of the single good ($t = 0, 1, \ldots, T$). While there are $T + 1$ consumption periods, there is just a single trading period, at $t = 0$.

For simplicity, assume first that the individual can borrow and lend at a simple interest rate of i per period.

As before, the standard axioms allow the individual's consumption problem to be reduced to the maximization of an inter-temporal utility function, say,

$$U(c_0, c_1, \ldots, c_T) = \sum_{t=0}^{T} \beta^t \ln c_t \tag{15.24}$$

where c_t is his consumption (of the single available good) in period t.

To simplify the statement of the budget constraint, define M to be the present value of the individual's income stream. As seen in Section 15.2.4, the present value of m_t payable t periods from now when the interest rate is i per period is $m_t/(1+i)^t$.

Thus the budget constraint is

$$\sum_{t=0}^{T} \frac{c_t}{(1+i)^t} = \sum_{t=0}^{T} \frac{m_t}{(1+i)^t} = M \tag{15.25}$$

The solution of this problem is left as an exercise; see Exercises 15.10 and 15.11.[3]

The parameter β in the utility function is effectively the personal discount factor of the consumer, while the market discount factor is $1/(1+i)$.

The assumption that there is a single constant interest rate for all transactions over T periods undoubtedly simplifies the notation and is intuitively appealing. However, we have seen in Section 12.6.5 that, in a world in which there are $T + 1$ variables in utility functions, T independent markets determining T relative prices are required to guarantee that equilibrium is Pareto optimal. Fixing the interest rate leaves only one independent price, so that markets are incomplete and the outcome is not necessarily Pareto optimal.

When markets are complete, there will be T independent discount factors, say, $p_1, p_2, \ldots, p_T$, where p_t denotes the price of date-t consumption in units of the numeraire, date-0 consumption ($p_0 = 1$). The full budget constraint is

$$\sum_{t=0}^{T} c_t p_t = \sum_{t=0}^{T} m_t p_t = M \tag{15.26}$$

From equilibrium discount factors, it is easy to find corresponding maturity-dependent discount rates $i_0, i_1, \ldots, i_T$, using

$$p_t = \frac{1}{1 + i_t} \tag{15.27}$$

$$i_t = \frac{1}{p_t} - 1 \tag{15.28}$$

where i_t represents the interest rate per period for a loan or investment beginning at date 0 and ending at date t.

The assumption of perfect competition ensures that the interest rate is the same for borrowers and investors; a more realistic model would relax this assumption.

15.4 Term structure of interest rates

15.4.1 Interest rate concepts

DEFINITION 15.4.1 A **bond** is a security that provides a regular (in practice, usually annual or semi-annual) stream of equal **coupon** (or **dividend**) payments and a principal repayment (the par value or **redemption value**) at the date of the last coupon (known as the **maturity date**).

DEFINITION 15.4.2 A **pure discount** or **zero-coupon bond** is a bond with no coupon payments.

The goods in the inter-temporal consumption problem of the previous section are effectively pure discount bonds of different maturities ($t = 1, 2, \ldots, T$).

DEFINITION 15.4.3 A **console** is a bond with maturity at infinity (no principal repayment).

DEFINITION 15.4.4 The notation $_{t_1}i_{t_2 t_3}$ denotes the annualized interest rate for a risk-free transaction for which

- t_1 is the **commitment date** (often dropped when it is clear from the context),
- t_2 is the **lending date** and
- t_3 is the **repayment date**.

DEFINITION 15.4.5 The **gross redemption yield** of a bond trading at a given price is the internal rate of return earned by buying the bond at that price and holding it to redemption.

There are five related interest rate concepts that are frequently encountered in the analysis of bonds. We will consider them in the case of discrete time, measuring time in years, say. Defining the corresponding concepts for continuously compounded rates is left as an exercise; see Exercise 15.13. These concepts are as follows:

1. the **spot rate**, which can be inferred from the price of a zero-coupon bond by making the appropriate substitutions in the standard equation (15.4)

$$i_{0t} = \left(\frac{1}{P_{0t}}\right)^{1/t} - 1 \quad (\times 100\%) \tag{15.29}$$

per annum, compounded annually, where P_{0t} is the price at time 0 of a zero-coupon bond maturing at time t with par value 1;
2. the **forward rate** for a loan taken out at time t_1 and repaid at time t_2, which can be inferred from the two spot rates i_{0t_1} and i_{0t_2} using the equation

$$i_{t_1 t_2} = {}^{t_2-t_1}\sqrt{\frac{(1+i_{0t_2})^{t_2}}{(1+i_{0t_1})^{t_1}}} - 1 \tag{15.30}$$

3. the **current yield** of a bond, which is simply the annual coupon divided by the current market price, but which is of little significance from an economic point of view;
4. the **yield to maturity** of a bond, which – assuming six-monthly coupons – is found at a coupon payment date by solving (using (15.23)) for the internal rate of return per six months and doubling the result (or by similar equally arbitrary techniques); and
5. the **effective annual yield** on a bond, which is the true internal rate of return.

For pure discount bonds, the implicit spot rate and the effective annual yield are the same.

15.4.2   Describing the term structure

Term structure theory deals with the differences in the effective annual yields on pure discount bonds of different maturities (i.e. the differences in the relevant implicit spot rates), the constraints on such differences and the reasons behind them.

As well as varying with maturity at a fixed point in calendar time, spot rates can, of course, vary over calendar time. In practice, the term structure of interest rates for different currencies at a single point in time can differ substantially.

Coupon-bearing bonds of different maturities will always have different yields to maturity; term structure theory is not needed to explain this.

The term structure is usually presented as the **yield curve**, which is the graph of the spot rate i_{0t} against time t. If the vertical axis in a yield curve diagram is labelled "yield to maturity", it is implicitly assumed that the variable on this axis is the properly computed yield to maturity of a pure discount bond.

There are many other sequences of numbers, in addition to

$$i_{01}, i_{02}, i_{03}, \ldots \tag{15.31}$$

which can be used to describe the term structure (in discrete time) equally well, provided that the no-arbitrage principle holds. Most useful in practice are the discount factors,

$$\frac{1}{1+i_{01}}, \quad \left(\frac{1}{1+i_{02}}\right)^2, \quad \left(\frac{1}{1+i_{03}}\right)^3, \quad \ldots \tag{15.32}$$

as will be seen later. One year forward rates can also be used:

$$_0i_{01}, \; _0i_{12}, \; _0i_{23}, \; _0i_{34}, \; \ldots \tag{15.33}$$

15.4.3   Estimating the term structure

The term structure for a particular currency at a particular instant in calendar time is usually estimated using data on government bonds denominated in that currency but with different maturity dates and different coupon rates.

The value of a bond with redemption value 1, semi-annual coupon c_j and $T_j/2$ years to maturity is (measuring time in six-month units)

$$P_j = \frac{c_j}{(1+i_{01})} + \frac{c_j}{(1+i_{02})^2} + \cdots + \frac{1+c_j}{(1+i_{0T_j})^{T_j}} \tag{15.34}$$

Everything in this equation is known except for the discount factors $d_t \equiv 1/(1 + i_{0t})^t$ $(t = 1, 2, \ldots, T_j)$ and the equation is linear in the discount factors. Given a sample of $n = T \equiv \max_{j=1}^n T_j$ independent bonds, it is possible to solve exactly for the T corresponding discount factors. To value another bond, one need only substitute the estimated discount factors into (15.34). Given $n > T$ bonds (n observations), we can incorporate an error term and use multiple regression to estimate the discount factors statistically.

If the no-arbitrage principle holds, if all bonds are genuinely risk-free (e.g. issued by reputable, solvent, sovereign governments) and if all bond prices are observed simultaneously and without error, the vector of OLS residuals from this estimation procedure will be $\mathbf{0}_n$. In practice, however, there are many empirical complications, ambiguities and approximations involved in estimating the term structure or yield curve. During the international financial crisis that began in 2008, differences in the yields on bonds issued by sovereign governments within the Euro Zone (e.g. Germany and Greece) made it clearer than ever that not all government debt is completely risk-free. Possibly the biggest source of ambiguity is in the treatment of **accrued interest**.

The coupon or dividend on a bond is paid to whomsoever was the registered owner of the bond on a nominated date, usually several weeks before the dividend payment date, known as the **ex-dividend date**. In the following definition, we distinguish between two concepts.

DEFINITION 15.4.6

(a) The **clean price** or **dealing price** of a bond, denoted P_d, is the benchmark price, which will only change if the yield curve moves.
(b) The **dirty price**, denoted P, is the price actually paid by the buyer to the seller in a bond transaction, and is adjusted for accrued interest. The dirty price will drift upwards linearly between ex-dividend dates and will drop sharply on the ex-dividend date.

The relationship between the clean and dirty prices is given by

$$P = P_d + \frac{\tau}{365} c \tag{15.35}$$

or

$$P = P_d + \frac{\tau}{360} c \tag{15.36}$$

depending on whether the relevant market uses a 360-day or 365-day convention, where c is the annual coupon and τ is the number of days to the relevant ex-dividend date. The number τ will be zero on the dividend payment date; negative after the ex-dividend date and before the dividend payment date; and positive after the dividend payment date and before the next ex-dividend date. Calculating the clean price is effectively extrapolating from current market conditions to find what the bond price would be in similar market conditions on the next (or the previous) ex-dividend date.

Figure 15.1 shows a graph of clean (flat) and dirty (sawtooth or zigzag) prices against calendar time, assuming a flat and stable yield curve.

When estimating the yield curve, the dirty price should be used.

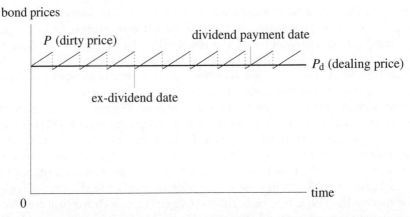

Figure 15.1 The relationship between clean and dirty bond prices and calendar time, assuming a flat and stable yield curve

Other difficulties that will be encountered in empirical estimation of the yield curve include the following:

- Quotes for different bonds are asynchronous, and there are usually only a handful of **benchmark** bonds in each currency that are very actively traded. Usually, there are not nearly enough reliable (active market) data available to make reliable estimation of discount factors feasible.
- It is difficult to distinguish between bid prices, ask prices and negotiated transactions prices within the bid–ask spread.
- Prices are usually discrete, with different tick sizes in different jurisdictions (US markets have worked with sixteenths and thirty-seconds; other jurisdictions might use two-thousandths or ten-thousandths of par value).
- Coupons are paid at different dates during the year.
- Some practitioners like to assume a 360-day year but others a 365-day year.

Estimates of the term structure (yield curve) are usually extremely sensitive to small variations in the way all the above problems are treated, but estimation is still a valuable and widely practised exercise.

If you believe your term-structure estimates, then you should buy any bonds that you consider under-valued and short-sell any that you consider over-valued. Alternatively, if you believe in the no-arbitrage principle but find that some bonds appear over-valued and others under-valued, then you might choose to disbelieve your estimates.

Alternative approaches involve assuming some general functional form for the yield curve, such as a cubic or exponential spline[4] or any similar function with just a few parameters to be estimated, and using the limited data available to estimate the parameters of this function.

The term structure and its movements over time can look very interesting in times and places where inflation is rampant and/or currencies are unstable, like the EU in the early 1990s, Russia in the mid-1990s and Brazil in 1999. Kahn (1990) discusses estimation of the term structure in the more stable environment of the US dollar.

Bond traders will also want to predict how bond prices are changing over time, which requires them first to predict how the term structure is changing over time. The next section considers various popular theories purporting to explain this.

15.4.4   Theories of the term structure

The pure expectations hypothesis, outlined in Section 13.10.2, is just one of a number of popular theories of the term structure. These are all unsatisfactory, if not just plain wrong.[5]

Pure expectations hypothesis

What is a good predictor of the future yield curve? One might argue, as the pure expectations hypothesis does, that the current set of forward rates would be

$$E_{t_0}[_{t_1}i_{t_1 t_2}] = {}_{t_0}i_{t_1 t_2} \tag{15.37}$$

But one could equally well argue that current forward prices for bonds should be a good predictor of future spot prices of bonds.

As seen in Section 13.10.2, Siegel's paradox tells us that we cannot simultaneously have both (and suggests that we probably have something in between, with the two versions of the pure expectations hypothesis providing upper and lower bounds).

In particular

$$E_{t_0}[_{t_1}i_{t_1 t_2}] = {}_{t_0}i_{t_1 t_2} \tag{15.38}$$

is incompatible with

$$E_{t_0}\left[\left(\frac{1}{1 + {}_{t_1}i_{t_1 t_2}}\right)^{t_2 - t_1}\right] = \left(\frac{1}{1 + {}_{t_0}i_{t_1 t_2}}\right)^{t_2 - t_1} \tag{15.39}$$

if there is any uncertainty at t_0 about the term structure that will prevail at t_1.

As noted in Section 13.10.2, the logarithmic expectations hypothesis is a more plausible and internally consistent alternative.

Segmented market hypothesis and preferred habitat hypothesis

These two hypotheses are essentially the same, although the latter is slightly more sophisticated.

Financial institutions have assets that will deliver payoffs at various times in the future and liabilities that will fall due at various times in the future. They like to reduce risk by matching the dates of payoffs as nearly as possible to the dates of liabilities. Thus, short-horizon rates are set by interactions between commercial banks managing chequing accounts, which generally involve short-horizon liabilities. Long-horizon rates are set by interactions between pension funds, insurance companies, large manufacturing companies building new plants, etc., which generally involve long-horizon liabilities. Risk aversion means that those with short-term liabilities tend to make short-term investments and vice versa in order to reduce the variance of cash flows in the distant future. Thus there is little interaction between long-term and short-term markets.

Liquidity premium hypothesis

This hypothesis just says that longer bonds should have higher returns to compensate for illiquidity, which arises because there are fewer investors and fewer opportunities for re-financing at the long end of the market.

As an upward sloping yield curve is more common,[6] this makes sense.

15.4.5   *Duration, volatility and convexity of bonds*

Related questions that are of interest to traders and investors in bond markets are:

1. What is the best bond portfolio to hold in order to meet a sequence of known future liabilities?
2. How will the price of a given bond react to a shift in the yield curve?
3. What effect will a shift in the yield curve have on the relative values of a sequence of liabilities and a portfolio of bonds held to meet those liabilities?

The answers to all of these questions are, perhaps surprisingly, closely related. If it is planned to sell bonds at market prices to meet known future liabilities, then a sudden fall in the value of the bonds will cause a cash-flow crisis. **Duration matching** is the art of balancing the cash flows arising from asset holdings with the cash flows due on liabilities.

Duration, volatility and convexity are measures that will be defined shortly and which have emerged to help answer the above questions. In order to address these questions, we will assume that there is a flat yield curve, or that there exists a discount rate i such that

$$i_{0t} = i \quad \forall t \tag{15.40}$$

This assumption has been relaxed in the general term structure literature, but will be maintained for the purposes of this book.

Under the assumption of a flat yield curve, the value of a bond with a periodic coupon of c and T periods to maturity is

$$P(1+i, c, T) = \frac{c}{(1+i)} + \frac{c}{(1+i)^2} + \cdots + \frac{1+c}{(1+i)^T} \tag{15.41}$$

Using the formula on p. xx, the present values of the various coupons in (15.41) may be summed, as they are the first T terms of a geometric progression with common ratio $r = 1 + i$ and initial (rightmost!) term $a = c/(1+i)^T$, to yield

$$
\begin{aligned}
P(1+i, c, T) &= \frac{c}{(1+i)^T} \frac{(1+i)^T - 1}{i} + \frac{1}{(1+i)^T} \\
&= \frac{c}{i}\left(1 - \frac{1}{(1+i)^T}\right) + \frac{1}{(1+i)^T} \\
&= \frac{c}{i} + \left(1 - \frac{c}{i}\right)\frac{1}{(1+i)^T}
\end{aligned}
\tag{15.42}
$$

A "period" will generally be a year or a half-year.

Expression (15.42) shows that, for positive i, the bond price will be equal to the redemption value of unity when $c = i$. The bond price will be greater than the redemption value when $c > i$ and less than the redemption value when $c < i$. (To see this, note that the second line of (15.42) expresses the bond price as a convex combination of 1 and c/i.)

DEFINITION 15.4.7 The **interest rate elasticity** of the bond whose value P is given by (15.41) is

$$\frac{1+i}{P(1+i, c, T)} \frac{\partial P}{\partial(1+i)} \tag{15.43}$$

i.e. the elasticity of P with respect to $1 + i$.

DEFINITION 15.4.8 The **duration**, $D(1 + i, c, T)$, of a bond is the weighted average time to receipt of cash flows, where the weight associated with each time is the contribution of the corresponding cash flow to $P(1 + i, c, T)$:

$$D(1 + i, c, T) \equiv \frac{\Sigma + T/(1 + i)^T}{P(1 + i, c, T)} \tag{15.44}$$

where

$$\Sigma \equiv \frac{c}{(1 + i)} + \frac{2c}{(1 + i)^2} + \frac{3c}{(1 + i)^3} + \cdots + \frac{Tc}{(1 + i)^T} \tag{15.45}$$

By evaluating the derivative in (15.43) and comparing the result with (15.44), it can be seen that (for any i) the duration of a bond is just the negative of its interest rate elasticity. Thus the duration of any coupon-bearing bond is the lifetime of a zero-coupon bond with the same interest rate elasticity. The duration of a zero-coupon bond is just its time to maturity.[7]

To calculate the numerical value of duration for a coupon-bearing bond, it is necessary to evaluate Σ numerically. There are two ways of deriving an expression for this sum. First, we can expand it as follows:

$$
\begin{aligned}
\Sigma = {} & \frac{c}{(1 + i)} + \frac{c}{(1 + i)^2} + \frac{c}{(1 + i)^3} + \cdots + \frac{c}{(1 + i)^T} \\
& + \frac{c}{(1 + i)^2} + \frac{c}{(1 + i)^3} + \cdots + \frac{c}{(1 + i)^T} \\
& \quad\quad + \frac{c}{(1 + i)^3} + \cdots + \frac{c}{(1 + i)^T} \\
& \quad\quad\quad\quad \ddots \quad\quad \vdots \\
& \quad\quad\quad\quad\quad\quad + \frac{c}{(1 + i)^T}
\end{aligned}
\tag{15.46}
$$

Since each of the T rows above (read from right to left) is just a geometric progression with common ratio $r = 1 + i$ and initial term $a = c/(1 + i)^T$, we have (again from the fact that $a + ar + ar^2 + ar^3 + \cdots + ar^{n-1} = a(r^n - 1)/(r - 1)$) that

$$
\begin{aligned}
\Sigma = {} & \frac{c}{(1 + i)^T} \frac{1}{i} \sum_{t=1}^{T} ((1 + i)^t - 1) \\
= {} & \frac{c}{(1 + i)^T} \frac{1}{i} \left((1 + i) \frac{((1 + i)^T - 1)}{i} - T \right) \\
= {} & \frac{c(1 + i)}{i^2} - \frac{c}{(1 + i)^{T-1}} \frac{1}{i^2} - \frac{c}{(1 + i)^T} \frac{1}{i} T
\end{aligned}
\tag{15.47}
$$

An alternative way to calculate Σ is to use the fact that

$$\Sigma - \frac{1}{1+i}\Sigma = \frac{c}{(1+i)} + \frac{2c}{(1+i)^2} + \frac{3c}{(1+i)^3} + \cdots + \frac{Tc}{(1+i)^T}$$

$$- \left(\frac{c}{(1+i)^2} + \frac{2c}{(1+i)^3} + \frac{3c}{(1+i)^4} + \cdots + \frac{Tc}{(1+i)^{T+1}} \right)$$

$$= \frac{c}{(1+i)} + \frac{c}{(1+i)^2} + \frac{c}{(1+i)^3} + \cdots + \frac{c}{(1+i)^T} - \frac{Tc}{(1+i)^{T+1}} \qquad (15.48)$$

which implies that

$$\frac{i}{1+i}\Sigma = \frac{c}{(1+i)^T}\frac{((1+i)^T - 1)}{i} - \frac{Tc}{(1+i)^{T+1}} \qquad (15.49)$$

Multiplying both sides by $(1+i)/i$ then yields

$$\Sigma = \frac{c(1+i)}{i^2} - \frac{c}{(1+i)^{T-1}}\frac{1}{i^2} - \frac{c}{(1+i)^T}\frac{1}{i}T \qquad (15.50)$$

as before. This simplifies to

$$\Sigma = \frac{c(1+i)}{i^2} - \frac{c}{(1+i)^T}\frac{1}{i^2}(1+i(T+1)) \qquad (15.51)$$

The numerator in the duration formula is obtained by adding the term corresponding to the redemption payment to Σ:

$$\Sigma + \frac{T}{(1+i)^T} = \frac{c(1+i)}{i^2} - \frac{c(1+i(T+1)) - i^2T}{i^2(1+i)^T} \qquad (15.52)$$

This formula can be used to construct a spreadsheet showing how duration depends on the relevant parameters c, i and T.

The interest rate elasticity of a bond measures the responsiveness of the price of a particular bond to changes in general market conditions (parallel shifts in the yield curve). In this respect, it is analogous to the β of an equity, which will be considered in Chapter 17. Financial institutions can reduce their exposure to shifts in the yield curve by attempting to keep the net duration of their assets and liabilities combined close to zero.

Duration depends on the yield i, yet to be of practical use it must be evaluated numerically at a particular value of i. One common convention is to use the semi-annual gross redemption yield of the bond in question, i.e. the i that solves

$$P_d = \sum_t \frac{c/2}{(1+i/2)^t} + \frac{1}{(1+i/2)^T} \qquad (15.53)$$

where P_d is the dealing price of the bond and the sum is over the times to all outstanding coupon payments.

DEFINITION 15.4.9 The **volatility** of a bond is

$$\frac{D(1+i,c,T)}{(1+i/2)} \tag{15.54}$$

Volatility is the preferred measure in the UK; duration is the preferred measure in the USA.

Elton *et al.* (2010, p. 561) note that there are at least a dozen different measures of duration. The simplest, illustrated above, is based on an assumption that the yield curve is flat. Different assumptions about the yield curve lead to different duration functions. The interested reader is referred to the discussion of these in Elton *et al.* (2010).

Note that the duration of a portfolio of bonds is the weighted average of the durations of the components. The weights, however, are the shares of the different components in the present value of the portfolio, *where cash flows are valued at the appropriate discount rate*. This is not in general the same as weighting by the cost of the different investments.

As with any elasticity, duration can be used to obtain a first-order approximation to the change in a bond price brought about by a shift in the (flat) term structure:

$$\%\Delta P \approx D(1+i,c,T) \times \%\Delta(1+i) \tag{15.55}$$

This linear approximation will work well for small changes in interest rates, but not for larger changes.

A second-order Taylor expansion can be used to obtain a quadratic approximation to $\%\Delta P$:

$$P(1+i+\Delta i,c,T)$$
$$\approx P(1+i,c,T) + \Delta i \frac{\partial P}{\partial(1+i)}(1+i,c,T) + \frac{(\Delta i)^2}{2}\frac{\partial^2 P}{\partial(1+i)^2}(1+i,c,T) \tag{15.56}$$

Rearranging yields

$$\frac{P(1+i+\Delta i,c,T) - P(1+i,c,T)}{P(1+i,c,T)}$$
$$\approx \Delta i \frac{\partial P(1+i,c,T)/\partial(1+i)}{P(1+i,c,T)} + \frac{(\Delta i)^2}{2}\frac{\partial^2 P(1+i,c,T)/\partial(1+i)^2}{P(1+i,c,T)}$$
$$= \%\Delta(1+i)D(1+i,c,T) + (\%\Delta(1+i))^2 \frac{(1+i)^2}{2}\frac{\partial^2 P(1+i,c,T)/\partial(1+i)^2}{P(1+i,c,T)} \tag{15.57}$$

The coefficient of the quadratic term

$$C(1+i,c,T) \equiv \frac{(1+i)^2}{2}\frac{\partial^2 P(1+i,c,T)/\partial(1+i)^2}{P(1+i,c,T)} \tag{15.58}$$

is called the **convexity** of the bond at i.

It is easily shown that

$$C(1+i,c,T) = \frac{1}{2P(1+i,c,T)}\left(\sum_{t=1}^{T}\frac{t(t+1)c}{(1+i)^t} + \frac{T(T+1)}{(1+i)^T}\right) \tag{15.59}$$

The details are left as an exercise; see Exercise 15.16.

EXERCISES

15.1 Calculate the present value at an interest rate of 5% per annum, compounded annually, of €1000 payable three years from now.

15.2 Calculate the continuously compounded annualized rate of return during 2010 and 2011 on a share whose closing price was 72 cent on 31 December 2009 and 108 cent on 31 December 2011.

15.3 Suppose an investment earns interest at time t at the variable simple interest rate $r(t)$ so that $P_{t+\Delta t} = (1+r(t)\Delta t)P_t$. Explore the average return on this investment from 0 to T.

15.4 If the interest rate for one-year deposits or loans is r_1 per annum compounded annually, the interest rate for two-year deposits or loans is r_2 per annum compounded annually and the forward interest rate for one-year deposits or loans beginning in one year's time is f_{12} per annum compounded annually, calculate the relationship that must hold between these three rates if there are to be no arbitrage opportunities.

15.5 Two university students meet in a local bank branch on 1 October.

(a) The first student has just returned from a summer working abroad and has managed to save €2000. She intends to live off her savings for the next academic year, withdrawing nine equal cash amounts at monthly intervals from 1 November until 1 July, when she intends to travel abroad again leaving a zero balance in her bank account. The bank manager offers her an interest rate of 1% per month, compounded monthly, on her €2000.

  (i) What is the present value on 1 October of the nine cash withdrawals, computed using this rate of interest? (Hint: recall the formula for the sum of a geometric series given on p. xx.)
  (ii) How much will she receive each month?

(b) The second student has just passed supplemental exams, for which he has been studying all summer, and needs to borrow €2000 to pay his fees and a deposit on a flat. He intends, despite the potential adverse effect on his academic progress, to work part-time during the year and pay off his loan in nine equal monthly instalments, from 1 November until 1 July, and to begin the following summer with no remaining debt. The bank manager demands an interest rate of 2% per month, compounded monthly.

  (i) What is the present value on 1 October of the nine repayments that he will make, computed using this borrowing rate?
  (ii) How much will he repay each month?

(c) Now suppose that the two students realize that they can bypass the banking system, and the first student offers to lend her savings to the second at an interest rate of 1.5% per month, compounded monthly. By how much, per month, will each student benefit if they agree to this arrangement, rather than using the bank as an intermediary?

15.6 Show, using a MacLaurin approximation, that the discount factor can be approximated using powers of the interest rate as follows:

$$\frac{1}{1+i} = 1 - i + i^2 - i^3 + i^4 - \cdots$$

Use this result to calculate first-order approximations to the present value of €10 000 payable in one year's time when the interest rate is (a) 1%, (b) 2%, (c) 10% and (d) 20% per annum, compounded annually in each case.

Now calculate the exact present value in each case, and compare your answers with the approximations previously obtained.

In what circumstances does the approximation $1 - i$ provide a useful substitute for the true discount factor $1/(1+i)$?

15.7 Suppose that in a rugby international between France and New Zealand the following are the best odds available:

Result	Fractional odds
France	5/4
New Zealand	2/1
draw	8/1

Show that an arbitrage opportunity exists and calculate the bets necessary to guarantee a profit of €10 000 irrespective of the outcome of the match.

15.8 Assuming a discount rate of $r > 0$ (per annum, compounded annually), find a general expression for today's (net present) value of a bond that promises to pay an annual coupon of €C on today's date in each of the next T years and an additional sum of €100 on today's date in the Tth year.

Show that the value of the bond exceeds €100 if and only if

$$r < \frac{C}{100}$$

Calculate the elasticity of the bond value with respect to r and show that it can be interpreted as a weighted average of the coupon payment dates, where each date is weighted by the share of the total present value of the bond that will be realized on that date.

15.9 Explore and comment on the additivity over time of a discrete rate of return that changes at the end of each compounding period from r_1 to r_2 to r_3 and so on.

15.10 Suppose that an individual lives for T periods, has income in each period of m_t $(t = 1, 2, \ldots, T)$, and can borrow and lend at a simple interest rate of i per period.

If the individual's preferences are represented by the inter-temporal utility function

$$U(c_1, c_2, \ldots, c_T) = \sum_{t=1}^{T} \ln c_t$$

where c_t is his consumption (of the single available good) in period t, what is his maximization problem? Draw an analogy with the single-period Cobb–Douglas case and, using this analogy, solve for optimal consumption. (Hint: this will be easier if you define M to be the present value of the individual's income and work with that.)

15.11 As in the previous exercise, suppose that an individual lives for T periods, has income in each period of m_t $(t = 1, 2, \ldots, T)$, and can borrow and lend at a simple interest rate of i per period.

Now consider the more general situation in which preferences are represented by the inter-temporal utility function

$$U(c_1, c_2, \ldots, c_T) = \sum_{t=1}^{T} \beta^t \ln c_t$$

The previous exercise considered the special case of $\beta = 1$. How does the solution change for $\beta \neq 1$? For what value of β will consumption be equal in each period? For what value of β will the same amount be set aside today to provide for consumption in each future period?

15.12 State and solve the inter-temporal utility-maximization problem for a consumer whose utility function is again given by $U(c_1, c_2, \ldots, c_T) = \sum_{t=1}^{T} \beta^t \ln c_t$, but who faces a price of p_t units of period-0 consumption for each unit of period-t consumption bought or sold $(t = 1, 2, \ldots, T)$.

What is the relationship between the price p_t and the interest rate i_t applicable to a loan running from date 0 to date t?

15.13 Write out expressions for spot rate, forward rate, yield to maturity and effective annual yield in terms of continuously compounded interest rates.

15.14 Derive the combined implications of Jensen's inequality and the pure expectations hypothesis for the relationship between the forward prices and future spot prices of pure discount bonds.

15.15 Suppose that forward continuously compounded interest rates are unbiased predictors of future spot continuously compounded interest rates. What are the implications of this hypothesis for the prices of pure discount bonds?

15.16 Assuming a flat yield curve with discount rate i, prove that the convexity of a bond with coupon c and T periods to maturity is given by

$$C(1+i, c, T) = \frac{1}{2P(1+i, c, T)} \left(\sum_{t=1}^{T} \frac{t(t+1)c}{(1+i)^t} + \frac{T(T+1)}{(1+i)^T} \right) \qquad (15.60)$$

where $P(1+i, c, T)$ denotes the implicit value of the bond.

15.17 The following table gives the prices quoted on 1 February 2009 for zero-coupon bonds with various maturity dates and par value of €1000:

Maturity date	Price (€)
1 February 2009	1000
1 February 2010	820
1 February 2011	725
1 February 2012	675
1 February 2013	600

(a) Calculate:

    (i) the annualized spot rates of interest for one-year, two-year, three-year and four-year loans taken out on 1 February 2009;

    (ii) the forward rates of interest for one-year loans to be taken out on 1 February 2010 and 1 February 2011; and

    (iii) the ex-dividend price on 1 February 2009 of a coupon-bearing bond with an annual dividend of €10 payable on 1 February each year, a par value of €100, and a maturity date of 1 February 2013.

(b) Plot the yield curve implied by the bond prices given in the table.

15.18 Four pure discount bonds, each with par value of €100, but maturing one year, two years, three years and four years from now, are trading at €$88\frac{8}{9}$, €80, €$72\frac{8}{11}$ and €$66\frac{2}{3}$, respectively.

(a) Calculate the implied annualized spot rates of interest for investments lasting one year, two years, three years and four years from today (to two decimal places) and plot the yield curve.

(b) Calculate the implied forward rates of interest for one-year investments starting in one year, in two years, and in three years from now.

(c) Given the information in this question, what would you pay for a four-year bond with annual coupons of €9.90 and a par value of €100?

15.19 It is often suggested that the expected value of the exchange rate between the euro and the US dollar in 30 days time should equal the corresponding 30-day forward exchange rate currently prevailing. Explain why this hypothesis is paradoxical if extended to all currency pairs. Suggest an alternative hypothesis that is based on the same intuition but is internally consistent. What are the implications of these observations for the theory of the term structure of interest rates?

15.20 Write out an expression for the value P of a bond with an annual coupon of c, T years to maturity and a par value of €1 when there is a flat yield curve corresponding to a discount rate of i per year compounded annually. Calculate the elasticity of the bond price P with respect to $1 + i$. Calculate the weighted average time to receipt of cash flows, where each time is weighted by the share of the present value of the corresponding cash flow in the total value of the bond. Show that the two expressions are equal.

15.21 Explain what is meant by the duration and the convexity of a stream of cash flows.

15.22 Calculate (as a function of the interest rate i) the duration of the following bonds:

(a) a five-year zero-coupon bond;
(b) a three-year bond with a 20% annual coupon; and
(c) a twenty-year bond with a 5% annual coupon.

15.23 Calculate the duration and convexity of a bond with an annual coupon of €10, par value of €100 and three years to maturity, using a discount rate of 10%. Assume that the bond has just gone ex-dividend.

15.24 Calculate the duration of a portfolio of €4000 divided equally between the four bonds in Exercise 15.18, first as a function of the discount rate i and second assuming a discount rate of 10%.

15.25 Suppose that you face liabilities of €2500 in one year's time, €5000 in two years' time and €5500 in three years' time. Assume a flat term structure with a discount rate of 5% per annum (compounded annually).

(a) Calculate the duration of your liabilities.

Suppose further that you can invest in a zero-coupon bond maturing in one year and/or in a bond with a 5% annual coupon and three years to maturity, and that the latter bond has just gone ex-dividend.

(b) Find a portfolio of these bonds with the same duration as your liabilities.
(c) Suppose that you can also invest in an ex-dividend bond with two years to maturity and a 10% annual coupon. Find (in terms of par value) a portfolio of the three bonds that exactly matches your liabilities. Give two other examples of portfolios of the three bonds with the same duration as your liabilities.
(d) Find (to the nearest cent) the market price of each bond, assuming a par value of €100.

15.26 Discuss briefly the factors that might influence the choice of duration of assets held by an investor (in particular an institutional investor) facing given future liabilities.

15.27 Calculate the gross redemption yield on a bond with two years to maturity, par value of €1000 and an annual coupon of 10%, which is selling for €800.

16 Single-period choice under uncertainty

16.1 Introduction

Section 13.3 introduced a selection of real-world situations in which consumers (or investors, as they might now be called) face uncertainty about the outcomes of their decisions, i.e. about the final values of their investments. Section 12.2.1 pointed out that the "goods" considered in the intervening material might be distinguished not by their intrinsic physical characteristics but by the state of nature in which they are consumed. This chapter brings these two concepts together to analyse in more detail the consumer's choice under uncertainty.

This chapter deals with choice under uncertainty exclusively in a single-period context. Trade takes place at the beginning of the period and uncertainty is resolved at the end of the period. Consumption can take place either at the end of the period only or both at the beginning of the period and at the end of the period. This framework is sufficient to illustrate the similarities and differences between the most popular approaches to analysing choice under uncertainty. For simplicity, we will assume throughout these final two chapters that there is just one single physical consumption good.

The chapter begins with a brief section providing initial motivation for what follows. The analysis of choice under uncertainty begins in Section 16.3 with the re-interpretation of the general equilibrium model in terms of the state-contingent claims introduced in Section 13.3.4. Economic theory has, over the years, used many different, sometimes overlapping, sometimes mutually exclusive, approaches to the analysis of choice under uncertainty. Among the approaches considered in this chapter are state-dependent and state-independent utility (Section 16.3), expected utility (Section 16.4), mean–variance utility (Section 16.8) and other non-expected-utility approaches (Section 16.9). The expected-utility approach leads on to analysis of risk aversion (Section 16.5) and risk neutrality (Section 16.6). That analysis in turn allows us to revisit and, in a sense, resolve Siegel's paradox (Section 16.7). The single-period portfolio choice problem, which could also be brought under the heading of "single-period choice under uncertainty", is of sufficient practical and mathematical importance that we will deal with it in a separate chapter (Chapter 17).

It will become clear in what follows that the theoretical lines of demarcation between consumers, investors and gamblers are much fuzzier than is usually acknowledged in everyday conversation.

16.2 Motivation

Table 13.1 presented a simple example of a choice under uncertainty, as well as raising the question of the economic valuation of a simple type of risky asset. Now consider another similar example, namely, a range of seven lotteries based on the toss of a fair coin. In these seven lotteries:

(a) with 50% probability you get nothing;
(b) with 50% probability you get, respectively,

    (i) €1,
   (ii) €10,
  (iii) €100,
  (iv) €1000,
   (v) €10 000,
  (vi) €100 000 or
 (vii) €1 000 000.

As in the example in Table 13.1, the reader is encouraged to consider how much (s)he would be willing to pay for a ticket in each of the seven lotteries described above and to keep a record of these amounts.

The nature of the uncertainty facing the consumer is different in different applications. In this case, there are just two equally likely states of nature, heads and tails. In the examples in Table 13.1, there are $^nC_6$ equally likely states of nature. In most real-world situations, there are infinitely many states of nature, usually a continuum of states, of which some are more likely than others.

Different approaches to the economic analysis of choice under uncertainty are better suited to different situations. The first approach considered here assumes that the underlying sample space comprises a finite number (S) of states of nature with different probabilities. A more thorough analysis of choice under uncertainty, allowing for infinite and continuous sample spaces and based on additional axioms of choice, follows later.

16.3　Pricing state-contingent claims

16.3.1　*Model structure*

Recall the definitions of state-contingent claims or Arrow–Debreu securities and of complex securitiescomplex security introduced in Section 13.3.4. Consider a world with a single physical consumption good, a sample space Ω containing S possible states of nature (distinguished by a first subscript, usually i), markets for N complex securities (distinguished by a second subscript, usually j) and H consumers (distinguished by a superscript, usually h).

We will let $\mathbf{Y}$ denote the $S \times N$ matrix whose jth column contains the payoffs of the jth complex security in each of the S states of nature, i.e.

$$\mathbf{Y} \equiv [\mathbf{y}_1 \ \mathbf{y}_2 \ \cdots \ \mathbf{y}_N] \tag{16.1}$$

where $\mathbf{y}_j = (y_{1j}, y_{2j}, \ldots, y_{Sj})$ is the associated $S \times 1$ payoff vector for security j.

We can assume either that individual endowments take the form of portfolios of the *elementary* state-contingent claims or that they take the form of portfolios of the *traded* complex securities for which markets exist; the significance of these alternative assumptions will become clear below.[1]

16.3.2 Types of utility function

In principle, each individual consumer's utility could be a completely arbitrary function of S variables, $x_1, x_2, \ldots, x_S$, representing the quantities of the single physical consumption good to be consumed in each of the S states of nature. The analysis becomes easier the more structure is imposed on individual preferences. For example, utility could be assumed to be additive across states and, within this context, utility might be state-dependent or state-independent.

Possible functional forms of individual h's utility function, for the case where all consumption takes place at the end of the period, include the following:

- arbitrary function

$$u^h(x_1, x_2, \ldots, x_S) \tag{16.2}$$

- additive across states

$$\sum_{i=1}^{S} u_i^h(x_i) \tag{16.3}$$

- state-independent, with subjective weights p_i^h, different for each individual consumer

$$\sum_{i=1}^{S} p_i^h v^h(x_i) \tag{16.4}$$

- state-independent, with weights given by objective probabilities π_i

$$\sum_{i=1}^{S} \pi_i v^h(x_i) \tag{16.5}$$

The function $v^h \colon \mathbb{R} \to \mathbb{R}$ can be thought of as h's **utility function on sure things**. We will return to this concept in Section 16.4.2. It will be assumed that v^h is strictly increasing, strictly concave and differentiable, so that standard results from optimization theory can be applied.

For simplicity of exposition, we will continue to assume for the time being that consumption decisions are made at date 0, before uncertainty is resolved, and that all consumption takes place at date 1, after uncertainty has been resolved. The model can be extended to allow for consumption to take place at both date 0 and date 1, just as the inter-temporal model of the previous chapter could include or exclude consumption at date 0.

16.3.3 Model equilibrium

If markets exist for all S elementary state-contingent claims, then prices are determined by the S market-clearing equations in a general equilibrium model, which can be reduced to

aggregate consumption in state i = aggregate endowment in state i $(i = 1, 2, \ldots, S)$

$$\tag{16.6}$$

We have seen in Section 12.5.5 that equilibrium exists in such a model, and in Section 12.6.3 that this equilibrium is Pareto optimal. Each individual will have an optimal consumption choice depending on endowments and preferences and conditional on the state of the world. We have three ways of viewing this optimal choice:

1. The optimal future consumption in the S possible states of the world for a particular investor can be viewed as an S-dimensional vector, denoted

$$\mathbf{x}^{*h} = \begin{bmatrix} x_1^{*h} \\ x_2^{*h} \\ \vdots \\ x_S^{*h} \end{bmatrix} \tag{16.7}$$

2. Equivalently, optimal future consumption can be viewed as a random variable $\tilde{x}^{*h}$ taking on the value x_i^{*h} in state i.
3. Finally, optimal future consumption can also be viewed as a portfolio of elementary state-contingent claims, comprising x_1^{*h} units of the state-1-contingent claim, x_2^{*h} units of the state-2-contingent claim, and so on.

If markets exist for N complex securities, then the investor will optimally hold a portfolio $\mathbf{w}^{*h} = \left(w_1^{*h}, w_2^{*h}, \ldots, w_N^{*h}\right)$ of these complex securities, such that

$$x_i^{*h} = \sum_{j=1}^{N} y_{ij} w_j^{*h}, \quad i = 1, 2, \ldots, S \tag{16.8}$$

or

$$\mathbf{x}^{*h} = \mathbf{Y}\mathbf{w}^{*h} \tag{16.9}$$

If the individual's initial endowment is a portfolio of elementary state-contingent claims $\mathbf{e}^h = (e_1^h, e_2^h, \ldots, e_S^h)$, then (s)he must purchase a vector of complex securities given by $\mathbf{t}^h = (t_1^h, t_2^h, \ldots, t_N^h)$, where

$$\mathbf{x}^{*h} - \mathbf{e}^h = \mathbf{Y}\mathbf{t}^h \tag{16.10}$$

Here $\mathbf{t}^h$ may have negative components representing sales of complex securities.

If the individual's initial endowment is a portfolio of traded complex securities $\mathbf{w}^h = (w_1^h, w_2^h, \ldots, w_N^h)$, then the optimal vector of net purchases is given by

$$\mathbf{t}^h = \mathbf{w}^{*h} - \mathbf{w}^h \tag{16.11}$$

THEOREM 16.3.1

(a) *If there are S complex securities ($S = N$) and the associated square payoff matrix $\mathbf{Y}$ is non-singular, then, for any initial endowment and any individual preferences, markets are complete and the equilibrium allocation is Pareto optimal.*

(b) *More generally, if each individual's vector of desired trades $\mathbf{x}^{*h} - \mathbf{e}^h$ is in the column space of the payoff matrix $\mathbf{Y}$, then, for this particular combination of endowment and preferences, the equilibrium allocation will be Pareto optimal.*

Proof:

(a) The matrix $\mathbf{Y}$ can be inverted to solve (16.10) for optimal trades in terms of complex securities:

$$\mathbf{t}^h = \mathbf{Y}^{-1}(\mathbf{x}^{*h} - \mathbf{e}^h) \tag{16.12}$$

(b) Even though $\mathbf{Y}$ may not be invertible, (16.10) can still be solved for $\mathbf{t}^h$ for certain combinations of preferences and endowments. When $\mathbf{Y}$ is singular, either the solution for $\mathbf{t}^h$ will not be unique or no solution for $\mathbf{t}^h$ will exist. $\square$

An $(N + 1)$th security would be redundant, while either a singular square matrix or less than N complex securities would result in incomplete markets, making the Pareto optimal allocation unattainable for some combinations of preferences and endowments.

If $\mathbf{Y}$ is invertible, it is also easy to work back from the (equilibrium) prices of complex securities, say $\mathbf{p} = (p_1, p_2, \ldots, p_N)$, to the individual (equilibrium) state prices, say $\boldsymbol{\phi} = (\phi_1, \phi_2, \ldots, \phi_S)$, since the no-arbitrage principle tells us that we must have

$$\mathbf{p} = \mathbf{Y}^\top \boldsymbol{\phi} \tag{16.13}$$

or

$$\boldsymbol{\phi}^\top = \mathbf{p}^\top \mathbf{Y}^{-1} \tag{16.14}$$

If individual endowments are portfolios of the N traded complex securities, then it can be seen from (16.9) that equilibrium allocations are Pareto optimal if and only if the Pareto optimal allocations are also portfolios of the N traded complex securities, i.e. if $\mathbf{x}^{*h}$ lies in the column space of $\mathbf{Y}$.

16.3.4 *Completion of markets using options*

In real-world markets, the relatively small number of linearly independent traded securities, N, is generally less than the very large number of states of nature, S. However, we will now show that options on existing securities may be sufficient to form complete markets, and thereby ensure Pareto optimality for arbitrary preferences.

To illustrate this, let us assume that there exists a **state index portfolio** yielding *different* non-zero payoffs in each state of nature (possibly one mimicking aggregate consumption; see Section 16.3.5). As usual, the payoffs of the state index portfolio can be viewed either as a random variable $\tilde{y}$ or as a vector of payoffs $\mathbf{y} = (y_1, y_2, \ldots, y_S)$. Without loss of generality, we can rank the states so that $y_i < y_j$ if $i < j$.

We now present some results, following Huang and Litzenberger (1988, Chapter 5), showing conditions under which trading in such a state index portfolio and in options on the state index portfolio can lead to the Pareto optimal complete markets equilibrium allocation. We will show that it is possible to achieve completion of markets using options on the state index portfolio by assuming that there exist $S - 1$ European call options on $\tilde{y}$ with exercise prices $y_1, y_2, \ldots, y_{S-1}$.

Here, the original state index portfolio and the $S - 1$ European call options yield the payoff matrix

$$
\begin{bmatrix}
y_1 & y_2 & y_3 & \cdots & y_S \\
0 & y_2 - y_1 & y_3 - y_1 & \cdots & y_S - y_1 \\
0 & 0 & y_3 - y_2 & \cdots & y_S - y_2 \\
\vdots & \vdots & \vdots & & \vdots \\
0 & 0 & 0 & \cdots & y_S - y_{S-1}
\end{bmatrix}^T
=
\begin{bmatrix}
\text{state index portfolio} \\
\text{call option 1} \\
\text{call option 2} \\
\vdots \\
\text{call option } S - 1
\end{bmatrix}^T
\tag{16.15}
$$

and as this triangular matrix is non-singular (since we have assumed different payoffs in each state in order to guarantee that the diagonal entries of the matrix are non-zero), we have constructed a complete market.

16.3.5   Security values and aggregate consumption

Instead of assuming, as in the previous section, that a state index portfolio exists, we can complete markets in a similar manner by assuming identical probability beliefs and a time-additive, state-independent utility function

$$
v_0^h(x_0) + \sum_{\omega \in \Omega} \pi(\omega) v_1^h(x_1(\omega))
\tag{16.16}
$$

where x_0 denotes consumption at date 0, $x_1(\omega)$ denotes consumption at date 1 if state ω materializes and $\pi(\omega)$ denotes the agreed probability that the state of the world is ω (effectively modifying (16.5) to allow for date-0 consumption).

The assumptions of identical probability beliefs and state-independent utility guarantee that equilibrium consumption decisions will be the same in any two states in which the aggregate consumption or endowment is the same, so we can let $f^h(k)$ denote individual h's equilibrium consumption in those states where aggregate consumption equals k. It is left as an exercise for the reader to prove this from the first-order conditions of the individual utility-maximization problems; see Exercise 16.4.

We will first consider the relationship between security values and aggregate consumption, which will be denoted $\tilde{C}$.

Let $C(\omega) \equiv$ aggregate consumption in state ω, and let $\Omega_k \equiv \{\omega \in \Omega : C(\omega) = k\}$. Let ϕ_k be the market price of the security with payoffs

$$
y_k(\omega) = \begin{cases} 1 & \text{if } C(\omega) = k \\ 0 & \text{otherwise} \end{cases}
\tag{16.17}
$$

This security is just a portfolio of the elementary state-contingent claims associated with all the states in which aggregate consumption takes the value k.

Let the agreed probability of the event Ω_k (i.e. the probability of aggregate consumption taking the value k) be

$$
\pi_k = \sum_{\omega \in \Omega_k} \pi(\omega)
\tag{16.18}
$$

Individual h's utility-maximization problem is

$$\max_{x_0, \{x_1(\omega) : \omega \in \Omega\}} v_0^h(x_0) + \sum_{\omega \in \Omega} \pi(\omega) v_1^h(x_1(\omega)) \qquad (16.19)$$

subject to the budget constraint

$$x_0 + \sum_{\omega \in \Omega} \phi(\omega) x_1(\omega) = M \qquad (16.20)$$

where date-0 consumption is treated as the numeraire, M denotes wealth and $\phi(\omega)$ is the price (in units of date-0 consumption) of the state-contingent claim paying off in state ω.

A risk-free security, paying off one unit of date-1 consumption in all states of the world, is just a portfolio of one of each state-contingent claim, so must, by the no-arbitrage principle, trade at a price of $\sum_{\omega \in \Omega} \phi(\omega)$ or, equivalently, $\sum_k \phi_k$. The risk-free rate of return can thus be easily derived from the state prices.

Eliminating the Lagrange multiplier, the first-order conditions for problem (16.19) simplify to

$$\phi(\omega) = \frac{\pi(\omega) v_1^{h\prime}(x(\omega))}{v_0^{h\prime}(x_0)} \quad \forall \, \omega \in \Omega \qquad (16.21)$$

The no-arbitrage principle implies that

$$\begin{aligned} \phi_k &= \sum_{\omega \in \Omega_k} \phi(\omega) \\ &= \sum_{\omega \in \Omega_k} \frac{v_1^{h\prime}(x(\omega))}{v_0^{h\prime}(x_0)} \pi(\omega) \\ &= \frac{v_1^{h\prime}(f^h(k))}{v_0^{h\prime}(x_0)} \sum_{\omega \in \Omega_k} \pi(\omega) \\ &= \frac{v_1^{h\prime}(f^h(k))}{v_0^{h\prime}(x_0)} \pi_k \end{aligned} \qquad (16.22)$$

since individual consumption is the same for all states in Ω_k.

Therefore, an arbitrary security with payoff $\tilde{x}$ taking the value $x(\omega)$ in state ω has value

$$\begin{aligned} \phi_{\tilde{x}} &= \sum_{\omega \in \Omega} \phi(\omega) x(\omega) \\ &= \sum_k \sum_{\omega \in \Omega_k} \phi(\omega) x(\omega) \end{aligned}$$

$$= \sum_k \frac{v_1^{h\prime}(f^h(k))}{v_0^{h\prime}(x_0)} \sum_{\omega \in \Omega_k} \pi(\omega) x(\omega)$$

$$= \sum_k \phi_k \sum_{\omega \in \Omega_k} \frac{\pi(\omega)}{\pi_k} x(\omega)$$

$$= \sum_k \phi_k E[\tilde{x} \mid \tilde{C} = k] \tag{16.23}$$

In other words, the value of an arbitrary security is given by a formula reminiscent of a familiar result from probability theory, but with state prices playing the role of probabilities. It is seen frequently in asset pricing models that the value of an asset is given, as it is here, by a formula similar to that for the statistical expectation of its payoff, but with probabilities (π_k in this case) replaced by an alternative measure derived from individual preferences (ϕ_k in this case).

Note also that a security whose payoff is independent of the value of aggregate consumption ($E[\tilde{x} \mid \tilde{C} = k] = E[\tilde{x}]$ for all k) will trade at $E[\tilde{x}] \sum_k \phi_k$, i.e. at the same price as a risk-free security paying off $E[\tilde{x}]$ with probability 1. Under the assumptions of this model, securities can trade at prices different from their discounted expected value only if their payoffs are dependent on aggregate consumption. The nature of that dependence will determine whether a security trades at a premium over or at a discount from that discounted expected value.

Thus this simple model provides one possible answer to a fundamental question common to most asset-pricing models, namely, "Why do risky securities sometimes trade at prices different from their simple discounted expected value?"

16.3.6 Replicating elementary claims with a butterfly spread

Let $\mathbf{x}_k$ be the vector of payoffs in the various possible states on a European call option on aggregate consumption with one period to maturity and exercise price k. For simplicity, let us assume first that the possible values of aggregate consumption $C(\omega)$ are just the integers $1, 2, \ldots, S$. Then payoffs are as given in Table 16.1.

Table 16.1 Payoffs for call options on aggregate consumption

$\tilde{C}$	$\mathbf{x}_0$	$\mathbf{x}_1$	$\mathbf{x}_2$	$\cdots$	$\mathbf{x}_{S-1}$
1	1	0	0	$\cdots$	0
2	2	1	0	$\cdots$	0
3	3	2	1	$\cdots$	0
$\vdots$	$\vdots$	$\vdots$	$\vdots$	$\vdots$	$\vdots$
S	S	$S-1$	$S-2$	$\cdots$	1

Elementary claims against aggregate consumption can be constructed as follows, using a **butterfly spread**: for example, for state 1,

$$(\mathbf{x}_0 - \mathbf{x}_1) - (\mathbf{x}_1 - \mathbf{x}_2) \tag{16.24}$$

yields the payoff

$$\left(\begin{bmatrix} 1 \\ 2 \\ 3 \\ \vdots \\ S \end{bmatrix} - \begin{bmatrix} 0 \\ 1 \\ 2 \\ \vdots \\ S-1 \end{bmatrix} \right) - \left(\begin{bmatrix} 0 \\ 1 \\ 2 \\ \vdots \\ S-1 \end{bmatrix} - \begin{bmatrix} 0 \\ 0 \\ 1 \\ \vdots \\ S-2 \end{bmatrix} \right) = \begin{bmatrix} 1 \\ 1 \\ 1 \\ \vdots \\ 1 \end{bmatrix} - \begin{bmatrix} 0 \\ 1 \\ 1 \\ \vdots \\ 1 \end{bmatrix} = \begin{bmatrix} 1 \\ 0 \\ 0 \\ \vdots \\ 0 \end{bmatrix} \qquad (16.25)$$

i.e. this replicating portfolio pays 1 if aggregate consumption is 1, and 0 otherwise. The prices of this, and the other elementary claims, must, by the no-arbitrage principle, equal the prices of the corresponding replicating portfolios.

This gets more complicated when the values that aggregate consumption can take on, $y_1, y_2, \ldots, y_S$, say, are not just a sequence of consecutive integers. We must work from the payoff matrix

$$\mathbf{Y} = \begin{bmatrix} y_1 & 0 & 0 & \cdots & 0 \\ y_2 & y_2 - y_1 & 0 & \cdots & 0 \\ y_3 & y_3 - y_1 & y_3 - y_2 & \cdots & 0 \\ \vdots & \vdots & \vdots & \ddots & \vdots \\ y_S & y_S - y_1 & y_S - y_2 & \cdots & y_S - y_{S-1} \end{bmatrix} \qquad (16.26)$$

to find a portfolio $\mathbf{w}_k = (w_{1k}, w_{2k}, \ldots, w_{Sk})$ of options whose payoffs satisfy

$$\mathbf{Y}\mathbf{w}_k = \mathbf{e}_k \qquad (16.27)$$

or, with δ_{jk} denoting the Kronecker delta,

$$\sum_{j=1}^{S} y_{ij} w_{jk} = \delta_{ik} \qquad (16.28)$$

or, letting $\mathbf{W}$ denote the matrix we are looking for, $\mathbf{Y}\mathbf{W} = \mathbf{I}$ or just $\mathbf{W} = \mathbf{Y}^{-1}$.

16.4 The expected-utility paradigm

16.4.1 Background

In the previous section, certain functional forms were proposed for utility functions in a world of uncertainty, and it was seen that assuming additional structure on the utility function allowed more powerful conclusions to be drawn. In this section, we investigate whether there might be some axiomatic grounding behind the assumption of such functional forms.

Consider a general lottery $\tilde{x}$ with n different possible payoffs $x_1, x_2, \ldots, x_n$ and associated probabilities $p_1, p_2, \ldots, p_n$.

Specifically, we would like to know whether we can represent the utility that someone derives from this lottery by an **expected-utility function**:

$$u(\tilde{x}) = p_1 v(x_1) + p_2 v(x_2) + \cdots + p_n v(x_n) = E[v(\tilde{x})] \qquad (16.29)$$

where v again represents a utility function on sure things. We will show that, under certain axioms about individual preferences, we can do so.

Related questions concern the individual response to an **actuarially fair gamble**, i.e. to a choice between

- a certain wealth, W, and
- a lottery, $\tilde{x}$, with expected value

$$W = p_1 x_1 + p_2 x_2 + \cdots + p_n x_n = E[\tilde{x}] \tag{16.30}$$

An individual who prefers the lottery to the certain amount is said to accept the actuarially fair gamble.

Under what assumptions do individuals rank risky investment opportunities by expected value? How are risky investment opportunities ranked, if not by expected value? The sections that follow will consider these questions.

16.4.2 Definition of expected utility

The objects of choice with which we are concerned in a world with uncertainty could still be called consumption plans, as in Section 12.2.3, but we will acknowledge the additional structure now described by referring to them instead as lotteries, as in Section 13.3.1.

If there are k physical commodities, a consumption plan must specify a k-dimensional vector, $\mathbf{x} \in \mathbb{R}^k$, for each time and state of the world. We assume a finite number of times, $t = 0, 1, 2, \ldots, T$, say. The possible states of the world are denoted by the set Ω. So a consumption plan or lottery is just a collection of k-dimensional random vectors ($T + 1$ of them), i.e. a vector stochastic process. For simplicity, we will assume from now on that $k = 1$, and that there are at most two time periods, $t = 0$ and $t = 1$.

Again, to distinguish the certainty and uncertainty cases, we let $\mathcal{L}$ denote the collection of lotteries under consideration; $\mathcal{X}$ will now denote the set of possible values of the lotteries in $\mathcal{L}$. Preferences are now described by a relation on $\mathcal{L}$. We will continue to assume that preference relations are complete, reflexive, transitive and continuous.

Although we have moved from a finite-dimensional to an infinite-dimensional problem by explicitly allowing a continuum of states of nature, it can be shown that the earlier theory of choice under certainty carries through to choice under uncertainty; in particular, a preference relation can always be represented by a continuous utility function on $\mathcal{L}$; see Theorem 16.4.1.

However, we would like utility functions to have a stronger property than continuity, namely, the expected-utility property.

DEFINITION 16.4.1 Let $\mathcal{L}$ denote the set of all random variables (or lotteries) with values in the (one-dimensional) consumption set $\mathcal{X}$ and let $u: \mathcal{L} \to \mathbb{R}$ be a utility function representing the preference relation $\succeq$.

Then $\succeq$ is said to have an **expected-utility representation** if there exists a utility function on sure things, $v: \mathcal{X} \to \mathbb{R}$, such that

$$u(\tilde{x}) = E[v(\tilde{x})]$$
$$= \int v(x) \, dF_{\tilde{x}}(x) \tag{16.31}$$

Such a representation will often be called a **von Neumann–Morgenstern** (or VNM) **utility function**, after its originators,[2] or just an **expected-utility function**.

Preferences with such a representation will be called **expected-utility preferences**.

The set $\mathcal{X}$ can be identified with a subset of $\mathcal{L}$, in that each *sure thing* in $\mathcal{X}$ can be identified with the trivial lottery that pays off that sure thing with probability 1. Similarly, the utility function on sure things v, with domain $\mathcal{X}$, can be identified with the utility function on lotteries u, with domain $\mathcal{L}$, since they have identical values on $\mathcal{X}$, where both are defined. In other words, v is just the restriction $u|_{\mathcal{X}}$. Henceforth, we will use a single letter to denote both functions.

Any strictly increasing transformation of a VNM utility function represents the same preferences. However, only strictly increasing affine transformations, i.e. transformations of the form

$$f(x) = a + bx \quad (b > 0) \tag{16.32}$$

retain the expected-utility property. Proof of this is left as an exercise; see Exercise 16.5.

16.4.3 Further axioms

In this section, we add three further axioms to the six listed in Section 12.3. The first of these axioms is motivated by the following thought experiment:

- Suppose you think a Lotto ticket is just worth €1.
- Which do you prefer:

  —tossing a fair €1 coin with the €1 coin as a prize if you call the outcome correctly, or
  —tossing a fair €1 coin with this Lotto ticket as a prize if you call the outcome correctly?

- Substitute your own value for the Lotto ticket and ask yourself the same question.

The substitution axiom (below) implies that you will be indifferent; see Exercise 16.6.

AXIOM 7 (SUBSTITUTION OR INDEPENDENCE AXIOM). *If $a \in (0, 1]$ and $\tilde{p} \succ \tilde{q}$, then*

$$a\tilde{p} \oplus (1-a)\tilde{r} \succ a\tilde{q} \oplus (1-a)\tilde{r} \quad \forall \tilde{r} \in \mathcal{L} \tag{16.33}$$

The next axiom is just a generalization of the continuity axiom.[3]

AXIOM 8 (ARCHIMEDEAN AXIOM). *If $\tilde{p} \succ \tilde{q} \succ \tilde{r}$ then*

$$\exists\, a, b \in (0, 1) \quad s.t. \ a\tilde{p} \oplus (1-a)\tilde{r} \succ \tilde{q} \succ b\tilde{p} \oplus (1-b)\tilde{r} \tag{16.34}$$

The last axiom is a further generalization of the substitution axiom.

AXIOM 9 (SURE-THING PRINCIPLE). *If probability is concentrated on a set of sure things that are preferred to $\tilde{q}$, then the associated consumption plan is also preferred to $\tilde{q}$. In other words, if $y \succeq \tilde{q}$ for all $y \in \mathcal{Y}$ and $\tilde{p} \in \mathcal{Y}$ with probability 1, then $\tilde{p} \succeq \tilde{q}$.*

16.4.4 The Allais paradox

Now let us consider the **Allais paradox**,[4] which shows that the above axioms (in particular, the substitution axiom) applied to particular (compound) lotteries do not always give the results suggested by intuition.

Before reading on, consider carefully which you would prefer from each of the following three pairs of lotteries, and make a note of your choices:

$$1 \times €1m \quad \text{or} \quad 0.1 \times €5m \oplus 0.89 \times €1m \oplus 0.01 \times €0? \quad (16.35)$$

$$1 \times €1m \quad \text{or} \quad \tfrac{10}{11} \times €5m \oplus \tfrac{1}{11} \times €0? \quad (16.36)$$

$$0.11 \times €1m \oplus 0.89 \times €0 \quad \text{or} \quad 0.1 \times €5m \oplus 0.9 \times €0? \quad (16.37)$$

The reader, having first noted his or her instinctive preferences, is encouraged then to investigate the mean, variance, skewness and kurtosis of the payoffs for each of the five lotteries in the Allais paradox; see Exercise 16.7. Experiments based on these and similar choices generally find that the preferences of a significant proportion of the population violate the substitution axiom.

We will now show that the substitution axiom (assuming that it holds) and the answer to the previous question are sufficient in each case above to determine the answer to the next question. Suppose that, in the case of (16.35), an individual prefers the lottery with the higher expected payoff:

$$0.1 \times €5m \oplus 0.89 \times €1m \oplus 0.01 \times €0 \succ 1 \times €1m \quad (16.38)$$

Note that

$$0.11 \times \left(\frac{10}{11} \times €5m \oplus \frac{1}{11} \times €0 \right) \oplus 0.89 \times €1m$$

$$= 0.1 \times €5m \oplus 0.89 \times €1m \oplus 0.01 \times €0 \quad (16.39)$$

and

$$0.11 \times (1 \times €1m) \oplus 0.89 \times €1m = 1 \times €1m \quad (16.40)$$

Thus the substitution axiom implies that an individual prefers the uncertain payoff in the case of (16.35) if and only if (s)he also prefers the uncertain payoff in the case of (16.36); see Exercise 16.8.

Finally, note that each lottery in (16.37) represents an 11% chance of the corresponding lottery in (16.36) and an 89% chance of nothing. Thus, by the substitution axiom again, for the individual who prefers the uncertain payoffs in the first two situations,

$$0.1 \times €5m \oplus 0.9 \times €0 \quad \succ \quad 0.11 \times €1m \oplus 0.89 \times €0 \quad (16.41)$$

While the substitution axiom ensures that an individual who prefers the higher expected payoff in any one of these situations also does so in the other two situations, empirical trials have generally found that many individuals prefer the higher expected payoff in one or two cases, but not in all three.

If the restrictions implied by the substitution axiom appear counter-intuitive, then the entire theory of expected utility based on it should be treated with caution. However, it is still a useful starting point and a basis for comparative analysis of competing theories.[5]

16.4.5 Existence of expected-utility functions

We will now consider necessary and sufficient conditions on preference relations for an expected-utility representation, as described in Definition 16.4.1, to exist.

THEOREM 16.4.1 *If $\mathcal{X}$ contains only a finite number of possible values, then the substitution and Archimedean axioms are necessary and sufficient for a preference relation to have an expected-utility representation.*

Proof: We will just sketch the proof that the axioms imply the existence of an expected-utility representation. The proof of the converse is left as an exercise; see Exercise 16.9. For full details, see Huang and Litzenberger (1988, Sections 1.9 and 1.10 and Exercises 1.3 and 1.4).

Since $\mathcal{X}$ is finite, and unless the consumer is indifferent among all possible choices, there must exist maximal and minimal sure things, say, p^+ and p_-, respectively. By the substitution axiom, and a simple inductive argument, these are maximal and minimal in $\mathcal{L}$ as well as in $\mathcal{X}$. (If $\mathcal{X}$ is not finite, then an inductive argument can no longer be used and the sure-thing principle is required.)

From the Archimedean axiom, it can be deduced that, for every other lottery, $\tilde{p}$, there exists a unique $V(\tilde{p})$ such that

$$\tilde{p} \sim V(\tilde{p})p^+ \oplus (1 - V(\tilde{p}))p_- \tag{16.42}$$

It is easily seen that V then represents $\succeq$ in the sense defined by (12.1).

We leave it as an exercise to deduce from the axioms that, if $\tilde{x} \sim \tilde{y}$ and $\tilde{z} \sim \tilde{t}$, then, for all $\pi \in [0, 1]$,

$$\pi\tilde{x} \oplus (1 - \pi)\tilde{z} \sim \pi\tilde{y} \oplus (1 - \pi)\tilde{t} \tag{16.43}$$

see Exercise 16.10. It remains to show that V is linear in probabilities.

Define $\tilde{z} \equiv \pi\tilde{x} \oplus (1 - \pi)\tilde{y}$. Then, using the definitions of $V(\tilde{x})$ and $V(\tilde{y})$ and (16.43),

$$\begin{aligned}
\tilde{z} &\sim \pi\tilde{x} \oplus (1 - \pi)\tilde{y} \\
&\sim \pi(V(\tilde{x})p^+ \oplus (1 - V(\tilde{x}))p_-) \oplus (1 - \pi)(V(\tilde{y})p^+ \oplus (1 - V(\tilde{y}))p_-) \\
&= (\pi V(\tilde{x}) + (1 - \pi)V(\tilde{y}))p^+ \oplus (\pi(1 - V(\tilde{x})) + (1 - \pi)(1 - V(\tilde{y})))p_-
\end{aligned} \tag{16.44}$$

Also

$$\tilde{z} \sim V(\tilde{z})p^+ \oplus (1 - V(\tilde{z}))p_- \tag{16.45}$$

It follows by uniqueness of $V(\tilde{z})$ that

$$V(\tilde{z}) = V(\pi\tilde{x} \oplus (1 - \pi)\tilde{y}) = \pi V(\tilde{x}) + (1 - \pi)V(\tilde{y}) \tag{16.46}$$

This shows linearity for compound lotteries with only two possible outcomes: by an inductive argument, every lottery can be reduced recursively to a two-outcome lottery when there is only a finite number of possible outcomes altogether.   □

THEOREM 16.4.2 *For more general $\mathcal{L}$, to these conditions must be added some technical conditions and the sure-thing principle.*

Proof: A proof of this more general theorem can be found in Fishburn (1970).   □

Note that expected utility depends only on the distribution function of the consumption plan. Two consumption plans having very different consumption patterns across states of nature but the same probability distribution give the same utility, e.g. if wet days and dry days are equally likely, then an expected-utility maximizer is indifferent between any consumption plan and the plan formed by switching consumption between wet and dry days.

The basic objects of choice under expected utility are not consumption plans themselves but classes of consumption plans with the same cumulative distribution function.

16.4.6   Common expected-utility functions

Some commonly used functional forms for the expected utility derived from a wealth of w include the following:

- affine utility

$$u(w) = a + bw, \quad b > 0 \tag{16.47}$$

- quadratic utility

$$u(w) = w - \frac{b}{2}w^2, \quad b > 0 \tag{16.48}$$

- logarithmic utility

$$u(w) = \ln(aw^b) = \ln a + b \ln w, \quad b > 0 \tag{16.49}$$

- negative-exponential utility

$$u(w) = -e^{-cw}, \quad c > 0 \tag{16.50}$$

- narrow-power utility

$$u(w) = \frac{B}{B-1} w^{1-1/B}, \quad w > 0, \ B > 0, \ B \neq 1 \tag{16.51}$$

- extended-power utility

$$u(w) = \frac{1}{B-1}(A + Bw)^{1-1/B}, \quad B > 0, \ A \neq 0, \ w > \max\left\{-\frac{A}{B}, 0\right\} \tag{16.52}$$

Note that increasing affine transformations can be applied to any of these functions to obtain alternative representations of the same underlying preferences, while retaining the expected-utility property. In particular, in different circumstances, the extended-power utility function may be more conveniently represented by $[1/(C+1)B](A+Bw)^{C+1}$; see Exercise 16.18 and Theorem 17.3.2.

EXAMPLE 16.4.1 Let us consider the problem faced by an investor with the negative-exponential expected-utility function (16.50) who can divide his initial wealth W_0 between a risk-free asset with return r_f and a risky asset with normally distributed return $\tilde{r} \sim N(\mu, \sigma^2)$. If he invests b (euro) in the risky asset, then his final wealth will be $W_0 r_f + b(\tilde{r} - r_f)$. So he will choose b to maximize

$$E[-\exp(-c(W_0 r_f + b(\tilde{r} - r_f)))]$$

$$= -\frac{1}{\sqrt{2\pi}\sigma} \int_{-\infty}^{\infty} \exp(-c(W_0 r_f + b(r - r_f))) \exp\left(\left(-\frac{1}{2}\right)\left(\frac{r-\mu}{\sigma}\right)^2\right) dr \quad (16.53)$$

using the normal pdf given in (13.6). We will not re-evaluate this integral, since we know that for any normally distributed random variable, $\tilde{x}$,

$$E[e^{\tilde{x}}] = e^{E[\tilde{x}]+0.5\text{Var}[\tilde{x}]} \quad (16.54)$$

see Exercise 13.14. Thus the expression to be maximized in this case is

$$- \exp(-c(W_0 r_f + b(\mu - r_f)) + 0.5c^2 b^2 \sigma^2) \quad (16.55)$$

This is a decreasing transformation of a quadratic function of b, so has its maximum at the turning point of the quadratic, which is

$$b = \frac{c(\mu - r_f)}{c^2\sigma^2} = \frac{\mu - r_f}{c\sigma^2} \quad (16.56)$$

Note that the optimal investment in the risky asset is independent of initial wealth W_0. ◇

For another example of the maximization of expected utility, see Exercise 16.14.

16.5 Risk aversion

We have already looked at several examples of individual choice between two investment strategies with uncertain outcomes, and of choice between a strategy with a certain outcome and one with an uncertain outcome. In this section, we introduce a formal classification of the possible responses to such choices.

DEFINITION 16.5.1

(a) An individual is **risk-averse** if he or she is unwilling to accept, or indifferent to, **any** actuarially fair gamble.
(b) An individual is **strictly risk-averse** if he or she is unwilling to accept any actuarially fair gamble.

(c) An individual is **risk-neutral** if he or she is indifferent to any actuarially fair gamble.
(d) An individual is **risk-loving** if he or she is willing to accept, or indifferent to, any actuarially fair gamble.
(e) An individual is **strictly risk-loving** if he or she is willing to accept any actuarially fair gamble.

We will also use the terms **risk aversion** and **risk neutrality** to describe the corresponding preferences and the resulting behaviour.

These definitions apply independently of the expected-utility axioms, but the remainder of the analysis here will assume these axioms. Let us consider an actuarially fair gamble involving a lottery with two possible outcomes

$$px_1 \oplus (1-p)x_2 \tag{16.57}$$

1. For a risk-averse individual with expected-utility function v,

$$v(px_1 + (1-p)x_2) \geq pv(x_1) + (1-p)v(x_2) \quad \forall\, x_1, x_2 \in \mathbb{R}, \ p \in (0,1) \tag{16.58}$$

i.e. the expected-utility function of a risk-averse individual is a *concave* function; see Figure 16.1.

2. For a risk-loving individual with expected-utility function u,

$$u(px_1 + (1-p)x_2) \leq pu(x_1) + (1-p)u(x_2) \quad \forall\, x_1, x_2 \in \mathbb{R}, \ p \in (0,1) \tag{16.59}$$

i.e. the expected-utility function of a risk-loving individual is a *convex* function.

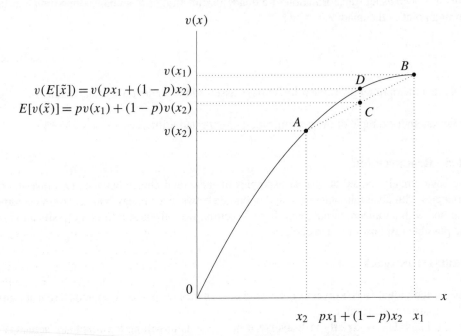

Figure 16.1 Concave expected utility function of a risk-averse individual

3. For a risk-neutral individual with expected-utility function f,

$$f(px_1 + (1-p)x_2) = pf(x_1) + (1-p)f(x_2) \quad \forall \, x_1, x_2 \in \mathbb{R}, \; p \in (0,1) \qquad (16.60)$$

i.e. the expected-utility function of a risk-neutral individual is an affine function.
4. A risk-neutral investor will pay exactly the expected value for any gamble; assuming that utility is increasing in wealth, a risk-averse investor will pay less; a risk-loving investor will pay more.

By appealing to Jensen's inequality (see Section 13.10), we can assert the converse of each of the above statements. For example, an individual with a concave expected-utility function will always exhibit risk-averse behaviour, and so on. Similarly, an individual with expected-utility preferences exhibits *strictly* risk-averse behaviour if and only if his VNM utility function is *strictly* concave, and so on. Note that increasing affine transformations of expected-utility functions preserve not only the expected-utility property of the original function but also its convexity or concavity.

Risk-neutral preferences can be represented by any expected-utility function of the form $f(\tilde{w}) = E[a + b\tilde{w}]$. The same preferences can be represented by any increasing affine transformation of this function, in particular by $g = -(a/b) + (1/b)f$ or by

$$g(\tilde{w}) = -\frac{a}{b} + \frac{1}{b}f(\tilde{w})$$

$$= -\frac{a}{b} + \frac{1}{b}E[a + b\tilde{w}]$$

$$= E[\tilde{w}] \qquad (16.61)$$

In other words, an individual has risk-neutral preferences if and only if his or her preferences rank all available options by expected value.

On the other hand, the utility functions representing different types of risk-averse (risk-loving) preferences can take any of the countless functional forms that are concave (convex) but are not increasing affine transformations of each other.

Thus, the term risk-neutral immediately specifies the functional form of the underlying expected-utility function; but the terms risk-averse and risk-loving each allow for many inherently different underlying preferences and expected-utility functions.

Our knowledge of the properties of concave and convex functions allows us to determine attitudes to risk by considering the second derivative of the underlying expected-utility function.

Most functions do not fall into any of these categories, and represent behaviour that is *locally risk-averse* at some wealth levels and *locally risk-loving* at other wealth levels.

To see this, just consider the obvious examples in everyday life of individuals who simultaneously gamble and buy insurance. Gambling in general involves the consumer in accepting an actuarially fair (or, more often, unfavourable) gamble, which is risk-loving behaviour; insurance involves the insurance company accepting a gamble that is actuarially favourable to it and thus involves the consumer in declining a gamble that is actuarially fair (or, more often, favourable) to him, which is risk-averse behaviour.

However, in most of what follows we will find it convenient to assume that individuals are globally risk-averse. Individuals who are globally risk-averse will never gamble, in the sense that they will never have a bet unless they believe that the expected return on the bet is positive. Thus assuming global risk aversion (and rational expectations) rules out the existence of national and state lotteries (except with large rollovers) and most other forms of betting and gaming.

We can distinguish more formally between local and global risk aversion.

DEFINITION 16.5.2 An individual with VNM utility function u is **locally risk-averse** at w if $u''(w) < 0$ and **globally risk-averse** if $u''(w) < 0$ for all w.

Individuals who gamble are not globally risk-averse but may still be locally risk-averse around their current wealth level. We will return to these concepts in Section 17.3.2. Some people (i.e. some preference relations) are more risk-averse than others; some functions are more concave than others. We will now introduce one method of measuring these differences; see Varian (1992, Sections 11.5–11.7) for more details.

The importance and usefulness of the Arrow–Pratt measures of risk aversion that we now define will become clearer as we proceed, in particular from the analysis of the portfolio choice problem.[6]

DEFINITION 16.5.3 The **Arrow–Pratt coefficient of absolute risk aversion** associated with expected-utility preferences represented by the expected-utility function u is

$$R_A(w) = -u''(w)/u'(w) \qquad (16.62)$$

which is the same for u and $au + b$, i.e. which is invariant under any affine transformation of u (increasing or decreasing).

Note that absolute risk aversion varies with the level of wealth. The second derivative, $u''(w)$, alone is meaningless as a measure of risk aversion, since it is not invariant under increasing affine transformations of u: u (and, hence, u' and u'') can be multiplied by any positive constant and still represent the same preferences. However, the above ratio is independent of the expected-utility function chosen to represent the underlying preferences.

DEFINITION 16.5.4 The **Arrow–Pratt coefficient of relative risk aversion** is

$$R_R(w) = w R_A(w) \qquad (16.63)$$

DEFINITION 16.5.5 The utility function u exhibits **increasing (constant, decreasing) absolute risk aversion** (IARA, CARA, DARA) if and only if

$$R'_A(w) > (=, <) 0 \quad \forall w \qquad (16.64)$$

DEFINITION 16.5.6 The utility function u exhibits **increasing (constant, decreasing) relative risk aversion** (IRRA, CRRA, DRRA) if and only if

$$R'_R(w) > (=, <) 0 \quad \forall w \qquad (16.65)$$

Note that:

- CARA or IARA $\Rightarrow$ IRRA;
- CRRA or DRRA $\Rightarrow$ DARA.

Note again that many expected-utility functions will not belong to any of these six categories, although all the examples that we have considered so far do.

Some examples of utility functions and their risk measures follow; see Exercise 16.18 for others.

EXAMPLE 16.5.1 For the negative-exponential expected-utility function, we have the following:

$$u(w) = -e^{-bw}, \quad b > 0 \tag{16.66}$$

$$u'(w) = be^{-bw} > 0 \tag{16.67}$$

$$u''(w) = -b^2 e^{-bw} < 0 \tag{16.68}$$

$$R_A(w) = b \tag{16.69}$$

$$R_R(w) = bw \tag{16.70}$$

$$R'_A(w) = 0 \tag{16.71}$$

$$R'_R(w) = b \tag{16.72}$$

In other words, these preferences exhibit CARA and IRRA. Because it exhibits constant absolute risk aversion and also because the corresponding utility-maximization problem is easily solved when wealth is normally distributed (see Example 16.4.6), the negative-exponential expected-utility function has been used as the basis of various empirical studies of risk aversion. It was also used by Grossman and Stiglitz (1989) in demonstrating the impossibility of informationally efficient markets. ◇

EXAMPLE 16.5.2 For the narrow-power expected-utility function, we have the following:[7]

$$u(w) = \frac{B}{B-1} w^{1-1/B}, \quad w > 0, \ B > 0, \ B \neq 1 \tag{16.73}$$

$$u'(w) = w^{-1/B} \tag{16.74}$$

$$u''(w) = -\frac{1}{B} w^{-1/B-1} \tag{16.75}$$

$$R_A(w) = \frac{1}{B} w^{-1} \tag{16.76}$$

$$R_R(w) = \frac{1}{B} \tag{16.77}$$

$$R'_A(w) = -\frac{1}{B} w^{-2} < 0 \tag{16.78}$$

$$R'_R(w) = 0 \tag{16.79}$$

In other words, these preferences exhibit CRRA and DARA. ◇

By integrating back from the defining equations

$$u''(w) = -cu'(w) \tag{16.80}$$

and

$$wv''(w) = -kv'(w) \tag{16.81}$$

respectively, it is possible to specify all the VNM utility functions exhibiting CARA and all those exhibiting CRRA; see Exercise 16.23.

16.6   Arbitrage, risk neutrality and the efficient markets hypothesis

In Section 12.5.6, we referred to the strong parallels between the no-arbitrage principle, the risk-neutral world and the efficient markets hypothesis. We are now in a position to define precisely the second of these concepts and, as promised in Section 13.10, to attempt to make the concept of the EMH more rigorous, so enabling us to spell out more clearly the parallels between the three concepts.

In a simple world of certainty, the no-arbitrage principle is a very useful tool for making predictions about the relationships between different rates of return. More generally, the no-arbitrage principle says that *if even one market participant has preferences exhibiting local non-satiation*, then securities with equal payoffs at all times and in all states of nature will have the same price.

If $\widetilde{P}_{it}$ denotes the value of security i at time t (omitting the tilde if the value is known with certainty, for example, at $t = 0$), then the no-arbitrage principle says that

$$\widetilde{P}_{it} = \widetilde{P}_{jt} \quad \text{with probability } 1 \; \forall \, t > 0 \; \Rightarrow \; P_{i0} = P_{j0} \tag{16.82}$$

In an equally simple world of uncertainty in which unlimited short-selling is possible, it is easy to derive a similar result, namely, that *if even one market participant is risk-neutral*, then all securities will have equal expected returns, or, in terms of security prices,

$$E[\widetilde{P}_{it}] = E[\widetilde{P}_{jt}] \quad \forall \, t > 0 \Rightarrow P_{i0} = P_{j0} \tag{16.83}$$

To see this, note that if two securities have different expected returns, then any risk-neutral individual will seek to exploit the discrepancy in expected returns on an infinite scale, by short-selling the asset with the lower expected return and investing the proceeds in the asset with the higher expected return, thereby increasing his or her expected future wealth, and thus his or her utility, without bound.

As in the case of an arbitrage opportunity, the risk-neutral individual's demand for the securities with different expected returns is not well-defined, i.e. not finite. If even one individual in such an economy has risk-neutral preferences, it follows that prices that permit differences in expected returns will not allow markets to clear. Conversely, if such an economy is in equilibrium, then markets must clear, demand for all securities must be finite and prices will adjust in equilibrium so that all securities have equal expected returns.

In summary, then, in equilibrium we must have at least one of the following:

- no risk-neutral individual; or
- restrictions on short-selling; or
- all expected returns equal.

We will refer to the hypothetical economy satisfying only the last of these conditions as the **risk-neutral world**. In equilibrium in such a world, the risk-neutral investor will be indifferent between all possible investment positions. One possible equilibrium allocation would be to share whatever was not demanded by non-risk-neutral investors on a *pro rata* basis among the risk-neutral investors according to the equilibrium values of their endowments.

While the risk-neutral world clearly does not correspond to the real world of apparently widely varying expected returns that we see around us, it (like other hypothetical but unrealistic concepts such as perfect competition) remains a very useful pedagogic benchmark against which to compare other economic models.

While the risk-neutral world does not in any way represent the behaviour of real-world long-horizon markets, it more closely represents what happens in short-horizon markets, such as betting markets, where security prices must converge quickly to their fundamental, predetermined maturity values.

In the more complicated world in which we live, the efficient markets hypothesis plays a role somewhat akin to those of the no-arbitrage principle and the risk-neutral world. The EMH can be looked at from either a comparative static or a dynamic perspective. We will begin with the former. One interpretation of the EMH from this perspective is that all securities will have the same expected returns *after adjustment for risk*. In other words, there may exist a risk-neutral measure with expectation operator E^* such that, in terms of security prices,

$$E^*[\widetilde{P}_{it}] = E^*[\widetilde{P}_{jt}] \; \forall \, t > 0 \; \Rightarrow \; P_{i0} = P_{j0} \tag{16.84}$$

The adjustment for risk depends on assumptions made about investor preferences and about the investment opportunity set, so it is impossible to test the EMH independently of the model of equilibrium underlying the risk-neutral measure; see Campbell et al. (1997, Section 1.5).

One such model of equilibrium is the capital asset pricing model (CAPM), which will be introduced in Section 17.5. If the assumptions of this model held, then, as we will see, the risk-neutral measure would be defined by

$$E^*[\widetilde{P}_{it}] = \frac{E[\widetilde{P}_{it}]}{1 + r_{\mathrm{f}} + \beta_{im} E[\widetilde{r}_m - r_{\mathrm{f}}]} \tag{16.85}$$

see Exercise 17.5.

The EMH, as viewed from a dynamic perspective, has also been interpreted in a multitude of different ways, all of which say something about the predictability of asset returns. For a full discussion, see Campbell et al. (1997, Chapter 2). If asset prices at time t fully reflect all available information, as proposed by the EMH, then future values of the stochastic process $\{\widetilde{P}_t\}$, say, followed by an asset price should not be predictable on the basis of its own past values. As mentioned in Sections 3.4, 8.3 and 14.4.1, this hypothesis has been made more precise by arguing that, and testing whether, the price process follows a random walk with iid increments, or some other process with a unit root. (Recall that $\widetilde{P}_t$ follows a random walk if

$$\widetilde{P}_t = \widetilde{P}_{t-1} + \widetilde{\varepsilon}_t \tag{16.86}$$

where $\widetilde{\varepsilon}_t \sim \mathrm{iid}(0, \sigma^2)$.)

16.7   Uncovered interest rate parity: Siegel's paradox revisited

Just as the no-arbitrage principle leads to the widely accepted principle of covered interest rate parity (Example 12.5.6), so we will now demonstrate that in the risk-neutral world (whatever about the real world) the more debatable principle of **uncovered interest rate parity** holds.

Hitherto, there has been no ambiguity as to the meaning of risk neutrality, as we have assumed the existence of a common unit of account that can be used as a numeraire in which to measure the initial prices and final values of all assets. We have not made it explicit so far that the idea of the risk-neutral world is based on the assumption that an individual is risk-neutral *in a specific currency* or, more generally, with respect to a specific numeraire. We will now show the relevance of the **numeraire currency** in this situation. An investor is risk-neutral with respect to a given numeraire currency if his objective is to maximize the expected value of his final wealth measured in terms of that numeraire. We referred to the numeraire initially as the "euro", implying that we were dealing with an unambiguous single-currency world. Now let us consider instead a sterling-denominated risk-neutral world, in which future exchange rates are uncertain, so that the euro is a risky asset.

Suppose, as in the case of covered interest rate parity, that an individual has access to one-period risk-free investments in both EUR and GBP at interest rates, respectively, of i_{EUR} and i_{GBP}. Let S_t denote the current spot GBP/EUR exchange rate and $\widetilde{S}_{t+1}$ the unknown spot GBP/EUR exchange rate next period.

A sterling payoff next period can be engineered from a sterling investment – of, say, one pound – in this period in either of two ways:

1. invest the principal in the risk-free sterling investment, for a return of $1 + i_{GBP}$; or
2. convert the principal to EUR $1/S_t$, invest the proceeds in the risk-free euro investment for a euro return of $(1 + i_{EUR})/S_t$ and convert this euro return back to sterling for a sterling return of $(1 + i_{EUR})\widetilde{S}_{t+1}/S_t$.

In a risk-neutral world (where at least one individual has risk-neutral preferences *for sterling payoffs*), these two strategies must have the same expected payoff, so that

$$E[\widetilde{S}_{t+1}] = \frac{1 + i_{GBP}}{1 + i_{EUR}} S_t \tag{16.87}$$

Like the theory of covered interest rate parity, the theory of uncovered interest rate parity says that apparent gains in the money market when interest rates differ across currencies will be eroded by losses in the foreign exchange market, or, in other words, a currency with a higher interest rate will depreciate against one with a lower interest rate, at least insofar as *expected changes in spot rates are concerned*.

Combining the theories of covered and uncovered interest rate parity, it follows that when both hold

$$E[\widetilde{S}_{t+1}] = F_t = \frac{1 + i_{GBP}}{1 + i_{EUR}} S_t \tag{16.88}$$

where F_t again denotes the one-period forward GBP/EUR exchange rate.

Note that we could have based the preceding example on any two currencies. The critical assumption was that at least one investor had risk-neutral preferences *when payoffs were expressed in sterling*.

The counter-hypothesis that the forward EUR/GBP exchange rate equals the expected future spot EUR/GBP exchange rate holds in a euro-denominated risk-neutral world (in which sterling is a risky asset). Siegel's paradox (Theorem 13.10.2) is then seen to be no more than the statements that

- the world cannot simultaneously be risk-neutral from the perspective of both euro and sterling investors; or
- uncovered interest rate parity cannot hold in both currencies simultaneously.

Another way of saying this is that no equilibrium exists in a world with unlimited borrowing and short-selling and risk-neutral investors in more than one currency. Since risk-neutral preferences are not *strictly* convex, such a world does not satisfy the assumptions underlying Theorem 12.5.7, so Siegel's paradox is not such a puzzle as it seemed after all.

We conclude this section with an example to illustrate how the theory of uncovered interest rate parity unravels in a simple two-investor, two-asset, two-currency world in which there is *no borrowing or short-selling*.

EXAMPLE 16.7.1 Suppose that there are two investors living in two different currency zones with both currencies available for investment to both investors. They live in a single-period world where they can trade their endowments at $t = 0$ and consume their final wealth at $t = 1$. The payoffs at $t = 1$ on the two investments in their own respective currencies are known at $t = 0$ to be r_1 and r_2 but the exchange rate at $t = 1$ is unknown and will be determined exogenously. Thus each investor faces a choice between one asset that is risk-free in his own currency and one asset that is risky in his own currency.

We will assume that 1 unit of currency 1 can be exchanged for s units of currency 2 at $t = 0$.

To preserve symmetry in this new two-currency world, we will drop the assumption that a common numeraire currency exists, and will assume instead that each investor uses a different numeraire, namely one unit of his own currency. Furthermore, we will assume that each investor is risk-neutral with respect to his final wealth measured in terms of the relevant numeraire.

We assume first for complete generality that the two investors have their own beliefs concerning the exchange rate at $t = 1$, with $\tilde{s}_2$ denoting the final value of 1 unit of currency 2 from investor 1's perspective and vice versa for $\tilde{s}_1$.

Our objective now, following our approach in Section 12.6.1, is to construct an Edgeworth box, shown in Figure 16.2, to illustrate the nature of equilibrium in this world.

If individual 1 has endowments of e_1^1 and e_1^2, respectively, of the two assets and decides to hold a portfolio comprising x_1^1 and x_1^2, respectively, of the two assets, then the latter quantities will be chosen to maximize

$$E[x_1^1 r_1 + x_1^2 r_2 \tilde{s}_2] \tag{16.89}$$

subject to non-negativity constraints and the budget constraint

$$s x_1^1 + x_1^2 = s e_1^1 + e_1^2 \tag{16.90}$$

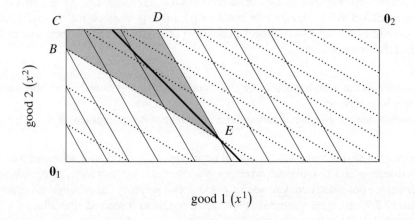

good 1 (x^1)

Figure 16.2 Edgeworth box diagram for Example 16.7

His indifference curves in the Edgeworth box are parallel lines of the form

$$x_1^1 r_1 + x_1^2 r_2 E[\tilde{s}_2] = k \qquad (16.91)$$

with common slope $-r_1/(r_2 E[\tilde{s}_2])$.

Similarly, individual 2's indifference curves in the Edgeworth box will be parallel lines of the form

$$x_2^1 r_1 E[\tilde{s}_1] + x_2^2 r_2 = k \qquad (16.92)$$

with common slope $-(r_1 E[\tilde{s}_1])/r_2$. (The slope will be the same whether calculated from the perspective of individual 1's origin or from the perspective of individual 2's origin.)

The budget lines for both individuals have slope $-s$.

The two individuals will have the same set of indifference curves if and only if

$$-\frac{r_1}{r_2 E[\tilde{s}_2]} = -\frac{r_1 E[\tilde{s}_1]}{r_2} \qquad (16.93)$$

or

$$E[\tilde{s}_1]E[\tilde{s}_2] = 1 \qquad (16.94)$$

Figure 16.2 is drawn on the assumption that

$$E[\tilde{s}_1]E[\tilde{s}_2] > 1 \qquad (16.95)$$

Suppose E in Figure 16.2 represents the endowment point, the dotted parallel lines represent individual 1's indifference curves, the normal parallel lines represent individual 2's indifference curves and the single thick line represents a common budget constraint. Then any point in the interior of the shaded region $BCDE$ Pareto dominates E. But all points in the interior of $BCDE$ are in turn Pareto dominated by some point on the boundary BCD.

It can be seen from Figure 16.2 that either there will be a corner equilibrium of one of four types or multiple interior equilibria. If (16.95) holds, as shown, then either:

(a) individual 1 will hold the whole aggregate endowment of asset 2 (solution on the top of the Edgeworth box); or
(b) individual 1 will hold none of asset 1 (solution on the left-hand side of the Edgeworth box).

If the opposite strict inequality to (16.95) holds, then either:

(c) individual 1 will hold none of asset 2 (solution along the bottom of the Edgeworth box); or
(d) individual 1 will hold the whole aggregate endowment of asset 1 (solution on the right-hand side of the Edgeworth box).

If (16.94) holds, then:

(e) the budget line and both individuals' indifference curves will have the same slope so that any point on the budget line maximizes both individuals' utility simultaneously.

If the two individuals have common beliefs, then we have $\tilde{s}_1 = 1/\tilde{s}_2$ and by Jensen's inequality

$$E[\tilde{s}_1] = E[\frac{1}{\tilde{s}_2}] > \frac{1}{E[\tilde{s}_2]} \tag{16.96}$$

Thus individual 2's indifference curves have the steeper slopes, as illustrated.[8]

In fact, the contract curve for the case illustrated comprises the left and top edges of the Edgeworth box. Given the endowment point E, there are multiple equilibria, any point on *BCD* being attainable given the appropriate budget line. Any budget line with slope steeper than individual 1's indifference curves and less steep than individual 2's indifference curves will yield an equilibrium somewhere on *BCD*. Thus the equilibrium exchange rate must satisfy

$$\frac{r_1}{r_2 E[\tilde{s}_2]} \leq s \leq \frac{r_1 E[\tilde{s}_1]}{r_2} \tag{16.97}$$

These limiting exchange rates are precisely those implied by uncovered interest rate parity for each individual, as given (in different notation) in (16.87). By (16.96), the upper and lower bounds for the equilibrium exchange rate in (16.97) are different.

Note also that individual 1, being risk-neutral in currency 1, will prefer currency 2 (his risky asset) to currency 1 (his risk-free asset) if and only if the former has the higher expected return, or

$$r_2 s E[\tilde{s}_2] \geq r_1 \tag{16.98}$$

Similarly, individual 2, being risk-neutral in currency 2, will prefer currency 1 (his risky asset) to currency 2 (his risk-free asset) if and only if

$$\frac{r_1 E[\tilde{s}_1]}{s} \geq r_2 \tag{16.99}$$

Inequalities (16.98) and (16.99), after rearrangement, are precisely the inequalities in the equilibrium condition (16.97), thus the exchange rate s supports an equilibrium if and only if it causes both investors simultaneously to view the foreign currency as having the same or higher expected return. Consequently, in equilibrium, each risk-neutral investor will seek to hold his entire wealth in foreign currency (subject to feasibility constraints). Thus the individual whose wealth is greater than the aggregate endowment of the foreign currency will hold the surplus in his own currency. The top-left corner of the Edgeworth box, where neither individual holds his own currency, represents the cross-over between these two types of equilibria.

We now have another interpretation of Siegel's paradox. If uncovered interest rate parity held simultaneously in both currencies, then the budget line would have to be simultaneously parallel to both BE and DE, which is impossible unless (16.94) holds. But (16.96) shows that this is inconsistent with rational expectations. What we have shown here is that the equilibrium exchange rate must lie between the two extremes given by uncovered interest rate parity in the two currencies.

The fact that the equilibrium in this example is always a corner solution coincides with the conclusion above that there is no equilibrium in a similar world with unlimited borrowing and short-selling. Allowing limited borrowing and short-selling would just increase the dimensions of the Edgeworth box in Figure 16.2. Allowing unlimited borrowing and short-selling corresponds to allowing the boundary of the Edgeworth box, and thus demands, to extend to infinity.   ◇

16.8 Mean–variance paradigm

In the preceding sections, we analysed in some detail the behaviour of individuals who care only about the expected value of final wealth and the implications for equilibrium of that behaviour. The conclusion was that such behaviour does not very closely approximate to the real world.

A more plausible hypothesis is that an individual may care about both the expected value and the variance of his final wealth. This hypothesis is one possible way of reducing the problem of choice under uncertainty from a large dimension, namely, the number of risky assets N, to a small dimension, in this case two. Such parsimonious theories are more practical than general utility-maximization approaches. We will say that such an individual has **mean–variance preferences**.

Three more detailed arguments are commonly used to motivate the mean–variance framework for analysis of choice under uncertainty, in particular, for analysis of the portfolio choice problem to be considered in Chapter 17. These are as follows:

• quadratic expected utility

$$
\begin{aligned}
E[u(\widetilde{W})] &= E[\widetilde{W}] - \frac{b}{2} E[\widetilde{W}^2] \\
&= E[\widetilde{W}] - \frac{b}{2} ((E[\widetilde{W}])^2 + \mathrm{Var}[\widetilde{W}]) \\
&= u(E[\widetilde{W}]) - \frac{b}{2} \mathrm{Var}[\widetilde{W}]
\end{aligned}
\tag{16.100}
$$

- normally distributed wealth (or multivariate normally distributed asset returns)[9]

$$E[u(\widetilde{W})] = \frac{1}{\sqrt{2\pi \operatorname{Var}[\widetilde{W}]}} \int_{-\infty}^{\infty} u(w)\, e^{-0.5(w-E[\widetilde{W}])^2/\operatorname{Var}[\widetilde{W}]}\, dw \qquad (16.101)$$

- Taylor-approximated expected-utility functions (see Sections 9.6 and 13.9)

$$u(\widetilde{W}) = u(E[\widetilde{W}]) + u'(E[\widetilde{W}])(\widetilde{W} - E[\widetilde{W}])$$
$$+ \frac{1}{2}u''(E[\widetilde{W}])(\widetilde{W} - E[\widetilde{W}])^2 + \widetilde{R}_3 \qquad (16.102)$$

$$\widetilde{R}_3 = \sum_{n=3}^{\infty} \frac{1}{n!} u^{(n)}(E[\widetilde{W}])(\widetilde{W} - E[\widetilde{W}])^n \qquad (16.103)$$

which implies

$$E[u(\widetilde{W})] = u(E[\widetilde{W}]) + \frac{1}{2}u''(E[\widetilde{W}])\operatorname{Var}(\widetilde{W}) + E[\widetilde{R}_3] \qquad (16.104)$$

where

$$E[\widetilde{R}_3] = \sum_{n=3}^{\infty} \frac{1}{n!} u^{(n)}(E[\widetilde{W}]) m^n[\widetilde{W}] \qquad (16.105)$$

This expansion (which assumes that utility functions are continuously differentiable) yields further insights into the properties of utility functions in general and polynomial utility functions in particular. Dropping the remainder term in (16.104) gives a Taylor approximation for expected utility involving only the mean and variance of final wealth.

More generally, it follows that the sign of the nth derivative of the utility function at expected wealth determines the direction of preference for the nth central moment of the probability distribution of terminal wealth, *ceteris paribus*.[10] In fact, expected utility is monotone (or constant) in all central moments, *ceteris paribus*. This is true since it follows from Taylor's expansion that, for $n > 1$, expected utility is an increasing (decreasing, constant) function of the nth central moment of terminal wealth, *ceteris paribus*, whenever the nth derivative of the utility function at expected wealth is positive (negative, zero). In particular, risk aversion implies concave utility, which implies a negative second derivative, which implies variance aversion, and vice versa.

Similarly, a positive third derivative implies a preference for greater skewness. It can be shown fairly easily that an increasing utility function that exhibits non-increasing absolute risk aversion has a non-negative third derivative; see Exercise 16.20.

Note that the expected-utility axioms are neither necessary nor sufficient to guarantee that the Taylor approximation to the first few moments is an exact representation of the utility function.

Mean–variance preferences can be represented by indifference maps in mean–variance or mean–standard deviation space, just as preferences in a two-good economy are represented by similar indifference maps in commodity space. The mean is normally shown on the vertical axis and the variance (or standard deviation) on the horizontal axis; see Sections 17.4.2 and 17.4.4.

Risk-neutral preferences can also be represented by an indifference map in mean–variance space. Since the risk-neutral individual ranks all options by expected value or mean, his indifference curves will be a set of parallel horizontal lines in mean–variance (or mean–standard deviation) space.[11]

16.9   Other non-expected-utility approaches

Justifying the use of mean–variance preferences by assuming either quadratic utility or normally distributed wealth is not very satisfactory. Quadratic utility implies the existence of a bliss point and exhibits increasing absolute risk aversion (see Exercise 16.18), which, as we shall see in Section 17.3.2, also has unrealistic implications. Multivariate normally distributed asset returns are not possible since real-world asset returns are bounded below by zero, but the normal distribution is unbounded. (However, the theoretical device of allowing unlimited short sales somewhat deflates this argument.) Another argument against the normal distribution is that empirical research has shown that the skewness of returns is non-zero,[12] although zero skewness is an implication of normality, or of any symmetric distribution of returns. Thus, intuition and empirical evidence both suggest that higher moments than mean and variance are relevant.[13]

Those not happy with the explanations of choice under uncertainty provided within the expected-utility paradigm have proposed various other alternatives in recent years. As well as the extremely rigorous work of Machina (1982) cited in note 5 of this chapter, these include both qualitative approaches talking about fun and addiction and more formal approaches looking at other factors such as maximum or minimum possible payoffs and irrational expectations. They also include concepts such as state-dependent utility, discussed in Section 16.3.2. However, they are beyond the scope of this book.

To conclude this chapter, we remind readers to consider again their personal valuations of the various lotteries discussed in Sections 13.3.1 and 16.2 in the light of the various theories of choice under uncertainty subsequently discussed.

EXERCISES

16.1 Consider a two-period world in which three states of nature, with probabilities $\pi_1 = 0.2$, $\pi_2 = 0.3$ and $\pi_3 = 0.5$, respectively, might arise at date 1. Call the unit of consumption in this world a **euro** and suppose that the corresponding equilibrium state-contingent claim prices are, respectively, $\phi_1 = 0.1$, $\phi_2 = 0.3$ and $\phi_3 = 0.6$ (using date-0 consumption as the numeraire).

(a) What would be the equilibrium price of a risk-free asset if it were also traded?
(b) Consider an individual whose utility from consuming x_0 at date 0 and the prospect of consuming x_1 at date 1 if state 1 materializes, x_2 if state 2 materializes and x_3 if state 3 materializes is

$$u(x_0, x_1, x_2, x_3) = 1.1 \ln x_0 + \pi_1 \ln x_1 + \pi_2 \ln x_2 + \pi_3 \ln x_3$$

This individual now has €100 and will receive a further income of €1000 at date 1, but only if state 2 occurs. Find his optimal consumption pattern and the trades that he would have to make to achieve this optimum.

(c) How would this optimal consumption pattern change if there were no market for the state-contingent claim paying off in state 2, but the other state-contingent claim prices were the same?

(d) Now suppose that there are no state-contingent claims markets, but that call options can be bought and written on a state index portfolio that pays off €10 in state 1, €5 in state 2 and €20 in state 3. Find the option portfolio that is equivalent to the endowment in (b) and the option trades that would give the same optimal allocation across states as you calculated in (b).

16.2 Consider a two-period world in which two states of nature might arise at date 1, the "up" state, which has probability $p = 0.4$, or the "down" state. Two complex securities are traded in this world:

- a risk-free security with a payoff of $1 + r$ in both future states; and
- a risky security with a payoff of uS in the "up" state and a payoff of dS in the "down" state.

Suppose that the risk-free security has an equilibrium price of 1 and the risky security an equilibrium price of S.

(a) Use the no-arbitrage principle to deduce the equilibrium state prices.
(b) Calculate the payoffs in each state at date 1 on an option to buy the risky security at that date for a price of K. (Hint: consider separately the cases of $K \leq dS$, $dS \leq K \leq uS$ and $uS \leq K$.)
(c) Calculate the value at date 0 of such an option.
(d) How would the value of the option change if p changed to 0.7?
(e) What happens if $p = 1$?

16.3 Consider a world with n states of nature described by the sample space $\Omega = \{\omega_1, \omega_2, \ldots, \omega_n\}$. Suppose that the only securities traded are a state index portfolio with payoff y_i in state ω_i, $i = 1, 2, \ldots, n$, $y_1 < y_2 < \cdots < y_n$, and options on that state index portfolio with strike prices $y_1, y_2, \ldots, y_{n-1}$.

Find a portfolio of these traded securities that has the same payoffs as the state i state-contingent claim.

16.4 Show that, in the context of Section 16.3.5, equilibrium consumption decisions will be the same in any two states in which the aggregate consumption or endowment is the same.

16.5 In general, a strictly increasing transformation of a utility function represents the same underlying preferences.

(a) Show that a strictly increasing affine transformation of a von Neumann–Morgenstern utility function also has an expected-utility representation.
(b) Show that any other strictly increasing transformation of a VNM utility function loses the expected-utility representation.
(c) Show that a strictly increasing affine transformation of the underlying utility function for sure things gives another VNM utility function representing the same underlying preferences.

(d) Show that any other strictly increasing transformation of the underlying utility function for sure things gives a VNM utility function that represents different underlying preferences.

16.6 Deduce the following from the substitution axiom and the Archimedean axiom:

(a) $\tilde{p} \succ \tilde{q}$ and $0 \leq a < b \leq 1$ imply that $b\tilde{p} \oplus (1-b)\tilde{q} \succ a\tilde{p} \oplus (1-a)\tilde{q}$.
(b) $\tilde{p} \succeq \tilde{q} \succeq \tilde{r}$ and $\tilde{p} \succ \tilde{r}$ imply that there exists a unique $a^* \in [0, 1]$ such that $\tilde{q} \sim a^*\tilde{p} \oplus (1-a^*)\tilde{r}$.
(c) $\tilde{p} \succ \tilde{q}$ and $\tilde{r} \succ \tilde{s}$ and $a \in [0, 1]$ imply that $a\tilde{p} \oplus (1-a)\tilde{r} \succ a\tilde{q} \oplus (1-a)\tilde{s}$.
(d) $\tilde{p} \sim \tilde{q}$ and $a \in [0, 1]$ imply that $\tilde{p} \sim a\tilde{p} \oplus (1-a)\tilde{q}$.
(e) $\tilde{p} \sim \tilde{q}$ and $a \in [0, 1]$ imply that $a\tilde{p} \oplus (1-a)\tilde{r} \sim a\tilde{q} \oplus (1-a)\tilde{r}$, for all $\tilde{r} \in \mathcal{L}$.

16.7 Calculate the means, variances, standard deviations, skewness coefficients and kurtosis coefficients (as defined in Section 13.6) of the various lotteries involved in the Allais paradox, set out in Section 16.4.4.

Did you originally rank the lotteries by expected value?
If not, did you prefer the lottery with lower variance?

16.8 Assuming that the substitution axiom holds, show that

$$0.1 \times \text{€}5\text{m} \oplus 0.89 \times \text{€}1\text{m} \oplus 0.01 \times \text{€}0 \succ 1 \times \text{€}1\text{m}$$

if and only if

$$\frac{10}{11} \times \text{€}5\text{m} \oplus \frac{1}{11} \times \text{€}0 \succ 1 \times \text{€}1\text{m}$$

16.9 Show that, if the preference relation denoted by $\succeq$ has an expected-utility representation, then it must satisfy the substitution axiom and the Archimedean axiom.

16.10 Deduce from the Archimedean and substitution axioms that, if $\tilde{x} \sim \tilde{y}$ and $\tilde{z} \sim \tilde{t}$, then, for all $\pi \in [0, 1]$,

$$\pi\tilde{x} \oplus (1-\pi)\tilde{z} \sim \pi\tilde{y} \oplus (1-\pi)\tilde{t}$$

16.11 A coin, which has probability p of landing heads up, is tossed repeatedly. You are offered a lottery ticket that will pay you $\text{€}2^j$ if the first head occurs at the jth flip.

(a) What is the maximum amount of money that you personally would be willing to pay for this lottery ticket when $p = 1/2$?
(b) What is the expected value of this lottery ticket when $p = 1/2$?
(c) Consider an individual whose expected-utility function is $u(x) = \ln x$. Express the utility of this game to her as a sum.
(d) Evaluate the sum in part (c).
(e) What is the maximum amount of money that the individual with logarithmic utility would be willing to pay to participate in this game?

(This exercise describes the so-called **St Petersburg paradox**.)

16.12 Consider the following four lotteries:

L_1: €50 with 60% probability
L_2: €100 with 50% probability
L_3: €50 with 40% probability
L_4: €100 with 30% probability

Suppose an investor prefers L_1 to L_2 and L_4 to L_3. Are his preferences consistent with the axioms of expected utility? If so, why? If not, why not?

16.13 A person has an expected-utility function of the form $u(x) = \sqrt{x}$. She initially has wealth of €4. She has a lottery ticket that will be worth €12 with probability $1/2$ and will be worth nothing with probability $1/2$. What is her expected utility? What is the lowest price at which she would part with the ticket?

16.14 A consumer has a VNM expected-utility function given by

$$u(w) = \ln w$$

He is offered the opportunity to bet on the toss of a coin that has a probability π of coming up heads. If he bets €x, he will have €$(w + x)$ if a head comes up and €$(w - x)$ if tails comes up.
  Solve for his optimal choice of x as a function of π and w.
  What is the optimal choice of x when the coin is fair ($\pi = 0.5$)?

16.15 Assume that your preferences have an expected-utility representation, based on the utility function for sure things $u(w) = \ln w$, and that your current level of wealth is €5000.

(a) Suppose that you are exposed to a situation that results in a 50/50 chance of winning or losing €1000. If you can buy insurance that completely removes the risk for a premium of €125, will you buy it or take the gamble? Explain your answer.
(b) Suppose that you accept the gamble outlined in (a) and lose, so that your wealth is reduced to €4000. If you are faced with the same gamble again in the next period and have the same offer of insurance as before, will you buy the insurance the second time around? Explain your answer.

16.16 A consumer has an expected-utility function of the form

$$u(w) = -\frac{1}{w}$$

He is offered a gamble that gives him wealth of w_1 with probability p and w_2 with probability $1 - p$. What wealth would he need now to be just indifferent between keeping his current wealth or accepting this gamble?

16.17 Explain the implications of Jensen's inequality for the relationship between the risk aversion (or otherwise) of an investor and the concavity (or otherwise) of an expected-utility function representing her preferences.

16.18 Calculate the coefficients of absolute and relative risk aversion for each of the following utility functions:

(a) the quadratic utility function given by

$$u(w) = w - \frac{B}{2}w^2$$

(b) the extended power utility function given by

$$u(w) = \frac{1}{(C+1)B}(A + Bw)^{C+1}$$

(c) the logarithmic utility function given by

$$u(w) = \ln w$$

Determine for each utility function the restrictions on the parameters A and/or B and/or C and on the possible values of wealth w that are necessary to guarantee that it is a well-defined, strictly increasing, strictly concave, single-valued, real-valued function.

Finally, determine whether absolute and relative risk aversion are increasing, decreasing or constant for these functions.

16.19 Investigate carefully the limiting behaviour of the utility function in Exercise 16.18(b) and of the associated risk aversion measures at the "singularities" $B = 0$ and $C = -1$.

16.20 Show that a utility function that is increasing and exhibits non-increasing absolute risk aversion must have a non-negative third derivative.

Hence, show that an investor with such a utility function prefers the distribution of wealth to have higher skewness, *ceteris paribus*.

16.21 Consider two risky assets with gross returns represented by the random variables $\tilde{r}_1$ and $\tilde{r}_2 \equiv a\tilde{r}_1 + b$, where a and b are scalars with $a > 0$ and $a \neq 1$.

(a) Show that the returns on these assets are perfectly correlated.
(b) Show that the variances of the returns on the two assets are different.
(c) Find the weights with which these assets would have to be combined in a portfolio to attain a riskless return.
(d) Suppose investors are not allowed to short-sell any asset for more than the fraction γ of their initial wealth. Find the relation between a and γ that must be satisfied for the riskless portfolio above to be available.
(e) What is the equilibrium (arbitrage-free) rate of return on a riskless asset in this economy if there are no restrictions on short-selling?

16.22 Suppose that you are offered the chance to participate in a game where you can bet repeatedly on the toss of a fair coin and can stop at any time you want.

(a) What is the expected profit from the strategy of betting €2^{j-1} on heads on the jth toss, stopping after the first win?
(b) What is the variance of this profit?
(c) What is the expected number of tosses before you stop?
(d) Why would you or would you not use this strategy in practice?

16.23 Find all utility functions exhibiting constant absolute risk aversion and all those exhibiting constant relative risk aversion.

17 Portfolio theory

17.1 Introduction

Portfolio theory is an important topic in the theory of choice under uncertainty. It deals with the portfolio choice problem facing an investor who must decide how to distribute an initial wealth of, say, W_0 among a number of single-period investments. The choice of portfolio will depend on both the investor's preferences and his beliefs about the uncertain payoffs of the various securities.

The chapter begins by considering some issues of definition and measurement. Section 17.3 then looks at the portfolio choice problem in a general expected-utility context. Section 17.4 considers the same problem from a mean–variance perspective. This leads on to a discussion of the properties of equilibrium security returns in Section 17.5.

17.2 Preliminaries

The investment opportunity set for the portfolio choice problem will generally consist of N risky assets. From time to time, we will add an $(N + 1)$th, risk-free asset. The notation used throughout this chapter is set out in Table 17.1. In Section 17.5, we will occasionally have to add a subscript i to identify the investments of the typical or ith investor.

The investor's date-0 investments are:

- b_j (euro) in the jth risky asset, $j = 1, 2, \ldots, N$; and
- $(W_0 - \sum_j b_j)$ in the risk-free asset, if it exists.

The investor's date-1 payoffs are:

- $b_j \tilde{r}_j$ from the jth risky asset; and
- $(W_0 - \sum_j b_j) r_f$ from the risk-free asset.

While we have suggested for clarity that returns are calculated from the euro values of assets, the theory applies equally well to any currency that we might choose to use as numeraire. The euro itself is not considered as an asset within this model. When we assume the existence of a risk-free asset, a euro investment in the risk-free asset earns a risk-free return. When we assume that all assets are risky, the euro is purely a unit of account and all wealth must be held in the form of risky assets.

Note that we have not proved that, in the presence of randomly varying exchange rates, a preference relation that has the expected-utility property with respect to values in euro retains that property with respect to values in other currencies. Likewise, an individual whose

Table 17.1 Notation for the portfolio choice problem

Symbol	Interpretation
W_0	investor's initial wealth
μW_0	investor's (desired) expected final wealth
N	number of risky assets
I	number of investors
r_f	gross return on the risk-free asset
$\tilde{r}_j \in \mathbb{R}$	gross return on jth risky asset
$\tilde{\mathbf{r}} \in \mathbb{R}^N$	$(\tilde{r}_1, \tilde{r}_2, \dots, \tilde{r}_N)$
$\mathbf{e} \equiv (e_1, e_2, \dots, e_n) \equiv E[\tilde{\mathbf{r}}] \in \mathbb{R}^N$	vector of expected returns
$\mathbf{V} \equiv \text{Var}[\tilde{\mathbf{r}}] \in \mathbb{R}^{N \times N}$	variance–covariance matrix of returns
$\mathbf{1}$	$(1, 1, \dots, 1)$, the N-dimensional vector of 1s
$a_j \in \mathbb{R}$	proportion of wealth invested in jth risky asset (weight of the jth asset in the portfolio)
$\mathbf{a} = (a_1, a_2, \dots, a_n) \in \mathbb{R}^N$	portfolio weight vector
$b_j \equiv a_j W_0 \in \mathbb{R}$	(euro) amount invested in jth risky asset
$\mathbf{b} = (b_1, b_2, \dots, b_n) \in \mathbb{R}^N$	portfolio vector
$\tilde{r}_\mathbf{a} = \mathbf{a}^\top \tilde{\mathbf{r}}$	gross return on the portfolio $\mathbf{a}$
$\tilde{W}_1 = \tilde{r}_\mathbf{a} W_0$	investor's actual final wealth
$\mu \equiv E[\tilde{r}_\mathbf{a}] \equiv E[\tilde{W}_1 / W_0]$	investor's (desired) expected gross return

preferences have the mean–variance property with respect to values in euro may not retain that property with respect to values in other currencies. We will return to these questions in Section 17.6.

The presentation here is in terms of a single-period problem, and the unconditional distribution of returns. The analysis of the multi-period, infinite-horizon, discrete-time problem, concentrating on the conditional distribution of the next period's returns given this period's, is quite similar, but is beyond the scope of this book.

DEFINITION 17.2.1 The portfolio vector $\mathbf{b}$ is said to be a **unit-cost** or **normal portfolio** if its components sum to unity $(\mathbf{b}^\top \mathbf{1} = 1)$.

The portfolio held by an investor with initial wealth W_0 can be thought of either as a portfolio vector $\mathbf{b}$ with $\mathbf{b}^\top \mathbf{1} = W_0$ or as the corresponding normal portfolio or portfolio weight vector, $\mathbf{a} = (1/W_0)\mathbf{b}$. It should be clear from the context which meaning of 'portfolio' is intended.

DEFINITION 17.2.2 The portfolio vector $\mathbf{b}$ is said to be a **zero-cost** or **hedge portfolio** if its components sum to zero $(\mathbf{b}^\top \mathbf{1} = 0)$.

The vector of net trades carried out by an investor moving from the portfolio $\mathbf{b}_0$ to the portfolio $\mathbf{b}_1$ can be thought of as the hedge portfolio $\mathbf{b}_1 - \mathbf{b}_0$.

The set of all possible portfolio vectors will be called the **portfolio space**. It is an N-dimensional real vector space, i.e. it is just Euclidean space, $\mathbb{R}^N$. In the presence of a risk-free asset, the investor can borrow or invest risklessly in order to hold any risky asset portfolio in $\mathbb{R}^N$. In the absence of a risk-free asset, only portfolios on the hyperplane $\mathbf{b}^\top \mathbf{1} = W_0$ are attainable.

The mapping $\mathbf{b} \mapsto \mathbf{b}^\top \tilde{\mathbf{r}}$ associates a random variable, namely, the payoff of the portfolio, with each vector in the portfolio space. In this sense, the portfolio space is equivalent to the vector space of random variables considered in Section 13.4. In this chapter, however, we will view portfolios primarily as Euclidean vectors rather than as random variables.

In the Euclidean interpretation, the set of unit-cost portfolios and the set of zero-cost portfolios are parallel hyperplanes in the portfolio space, both normal to the vector $\mathbf{1}$. We will call the set of unit-cost portfolios the **portfolio weight hyperplane**.

When short-selling is allowed, $\mathbf{b}$ can have negative components; if short-selling is not allowed, then the portfolio choice problem will have non-negativity constraints, $b_j \geq 0$ for $j = 1, 2, \ldots, N$. In the latter case, without a risk-free asset, the set of portfolios available to the investor with initial wealth W_0 is the simplex

$$\{\mathbf{b} \in \mathbb{R}^N : \mathbf{b}^\top \mathbf{1} = W_0; \ b_j \geq 0, \ j = 1, 2, \ldots, N\} \tag{17.1}$$

DEFINITION 17.2.3 The **excess return** on the jth risky asset (or on a portfolio) is its return in excess of the risk-free rate, $\tilde{r}_j - r_f$.

DEFINITION 17.2.4 The **risk premium** on an asset or portfolio is its expected excess return.

A number of further comments are in order at this stage.

1. The derivation of the mean–variance frontier is generally presented in the literature (e.g. Huang and Litzenberger (1988, Chapter 3)) in terms of unit-cost portfolios or portfolio weight vectors or, equivalently, with initial wealth normalized to unity ($W_0 = 1$). This assumption is not essential and will be avoided here, as the development of the theory is more elegant if presented in terms of arbitrary W_0.
2. Since we will be dealing on occasion with hedge portfolios, we will in future avoid the concepts of *rate of return* and *net return*, which are usually thought of as the ratio of profit to initial investment. These terms are meaningless for a hedge portfolio as the denominator is zero. Instead, we will speak of the **gross return** on a portfolio or the **portfolio payoff**; see also Section 15.2. This can be defined unambiguously as follows. There is no ambiguity about the payoff on one of the original securities, which is just the gross return per euro invested. The payoff on a unit-cost or normal portfolio $\mathbf{b}$ is equivalent to the gross return. It is just $\mathbf{b}^\top \tilde{\mathbf{r}}$. The payoff on a zero-cost portfolio, $\mathbf{b}$, can also be defined as $\mathbf{b}^\top \tilde{\mathbf{r}}$.
3. The excess return and risk premium will be the same whether they are calculated on a gross return basis or on a net return basis.
4. If individual asset returns are multivariate normally distributed, $\tilde{\mathbf{r}} \sim \text{MVN}(\cdot, \cdot)$, then all portfolio returns are normally distributed; see Exercise 13.16.

17.3 Single-period portfolio choice problem

17.3.1 Canonical portfolio choice problem

We will solve the portfolio choice problem first in a general expected-utility context. Unless otherwise stated, we assume throughout this section that individuals:

1. have von Neumann–Morgenstern (VNM) utilities, i.e. preferences have the expected-utility representation

$$v(\widetilde{W}) = E[u(\widetilde{W})] = \int u(W) \, dF_{\widetilde{W}}(W) \tag{17.2}$$

where v is the utility function for random variables (gambles, lotteries) and u is the utility function for sure things;

2. prefer more to less (or are greedy), i.e. u is increasing or

$$u'(W) > 0 \quad \forall\, W \tag{17.3}$$

and

3. are (strictly) risk-averse, i.e. u is strictly concave or

$$u''(W) < 0 \quad \forall\, W \tag{17.4}$$

It is assumed here that there are no constraints on short-selling or on borrowing (which is the same as short-selling the risk-free security).

This canonical portfolio choice problem is solved by finding the values of b_j which maximize the expected utility of date-1 wealth,

$$\widetilde{W} = \left(W_0 - \sum_j b_j \right) r_{\mathrm{f}} + \sum_j b_j \tilde{r}_j$$

$$= W_0 r_{\mathrm{f}} + \sum_j b_j (\tilde{r}_j - r_{\mathrm{f}}) \tag{17.5}$$

i.e. by solving the unconstrained maximization problem

$$\max_{\{b_j\}} \; f(b_1, b_2, \dots, b_N) \equiv E\left[u\left(W_0 r_{\mathrm{f}} + \sum_j b_j (\tilde{r}_j - r_{\mathrm{f}}) \right) \right] \tag{17.6}$$

The first-order conditions are

$$E[u'(\widetilde{W})(\tilde{r}_j - r_{\mathrm{f}})] = 0 \quad \forall\, j \tag{17.7}$$

since we can pass the differentiation operator through the expectation operator and use the chain rule; see Section 13.6.2.

The Hessian matrix of the objective function is

$$\mathbf{A} \equiv E[u''(\widetilde{W})(\tilde{\mathbf{r}} - r_{\mathrm{f}}\mathbf{1})(\tilde{\mathbf{r}} - r_{\mathrm{f}}\mathbf{1})^{\top}] \tag{17.8}$$

Since we have assumed that investor behaviour is strictly risk-averse, $u''(\widetilde{W}) < 0$ and thus, provided that the variance–covariance matrix, $\mathbf{V}$, is positive-definite, $\mathbf{h}^{\top}\mathbf{A}\mathbf{h} < 0$ for all $\mathbf{h} \neq \mathbf{0}_N$. Hence, $\mathbf{A}$ is a negative definite matrix and, by Theorem 10.2.5, the objective function f is a strictly concave (and, hence, strictly quasi-concave) function. Thus, under the present assumptions, Theorems 10.3.3 and 10.3.5 guarantee that the first-order conditions have a unique solution. The trivial case, in which the random returns are not really random at all, can be ignored.

Note that there is no guarantee that the portfolio choice problem has any finite or unique solution if the expected-utility function is not concave.

Other ways of writing (17.7) are

$$E[u'(\widetilde{W})\tilde{r}_j] = E[u'(\widetilde{W})]r_f \quad \forall j \tag{17.9}$$

or

$$\text{Cov}[u'(\widetilde{W}), \tilde{r}_j] + E[u'(\widetilde{W})]E[\tilde{r}_j] = E[u'(\widetilde{W})]r_f \quad \forall j \tag{17.10}$$

or

$$E[\tilde{r}_j - r_f] = -\frac{\text{Cov}[u'(\widetilde{W}), \tilde{r}_j]}{E[u'(\widetilde{W})]} \quad \forall j \tag{17.11}$$

In other words, the risk premiums on the risky assets are (negatively) proportional to the covariance of their returns with optimal marginal utility. (Since terminal wealth is random, so too at the time of decision-making is marginal utility evaluated at optimal terminal wealth.)

There is one final way of writing (17.7), which leads to another useful interpretation of the first-order conditions. Suppose p_j is the price of the random payoff $\tilde{x}_j$. Then $\tilde{r}_j = \tilde{x}_j/p_j$ and

$$p_j = E\left[\frac{u'(\widetilde{W})}{E[u'(\widetilde{W})]r_f}\tilde{x}_j\right] \quad \forall j \tag{17.12}$$

In other words, securities are valued by taking the expected present values of their payoffs, using the **stochastic discount factor** $u'(\widetilde{W})/(E[u'(\widetilde{W})]r_f)$. This valuation method, at the optimum, gives the same security values for all investors, although they may have different underlying utility functions. Practical corporate finance and theoretical asset pricing models to a large extent are (or should be) concerned with analysing this discount factor.

Exercise 17.2 asks the reader to analyse the portfolio choice problem for the special case of quadratic expected utility and $N = 2$.

17.3.2 Risk aversion and portfolio composition

Before proceeding, the reader might now like to review the material in Section 16.5.

For the moment, assume only one risky asset ($N = 1$). In this case, the subscript identifying the asset number can be dropped. In other words, we now deal with the basic analysis of the choice between one risk-free and one risky asset, following Huang and Litzenberger (1988, Chapter 1).

Such an example is sufficient to illustrate several useful principles, including the following:

1. The investment decision depends on the investor's degree of risk aversion.
2. Even risk-averse investors are locally risk-neutral at the margin.

We first consider the concept of **local risk neutrality**.

The optimal investment in the risky asset is positive if and only if the objective function is increasing at $b = 0$ if and only if

$$f'(0) > 0 \tag{17.13}$$

if and only if

$$E[u'(W_0 r_{\mathrm{f}})(\tilde{r} - r_{\mathrm{f}})] > 0 \tag{17.14}$$

if and only if

$$u'(W_0 r_{\mathrm{f}}) E[\tilde{r} - r_{\mathrm{f}}] > 0 \tag{17.15}$$

if and only if

$$E[\tilde{r}] > E[r_{\mathrm{f}}] = r_{\mathrm{f}} \tag{17.16}$$

(since we continue to assume that utility functions are strictly increasing).

This is the property of local risk neutrality – a greedy, risk-averse investor will always prefer a little of a risky asset paying an expected return higher than r_{f} to none of the risky asset.

We denote the wealth elasticity of demand for the risky asset by[1]

$$\eta \equiv \frac{W_0}{b} \frac{db}{dW_0} \tag{17.17}$$

Then we have

$$\frac{da}{dW_0} = \frac{d(b/W_0)}{dW_0} = \frac{W_0(db/dW_0) - b}{W_0^2}$$

$$= \frac{b}{W_0^2}(\eta - 1) \tag{17.18}$$

Note that, when there is a positive risk premium on the risky asset, b is positive (by local risk neutrality) and thus

$$\mathrm{sign}\left(\frac{da}{dW_0}\right) = \mathrm{sign}(\eta - 1) \tag{17.19}$$

Knowledge of the risk-aversion properties of utility functions set out in Definitions 16.5.5 and 16.5.6 allows us to sign the relationship between the optimal risky asset investment and initial wealth, as follows.

THEOREM 17.3.1 *Again assuming a positive risk premium on the risky asset:*

- *DARA $\Rightarrow$ the risky asset is a normal good at all wealth levels ($db/dW_0 > 0$);*
- *CARA $\Rightarrow$ $db/dW_0 = 0$;*
- *IARA $\Rightarrow$ the risky asset is an inferior good at all wealth levels ($db/dW_0 < 0$);*
- *DRRA $\Rightarrow$ an increasing proportion of wealth is invested in the risky asset ($da/dW_0 > 0$ or $\eta > 1$);*
- *CRRA $\Rightarrow$ a constant proportion of wealth is invested in the risky asset ($da/dW_0 = 0$ or $\eta = 1$);*
- *IRRA $\Rightarrow$ a decreasing proportion of wealth is invested in the risky asset ($da/dW_0 < 0$ or $\eta < 1$).*

Proof: We will prove the first of these properties only. The second follows directly from Example 16.4.6 and Exercise 16.23. The other results are proved similarly; see Exercise 17.3.

Writing the first-order condition (17.7) as

$$E[u'(W_0 r_f + b(\tilde{r} - r_f))(\tilde{r} - r_f)] = 0 \tag{17.20}$$

then differentiating with respect to W_0 to obtain

$$E\left[u''(\widetilde{W})(\tilde{r} - r_f)\left(r_f + \frac{db}{dW_0}(\tilde{r} - r_f)\right)\right] = 0 \tag{17.21}$$

and rearranging, we have

$$\frac{db}{dW_0} = \frac{E[u''(\widetilde{W})(\tilde{r} - r_f)]r_f}{-E[u''(\widetilde{W})(\tilde{r} - r_f)^2]} \tag{17.22}$$

By concavity, the denominator is positive and $r_f > 0$. Therefore

$$\text{sign}(db/dW_0) = \text{sign}(E[u''(\widetilde{W})(\tilde{r} - r_f)]) \tag{17.23}$$

We will show that both are positive.

For decreasing absolute risk aversion,

$$\tilde{r} > r_f \Rightarrow R_A(\widetilde{W}) < R_A(W_0 r_f) \tag{17.24}$$

and

$$\tilde{r} \leq r_f \Rightarrow R_A(\widetilde{W}) \geq R_A(W_0 r_f) \tag{17.25}$$

Recalling the definition of absolute risk aversion, Definition 16.5.3, and multiplying both sides of each inequality by $-u'(\widetilde{W})(\tilde{r} - r_f)$ gives, respectively,

$$u''(\widetilde{W})(\tilde{r} - r_f) > -R_A(W_0 r_f)u'(\widetilde{W})(\tilde{r} - r_f) \tag{17.26}$$

in the event that $\tilde{r} > r_f$, and

$$u''(\widetilde{W})(\tilde{r} - r_f) \geq -R_A(W_0 r_f)u'(\widetilde{W})(\tilde{r} - r_f) \tag{17.27}$$

(effectively the same result) in the event that $\tilde{r} \leq r_f$.

Integrating over the events $\tilde{r} > r_f$ and $\tilde{r} \leq r_f$ implies that

$$E[u''(\widetilde{W})(\tilde{r} - r_f)] > -R_A(W_0 r_f)E[u'(\widetilde{W})(\tilde{r} - r_f)] \tag{17.28}$$

provided that $\tilde{r} > r_f$ with positive probability.

The right-hand side of inequality (17.28) is 0 at the optimum, by the first-order condition, hence the left-hand side is positive as claimed. $\square$

17.3.3 *Mutual fund separation*

A **mutual fund** is a special type of (managed) portfolio.

Commonly, investors delegate portfolio choice to mutual fund operators or managers. We are interested in conditions under which large groups of investors will agree on portfolio composition. For example, all investors with similar utility functions might choose the same portfolio, or all investors with similar probability beliefs might choose the same portfolio. More realistically, we may be able to define a group of investors whose portfolio choices all lie in a subspace of small dimension (say, two) of the N-dimensional portfolio space. The first such result, Theorem 17.3.2, is due to Cass and Stiglitz (1970).

We begin with a formal definition.

DEFINITION 17.3.1 Two-fund monetary separation is said to exist when a group of agents with different wealths (but the same increasing, strictly concave, VNM utility) all hold the same risky unit-cost portfolio, $\mathbf{a}^*$, say. The mix of the risk-free asset and this risky portfolio may differ between investors.

Formally, two-fund monetary separation exists when there is a portfolio $\mathbf{a}^*$ such that, for all other portfolios $\mathbf{b}$ and wealths W_0, there exists λ such that

$$E[u(W_0 r_{\mathrm{f}} + \lambda \mathbf{a}^{*\top}(\tilde{\mathbf{r}} - r_{\mathrm{f}}\mathbf{1}))] \geq E[u(W_0 r_{\mathrm{f}} + \mathbf{b}^{\top}(\tilde{\mathbf{r}} - r_{\mathrm{f}}\mathbf{1}))] \tag{17.29}$$

THEOREM 17.3.2 *Two-fund monetary separation exists if and only if*

- *risk tolerance* $(1/R_A(W))$ *is linear in wealth (including constant),*
- *i.e. there exists hyperbolic absolute risk aversion (HARA, including CARA),*
- *i.e. the utility function is of one of these types (see Exercise 17.4):*

  – *extended power*

$$u(W) = \frac{1}{(C+1)B}(A + BW)^{C+1} \tag{17.30}$$

  – *logarithmic*

$$u(W) = \frac{1}{B}\ln(A + BW) \tag{17.31}$$

  – *negative exponential*

$$u(W) = \frac{A}{B}\exp(BW) \tag{17.32}$$

  where A, B and C are chosen to guarantee that $u' > 0$ and $u'' < 0$,
- *i.e. marginal utility satisfies*

$$u'(W) = (A + BW)^{C} \quad or \quad u'(W) = A\exp(BW) \tag{17.33}$$

  where A, B and C are again chosen to guarantee that $u' > 0$ and $u'' < 0$.

Proof: The proof that these conditions are necessary for two-fund separation is difficult and tedious. The interested reader is referred to Cass and Stiglitz (1970).

We will show that $u'(W) = (A + BW)^C$ is sufficient for two-fund separation.

The optimal euro investments b_i constitute the unique solution to the first-order conditions

$$
\begin{aligned}
0 &= E\left[u'(\widetilde{W})\frac{\partial \widetilde{W}}{\partial b_i}\right] \\
&= E[(A + B\widetilde{W})^C(\tilde{r}_i - r_f)] \\
&= E\left[\left(A + BW_0 r_f + \sum_{j=1}^{N} Bb_j(\tilde{r}_j - r_f)\right)^C (\tilde{r}_i - r_f)\right]
\end{aligned}
\tag{17.34}
$$

$i = 1, 2, \ldots, N$, or, equivalently, dividing across by $A + BW_0 r_f$, to the system of equations

$$
E\left[\left(1 + \sum_{j=1}^{N} \frac{Bb_j}{A + BW_0 r_f}(\tilde{r}_j - r_f)\right)^C (\tilde{r}_i - r_f)\right] = 0
\tag{17.35}
$$

or

$$
E\left[\left(1 + \sum_{j=1}^{N} x_j(\tilde{r}_j - r_f)\right)^C (\tilde{r}_i - r_f)\right] = 0
\tag{17.36}
$$

where $x_i \equiv Bb_i/(A + BW_0 r_f)$, for $i = 1, 2, \ldots, N$.

The unique solutions for x_i are independent of W_0, which does not appear in (17.36). Since A and B do not appear either, the unique solutions for x_i are also independent of those parameters. However, they do depend on C. But the risky portfolio weights are

$$
\begin{aligned}
a_i &= \frac{b_i}{\sum_{j=1}^{N} b_j} = \frac{Bb_i/(A + BW_0 r_f)}{\sum_{j=1}^{N} Bb_j/(A + BW_0 r_f)} \\
&= \frac{x_i}{\sum_{j=1}^{N} x_j}
\end{aligned}
\tag{17.37}
$$

and so are also independent of initial wealth and of A and B.

Since the euro investment in the ith risky asset satisfies

$$
b_i = x_i\left(\frac{A}{B} + W_0 r_f\right)
\tag{17.38}
$$

we also have in this case that the euro investment in the common risky portfolio is a linear function of the initial wealth. The other sufficiency proofs are similar and are left as exercises; see Exercise 17.6. $\qquad\square$

A portfolio separation result like this allows us to assert that the equilibrium outcome is Pareto efficient even in an incomplete market in which the only two markets are for a risk-free asset and for the relevant portfolio of risky assets.

17.4   Mathematics of the portfolio frontier

17.4.1   Portfolio frontier in $\mathbb{R}^N$: risky assets only

The portfolio frontier

DEFINITION 17.4.1 The (**mean–variance**) **portfolio frontier** is the set of solutions to the **mean–variance portfolio choice problem** that is faced by an investor with an initial wealth of W_0 who desires an expected final wealth of $W_1 \equiv \mu W_0$ (or, equivalently, an expected rate of return of μ), but with the smallest possible variance of final wealth. There is a solution for each (W_0, μ) or, equivalently, each (W_0, W_1) pair.

Not all individuals with mean–variance preferences will necessarily choose a portfolio on the mean–variance portfolio frontier. It will become clear later that an individual with mean–variance preferences represented by indifference curves that are convex and upward-sloping in variance–mean space will generally do so.

The mean–variance frontier can also be called the **two-moment portfolio frontier**, in recognition of the fact that the same approach can be extended (with difficulty) to higher moments.[2]

The mean–variance portfolio frontier is a subset of the portfolio space: we will show later that it is in fact a vector subspace of the portfolio space. However, introductory treatments generally present it (without proof) as the envelope function, in mean–variance space or mean–standard deviation space ($\mathbb{R}_+ \times \mathbb{R}$), of the variance-minimization problem. We will come to this representation in Section 17.4.2.

DEFINITION 17.4.2 The portfolio vector **b** is called a **frontier portfolio** if its return has the minimum variance among all portfolios that have the same cost, $\mathbf{b}^\top \mathbf{1}$, and the same expected payoff, $\mathbf{b}^\top \mathbf{e}$.

We will begin by supposing that all assets are risky. Formally, the frontier portfolio corresponding to initial wealth W_0 and expected return μ (expected terminal wealth μW_0) is the solution to the quadratic programming problem

$$\min_{\mathbf{b}} \mathbf{b}^\top \mathbf{V} \mathbf{b} \tag{17.39}$$

subject to the linear constraints

$$\mathbf{b}^\top \mathbf{1} = W_0 \tag{17.40}$$

and

$$\mathbf{b}^\top \mathbf{e} = W_1 = \mu W_0 \tag{17.41}$$

The first constraint is just the budget constraint, while the second constraint states that the expected rate of return on the portfolio is the desired mean return μ.

The frontier in the risky-assets-only case is the set of solutions for all values of W_0 and W_1 (or μ) to this variance-minimization problem, or to the equivalent maximization problem:

$$\max_{\mathbf{b}} -\mathbf{b}^\top \mathbf{V} \mathbf{b} \tag{17.42}$$

subject to the same linear constraints (17.40) and (17.41).

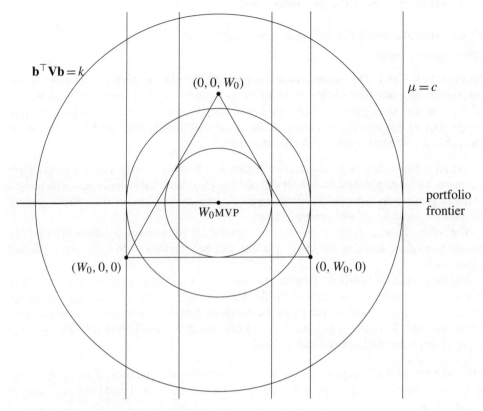

Figure 17.1 Portfolio frontier in $\mathbb{R}^3$

The variance minimization problem, like any constrained optimization problem, has a dual problem, in this case an expected value maximization problem, subject to a variance constraint. Indeed, the latter is the problem of practical interest, but the former approach provides more useful insights at this stage.

Figure 17.1 illustrates the construction of the portfolio frontier for $N = 3$ in the simple case where all assets share a common variance and common pairwise covariances. The set of portfolios costing W_0 is a simplex if there is no short-selling, but is the entire affine hyperplane containing the simplex if short-selling is allowed. Figure 17.1 shows the level sets of the expected return and variance functions in this affine hyperplane. The iso-variance curves (with equations of the form $\mathbf{b}^\top \mathbf{V} \mathbf{b} = k$) are concentric circles (or ellipses for a more general variance–covariance matrix). The iso-mean curves (with equations of the form $\mu = \mathbf{b}^\top \mathbf{e} = c$) are parallel lines, and the solutions to the variance-minimization problem for different μ (or W_1) values are the tangency points between these ellipses and lines.[3] The centre of the concentric ellipses is at the global minimum-variance portfolio corresponding to W_0, marked W_0MVP. In the simple case illustrated in Figure 17.1, variance is minimized at the centroid of the relevant simplex. A similar geometric interpretation can be applied in higher dimensions.

The properties of this two-moment frontier are well known, and can be found, for example, in Merton (1972) or Roll (1977). The basic notation here follows Huang and Litzenberger (1988, Chapter 3).

The solution

Problem (17.42) is a variant of the canonical quadratic programming problem dealt with in Section 14.3, except that it has equality constraints.

To avoid degeneracies, we require that:

1. not every portfolio has the same expected return, i.e.

$$\mathbf{e} \neq E[\tilde{r}_1]\mathbf{1} \tag{17.43}$$

   and in particular that $N > 1$; and
2. the variance–covariance matrix, $\mathbf{V}$, is positive definite. We already know that any variance–covariance matrix must be positive semi-definite, but we require this slightly stronger condition as we also did to guarantee that the matrix $\mathbf{A}$ in (17.8) was negative definite. To see why, suppose

$$\exists\, \mathbf{b} \neq \mathbf{0}_N \quad \text{s.t. } \mathbf{b}^\top \mathbf{V} \mathbf{b} = 0 \tag{17.44}$$

   Then there exists a portfolio whose return $\mathbf{b}^\top \tilde{\mathbf{r}} = \tilde{r}_{\mathbf{b}}$ has zero variance. This implies that $\tilde{r}_{\mathbf{b}} = r_0$, say, with probability 1 or, essentially, that this portfolio is risk-free. Arbitrage will force the returns on all risk-free assets to be equal in equilibrium, so this situation is equivalent economically to the introduction of a risk-free asset, which we will come to in Section 17.4.3.

   In the present problem, the place of the matrix $\mathbf{A}$ in the canonical quadratic programming problem of Section 14.3 is taken by the (symmetric) negative definite matrix, $-\mathbf{V}$, which is just the negative of the variance–covariance matrix of asset returns; $\mathbf{g}^1 = \mathbf{1}^\top$ and $\alpha_1 = W_0$; and $\mathbf{g}^2 = \mathbf{e}^\top$ and $\alpha_2 = W_1$. Assumption (17.43) guarantees that the $2 \times N$ matrix

$$\mathbf{G} = \begin{bmatrix} \mathbf{1}^\top \\ \mathbf{e}^\top \end{bmatrix} \tag{17.45}$$

is of full rank 2.

Making the appropriate substitutions in the generic solution (14.38) yields

$$\mathbf{b} = \mathbf{V}^{-1}\mathbf{G}^\top (\mathbf{G}\mathbf{V}^{-1}\mathbf{G}^\top)^{-1} \begin{bmatrix} W_0 \\ W_1 \end{bmatrix} \tag{17.46}$$

This says that the optimal $\mathbf{b}$ is a linear combination of the two columns of the $N \times 2$ matrix $\mathbf{V}^{-1}\mathbf{G}^\top (\mathbf{G}\mathbf{V}^{-1}\mathbf{G}^\top)^{-1}$. In this linear combination, the first and second columns of the matrix are weighted by initial wealth, W_0, and expected final wealth, W_1, respectively.

Let us define:

$$A \equiv \mathbf{1}^\top \mathbf{V}^{-1}\mathbf{e} = \mathbf{e}^\top \mathbf{V}^{-1}\mathbf{1} \tag{17.47}$$

$$B \equiv \mathbf{e}^\top \mathbf{V}^{-1}\mathbf{e} > 0 \tag{17.48}$$

$$C \equiv \mathbf{1}^\top \mathbf{V}^{-1}\mathbf{1} > 0 \tag{17.49}$$

$$D \equiv BC - A^2 \tag{17.50}$$

The inequalities in (17.48) and (17.49) follow from the fact that $\mathbf{V}^{-1}$ (like $\mathbf{V}$) is positive definite.

Then we can write

$$\mathbf{V}^{-1}\mathbf{G}^{\top}(\mathbf{G}\mathbf{V}^{-1}\mathbf{G}^{\top})^{-1} = \mathbf{V}^{-1}\mathbf{G}^{\top}\begin{bmatrix} C & A \\ A & B \end{bmatrix}^{-1}$$

$$= \frac{1}{D}\mathbf{V}^{-1}\mathbf{G}^{\top}\begin{bmatrix} B & -A \\ -A & C \end{bmatrix}$$

$$= \frac{1}{D}\mathbf{V}^{-1}[\mathbf{1} \quad \mathbf{e}]\begin{bmatrix} B & -A \\ -A & C \end{bmatrix}$$

$$= \frac{1}{D}\mathbf{V}^{-1}[B\mathbf{1} - A\mathbf{e} \quad C\mathbf{e} - A\mathbf{1}] \tag{17.51}$$

If we define

$$\mathbf{g} \equiv \frac{1}{D}\mathbf{V}^{-1}(B\mathbf{1} - A\mathbf{e}) \tag{17.52}$$

and

$$\mathbf{h} \equiv \frac{1}{D}\mathbf{V}^{-1}(C\mathbf{e} - A\mathbf{1}) \tag{17.53}$$

respectively, then we can write the solution (17.46) as

$$\mathbf{b} = W_0\mathbf{g} + W_1\mathbf{h} = W_0(\mathbf{g} + \mu\mathbf{h}) \tag{17.54}$$

Thus the set of solutions to this quadratic programming problem for all possible (W_0, W_1) combinations (including negative W_0) is the two-dimensional vector subspace of the portfolio space that is generated by the vectors $\mathbf{g}$ and $\mathbf{h}$, which constitute a basis for the frontier.

The components of $\mathbf{g}$ and $\mathbf{h}$ are functions solely of the means, variances and covariances of security returns, given by the vector $\mathbf{e}$ and the matrix $\mathbf{V}$. Thus the vector of optimal portfolio proportions,

$$\mathbf{a} = \frac{1}{W_0}\mathbf{b} = \mathbf{g} + \mu\mathbf{h} \tag{17.55}$$

is independent of the initial wealth W_0. This is another version of the mutual fund separation result: all investors choosing frontier portfolios, regardless of their initial wealth, will choose (linear) combinations of the two mutual funds $\mathbf{g}$ and $\mathbf{h}$.

Like any two-dimensional vector space, the portfolio frontier has many alternative bases; indeed, any pair of linearly independent frontier portfolios constitutes a basis, or a pair of mutual funds with respect to which the separation result can be restated. We will now list and discuss the properties of four commonly used bases for the frontier. The relevant basis vectors can also be described as **basis portfolios**.

Basis 1: The vectors **g** and **h**.

It is easy to see the economic interpretation of **g** and **h**:

- Vector **g** is the frontier portfolio corresponding to $W_0 = 1$ and $W_1 = 0$. In other words, it is the normal portfolio that would be held by an investor whose objective was to (just) go bankrupt with minimum variance.
- Similarly, vector **h** is the frontier portfolio corresponding to $W_0 = 0$ and $W_1 = 1$. In other words, it is the hedge portfolio that would be purchased by a variance-minimizing investor in order to increase his expected final wealth by one unit.

Basis 2: The vectors $\mathbf{V}^{-1}\mathbf{1}$ and $\mathbf{V}^{-1}\mathbf{e}$.

Equation (14.35) implies that the optimal **b** is a linear combination of the two columns of the $N \times 2$ matrix

$$\tfrac{1}{2}\mathbf{V}^{-1}\mathbf{G}^{\top} = \left[\tfrac{1}{2}\mathbf{V}^{-1}\mathbf{1} \quad \tfrac{1}{2}\mathbf{V}^{-1}\mathbf{e}\right] \tag{17.56}$$

with columns weighted by the Lagrange multipliers corresponding to the two constraints. For convenience, we will denote the Lagrange multipliers $2\gamma/C$ and $2\lambda/A$, respectively. This allows the solution to be written as

$$\mathbf{b} = \frac{\gamma}{C}\mathbf{V}^{-1}\mathbf{1} + \frac{\lambda}{A}\mathbf{V}^{-1}\mathbf{e} \tag{17.57}$$

It is easily shown that $(1/C)\mathbf{V}^{-1}\mathbf{1}$ and $(1/A)\mathbf{V}^{-1}\mathbf{e}$ are both unit portfolios, so the total cost of **b** is $\gamma + \lambda = W_0$.

Basis 3: The vectors MVP and **h**, where MVP denotes the **global minimum-variance unit-cost portfolio**, i.e. the unit-cost portfolio that minimizes the variance of the payoff, regardless of expected final wealth.

We know that the expected final wealth constraint is non-binding if and only if the corresponding Lagrange multiplier $\lambda = 0$. Thus we can see from (17.57) that $(\gamma/C)\mathbf{V}^{-1}\mathbf{1}$ is the global minimum-variance portfolio with cost W_0 (which, in fact, equals γ in this case). Setting $W_0 = \gamma = 1$ shows that, in terms of Basis 2,

$$\mathrm{MVP} = \frac{1}{C}\mathbf{V}^{-1}\mathbf{1} \tag{17.58}$$

We can also find the MVP in terms of Basis 1, as follows. Recalling that $\mathrm{Var}[\tilde{r}_\mathbf{b}] = \mathbf{b}^{\top}\mathbf{V}\mathbf{b}$ and substituting $\mathbf{g} + \mu\mathbf{h}$ for **b**, the variance of the generic frontier portfolio with expected return μ is given by the quadratic expression

$$\mathrm{Var}[\tilde{r}_{\mathbf{g}+\mu\mathbf{h}}] = \mathbf{g}^{\top}\mathbf{V}\mathbf{g} + 2\mu(\mathbf{g}^{\top}\mathbf{V}\mathbf{h}) + \mu^2(\mathbf{h}^{\top}\mathbf{V}\mathbf{h}) \tag{17.59}$$

which has its minimum at

$$\mu = -\frac{\mathbf{g}^{\top}\mathbf{V}\mathbf{h}}{\mathbf{h}^{\top}\mathbf{V}\mathbf{h}} \tag{17.60}$$

It can be shown that the latter expression reduces to A/C and that the minimum value of the variance is $1/C$; see Exercise 17.12. Thus the MVP is the unit-cost frontier portfolio with expected return A/C and, in terms of Basis 1,

$$\text{MVP} = \mathbf{g} + \frac{A}{C}\mathbf{h} \tag{17.61}$$

Now we can combine (17.54) and (17.61) and write the generic frontier portfolio in terms of Basis 3 as

$$\mathbf{b} = W_0\left(\text{MVP} + \left(\mu - \frac{A}{C}\right)\mathbf{h}\right) \tag{17.62}$$

see Exercise 17.12 again. Furthermore,

$$\begin{aligned}
\text{Cov}[\tilde{r}_{\mathbf{h}}, \tilde{r}_{\text{MVP}}] &= \mathbf{h}^\top \mathbf{V}\left(\mathbf{g} - \frac{\mathbf{g}^\top \mathbf{V}\mathbf{h}}{\mathbf{h}^\top \mathbf{V}\mathbf{h}}\mathbf{h}\right) \\
&= \mathbf{h}^\top \mathbf{V}\mathbf{g} - \frac{\mathbf{g}^\top \mathbf{V}\mathbf{h} \times \mathbf{h}^\top \mathbf{V}\mathbf{h}}{\mathbf{h}^\top \mathbf{V}\mathbf{h}} \\
&= 0 \tag{17.63}
\end{aligned}$$

i.e. the returns on the portfolio with weights $\mathbf{h}$ and on the minimum-variance portfolio are uncorrelated. The basis portfolios MVP and $\mathbf{h}$ are in this sense orthogonal.

The global MVP has another interesting property. If $\mathbf{a}$ is any unit-cost portfolio, frontier or not, then the MVP must by definition be the minimum-variance affine combination of itself and $\mathbf{a}$, i.e. $\beta = 0$ solves

$$\min_{\beta} \tfrac{1}{2} \text{Var}[\tilde{r}_{\beta\mathbf{a}+(1-\beta)\text{MVP}}] \tag{17.64}$$

which has necessary and sufficient first-order condition

$$\beta \, \text{Var}[\tilde{r}_{\mathbf{a}}] + (1 - 2\beta)\, \text{Cov}[\tilde{r}_{\mathbf{a}}, \tilde{r}_{\text{MVP}}] - (1 - \beta)\, \text{Var}[\tilde{r}_{\text{MVP}}] = 0 \tag{17.65}$$

Hence, setting $\beta = 0$,

$$\text{Cov}[\tilde{r}_{\mathbf{a}}, \tilde{r}_{\text{MVP}}] - \text{Var}[\tilde{r}_{\text{MVP}}] = 0 \tag{17.66}$$

and the covariance of any unit-cost portfolio with the MVP is $1/C$.

Basis 4: Any two frontier portfolios with uncorrelated returns.
We will return to this idea shortly.

As shown in Figure 17.2, the set of unit-cost frontier portfolios in $\mathbb{R}^N$ is the line L passing through $\mathbf{g}$, parallel to $\mathbf{h}$. Some writers consider this set to be the portfolio frontier. It follows immediately that the set of unit-cost frontier portfolios (like any straight line in $\mathbb{R}^N$) is an affine set, and can be generated by affine combinations of any pair of frontier portfolios with weights of the form β and $(1 - \beta)$. This is just another way of restating the two-fund separation result, this time in terms of two unit-cost portfolios.

Before proceeding further, the reader is advised to review the material on scalar product spaces in Section 5.4.9 and on metric spaces in Section 7.5.

Orthogonal decomposition of portfolios

We have already pointed out that the MVP and **h** are portfolios with uncorrelated returns, and have called them orthogonal. We now make this idea more precise by introducing a scalar product on the portfolio space, namely, that based on the variance–covariance matrix **V**. Since **V** is a positive definite matrix, and therefore non-singular, it defines a well-behaved scalar product and all the standard results on orthogonal projection, etc., from linear algebra are valid, as discussed in Section 13.6.3.

Two portfolios **b**$_1$ and **b**$_2$ are orthogonal with respect to this scalar product if and only if

$$\mathbf{b}_1^\top \mathbf{V} \mathbf{b}_2 = 0 \tag{17.67}$$

if and only if

$$\mathrm{Cov}[\mathbf{b}_1^\top \tilde{\mathbf{r}}, \mathbf{b}_2^\top \tilde{\mathbf{r}}] = 0 \tag{17.68}$$

if and only if the random variables representing the returns on the portfolios are uncorrelated.

Thus, the terms "orthogonal" and "uncorrelated" may legitimately, and shall, be applied interchangeably to pairs of portfolios. Furthermore, the squared length of a portfolio vector corresponds to the variance of its payoff.

Indeed, as for any scalar product space, there exists a basis for the full N-dimensional portfolio space consisting of N portfolios whose returns are uncorrelated, with each having unit variance, i.e. an orthonormal basis. The mathematics of the portfolio frontier could be developed in terms of the unit-cost versions of these uncorrelated portfolios. The details are left as an exercise; see Exercise 17.11.

This scalar product structure allows the equations of the portfolio frontier in mean–variance and mean–standard deviation space to be derived heuristically using the stylized diagram illustrating the portfolio decomposition in Figure 17.2, which we will now explain.

The line L' in this figure denotes the set of all zero-cost frontier portfolios, while the line L denotes the set of all unit-cost frontier portfolios. Two zero-cost frontier portfolios are marked on L', namely the zero portfolio, **0**, and the portfolio **h**, which would be purchased by a variance-minimizing investor in order to increase his expected final wealth by one unit. Five unit-cost frontier portfolios are marked on L. These include the special portfolios **g** and MVP, which have already been discussed.

Figure 17.2 also shows that, for any frontier portfolio such as **p** (apart from MVP), there is a unique unit-cost frontier portfolio, **z**$_\mathbf{p}$, which is orthogonal to **p**, called the **zero-covariance frontier portfolio** of **p**. The relationship between $\mu \equiv E[\tilde{r}_\mathbf{p}]$ and $\mu_\mathbf{z} \equiv E[\tilde{r}_{\mathbf{z}_\mathbf{p}}]$ can be worked out by solving

$$\mathrm{Cov}[\tilde{r}_{\mathbf{g}+\mu\mathbf{h}}, \tilde{r}_{\mathbf{g}+\mu_\mathbf{z}\mathbf{h}}] = 0 \tag{17.69}$$

or, equivalently, since $\tilde{r}_\mathbf{h}$ and $\tilde{r}_{\mathrm{MVP}}$ are uncorrelated, as shown in (17.63), by solving

$$\mathrm{Var}[\tilde{r}_{\mathrm{MVP}}] + (\mu - E[\tilde{r}_{\mathrm{MVP}}])(\mu_\mathbf{z} - E[\tilde{r}_{\mathrm{MVP}}])\mathrm{Var}[\tilde{r}_\mathbf{h}] = 0 \tag{17.70}$$

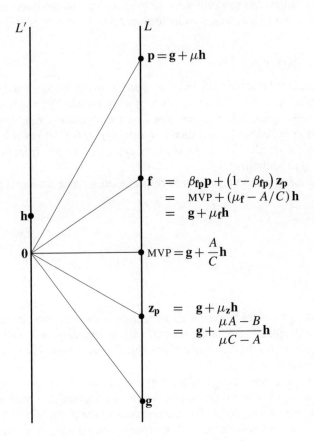

Figure 17.2 The portfolio decomposition

To make this true, we must have

$$(\mu - E[\tilde{r}_{MVP}])(\mu_z - E[\tilde{r}_{MVP}]) < 0 \tag{17.71}$$

or μ and μ_z on opposite sides of $E[\tilde{r}_{MVP}]$, or equivalently $\mathbf{p}$ and $\mathbf{z_p}$ on opposite sides of the MVP as shown in Figure 17.2.

Some simple manipulations will reveal that

$$\mu_z = \frac{\mu A - B}{\mu C - A} \tag{17.72}$$

where A, B and C are as defined in (17.47)–(17.49); see Exercise 17.12. Note that the denominator is zero when $\mu = A/C$, i.e. when $\mathbf{p} = \text{MVP}$. However, we have already seen that the MVP and $\mathbf{h}$ are orthogonal frontier portfolios; the difference is that in this case $\mathbf{h}$ is a zero-cost portfolio, but in all other cases $\mathbf{z_p}$ is a unit-cost portfolio.

Finally, Figure 17.2 shows that, given a fixed unit-cost frontier portfolio **p** (other than the MVP), any other unit-cost frontier portfolio, such as **f**, can be decomposed in the form

$$\mathbf{f} = \beta_{\mathbf{fp}}\mathbf{p} + (1 - \beta_{\mathbf{fp}})\mathbf{z_p} \tag{17.73}$$

i.e. written as an affine combination of **p** and $\mathbf{z_p}$.

We now turn our attention to the decomposition of non-frontier portfolios. We have the following theorem.

THEOREM 17.4.1 *Let* **u** *be any portfolio. Then* **u** *is uncorrelated with all frontier portfolios if and only if* **u** *is a zero-mean hedge portfolio.*

Proof: We know from (17.57) that we can write any frontier portfolio as

$$\mathbf{b} = \frac{\gamma}{C}\mathbf{V}^{-1}\mathbf{1} + \frac{\lambda}{A}\mathbf{V}^{-1}\mathbf{e} \tag{17.74}$$

Portfolios **b** and **u** are uncorrelated if and only if

$$\mathbf{u}^\top \mathbf{V} \mathbf{b} = 0 \tag{17.75}$$

if and only if

$$\frac{\gamma}{C}\mathbf{u}^\top \mathbf{1} + \frac{\lambda}{A}\mathbf{u}^\top \mathbf{e} = 0 \tag{17.76}$$

But (17.76) holds for all γ, λ (i.e. for all frontier portfolios) if and only if

$$\mathbf{u}^\top \mathbf{1} = \mathbf{u}^\top \mathbf{e} = 0 \tag{17.77}$$

if and only if **u** is a zero-mean hedge portfolio. $\square$

Similarly, since the MVP is collinear with $\mathbf{V}^{-1}\mathbf{1}$, it is orthogonal to all portfolios **w** for which $\mathbf{w}^\top \mathbf{V} \mathbf{V}^{-1}\mathbf{1} = 0$, or in other words to all portfolios for which $\mathbf{w}^\top \mathbf{1} = 0$. But these are precisely all hedge portfolios.

Similarly again, any portfolio collinear with $\mathbf{V}^{-1}\mathbf{e}$ is orthogonal to all portfolios with zero expected return, since for any such portfolio, $\mathbf{w}^\top \mathbf{V} \mathbf{V}^{-1}\mathbf{e} = \mathbf{w}^\top \mathbf{e} = 0$; in particular, **g** and $\mathbf{V}^{-1}\mathbf{e}$ are orthogonal, so that $\mathbf{z_g} = (1/A)\mathbf{V}^{-1}\mathbf{e}$.

In fact, what Theorem 17.4.1 shows is that the portfolio space can be decomposed into the two-dimensional portfolio frontier and its $(N - 2)$-dimensional orthogonal complement, the set of zero-mean, zero-cost portfolios. Equivalently, the portfolio space is the direct sum of the portfolio frontier and the set of zero-mean, zero-cost portfolios. Any (frontier or non-frontier) portfolio **q** with non-zero cost W_0 can be written in the form $\mathbf{f_q} + \mathbf{u_q}$, where

$$\begin{aligned}\mathbf{f_q} &\equiv W_0(\mathbf{g} + E[\tilde{r}_\mathbf{q}]\mathbf{h}) \\ &= W_0(\beta_{\mathbf{qp}}\mathbf{p} + (1 - \beta_{\mathbf{qp}})\mathbf{z_p}) \quad \text{(say)}\end{aligned} \tag{17.78}$$

is the frontier portfolio costing W_0 with expected return $E[\tilde{r}_\mathbf{q}]$ and

$$\mathbf{u_q} \equiv \mathbf{q} - \mathbf{f_q} \tag{17.79}$$

is a hedge portfolio with zero expected return. Theorem 17.4.1 shows that any portfolio sharing these properties of $\mathbf{u_q}$ is uncorrelated with all frontier portfolios.

Geometrically, this decomposition is equivalent to the orthogonal projection of $\mathbf{q}$ onto the frontier. In fact, the three components in the decomposition of $\mathbf{q}$ (i.e. the vectors $\mathbf{p}$, $\mathbf{z_p}$ and $\mathbf{u_q}$) are mutually orthogonal.

The portfolio frontier is a (two-dimensional) plane in $\mathbb{R}^N$; it is not an affine hyperplane for $N > 3$. Thus, in the context of the discussion in Section 7.4.1, it should be noted that (for $N > 3$) it is possible to travel around the frontier without passing through it. In fact, the space surrounding the frontier is filled with non-frontier portfolios, in the same way as a point in the plane or a line in 3-space is surrounded by a higher number of dimensions.

Furthermore, it should now be evident that not only is $\mathbf{z_{f_q}}$ the unique unit-cost frontier portfolio that is orthogonal to $\mathbf{f_q}$, but it is also the unique unit-cost frontier portfolio that is orthogonal to $\mathbf{q}$ itself, and so we can extend the notion of zero-covariance frontier portfolios, originally defined for frontier portfolios only, to any portfolio. In fact, all portfolios with expected return μ are orthogonal to the same unique unit-cost frontier portfolio (with expected return μ_z), which we will henceforth denote $\mathbf{z}_\mu$ rather than $\mathbf{z_p}$ where appropriate.

Astute readers will have realized that the choice of the symbol β (rather than the normal λ) to describe affine combinations of the unit-cost frontier portfolios $\mathbf{p}$ and $\mathbf{z_p}$ is deliberate (and not just a means of avoiding confusion with the Lagrange multiplier λ). We will now indicate the reasons for this choice of notation.

If $\mathbf{q}$ is any unit-cost portfolio, we can rewrite the decomposition as

$$\mathbf{q} = \mathbf{f_q} + \mathbf{u_q} = \beta_{qp}\mathbf{p} + (1 - \beta_{qp})\mathbf{z_p} + \mathbf{u_q} \tag{17.80}$$

Since $\mathrm{Cov}[\tilde{r}_{\mathbf{u_q}}, \tilde{r}_{\mathbf{p}}] = \mathrm{Cov}[\tilde{r}_{\mathbf{z_p}}, \tilde{r}_{\mathbf{p}}] = 0$, taking covariances of returns with $\tilde{r}_{\mathbf{p}}$ in (17.80) gives

$$\mathrm{Cov}[\tilde{r}_{\mathbf{q}}, \tilde{r}_{\mathbf{p}}] = \mathrm{Cov}[\tilde{r}_{\mathbf{f_q}}, \tilde{r}_{\mathbf{p}}] = \beta_{qp}\mathrm{Var}[\tilde{r}_{\mathbf{p}}] \tag{17.81}$$

or

$$\beta_{qp} = \frac{\mathrm{Cov}[\tilde{r}_{\mathbf{q}}, \tilde{r}_{\mathbf{p}}]}{\mathrm{Var}[\tilde{r}_{\mathbf{p}}]} \tag{17.82}$$

Thus β in (17.78) has its usual definition from probability theory, given by (13.41), which stated that the β of the random variable $\tilde{y}$ with respect to the random variable $\tilde{x}$ is

$$\beta = \frac{\mathrm{Cov}[\tilde{x}, \tilde{y}]}{\mathrm{Var}[\tilde{x}]} \tag{17.83}$$

We will sometimes refer to $\mathbf{z_p}$, which we have heretofore called the zero-covariance frontier portfolio of $\mathbf{p}$, as the **zero-beta frontier portfolio** of p.

Reversing the roles of $\mathbf{p}$ and $\mathbf{z_p}$, it can be seen that

$$\beta_{q\mathbf{z_p}} = 1 - \beta_{qp} \tag{17.84}$$

We can extend the orthogonal portfolio decomposition (17.80) to cover not only

- the original portfolio proportions (viewed as mutually orthogonal vectors);

but also

- portfolio proportions (viewed as scalars or components)

$$q_i = \beta_{\mathbf{qp}} p_i + (1 - \beta_{\mathbf{qp}}) z_i + u_i(\mathbf{q}) \tag{17.85}$$

where q_i, p_i, z_i and $u_i(\mathbf{q})$ denote, respectively, the ith components of $\mathbf{q}$, $\mathbf{p}$, $\mathbf{z_p}$ and $\mathbf{u_q}$;
- returns (mutually uncorrelated random variables)

$$\tilde{r}_{\mathbf{q}} = \beta_{\mathbf{qp}} \tilde{r}_{\mathbf{p}} + (1 - \beta_{\mathbf{qp}}) \tilde{r}_{\mathbf{z_p}} + \tilde{r}_{\mathbf{u_q}} \tag{17.86}$$

and
- expected returns (numbers)

$$E[\tilde{r}_{\mathbf{q}}] = \beta_{\mathbf{qp}} E[\tilde{r}_{\mathbf{p}}] + (1 - \beta_{\mathbf{qp}}) E[\tilde{r}_{\mathbf{z_p}}] \tag{17.87}$$

since the expected value of the disturbance term $E[\tilde{r}_{\mathbf{u_q}}]$ is zero.

We could even write down a similar decomposition of the variance of portfolio returns; however, we will derive a more useful variance decomposition in (17.103).

Equation (17.87), or the equivalent

$$E[\tilde{r}_{\mathbf{q}}] - E[\tilde{r}_{\mathbf{z_p}}] = \beta_{\mathbf{qp}}(E[\tilde{r}_{\mathbf{p}}] - E[\tilde{r}_{\mathbf{z_p}}]) \tag{17.88}$$

may be familiar to some readers from earlier courses in financial economics. It is important to note that these equations are quite general and require neither asset returns to be normally distributed nor any assumptions about preferences.

The next theorem states conditions on distributions and preferences, which together ensure that investors hold frontier portfolios.

THEOREM 17.4.2 *All risk-averse investors with expected-utility preferences prefer the frontier component of the portfolio* $\mathbf{q}$ *(i.e.* $\mathbf{f_q}$*) to* $\mathbf{q}$ *itself, for all portfolios* $\mathbf{q}$*, if and only if*

$$E[\tilde{r}_{\mathbf{u_q}} \mid \tilde{r}_{\mathbf{f_q}}] = 0 \quad \forall \mathbf{q} \tag{17.89}$$

Note the subtle distinction between uncorrelated returns (in the definition of the decomposition) and independent returns (in this theorem). It is a mathematical fact that $\mathrm{Corr}[\tilde{r}_{\mathbf{u_q}}, \tilde{r}_{\mathbf{f_q}}] = 0$ for all $\mathbf{q}$, whatever the probability distribution of asset returns. Equation (17.89), on the other hand, is true only when asset returns are normally distributed or follow a related distribution.

Proof:

(a) First suppose that all risk-averse investors with expected-utility preferences prefer $\mathbf{f_q}$ to $\mathbf{q}$, for all portfolios $\mathbf{q}$.

Then for any frontier portfolio $\mathbf{p}$ and any zero-cost zero-mean portfolio $\mathbf{u}$, such investors prefer $\mathbf{p}$ to $\mathbf{p} + k\mathbf{u}$ for $k \neq 0$. Thus, for any initial wealth W_0 and any concave utility function v, $k = 0$ must solve the utility-maximization problem

$$\max_{k \in \mathbb{R}} E[v(W_0(\tilde{r}_\mathbf{p} + k\tilde{r}_\mathbf{u}))] \tag{17.90}$$

which has first-order condition

$$E[v'(W_0(\tilde{r}_\mathbf{p} + k\tilde{r}_\mathbf{u}))\tilde{r}_\mathbf{u}] = 0 \tag{17.91}$$

or, setting $k = 0$,

$$E[v'(W_0\tilde{r}_\mathbf{p})\tilde{r}_\mathbf{u}] = 0 \tag{17.92}$$

We will now prove (17.89) by contradiction.

Suppose (17.89) does not hold, i.e.

$$m(r) \equiv E[\tilde{r}_\mathbf{u} \mid \tilde{r}_\mathbf{p} = r] \neq 0 \tag{17.93}$$

for some value of r, for some $\mathbf{p}$ and $\mathbf{u}$. (Since $\mathbf{p}$ and $\mathbf{u}$ are arbitrary, $\mathbf{p} + \mathbf{u}$ could be any portfolio.)

Since $\mathbf{u}$ is a zero-mean portfolio,

$$\begin{aligned} E[\tilde{r}_\mathbf{u}] &= E[E[\tilde{r}_\mathbf{u} \mid \tilde{r}_\mathbf{p}]] \\ &= E[m(\tilde{r}_\mathbf{p})] \\ &= \int_{-\infty}^{\infty} m(r)\, dF_{\tilde{r}_\mathbf{p}}(r) \\ &= 0 \end{aligned} \tag{17.94}$$

But, by our hypothesis (17.93), for some r^*,

$$c \equiv \int_{-\infty}^{r^*} m(r)\, dF_{\tilde{r}_\mathbf{p}}(r) = -\int_{r^*}^{\infty} m(r)\, dF_{\tilde{r}_\mathbf{p}}(r) \neq 0 \tag{17.95}$$

see Exercise 17.17.

The first-order condition (17.92) holds for any concave function v and any initial wealth W_0, in particular for the piecewise linear utility function defined by

$$v(W) = \begin{cases} k_1 W & \text{if } W \leq W_0 r^* \\ k_1 W_0 r^* + k_2(W - W_0 r^*) & \text{if } W \geq W_0 r^* \end{cases} \tag{17.96}$$

where $k_2 < k_1$. For this utility function,

$$v'(W) = \begin{cases} k_1 & \text{if } W \leq W_0 r^* \\ k_2 & \text{if } W \geq W_0 r^* \end{cases} \tag{17.97}$$

Thus

$$
\begin{aligned}
0 &= E[v'(W_0\tilde{r}_{\mathbf{p}})\tilde{r}_{\mathbf{u}}] \\
&= E[E[v'(W_0\tilde{r}_{\mathbf{p}})\tilde{r}_{\mathbf{u}} \mid \tilde{r}_{\mathbf{p}}]] \\
&= E[v'(W_0\tilde{r}_{\mathbf{p}})E[\tilde{r}_{\mathbf{u}} \mid \tilde{r}_{\mathbf{p}}]] \\
&= E[v'(W_0\tilde{r}_{\mathbf{p}})m(\tilde{r}_{\mathbf{p}})] \\
&= \int_{-\infty}^{r^*} k_1 m(r)\,dF_{\tilde{r}_{\mathbf{p}}}(r) + \int_{r^*}^{\infty} k_2 m(r)\,dF_{\tilde{r}_{\mathbf{p}}}(r) \\
&= k_1 c - k_2 c \\
&\neq 0
\end{aligned}
\tag{17.98}
$$

which is a contradiction, and hence

$$
m(r) \equiv E[\tilde{r}_{\mathbf{u}} \mid \tilde{r}_{\mathbf{p}} = r] = 0 \quad \forall\, r, \mathbf{p}, \mathbf{u}
\tag{17.99}
$$

(b) Now suppose that (17.89) holds.
   Then for any concave function v and any initial wealth W_0,

$$
\begin{aligned}
E[v(W_0\tilde{r}_{\mathbf{q}})] &= E[v(W_0(\tilde{r}_{\mathbf{f_q}} + \tilde{r}_{\mathbf{uq}}))] \\
&= E[E[v(W_0(\tilde{r}_{\mathbf{f_q}} + \tilde{r}_{\mathbf{uq}})) \mid \tilde{r}_{\mathbf{f_q}}]] \\
&\leq E[v(E[W_0(\tilde{r}_{\mathbf{f_q}} + \tilde{r}_{\mathbf{uq}}) \mid \tilde{r}_{\mathbf{f_q}}])] \\
&= E[v(W_0\tilde{r}_{\mathbf{f_q}})]
\end{aligned}
\tag{17.100}
$$

where the inequality follows from Jensen's inequality and the final step from (17.89). But this is precisely what we set out to prove.

The above proof is based on Huang and Litzenberger (1988, pp. 85–8). $\square$

COROLLARY 17.4.3 *If condition (17.89) holds, then all risk-averse investors with expected-utility preferences choose frontier portfolios.*

Proof: By the previous theorem, no such investor will choose a non-frontier portfolio $\mathbf{q}$ over its frontier component $\mathbf{f_q}$. $\square$

17.4.2 *Portfolio frontier in mean–variance space: risky assets only*

We now move on to consider the mean–variance and (equivalent) mean–standard deviation relationships along the line in $\mathbb{R}^N$ containing the unit-cost frontier portfolios. In other words, we wish to graph the envelope function for the portfolio variance minimization problem (17.39). In general, the envelope function gives the minimum variance achievable as a function of all the exogenous parameters of the problem, $\sigma^2(\mu, W_0, \mathbf{e}, \mathbf{V})$. In this section, we are mainly interested in the relationship between the minimum variance achievable and a single exogenous parameter, the desired expected rate of return, μ. In what follows, we will ignore the relationship between σ^2 and the other exogenous parameters, W_0, $\mathbf{e}$ and $\mathbf{V}$.

As mentioned in Section 16.8, it is conventional to plot the mean return on the vertical axis and the variance or standard deviation of returns on the horizontal axis, although both the conventional presentation of the envelope theorem and the conventional interpretation of "mean–variance space" or "xy space" would suggest otherwise.

Using (17.62), we can write the typical unit-cost frontier portfolio $\mathbf{p}$ as

$$\mathbf{p} = \text{MVP} + \left(\mu - \frac{A}{C}\right)\mathbf{h} \tag{17.101}$$

or, in notation that perhaps better reflects the inherent structure of the frontier,

$$\mathbf{p} = \text{MVP} + (\mu - E[\tilde{r}_{\text{MVP}}])\mathbf{h} \tag{17.102}$$

since $E[\tilde{r}_{\text{MVP}}] = A/C$.

Taking variances on both sides of the orthogonal decomposition in (17.102), which is effectively applying Pythagoras's theorem to the right-angled triangle in Figure 17.2 with vertices at $\mathbf{0}$, $\mathbf{p}$ and MVP, yields

$$\sigma^2 \equiv \text{Var}[\tilde{r}_{\mathbf{p}}] = \text{Var}[\tilde{r}_{\text{MVP}}] + (\mu - E[\tilde{r}_{\text{MVP}}])^2 \text{Var}[\tilde{r}_{\mathbf{h}}] \tag{17.103}$$

Recall from the coordinate geometry of conic sections (Section 4.2, in particular (4.4)) that (17.103) or the equivalent

$$V(\mu) = \frac{1}{C} + \frac{C}{D}\left(\mu - \frac{A}{C}\right)^2 \tag{17.104}$$

which is a quadratic equation in μ, is the equation of the parabola in mean–variance space with vertex at

$$V(\mu) = \text{Var}[\tilde{r}_{\text{MVP}}] = \frac{1}{C} \tag{17.105}$$

$$\mu = E[\tilde{r}_{\text{MVP}}] = \frac{A}{C} \tag{17.106}$$

Thus in mean–variance space, the frontier is a parabola; see Figure 17.3.

Similarly, in mean–standard deviation space, the frontier is a hyperbola, or at least the part of a hyperbola lying in the half-plane where $\sigma \geq 0$. To see this, recall by comparison with (4.16) that (17.103) is the equation of the hyperbola with vertex at

$$\sigma = \sqrt{\text{Var}[\tilde{r}_{\text{MVP}}]} = \sqrt{\frac{1}{C}} \tag{17.107}$$

$$\mu = E[\tilde{r}_{\text{MVP}}] = \frac{A}{C} \tag{17.108}$$

centre at $\sigma = 0$, $\mu = A/C$ and asymptotes as indicated in Figure 17.4. The other half of the hyperbola ($\sigma < 0$) has no economic meaning.

Recall that (17.103) could also represent two other types of conic sections. In the case of $\text{Var}[\tilde{r}_{\mathbf{h}}] < 0$ (which, of course, is impossible) it represents an ellipse with centre $(0, A/C)$.

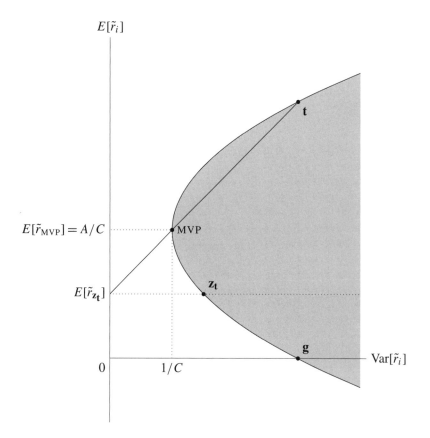

Figure 17.3 Portfolio frontier in mean–variance space: risky assets only

In the more practical case of $\text{Var}[\tilde{r}_{\text{MVP}}] = 0$, which essentially represents the presence of a risk-free asset, the square root can be taken on both sides:

$$\sigma = \pm(\mu - E[\tilde{r}_{\text{MVP}}])\sqrt{\text{Var}[\tilde{r}_{\mathbf{h}}]} \qquad (17.109)$$

In other words, the hyperbola then becomes the pair of lines that are its asymptotes otherwise.

Note that, when $N > 3$, there is a unique unit-cost portfolio in portfolio space corresponding to each point on the mean–variance or mean–standard deviation frontier, but that there are infinitely many points in portfolio space corresponding to each point inside the mean–variance or mean–standard deviation frontier.

When $N = 2$, all portfolios are frontier portfolios and there are no portfolios at all corresponding to points inside the mean–variance or mean–standard deviation frontiers.

When $N = 3$, the orthogonal complement of the portfolio frontier is a one-dimensional subspace of the portfolio space, spanned by any zero-cost, zero-mean portfolio, say, the unit-variance portfolio $\mathbf{u}$. Thus, there are exactly two points in portfolio space corresponding to each (σ^2, μ) or (σ, μ) point inside the mean–variance or mean–standard deviation frontier,

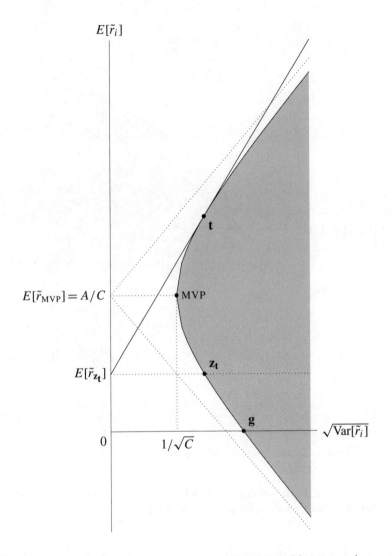

Figure 17.4 Portfolio frontier in mean–standard deviation space: risky assets only

respectively. They are

$$\text{MVP} + (\mu - E[\tilde{r}_{\text{MVP}}])\mathbf{h} \pm \sqrt{\sigma^2 - \text{Var}[\tilde{r}_{\text{MVP}}] - (\mu - E[\tilde{r}_{\text{MVP}}])^2 \text{Var}[\tilde{r}_{\mathbf{h}}]}\,\mathbf{u} \qquad (17.110)$$

Frontier portfolios on which the expected return, μ, exceeds $E[\tilde{r}_{\text{MVP}}]$ are termed **efficient**, since they maximize expected return given variance; other frontier portfolios minimize expected return given variance and are termed **inefficient**.

In other words, a frontier portfolio is an efficient portfolio if and only if its expected return exceeds the minimum-variance expected return $A/C = E[\tilde{r}_{\text{MVP}}]$.

The set of efficient unit-cost portfolios in $\mathbb{R}^N$, known as the **efficient frontier** or sometimes as the **Markowitz frontier**,[4] is the half-line emanating from the MVP in the direction of **h**, and, hence, like the set of all unit-cost frontier portfolios, is also a convex set (but not an affine set). While all affine combinations of frontier portfolios are frontier portfolios, only convex combinations of efficient portfolios are guaranteed to be efficient.

Along the efficient frontier, the dual problem to the variance minimization problem is that of expected return maximization; along the inefficient part of the frontier, the dual problem is that of expected return minimization.

We now consider zero-covariance (zero-beta) portfolios. There are two neat tricks that allow zero-covariance portfolios to be plotted in mean–standard deviation and mean–variance space, respectively. We begin with mean–standard deviation space.

Implicit differentiation with respect to σ of the μ–σ relationship along the frontier (17.103) reveals that the frontier in mean–standard deviation space has slope

$$\frac{d\mu}{d\sigma} = \frac{\sigma}{(\mu - E[\tilde{r}_{\mathrm{MVP}}]) \operatorname{Var}[\tilde{r}_{\mathbf{h}}]} \tag{17.111}$$

The tangent to this frontier at (σ, μ) intercepts the μ axis at

$$
\begin{aligned}
\mu - \sigma \frac{d\mu}{d\sigma} &= \mu - \frac{\sigma^2}{(\mu - E[\tilde{r}_{\mathrm{MVP}}]) \operatorname{Var}[\tilde{r}_{\mathbf{h}}]} \\
&= \mu - \frac{\operatorname{Var}[\tilde{r}_{\mathrm{MVP}}]}{(\mu - E[\tilde{r}_{\mathrm{MVP}}]) \operatorname{Var}[\tilde{r}_{\mathbf{h}}]} - (\mu - E[\tilde{r}_{\mathrm{MVP}}]) \\
&= E[\tilde{r}_{\mathrm{MVP}}] - \frac{\operatorname{Var}[\tilde{r}_{\mathrm{MVP}}]}{(\mu - E[\tilde{r}_{\mathrm{MVP}}]) \operatorname{Var}[\tilde{r}_{\mathbf{h}}]}
\end{aligned}
\tag{17.112}
$$

where we have substituted for σ^2 from the equation of the frontier (17.103).

A little rearrangement of (17.70) shows that the expression on the right-hand side of (17.112) is just the expected return on the zero-covariance frontier portfolio of any portfolio with expected return μ. This geometric construction is illustrated in Figure 17.4.

To find $\mathbf{z}_\mu$ in mean–variance space, note that the line joining (σ^2, μ) to the MVP intercepts the μ axis at

$$\mu - \sigma^2 \frac{\mu - E[\tilde{r}_{\mathrm{MVP}}]}{\sigma^2 - \operatorname{Var}[\tilde{r}_{\mathrm{MVP}}]} = \mu - \sigma^2 \frac{\mu - E[\tilde{r}_{\mathrm{MVP}}]}{(\mu - E[\tilde{r}_{\mathrm{MVP}}])^2 \operatorname{Var}[\tilde{r}_{\mathbf{h}}]} \tag{17.113}$$

After cancellation, this is exactly the expression for the zero-covariance return that we had in the first line of (17.112). This geometric construction is illustrated in Figure 17.3.

17.4.3 *Portfolio frontier in* $\mathbb{R}^N$*: risk-free and risky assets*

We now consider the mathematics of the portfolio frontier when there is a risk-free asset. The investor now wishes to choose the portfolio **b** to minimize the variance of date-1 wealth, $\widetilde{W}_1 = \mathbf{b}^\top \tilde{\mathbf{r}} + (W_0 - \mathbf{b}^\top \mathbf{1}) r_{\mathrm{f}}$, subject to attaining an expected date-1 wealth of at least μW_0.

The risk-free rate is unique by the no-arbitrage principle, since otherwise a greedy investor would borrow an infinite amount at the lower rate and invest it at the higher rate, which is impossible in equilibrium. Similarly, as already mentioned on p. 459, the no-arbitrage principle allows us to rule out variance–covariance matrices for risky assets that permit the construction of portfolios with zero return variance, i.e. synthetic risk-free assets.

In the presence of a risk-free asset, the frontier portfolio solves another instance of the canonical quadratic programming problem of Section 14.3, namely,

$$\min_{\mathbf{b}} \ \mathbf{b}^\top \mathbf{V} \mathbf{b} \tag{17.114}$$
$$\text{s.t. } \mathbf{b}^\top \mathbf{e} + (W_0 - \mathbf{b}^\top \mathbf{1}) r_{\mathrm{f}} \geq \mu W_0$$

There is no longer a restriction on portfolio weights, and whatever is not invested in the N risky assets is assumed to be invested in the risk-free asset.

The solution (see Exercise 17.18) is obtained by a method similar to the case where all assets were risky and is

$$\mathbf{b}^* = \frac{W_0(\mu - r_{\mathrm{f}})}{H} \mathbf{V}^{-1}(\mathbf{e} - r_{\mathrm{f}}\mathbf{1}) \tag{17.115}$$

where

$$H = (\mathbf{e} - r_{\mathrm{f}}\mathbf{1})^\top \mathbf{V}^{-1}(\mathbf{e} - r_{\mathrm{f}}\mathbf{1}) = B - 2Ar_{\mathrm{f}} + Cr_{\mathrm{f}}^2 \tag{17.116}$$

It can be shown that $H > 0$ for all r_{f}; see Exercise 17.19.

The unit-cost portfolio corresponding to this optimal solution will be denoted

$$\mathbf{t} \equiv \frac{1}{A - r_{\mathrm{f}}C} \mathbf{V}^{-1}(\mathbf{e} - r_{\mathrm{f}}\mathbf{1}) \tag{17.117}$$

so that

$$\mathbf{b}^* = \frac{W_0(\mu - r_{\mathrm{f}})(A - r_{\mathrm{f}}C)}{H} \mathbf{t} \tag{17.118}$$

The exception to this is when $r_{\mathrm{f}} = A/C$, in which case the optimal solution is to invest all of initial wealth in the risk-free security, and $\mathbf{b}^*$, still defined by (17.115), is a zero-cost, hedge portfolio of risky assets, held in order to increase (or, potentially, reduce) expected return. Assuming that $r_{\mathrm{f}} \neq A/C$ and pre-multiplying (17.117) by $\mathbf{e}^\top$ gives

$$E[\tilde{r}_{\mathbf{t}}] = \frac{B - r_{\mathrm{f}}A}{A - r_{\mathrm{f}}C} \tag{17.119}$$

The vector $\mathbf{t}$ must lie on the frontier of risky assets, as it is a linear combination of $\mathbf{V}^{-1}\mathbf{1}$ and $\mathbf{V}^{-1}\mathbf{e}$, which constitute a basis for that frontier. Comparing (17.119) with (17.72) reveals that $\mathbf{t}$ is nothing other than the zero-beta portfolio of any risky-asset portfolio with expected return equal to the risk-free rate r_{f}.

Note from (17.118) that the sign of the optimal holding of the portfolio $\mathbf{t}$ depends on the signs of $\mu - r_{\mathrm{f}}$ and $A - r_{\mathrm{f}}C$, i.e. on where the risk-free rate lies in relation to (a) the desired expected return, μ, and (b) the expected return on the MVP of risky assets only, A/C. These relationships will become clearer in the next section.

In the N-dimensional portfolio space, the portfolio frontier is now the one-dimensional vector subspace generated by the portfolio $\mathbf{t}$. In the absence of the risk-free asset, the optimal investment strategy could have been described as dividing wealth between the orthogonal portfolios $\mathbf{t}$ (with expected return $(B - r_{\mathrm{f}}A)/(A - r_{\mathrm{f}}C)$) and $\mathbf{z}_{\mathbf{t}}$ (with expected return r_{f}). In the presence of the risk-free asset, the optimal investment strategy is now to divide wealth between the portfolio **t** and the risk-free asset (with fixed return r_f).

Finally, it can be shown that

$$\text{Var}[\tilde{r}_t] = \frac{H}{(A - r_f C)^2} \tag{17.120}$$

and

$$\text{Var}[\tilde{r}_{b*}] = \frac{W_0^2 (\mu - r_f)^2}{H} \tag{17.121}$$

see Exercise 17.20.

17.4.4 *Portfolio frontier in mean–variance space: risk-free and risky assets*

We can now establish the shape of the frontiers in mean–standard deviation and mean–variance space in the presence of a risk-free asset. We begin with the former. Three alternative derivations follow.

1. The first derivation begins by setting $W_0 = 1$ in (17.121), which implies that the relationship between desired expected return μ and minimum attainable standard deviation σ along the frontier in this case is

$$\sigma = \frac{|\mu - r_f|}{\sqrt{H}} \tag{17.122}$$

Thus, in mean–standard deviation space, the frontier is a pair of straight lines crossing the vertical (mean) axis at $\mu = r_f$ and with slopes $\pm\sqrt{H}$; see Figure 17.5.

2. The second derivation reaches the same conclusion graphically by inspection of Figure 17.4. Three separate cases need to be considered.

(a) $r_f < E[\tilde{r}_{\text{MVP}}]$. It follows that

$$E[\tilde{r}_t] > E[\tilde{r}_{\text{MVP}}] > E[\tilde{r}_{z_t}] = r_f \tag{17.123}$$

Recalling the geometrical technique above for determining the locations of pairs of zero-beta portfolios, it can be seen that the tangent from $(0, r_f)$ to the efficient mean–standard deviation frontier of risky assets touches that frontier at **t**, and thus we call **t** the **tangency portfolio**.

(b) $r_f > E[\tilde{r}_{\text{MVP}}]$. The inequalities are now reversed:

$$E[\tilde{r}_t] < E[\tilde{r}_{\text{MVP}}] < E[\tilde{r}_{z_t}] = r_f \tag{17.124}$$

The tangency portfolio **t** in this case lies on the lower, inefficient, part of the mean–standard deviation frontier of risky assets.

(c) $r_f = E[\tilde{r}_{\text{MVP}}]$. As r_f approaches $E[\tilde{r}_{\text{MVP}}]$ from either direction, the mean and variance of the tangency portfolio tend to infinity but the net investment in this portfolio tends to zero. We have already seen that in the limiting case the optimal holding of risky assets

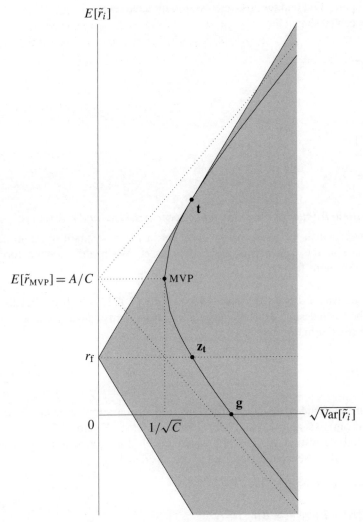

Figure 17.5 Portfolio frontier in mean–standard deviation space: risk-free and risky assets

is a zero-cost portfolio, which therefore cannot be represented on the mean–standard deviation diagram, as this diagram shows only unit-cost portfolios.

3. The third derivation considers what happens when we combine any portfolio **p** with the risk-free asset in proportions a and $(1 - a)$, respectively. This gives a portfolio with expected return

$$aE[\tilde{r}_\mathbf{p}] + (1 - a)r_\mathrm{f} = r_\mathrm{f} + a(E[\tilde{r}_\mathbf{p}] - r_\mathrm{f}) \tag{17.125}$$

and standard deviation of returns

$$a\sqrt{\mathrm{Var}[\tilde{r}_\mathbf{p}]} \tag{17.126}$$

Graphically, as we will see later in Figure 17.7, in mean–standard deviation space ($\sigma\mu$ space), these portfolios trace out the ray emanating from $(0, r_{\mathrm{f}})$, where $a = 0$, and passing through **p**, where $a = 1$. For each σ the highest return attainable is along the ray from r_{f} that is tangent to the frontier generated by the risky assets. On this ray, the risk-free asset is held in combination with the tangency portfolio **t**. Again, this only makes sense for $r_{\mathrm{f}} < A/C = E[\tilde{r}_{\mathrm{MVP}}]$.

Note that, in order to obtain an expected return above that on **t**, the investor must place a negative weight on the risk-free asset, i.e. borrow funds at the risk-free rate for investment, along with his initial wealth, in the tangency portfolio **t**.

Finally, we consider the shape of the frontier in mean–variance space in the presence of a risk-free asset. Equation (17.121) implies that the relationship between desired expected return μ and minimum attainable variance σ^2 along the frontier in this case is

$$\sigma^2 = \frac{(\mu - r_{\mathrm{f}})^2}{H} \tag{17.127}$$

Thus, in mean–variance space, as it was with risky assets only, the frontier is a parabola, in this case with vertex at $(0, r_{\mathrm{f}})$.

The tangency portfolio, **t**, is once again the only point that lies on both the frontier of all assets and the frontier of risky assets only. Thus, in mean–variance space, the tangency portfolio is the point of tangency between the former (outer) parabola and the latter (inner) parabola; see Figure 17.6.

Limited borrowing

Unlimited borrowing and lending at a single risk-free interest rate, as allowed in the preceding analysis, is unrealistic. Figures 17.7 and 17.8 show what happens, respectively, with

1. margin constraints on borrowing, and
2. a different borrowing rate (r_b) and lending rate (r_l).

In the former case, the proportion, β, of initial wealth invested in risky assets may be restricted to no more than, say, 1.25 or 1.5; the latter case is illustrated in Figure 17.7. The frontier is then the envelope of all the finite rays from the risk-free asset through risky portfolios, such as **p** in Figure 17.7, extending as far as the borrowing constraint allows. As Figure 17.7 illustrates, margin constraints on borrowing have less impact on the minimum attainable variance than on the portfolio composition used to attain the minimum variance, say when the desired expected return is $r_{\mathrm{f}} + 1.5(E[\tilde{r}_{\mathbf{p}}] - r_{\mathrm{f}})$ as illustrated.

In the latter case, there are two tangency portfolios, $\mathbf{t}_b$ and $\mathbf{t}_l$, say, with $E[\tilde{r}_{\mathbf{z}_{\mathbf{t}_b}}] = r_b$ and $E[\tilde{r}_{\mathbf{z}_{\mathbf{t}_l}}] = r_l$.

- There is a range of expected returns from $E[\tilde{r}_{\mathbf{t}_l}]$ up to $E[\tilde{r}_{\mathbf{t}_b}]$ over which a pure risky strategy provides minimum variance.
- Lower expected returns are achieved by a combination of risk-free lending at r_l and an investment in the risky asset portfolio $\mathbf{t}_l$.
- Higher expected returns are achieved by risk-free borrowing at r_b to fund an investment of more than the initial wealth in the risky asset portfolio $\mathbf{t}_b$.

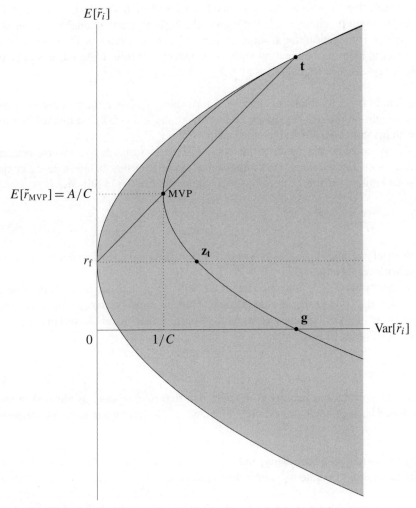

Figure 17.6 Portfolio frontier in mean–variance space: risk-free and risky assets

17.5   Market equilibrium and the capital asset pricing model

17.5.1   *Pricing assets and predicting security returns*

Up to this point, this chapter has dealt solely with the search for optimal investment strategies, which is itself a very important but relatively straightforward aspect of financial economics. The portfolio theory that we have learned so far will be of great assistance in addressing two other aspects of financial economics, which are perhaps of greater popular interest, namely:

- explaining the relationship between contemporaneous equilibrium returns on different assets, asset classes and financial markets; and
- predicting future asset returns.

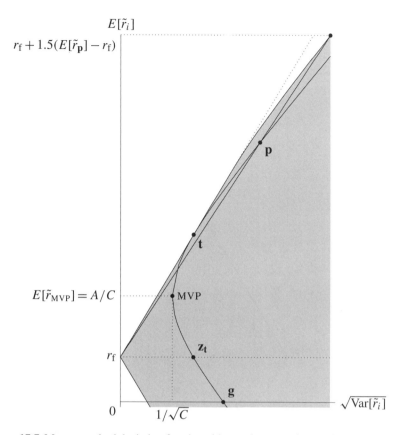

$E[\tilde{r}_i]$

$r_{\rm f} + 1.5(E[\tilde{r}_{\bf p}] - r_{\rm f})$

p

t

$E[\tilde{r}_{\rm MVP}] = A/C$ .... MVP

$\bf z_t$

$r_{\rm f}$

g

$\sqrt{\mathrm{Var}[\tilde{r}_i]}$

0     $1/\sqrt{C}$

Figure 17.7 Mean–standard deviation frontier with margin constraints on borrowing

In general, addressing either of the latter problems requires realistic assumptions concerning investor preferences and the probability distributions of future asset values, which in turn lead to useful and parsimonious asset pricing models, with empirically testable predictions.

The best-known such assumptions are those of the **capital asset pricing model (CAPM)**. At a very basic level, the CAPM is based on the assumption that *each investor holds a mean–variance frontier portfolio*. It has been known since early in the development of portfolio theory that this assumption is valid if, for example, preferences are quadratic or the probability distribution of asset returns is normal.[5]

17.5.2 Properties of the market portfolio

By the **market portfolio** we mean the aggregate of the portfolios held by all I individuals at a given point in time.

Given security prices, let m_j be the weight of security j in the market portfolio $\bf m$, W_{0i} (> 0) be individual i's initial wealth, and a_{ji} be the proportion of individual i's wealth invested in security j. Then total initial wealth is defined by

$$W_{0m} \equiv \sum_{i=1}^{I} W_{0i} \tag{17.128}$$

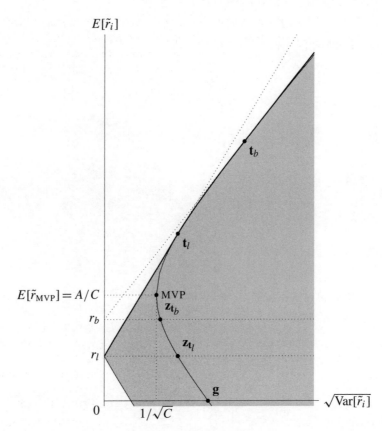

Figure 17.8 Mean–standard deviation frontier with different borrowing and lending rates

and in equilibrium the relation

$$\sum_{i=1}^{I} a_{ji} W_{0i} = m_j W_{0m} \tag{17.129}$$

must hold for all assets j. Dividing by W_{0m} yields

$$\sum_{i=1}^{I} a_{ji} \frac{W_{0i}}{W_{0m}} = m_j \quad \forall \, j \tag{17.130}$$

and thus in equilibrium the market portfolio **m** is a convex combination of the individual portfolios $\mathbf{a}_1, \mathbf{a}_2, \ldots, \mathbf{a}_I$, with weights $W_{01}/W_{0m}, W_{02}/W_{0m}, \ldots, W_{0I}/W_{0m}$, respectively, which, by definition, are positive and sum to unity. This simple observation leads to Theorem 17.5.1, the Black (1972) zero-beta version of the CAPM.

17.5.3   *Zero-beta capital asset pricing model*

THEOREM 17.5.1 (ZERO-BETA CAPITAL ASSET PRICING MODEL). *If each investor holds a mean–variance frontier portfolio (for example, if (17.89) holds and all investors are risk-averse expected-utility maximizers), then the market portfolio, **m**, is a mean–variance frontier portfolio, and, hence, by (17.87), for all portfolios **q**, the **CAPM equation***

$$E[\tilde{r}_{\mathbf{q}}] = (1 - \beta_{\mathbf{qm}}) E[\tilde{r}_{\mathbf{z_m}}] + \beta_{\mathbf{qm}} E[\tilde{r}_{\mathbf{m}}] \tag{17.131}$$

holds, where

$$\tilde{r}_{\mathbf{m}} = \sum_{j=1}^{N} m_j \tilde{r}_j \tag{17.132}$$

and

$$\beta_{\mathbf{qm}} = \frac{\mathrm{Cov}[\tilde{r}_{\mathbf{q}}, \tilde{r}_{\mathbf{m}}]}{\mathrm{Var}[\tilde{r}_{\mathbf{m}}]} \tag{17.133}$$

While we already know that it is possible to compute the β of any random variable with respect to any other random variable (see (13.41)), or of any portfolio with respect to any frontier portfolio (see (17.73)), the term **beta** or, more properly, **market beta** is commonly used to refer to the β of an asset or portfolio with respect to the market portfolio. Note that the market beta of the market portfolio itself is, by definition, equal to 1.

As a corollary to Theorem 17.5.1, we have another two-fund separation result: under the assumptions of the zero-beta CAPM theorem, every investor effectively holds a combination of the market portfolio, **m**, and its zero-beta frontier portfolio, $\mathbf{z_m}$. Equivalently, we could say that every investor holds a combination of two arbitrarily fixed frontier portfolios.

The zero-beta CAPM implies, for any particular individual security, say, the jth, that

$$E[\tilde{r}_j] = (1 - \beta_{jm}) E[\tilde{r}_{\mathbf{z_m}}] + \beta_{jm} E[\tilde{r}_{\mathbf{m}}] \tag{17.134}$$

since an individual security is just a portfolio with a weight of 1 on one asset and weights of 0 on all other assets.

Equation (17.134) says that, when CAPM holds, the only parameters of the potentially complex multivariate probability distribution of future asset values that influence expected returns are the N market betas of the individual assets and the expected returns on the market portfolio and its zero-beta portfolio.

17.5.4   *Traditional capital asset pricing model*

Now we can derive the traditional CAPM by adding the risk-free asset, which in turn determines the tangency portfolio, **t**. Note that, by construction,

$$r_{\mathrm{f}} = E[\tilde{r}_{\mathbf{z_t}}] \tag{17.135}$$

Normally in equilibrium there is zero aggregate supply of the risk-free asset, but the traditional CAPM holds regardless of whether or not this is the case.

The two theorems that follow should be self-evident from what has gone before.

THEOREM 17.5.2 (SEPARATION THEOREM). *In the presence of a risk-free asset, the risky asset holdings of all investors who hold mean–variance frontier portfolios are in the proportions given by the tangency portfolio,* **t**.

THEOREM 17.5.3 (TRADITIONAL CAPITAL ASSET PRICING MODEL). *If each investor holds a mean–variance frontier portfolio, then the market portfolio of risky assets,* **m**, *is the tangency portfolio,* **t**, *and, hence, for all portfolios* **q**, *the traditional CAPM equation*

$$E[\tilde{r}_{\mathbf{q}}] = (1 - \beta_{\mathbf{qm}})r_f + \beta_{\mathbf{qm}}E[\tilde{r}_{\mathbf{m}}] \tag{17.136}$$

holds.

Theorem 17.5.3 is sometimes known as the Sharpe–Lintner theorem.[6]

As with the zero-beta CAPM equation, the traditional CAPM equation applies to portfolios consisting of a single asset, so can also be written

$$E[\tilde{r}_j] = (1 - \beta_{j\mathbf{m}})r_f + \beta_{j\mathbf{m}}E[\tilde{r}_{\mathbf{m}}] \tag{17.137}$$

It can also be written in terms of risk premiums as

$$E[\tilde{r}_j - r_f] = \beta_{j\mathbf{m}}E[\tilde{r}_{\mathbf{m}} - r_f] \tag{17.138}$$

There are many important implications of this equation. Provided that the market portfolio is on the efficient part of the mean–variance frontier, it implies that assets with higher market betas have higher risk premiums. Similarly, an asset with a market beta of zero returns only the risk-free rate. Finally, an asset with a negative market beta returns less than the risk-free rate.

Recall at this point the discussion on p. 408. Like the duration of a bond, the beta of a risky asset measures the responsiveness of the price of the asset to changes in general market conditions.

There are two commonly encountered graphical representations of the traditional CAPM equation (17.136).

1. The **capital market line** is the line in mean–standard deviation space connecting the risk-free asset to the market portfolio, illustrating the expected return–standard deviation relationship on frontier portfolios when the CAPM holds; see Figure 17.9. It is just the mean–standard deviation portfolio frontier, with the additional insight that it passes through the market portfolio under the CAPM assumptions.
2. The security market line is the line in market beta-expected return space connecting the risk-free asset at $(0, r_f)$ (or $\mathbf{z}_{\mathbf{m}}$ if there is no risk-free asset) to the market portfolio at $(1, E[\tilde{r}_{\mathbf{m}}])$, and thus having equation

$$\mu = r_f + (E[\tilde{r}_{\mathbf{m}}] - r_f)\beta \tag{17.139}$$

As in mean–standard deviation and mean–variance analysis, it is conventional to plot expected return on the vertical axis when drawing the security market line; see Figure 17.10. If the traditional CAPM holds, then, by (17.136), the *population* means and betas for all portfolios and individual securities (such as the illustrated security i) must lie on the security market line.

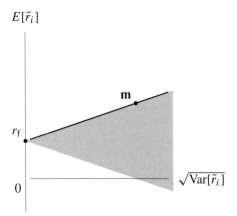

Figure 17.9 Capital market line

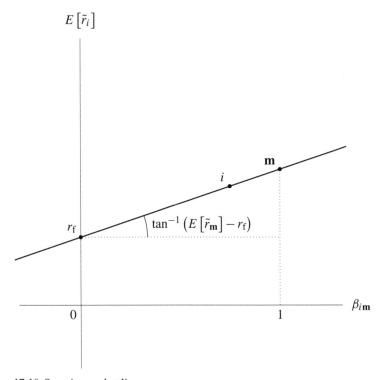

Figure 17.10 Security market line

One basic graphical empirical test of the CAPM is to plot *sample* mean returns against *sample* market betas and see whether the resulting scatterplot is close to being a straight line.

More rigorously, one could run a cross-sectional linear regression of sample mean returns against sample market betas and test the hypotheses that the intercept equals the risk-free rate and that the slope equals the risk premium on the market portfolio, as in (17.139). A whole branch of econometric analysis has grown up around the empirical testing of the CAPM, but it is outside the scope of this book; the interested reader is referred to Huang and Litzenberger (1988, Chapter 10) and Campbell et al. (1997, Chapter 5).

We can also think about what happens to the CAPM if there are different risk-free borrowing and lending rates as considered in Figure 17.8; see Elton *et al.* (2010, p. 88). If all individuals face this situation in equilibrium, realism demands that both risk-free assets are in zero aggregate supply and, hence, that all investors hold risky assets only.

17.5.5   *Risk premium of the market portfolio*

When a risk-free asset exists, we know that by construction it must lie on the opposite side of the MVP from the tangency portfolio, which under the CAPM assumptions must be the market portfolio of risky assets. Nothing so far has indicated whether or not the equilibrium expected return on the market portfolio of risky assets will exceed the risk-free rate of return. The next theorem gives sufficient conditions for the mean–variance efficiency of the market portfolio.

THEOREM 17.5.4 *If*

(a) *the CAPM assumptions are satisfied,*
(b) *risky assets are in strictly positive supply, and*
(c) *investors have strictly increasing and concave utility functions,*

then the market (i.e. tangency) portfolio is efficient, with

$$r_f < E[\tilde{r}_{\text{MVP}}] < E[\tilde{r}_{\mathbf{m}}] \tag{17.140}$$

Proof: The risk-free asset dominates any portfolio whose return $\tilde{r}$ satisfies

$$E[\tilde{r}] < r_f \tag{17.141}$$

since, by Jensen's inequality and concavity and monotonicity of the utility function, respectively,

$$E[u(W_0\tilde{r})] \leq u(E[W_0\tilde{r}])$$
$$< u(W_0 r_f) \tag{17.142}$$

Hence, the expected returns on all individuals' optimal equilibrium portfolios exceed r_f. It follows by a convexity argument that the expected return on the market portfolio must exceed r_f, provided that risky assets are in strictly positive supply.   □

This result says that the risk premium of the market portfolio is positive; we can go much further than this. The CAPM gives a relation between the risk premiums on individual assets and the risk premium on the market portfolio. The risk premium on the market portfolio must in turn adjust in equilibrium to give market-clearing. In some situations, the risk premium on the market portfolio can be written in terms of investors' utility functions.

For example, assume that there is a risk-free asset and that returns are multivariate normal. Recall the first-order condition (17.7) for the canonical portfolio choice problem for investor i and asset j:

$$
\begin{aligned}
0 &= E[u_i'(\widetilde{W}_{1i})(\tilde{r}_j - r_{\mathrm{f}})] \\
&= E[u_i'(\widetilde{W}_{1i})]E[\tilde{r}_j - r_{\mathrm{f}}] + \mathrm{Cov}[u_i'(\widetilde{W}_{1i}), \tilde{r}_j] \\
&= E[u_i'(\widetilde{W}_{1i})]E[\tilde{r}_j - r_{\mathrm{f}}] + E[u_{1i}''(\widetilde{W}_{1i})]\mathrm{Cov}[\widetilde{W}_{1i}, \tilde{r}_j]
\end{aligned}
\tag{17.143}
$$

using the definition of covariance and Stein's lemma (Theorem 13.7.1) for MVN distributions. Rearranging gives

$$
\frac{E[\tilde{r}_j - r_{\mathrm{f}}]}{\theta_i} = \mathrm{Cov}[\widetilde{W}_{1i}, \tilde{r}_j]
\tag{17.144}
$$

where

$$
\theta_i \equiv \frac{-E[u_i''(\widetilde{W}_{1i})]}{E[u_i'(\widetilde{W}_{1i})]}
\tag{17.145}
$$

is the ith investor's **global absolute risk aversion**. Since

$$
\widetilde{W}_{1i} = W_{0i}\left(r_{\mathrm{f}} + \sum_{k=1}^{N} a_{ki}(\tilde{r}_k - r_{\mathrm{f}})\right)
\tag{17.146}
$$

we have, by taking the covariance of each side with $\tilde{r}_j$ and dropping non-stochastic terms,

$$
\mathrm{Cov}[\widetilde{W}_{1i}, \tilde{r}_j] = \mathrm{Cov}\left[W_{0i}\sum_{k=1}^{N} a_{ki}\tilde{r}_k, \tilde{r}_j\right]
\tag{17.147}
$$

Hence,

$$
\frac{E[\tilde{r}_j - r_{\mathrm{f}}]}{\theta_i} = \mathrm{Cov}\left[W_{0i}\sum_{k=1}^{N} a_{ki}\tilde{r}_k, \tilde{r}_j\right]
\tag{17.148}
$$

Summing over investors, this gives (since we have $\sum_i W_{0i}a_{ki} = W_{0m}m_k$ by market-clearing and $\sum_k m_k\tilde{r}_k = \tilde{r}_{\mathbf{m}}$ by definition)

$$
E[\tilde{r}_j - r_{\mathrm{f}}]\left(\sum_{i=1}^{I} \theta_i^{-1}\right) = W_{0m}\,\mathrm{Cov}[\tilde{r}_{\mathbf{m}}, \tilde{r}_j]
\tag{17.149}
$$

or

$$
E[\tilde{r}_j - r_{\mathrm{f}}] = \left(\sum_{i=1}^{I} \theta_i^{-1}\right)^{-1} W_{0m}\,\mathrm{Cov}[\tilde{r}_{\mathbf{m}}, \tilde{r}_j]
\tag{17.150}
$$

i.e. in equilibrium, the risk premium on the jth asset is the product of the **aggregate relative risk aversion** of the economy and the covariance between the return on the jth asset and the return on the market.

Now take the average over j weighted by market portfolio weights:

$$E[\tilde{r}_{\mathbf{m}} - r_{\mathrm{f}}] = \left(\sum_{i=1}^{I} \theta_i^{-1}\right)^{-1} W_{0m} \operatorname{Var}[\tilde{r}_{\mathbf{m}}] \qquad (17.151)$$

i.e. in equilibrium, the risk premium on the market is the product of the aggregate relative risk aversion of the economy and the variance of the return on the market. Equivalently, the return to variability of the market equals the aggregate relative risk aversion.

We conclude with some examples.

EXAMPLE 17.5.1 Negative-exponential utility:

$$u_i(W) = -e^{-\alpha_i W}, \quad \alpha_i > 0 \qquad (17.152)$$

implies (see Example 16.5)

$$\left(\sum_{i=1}^{I} \theta_i^{-1}\right)^{-1} = \left(\sum_{i=1}^{I} \alpha_i^{-1}\right)^{-1} > 0 \qquad (17.153)$$

and, hence, the market portfolio is efficient.   ◇

EXAMPLE 17.5.2 Quadratic utility:

$$u_i(W) = W - \frac{\alpha_i}{2} W^2, \quad \alpha_i > 0 \qquad (17.154)$$

implies (see Exercise 16.18(a))

$$\left(\sum_{i=1}^{I} \theta_i^{-1}\right)^{-1} = \left(\sum_{i=1}^{I} \left(\frac{1}{\alpha_i} - E[\tilde{W}_i]\right)\right)^{-1} \qquad (17.155)$$

Recall that the quadratic utility function has a bliss point at $W = 1/\alpha_i$. Thus, aggregate relative risk aversion is positive, and the market portfolio efficient, provided that the average investor's expected wealth is below his bliss point. Of course, the whole derivation is based on the assumption of increasing utility and could break down if it were possible for investors to attain bliss points.   ◇

17.5.6   Capital asset pricing model and asset valuation

Suppose, as in (17.12), that p_j is the price of the random payoff $\tilde{x}_j$, but now suppose also that the traditional CAPM holds and the covariance of the payoff $\tilde{x}_j$ with the return $\tilde{r}_{\mathbf{m}}$ on the market portfolio is known.

Then $\tilde{r}_j = \tilde{x}_j / p_j$ and CAPM states that

$$E[\tilde{r}_j] = r_f + \beta_{jm} E[\tilde{r}_m - r_f]$$
$$= r_f + \frac{\mathrm{Cov}[\tilde{x}_j, \tilde{r}_m]}{p_j \, \mathrm{Var}[\tilde{r}_m]} E[\tilde{r}_m - r_f] \tag{17.156}$$

Multiplying across by p_j yields

$$E[\tilde{x}_j] = r_f p_j + \frac{\mathrm{Cov}[\tilde{x}_j, \tilde{r}_m]}{\mathrm{Var}[\tilde{r}_m]} E[\tilde{r}_m - r_f] \tag{17.157}$$

Solving for p_j yields

$$p_j = \frac{1}{r_f}\left(E[\tilde{x}_j] - \frac{\mathrm{Cov}[\tilde{x}_j, \tilde{r}_m]}{\mathrm{Var}[\tilde{r}_m]} E[\tilde{r}_m - r_f]\right) \tag{17.158}$$

Thus:

- regardless of the efficiency of the market portfolio, a payoff that is uncorrelated with the return on the market portfolio will trade at its discounted expected value (discounted at the risk-free rate); and
- provided that the market portfolio is mean–variance efficient,
  - a payoff that is negatively correlated with the market return will trade at a price higher than its discounted expected value (i.e. will attract a return lower than the risk-free rate); and
  - a payoff that is positively correlated with the market return will trade at a price lower than its discounted expected value (i.e. will attract a return higher than the risk-free rate).

17.6 Multi-currency considerations

The theory of choice under uncertainty is generally presented without reference to the importance of the choice of numeraire currency. We have already begun to see the significance of this choice in the discussion of Siegel's paradox and uncovered interest rate parity in Sections 13.10.2 and 16.7. The portfolio theory literature generally refers to the currency as the "dollar", for no better reason than that most of it has been written in the USA. We have taken a deliberate decision to refer to the unit of currency in this book as the "euro" for a number of reasons, and not solely because the book was written within the Euro Zone. More importantly, in this concluding section we wish to highlight again the place in the theory of equilibrium presented here of various implicit assumptions.

In the literature generally, propositions are made such as:

- the world is risk-neutral *with respect to the numeraire currency*; or
- preferences have the expected-utility property *with respect to the numeraire currency*; or
- individuals choose portfolios that are on the mean–variance frontier *with respect to the numeraire currency*.

Usually, however, the phrase in *italics* is omitted. The question arises as to the conditions under which such propositions are, or are not, invariant under changes of numeraire currency.

Under a fixed exchange rate regime, the choice of numeraire currency would make no difference to any result. However, several decades after the collapse in 1971 of the Bretton Woods system of fixed exchange rates and in the present era of increasing globalization, the assumption of a fixed exchange rate regime is no longer of any practical use. Even under the almost equally unrealistic assumption that exchange rates are statistically independent of the returns on the assets available for investment, it is difficult to derive any useful invariance results.

We have already seen in Section 16.7 that the properties of the risk-neutral world are not invariant under a change of numeraire currency, or more precisely if different individuals are risk-neutral with respect to different currencies. As we saw in that case, models of financial market equilibrium can be sensitive to the currency with respect to which they are described. A world that is risk-neutral from the perspective of the euro cannot also be risk-neutral from the perspective of sterling, or of any other currency. A similar caveat applies to the mean–variance world behind the CAPM, which has been described in this chapter.

If the portfolios chosen by all investors are mean–variance efficient frontier portfolios with respect to one currency, it does not follow automatically that they are also mean–variance efficient frontier portfolios with respect to a different currency. We have shown that, in a world of uncertain exchange rates, investors with mean–variance preferences with respect to their own currency will hold a combination of the asset that is risk-free in their own currency and the corresponding tangency portfolio of risky assets (which might include assets whose payoff in some other currency is risk-free, but which are rendered risky by exchange rate uncertainty).

Just as the risk-neutral equilibrium unravels if there are investors with risk-neutral preferences in two or more different currencies, so the two-fund-separation equilibrium could unravel if there are investors with mean–variance preferences in two or more different currencies; see Exercise 17.22.

EXERCISES

17.1 Show that the constrained maximization problem

$$\max_{\{a_0, a_1, \ldots, a_N\}} \quad E\left[u\left(a_0 r_{\mathrm{f}} + \sum_{i=1}^{N} a_i \tilde{r}_i\right)\right]$$

subject to

$$\sum_{i=0}^{N} a_i = W_0$$

has the same solutions for $a_1, a_2, \ldots, a_N$ as the unconstrained maximization problem

$$\max_{\{a_1, a_2, \ldots, a_N\}} \quad E\left[u\left(W_0 r_{\mathrm{f}} + \sum_{i=1}^{N} a_i (\tilde{r}_i - r_{\mathrm{f}})\right)\right]$$

17.2 Calculate the optimal portfolio holdings for an individual with initial wealth $W_0 = 1$ and VNM expected-utility function $u(w) = w - (b/2)w^2$ when $b = 0.001$, faced with a risk-free interest rate of 8% and two risky investment opportunities with expected returns given by the vector

$$\mathbf{e} = \begin{bmatrix} 0.12 \\ 0.15 \end{bmatrix}$$

and variances and covariances given by the matrix

$$\mathbf{V} = \begin{bmatrix} 0.02 & -0.01 \\ -0.01 & 0.04 \end{bmatrix}$$

17.3 Consider the problem of allocating initial wealth between a risk-free asset and a single risky asset in such a way as to maximize expected utility. Recall that decreasing absolute risk aversion (DARA) implies that the risky asset is a normal good if its expected return exceeds the risk-free rate.

Prove the analogous results for CARA, IARA, DRRA, CRRA and IRRA.

17.4 Find all utility functions exhibiting hyperbolic absolute risk aversion (HARA).

17.5 Derive equation (16.85).

17.6 Show that $u'(w) = Ae^{Bw}$ is a sufficient condition for two-fund monetary separation.

17.7 An investor with an initial wealth of W_0 can divide that wealth among N assets with uncertain (gross) rates of return, given by the random vector $\tilde{\mathbf{r}} \equiv (\tilde{r}_1, \tilde{r}_2, \ldots, \tilde{r}_N)$.

The vector (or portfolio) $W_0 \mathbf{b} \equiv (b_1 W_0, b_2 W_0, \ldots, b_N W_0)$ of investments is chosen such that its overall rate of return has minimum variance, subject to the condition that the expected (gross) rate of return on the portfolio is at least μ. Assume that negative holdings of assets (short-selling, or $b_i < 0$) *are* allowed.

Let $\mathbf{e}$ be the $N \times 1$ vector whose ith component is $E[\tilde{r}_i]$ and let $\mathbf{V}$ be the $N \times N$ matrix whose ijth element is $\text{Cov}[\tilde{r}_i, \tilde{r}_j]$, assuming that all these expectations and covariances are finite.

(a) Using the basic properties of expectations, variances and covariances, show that the expectation of the (gross) rate of return on the portfolio $W_0 \mathbf{b}$ can be written in matrix notation as $\mathbf{b}^\top \mathbf{e}$ and its variance as $\mathbf{b}^\top \mathbf{V} \mathbf{b}$.

(b) Show further that $\mathbf{V}$ is a symmetric positive semi-definite matrix.

(c) Formulate the investor's problem as a Kuhn–Tucker optimization problem.

(d) Write down the first-order conditions for the optimization problem, and investigate whether the standard second-order conditions are satisfied.

(e) Assuming that $\mathbf{V}$ is an invertible matrix and that the constraints are binding at the optimum, solve for the vector of optimal investments in terms of μ, $\mathbf{e}$, $\mathbf{V}$ and the Kuhn–Tucker multipliers.

(f) Substitute this expression for the vector of optimal investments in each of the constraint equations in turn in order to derive two simultaneous linear equations in the Kuhn–Tucker multipliers and, hence, find an expression for the optimal **b** involving only the exogenous parameters and not the Kuhn–Tucker multipliers.

17.8 What form does the portfolio frontier take when the N risky assets available to investors are mutually uncorrelated?

17.9 Now consider Exercise 17.7 again, but assuming that there are only three assets, that initial wealth is €1m, but that short-selling is **not** allowed.

(a) Show that the investor's problem can be formulated as a Kuhn–Tucker maximization problem with inequality constraints, involving either

   (i) three choice variables and six inequality constraints, or
   (ii) two choice variables and four inequality constraints.

(b) Now consider the specific example in which the expectation of the (net) return vector is $(0.04, 0.08, 0.12)$ and its variance–covariance matrix is

$$
\begin{bmatrix}
0.03 & -0.01 & -0.01 \\
-0.01 & 0.03 & -0.01 \\
-0.01 & -0.01 & 0.03
\end{bmatrix}
$$

Calculate the optimal investment proportions for desired expected returns of 4%, 5.5%, 6%, 8%, 10%, 11.5% and 12%. What are the signs of the Kuhn–Tucker multipliers in each case (strictly positive, zero, or strictly negative)?

(c) Using the envelope theorem, calculate the rate of change of the minimized variance with respect to the desired expected return in the situation where only the expected return constraint and the budget constraint are binding.

(d) Sketch rough graphs of the feasible set both in the $b_1 b_2$ plane (i.e. using the two-variable, four-constraint approach) and in the plane $b_1 + b_2 + b_3 = 1$ (i.e. using the three-variable, six-constraint approach) for two possible values of μ, indicating on each graph the indifference curves of the objective function.

(e) Over what range of desired expected returns is each short-selling constraint binding? Using this information, plot the envelope function for this problem.

(f) How would your answers in part (b) above change if all the covariances were 0.01 instead of -0.01? Explain.

17.10 Suppose there are two stocks available to investors with equilibrium expected returns of 14% and 8% and standard deviations of returns of 6 and 3 percentage points, respectively. Suppose further that the correlation between the returns on the two stocks is 1.

(a) Write down the correlation matrix for these two stocks.
(b) Find a riskless portfolio of the two stocks.
(c) Plot the portfolio frontier. (Note that the usual method does not work because of the existence of a riskless portfolio of the two stocks.)
(d) If there exists a risk-free security in equilibrium, what rate of return will it offer?

17.11 Show that there exists a basis for the portfolio space consisting of N portfolios whose returns are uncorrelated and each has unit variance. Develop the mathematics of the portfolio frontier in terms of the unit-cost versions of these uncorrelated portfolios.

17.12 The notation in this question is as defined in the text.

(a) For the returns on the hedge portfolio **h** and on the unit portfolios **g**, $(1/C)\mathbf{V}^{-1}\mathbf{1}$ and $(1/A)\mathbf{V}^{-1}\mathbf{e}$, calculate means, variances and all pairwise covariances.
(b) Show that $D \equiv BC - A^2$ is positive.
(c) Show that, for all unit portfolios **p**,

$$\text{Cov}[\tilde{r}_{\text{MVP}}, \tilde{r}_{\mathbf{p}}] = \frac{1}{C}$$

(d) Derive equation (17.62), i.e. show that any frontier portfolio can be written in the form

$$\mathbf{b} = W_0\left(\text{MVP} + \left(\mu - \frac{A}{C}\right)\mathbf{h}\right)$$

(e) Show that the decomposition of portfolio weight vectors

$$\mathbf{a} \equiv \text{MVP} + \left(E[\tilde{r}_{\mathbf{a}}] - \frac{A}{C}\right)\mathbf{h} + \mathbf{u}$$

extends naturally to a decomposition of the random variables representing portfolio returns.
(f) Work out the relationship between $E[\tilde{r}_{\mathbf{p}}]$ and $E[\tilde{r}_{\mathbf{z}_{\mathbf{p}}}]$ by solving (17.69).

17.13 Compute and graph the portfolio frontier (in both portfolio space and mean–variance space) when

$$\mathbf{e} = \begin{bmatrix} 1.03 \\ 1.08 \end{bmatrix} \quad \text{and} \quad \mathbf{V} = \begin{bmatrix} 0.02 & -0.01 \\ -0.01 & 0.05 \end{bmatrix}$$

17.14 What form does the portfolio frontier take when all assets have the same expected return? Find the proportions of the minimum-variance portfolio of two assets with the same expected return and variance–covariance matrix

$$\mathbf{V} \equiv \begin{bmatrix} \sigma_1^2 & \sigma_{12} \\ \sigma_{12} & \sigma_2^2 \end{bmatrix}$$

17.15 Find from first principles the weights of, expected return on, and variance of the return on the global minimum-variance portfolio.

17.16 Suppose that there are three risky assets available for investment, whose returns have equal variances and whose expected returns are 1%, 2% and 3%, respectively.
    Assume that the first and third assets are uncorrelated mean–variance frontier portfolios.

(a) Derive the full 3×3 correlation matrix of the asset returns.

(b) Hence, or otherwise, calculate the proportions of, and the expected return on, the global minimum-variance portfolio, and its beta with respect to each of the three underlying assets.

(c) How much of the second asset will be held by an investor seeking a mean–variance efficient portfolio?

(d) Finally, what proportions are held in each asset in the frontier portfolios that have betas of 0.25 and 0.75 with respect to asset 3?

17.17 Suppose $E[\tilde{x}] = 0$ but $m(y) = E[\tilde{x} \mid \tilde{y} = y] \neq 0$ for some y.
Show that there exists y^* such that

$$c \equiv \int_{-\infty}^{y^*} m(y) \, dF_{\tilde{y}}(y) = -\int_{y^*}^{\infty} m(y) \, dF_{\tilde{y}}(y) \neq 0$$

17.18 Solve problem (17.114), namely,

$$\min_{\mathbf{b}} \ \mathbf{b}^{\top} \mathbf{V} \mathbf{b}$$
$$\text{s.t.} \ \mathbf{b}^{\top} \mathbf{e} + (W_0 - \mathbf{b}^{\top} \mathbf{1}) r_f \geq \mu W_0$$

17.19 Show that H, as defined in equation (17.116), is positive for all values of r_f.

17.20 Derive the expressions for $\mathrm{Var}[\tilde{r}_t]$ and $\mathrm{Var}[\tilde{r}_{b*}]$ given in equations (17.120) and (17.121), respectively.

17.21 Consider the limiting behaviour of the variance of the return on an equally weighted portfolio as the number of securities included goes to infinity. Show that, if securities are added in such a way that the average of the variance terms and the average of the covariance terms are stable, then the portfolio variance approaches the average covariance as a lower bound.

17.22 Consider a world with N countries where the only traded securities are N risk-free assets, one denominated in each currency. Suppose that exchange rates in this world are uncertain and exogenously determined. Suppose that investors in each country choose portfolios of these N assets that are mean–variance efficient from the perspective of their home currency.

Explore the properties of the optimal portfolio choice and of market equilibrium in this world.

Notes

Notation and preliminaries

1 Solow (2009) is a good introduction to the important concepts related to formal proofs.
2 Venn diagrams, showing sets and their intersections and unions, were conceived around 1880 by the English logician and philosopher John Venn (1834–1923).
3 The Cartesian product and Cartesian coordinates are both named after the French mathematician René Descartes (1596–1650), who was the first to use coordinates in this way. The modern convention – which we follow from Section 5.3 onwards – of referring to $\mathbb{R}^n$ as Euclidean (n-)space, in honour of the ancient mathematician Euclid of Alexandria ($c.325$–$c.265$BC), is actually questionable, and it would be equally justifiable to refer to it as Cartesian space. The Euclidean norm (Definition 5.2.11) is also named after Euclid.
4 De Moivre's theorem is named after the French Huguenot mathematician Abraham de Moivre (1667–1754), who first published it in 1722; for further details, see, for example, Sydsæter *et al.* (2008, Appendix B.3).
5 See Simmons (1963, Sections 1.5 and 1.8) for more details on binary relations.
6 L'Hôpital's rule was first published in a 1696 book by the French mathematician Guillaume François Antoine, Marquis de l'Hôpital (1661–1704).
7 Binmore (1982, Chapters 7–16) is one of many texts that cover univariate calculus well, with the added advantage that the author is a leading economist. Good alternatives include Simon and Blume (1994, Chapters 2–4) or Stewart (2008, Chapters 1–8).
8 See, for example, Chiang and Wainwright (2005, Section 10.2).

1 Systems of linear equations and matrices

1 However, see Exercise 12.2 for an illustration of the peculiarities of linear demand functions.
2 Keynesian models are named after the English economist John Maynard Keynes (1883–1946), whose macroeconomic theories continue to be the subject of debate and dispute to the present day.
3 Chapter 14 discusses in more detail the use of random disturbance terms in linear economic models.
4 The German-born economist Wassily Wassilyovich Leontief (1905–1999), who grew up in Russia and later settled in America, was awarded the Nobel Memorial Prize in Economic Sciences in 1973 for the development of the input–output method and for its application to important economic problems.
5 The Kronecker delta, and also the Kronecker product introduced in Section 1.5.15, are called after the Prussian-born mathematician Leopold Kronecker (1823–1891).
6 An alternative form of this result exists but its derivation is left as part of Exercise 1.23.

2 Determinants

1 A result proved by induction is true for all finite positive integers, but may not be true at infinity. For example, a result proved by induction on the dimension of finite-dimensional vector spaces may not hold in infinite-dimensional spaces; see Chapter 5. Similarly, a result proved for discrete random variables by induction on the number of states of nature in the underlying sample space may not hold for continuous random variables; see Chapter 13.

2 These results are named after the Swiss mathematician Gabriel Cramer (1704–1752). Cramer (1750) presented the result that now bears his name as a method for solving five linear equations in five unknowns.

3 Gaussian elimination, Gauss–Jordan elimination, the Gaussian distribution (p. 337) and the Gauss–Markov theorem (Theorem 14.3.1) are all named after the German mathematician Johann Carl Friedrich Gauss (1777–1855). Gaussian elimination was, however, known to the ancient Chinese, as *fangcheng*, at least 1500 years before Gauss; see, for example, Hart (2010). Gauss–Jordan elimination is named after Gauss and the German geodesist Wilhelm Jordan (1842–1899).

3 Eigenvalues and eigenvectors

1 In Section 8.4.2, an alternative form for complex numbers, using polar coordinates, is also employed. Readers requiring some revision of the subject of complex numbers are referred to Simon and Blume (1994, Appendix A3.1) or Sydsæter *et al.* (2008, Appendix B.3).

2 We will shortly give a formal definition of this kind of relationship between the equations, which is known as linear dependence; see Definition 3.6.2.

3 We shall consider the geometry of orthogonal vectors, and vectors generally, in Chapter 5.

4 An $m \times n$ matrix $\mathbf{Q}$ with the property that $\mathbf{Q}\mathbf{Q}^\top = \mathbf{I}_m$ or $\mathbf{Q}^\top\mathbf{Q} = \mathbf{I}_n$ is sometimes called an orthonormal matrix. These properties can hold only if $m \leq n$ or $m \geq n$, respectively. Thus both properties can hold simultaneously only when $\mathbf{Q}$ is square; see Exercise 5.20.

5 Here "equivalence" means that any one statement implies the others.

6 This step can be undertaken using a procedure called the Gram–Schmidt process, which is explained in Section 5.4.7.

7 These formulas were first discovered by the French Huguenot mathematician François Viète or Franciscus Vieta (1540–1603), known as the father of algebra.

4 Conic sections, quadratic forms and definite matrices

1 This section draws heavily on Tranter (1953, Chapter 17).

2 The concept of matrix rank will be examined in more detail in Section 5.4.6.

3 The Cholesky decomposition or factorization is named after the French military officer and mathematician Major André-Louis Cholesky (1875–1918). Cholesky's method was published posthumously in 1924 by one of his fellow officers, Cholesky himself having been killed in World War I.

4 For further details of the LU-decomposition see, for example, Anton and Rorres (2011, p. 480).

5 Vectors and vector spaces

1 The similarity to a 1×2 or 2×1 matrix is obvious.

2 Infinite-dimensional vector spaces will be mentioned again in, for example, Sections 6.2, 9.6 and 13.4.

3 Recall the earlier definitions of row rank and column rank in Definition 4.4.2.

4 The Gram–Schmidt process is named after the Danish mathematician Jorgen Pedersen Gram (1850–1916) and the Estonian-born mathematician Erhard Schmidt (1876–1959), although, as in many such cases, others had used the process before them.

5 An orthogonal matrix was defined in Chapter 1 and first used in Chapter 3 in our treatment of eigenvalues, eigenvectors and diagonalization.

6 The specification of a scalar product in Definition 5.4.12 has been generalized by relaxing the symmetry and/or non-negativity conditions. For example, the **Minkowski inner product** introduced by the Lithuanian-born mathematician Hermann Minkowski (1864–1909) and used in relativity does not satisfy the non-negativity condition. All scalar products considered in this book will, however, be both symmetric and non-negative.

6 Linear transformations

1 It should be noted, however, that the vector space of differentiable functions is different from the vector space of integrable functions: not all differentiable functions are integrable, and not all integrable functions are differentiable. The vector space of functions that are both differentiable and integrable is another, distinct vector space.

7 Foundations for vector calculus

1 Note that the context makes it clear that (a_1, a_2) denotes a 2-vector but that (a_1, b_1) denotes an open interval in the real line, and so on.

2 Some writers, e.g. Simmons (1963) and Spivak (1965), simply use the words "cube" and "rectangle" in higher dimensions where we use "hypercube" and "hyperrectangle".

3 The Cauchy–Schwarz inequality is named after the French mathematician Augustin Louis Cauchy (1789–1857) and the German mathematician Karl Hermann Amandus Schwarz (1843–1921), who rediscovered it. The theorem to which we refer as Young's theorem (Theorem 9.7.2) is also sometimes known as Schwarz's theorem; see note 10 of Chapter 9.

4 This theorem is named after the German mathematician Heinrich Eduard Heine (1821–1881) and the Russian-born mathematician Georg Ferdinand Ludwig Phillip Cantor (1845–1918).

5 Some writers (e.g. Berge (1997)) use the term **semi-continuous correspondence** rather than the term hemi-continuous correspondence; Hildenbrand and Kirman (1988, pp. 260–1) argue in favour of the latter term.

6 For the purposes of this book, we are interested in the continuity properties of functions between real vector spaces only. These properties all generalize elegantly to general metric and topological spaces, which are the most natural and general setting for continuity. For further reading on various alternative but equivalent definitions of continuity, see Simmons (1963, p. 76). For further reading on continuity of correspondences, the interested reader is referred to de la Fuente (2000, Chapter 2) or Hildenbrand and Kirman (1988, Mathematical Appendix III). Hildenbrand (1974) gives a more complete treatment of the subject.

8 Difference equations

1 The polynomials in the lag operator that appear in properties 7, 8 and 9 are important in certain applications, as we shall see later. The sums of the infinite series that appear in properties 8 and 9 should be familiar results; but see equation (8.39) and Exercise 8.5.

2 See, for example, the discussion of the efficient markets hypothesis and the random walk in Section 16.6.

3 The idea of a unit eigenvalue or unit root of the characteristic polynomial associated with a given square matrix was referred to in Section 3.4. Unit roots may also arise from solution of a related characteristic equation associated with a form of higher-order difference equation, which will be introduced in Section 8.4. Unit roots, and the concept of co-integration, will be referred to again in Section 14.4.1.

4 Schur's theorem is named after the Russian-born mathematician Issai Schur (1875–1941).

5 This proof is based on a proof by Barankin (1945). For an alternative proof, see Woods (1978, Theorem 67).

9 Vector calculus

1 The Jacobian matrix is named after the Prussian-born mathematician Carl Gustav Jacob Jacobi (1804–1851), who wrote a long memoir devoted to the subject in 1841.

2 The Hessian matrix is named after the Prussian-born mathematician Ludwig Otto Hesse (1811–1874), a student of Jacobi.

3 Leibniz's law and Leibniz's integral rule (Theorems 9.7.4 and 9.7.5) take their names from the German-born mathematician Gottfried Wilhelm Leibniz (1646–1716).

4 According to Neary (1997, p. 102), the Cobb–Douglas function was first developed by the Swedish mathematician and economist Johan Gustaf Knut Wicksell (1851–1926), but it takes its name from the American economists Charles Wiggans Cobb (1875–1949) and Senator Paul Howard Douglas (1892–1976), who first tested it against statistical evidence.

5 Some writers define the directional derivative of f at $\mathbf{x}'$ only in the direction of a unit vector, for example, $\mathbf{u} \equiv (\mathbf{x} - \mathbf{x}')/\|\mathbf{x} - \mathbf{x}'\|$, as $f'(\mathbf{x}')\mathbf{u}$. Definition 9.5.1 extends this definition to cover any $\mathbf{u}$, even if $\|\mathbf{u}\| \neq 1$.

6 Taylor's theorem and Taylor's expansion take their names from the English mathematician Brook Taylor (1685–1731), who first mentioned the result now named after him in a letter written in 1712.

7 This theorem is named after the French mathematician Michel Rolle (1652–1719), who published it in 1691.

8 Maclaurin's series and the associated approximation are named after the Scottish mathematician Colin Maclaurin (1698–1746), who used it in Maclaurin (1742).
9 The standard reference for this theorem, credited to the Italian mathematician Guido Fubini (1879–1943), is to Fubini's posthumously published selected works, where the result appears in Italian under the title "Sugli integrali multipli" (Fubini, 1958, pp. 243–9).
10 For the history of this theorem, see Higgins (1940). The French mathematician Alexis Claude Clairaut (1713–1765) appears to have been the first person to attempt to establish equality of mixed partials; the German mathematician Karl Hermann Amandus Schwarz (1843–1921), who was also jointly responsible – see note 3 of Chapter 7 – for the Cauchy–Schwarz inequality (7.29), produced the first acceptable proof; and the English mathematician William Henry Young (1863–1942) provided a weaker set of sufficient conditions.
11 The integrand is undefined at $y = 0$; it is assumed to take the value 1, which is its limit as $y \to 0$ (see p. 160).

10   Convexity and optimization

1 This proof is based on Roberts and Varberg (1973, p. 98).
2 Lagrange multipliers and the Lagrangian are named after the Italian-born mathematician Joseph-Louis Lagrange (1736–1813).
3 More precisely, the constraints in the Lagrange multiplier conditions could be described as locally linearly independent.
4 As defined in Notation 1.2.3, a round-bracketed row of numbers separated by commas is used to denote a column vector.
5 Some writers, e.g. de la Fuente (2000, Chapter 7), call the envelope function the **value function**, a term usually reserved for the equivalent in dynamic optimization, which we do not consider in this book. Note also that, while we generally denote functions, including vector-valued functions, by italic letters, f, g, etc., we will generally denote the optimal response function by the same boldface letter ($\mathbf{x}$) used to denote the vector of choice variables. In the applications, we will also sometimes use a boldface letter to denote a function defined in terms of an optimal response function, for example an excess demand function.
6 Kuhn–Tucker multipliers are named after the Canadian mathematician Albert William Tucker (1905–1995) and his American graduate student Harold William Kuhn (b. 1925). Kjeldsen (2000) and others argue, however, that credit for the result should go to Karush (1939), whose work pre-dated that of Kuhn and Tucker (1950).
7 By calling G a non-empty, compact-valued correspondence, we mean that the set $G(\boldsymbol{\alpha})$ is a non-empty and compact set for all $\boldsymbol{\alpha} \in \mathbb{R}^q$.
8 The version of the theorem of the maximum given (without proof) by Varian (1992, p. 506) assumes that the objective function has "a compact range". This restriction is impracticable for the principal applications in economics, which are to the unbounded utility functions of consumers with strongly monotonic preferences; see Section 12.3.

11   Macroeconomic applications

1 Woods (1978) analyses the properties of such matrices in considerable detail.

12   Single-period choice under certainty

1 While all trading takes place simultaneously in the model, consumption can be spread over many periods.
2 This function takes its name from the English economist and Nobel Laureate Sir John Richard Nicholas Stone (1913–1991) and the Irish statistician Robert Charles (Roy) Geary (1896–1983). For a full history of how this utility function came to be labelled "Stone–Geary", see Neary (1997). The function was indeed mentioned by Geary (1950) and Stone (1954), but both Geary and Samuelson (1947) mentioned it only in comments on an earlier publication by Klein and Rubin (1947), who perhaps deserve some of the credit.
3 The Japanese economist Ken-Ichi Inada (1925–2002) set out similar conditions in a different context, for a production function (Inada, 1963).

4 The Walrasian auctioneer, Walrasian equilibrium and Walras's law, all introduced in this chapter, are named after the French mathematical economist Marie-Esprit-Léon Walras (1834–1910).

5 The Marshallian demand function takes its name from the English economist Alfred Marshall (1842–1924) who popularized its use.

6 A real-world example of a good with a negative price is an elderly motor vehicle for which the cost of environmentally acceptable scrappage may become so large as to render the market value of the vehicle negative.

7 Hicksian demand functions are named after the English economist Sir John Richard Hicks (1904–1989).

8 This version of Shephard's lemma and an equivalent result in production theory are named after the American mathematician, economist and engineer Ronald William Shephard (1912–1982). The result in production theory appeared in Shephard (1953). Takayama (1994, p. 135) suggests that Theorem 12.4.3 should be known as the Shephard–McKenzie lemma, as the equivalent result in consumer theory first appeared in McKenzie (1957, p. 188). Similar results were known before 1953, and Shephard's real contribution was to show that they could be derived without reference to the underlying utility function or production function. Subsequently, the simpler proof used here, based on the envelope theorem, has become the more popular approach to Shephard's lemma.

9 Roy's identity is named after the French economist René François Joseph Roy (1894–1977) (Roy, 1947).

10 These results were first published, in Italian (Slutsky, 1915), by the Russian mathematical statistician and economist Evgenii Evgen'evich Slutsky (or Slutskii) (1880–1948). According to Barnett (2004, p. 6), Slutsky's work on consumer behaviour was first brought to the attention of English-speaking economists by Allen (1936), who had independently developed some of the same ideas, many years after Slutsky, but before discovering the latter's work.

11 Giffen goods are named after the Scottish statistician and economist Sir Robert Giffen (1837–1910). As noted by Mason (1989), Alfred Marshall, writing in 1895, attributed the possibility of such goods to Giffen, but exhaustive subsequent research has failed to identify any passage in Giffen's writings where he pointed this out.

12 Brouwer's fixed-point theorem takes its name from the Dutch mathematician Luitzen Egbertus Jan Brouwer (1881–1966).

13 Kakutani's fixed-point theorem is named after the Japanese mathematician Shizuo Kakutani (1911–2004).

14 The no-arbitrage principle is also known as the *no-free-lunch principle* or the *law of one price*.

15 For this and similar examples, the three-letter alphabetic codes for currencies set out in the relevant international standard (ISO 4217:2008) published by the International Organization for Standardization (ISO) will be used. EUR denotes the euro and GBP the pound sterling.

16 The concept of the Edgeworth box was introduced (Edgeworth, 1881) by the Irish economist and statistician Francis Ysidro Edgeworth (1845–1926). It was later popularized (Bowley, 1924) by the English economist and statistician Sir Arthur Lyon Bowley (1869–1957), whose name is sometimes appended to the concept.

17 Pareto optimality, Pareto dominance and Pareto efficiency are all called after the French-born Italian engineer, sociologist, economist, and philosopher, Vilfredo Federico Damaso Pareto (1848–1923).

18 For an equivalent discussion in the context of two-period choice under uncertainty, see Huang and Litzenberger (1988, Chapter 5).

13 Probability theory

1 Bernoulli trials are named after Jacob Bernoulli (1654–1705), the eldest of a dynasty of Swiss mathematicians.

2 This distribution was introduced by and named after the French mathematician, geometer and physicist Siméon-Denis Poisson (1781–1840).

3 The American Kenneth Joseph Arrow (b. 1921) and the French-born Gerard Debreu (1921–2004) were separately awarded the Nobel Memorial Prize in Economic Sciences in 1972 and 1983, respectively, for their work in general equilibrium theory. Debreu's contributions have already been cited in Chapter 12. Another of Arrow's contributions to the economics of uncertainty, which are summarized by Greenberg and Lowrie (2010), will be mentioned in Section 16.5.

4 Note the similarity to the arguments about linear independence of solutions to difference equations in Section 8.4.2.

5 It might be argued that, rather than denoting the typical element merely by **a**, we should denote it by $\tilde{\mathbf{a}}$, since it is a random variable, or by $\tilde{a}$, since its realizations are real numbers, not vectors.

6 For more on infinite-dimensional vector spaces of random variables, the interested reader is referred to Billingsley (1995, Section 18).

7 This is not the same as assuming statistical independence, except in special cases, such as bivariate normality; see Section 13.7.

8 In the special case in which $\tilde{x}$ and $\tilde{y}$ are bivariate normally distributed, the conditional mean of the disturbance term is zero; see Section 13.7.

9 These results can be traced to the work of Charles M. Stein (b. 1920), in particular Stein (1973) and Stein (1981). The lemma was first used and proved indirectly in the finance literature by Rubinstein (1973), whose work was contemporaneous with but independent of Stein's. Ingersoll (1987, pp. 13–14) also proves the result without naming it Stein's lemma. The lemma is also used by Huang and Litzenberger (1988, p. 101 and p. 116) and by Cochrane (2005, p. 163). A multivariate extension of Stein's lemma appears in Balvers and Huang (2009, Appendix B), where it is called Stein's extended lemma.

10 This result was first proved (Jensen, 1906) by the Danish mathematician Johan Ludwig William Valdemar Jensen (1859–1925).

11 Jensen's original paper does not appear to acknowledge that his proof works only for non-zero y_i.

12 Siegel's paradox takes its name from the American financial economist Jeremy J. Siegel (b. 1945), who drew attention to the phenomenon during a period of intense research into the determination of exchange rates in the aftermath of the collapse in 1971 of the Bretton Woods system of fixed exchange rates (Siegel, 1972).

14   Quadratic programming and econometric applications

1 It should be noted that, in econometrics texts, the problem is often formulated using different notation and $\hat{\boldsymbol{\beta}}$ is used to denote the OLS solution.

2 A possible fix is to let the Kuhn–Tucker multipliers be defined by $\boldsymbol{\lambda}^* \equiv \max\{\mathbf{0}_m, -2(\mathbf{G}\mathbf{A}^{-1}\mathbf{G}^\top)^{-1}\boldsymbol{\alpha}\}$, where the max operator denotes component-by-component maximization. The effect of this is to knock out the non-binding constraints (those with negative Lagrange multipliers) from the original problem and the subsequent analysis.

3 The Gauss–Markov theorem is named after the German mathematician Johann Carl Friedrich Gauss (1777–1855), already mentioned in note 3 of Chapter 2, and the later Russian mathematician Andrei Andreyevich Markov (1856–1922). Gauss published the result in the 1820s and Markov almost a century later. Each name was associated separately with the theorem until the 1950s, since when the joint attribution has become standard.

4 The adaptation of difference equations to time-series econometric models is dealt with well by Enders (2010, Chapter 1).

5 With $\phi_0 = 0$, such a process is often used to model autocorrelation in the stochastic disturbance term in single-equation econometric (regression) models.

6 Given T consecutive values of a random variable, $\tilde{y}_t$, generated by a weakly stationary AR(1) process, it is relatively easy to show that the variance–autocovariance matrix $E[(\tilde{\mathbf{y}} - E[\tilde{\mathbf{y}}])(\tilde{\mathbf{y}} - E[\tilde{\mathbf{y}}])^\top]$, where $\tilde{\mathbf{y}} = (\tilde{y}_1, \tilde{y}_2, \ldots, \tilde{y}_T)$, is a band matrix. It follows that the autocorrelation matrix is also a band matrix; see Exercise 14.12.

7 See Section 16.6.

8 Equation (14.93) is essentially the characteristic equation used in the solution of the eigenvalue problem in Chapter 3. Indeed, as the roots of $\boldsymbol{\Phi}_1(L)$ are the reciprocals of the eigenvalues of the matrix $\boldsymbol{\Phi}_1$, an equivalent condition for stationarity is that the eigenvalues of $\boldsymbol{\Phi}_1$ must lie inside the unit circle. For the restricted VAR process currently under consideration, the eigenvalues are $\phi_{1_{11}}$ and $\phi_{1_{22}}$.

9 Details of the solution of (14.98) for the general VAR(p) process may be found in Hamilton (1994, pp. 264–6).

15 Multi-period choice under certainty

1 Good background reading for this section is Jacques (2009, Chapter 3) or Hirshleifer (1970, pp. 41–4).
2 These conditions are discussed in Hirshleifer (1970, pp. 51–6).
3 Note that the analysis is essentially the same whether t starts at 0 or at 1.
4 A **spline** is a special function defined piecewise by polynomials and widely used in numerical analysis.
5 Good background reading for this section is Elton *et al.* (2010, pp. 514–23).
6 One notable exception was in the run-up to European Monetary Union, as the interest rates for the various participating currencies converged.
7 Hull (2009, Sections 4.5–4.7) and Elton *et al.* (2010, Chapter 22) both cover duration.

16 Single-period choice under uncertainty

1 The remainder of Section 16.3 draws on Hirshleifer (1970, Chapter 8, Section C), Fama and Miller (1972, Chapter 4, Section III.C) and Huang and Litzenberger (1988, Chapter 5).
2 The VNM utility function was just one of countless contributions to the advancement of knowledge in fields ranging from statistics to computer science by the Hungarian polymath John von Neumann (1903–1957), working on this occasion (von Neumann and Morgenstern, 1944) with a German-born co-author, Oskar Morgenstern (1902–1977). For background information, see Leonard (1995, 2010). For von Neumann's other contributions, see Heims (1980).
3 von Neumann and Morgenstern (1944) appear to have named this axiom after a similar axiom in pure mathematics, which in turn is named after the ancient mathematician Archimedes of Syracuse (287–212BC). They say the axiom expresses what is known in geometrical axiomatics as the *Archimedean property*.
4 The French economist and Nobel Laureate Maurice Félix Charles Allais (1911–2010) was the first to propose this paradox (Allais, 1953, p. 527).
5 Machina (1982) has developed a theory of choice under uncertainty without the substitution axiom. He shows that, while preferences no longer have an expected-utility representation, his alternative theory can explain the Allais paradox and similar phenomena, as well as generating results similar to many of those derived in the remainder of this book. One advantage of the expected-utility approach is that the mathematics involved is much simpler than that required for Machina's approach.
6 The Arrow–Pratt coefficients are named after the American economist Kenneth Joseph Arrow (b. 1921), already mentioned in note 3 of Chapter 13, and the American mathematician and statistician John Winsor Pratt (b. 1931). They developed their ideas independently but almost contemporaneously (Arrow, 1965; Pratt, 1964).
7 Note that, in the case of narrow-power utility, $u'(w) \to 1/w$ as $B \to 1$, so that $u(w) \to \ln w$. In other words, the logarithmic utility function, and its risk aversion measures, are the limiting case of narrow-power utility as $B \to 1$.
8 This result does not conflict with the obvious symmetry of the model, for if we relabelled the currencies, or equivalently rotated the Edgeworth box through 90° and then took a mirror image, then individual 1's indifference curves would have the steeper slopes.
9 Chamberlain (1983) has determined more precisely the set of probability distributions of asset returns yielding mean–variance utility, of which multivariate normality is a special case.
10 The Latin phrase *ceteris paribus* is frequently used by economists as a shorthand for the assumption that all other variables are assumed to be held constant – in this case, that all the central moments apart from the nth are held constant.
11 Traditionally, mean–variance indifference curves in $\mu\sigma^2$ space are depicted as if the mean is an increasing and convex function of the variance (or, equivalently, as if the variance is an increasing and concave function of the mean). However, while Feldstein (1969) showed that, when utility is logarithmic and wealth is lognormally distributed, then the investor has mean–variance preferences and indifference curves are upward sloping, he also showed that the indifference curves change from convex to concave when they cross the line $\mu = \sqrt{2}\sigma$. Thus, without restricting the probability distribution of wealth, the common trend of drawing such indifference curves as convex everywhere is fallacious.
12 See, for example, Campbell and Hentschel (1992).
13 See Waldron (1991).

17   Portfolio theory

1   As b depends on other parameters of the problem besides W_0, the elasticity could also have been written as $(W_0/b)\partial b/\partial W_0$.

2   See Waldron (1991).

3   We will prove later that these tangency points themselves lie on a line that is itself orthogonal (in a sense to be defined) to the iso-mean lines.

4   The mathematics of the portfolio frontier was initially developed (without the use of matrix notation!) by the American economist Harry Max Markowitz (b. 1927) (Markowitz, 1952). Markowitz shared the 1990 Nobel Memorial Prize in Economic Sciences for his pioneering work in the theory of financial economics.

5   Chamberlain (1983) and Huang and Litzenberger (1988) have fully generalized these distributional conditions, but their results are beyond the scope of this book. Other recommended background reading for this section is Merton (1972), Roll (1977), Markowitz (1991) and Markowitz and Todd (2000).

6   The American economist William Forsyth Sharpe (b. 1934) shared the 1990 Nobel Memorial Prize in Economic Sciences for his pioneering work in the theory of financial economics. Sharpe (1964) developed the CAPM more or less independently from, but contemporaneously with, another American economist, John Virgil Lintner, Jr. (1916–1983) (Lintner, 1965).

References

Allais, M. F. C., 1953. "Le comportement de l'homme rationnel devant le risque: critique des postulats et axiomes de l'école Americaine", *Econometrica* 21(4): 503–546.

Allen, R. G. D., 1936. "Professor Slutsky's theory of consumers' choice", *Review of Economic Studies* 3(2): 120–129.

Anton, H. and Rorres, C., 2011. *Elementary Linear Algebra: With Supplemental Applications*, 10th edn. John Wiley, Hoboken, NJ.

Arrow, K. J., 1953. "Le rôle des valeurs boursières pour la repartition la meilleure des risques", *Econometrie* 11: 41–48.

Arrow, K. J., 1965. "The theory of risk aversion", *Aspects of the Theory of Risk Bearing – Yrjö Jahnsson Lectures*. Yrjö Jahnsson Saatio, Helsinki.

Balvers, R. J. and Huang, D., 2009. "Money and the C-CAPM", *Journal of Financial and Quantitative Analysis* 44(2): 337–368.

Barankin, E. W., 1945. "Bounds for the characteristic roots of a matrix", *Bulletin of the American Mathematical Society* 51(10): 767–770.

Barnett, V., 2004. "E. E. Slutsky: mathematical statistician, economist, and political economist?' *Journal of the History of Economic Thought* 26(1): 5–18.

Bass, T. A., 1999. *The Predictors*. Henry Holt, New York.

Berge, C., 1959. *Espaces Topologiques. Fonctions Multivoques*, Vol. III. of Collection Universitaire de Mathématiques. Dunod, Paris.

Berge, C., 1997. *Topological Spaces* (transl. of Berge (1959) by E. M. Patterson). Dover, Mineola, NY.

Berger, J. O., 1993. *Statistical Decision Theory and Bayesian Analysis* (corrected 3rd printing). Springer Series in Statistics, Springer, New York.

Bernstein, P. L., 2007. *Capital Ideas Evolving*. John Wiley, Hoboken, NJ.

Billingsley, P., 1995. *Probability and Measure*, 3rd edn. John Wiley, New York.

Binmore, K. G., 1982. *Mathematical Analysis: A Straightforward Approach*, 2nd edn. Cambridge University Press, Cambridge.

Black, F., 1972. "Capital market equilibrium with restricted borrowing", *Journal of Business* 45: 444–454.

Bowley, A. L., 1924. *The Mathematical Groundwork of Economics: An Introductory Treatise*. Clarendon Press, Oxford.

Campbell, J. Y. and Hentschel, L. N., 1992. "No news is good news: an asymmetric model of changing volatility in stock returns", *Journal of Financial Economics* 31: 281–318.

Campbell, J. Y., Lo, A. W. and MacKinlay, A. C., 1997. *The Econometrics of Financial Markets*. Princeton University Press, Princeton, NJ.

Cass, D. and Stiglitz, J. E., 1970. "The structure of investor preferences and asset returns, and separability in portfolio allocation: a contribution to the pure theory of mutual funds", *Journal of Economic Theory* 2: 122–160.

Chamberlain, G., 1983. "A characterization of the distributions that imply mean-variance utility functions", *Journal of Economic Theory* 29: 185–201.

Chiang, A. C.-I. and Wainwright, K., 2005. *Fundamental Methods of Mathematical Economics*, 4th edn. McGraw-Hill, New York.

Chu, K. H., 2005. "Solution to the Siegel paradox", *Open Economies Review* 16(4): 399–405.

Cochrane, J. H., 2005. *Asset Pricing*, rev. edn. Princeton University Press, Princeton, NJ.

Cramer, G., 1750. *Introduction á l'Analyse des Lignes Courbes Algébraique*. Frères Cramer & Cl. Philibert, Geneva.

Davidson, C., 1996. "Christine Downton's brain", *Wired* 4(12): 170–183.

de la Fuente, A., 2000. *Mathematical Methods and Models for Economists*. Cambridge University Press, Cambridge.

Debreu, G., 1959. *Theory of Value: An Axiomatic Analysis of Economic Equilibrium*. Yale University Press, New Haven, CT.

Debreu, G., 1964. "Continuity properties of Paretian utility", *International Economic Review* 5: 285–293.

Edgeworth, F. Y., 1881. *Mathematical Psychics: An Essay on the Application of Mathematics to the Moral Sciences*. C. K. Paul, London.

Elton, E. J., Gruber, M. J., Brown, S. J. and Goetzmann, W. N., 2010. *Modern Portfolio Theory and Investment Analysis*, 8th edn. John Wiley, Hoboken, NJ.

Enders, W., 2010. *Applied Econometric Time Series*, 3rd edn. John Wiley, Hoboken, NJ.

Fama, E. F., 1970. "Efficient capital markets: a review of theory and empirical work", *Journal of Finance* 25: 383–423.

Fama, E. F. and Miller, M. H., 1972. *The Theory of Finance*. Dryden Press, Hinsdale, IL.

Feldstein, M. S., 1969. "Mean-variance analysis in the theory of liquidity preference and portfolio selection", *Review of Economic Studies* 36: 5–12.

Fishburn, P. C., 1970. *Utility Theory for Decision Making*. John Wiley, Chichester.

Fubini, G., 1958. *Opere Scelte*, Vol. 2. Edizioni Cremonese, Rome.

Geary, R. C., 1950. "A note on 'A constant-utility index of the cost of living' ", *Review of Economic Studies* 18(1): 65–66.

Goursat, É., 1959. *Derivatives and Differentials; Definite Integrals; Expansion in Series; Applications to Geometry*, Vol. I of *A Course in Mathematical Analysis* (new Dover edn of 1904 edn, transl. E. R. Hedrick). Dover, New York.

Greenberg, M. and Lowrie, K., 2010. "Kenneth J. Arrow: understanding uncertainty and its role in the world economy", *Risk Analysis* 30(6): 887–880.

Grossman, S. J. and Stiglitz, J. E., 1989. "On the impossibility of informationally efficient markets", in S. J. Grossman, ed., *The Informational Role of Prices*, pp. 91–116. MIT Press, Cambridge, MA.

Gujarati, D. N., 2003. *Basic Econometrics*, 4th edn. McGraw-Hill, Boston.

Hamilton, J., 1994. *Time Series Analysis*. Princeton University Press, Princeton, NJ.

Hart, R., 2010. *The Chinese Roots of Linear Algebra*. Johns Hopkins University Press, Baltimore.

Heims, S. J., 1980. *John von Neumann and Norbert Wiener: From Mathematics to the Technologies of Life and Death*. MIT Press, Cambridge, MA.

Higgins, T. J., 1940. "A note on the history of mixed partial derivatives", *Scripta Mathematica* 7: 59–62.

Hildenbrand, W., 1974. *Core and Equilibria of a Large Economy*. Princeton University Press, Princeton, NJ.

Hildenbrand, W. and Kirman, A. P., 1988. *Equilibrium Analysis: Variations on Themes by Edgeworth and Walras*. North-Holland, Amsterdam.

Hirshleifer, J., 1970. *Investment, Interest and Capital*. Prentice Hall, Englewood Cliffs, NJ.

Hogg, R. V. and Craig, A. T., 1978. *Introduction to Mathematical Statistics*, 4th edn. Collier Macmillan, London.

Huang, C.-F. and Litzenberger, R. H., 1988. *Foundations for Financial Economics*. North-Holland, New York.

Hull, J. C., 2009. *Options, Futures, and Other Derivatives*, 7th edn. Pearson Prentice-Hall, Upper Saddle River, NJ.

Inada, K.-I., 1963. "On a two-sector model of economic growth: comments and a generalization", *Review of Economic Studies* 30(2): 119–127.

Ingersoll, J. E., 1987. *Theory of Financial Decision Making*. Rowman and Littlefield, Totowa, NJ.

Jacques, I., 2009. *Mathematics for Economics and Business*, 6th edn. Financial Times Prentice-Hall, Harlow.

Jensen, J. L. W. V., 1906. "Sur les fonctions convexes et les inégalités entre les valeurs moyennes", *Acta Mathematica* 30(1): 175–193.

Johnston, J. and DiNardo, J., 1997. *Econometric Methods*, 4th edn. McGraw-Hill, New York.

Kahn, R. N., 1990. "Estimating the U.S. Treasury term structure of interest rates", in Fabozzi, F. J., ed., *The Handbook of U.S. Treasury and Government Agency Securities Instruments, Strategies and Analysis*, Chapter 9, pp. 179–189. Probus Publishing, Chicago.

Karush, W., 1939. "Minima of Functions of Several Variables with Inequalities as Side Conditions", Masters Thesis, University of Chicago.

Kelly, K., 1994. *Out of Control: The New Biology of Machines, Social Systems, and the Economic World*. Perseus Books, Cambridge, MA.

Kjeldsen, T. H., 2000. "A contextualized historical analysis of the Kuhn–Tucker theorem in nonlinear programming: the impact of World War II". *Historia Mathematica* 37(4): 331–361.

Klein, L. R. and Goldberger, A. S., 1955. *An Econometric Model of the United States, 1929–1952*. North-Holland, Amsterdam.

Klein, L. R. and Rubin, H., 1947. "A constant-utility index of the cost of living", *Review of Economic Studies* 15(2): 84–87.

Kuhn, H. W. and Tucker, A. W., 1950. "Nonlinear programming", in Neyman, J., ed., *Proceedings of the Second Berkeley Symposium on Mathematical Statistics and Probability*, pp. 481–492. Berkeley, CA.

Leonard, R. J., 1995. "From parlor games to social science: Von Neumann, Morgenstern, and the creation of game theory 1928–1944", *Journal of Economic Literature* 33(2): 730–761.

Leonard, R. J., 2010. *Von Neumann, Morgenstern, and the Creation Of Game Theory: From Chess to Social Science, 1900–1960, Historical Perspectives on Modern Economics*, Cambridge University Press, Cambridge.

Lintner, J., 1965. "The valuation of risk assets and the selection of risky investments in stock portfolios and capital budgets", *Review of Economics and Statistics* 47: 13–37.

Machina, M. J., 1982. " 'Expected utility' analysis without the independence axiom", *Econometrica* 50(2): 277–323.

Maclaurin, C., 1742. *A Treatise of Fluxions in Two Books*. T. W. and T. Ruddimans, Edinburgh.

Markowitz, H. M., 1952. "Portfolio selection", *Journal of Finance* 7(1): 77–91.

Markowitz, H. M., 1991. *Portfolio Selection: Efficient Diversification of Investments*, 2nd edn. Blackwell, Oxford.

Markowitz, H. M. and Todd, G. P., 2000. *Mean-Variance Analysis in Portfolio Choice and Capital Markets*. Frank J. Fabozzi Associates, New Hope, PA.

Mason, R. S., 1989. *Robert Giffen and the Giffen Paradox*. Philip Allen, Deddington.

McKenzie, L., 1957. "Demand theory without a utility index", *Review of Economic Studies* 24(3): 185–189.

Mendelson, B., 1975. *Introduction to Topology*, 3rd edn. Allyn and Bacon, Boston.

Merton, R., 1972. "An analytic derivation of the efficient portfolio frontier", *Journal of Financial and Quantitative Analysis* 7: 1851–1872.

Mood, A. M., Graybill, F. A. and Boes, D. C., 1974. *Introduction to the Theory of Statistics*, 3rd edn. McGraw-Hill Series in Probability and Statistics. McGraw-Hill, Singapore.

Neary, J. P., 1997. "R. C. Geary's contributions to economic theory", in D. Conniffe, ed., *Roy Geary 1896–1983: Irish Statistician. Centenary Lecture by John E. Spencer and Associated Papers*. Economic and Social Research Institute, Oak Tree Press, Dublin.

Poundstone, W., 2005. *Fortune's Formula: The Untold Story of the Scientific Betting System that Beat the Casinos and Wall Street*. Hill and Wang, New York.

Pratt, J. W., 1964. "Risk aversion in the small and in the large", *Econometrica* 32(1/2): 122–136.

Purfield, C. and Waldron, P., 1997. "Extending the Mean-Variance Framework to Test the Attractiveness of Skewness in Lotto Play", Trinity Economic Papers Technical Paper 97/4, Trinity College, Dublin.

Ridley, M., 1993. "The mathematics of markets: a survey of the frontiers of finance", *The Economist* (9 October): 1–20.

Roberts, A. W. and Varberg, D. E., 1973. *Convex Functions*. Academic Press, New York.

Rockafellar, R. T., 1970. *Convex Analysis*. Princeton University Press, Princeton, NJ.

Roll, R., 1977. "A critique of the asset pricing theory's tests – Part 1: On past and potential testability of the theory", *Journal of Financial Economics* 4: 129–176.

Roy, R. F. J., 1947. "La distribution du revenu entre les divers biens", *Econometrica* 15(3): 205–225.

Rubinstein, M. E., 1973. "A comparative statics analysis of risk premiums", *Journal of Business* 46(4): 605–615.

Samuelson, P. A., 1947. "Some implications of 'Linearity' ", *Review of Economic Studies* 15(2): 88–90.

Sharpe, W., 1964. "Capital asset prices: a theory of capital market equilibrium under conditions of risk", *Journal of Finance* 19: 425–442.

Shephard, R. W., 1953. *Cost and Production Functions*. Princeton University Press, Princeton, NJ.

Siegel, J. J., 1972. "Risk, interest rates and the forward exchange", *Quarterly Journal of Economics* 86: 303–309.

Simmons, G. F., 1963. *Introduction to Topology and Modern Analysis*. McGraw-Hill, Singapore.

Simon, C. P. and Blume, L., 1994. *Mathematics for Economists*. Norton, New York.

Sims, C., 1980. "Macroeconomics and reality", *Econometrica* 48: 1–48.

Slutsky, E. E., 1915. "Sulla teoria del bilancio del consumatore", *Giornale degli Economisti* 51: 1–26.

Solow, D., 2009. *How to Read and Do Proofs: An Introduction to Mathematical Thought Processes*, 5th edn. John Wiley, Hoboken, NJ.

Spivak, M., 1965. *Calculus on Manifolds: A Modern Approach to Classical Theorems of Advanced Calculus*. Mathematics Monograph Series. Benjamin/Cummings, Menlo Park, CA.

Stein, C., 1973. "Estimation of the mean of a multivariate normal distribution", *Proceedings of the Prague Symposium on Asymptotic Statistics*, Vol. II. pp. 345–381.

Stein, C. M., 1981. "Estimation of the mean of a multivariate normal distribution", *Annals of Statistics* 9(6): 1135–1151.

Stewart, J., 2008. *Calculus (Early Transcendentals)*, 6th edn. Brooks-Cole, Albany.

Stone, J. R. N., 1954. "Linear expenditure systems and demand analysis: an application to the pattern of British demand", *Economic Journal* 64: 511–527.

Sydsæter, K., Hammond, P., Seierstad, A. and Strøm, A., 2008. *Further Mathematics for Economic Analysis*, 2nd edn. Financial Times Prentice-Hall, Harlow.

Takayama, A., 1994. *Analytical Methods in Economics*. Harvester Wheatsheaf, New York.

Tinbergen, J., 1937. *An Econometric Approach to Business Cycle Problems*. Hermann, Paris.

Tinbergen, J., 1939. *Statistical Testing of Business Cycle Theories*, Vols. I and II. League of Nations, Geneva.

Tranter, C. J., 1953. *Advanced Level Pure Mathematics*. Physical Science Texts. English Universities Press, London.

Varian, H. R., 1992. *Microeconomic Analysis*, 3rd edn. W. W. Norton, New York.

von Neumann, J. and Morgenstern, O., 1944. *Theory of Games and Economic Behavior* (60th anniversary edn, 2004). Princeton University Press, Princeton, NJ.

Waldron, P., 1991. "Essays in Financial Economics", Ph.D. thesis, University of Pennsylvania.

Woods, J. E., 1978. *Mathematical Economics*. Longman, London.

Index

Page numbers in boldface in this index refer to the page on which mathematical or technical terms are first introduced or defined in the text.